PEACEFUL USES OF AUTOMATION IN OUTER SPACE

PEACEFUL USES OF AUTOMATION IN OUTER SPACE

Proceedings of the First IFAC Symposium on
Automatic Control in the Peaceful Uses of Space,
held June 21-24, 1965, in
Stavanger, Norway

Edited by

JOHN A. ASELTINE

TRW Systems, Houston, Texas

Springer Science+Business Media, LLC
1966

Library of Congress Catalog Card No. 65-22182

Originally published by International Federation of Automatic Control
Düsseldorf, Germany in 1966
Softcover reprint of the hardcover 1st edition 1966

ISBN 978-1-4899-6203-4 ISBN 978-1-4899-6411-3 (eBook)
DOI 10.1007/978-1-4899-6411-3

TABLE OF CONTENTS

PANEL DISCUSSION

LAUNCH & MIDCOURSE GUIDANCE

SPECIFIC SYSTEMS

ATTITUDE CONTROL

GROUND STATIONS & TRACKING

COMPONENTS & TECHNIQUES

ENTRY & LANDING

GENERAL TOPICS

FOREWORD

As part of its program of services to the English-speaking technical community, the Instrument Society of America (ISA), with the cooperation of the International Federation of Automatic Control (IFAC), is pleased to serve as the publisher of the Proceedings of the First IFAC Symposium on Automatic Control in the Peaceful Uses of Space. It is hoped that the documentation of the information exchange resulting from the presentations and discussions by international authorities will stimulate others to pursue and expand programs of peaceful space research.

Based on the enthusiastic acceptance of this First Symposium, it is hoped that the IFAC Theory and Applications Committees will periodically review progress in this area and consider the possibilities for programming future international meetings on space research activities and their peaceful applications.

Emil J. Minnar, Manager
Technical Operations
Instrument Society of America

PREFACE

Plans for the first IFAC Symposium on Automatic Control in the Peaceful Uses of Space began in Basel at the 1963 IFAC Congress. The idea of a symposium in this rapidly expanding field which combines theory, applications, and components appealed to some of us. J. C. Lozier and I from the U.S.A., A. M. Letov and N. B. Petrov of the U.S.S.R., and J. G. Balchen of Norway agreed that such a meeting should be sponsored by IFAC. With the support of J. G. Truxal and W. E. Miller, Chairmen of the Theory and Applications Committees, the Symposium was proposed in the name of the Norwegian National Committee of IFAC by Balchen, and accepted by the IFAC Executive Committee.

The Secretary, Lars Monrad-Krohn, arranged for the facilities of Stavanger for the meeting, and plans were made for about 100 delegates. We decided the papers should present practical results wherever possible, as well as theory, and arranged the program accordingly.

In June of 1965 the meeting began at Stavanger where about 120 delegates heard the 44 papers appearing in this volume. Of these papers, 9 describe results of space experiments (Mariner and Vostok are examples); 5 describe planned flights; 13 report on designs and research projects; and 17 are theoretical.

The delegates represented 9 countries. Although the majority were from the U.S.A. and U.S.S.R., papers from France, Germany, Japan, Belgium, Denmark, Norway, and the United Kingdom are an indication that interest and participation are truly international.

Opening remarks by several persons are included in this volume. Discussions of papers were taped during the Symposium and edited versions follow many of the papers. A panel discussion by distinguished U.S.A. and U.S.S.R. scientists is also included.

This Symposium represents a trend in IFAC toward subjects crossing boundaries between theory, applications, and components. It also represented the response of IFAC members to the challenge of a new and rapidly expanding field. As this is written only a little over a month has passed since the meeting, but much has happened. Mariner, described in this volume, has returned the first pictures of the planet Mars; the Soviet spacecraft ZOND 3 has sent back the best pictures so far of the far side of the Moon. These are exciting times, and this volume contains much evidence of the excitement and many directions for things to come.

J. A. Aseltine
Program Chairman

Houston, Texas
August 1965

ACKNOWLEDGMENTS

International Federation of Automatic Control
J. F. Coales, United Kingdom
President of IFAC

Dr. G. Ruppel, Germany
Honorary Secretary of IFAC

Professor John Truxal, U.S.A.
Chairman, IFAC Technical Committee on Theory

Mr. W. E. Miller, U.S.A.
Chairman, IFAC Technical Committee on Applications

Mr. Chr. Rand, Norway
Chairman, Norsk Forening for Automatisering
(Norwegian National Committee of IFAC)

Mr. K. Holberg, Norway
Chairman, Symposium Organizing Committee

Mr. L. Monrad-Krohn, Norway
Symposium Secretary

Dr. John Aseltine, U.S.A.
Chairman, Symposium Program Committee

Professor Academician M. Letov, U.S.S.R.
Member, Symposium Program Committee

WELCOME

Karl Holberg
Royal Norwegian Council for Scientific
and Industrial Research

On behalf of the Norwegian National Committee of IFAC, the Royal Norwegian Council for Scientific and Industrial Research and the Organizing Committee I have the pleasure of welcoming you all to Norway and to the IFAC Symposium "Automatic Control in the Peaceful Uses of Space".

This Symposium is the first in IFAC's history to encompass an entire field of modern science and engineering and cover both theory, application and components, and I sincerely do hope that it will fulfill the long expressed need for free exchange of information in this for all nations so important work.

I am glad that Norway with almost no space programs of its own has had the opportunity to be instrumental in bringing the most noted experts together here in Stavanger.

With these few words I declare the IFAC Symposium "Automatic Control in the Peaceful Uses of Space" for opened and will call upon our first speaker this morning, Dr. Odd Dahl.

OPENING ADDRESS

Dr. O. Dahl
Royal Norwegian Council for Scientific
and Industrial Research

This is a significant symposium, not only within the professions, but in the departments of government and among the general public as well. One can see many reasons for this. Rockets and satellites have opened new perspectives in research. Large commercial and industrial interests are involved, and the public can easily appreciate this research in the form of long distance picture transmission, for example. And we must not forget space activity appeals to the imagination. In the face of this very favorable wind blowing today, one may raise the question, "Is not space activity over-supported? Should they not give a similar or even higher priority to more earthbound problems, such as technology in medicine and in the humanities?" Perhaps we should, but in the absence of master plans, the fact remains: space activity is strongly supported. We may hope that it will serve as a precedent for other large scale activities in the interests of science and humanity.

Space research activity is not a narrow specialized field. It embraces front line knowledge of many fields and few natural scientists or engineers may leave space alone today if they want to apply an international yardstick to themselves. For small nations this presents a dilemma, only partly to be resolved through international cooperation. All are confronted with the fact that knowledge costs very much today, and the price is rising steadily. In our quest for knowledge, society has to decide whether it is willing to, or considers itself able to, pay the price. Therefore, more than ever, it is part of our duty to show that we can never have too much knowledge. Sooner or later it will benefit society and in no uncertain way. When the wise man of olden days, Solomon, saw the human population as the undisturbed ocean, he compared the teachers to the winds, saying that the ocean would remain at rest if not distrubed by the wind. A long time ago, a countryman of mine, an explorer and humanist, gave his doctorate acceptance speech to the students of the St. Andrew's University, in which he talked about the spirit of adventure as forcing us to seek knowledge in all walks of life, not just in one's particular field. "Adventure comes to us in many disguises", he said, "tempting us away from the road we believe to be right when it becomes dull and long. But the true spirit of adventure never lags because it can't see the end of the road. There is no retreating because we've burned all the bridges".

In his days, geographic exploration was given support in all countries. It was the space research of those days and cost relatively the same as today. We did not need the North Pole but it was there, and our knowledge about the Arctic has given us a foundation for weather forecasting and an understanding of the interplay between Arctic and equatorial waters. Thus, through better knowledge of current systems in the ocean, shipping is made safer and fishing more predictable. We did not need nuclear physics, and, as we know, its large scale support has been criticized in many countries. But nuclear physics led the way to atomic energy which is still a controversial subject. Still, we can all agree that we would have had a vastly different world without it. With it, we may be able to shape the world to our needs.

This kind of knowledge is anchored to the basic studies of the laws of nature. New knowledge is brought forth at an increasing rate through the adventuring mind setting intellectual standards, which we also must realize, will affect every frontier of the mind. Space research gave us tangible results much sooner than expected: we must anticipate that the concepts of reliability so closely tied up with space activity will permeate all branches of technology and perhaps even change our behavior.

If we go far back in time, and look at the world today, we might say we owe it all to the star gazers, even our declaration of human rights. So, I believe we may continue our quest for knowledge in good conscience, and through our symposium try to improve upon a trail recently laid by pioneers (some of whom are with us today) who gave guiding instructions and directions.

Control itself is nothing new, and it has a rather clear meaning in society. The laws of nature exert "control", and we also introduce other various "controlling elements" in dealing with each other and between groups and nations. Such controls have become increasingly more stringent, relating cause and effect as the result of observations, and lately by more systematic analyses of such close examinations. These general problems of control are sadly complicated by the human element. It is also difficult to envision reliability built into such automatic controls, because we do not always know what we are seeking. Maybe the control engineers should apply themselves a little more to these large scale and diffuse control problems.

The control problems of space are, of course, rather special in formulation, but with general implications. In addition, automatic control is imperative in space activity since solutions to control problems of space must be found, a pace is set for adoption of automatic control to processes where such control is not an absolute necessity, but where it will increase efficiency and productivity in industry. I am, of course, not implying that industry would not have discovered automation without the space effort, but they will learn much from it.

I should now like to say a few words on the non-technical aspects of a symposium like this. Here we are, people from many countries, talking a common language, the language of exact knowledge, and assembled in a small town in Norway. It could have been in a large city, in a large country, but it is significant that this small country can be the host to such an important gathering of experts. It's in the real spirit of international cooperation. Perhaps it is one of the too-many meetings which leave no time for us to work. I do not believe so. When a field is moving fast, it is very necessary to come together like this. We must inform each other and try to adjust levels so the pleateau of knowledge may show a uniform buildup, while showing up depressions which should be filled out and humps which do not need to go higher just now.

I never liked the prefix "peaceful" which nowadays seems to be required when important topics are under discussion. I am not against peace of course, but it carries with it an implication that some aspect of our work is not respectable. It poses problems for all of us, that our knowledge may see non-constructive uses. But as long as your work is anchored to the laws of nature and truth I hope that you will prevail in your work, and push forward with a not too peaceful attitude.

With these scattered remarks and comments, I've tried to place our symposium in a larger perspective, wishing to impress upon you that the days you are spending together have a meaning beyond the restricted purpose of this symposium itself. As a Norwegian, I'm glad to see with us guests from many countries, not as strangers, but as friends, and I wish you success. Space is still wide open.

WELCOMING ADDRESS ON BEHALF OF IFAC

Professor John G. Truxal
Polytechnic Institute of Brooklyn

On behalf of the Theory Committee of IFAC, I'm delighted to welcome you as delegates to this symposium. I also wish to convey to you greetings from Mr. William Miller, Chairman of the Applications Committee of IFAC. These two committees are, of course, sponsors of this symposium. Finally, I wish to express the gratitude of IFAC President, John Coales, on behalf of all elements of the IFAC organization, to the Norwegian National Committee and the Royal Norwegian Council for Scientific and Industrial Research for their willingness to hold this symposium in a country which certainly matches any other IFAC member-nation in the warmth of its hospitality and the beauty of its land. I come from a part of the United States where we have a severe water shortage at present, so it's delightful to come to this country where there seems to be a superabundance of water from all directions.

The two IFAC sponsoring committees are particularly pleased with the central theme of the meeting. The exploration of space and the development of the technology which permits space travel have really captured the imagination and enthusiasm of people everywhere, particularly of the young people; both the student in our universities and the productive young engineer or scientist who's not reached obsolescence. By satisfying man's love for exploration of the unknown and man's insatiable enthusiasm for establishing new records of accomplishment, the space program has provided a peaceful focus around which technology and science can advance. Perhaps a century hence, when our great great grandchildren read in their history books about this century as the period during which mankind moved into space, the major contribution of this space age will not be the vehicles or the travel, but rather will be the demonstration that it is possible for society to marshal the resources of technology for an attack on a peaceful goal.

I would speak next only for the United States, but certainly there the past decade of the space program has changed entirely our governmental and our public attitude toward technology. In 1955, a decade ago, social and political forces viewed the technological age with distrust, with fear, and with apathy and antipathy. Technology had created military weapons against which defense was impossible. Technology in automating production was leaving in its wake masses of unskilled, uneducated workers, who were unemployable. Technology was promising computers that would replace man or compete with man, devices which only the technologist could control. It seems to me this was the picture in 1955.

Today, nearly a decade later, and thanks entirely, I think, to the space age, society is focusing instead on the potential benefits to be derived from the development of a technology which can be harnessed and controlled. Technological resources are being focused on the development, for example, of artifical replaceable parts of the human body; focused on the understanding of human sensing mechanisms; focused on the evolution of logical mass-transportation systems to serve our industrial and urban population; and focused on the economic development of water resources. Not only has the space program demonstrated the possibility of such technological accomplishments, but we could also document in detail the direct impact of space research and development in these seemingly unrelated areas.

Within the United States, perhaps the best example of this national enthusiasm which is sweeping our country, where we have marshalled the resources of technology, is the rapidly growing interest in the field of ocean engineering. (The opposite of space, I might add.) Oceanography comprises mining and exploration of the continental shelf, extraction of elements from the sea, and other phases of deep ocean engineering. Perhaps this particular problem illustrates the very basic reason for the importance which IFAC associates with the symposium we're now holding. The space program has, during its first decade, been characterized by regional rivalries; rivalries which evolved naturally from the geo-political scene of a decade ago. It is certainly the fervent hope of IFAC that international cooperation and willingness to exchange technical information, which will mark these four days, will help to establish a pattern of international technological collaboration encouraging the launching of technologies' next programs on a truly non-national basis. I certainly wish you the most profitable and delightful four days at this symposium.

SYMPOSIUM OPENING REMARKS

Academician B.N. Petrov
Soviet Academy of Science

I would like to welcome the members of this symposium in the name of the Soviet Academy of Science and the Soviet delegation, and to wish them good work. We are meeting here to discuss problems of automatic control for peaceful conditions. Today we should remember the words of one of the founders of astronautics, Konstantin Tsiolkovsky, "Earth is the cradle of humanity, but one cannot live in the cradle forever."

In 1957 the first artificial satellite of the earth was launched. In 1961 Yuri Garagin became the first man to enter outer space. In 1965, in the real meaning of the word, a man opened the door to space and entered it. The first man to do this was A. Leonov and since then we have heard about the walk in space by the American astronaut, Edward H. White. Everyone knows the achievements of American science and technology in this field for peaceful purposes. I am quite certain that the work of our symposium will be successful and will set high standards for space technology in its peaceful applications in the interest of all mankind.

I would like to express my thanks to the Norwegian National Committee of IFAC for arranging this symposium and to the Technical Committees of IFAC on Theory and Applications for taking the initiative in preparing this symposium. Finally I would like to present Professor Holberg, Chairman, Royal Norwegian Council for Scientific and Industrial Research, Automatic Control Committee, a souvenir of this symposium, a model of the first Soviet artificial satellite.

Panel Discussion

PANEL DISCUSSION

on

"The Problems of Control of the Space Flight in the Future"

Chairman, Professor A. M. Letov

Taking this chair, I feel rather awkward and anxious because in front of me sits an audience composed of very highly qualified experts from many countries. I'm particularly anxious because I've seen who the speakers for this evening are. I wish to thank the speakers who accepted a part in this discussion. When I introduce these speakers, you will understand why I am anxious.

Professor Lur'e is a Corresponding Member of the Academy of Sciences of the U.S.S.R. He also holds the Chair of Mechanics of the Leningrad Polytechnic Institute, and is well known as a theorist working in the domain of modern mechanics. He is one of my former teachers and oldest friends, and I have great respect for him. Professor Lur'e certainly will not hesitate to punish me as a small boy if I fail to achieve the scientific level expected of me.

Professor Charles Draper is the Chairman of the Department of Aeronautics and Astronautics of the Massachusetts Institute of Technology and Director of its Instrumentation Laboratory. He is the creator of the well known floated gyroscope which is very highly accurate. I had the pleasure of meeting Professor Draper in Moscow at the 1960 IFAC Congress, as well as at the Symposium on Nonlinear Oscillations in Kiev. In my opinion Professor Draper is a very brilliant lecturer and scientist. During a reception by the President of the Ukrainian Academy of Sciences, Professor Draper and I were fortunate enough to taste the famous Popov Vodka produced in the Ukraine. Both Professor Draper and I were very pleased with this beverage. I mention this fact only to create in Professor Draper's mind agreeable memories which should serve here as a plea for mercy.

Unfortunately, I can't recount personal experiences concerning the other two speakers, Dr. Gillette and Dr. Fedosov. They are both younger and I believe they will contribute valuable information for scientific discussions. They are beginning to attain great reknown in the scientific world, but have not yet achieved that of Professor Draper and Professor Lur'e. I am certain they will find an opportunity to show their talents and forensic capabilities this evening, and this is another reason that I am apprehensive about the dangers I am facing.

Dr. Dean Gillette has nothing in common with the famous inventor of the safety blade. He's the head of a large laboratory in the Bell Telephone Laboratories, and although he's young, he's very experienced in the mathematical study of the problems we are discussing.

Dr. Fedosov is a Professor in the Moscow Polytechnical Institute. He has devoted many years to the study of mathematical problems linked with the flight of rockets and satellites, and also of the use of computers in this area. He is particularly interested in the statistical aspects of these problems.

Now, after having introduced these main speakers, I will say only a few words about myself. I am the director of a small laboratory in the Moscow Institute of Automatics and Telemechanics of the Academy of Sciences of U.S.S.R. My department is responsible for the mathematical treatment of certain problems of automatic control. It seems that people say I have succeeded in obtaining some results in this area. What will happen to me after the end of this discussion I am not capable of stating. I am personally afraid that after this discussion there will be an execution reminiscent of the popular Russian expression "a cat in a bag".

Now I will attempt to illustrate the problems which we will deal with here. The speakers and I had a preliminary meeting and arranged to develop the discussion in the following direction. I propose that we consider the problems which we might encounter if we all decided to go into space this evening. (I personally never considered going into space and I am not yet prepared to do such a thing. But I would like to state that Professor Draper has a totally different opinion. However, science and technology are developing at such a rate that even my opinion might change this evening.) The first problem to arise concerns forces acting on our space vehicle. These include the gravitational fields of Isaac Newton, the magnetic fields studied by Maxwell, and the fields of sun radiation. The first question to be raised in connection with the resultant field is, "What is the total force acting on the vehicle and the resulting motion?" It seems that we are not sufficiently informed about these different fields and every effort should be made to obtain this necessary information. We must learn how to use this vector force and the resulting motion in order to control and command the vehicle. The second problem is related to rendezvous, that is, the meeting of two or three or more vehicles in space.

We are also concerned with the problem of optimality in space. Because our personal abilities and resources are necessarily limited, we want to organize our flight in the best way. Therefore, the third problem with which we will be immediately concerned with is to determine the philosophical, mathematical, and engineering aspects of this optimality.

The last problem which I would like to bring to your attention concerns quantity. "What is the quantity of information we should place on board the vehicle in order to guide it in the proper manner?" These four problems were outlined in the preliminary meeting of the main speakers, but it was also agreed to mention the problem of the reactive "gun" with which the astronauts are able to direct their own movement when they leave their vehicle.

All these problems have, of course, theoretical aspects, but they also have aspects which are linked with their practical achievement. The speakers of this meeting represent in a very brilliant way both of these aspects. Therefore I will first ask Professor Lur'e to kindly give his opinion on the theoretical aspects. I will ask, in turn, Professor Charles Draper to speak on the creation of the necessary instruments for guidance and flight information, and to give some figures on the accuracy of these instruments. I would also like him to speak about the servomotors and actuators needed to provide for the necessary maneuvers. Of course the problem wouldn't be fully discussed if we omitted the disposition of the scientific information received; this problem is called data processing. I ask Dr. Fedosov and Dr. Gillette to take part in the discussion of this aspect.

Now everybody present, including the speakers, can easily disregard my proposals and speak about anything he likes. I apologize for this long introduction and thank you for your attention. Is my proposal acceptable to the main speakers and audience? Your applause gives me reason for hopeful expectation.

Dr. Charles Draper

It is difficult to follow a speaker who has been so complete as the one who has just given us the theoretical side of the story. I represent the field of technology. I feel that I must not alone consider the problems and imagine what should be done, I must do the problems with actual working equipment.

I feel that one does well to start with the general situation and proceed to special situations, so I will imagine a general travel in space. Vehicles will move from the earth to anyplace in space, and I admit that now this means the solar system, but any place in space to which there may be a reason to go. I am not thinking of a demonstration or a stunt; I am thinking of jobs that have to be done reliably and as a routine matter. This means vehicle equipment for guidance and information handling that will operate without reliance on information from outside the vehicle. This does not mean that if information from outside sources is available it should not be used; quite the contrary. It implies that the equipment on board should be able to take advantage of all possible sources of information. This is very similar to the situation of the captain of an ocean going vessel or an airplane today.

Also, I feel that the equipment must be suitable for either manned or unmanned vehicles. If the craft carries men, I am sure that the equipment should not use a human being as a component; the man should not carry out any routine task. Any matter that may be programmed should be put into the computers of the system without requiring special attention from a human pilot. This means in essence that the system should be automatic for routine and programmable matters. On the other hand the man is required for exercising judgment, making wise decisions and applying wisdom to situations that are not routine and cannot be programmed. The system should be designed so it will operate under any circumstances that may appear.

The guidance problem is basically a problem in geometry. It involves a vehicle space and its coordinates rigidly attached to the vehicle; it also involves one or more reference spaces that are outside the vehicle and within which the path of the vehicle is important. For example, operations near the earth may use earth space as reference space. Operations beyond the earth would use perhaps celestial space as a reference.

For purposes of accurate and continuous geometrical control during periods of acceleration, launching, injection into orbit, leaving orbit, midcourse corrections, landing on planets, an artificial space inside of the vehicle and mechanized so it may be matched to the outside space is required. Means for aligning this artificial mechanized internal space with the outer space in which the vehicle moves must be provided. This means that optical links, radio links, radar links, all radiation means available to connect the internal and external spaces must be provided.

Computers to provide for geometrical transformations among the spaces involved are absolutely necessary. Storage for programs, instructions, means of operation, plans, all these things must be provided for the computer and the difference from one mission to another mission depends entirely on what this stored information may be.

For manned flights, readout systems are required that will inform the man completely and continuously as to geometrical and dynamical situations. The man must be continuously informed as to the situation. The means must be provided by which the man can command the machine; the man must be able to command the machine to follow his will. The man, in fact, does not need to be aboard the vehicle; he may be any other place that affords him communications with the vehicle.

The elements required to bring such systems into existance all are ready today. You will hear about some of these tomorrow. And the performance is adequate and the reliability I believe is surprisingly good today. Cost in the future will be reduced, the size will become less, the weight will become less. I believe the mechanization of guidance will not be the limiting factor in any space mission of the future. If the mission is not a complete overall mission, but is simpler and does not require all the elements it is always possible to reduce the complete system to a simpler form that will do the more specialized mission; this is a matter for judgment.

Professor A. I. Lur'e

I am a specialist in the theoretical aspects of these problems, and I believe that the problems of space will expose a very large area for the theoretician. Now I will formulate some questions, but I won't always be able to give answers. Of course, on several questions I can give my personal opinion. Every control system is composed of a controlled plant, control signals, servomotors, and a measuring instrument. Now I will speak about the different elements of the system.

I understand the term control plant to mean, not the vehicle itself by which we will all go into space, but the system of equations which define it. In the equation of motion there is a left hand side and a right hand side. Take first the left hand side of the equation: is this side sufficiently determined for practical realization? It seems there is no definite answer to this question. First of all, if all the data is known for a given plant we can say that the left side is determined. However, if we imagine a future vehicle, this calculation must be made on the basis of the task to be done by that vehicle, and this decreases the deterministic character of the data. For example, the man who has gone out in space, to have a walk for instance, should be considered to a given extent as an automated system. I'm convinced it would not be sufficient for the astronaut to have simply a reaction gun. Therefore, this left hand side of the equation isn't fully determined, because we can't provide for all the movements the astronaut will have to make to fulfill a given task.

The Chairman has already stated that the right hand side of the equation is determined by the force fields which are acting on the vehicle. Therefore, the problems are to know this field and to know to what degree of accuracy we require this knowledge.

It seems there are four different fields to be considered. The easiest problem concerns the gravitational field, as the calculation of the force vector and momentum do not present any great difficulty. If we consider the force vector, we can view it as either the source of disturbances for the vehicle or the means for passive stabilization. The passive methods which were considered today in a series of very interesting papers were devoted mainly to a satellite near the earth. The idea of using the gradiant of the gravitational field for the passive stabilization of an earth satellite is quite logical, but I question whether the same idea could be used for a vehicle sent to the moon or to other planets.

The aerodynamical problems are related to free molecular flow. Here the problem of computing the force vector and the momentum is very complicated, although it has been studied by a number of authors. This problem is particularly complicated for satellites not having convex form, which is the case for satellites using solar cells. I don't want to enter into details, but the fact is, for satellites or bodies having this form, the basic hypotheses of free molecular flow are not always fulfilled. Determination of the vector force and momentum becomes more complicated here, because there is a dependence on the angles involved. If the gravitational field was clear, the aerodynamical field can still use much investigation, particularly in the area of changing density, where the density data varies with different authors.

The determination of the sun radiation pressure field involves knowing whether the surface is fully reflecting, fully absorbing, or has some peculiar properties. Here there are several serious problems. Sun radiation pressure arises from a potential field derived from the sun. Therefore, this field doesn't provide means for damping the oscillation which would arise if a body were to travel there. Would it be possible to use this field for damping by changing vehicle reflecting ability?

In conclusion I would like to speak about the problem of optimization. If we know the initial conditions and would like to obtain a definite final condition, we must find the control which provides the optimum. This gives us a control law of limited possibilities. Should we stick to it? It is difficult to say how to evaluate the damage resulting from departure from an optimum. Or should we put the problem of optimization in quite a different way, continuously using all the information coming in?

I ask many questions; of course one man can ask more than any fifty men could answer, but I hope that even in this limited discussion you will receive a reply to part of these questions and obtain some positive results.

Dr. Dean Gillette

The Chairman has described me as being young, argumentative, and I think clean-shaven. I thank him for his reassurances on the former, I can assure you on the latter, and I think you will find the argumentative part is true too. Now the distinction between the two sides of the table up here isn't really what you think. I believe that you will find that the first two speakers are more interested in the theoretical aspects, and the last two are the engineers. So my arguments are probably going to be with Professor Draper.

The Chairman has asked me to talk about data processing and I would like to stay very close to what Dr. Draper has said, and talk about data processing by the combination of men and machines. Now machines, as I'm sure you all know, are quite good for reliable storage of large amounts of pre-established information and for the most part they can recall this storage upon command. They're also good for accurate and rapid computation of routine problems. Now I have problems myself with accurate calculations, particularly with checkbooks, and with exact recall of data. Men are not as good as machines at recalled detailed information and routine calculation.

But men are awfully good at what is broadly called pattern recognition. When Dr. Draper speaks of understanding a situation, this falls within the category that I refer to as pattern recognition, and men are awfully good at this. Men are also good at correlating widely differing classes of information. This may be what is meant by "experience". This fundamental dichotomy, I think, should be used in asking of any particular problem at hand whether a man or a machine should undertake it. This is my first point, almost in direct conflict with Dr. Draper. I believe that the equipment and the definition of objectives for parts of the equipment should be chosen as a result of analysis of the task at hand and not laid out generally in advance.

As a specific example, and it strictly is in the realm of opinion, I believe that approach to a planet or takeoff from a planet, earth or the moon, should if possible be carried out from the planet. This can probably be translated technically to a statement that I believe in radio guidance. Dr. Draper used the analogy of a captain at sea being essentially autonomous. I would like to mention a parallel problem, that of a captain of a large passenger liner, essentially autonomous, but there are times when I believe that he would just love to have an automatic finding system.

In looking at these problems and in the selection of what kind of data processor you should use, I think one of the fundamental approaches that is going to help in system reliability is what I call "safety oriented". You can see an example of this in the recent Gemini 4 mission. The landing was to be carried out by a combination of man and computer on board. With the onboard computer, the astronauts hoped to be able to land on the flight deck or right next to the recovery carrier. The machine aboard was not operable; fortunately a safety precaution was available and the men could come down to a safe landing but not a precise one without the computer. The type of design approach that leads to reliability, I believe, is one in which you start out with the system that will give you the minimum required capability, in this case, safe return. In order to get a more precise result or to fulfill more complex requirements, one may then add an overlay of more sophisticated equipment. My approach I'm suggesting then, the safety oriented approach, is one in which you build first for the simplest system that will meet minimum requirements, then add to it for more precision. I think in this way you will get more reliability.

Now this will probably lead you away from optimal solutions, theoretically optimal solutions. We heard an example today of a simple scheme for orientation of the Vostok vehicle. It was probably just a slight departure from optimal in the sense of minimum fuel usage. Now in his opening remarks, the Chairman pointed out the need for optimization, because of the limited resources that are available. One of my predictions for the future, and I hope it's more than simply wishful thinking, is that we will not be as limited as we are now with respect to propulsion, and can depart from optimality. I think this is going to be particularly important in the future because as we clearly see more complex missions, as we see longer time of flight in space, we are going to have to press much harder for reliability.

So my point of controversy with Dr. Draper is that I think each class of mission should be examined by itself when asking how to match the man/machine difference in data processing so as to obtain the most in reliability from system design. The safety oriented approach that I have suggested may turn out to be a useful one.

Dr. E. A. Fedosov

Dr. Lur'e in his talk has touched a very interesting question, namely, the coming out of the astronauts, leaving the vehicle and going into space. In this case the astronaut is a system with a variable inertia, because when he is working with various tools for accomplishing different tasks his moment of inertia is variable. We have met already with such problems in the works of our American colleagues. I mean the works of Sammuels and Berkin. But these works are limited to the so-called white noise coefficients of the equations. Whereas, the problem of automatization of the work of the astronaut requires concentration of the equations where the coefficients are not white noise. Some works concerning such problems have been published in Soviet literature, for instance by Sinitsin and others. We would be very interested to hear from our American colleagues, in particular from Professor Truxal, about their work on these problems.

The second question I would like to consider is the problem of digital computers. Professor Draper and Dr. Dean Gillette have taken into consideration the problem of computers, and they have put many questions concerning the use of these computers. I will repeat some of these problems. The first is the solution of current navigational problems; the second is the problem of working with different scientific data and the problem of recognition of patterns. And third is the control of the means for commanding the vehicle itself. These three classes of problems can be solved by means of very different algorithms. That's why there is a problem of how to build the on-board computer.

Two solutions can be considered: first a central universal computer which will be the brain of the vehicle and able to solve all the three classes of problems. Or a system can be considered consisting of at least three computers, each specialized in a certain category of problems, analogous to the nervous system. These two ultimate solutions are in seeming contradiction.

Dr. Dean Gillette seems to be in favor of the second solution, that is, construction of narrowly specialized computers, insisting upon high reliability. Although I agree entirely as far as reliability and safety is concerned, I can't agree on account of the sizes and of the required energy to run these computers. It seems to me that the construction of such a system of specialized computers will require connection between computers speaking different languages, and this will complicate the communication channels and transformations from machine to machine. This apparently can make us lose the reliability of the system. Besides, such a de-centralized system is very difficult to control, in particular, the channels of communication are complicated.

In addition, the programming represents quite a problem. I think that such a de-centralized system would require the existance on-board of a certain kind of dispatcher, and this dispatcher would control the relations between these computers. It seems that what I used as an argument in favor becomes also criticism of the de-centralized system for computers which will be aboard the satellite - the fact that they use very different algorithms. That's why I suppose the future belongs to a centralized unique computer.

In conclusion I would like to unveil a secret of the Soviet Union: the point of view concerned with a centralized computer system is only my personal point of view and there is quite a wide circle of Soviet scientists who are rather in favor of the point of view of Dr. Dean Gillette.

Comments by Panel and Audience

Letov - I think that I was certainly correct in estimating the battle fever of the younger opponents. They have started quite a fight, which is very good for our discussion. Is anybody in the audience willing to speak on behalf of the subject touched by the principal speakers or open a new field of discussion?

Truxal - I don't want to argue with anybody. I heard my name mentioned in the middle of the last speech. My pattern recognition properties were not as effective as the computer might have been, but I think Professor Fedosov was referring to the stochastic control problem that work which falls into that category. It's my impression that the stochastic control problems which can be solved are so specialized and so few and so simple that we're still a great distance from solving any realistic problems on the very simplified missions and flights to which Dr. Draper referred. It is also my feeling that stochastic control theory as it has developed recently in the Soviet Union, and perhaps to a lesser extent in the United States, is still far from a useful tool in the design of specific systems. I really know of no stochastic control results which lead to algorithms which are manageable on the computing facilities which we have available.

Lawton - Dr. Gillette made the statement that the best way to make an approach to one of the planets is to be guided down from that planet. This may be a little hard to realize.

Gillette - It is very difficult the first time on the moon, I grant you that.

Lawton - The first time anywhere!

Raouschenbakh - I join the opinion of Dr. Gillette. I probably don't belong to the Soviet scientists who are of the opinion of Dr. Draper. Space systems, in my opinion, should be very simple and very reliable. Therefore, complex systems which are obtained as the result of solving variational problems have two purposes. The first practical purpose is for the author because it's a good thesis. But the second purpose arises from the fact that it is difficult to evaluate what we are losing when we don't make such systems.

King - I agree with Professor Fedosov. In my opinion people too often associate themselves with particular parts of a system which leads away from an integrated system which may be better.

Letov - Is your viewpoint also a secret of the United States?

King - Yes.

Uditski - On the point of view of universalization of specialization of devices I think that Dr. Gillette is right. In what concerns the universalization of the commands, I am of the opinion of Dr. Fedosov.

Petrov - One of the speakers has expressed the thought that the satellite should have a large number of channels of communication with earth or other satellites. Although the creation of an autonomous system is most interesting, it is impossible to avoid communication links with the external world. The question I would like to ask in this connection is: how does the audience visualize the use of the Laser or devices of a similar type for this purpose?

Kershner - It has been announced but it may not have been widely noticed that there have been over the past 6 months very successful ranging and direction determinations of satellites in orbit by Laser. These experiments were made under very difficult conditions from earth where the whole problem of atmospheric absorption is present. The problem of modulation is reasonably well in hand, so I think that the extrapolation to communication techniques from one satellite to another or in any airless environment is not a very big extrapolation at all. It very clearly is coming very soon.

Letov - If there are no other comments, I will summarize the results of the discussion. Professor Lur'e has given an outline of what we should know about the force fields which are acting on a vehicle which is moving in these fields. According to an old rule which I think is attributed to Lagrange a good formulation of the problem is already half a success. Professor Draper has predicted a promising future for achievement of accurate instruments for future space flights. Dr. Gillette and Dr. Fedosov have given their respective points of view on the possibilities of using computers in order to create such systems of control.

It is quite natural that some points of view are different. Life will show in the future who is right. In any case, I don't consider in such a hopeless way my own possibilities of going into space as I did at the beginning of this meeting.

I would like to thank everyone for his active participation in the discussion. I believe that this has been the first opportunity for such a discussion in the framework of the International Federation of Automatic Control. The success was so great that there was an exchange of state secrets between members of the audience. Of course I hope what has been said is not for the record in the newspapers. One of my personal opinions has already been misinterpreted and published in the Norwegian press as if the Soviet Union were prepared to send five astronauts into space.

Launch and Midcourse Guidance

THRUST PROGRAMMING IN A CENTRAL GRAVITATIONAL FIELD

by A. I. Lur'e

ABSTRACT

Programming thrust in a central force field for satisfaction of one optimality condition or another (response, minimal fuel consumption, etc.) leads to discussion of the Mayer-Bolza variational problem. In this report we will present a survey of methods of reducing this problem to a boundary-value problem for a system of nonlinear differential equations, prove that it is possible to reduce the order of this system by using its first integrals, and consider the case of motion in plane and spherical orbits.

The first section contains the equations of motion and stationarity conditions for three types of thrust devices. In the second section we will show that use of scalar and vector integrals of these equations reduces investigation of the Lagrange vector $\underline{\lambda}$ to a first-order differential equation for which one integral is known. Criteria for classifying types of motion are formulated in terms of a constant vector $\underline{a}$ in this differential equation. In the discussion we introduce orbital axes and write the differential equations of the problem in terms of these axes.

Boundary-value problems are stated in the third section. Here we will show that our method of determining the vector $\underline{\lambda}$ or motion along a Kepleristic arc reduces to use of well-known solutions of variational equations for the differential equations of Kepleristic motion.

The system of differential equations for our optimization problem admits a class of particular solutions that determine motion along spherical orbits; we will discuss these equations in the fourth section. In the fifth section we will show that motions (including plane and spherical) with a bounded-power thrust device and minimum fuel consumption are determined by the solution to the variational problem of minimizing the integral of the square of the acceleration due to thrust. We will also establish that it is possible to use the Ritz method for solution of this class of problems. In the sixth section we will consider motions in singular-control regimes in the presence of thrust with bounded absolute value.

SECTION 1. THE EQUATIONS OF MOTION OF A SPACE VEHICLE IN A CENTRAL GRAVITATIONAL FORCE FIELD IN THE PRESENCE OF THRUST

1.1 The vector form of the equations of motion. We will denote the vector-radius of the center of inertia of a space vehicle with origin at the center of attraction by $\underline{r}$, and we will denote the velocity vector by $\underline{v}$. These vectors determine the instantaneous plane of an orbit π. The vector momentum per unit mass is

$$\underline{k} = \underline{r} \times \underline{v} = k\underline{n} \tag{1.1}$$

and is directed along the normal ($\underline{n}$ is its unit vector) to this plane.

The differential equations of motion of the center of inertia can be written in the form

$$\dot{\underline{r}} = \underline{v}, \quad \dot{\underline{v}} = -\frac{\mu}{r^3}\underline{r} + \underline{w}, \tag{1.2}$$

Here is a constant equal to the product of the mass at the gravitational center and the gravitational constant, and w is the acceleration due to thrust,

$$\underline{w} = \frac{cq}{m}\underline{e}, \tag{1.3}$$

where c denotes the flow speed and q is the mass-flow rate, which is related to the mass-flow rate by the consumption equation

$$m' = -q \tag{1.4}$$

Finally, $\underline{e}$ is the unit vector determining the thrust direction

$$1 - \underline{e}\cdot\underline{e} = 0 \tag{1.5}$$

We should also recall that, by the angular momentum theorem,

$$\dot{\underline{k}} = \underline{r} \times \underline{w} \tag{1.6}$$

1.2 Statement of the problem. We will call the vectors $\underline{r}$ and $\underline{v}$ and the mass m the "coordinates" of the system, while we will call $\underline{e}$, c, and q the "controls;" we will subject the controls to certain conditions determined by the type of thrust device (see 1.3).

We will attempt to find the coordinates in the class of continuous functions of time in the interval $(0, t_1)$. Henceforth, we will assume that their initial values are

$$t=0 \quad \underline{r}=\underline{r}^0, \quad \underline{v}=\underline{v}^0, \quad m=m^0, \tag{1.7}$$

while we will assume that the values r^1, v^1, and m^1 at the right end of the time interval are subject to the following conditions: 1) the vectors r^1 and v^1 are related by expressions of the form

$$t=t_1, \quad \varphi_\ell(\underline{r}^1, \underline{v}^1)=0, \quad \ell=1,\dots, r\le 6 \tag{1.8}$$

where t_1 is either previously given ($t_1 = t_1^*$) or unknown; 2) the final mass is preset

$$t=t_1 \quad , \quad m=m^1=m_*^1 \tag{1.9}$$

or is required to ensure minimum total mass consumption

$$J=m^0-m^1=Min; \tag{1.10}$$

3) requirement (1.10) can be replaced by a minimum-time condition (the response problem)

$$J=t_1=Min, \tag{1.11}$$

where, in this case, m^1 is either previously given or unknown.

The problem consists in programming the thrust -- finding functional relationships between the controls e, c, q, and time (in the class of piece wise-continuous functions) ensuring satisfaction of the conditions stated above.

1.3 Thrust devices. We will consider thrust devices of three types.

1. A device that ensures that the acceleration vector due to thrust (w = const) has constant absolute value, where the flow speed c is constant or given as a function of time. The consumption equation can be integrated,

$$m=m^0\exp\left(-w\int_0^t \frac{dt}{c}\right),$$

and we will eliminate the mass m from further discussion. The response requirement ensures minimum total mass consumption. The vector $\underline{e}$ serves as the control.

2. Motors with bounded power N(t) -- plasma or ion motors. The controls include c and q, which are related by the expression

$$N(t)=\frac{1}{2}c^2q=\frac{1}{2}\frac{m^2w^2}{q}, \quad N_{min}\le N(t)\le N_{max} \tag{1.12}$$

If we adopt the same assumptions as those used in (1, 2), these inequalities can be written in the form of an equation:

$$[N_{max}-N(t)][N(t)-N_{min}]-\nu_2^2=0 \tag{1.13}$$

where ν_2 is treated as an auxiliary "control."

3. Chemical-fuel motors with bounded mass consumption q. The flow speed is assumed to be constant

$$c=const, \quad 0\le q(t)\le q_{max} \tag{1.14}$$

The "controls" are q and ν_3:

$$q(t)\left[q_{max}-q(t)\right]-\nu_3^2=0 \tag{1.15}$$

1.4 The Mayer-Bolza problem. The fundamental papers (1-5) stated a problem that can be interpreted within the realm of the Mayer-Bolza problem of variational calculus. This problem has been most completely discussed in Bliss's "lectures" (6), although in recent years a large number of other papers have also appeared on this problem (7, 8). The monograph (9) is devoted to optimal flight-control problems.

We now introduce three types of Lagrange multipliers into the discussion: 1) the vectors $\underline{\lambda}_r, \underline{\lambda}_v$, and the scalar λ_m for thrust devices of the second and third types. We use these multipliers in the right sides of the differential equations of motion to compose the "Hamiltonian"

$$H_\lambda=\underline{\lambda}_r\cdot\underline{v}-\underline{\lambda}_v\cdot\underline{r}\frac{\mu}{r^3}+\underline{\lambda}_v\cdot\underline{e}\,w \tag{1.16$_1$}$$

$$H_\lambda=\underline{\lambda}_r\cdot\underline{v}-\underline{\lambda}_v\cdot\underline{r}\frac{\mu}{r^3}+\lambda_v\cdot e\frac{cq}{m}-\lambda_m q \tag{1.16$_{2,3}$}$$

2) The scalar factors μ_e and μ_s are introduced to allow for the terminal relations imposed on the "controls." The second term of the Hamiltonian can be expressed in terms of them:

$$H_\mu=\mu_e(1-\underline{e}\cdot\underline{e})=0 \tag{1.17$_1$}$$

$$H_\mu=\mu_e(1-\underline{e}\cdot\underline{e})+\mu_1(c^2q-2N)+\mu_2\left[(N_{max}-N)(N-N_{min})-\nu_2^2\right]=0 \tag{1.17$_2$}$$

$$H_\mu=\mu_e(1-\underline{e}\cdot\underline{e})+\mu_3\left[q(q_{max}-q)-\nu_3^2\right]=0 \tag{1.17$_3$}$$

3) The constant scalars ζ_ℓ are introduced in order to write out the boundary conditions; they are used to write expressions for a "master function" θ:

$$\theta=J+\theta_1, \quad \theta_1=\sum_{\ell=1}^{r}\zeta_\ell\,\varphi_\ell(\underline{r}^1,\underline{v}^1)+\zeta_t(t_1-t_1^*)+\zeta_m(m^1-m_*^1)=0 \tag{1.18}$$

Here J is functional (1.10) or (1.11); if t_1 and (or) m^1 is not previously given, then ζ_t and (or) ζ_m is assumed to be equal to zero.

Moreover, if we set the variation of the functional

$$I=\theta+\int_0^{t_1}\left(\underline{\lambda}_r\cdot\dot{\underline{r}}+\underline{\lambda}_v\cdot\dot{\underline{v}}+\lambda_m\dot{m}-H_\lambda-H_\mu\right)dt \tag{1.19}$$

to zero, we obtain three groups of equations: stationarity conditions, Erdmann-Weierstrass conditions at the points t_* of possible discontinuities in the "controls," and boundary conditions for the Lagrange multipliers $\underline{\lambda}_r$, $\underline{\lambda}_v$, λ_m and the

Hamiltonian H . The Weierstrass minimum criterion, which is of cardinal value, can be added to them. We will now write out the value of E for the above three types of thrust devices:

$$E = \underline{\lambda}_v \cdot (\underline{e} - \underline{e}^*) \qquad (1.20)_1$$

$$E = \frac{1}{m}\underline{\lambda}_v \cdot (cq\underline{e} - c^*q^*\underline{e}^*) - \lambda_m (q - q^*) \qquad (1.20)_2$$

$$E = \frac{c}{m}\, \underline{\lambda}_v \cdot (q\underline{e} - q^*\underline{e}^*) - \lambda_m (q - q^*) \qquad (1.20)_3$$

where the stars denote admissible (i.e., satisfying the imposed relations) controls. The Weierstrass minimum criterion consists in requiring that the quantity E be non-negative:

$$E \geq 0 \qquad (1.21)$$

1.5 Stationarity conditions. These determine a system of differential equations and terminal relations for Lagrange multipliers of the first and second kinds:

$$\dot{\underline{\lambda}}_v = -\mathrm{grad}_{\underline{v}} H_\lambda = -\underline{\lambda}_r \qquad (1.22)$$

$$\dot{\underline{\lambda}}_r = -\mathrm{grad}_{\underline{r}} H_\lambda = \frac{\mu}{r^3}\left(\underline{\lambda}_v - 3\frac{\underline{\lambda}_v \cdot \underline{r}}{r^2}\underline{r}\right) \qquad (1.23)$$

$$\dot{\lambda}_m = -\frac{\partial H_\lambda}{\partial m} = \underline{\lambda}_v \cdot \underline{e}\ \frac{cq}{m^2} \qquad (1.24)_{2,3}$$

$$\mathrm{grad}_e (H_\lambda + H_\mu) = w\underline{\lambda}_v - 2\mu_e \underline{e} = 0 \qquad (1.25)_1$$

$$\mathrm{grad}_e (H_\lambda + H_\mu) = c/m\,\underline{\lambda}_v - 2\mu_e \underline{e} = 0 \qquad (1.25)_{2,3}$$

$$\left.\begin{aligned} \frac{\partial (H_\lambda + H_\mu)}{\partial q} &= \underline{\lambda}_v \cdot e\frac{cq}{m} - \lambda_m + c^2\mu_1 = 0, \\ \frac{\partial H_\lambda + H_\mu}{\partial c} &= q\left(\frac{1}{m}\underline{\lambda}_v \cdot e + 2\mu_1 c\right) = 0 \\ \frac{\partial H_\mu}{\partial N} &= -2\mu_1 + \mu_2 (N_{MAX} + N_{MIN} - 2N) = 0 \\ \frac{\partial H_\mu}{\partial \nu_2} &= -\nu_2 \mu_2 = 0 \end{aligned}\right\} \qquad (1.26)_2$$

$$\frac{\partial (H_\lambda + H_\mu)}{\partial q} = \underline{\lambda}_v \cdot \underline{e}\frac{c}{m} - \lambda_m + \mu_3 (q_{max} - 2q) = 0, \ \frac{\partial H_\mu}{\partial \nu_3} = -\nu_3 \mu_3 = 0 \qquad (1.26)_3$$

1.6 Equations common to all three types of thrust devices. These are the stationarity conditions (1.25), (1.22), and (1.23). It follows from (1.25) that the vectors $\underline{\lambda}_v$ and $\underline{e}$ are parallel; by using the Weierstrass criterion, we can easily show that these vectors are codirectional. Indeed, since the desired controls belong to the class of admissible controls, these criteria must be satisfied when c = c* and q = q*. For all types of thrust devices under discussion, therefore, we have the same inequality $(1.20)_1$. If we now set $\underline{\lambda}_v = \tilde{\lambda}_v \underline{e}$ where $\tilde{\lambda}_v = \pm\lambda_v = \pm|\underline{\lambda}_v|$, we find that

$$E = \underline{\lambda}_v \cdot (\underline{e} - \underline{e}^*) = \tilde{\lambda}_v (1 - \underline{e} \cdot \underline{e}^*) \geq 0, \quad \tilde{\lambda}_v = \lambda_v$$

which it was required to prove. Thus,

$$\underline{\lambda}_v = \lambda_v \underline{e} \qquad (1.27)$$

eliminating $\underline{\lambda}_r$ from (1.22) and (1.23) and, for brevity, substituting $\underline{\lambda}$ for $\underline{\lambda}_v$, we are led to a second-order differential equation for the vector $\underline{\lambda}$:

$$\ddot{\underline{\lambda}} = \frac{\mu}{r^3}\left(3\,\frac{\underline{\lambda} \cdot \underline{r}}{r^2}\underline{r} - \underline{\lambda}\right) \qquad (1.28)$$

These conditions exhaust the set of stationarity conditions for the first type of device.

1.7 Additional stationarity conditions. Thrust devices of the second type. We assume that the flow rate remains finite; then, by (1.12), the consumption q = 0; after we eliminate μ_1 from $(1.26)_2$ and account for (1.27) and $(1.24)_2$, we find that

$$\lambda_m = \frac{c\lambda}{2m}, \quad \frac{\dot{\lambda}_m}{\lambda_m} = \frac{2q}{m}, \quad \lambda_m = \frac{A}{m^2}, \quad \lambda = \frac{2A}{mc}, \qquad (1.29)$$

where A is a constant of integration (A > 0). We now have

$$c = \frac{2A}{m\lambda}, \quad N = \frac{1}{2}c^2 q = \frac{2A^2 q}{\lambda^2 m^2}, \quad w = \frac{cq}{m} = \frac{2A}{\lambda m^2}q$$

so that, by Equation (1.4),

$$\lambda = A\frac{w}{N}, \quad \frac{q}{m^2} = \left(\frac{1}{m}\right)^{\cdot} = \frac{w^2}{2N}, \quad \frac{1}{m} - \frac{1}{m_0} = \frac{1}{2}\int_0^t \frac{w^2}{N}dt$$

After elimination of μ_1, the third equation of $(1.26)_2$ takes the form

$$\frac{\lambda}{mc} + \mu_2 (N_{max} + N_{min} - 2N) = 0$$

and the assumption the $\mu_2 = 0$ leads to $\lambda = 0$, $\underline{\lambda} = 0$, $\dot{\underline{\lambda}} = 0$, $\lambda_m = 0$. Below we will show that this leads to a contradiction with the boundary conditions of the problem, both in the problem of minimizing the total consumption of mass and in the problem of minimizing t_1. The assumption $\mu_2 = 0$ must be rejected, so $\nu_2 = 0$ and $N = N_{max}$ or $N = N_{min}$. But consideration of the Weierstrass criterion now shows that only the first possibility remains. Indeed, using $(1.20)_2$, (1.27), and (1.29), we find that

$$\frac{1}{2}c(q + q^*) - c^*q^* \geq 0 \quad \text{or} \quad \frac{N}{c}\left(1 + \frac{N^*}{N}\right) - \frac{2N^*}{c^*} > 0$$

where, of the three admissible controls c*, q*, and N*, two are independent and one of them may be assigned arbitrary values; setting c = c*, the inequality in the second form yields N − N* ≥ 0, while if we set q = q*, the first inequality yields c ≥ c*, i.e., N ≥ N* again. But N* can, without forcing inequalities (1.12) to be false, be chosen arbitrarily close to N_{max}, so the inequalities we have found require that $N = N_{max}$.

Thus, for the type of thrust device under discussion, the vector λ is the only constant factor different from w.

$$\underline{\lambda} = \frac{A}{N_{max}}\underline{w} = \frac{A}{N_{max}}\left(\dot{\underline{v}} + \frac{\mu}{r^3}\underline{r}\right) \qquad (1.30)$$

while the mass is given by the equation

$$\frac{1}{m} - \frac{1}{m_0} = \frac{1}{2N_{max}}\int_0^t \left|\dot{\underline{v}} + \frac{\mu}{r^3}\underline{r}\right|^2 dt \qquad (1.31)$$

1.8 Additional stationarity conditions. Thrust devices of the third type. In our discussion it is desirable to replace λ_m by the quantity η -- a criterion for distinguishing regimes -- defined by the expression

$$\eta = \frac{c}{m}\lambda - \lambda_m^* \qquad (1.32)$$

It follows from $(1.24)_3$ that this criterion satisfies the differential equation

$$\dot{\eta} = \frac{c}{m}\dot{\lambda} \qquad (1.33)$$

Conditions $(1.26)_3$ now take the form

$$\eta + \mu_3(q_{max} - 2q) = 0, \quad \mu_3 \nu_3 = 0$$

and it follows from (1.15) and the assumption that $\nu_3 = 0$ that the fuel consumption may take only one of its limit values, $q = 0$ or $q = q_{max}$. Weierstrass criterion $(1.20)_3$ now takes the form

$$\eta(q - q^*) \geq 0$$

and it becomes possible to realize two regimes: the maximum thrust regime,

$$q = q_{max} \quad \text{for} \quad \eta > 0 \qquad (1.34)_a$$

and the regime of motion along a Kepleristic arc,

$$q = 0 \quad \text{for} \quad \eta < 0 \qquad (1.34)_{aa}$$

According to the Weierstrasse condition, at the point at which the regimes shift the Lagrange multipliers and, consequently, η and $\dot{\eta}$, remains continuous. Thus,

$$\eta(t_*) = 0 \qquad (1.35)$$

where $\dot{\eta}(t_* - 0) > 0$ upon transition from a Kepleristic arc to the maximal thrust regime and $\dot{\eta}(t_* - 0) < 0$ when the opposite transition occurs. The case $\dot{\eta}(t_*) = 0$ remains in doubt because the sign of the second derivative $\ddot{\eta}(t_* \mp 0)$, which does not remain continuous, provides no basis for deciding what the next regime will be. None of the three possibilities are excluded: the previous regime continues if, during further solution, it is discovered that $\eta(t)$ does not change sign, transition from the maximal thrust regime to motion along a Kepleristic arc occurs, when $\eta(t)$ changes from positive to negative, or the opposite transition occurs, when $\eta(t)$ changes from negative to positive.

All of what we have said is based on the assumption that $\nu_3 = 0$; it is, however, not permissible to eliminate the possibility that $\nu_3 \neq 0$ in some time interval (t_*, t_{**}). Then

$$0 \leq q(t) \leq q_{max}, \quad \eta \equiv 0 \qquad (1.36)$$

In this "singular control regime" with variable thrust the consumption $q(t)$ is the desired functior of time, is determined by the presence of the second of Equations (1.36), and, in this case, we need not carry out any new integrations of the equations of motion to show that

$$q = \frac{m}{c}\left(\dot{\underline{v}} \cdot \underline{e} + \frac{\mu}{r^3}\underline{r} \cdot \underline{e}\right) \qquad (1.37)$$

Thus, when $\dot{\eta}(t_*) = 0$ a fourth possibility is not excluded: the desired solution includes an arc with a singular control regime if $\underline{r}$ and $\underline{v}$ can remain continuous. In the singular control regime $\dot{\lambda} = 0$ and $\lambda = \frac{m\lambda_m}{c}$ is a nonzero constant because the assumption that $\lambda = 0$ causes all of the Lagrange multipliers ($\underline{\lambda}$, $\dot{\underline{\lambda}}$, λ_m) to vanish is not permitted in solutions to the Mayer-Bolza problem. The constant λ is determined by means of the condition that the adjoint be continuous at t_*; it is also necessary to verify continuity of the vectors $\underline{\lambda}$, $\dot{\underline{\lambda}}$, and the Hamiltonian H_λ (of course, $|\dot{\underline{\lambda}}| \neq \dot{\lambda} = 0$). Since $\lambda \neq 0$, this reduces to the requirement that the vectors $\underline{e}$ and $\dot{\underline{e}}$ be continuous.

The time $t_{**} - 0$ at which the singular regime ends remains unknown; this makes it necessary for $q(t)$ to leave the boundaries $q = 0$ or $q = q_{max}$, because q may not remain continuous. Apparently, t_{**} must be retained as a new unknown in the problem under discussion, and, in the last analysis, must be determined from the boundary conditions. See also 6.21.

SECTION 2. THE INTEGRALS OF THE SYSTEM OF EQUATIONS

2.1 Scalar integrals. The equations of motion, consumption, and stationarity admit a first integral expressing the constancy of the Hamiltonian

$$H_\lambda = h \qquad (2.1)$$

For thrust devices of the first type, this integral, by $(1.16)_1$, (1.22), and (1.27), can be written in the form

$$-\dot{\underline{\lambda}} \cdot \underline{v} - \underline{\lambda} \cdot \underline{r}\frac{\mu}{r^3} + w\lambda = h \qquad (2.2)_1$$

For thrust devices of the second type, we find that, when we account for $(1.16)_2$ and (1.29)

$$-\dot{\underline{\lambda}} \cdot \underline{v} - \underline{\lambda} \cdot \underline{r}\frac{\mu}{r^3} + \frac{1}{2}\lambda w = h \qquad (2.2)_2$$

Here the vector $\underline{\lambda}$ can be replaced by the vector $\underline{w}$

$$-\dot{\underline{w}} \cdot \underline{v} - \underline{w} \cdot \underline{r}\frac{\mu}{r^3} + \frac{1}{2}w^2 = h_1, \left(h_1 = \frac{N_{max}}{A}h\right)$$

which is proportional to it. Finally, by using regime criterion (1.32) we can show that, for thrust devices of the third type,

$$-\dot{\underline{\lambda}} \cdot \underline{v} - \underline{\lambda} \cdot \underline{r}\frac{\mu}{r^3} + \eta q = h \qquad (2.2)_3$$

where q is constant (q_{max} or 0) when $\eta \neq 0$, while $q = 0$ when $\eta \equiv 0$.

A check of the differentiation shows that (2.1) is actually the integral of the equations we have listed. The calculation is carried out analogously for all three cases, so we will limit discussion to the first; as a preliminary, we should note the following relations:

$$\dot{\underline{\lambda}} = \dot{\lambda}\underline{e} + \lambda\dot{\underline{e}}, \quad \underline{e} \cdot \dot{\underline{e}} = 0, \quad \dot{\underline{\lambda}} \cdot \underline{w} = \dot{\lambda} w$$

The result of differentiating $(2.2)_1$ can therefore be represented in the form

$$(\ddot{\lambda} + \frac{\mu}{r^3}\lambda - \frac{3\mu}{r^5}\lambda\cdot r\, r)\cdot v + \dot{\lambda}\cdot(\dot{v} + \frac{\mu}{r^3}r - w) = 0 \quad (2.3)$$

and the vectors in parentheses vanish as a result of the equations of motion and stationarity equation (1.28). We will use formula (2.3) in 2.3.

2.2 Vector integrals. Under condition (1.27), the system of equations (1.2) and (1.28) also admits a vector integral. Multiplying the expressions in the left and right sides of (1.28) vectorially by r, and using (1.2) and (1.27), we find that

$$\ddot{\lambda}\times r + \lambda\times\frac{\mu}{r^3}r = \ddot{\lambda}\times r - \lambda\times\dot{v} = \frac{d}{dt}(\dot{\lambda}\times r - \lambda\times v) = 0$$

so that

$$\dot{\lambda}\times r - \lambda\times v = a = a\sigma \quad (2.4)$$

where a is a constant vector, $a = |a|$, and σ is a constant unit vector.

Henceforth, all constructions in this article will be based on the scalar and vector integrals. The existence of the first is well known, although the vector integral is mentioned in neither (1-5) of the later paper (10); the existence of this integral was noted in (11) for the case of plane motion, but it was not applied there.

2.3 Equivalence of systems of equations. Differentiating Equation (2.4) and using the equations of motion (1.2), we obtain the expression

$$(\ddot{\lambda} + \frac{\mu}{r^3}\lambda)\times r = 0 \quad (2.5)$$

It follows from this equation that the vectors in parentheses are colinear with r and can be represented in the form br, where b is a scalar. Substitution into (2.3) now yields

$$(b - \frac{3\mu}{r^5}\lambda\cdot r)\, r\cdot v = 0,$$

because, by (1.2), there is no second term. Thus,

$$r\cdot v \neq 0,\quad b = \frac{3\mu}{r^5}\lambda\cdot r,\quad \ddot{\lambda} + \frac{\mu}{r^3}\lambda = \frac{3\mu}{r^5}\lambda\cdot r\, r$$

Thus, when the system of equations of motion (1.2) is taken into account, the second-order equation (1.28) is equivalent to the two first-order equations (2.2) and (2.4); the latter contains four constants: h and the vector a. The case $r\cdot v = 0$ corresponds to motion in spherical orbits and will be the subject of a special discussion.

2.4 A first-order equation for the vector λ. Successively taking the scalar products of (2.4) and r, v, and k, we obtain, when (1.1) is taken into account, the equations

$$\lambda\cdot k = a\sigma\cdot r \quad (2.6)$$

$$\dot{\lambda}\cdot k = a\sigma\cdot v \quad (2.7)$$

$$(\dot{\lambda}\cdot r + \lambda\cdot v)\, r\cdot v - (\dot{\lambda}\cdot v\, r^2 + \lambda\cdot r\, v^2) = a\sigma\cdot k \quad (2.8)$$

where, in virtue of (1.3), (1.6), and (1.27), Equation (2.6) is the integral of (2.7).

When $r \cdot v \neq 0$, it is possible to use Equations (2.7), (2.8), and (2.2) to determine $\dot{\lambda}\cdot r$, $\dot{\lambda}\cdot v$, and $\dot{\lambda}\cdot k$ -- the covariant components of $\dot{\lambda}$ in the vector basis (r, v, k), and then to use them to compose an expression for this vector:

$$\dot{\lambda} = \frac{1}{k^2}(\dot{\lambda}\cdot r\, v\times k + \dot{\lambda}\cdot v\, k\times r + \dot{\lambda}\cdot k\, k)$$

This expression is the desired differential equation; after substitution for $\dot{\lambda}\cdot r$, $\dot{\lambda}\cdot v$, and $\dot{\lambda}\cdot k$ this equation reduces to the form

$$\dot{\lambda} = \frac{1}{r\cdot v}\Big[\lambda\cdot(v\times n)\, v\times n - \frac{\mu}{r^3}\lambda\cdot r\, r + a(\sigma\cdot n\, v\times n + \frac{r\cdot v}{k}\sigma\cdot v\, n) + r\vartheta\Big] \quad (2.9)$$

where

$$\vartheta = w\lambda - h \quad (2.10)_1$$

$$\vartheta = \frac{1}{2}w\lambda - h \quad (2.10)_2$$

$$\vartheta = nq - h \quad (2.10)_3$$

for the first, second, and third types of thrust devices, respectively.

2.5 Orbital axes. We define an orbital trihedron by the following triple of unit vectors: e_r and e in the instantaneous plane Π of the orbit, and n, its perpendicular,

$$e_r = \frac{r}{r},\quad e_\varphi = n\times e_r,\quad n = \frac{1}{k}r\times v \quad (2.11)$$

We denote the angular-velocity vector of the trihedron by ω,

$$\omega = e_r\omega_r + e_\varphi\omega_\varphi + n\,\omega_n$$

and it is required to determine its projections on the orbital axes. Using the well-known kinematic formulas

$$\dot{e}_r = \omega\times e_r = -n\omega_\varphi + e_\varphi\omega_n$$

$$\dot{e}_\varphi = \omega\times e_\varphi = -e_r\omega_n + n\omega_r$$

$$\dot{n} = \omega\times n = -e_\varphi\omega_r + e_r\omega_\varphi$$

we obtain the following expressions for the vectors v and k:

$$v = (re_r)^{\cdot} = \dot{r}e_r + r(-n\omega_\varphi + e_\varphi\omega_n)$$

$$k = r\times v = r^2(e_\varphi\omega_\varphi + n\omega_n)$$

But v is located in the plane Π and k is perpendicular to it: thus,

$$\omega_\varphi = 0,\quad \omega_n = \frac{k}{r^2},\quad v = v_r e_r + \frac{k}{r}e_\varphi,\quad v_r = \dot{r} \quad (2.12)$$

It remains to find ω_r; in order to do so, we substitute the expression that we have obtained for v into Equation (1.2):

$$\dot{v} = \ddot{r}e_r + \dot{r}\frac{k}{r^2}e_\varphi + (\frac{\dot{k}}{r} - \dot{r}\frac{k}{r^2})e_\varphi + \frac{k}{r}(-e_r\frac{k}{r^2} + n\omega_r) = -\frac{\mu}{r^2}e_r + we, \quad (2.13)$$

By α_1, α_2, and α_3 we denote the cosines of the angles between the thrust vector and the axes of the orbital trihedron:

$$\mathbf{e} = \alpha_1 \mathbf{e}_r + \alpha_2 \mathbf{e}_\varphi + \alpha_3 \mathbf{n} \tag{2.14}$$

where, of course,

$$\alpha_1^2 + \alpha_2^2 + \alpha_3^2 = 1 \tag{2.15}$$

Now, returning to (2.13), we obtain, first of all, the equations of motion in the instantaneous orbital plane,

$$\dot{r} = v_r, \quad \dot{v}_r = \frac{k^2}{r^3} - \frac{\mu}{r^2} + w\alpha_1, \quad \dot{k} = r w \alpha_2 \quad (w = cg/m) \tag{2.16}$$

and, second of all, an expression for the vector $\boldsymbol{\omega}$:

$$\boldsymbol{\omega} = \frac{k}{r^2}\mathbf{n} + w\frac{r}{k}\alpha_3 \mathbf{e}_r \tag{2.17}$$

The differential equations of rotation for the orbital trihedron take the form

$$\dot{\mathbf{e}}_r = \frac{k}{r^2}\mathbf{e}_\varphi, \quad \dot{\mathbf{e}}_\varphi = -\frac{k}{r^2}\mathbf{e}_r + w\frac{r}{k}\alpha_3\mathbf{n}, \quad \dot{\mathbf{n}} = -w\frac{r}{k}\alpha_3\mathbf{e}_\varphi \tag{2.18}$$

We denote the vector whose projections on the axes of the orbital trihedron are equal to the projections of $\mathbf{e}$ on these axes by $\mathbf{e}^*$. Then

$$\dot{\mathbf{e}} = \mathbf{e}^* + \boldsymbol{\omega}\times\mathbf{e} = \left(\dot{\alpha}_1 - \frac{k}{r^2}\alpha_2\right)\mathbf{e}_r + \left(\dot{\alpha}_2 + \frac{k}{r^2}\alpha_1 - \frac{r}{k}w\alpha_3^2\right)\mathbf{e}_\varphi + \left(\dot{\alpha}_3 + \frac{r}{k}w\alpha_2\alpha_3\right)\mathbf{n} \tag{2.19}$$

We now consider one more vector, $\boldsymbol{\nu}$, the angular velocity of the vector $\mathbf{e}$; of course, we can set $\boldsymbol{\nu}\cdot\mathbf{e} = 0$. It is related to the vectors $\boldsymbol{\omega}$ and $\mathbf{e}$ by the expressions

$$\dot{\mathbf{e}} = \boldsymbol{\nu}\times\mathbf{e} = \mathbf{e}^* + \boldsymbol{\omega}\times\mathbf{e}, \quad \boldsymbol{\nu} = \mathbf{e}\times(\boldsymbol{\nu}\times\mathbf{e}) = \boldsymbol{\omega} + \mathbf{e}\times\mathbf{e}^* - \boldsymbol{\omega}\cdot\mathbf{e}\,\mathbf{e} \tag{2.20}$$

The vector $\boldsymbol{\sigma}$ in the expression for the vector integral has constant direction, so

$$\dot{\boldsymbol{\sigma}} = \boldsymbol{\sigma}^* + \boldsymbol{\omega}\times\boldsymbol{\sigma} = 0$$

and, if we denote the projections of this vector on the axes of the orbital trihedron by σ_s, we obtain the differential equations

$$\dot{\sigma}_1 = \frac{k}{r^2}\sigma_2, \quad \dot{\sigma}_2 = -\frac{k}{r^2}\sigma_1 + \frac{r}{k}w\alpha_3\sigma_3, \quad \dot{\sigma}_3 = -\frac{r}{k}w\alpha_3\sigma_2 \tag{2.21}$$

which admit the obvious integral

$$\sigma_1^2 + \sigma_2^2 + \sigma_3^2 = 1 \tag{2.22}$$

if we assume that the initial values σ_s^0 also satisfy the relation

$$\sigma_1^{0\,2} + \sigma_2^{0\,2} + \sigma_3^{0\,2} = 1 \tag{2.23}$$

For the axes of constant direction we take the orbital axes at the starting time: $\mathbf{e}_r^0$, $\mathbf{e}_\varphi^0$, $\mathbf{n}^0$. The orbital trihedron may be oriented relative to these axes by means of the three Euler angles, Ω, i, u, the longitude of the ascending node, the inclination of the instantaneous orbital plane Π to the plane Π^0 (the vectors $\mathbf{e}_r^0$ and $\mathbf{e}_\varphi^0$), and the angle in the plane Π between the directions to the ascending node (the unit vector along the line of nodes is denoted by $\mathbf{m}$) and the radius vector $\mathbf{r}$. The angular-velocity vector may be represented in terms of the derivatives of the Euler angles by the relation

$$\boldsymbol{\omega} = \dot{\Omega}\mathbf{n}^0 + \frac{di}{dt}\mathbf{m} + \dot{u}\mathbf{n} \tag{2.24}$$

and comparison with (2.17) leads to the following system of differential equations for the Euler angles:

$$\frac{di}{dt} = \frac{r}{k}w\alpha_3\cos u, \quad \dot{\Omega} = \frac{r}{k}w\alpha_3\frac{\sin u}{\sin i}, \quad \dot{u} = \frac{k}{r^2} - \frac{r}{k}w\alpha_3\sin u\,\mathrm{ctg}\,i \tag{2.25}$$

Integration of this system does not lead to any new constants, so, by hypothesis, the initial values of the Euler angles are equal to zero.

These equations take a more symmetric form when Cayley-Klein parameters α and β are introduced:

$$\dot{\alpha} = \frac{i}{2}\left(\frac{k}{r^2}\alpha + w\frac{r}{k}\alpha_3\beta\right), \quad \dot{\beta} = \frac{i}{2}\left(-\frac{k}{r^2}\beta + w\frac{r}{k}\alpha_3\alpha\right) \tag{2.26}$$

where*

$$\alpha = \cos\frac{i}{2}\,e^{\frac{i}{2}(\Omega+u)}, \quad \beta = i\sin\frac{i}{2}\,e^{\frac{i}{2}(\Omega-u)} \tag{2.27}$$

*There should be no confusion between $i = \sqrt{-1}$ and the angle i, which appears only as an argument of a trigonometric function.

Equation (2.26) integrates under the initial conditions $\alpha = 1$, $\beta = 0$. We should also note that the projections σ_s of the vector $\boldsymbol{\sigma}$ can be expressed in terms of their initial values and the Cayley-Klein parameters by means of the formulas

$$\begin{aligned}\sigma_1 + i\sigma_2 &= \bar{\alpha}^2(\sigma_1^0 + i\sigma_2^0) - \beta^2(\sigma_1^0 - i\sigma_2^0) + 2\bar{\alpha}\beta\sigma_3^0\\ \sigma_1 - i\sigma_2 &= -\bar{\beta}^2(\sigma_1^0 + i\sigma_2^0) + \alpha^2(\sigma_1^0 - i\sigma_2^0) + 2\alpha\bar{\beta}\sigma_3^0\\ \sigma_3 &= -\bar{\alpha}\bar{\beta}(\sigma_1^0 + i\sigma_2^0) - \alpha\beta(\sigma_1^0 - i\sigma_2^0) + (\alpha\bar{\alpha} - \beta\bar{\beta})\sigma_3^0\end{aligned} \tag{2.28}$$

where the bars denote complex conjugates.

2.6 Differential equations in orbital axes. Using formulas (1.27) and (2.19), we can replace differential equation (2.9) by the system

$$\frac{\dot{\lambda}}{\lambda} = \frac{1}{\lambda}\dot{\boldsymbol{\lambda}}\cdot\mathbf{e} = \frac{1}{\mathbf{r}\cdot\mathbf{v}}\left\{|\mathbf{e}\cdot(\mathbf{v}\times\mathbf{n})|^2 - \frac{\mu}{r}\alpha_1^2 + \frac{a}{\lambda}\left[\sigma_3\,\mathbf{e}\cdot(\mathbf{v}\times\mathbf{n}) + \frac{\mathbf{r}\cdot\mathbf{v}}{k}\boldsymbol{\sigma}\cdot\mathbf{v}\,\alpha_3\right] + r\alpha_1\frac{g}{\lambda}\right\} \tag{2.29}$$

$$\begin{aligned}\mathbf{e}^* + \boldsymbol{\omega}\times\mathbf{e} = \frac{1}{\mathbf{r}\cdot\mathbf{v}}\Big\{&\mathbf{e}\cdot(\mathbf{v}\times\mathbf{n})\,\mathbf{e}\times[(\mathbf{v}\times\mathbf{n})\times\mathbf{e}] - \frac{\mu}{r^2}\alpha_1\mathbf{e}\times(\mathbf{r}\times\mathbf{e}) +\\ &+ \frac{a}{\lambda}\left[\sigma_3\mathbf{e}\times\{(\mathbf{v}\times\mathbf{n})\times\mathbf{e}\} + \frac{\mathbf{r}\cdot\mathbf{v}}{k}\boldsymbol{\sigma}\cdot\mathbf{v}\,\mathbf{e}\times(\mathbf{n}\times\mathbf{e})\right] +\\ &+ \mathbf{e}\times(\mathbf{r}\times\mathbf{e})\frac{g}{\lambda}\Big\}\end{aligned} \tag{2.30}$$

where, by (2.6),

$$\lambda = \frac{ar\sigma_1}{k\alpha_3} \tag{2.31}$$

when $\alpha_3 = 0$, which, as we will see below, corresponds to the case of plane motion, this relation drops out (we then also have $\sigma_1 = 0$, $\sigma_2 = 0$).

SECTION 3. BOUNDARY CONDITIONS. TYPES OF MOTION.

3.1 The general boundary-value problem. The position and velocity of a space vehicle are given at time t_1, i.e., the vectors $\underline{r}' = \underline{r}'_*$, $\underline{v}' = \underline{v}'_*$. Known methods can be applied to use them for determination of r', v_r', k, and the position of the orbital trihedron -- the vectors $\underline{e}_r'$, $\underline{e}_\varphi'$, and $\underline{n}^1$ (or the Euler angles, or the Cayley-Klein parameters, the Hamilton-Rodrigues function, etc.). The term θ_1 in master function (1.18) is representable in the form

$$\theta_1 = \underline{\rho}_1 \cdot (\underline{r}' - \underline{r}'_*) + \underline{\rho}_2 \cdot (\underline{v}' - \underline{v}'_*) + \rho_t (t_1 - t_1^*) + \rho_m (m' - m'_*) \tag{3.1}$$

where $\underline{\rho}_1$ and $\underline{\rho}_2$ are vector Lagrange multipliers. Since the Lagrange multipliers and the boundary value of the Hamiltonian H^1 are determined by the relations

$$\underline{\lambda}' = -\mathrm{grad}_{\underline{v}'}\theta_1, \quad \dot{\underline{\lambda}}' = \mathrm{grad}_{\underline{r}'}\theta_1, \quad \lambda_m' = -\frac{\partial\theta}{\partial m'}, \quad H_\lambda' = \frac{\partial\theta}{\partial t_1} = h \tag{3.2}$$

we have $\underline{\lambda}' = -\underline{\rho}_2$, $\dot{\underline{\lambda}}' = \underline{\rho}_1$, and no information can be obtained about the boundary values of $\underline{\lambda}$ and $\dot{\underline{\lambda}}$. When (1.10) is given, the functional to be minimized, ρ_m, is equal to 0 and, therefore,

$$\lambda_m' = 1, \quad h = \rho_t \; ; \tag{3.3}$$

i.e., h remains unknown if t_1 is fixed, and equal to zero when it is not given.

If, however, the time is minimized -- the functional is determined with (1.11) -- then $\rho_t = 0$ and

$$h = 1, \quad \lambda_m' = \rho_m$$

i.e., λ^1_m is unknown when m^1 is given, and otherwise equal to zero.

3.2 Rotation of the orbital plane. The unit vector of the normal $\underline{n}' = \underline{n}'_*$ to the instantaneous plane Π' at time t_1 is given, so that

$$\theta_1 = \underline{\rho} \cdot (\underline{n}' - \underline{n}'_*) = \underline{\rho}\left(\frac{\underline{r}' \times \underline{v}'}{|r' \times v'|} - \underline{n}'_*\right)$$

By (3.2),

$$\dot{\underline{\lambda}}' = \frac{1}{|\underline{r}' \times \underline{v}'|}\,\mathrm{grad}_{\underline{r}'}\,\underline{r}' \cdot (\underline{v}' \times \underline{\rho}) - \frac{\underline{\rho} \cdot (\underline{r}' \times \underline{v}')}{|\underline{r}' \times \underline{v}'|^2}\,\mathrm{grad}\,|r' \times v'| =$$

$$= \frac{1}{|\underline{r}' \times \underline{v}'|^3}\left[\underline{v}' \times \underline{\rho}\,|\underline{r}' \times \underline{v}'|^2 - \underline{\rho} \cdot (\underline{r}' \times \underline{v}')\,\underline{v}' \times (\underline{r}' \times \underline{v}')\right]$$

and, setting $\underline{\rho}^* = \frac{1}{k'}\underline{\rho}$, we find that

$$\dot{\lambda}' = \underline{v}' \times \underline{\rho}^* - \underline{\rho}^* \cdot \underline{n}'\,\underline{v}' \times \underline{n}' = \underline{v}' \times [\underline{n}' \times (\underline{\rho}^* \times \underline{n}')]$$

We can compute $\dot{\underline{\lambda}}$ similarly. We obtain

$$\dot{\underline{\lambda}}' = \underline{v}' \times \underline{b}, \quad \underline{\lambda}_1 = \underline{r}' \times \underline{b}, \quad \underline{b} = \underline{n}' \times (\underline{\rho}^* \times \underline{n}') \tag{3.4}$$

where b is a vector in the plane Π' , because $\underline{b} \cdot \underline{n}' = 0$. Each of the vectors $\dot{\underline{\lambda}}^1$ and $\underline{\lambda}^1$ is perpendicular to two vectors ($\underline{v}^1$ and $\underline{b}$ or $\underline{r}^1$ and b) located in the plane Π' ; thus, $\underline{\lambda}^1$ and $\dot{\underline{\lambda}}^1$ (and, consequently, $\underline{e}^1$ and $\dot{\underline{e}}^1$) are colinear with $\underline{n}^1$, so that

$$\alpha_1' = 0, \quad \alpha_2' = 0, \quad \alpha_3' = \varepsilon = \pm 1 \; ; \tag{3.5}$$

and by (2.19) and (2.15),

$$\dot{\alpha}_1' = 0, \quad \dot{\alpha}_2' = \frac{r'}{k'}w', \quad \dot{\alpha}_3' = 0.$$

Returning to vector integral (2.4), we find that

$$a\underline{\sigma} = (\underline{v}' \times \underline{b}) \times \underline{r}' - (\underline{r}' \times \underline{b}) \times \underline{v}' = \underline{b} \times (\underline{r}' \times \underline{v}') = k\,\underline{b} \times \underline{n}' \tag{3.6}$$

so that since $a \neq 0$ when the motion is not plane, a situation that we will discuss below), the vector $\underline{\sigma}$ is located in the plane Π' and is perpendicular to $\underline{b}$. Thus,

$$\underline{\sigma} \cdot \underline{n}' = 0, \quad \sigma_3' = 0, \quad \sigma_1' = \sin\mu, \quad \sigma_2' = \cos\mu \tag{3.7}$$

where the angle μ is defined by the relation $\underline{b} = b(\underline{e}_r' \cos\mu + \underline{e}_\varphi' \sin\mu)$. If we now project $\underline{\sigma}$ onto the initial axes, we obtain the relations

$$\begin{aligned} \sigma_1^0 &= \cos\Omega' \sin(\mu + u') + \cos i' \sin\Omega' \cos(\mu + u') \\ \sigma_2^0 &= \sin\Omega' \sin(\mu + u') - \cos i' \cos\Omega' \cos(\mu + u') \\ \sigma_3^0 &= -\sin i' \cos(\mu + u') \end{aligned} \tag{3.8}$$

where Ω^1 and i^1 are given angles determining $\underline{n}^1$ in the initial system of orbital axes.

When we take (3.4) into account, it follows from (2.9) and (2.31) that we also have

$$\frac{\lambda'}{a} = \varepsilon\frac{r'}{k'}\sin\mu, \quad \vartheta' = 0 \tag{3.9}$$

whence, in particular, it follows that the signs of ε and $\sin\mu$ are the same.

3.3 Three-dimensional motions. When (2.31) is taken into account, the developed form of Equation (2.29) and (2.30) is

$$\frac{\dot{\lambda}}{\lambda} = \frac{1}{v_1}\left(\frac{k^2}{r^3} - \frac{\mu}{r^2}\right)\alpha_1^2 - 2\frac{k}{r^2}\alpha_1\alpha_2 + \frac{v_r}{r}(1 - \alpha_1^2) + \frac{k\alpha_3}{r^2\sigma_1}\left[\sigma_2\alpha_3 + \sigma_3\left(-\alpha_2 + \frac{k\alpha_1}{rv_r}\right)\right] + \frac{\alpha_1}{v_r}\vartheta\frac{k\alpha_3}{ar\sigma_1} \tag{3.10}$$

$$\dot{\alpha}_1 = \frac{1}{v_r}\left(\frac{k^2}{r^3} - \frac{\mu}{r^2}\right)\alpha_1(1 - \alpha_1^2) + 2\frac{k}{r^2}\alpha_1^2\alpha_2 - \frac{v_r}{r}(1 - \alpha_1^2)\alpha_1 + \frac{k\alpha_3}{r^2\sigma_1}\left[\alpha_1(\alpha_2\sigma_3 - \alpha_3\sigma_2) + \frac{k\sigma_3}{rv_r}(1 - \alpha_1^2)\right] + \frac{1 - \alpha_1^2}{v_r}\vartheta\frac{k\alpha_3}{ar\sigma_1} \tag{3.11}$$

$$\dot{\alpha}_2 = -\frac{1}{v_r}\left(\frac{k^2}{r^3} - \frac{\mu}{r^2}\right)\alpha_1^2\alpha_2 - 2\frac{k}{r^2}\alpha_1(1 - \alpha_2^2) + \frac{v_r}{r}\alpha_1^2\alpha_2 - \frac{k\alpha_3}{r^2\sigma_1}\left[\alpha_2\alpha_3\sigma_2 + \sigma_3(1 - \alpha_2^2) - \sigma_3\frac{k}{rv_r}\alpha_1\alpha_2\right] - \frac{\alpha_1\alpha_2}{v_r}\vartheta\frac{k\alpha_3}{ar\sigma_1} + \frac{r}{k}w\alpha_3^2 \tag{3.12}$$

$$\dot{\alpha}_3 = -\frac{1}{v_r}\left(\frac{k^2}{r^3}-\frac{\mu}{r^2}\right)\alpha_1^2\alpha_3 + 2\frac{k}{r^2}\alpha_1\alpha_2\alpha_3 + \frac{v_r}{r}\alpha_1^2\alpha_3 + \tag{3.13}$$
$$+ \frac{k\alpha_3}{r^2\sigma_1}\left[\sigma_2(1-\alpha_3^2)+\sigma_3\alpha_2\alpha_3 - \sigma_3\frac{k}{rv_r}\alpha_1\alpha_3\right] - \frac{\alpha_1\alpha_3}{v_r}\vartheta\frac{k\alpha_3}{ar\sigma_1} -$$
$$-\frac{r}{k}w\alpha_2\alpha_3 .$$

Differential equation (3.10) is a corollary of (2.31) and Equations (3.11) - (3.13) and (2.21); it is used in writing Equation (1.33) for the regime criteria.

In the most difficult case, the case of thrust devices of the third kind, the unknowns are

$$r,\ v,\ k,\ m,\ \alpha_1,\ \alpha_2,\ \alpha_3,\ \sigma_1,\ \sigma_2,\ \sigma_3,\ \eta \tag{3.14}$$

a total of 11. In order to find them, we can use the three differential equations of motion (2.16), consumption equation (1.4), Equations (3.11) - (3.13), and Equations (2.26) (or (2.25)), which define the Cayley-Klein parameters (or their angles), which can be used, along with formulas (2.28), to represent σ_s, and Equation (1.33), the regime criterion. For fixed r^0, v^0, k^0, and m^0, the Cauchy integral contains ten constants α_s^0, σ_s^0, η^0, k, and a; eight of them are independent, since α^0_s and σ_s^0 must satisfy relations (2.15) and (2.22).

In the general boundary-value problem, for example, when t_1 is minimized and the final mass is given, the constant h = 1, but t_1 is among the constants; seven of the quantities must be determined from the same number of conditions, which state that r^1, v^1, k^1, m^1 and the Euler angles Ω^1, i^1, and u^1 take preassigned values. If, however, the mass is not given, $\lambda_m^1 = 0$, and the seventh condition is given by the relation

$$\eta^1 = \frac{c}{m^1}\ \frac{ar\sigma_1^1}{k^1\alpha_3^1}$$

which is a consequence of (1.32) and (2.31).

Concerning integration of the equations on orbital sections following Kepleristic arcs and the conjugates of solutions at points at which regimes change, see (3.6).

In cases concerning thrust devices of the first type (w = const), the consumption equation and the regime criteria drop out and the time t_1 is minimized. Six unknown constants (α_s^0, σ_s^0, a, and t_1) must be found. In the problem of rotation in a plane orbit, the number of constants may be reduced to five by expressing σ_s^0 by means of (3.8) in terms of the one constant μ. The five equations required for determining these constants are: two of the three of Equations (3.5), two equations that state that Ω^1 and i^1 take given values, and, finally, the relationship

$$wa\sin\mu = \varepsilon\frac{k^1}{r^1}$$

which is obtained by means of (3.9) and $(2.10)_1$.

3.4 <u>Plane motion.</u> In motion in plane orbits, $\underline{n}$ is a constant vector, $\dot{\underline{n}} = 0$, and (by (2.18)), $\alpha_3 = 0$ -- the thrust is in the orbital plane; moreover, this plane contains the vector $\underline{\lambda}$, since, by (2.6) and (2.21), the vectors $\underline{\sigma}$ and $\underline{n}$ are colinear: $\sigma_1 = 0$, $\sigma_2 = 0$, $\sigma_3 = \pm 1$, and $\underline{a} = \tilde{a}\underline{n}$, where $\tilde{a} = \pm a$. Setting $\alpha_1 = \cos\psi$ and $\alpha_2 = \sin\psi$, where ψ is the angle between the vector $\underline{e}$ and the radius vector, we obtain two differential equations that now replace Equations (3.10) - (3.13); here we must keep in mind that (2.31) has dropped out, so that in these equations we must replace λ.

We find that

$$\frac{\dot{\lambda}}{\lambda} = -\frac{1}{v_r}\left(\frac{v_r^2}{r}+\frac{\mu}{r^2}-\frac{k^2}{r^3}\right)\cos^2\psi + \frac{v_r}{r} - 2\frac{k}{r^2}\sin\psi\cos\psi + \tag{3.15}$$
$$+ \frac{\tilde{a}}{\lambda r}\left(-\sin\psi + \frac{k}{rv_r}\cos\psi\right) + \frac{\cos\psi}{v_r}\frac{\vartheta}{\lambda}$$

$$\dot{\psi} = \frac{1}{v_r}\left(\frac{v_r^2}{r}+\frac{\mu}{r^2}-\frac{k^2}{r^3}\right)\sin\psi\cos\psi - 2\frac{k}{r^2}\cos^2\psi - \tag{3.16}$$
$$-\frac{\tilde{a}}{\lambda r}\left(\cos\psi + \frac{k}{rv_r}\sin\psi\right) - \frac{\sin\psi}{v_r}\frac{\vartheta}{\lambda}$$

Together with these equations, we must consider the equations of motion (2.16), the mass consumption equation (1.4), and the regime criterion (1.33). We obtain a seventh-order system and the general solution contains ten constants (including h and $\tilde{a}$). The order drops to five when thrust devices of the first type are considered. The problem becomes substantially simpler because it is not necessary to keep track of the sign of the regime criterion and there are no transitions from regime to regime.

The solution of the general boundary-value problem (the plane problem) for thrust devices of the third type reduces to determination of five constants (with fixed r^0, v^0, k^0, and m^0). Four equations provide the boundary conditions, which are determined by giving r^1, v^1, k^1, and the angle

$$\varphi^1 - \varphi^0 = \int_0^{t_1}\frac{k}{r^2}dt \tag{3.17}$$

the difference between the real anomalies. Let, for example, t_1 be given, and assume that it is required to minimize m^1; then the fifth equation takes the form

$$\eta^1 = \frac{c}{m^1}\lambda^1 - 1$$

3.5 <u>A special case of plane motion.</u> It occurs when $\sigma_s = 0$. Then the last equation of (2.21) implies that $\alpha_3 = 0$ or $\sigma_2 = 0$. In the first case, according to (3.4), we have $\sigma_1 = \sigma_2 = 0$, $\underline{a} = 0$, while in the second case, by (2.21), we have $\sigma_1 = 0$. In both cases a = 0, the constant $\tilde{a}$ drops out of Equations (3.15) - (3.16), and it is impossible to solve the general boundary-value problem. The problem is solvable when the scalar invariants $r^1, v^1, k^1, \vartheta^1$ are given at the right end instead of the vectors $\underline{r}^1$ and $\underline{v}^1$. Indeed, when the master function is of the form

$$\theta_1 = \theta_1(r^1, v^1; \underline{r}^1\cdot\underline{v}^1)$$

we have, by (3.2)

$$\dot{\lambda}^1 = \frac{\partial\theta_1}{\partial r^1}\frac{\underline{r}^1}{r^1} + \frac{\partial\theta_1}{\partial(\underline{r}^1\cdot\underline{v}^1)}\underline{v}^1, \quad \lambda^1 = -\frac{\partial\theta_1}{\partial v^1}\frac{\underline{v}^1}{v^1} - \frac{\partial\theta^1}{\partial(\underline{r}^1\cdot\underline{v}^1)}\underline{r}^1 ,$$

and substitution into the expression for vector integral (2.4) yields $\underline{a} = 0$.

Differential equations (3.15) reduce to the form

$$\frac{\dot\lambda}{\lambda} = \frac{v_r}{r} - \left(\dot\psi + \frac{2k}{r^2}\right)\operatorname{ctg}\psi, \quad \frac{\vartheta}{\lambda} = \left(\frac{v_r^2}{r} + \frac{\mu}{r^2} - \frac{k^2}{r^3}\right)\cos\psi - \frac{v_r}{\sin\psi}\left(\dot\psi + \frac{2k}{r^2}\cos^2\psi\right) \tag{3.18}$$

Elimination of λ from them proceeds in the same manner for each value of ϑ. We have

$$\left(\frac{\vartheta}{\lambda}\right)^{\cdot} = \frac{\dot\vartheta}{\lambda} - \frac{\vartheta}{\lambda}\frac{\dot\lambda}{\lambda} = \frac{\eta q}{\lambda} - \frac{\vartheta}{\lambda}\frac{\dot\lambda}{\lambda} = \left(\frac{cq}{m} - \frac{\vartheta}{\lambda}\right)\frac{\dot\lambda}{\lambda} = \left(w - \frac{\vartheta}{\lambda}\right)\frac{\dot\lambda}{\lambda}$$

Here ϑ is taken in the form $(2.10)_3$. Similarly, for the forms $(2.10)_1$ and $(2.10)_2$ we have

$$\left(\frac{\vartheta}{\lambda}\right)^{\cdot} = \frac{h}{\lambda}\frac{\dot\lambda}{\lambda} = \left(w - \frac{\vartheta}{\lambda}\right)\frac{\dot\lambda}{\lambda}$$

$$\left(\frac{\vartheta}{\lambda}\right)^{\cdot} = \left(\frac{1}{2}\frac{A}{N_{max}}\lambda - \frac{h}{\lambda}\right)^{\cdot} = \left(\frac{1}{2}\frac{A}{N_{max}}\lambda + \frac{h}{\lambda}\right)\frac{\dot\lambda}{\lambda} = \left(w - \frac{\vartheta}{\lambda}\right)\frac{\dot\lambda}{\lambda}$$

After substitution of values of $\dot\lambda/\lambda$ and ϑ/λ into these expressions, we obtain a first-order equation for ψ when we retain v_r and return to the equations of motion:

$$\ddot\psi + 2\left(\dot\psi + \frac{k}{r^2}\right)\left(\dot\psi + 2\frac{k}{r^2}\right)\operatorname{ctg}\psi + 3\frac{\mu}{r^3}\cos\psi\sin\psi + 2\frac{v_r}{r}\dot\psi + \frac{w}{r}\sin\psi = 0 \tag{3.19}$$

Sometimes it is preferable to consider the criterion

$$\zeta = \frac{\eta}{\lambda} \tag{3.20}$$

instead of η. Instead of (1.33) we obtain the differential equation

$$\dot\zeta = \left(\frac{c}{m} - \zeta\right)\frac{\dot\lambda}{\lambda} = \left(\frac{c}{m} - \zeta\right)\left[\frac{v_r}{r} - \left(\dot\psi + \frac{2k}{r^2}\right)\operatorname{ctg}\psi\right] \tag{3.21}$$

It is necessary to return to Equations (3.19) when the initial value of the radial velocity component v_r is equal to zero (application of thrust at perigee or apogee).

Assume, for example, that the problem of minimizing time is under consideration, and that a functional relationship of the form $\varphi(r', v') = 0$ is given at the right end. Then

$$\theta = t_1 + \rho\varphi(r', v')$$

so that, by (3.2), (1.27), and (2.20),

$$\dot{\underline{\lambda}}_1 = \lambda_1(\underline{e}_r'\cos\psi' + \underline{e}_\varphi'\sin\psi') = -\rho\frac{\partial\varphi}{\partial v'}\frac{\underline{v}'}{v'} = -\rho\frac{\partial\varphi}{\partial v'}(v_r'e_r' + \frac{k'}{r'}e_\varphi')$$

$$\dot{\underline{\lambda}}' = (\dot\lambda'\cos\psi' - \lambda'\nu_3'\sin\psi')\underline{e}_r' + (\dot\lambda'\sin\psi' + \lambda'\nu_3'\cos\psi')\underline{e}_\varphi' = \rho\frac{\partial\varphi}{\partial r'}\underline{e}_r'$$

where $h = 1$, $\nu_3' = \dot\psi' + \frac{k'}{r'^2}$. After $\dot\lambda'/\lambda'$ and ρ/λ' are eliminated from the four equations thus obtained, we are led to the desired three boundary conditions:

$$\operatorname{ctg}\psi' = \frac{v_r'r'}{k'}, \quad \dot\psi' = -\frac{k'}{r'^2}\left(1 + \frac{d\ln v'}{d\ln r'}\right), \quad \varphi(r', v') = 0$$

For example, when

$$\varphi = v'^2 - \frac{\alpha\mu}{r'}, \quad \frac{d\ln v'}{d\ln r'} = -\frac{1}{2}, \quad \dot\psi' = -\frac{1}{2}\frac{k'}{r'^2} = -\frac{1}{2}\dot\varphi'$$

Here $\alpha = 1$ in the problem of obtaining the first cosmic speed, while $\alpha = 2$ for the second.

In the problem of obtaining a circular orbit

$$\theta_1 = \rho_1\left(v'^2 - \frac{\mu}{r'}\right) + \rho_2\underline{r}'\cdot\underline{v}'$$

and after elimination of ρ_1/λ, ρ_2/λ, and $\dot\lambda/\lambda$ from the four equations obtained, we obtain the relationships

$$\dot\psi' = -\frac{v'}{2r'}(1 + 3\cos\psi'), \quad v'^2 = \frac{\mu}{r'}, \quad v_r' = 0$$

3.6 Motion along a Kepleristic arc. For the vector $\underline{\lambda}$, differential equation (1.28) is an equation in the variations of the system of equations of motion; indeed,

$$\delta\dot{\underline{r}} = \delta\underline{v}, \quad \delta\dot{\underline{v}} = -\delta\frac{\mu}{r^3}\underline{r} = \frac{\mu}{r^3}\left(\frac{3}{r^2}\underline{r}\cdot\delta\underline{r}\,\underline{r} - \delta\underline{r}\right)$$

and it is sufficient to set $\delta\underline{r} = \underline{\lambda}$ and $\delta\underline{v} = \dot{\underline{\lambda}}$. But the solution of the equations of motion is well known; the derivatives of the radius vector $\underline{r}$ with respect to the six constants in it (p, ε, t_0, Ω, i, ω) yield a system of linearly independent particular solutions of the equation in variations. By composing certain expressions that are linear with respect to these derivatives, we obtain the system of vectors (12)

$$\underline{q}_1 = \left[\frac{r}{p}(1-\varepsilon^2) - \frac{3n(t-t_0)}{2\sqrt{1-\varepsilon^2}}\varepsilon\sin\varphi\right]\underline{e}_r - \frac{3n(t-t_0)}{2\sqrt{1-\varepsilon^2}}(1+\varepsilon\cos\varphi)\underline{e}_\varphi$$

$$\underline{q}_2 = -\cos\varphi\,\underline{e}_r + \frac{2+\varepsilon\cos\varphi}{1+\varepsilon\cos\varphi}\sin\varphi\,\underline{e}_\varphi$$

$$\underline{q}_3 = \sin\varphi\,\underline{e}_r + \frac{2+\varepsilon\cos\varphi}{1+\varepsilon\cos\varphi}\cos\varphi\,\underline{e}_\varphi$$

$$\underline{q}_4 = \frac{r(1-\varepsilon^2)}{p}\underline{e}_\varphi$$

$$\underline{q}_5 = \frac{r(1-\varepsilon^2)}{p}\cos\varphi\,\underline{n}, \quad \underline{q}_6 = \frac{r(1-\varepsilon^2)}{p}\sin\varphi\,\underline{n}$$

and their derivatives

$$\frac{1}{n}\dot{\underline{q}}_1 = -\left[\frac{\varepsilon\sin\varphi}{2\sqrt{1-\varepsilon^2}} - \frac{3}{2}n(t-t_0)\frac{p^2}{(1-\varepsilon^2)^2r^2}\right]\underline{e}_r - \frac{1+\varepsilon\cos\varphi}{2\sqrt{1-\varepsilon^2}}\underline{e}_\varphi,$$

$$\frac{1}{n}\dot{\underline{q}}_2 = \frac{p}{r(1-\varepsilon^2)^{3/2}}\left(-\sin\varphi\,\underline{e}_r + \frac{\varepsilon+\cos\varphi}{1+\varepsilon\cos\varphi}\underline{e}_\varphi\right),$$

$$\frac{1}{n}\dot{\underline{q}}_3 = -\frac{p}{r(1-\varepsilon^2)^{3/2}}\left(\cos\varphi\,\underline{e}_r + \frac{\sin\varphi}{1+\varepsilon\cos\varphi}\underline{e}_\varphi\right),$$

$$\frac{1}{n}\dot{\underline{q}}_4 = \frac{1}{\sqrt{1-\varepsilon^2}}\left[-(1+\varepsilon\cos\varphi)\underline{e}_r + \varepsilon\sin\varphi\,\underline{e}_\varphi\right],$$

$$\frac{1}{n}\dot{\underline{q}}_5 = -\frac{1}{\sqrt{1-\varepsilon^2}}\sin\varphi\,\underline{n}, \quad \frac{1}{n}\dot{\underline{q}}_6 = \frac{1}{\sqrt{1-\varepsilon^2}}(\cos\varphi + \varepsilon)\underline{n}$$

The real anomaly (computed from the perigee of the ellipse) is denoted by t_*, and the expressions for r, v_r, and $k = r^2\dot\varphi$ have the well-known form

$$r = \frac{p}{1+\varepsilon\cos\varphi}, \quad v_r = \frac{np\varepsilon}{(1-\varepsilon^2)^{3/2}}\sin\varphi, \quad k = \frac{np^2}{(1-\varepsilon^2)^{3/2}}, \quad \left(n = \sqrt{\frac{\mu}{p^3}}(1-\varepsilon^2)^{3/2}\right)$$

The Kepler constants of the orbit and the real anomaly at the time t_* of transition can be expressed in terms of r_*, v_{r*}, and k_* by means of the formulas

$$p = \frac{k_*^2}{\mu}, \quad 1 - \frac{k_*^2}{r_*\mu} = \varepsilon\cos\varphi_*, \quad \frac{v_{r*}k_*}{\mu} = \varepsilon\sin\varphi_*$$

so that

$$\varepsilon = \sqrt{1 - 2\frac{k_*^2}{\mu^2}\left(\frac{\mu}{r_*} - \frac{v_*^2}{2}\right)} = \sqrt{1 + 2h_1\frac{k_*^2}{\mu^2}}$$

where h_1 is the energy constant of the Kepleristic orbit ($h_1 < 0$); the time t_0 of passage through perigee can be determined by means of well-known formulas or tables of solutions of the Kepler equation.

The solution of differential equation (1.28) can now be written in the form

$$\underline{\lambda} = \sum_{k=1}^{6} C_k \underline{q}_k, \quad \dot{\underline{\lambda}} = \sum_{k=1}^{6} C_k \dot{\underline{q}}_k \qquad (3.22)$$

and, in accordance with the Erdmann-Weierstrass conditions, the constants C_k must be subject to the requirements of continuous conjugate values of the vectors $\underline{\lambda}$ and $\dot{\underline{\lambda}}$ and the Hamiltonian H_λ at the point of discontinuity of control t_* (at it $\eta = 0$).

By $(2.2)_3$, on a Kepleristic arc (w = 0)

$$\sum_{k=1}^{6} C_k\left[\dot{\underline{q}}_k\cdot\left(\underline{e}_r v_r + \underline{e}_\varphi \frac{k}{r}\right) + \frac{n^2p^3}{(1-\varepsilon^2)^3 r^2}\underline{q}_k\cdot\underline{e}_r\right] = -h$$

Only the first term proves to be nonzero; we find that

$$C_1 = -2h\frac{1-\varepsilon^2}{n^2 p} = -\frac{h}{h_1}\frac{\mu}{2h_1} = -\frac{h}{h_1}\frac{p}{1-\varepsilon^2} \qquad (3.23)$$

If we carry out an analogous computation with vector integral (2.4), we find that

$$\frac{1}{2}(-3C_1 + \varepsilon C_2)\frac{np}{\sqrt{1-\varepsilon^2}}\underline{n} + \frac{np}{\sqrt{1-\varepsilon^2}}(C_5\underline{i}_1 + C_6\underline{i}_2) = a\underline{\sigma}$$

where $\underline{i}_1$ is a unit vector directed toward the perigee, $\underline{i}_2 = \underline{n}\times\underline{i}_1$. Thus,

$$a\sigma_3 = \frac{1}{2}(-3C_1+\varepsilon C_2)\frac{np}{\sqrt{1-\varepsilon^2}} \qquad (3.24)$$

$$a\sigma_1 = (C_5\cos\varphi + C_6\sin\varphi)\frac{np}{\sqrt{1-\varepsilon^2}}, \quad a\sigma_2 = (-C_5\sin\varphi + C_6\cos\varphi)\frac{np}{\sqrt{1-\varepsilon^2}} \qquad (3.25)$$

Naturally, σ_3 and the projections of $\underline{\sigma}$ onto the constant directions $\underline{i}_1$ and $\underline{i}_2$ are constant; the values of σ_1 and σ_2 in the text can be derived by starting with Equations (2.21) and introducing the independent variable φ ($d\varphi = k\,dt/r^2$).

The coefficients C_3 and C_4 dropped out during our discussion of the scalar and vector integrals. In order to find these coefficients, we can use the first formula of (3.22), from which we obtain the equations

$$\sum_{k=1}^{4} C_k \underline{q}_k\cdot\underline{e}_r = \lambda\alpha_1, \quad \sum_{k=1}^{4} C_k\dot{\underline{q}}_k\cdot\underline{e}_\varphi = \lambda\alpha_2,$$
$$C_5\underline{q}_5\cdot\underline{n} + C_6\underline{q}_6\cdot\underline{n} = \lambda\alpha_3 \qquad (3.26)$$

The last equation can be rewritten in the form

$$\lambda\alpha_3 = \frac{r}{p}(1-\varepsilon^2)(C_5\cos\varphi + C_6\sin\varphi) = \frac{r}{k}a\sigma_1$$

and leads to Equation (2.31), which we have used often above.

The law for changing the direction of the device used to produce thrust is given by formulas of the form

$$\alpha_1\sum_{k=1}^{4} C_k\underline{q}_k^*\cdot\underline{e}_r^* = \alpha_1^*\sum_{k=1}^{4} C_k\underline{q}_k\cdot\underline{e}_r, \quad \text{etc.}$$

The generally cumbersome formulas defining the coefficients C_k in terms of the quantities $a\sigma_k$, α_k, r, k, and v_r, at time t_* need not be introduced here; they can be computed without difficulty from equations we have cited. A considerable simplification ($C_1 = 0$) will occur when h = 0 -- in the problem of minimizing the total fuel consumption with t_1 not fixed.

Since the mass m on a Kepleristic arc is constant, it immediately follows from (1.33) that

$$\eta = \frac{c}{m_*}(\lambda - \lambda_*) \qquad (3.27)$$

A large part of (3) is devoted to determing the vector $\underline{\lambda}$ (in the problem of plane motion). It follows from what we have said above that determination of $\underline{\lambda}$ does not require integration.

3.61 The vector $\underline{\lambda}$ in the special case of plane motion with h = 0. It immediately follows from Equations (3.24) and (3.25) (since a = 0) that $C_1 = C_2 = C_5 = C_6$; the constants C_3 and C_4 can be found with Equations (3.26). They take the form

$$C_3\sin\varphi = \lambda\cos\varphi, \quad C_3\frac{2+\varepsilon\cos\varphi}{1+\varepsilon\cos\varphi} + C_4\frac{1-\varepsilon^2}{1+\varepsilon\cos\varphi} = \lambda\sin\varphi$$

After r and v_r have been used to replace cos φ and sin φ, we obtain the formulas

$$\operatorname{tg}\psi = \frac{r}{v_r}\left(\frac{k_*}{r^2} + \mathcal{B}\right), \quad \mathcal{B} = \sqrt{\frac{\mu}{p^3}}\left[-1+\varepsilon\sqrt{1-\varepsilon^2}\,\frac{C_4}{C_3}\right].$$

It is also possible to obtain these formulas by integrating Equation (3.16) with $\tilde{a} = 0$, $\vartheta = \eta\xi = 0$. The expressions for λ and the criterion η can also be composed without difficulty. Consideration of the criterion ζ leads to a more compact formula. Differential equation (3.21) can be written in the form

$$-\left[\ln\left(\frac{c}{m_*} - \zeta\right)\right]^{\cdot} = \left(\ln\frac{r}{\sin\psi}\right)^{\cdot} - \frac{2k}{r^2}\operatorname{ctg}\psi = \left(\ln\frac{r}{\sin\psi}\right)^{\cdot} - \left[\ln\frac{r^2}{k\mathcal{B}+r^2}\right]^{\cdot}$$

and after the constants of integration are determined, we can use the expression for tg ψ to obtain the expression

$$\zeta = \frac{c}{m_*}\left(1 - \frac{v_r}{v_r^*}\frac{\cos\psi_*}{\cos\psi}\right)$$

SECTION 4. ORBITS ON THE SURFACE OF A SPHERE.

4.1 Statements of the problem of spherical motion. The coordinate $\underline{r}$ of a motion along a spherical surface must satisfy the condition

$$\underline{r}\cdot\underline{r} = r^2 = \text{const} \qquad (4.1)$$

The additional term

$$\frac{1}{2}\int_0^{t_1}\nu(\underline{r}\cdot\underline{r} - r^2)\,dt$$

must now be introduced into functional (1.19), where ν is some Lagrange multiplier. This causes the term $\nu \underline{r}$ to enter stationarity condition (1.23), and Equation (1.28) must now take the form

$$\ddot{\underline{\lambda}} = \frac{\mu}{r^3}\left(3\frac{\underline{\lambda}\cdot\underline{r}}{r^2}\underline{r} - \underline{\lambda}\right) - \nu \underline{r} \qquad (4.2)$$

All of the remaining conditions remain the same. The existence of the term $\nu \underline{r}$, which is colinear with $\underline{r}$, in (4.2) does not, however, have any influence on the derivation of the vector integral of 2.2 or, of course, the scalar integral H_λ does not change). But a consequence of the vector integral and the equations of motion will be that the vectors $\ddot{\underline{\lambda}} + \frac{\mu}{r^3}\underline{\lambda}$ and $\underline{r}$, which we discussed in 2.3, are colinear. Thus, by using integrals (2.2) and (2.4) (in conjunction, of course, with the equations of motion, mass consumption, and the regime criterion) does not affect Equation (4.2); it can be satisfied by appropriate selection of ν. It would be an error to consider spherical motion on the basis of differential equation (1.28).

4.2 The differential equations of spherical motion. By hypothesis,

$$\underline{r}\cdot\underline{r} = r^2, \quad \underline{r}\cdot\underline{v} = 0, \quad v_r = 0 \qquad (4.3)$$

The equations of motion (2.16) can be written in the form

$$w\alpha_1 = \frac{\mu}{r^2} - \frac{k^2}{r^3}, \quad \dot{k} = rw\alpha_2, \quad \left(w = \frac{cq}{m}\right) \qquad (4.4)$$

We now return to Equations (3.10) - (3.13). If, in these equations, we retain only terms containing v_r in the numerator, we obtain the same relation from all four equations:

$$\left(\frac{k^2}{r^3} - \frac{\mu}{r^2}\right)\alpha_1 + \frac{k^2}{r^3}\frac{\sigma_3}{\sigma_1}\alpha_3 + \vartheta\frac{k\alpha_3}{ar\sigma_1} = 0 \qquad (4.5)$$

It is possible to use (3.12) and (3.13) to obtain a relation that does not contain

$$\dot{\alpha}_2 - \frac{\dot{\alpha}_3}{\alpha_3}\alpha_2 = -2\frac{k}{r^2}\alpha_1 - \frac{k}{r^2\sigma_1}(\sigma_2\alpha_2 + \sigma_3\alpha_3) + w\frac{r}{k}(1-\alpha_1^2) \qquad (4.6)$$

As we showed in (3.5), ϑ satisfies the differential equation

$$\left(\frac{\vartheta}{\lambda}\right)^{\cdot} = \left(w - \frac{\vartheta}{\lambda}\right)\frac{\dot{\lambda}}{\lambda} = \left(w - \frac{\vartheta}{\lambda}\right)\left(\frac{k}{r^2}\frac{\sigma_2}{\sigma_1} - \frac{r}{k}w\alpha_2 - \frac{\dot{\alpha}_3}{\alpha_3}\right), \qquad (4.7)$$

where, in the case of the third type of thrust device, this relationship represents another form of the differential equation for the regime criterion η; this last is expressed in terms of η by means of formula $(2.10)_3$.

The nine quantities k, m, α_s, σ_s, and ϑ are related by ten equations: the equations of motion (4.4), the mass consumption equation (1.4), the four equations (4.5) - (4.7), (2.15), and, finally, the three differential equations (2.21). But Equation (4.7) written, by means of (4.5) and (4.4), in the form

$$\left(w\alpha_1^2 - \frac{k^2}{r^3}\frac{\sigma_3}{\sigma_1}\alpha_3\right)^{\cdot} = \left[w(1-\alpha_1^2) + \frac{k^2}{r^3}\frac{\sigma_3}{\sigma_1}\alpha_3\right] \times \left(\frac{k}{r^2}\frac{\sigma_2}{\sigma_1} - \frac{r}{k}w\alpha_2 - \frac{\dot{\alpha}_3}{\alpha_3}\right) \qquad (4.8)$$

is satisfied because of other equations that can be verified by direct differentiation. In this case $\dot{\alpha}_1$ is determined by differentiation of the equations of motion (4.4),

$$\dot{\alpha}_1 = -\frac{\dot{q}}{m}\alpha_1 - \frac{2k}{r^2}\alpha_2 \qquad (4.9)$$

while $\dot{\alpha}_2$ and $\dot{\alpha}_3$ are determined by means of (4.6) and the relationship

$$\alpha_1\dot{\alpha}_1 + \alpha_2\dot{\alpha}_2 + \alpha_3\dot{\alpha}_3 = 0$$

Thus, Equation (4.7) can be eliminated from the discussion and (4.6) is only used to compute the regime criterion with the aid of $(2.10)_3$. We thus obtain eight equations for determination of the same number of unknowns k, m, α_s, and σ_s.

4.3 Spherical motion with w = const (devices of the first type. In order to simplify the notation, we introduce the dimensionless variables

$$\nu = \frac{k}{r}\sqrt{\frac{r}{\mu}}, \quad \sqrt{\frac{\mu}{r^3}}t = nt = \tau, \quad w\frac{r^2}{\mu} = \zeta \qquad (4.10)$$

Here ν is the ratio of the velocity v = k/r to the velocity on a circular Kepleristic orbit of the same radius, n is the angular velocity on such an orbit, and ζ is the ratio of the force of gravity to the thrust ("overload")*. Assuming that we are dealing with minimization of t_1, by $(2.10)_1$ we have

$$\frac{\vartheta}{\lambda} = w - \frac{h}{\lambda} = w - \frac{1}{\lambda} = w - \frac{v\alpha_s}{a\sigma_1}$$

and Equations (4.4) - (4.7) take the form (the prime denote differentiation with respect to)

$$\alpha_1 = \frac{1-\nu^2}{\zeta}, \quad \alpha_2 = \frac{\nu'}{\zeta} \qquad (4.4)_1$$

$$\alpha_3 = \frac{\sigma_1}{\nu\zeta}\,\frac{\zeta^2 - (1-\nu^2)^2}{\delta - \nu\sigma_3}, \quad \left(\delta = \frac{1}{an}\right) \qquad (4.5)_1$$

$$\alpha_2' - \frac{\alpha_3'}{\alpha_3}\alpha_2 = -2\nu\alpha_1 - \frac{\nu}{\sigma_1}(\sigma_2\alpha_2 + \sigma_3\alpha_3) + \frac{\zeta}{\nu}(1-\alpha_1^2), \qquad (4.6)_1$$

where one of the two last equations is a consequence of the preceding equations.

---

* There is no chance of confusing this notation with previously introduced notations for the regime criterion, since this criterion does not figure in the problem under discussion.

It is natural to retain $(4.5)_1$. Of course, the equations we have written must be supplemented by (2.21) and (2.15). The last, along with $(4.4)_1$ and $(4.5)_1$, can be used to derive the differential equation

$$\nu'^2 + (1-\nu^2)^2 + \frac{\sigma_1^2}{\nu^2}\frac{\left[\zeta^2 - (1-\nu^2)^2\right]^2}{(\delta - \nu\sigma_3)^2} = \zeta^2 \qquad (4.11)$$

In this equation σ_3 and σ_1 can be expressed by means of formulas (2.28) in terms of the Cayley-Klein parameters; these last are found by means of system (2.26), which now takes the form

$$\alpha' = \frac{i}{2}\left(\nu\alpha + \frac{\sigma_1}{\nu^2}\,\frac{\zeta^2 - (1-\nu^2)^2}{\delta - \nu\sigma_3}\beta\right),$$

$$\beta' = \frac{i}{2}\left(-\nu\beta + \frac{\sigma_1}{\nu^2}\,\frac{\zeta^2 - (1-\nu^2)^2}{\delta - \nu\sigma_3}\alpha\right), \qquad (4.12)$$

and the two equations for the conjugates $\bar{\alpha}$ and $\bar{\beta}$. They are integrated with the initial conditions $\alpha^0 = \bar{\alpha}^0 = 1$ and $\beta^0 = \bar{\beta}^0 = 0$, so we now have the integral

$$\alpha\bar{\alpha} + \beta\bar{\beta} = 1 \qquad (4.13)$$

The solution contains the constants ν^0, σ_3^0, and δ. In the general boundary-value problem, ν^0 and ν' are given, along with the vectors $\underline{e}_r'$ and $\underline{e}_\varphi'$, which determine the position of the center of inertia and the direction of its velocity vector at time t_1. With a fixed ν^0 at the right end, the conditions

$$\nu' = \nu(\sigma_3^0, \delta, t_1),$$

$$\alpha' = \alpha(\sigma_3, \delta, t_1),\ \bar{\alpha}' = \bar{\alpha}(\sigma_3^0, \delta, t_1),\ \beta' = \beta(\sigma_3^0, \delta, t_1),\ \bar{\beta}' = \bar{\beta}(\sigma_3^0, \delta, t_1)$$

must be satisfied; four of them are independent because the Cayley-Klein parameters satisfy (4.13). The number of independent constants σ_3^0, δ, and t_1 is also four. Of course, α', $\bar{\alpha}'$, β', and $\bar{\beta}'$ are determined when $\underline{e}_r'$, $\underline{e}_\varphi'$, and $\underline{n}' = \underline{e}_r' \times \underline{e}_\varphi'$ are given.

4.31 Rotation in a circular orbit. We will now consider a particular solution of the problem of the preceding paragraph under the condition that the force of gravity is colinear with the normal n to the instantaneous orbital plane:

$$\alpha_1 = 0,\quad \alpha_2 = 0,\quad \alpha_3 = \pm 1 = \varepsilon \qquad (4.14)$$

It follows from $(4.4)_1$ that under these conditions $\nu = 1$ -- the motion occurs with constant velocity equal to the velocity along a circular orbit of radius r. Returning to $(4.5)_1$ or (4.11), we obtain the equation

$$\varepsilon \zeta \sigma_1 + \sigma_3 = \delta \qquad (4.15)$$

into which the stationarity conditions degenerate in this simple case. Differential equations (2.21) now take the form

$$\sigma_1' = \sigma_2,\quad \sigma_2' = -\sigma_1 + \varepsilon\zeta\sigma_3,\quad \sigma_3' = -\varepsilon\zeta\sigma_2 \qquad (4.16)$$

In virtue of these equations, the derivative of the left side of (4.15) with respect to τ is equal to zero, so the last may be realized.

Equations (4.12) are a system of differential equations with constant coefficients:

$$\alpha' = \frac{i}{2}(\alpha + \varepsilon\zeta\beta),\quad \beta' = \frac{i}{2}(-\beta + \varepsilon\zeta\alpha) \qquad (4.17)$$

Their general integrals can be written in the form

$$\alpha = \alpha^0 \kappa_1 + \beta^0 \kappa_2,\quad \beta = \alpha_0 \rho_1 + \beta_0 \rho_2 \qquad (4.18)$$

We now introduce a system of solutions with a single matrix of initial conditions

$$\kappa_1(\tau) = \cos^2\frac{\chi}{2} e^{i\omega} + \sin^2\frac{\chi}{2} e^{-i\omega},\quad \rho_1(\tau) = \frac{1}{2}\varepsilon \sin\chi (e^{i\omega} - e^{-i\omega}),$$

$$\kappa_2(\tau) = \frac{1}{2}\varepsilon\sin\chi(e^{i\omega} - e^{-i\omega}),\quad \rho_2(\tau) = \sin^2\frac{\chi}{2} e^{i\omega} + \cos^2\frac{\chi}{2} e^{-i\omega}, \qquad (4.19)$$

where

$$\operatorname{tg}\chi = \zeta,\quad \omega = \frac{\tau}{2\cos\chi}. \qquad (4.20)$$

TABLE OF COSINES

	$\underline{e}_r^0$	$\underline{e}_\varphi^0$	$\underline{n}^0$
$\underline{e}_r$	$\sin^2\chi + \cos^2\chi\cos 2\omega$	$\cos\chi \sin 2\omega$	$\varepsilon\cos\chi\sin\chi(1-\cos 2\omega)$
$\underline{e}_\varphi$	$-\cos\chi\sin 2\omega$	$\cos 2\omega$	$\varepsilon\sin\chi\sin 2\omega$
$\underline{n}$	$\varepsilon\cos\chi\sin\chi(1-\cos 2\omega)$	$-\varepsilon\sin\chi\sin 2\omega$	$\cos^2\chi + \sin^2\chi\cos 2\omega$

The above table of cosines corresponds to $\alpha^0 = 1$, $\beta^0 = 0$. i.e., the transition from the initial position of the orbital trihedron. The equation of motion of the center of inertia can be written with the aid of the first row of this table:

$$\underline{r} = r\underline{e}_r = r[\underline{e}_r^0(\sin^2\chi + \cos^2\chi\cos 2\omega + \\ + \underline{e}_\varphi^0 \cos\chi\sin 2\omega + \\ + \underline{n}_0\varepsilon\cos\chi\sin\chi(1-\cos 2\omega)] \qquad (4.21)$$

It is easy to obtain the relationships

$$\underline{r}\cdot\underline{N} = r\sin\chi,\quad \underline{N} = \underline{e}_r^0\sin\chi + \varepsilon\underline{n}^0\cos\chi \qquad (4.22)$$

from which it follows that the orbit is the circle on a sphere that is generated by the intersection of a plane perpendicular to the vector $\underline{N}$ at a distance of r $\sin\chi$ from the center of attraction.

The solution we have constructed does not permit us to preassign a given direction of the vector n^1 of the normal to the instantaneous plane Π' -- this direction is given by two angles, Ω' and i^1, and we have found only one quantity, $2\omega_1 = \tau_1/\cos\chi$. It would therefore be justified to attempt to obtain a solution that would, after the appropriate time τ_*, reverse the thrust (transferring from $\alpha_3 = \varepsilon$ to $\alpha_3 = -\varepsilon$.

Then, by (4.18)

$$0 \le \tau \le \tau_* - 0 \quad \alpha_3 = \varepsilon \quad \alpha = \kappa_1^+(\tau) \quad \beta = \rho_1^+(\tau)$$

$$\tau_* + 0 \le \tau \le \tau_1 \quad \alpha_3 = -\varepsilon \quad \alpha = \alpha_*\kappa_1^-(\tau - \tau_*) + \beta_*\kappa_2^-(\tau - \tau_*)$$

$$\beta = \alpha_*\rho_1^-(\tau - \tau_*) + \beta_*\rho_2^-(\tau - \tau_*)$$

where κ_i^+ and ρ_i^+ are given by formulas (4.19), while κ_i^- and ρ_i^- are given by the same formulas with $-\varepsilon$ substituted for ε. We obtain

$$\alpha' = \alpha(\tau_1) = \kappa_1^+(\tau_*)\kappa_1^-(\tau_1 - \tau_*) + \rho_1^+(\tau_*)\kappa_2^-(\tau_1 - \tau_*),$$

$$\beta' = \beta(\tau_1) = \kappa_1^+(\tau_*)\rho_1^-(\tau_1 - \tau_*) + \rho_1^+(\tau_*)\rho_2^-(\tau_1 - \tau_*),$$

and after calculations with formulas (4.19) we obtain the equations

$$\alpha' = i\cos\chi\sin\omega_1 + \cos\omega_1 + \sin^2\chi\left[\cos(\omega_1 - 2\omega_*) - \cos\omega_1\right]$$

$$\beta' = \varepsilon\sin\chi\{\cos\chi\left[\cos(\omega_1 - 2\omega_*) - \cos\omega_1\right] - i\sin\chi\sin(\omega_1 - 2\omega_*)\} \quad (4.23)$$

$$\omega_1 = \frac{\tau_1}{2\cos\chi}, \quad \omega_* = \frac{\tau_*}{2\cos\chi}$$

The unknowns τ_* and τ_1 are the smallest positive roots of the system

$$\alpha'\bar{\alpha}' - \beta'\bar{\beta}' = \cos i', \quad 2\alpha'\beta' = i\sin i' e^{i\Omega_1} \quad (4.24)$$

When $0 \le \tau \le \tau_1 - 0$, the orbit is composed of two segments of circles that are at the same distance from the center of attraction; their common tangent at the point $r\underline{e}_r(\tau_*)$ is directed along the vector $\underline{e}_\varphi(\tau_*)$, while the normals to the planes of the circles are the vectors $\underline{N}^+$ and $\underline{N}^-$. When, at the point $\tau_1 + 0$, thrust is terminated, the Kepleristic orbit becomes a great circle whose plane has the required orientation (its normal is the vector $\underline{n}^1$).

In the problem under discussion, the stationarity conditions and the Weierstrass criterion, as we indicated above, led to the single requirement (4.15); it is separately satisfied in each of the transitions $\underline{e}_r^0 \to \underline{e}_r^*$ and $\underline{e}_r^* \to \underline{e}_r^1$, while for the process as a whole, $\underline{e}_r^0 \to \underline{e}_r^1$, it may be satisfied only when $\sigma_1(\tau_*) = 0$, since $\underline{\sigma}$ is a constant vector and its projections σ_s on the axes of the orbital trihedron must be continuous functions of τ. But, because $\alpha_1 = \alpha_2 = 0$ and $\alpha_3 = \varepsilon$, we have

$$\underline{\lambda} = \lambda\underline{n}\varepsilon = \frac{ar}{k}\sigma_1\underline{n} = \frac{r^2}{\delta\mu}\sigma_1\underline{n},$$

$$\lambda' = \frac{r^2}{\delta\mu}(\sigma_2\underline{n} - \varepsilon\zeta\sigma_1\underline{e}_\varphi)$$

and, moreover,

$$\underline{\lambda}(\tau_*) = 0, \quad \underline{\lambda}'(\tau_*) = \frac{r^2}{\delta\mu}\sigma_2\underline{n}^*, \quad (H_\lambda)_{\tau=\tau_*} = 0$$

This contradicts the condition $h = 1$, so that one of the Erdmann-Weierstrass conditions does not hold -- the transition $\underline{e}_r^0 \to \underline{e}_r^1$ is not optimal.

In the original statement of the problem it was assumed that the force of gravity may be arbitrarily oriented relative to the orbital axes; it seems that under this assumption the requirement of minimum time can be realized only in the presence of a gravitational component in the instantaneous orbital plane; the particular solution $\alpha_1 = \alpha_2 = 0$ is not desirable.

4.32 <u>Optimal rotation of the plane of a circular orbit.</u> We now change the statement of the problem as follows: It is assumed that the orbit is the spherical curve and that there is an acceleration due to thrust that is bounded in magnitude and must be colinear with the normal to the instantaneous orbital plane. Under these conditions, the velocity will have constant magnitude and be colinear with $\mathbf{e}_\varphi$, which follows from the equations of motion $(4.4)_1$. The remaining equations of motion reduce to Equations (2.18); in the dimensionless notation introduced above, they take the form

$$\underline{e}_r' = \underline{e}_\varphi, \quad \underline{e}_\varphi' = -\underline{e}_r^* + \varepsilon\zeta\underline{n}, \quad \underline{n}' = -\varepsilon\zeta\underline{e}_\varphi \quad (4.25)$$

where $|\varepsilon(t)| \le 1$. The selection of the "control" $\varepsilon(t)$ is subject to the requirement of minimum time for transition from the initial position of the orbital plane to the terminal position, which is determined by preassignment of the normal vector $\underline{n}' = \tilde{\underline{n}}'$. Thus, the master function is given by the expression

$$\theta = \tau_1 + \underline{\rho}\cdot(\underline{n}' - \tilde{\underline{n}}')$$

We now introduce three Lagrangian vectors $\underline{\lambda}_1$, $\underline{\lambda}_2$, and $\underline{\lambda}_3$ and write the Hamiltonian

$$H_\lambda = \underline{\lambda}_1\cdot\underline{e}_\varphi - \underline{\lambda}_2\cdot\underline{e}_r + \varepsilon\zeta(\underline{\lambda}_2\cdot\underline{n} - \underline{\lambda}_3\cdot\underline{e}_\varphi)$$

It follows from Pontryagin's principle of the maximum that $\varepsilon(t)$ can take only its extreme values:

$$\varepsilon = 1 \quad \text{FOR} \quad \underline{\lambda}_2\cdot\underline{n} - \underline{\lambda}_3\cdot\underline{e}_\varphi > 0$$

$$\varepsilon = -1 \quad \text{FOR} \quad \underline{\lambda}_2\cdot\underline{n} - \underline{\lambda}_3\cdot\underline{e}_\varphi < 0$$

The time τ_* at which the regimes change is given by the equation

$$\underline{\lambda}_2\cdot\underline{n} - \underline{\lambda}_3\cdot\underline{e}_\varphi = 0 \quad (4.26)$$

The stationarity conditions lead to a system of linear equations,

$$\underline{\lambda}_1' = \underline{\lambda}_2, \quad \underline{\lambda}_2' = -\underline{\lambda}_1 + \varepsilon\zeta\underline{\lambda}_3, \quad \underline{\lambda}_3' = -\varepsilon\zeta\underline{\lambda}_2 \quad (4.27)$$

which has the particular solution $\underline{\lambda}_1 = \underline{e}_r$, $\underline{\lambda}_2 = \underline{e}_\varphi$, $\underline{\lambda}_3 = \underline{n}$; its general solution, which contains nine arbitrary constants, can be written in the form

$$\underline{\lambda}_1 = \mathbf{A}\cdot\underline{e}_r, \quad \underline{\lambda}_2 = \mathbf{A}\cdot\underline{e}_\varphi, \quad \lambda_3 = \mathbf{A}\cdot\underline{n}$$

where **A** is a constant tensor of rank two. The boundary conditions can be written by means of the above-noted master function; they have the form

$$\underline{\lambda}_1' = 0, \quad \underline{\lambda}_2' = 0, \quad H_\lambda' = 1 \quad (4.28)$$

As a result, if we represent **A** in the form of the sum of three diads,

$$\mathbf{A} = \underline{a}_1\underline{e}_r' + \underline{a}_2\underline{e}_\varphi' + \underline{a}\,\underline{n}$$

we find that $\underline{a}_1 = 0$, $\underline{a}_2 = 0$, and the solution of Equations (4.27) with coundary conditions (4.28) can be written in the form

$$\underline{\lambda}_1 = \underline{a}\,\underline{e}_r\cdot\underline{n}', \quad \underline{\lambda}_2 = \underline{a}\,\underline{e}_\varphi\cdot\underline{n}', \quad \lambda_3 = \underline{a}\,\underline{n}\cdot\underline{n}'$$

where $\underline{a}$ is a constant vector that is colinear with the Lagrange multipliers $\underline{\lambda}_s$.

The scalar integral expressing the constancy of the Hamiltonian now takes the form

$$(\underline{n}'\times\underline{a})\cdot(\underline{n} + \varepsilon\zeta\underline{e}_r) = 1 \quad (4.29)$$

and it is sufficient to write $\underline{n}'\times\underline{a} = \underline{\sigma}/\delta$ to obtain an expression in the form (4.15); the vector

$\underline{\sigma}$ is located in the plane Π', which agrees with the boundary condition (3.7) that we found earlier for the general problem of rotation in the orbital plane. The time τ_* is determined by condition (4.26), which we write in the form

$$\mathbf{a}\cdot(\mathbf{n}^*\mathbf{e}_\varphi^*\cdot\mathbf{n}' - \mathbf{e}_\varphi^*\,\mathbf{n}^*\cdot\mathbf{n}') = \mathbf{a}\cdot[\mathbf{n}'\times(\mathbf{n}^*\times\mathbf{e}_\varphi^*)] =$$
$$= (\mathbf{a}\times\mathbf{n}')\cdot(\mathbf{n}^*\times\mathbf{e}_\varphi^*) = 0$$

or

$$\frac{1}{\delta}\boldsymbol{\sigma}\cdot\mathbf{e}_r^* = 0, \quad \sigma_1(\tau_*) = 0.$$

Nothing interferes with the assumption that this condition is satisfied. It also follows from (4.29) that

$$\varepsilon(\tau_1)\gamma\sigma_1(\tau_1) = \delta, \quad \varepsilon(\tau_1) = \operatorname{sgn}\sigma_1(\tau_1)$$

which determines the choice of the sign of $\varepsilon(\tau_1)$. This also confirms formula (3.9).

In the statement of the problem that we are using, the stationarity conditions and the Weierstrass criterion (the principal of the maximum) are satisfied, while, under condition (4.26), the form of the expressions for λ_s -- the vectors and the Hamiltonian -- leave no doubt that the Erdmann-Weierstrass conditions are also satisfied. The differential equations of motion (4.25) are considered above in the equivalent form (4.17). The orbit found is composed of two segments of circles, and is optimal for the statement of the problem adopted in this paragraph; the minimal time and switching time are determined by system (4.24). Of course, this minimum is larger than that achieved in the more general statement of the problem in which it is assumed that the thrust may have components in the instantaneous orbital plane.

The problem of rotation of the plane of a circular orbit was considered in (13, 14, 15).

SECTION 5. LIMITED-POWER ENGINES.

5.1 Differential equation of motion. We showed in 1.7 that for this type of thrust device the vector $\boldsymbol{\lambda}$ differs only by the constant factor A/N_{max} from the vector of acceleration due to thrust. As a result, if we substitute expression (1.30) for the vector $\boldsymbol{\lambda}$ into differential equation (1.28) we obtain a fourth-order differential equation for the vector $\mathbf{r}$; this equation takes the form

$$\mathbf{L}(\mathbf{r}) = \left(\ddot{\mathbf{r}} + \frac{\mu}{r^3}\mathbf{r}\right)^{\cdot\cdot} - \frac{\mu}{r^3}\left(3\frac{\ddot{\mathbf{r}}\cdot\mathbf{r}}{r^2}\mathbf{r} + 2\frac{\mu}{r^3}\mathbf{r} - \ddot{\mathbf{r}}\right) = 0 \tag{5.1}$$

We will assume that the total fuel consumption is minimized; then

$$\theta = -m' + \theta_1(\mathbf{r}', \dot{\mathbf{r}}') + \rho_t(t_1 - t_1^*) \tag{5.2}$$

where ρ_t = h and, if t_1 is not preassigned, h = 0. The boundary conditions at the right end are given in terms of the vector $\mathbf{r}$; according to (1.30) and (3.2), they can be written in the form

$$t = t_1, \quad \left(\ddot{\mathbf{r}} + \frac{\mu}{r^3}\mathbf{r}\right)^{\cdot}_{t_1} = \operatorname{grad}_{\mathbf{r}'}\theta_1, \tag{5.3}$$
$$\left(\ddot{\mathbf{r}} + \frac{\mu}{r^3}\mathbf{r}\right)_{t_1} = -\operatorname{grad}_{\dot{\mathbf{r}}'}\theta_1, \quad H_\lambda\big|_{t_1} = h$$

It is assumed that $\mathbf{r}^0$ and $\dot{\mathbf{r}}^0$ are given at the left end. The expression for H_λ is written below.

5.11 Scalar and vector integrals. After substitution of expression (1.30) into Equations $(2.2)_2$ and (2.4), these integrals take the forms

$$H_\lambda = -(\ddot{\mathbf{r}}\cdot\dot{\mathbf{r}})^{\cdot} + \frac{3}{2}\ddot{\mathbf{r}}\cdot\ddot{\mathbf{r}} - \frac{\mu}{r^3}\left[\frac{\mu}{2r} + \dot{\mathbf{r}}\cdot\dot{\mathbf{r}} - \frac{3}{r^2}(\mathbf{r}\cdot\dot{\mathbf{r}})^2\right] = h \tag{5.4}$$

$$(\ddot{\mathbf{r}}\times\mathbf{r})^{\cdot} - 2\left(\ddot{\mathbf{r}} + \frac{\mu}{r^3}\mathbf{r}\right)\times\dot{\mathbf{r}} = \mathbf{b} = b\boldsymbol{\sigma} \tag{5.5}$$

where b and h are constants. As in (2.3), it turns out that system (5.4) - (5.5), which already contains four constants, is equivalent to the initial equation (5.1) under the condition $\mathbf{r}\cdot\dot{\mathbf{r}} \neq 0$. Relationship (2.31), which can also be written in the form

$$w\alpha_3 k = \left|\ddot{\mathbf{r}} + \frac{\mu}{r^3}\mathbf{r}\right|\,|\mathbf{r}\times\dot{\mathbf{r}}|\,\alpha_3 = \left(\ddot{\mathbf{r}} + \frac{\mu}{r^3}\mathbf{r}\right)\cdot\mathbf{n}\,|\mathbf{r}\times\dot{\mathbf{r}}| =$$
$$= \left(\ddot{\mathbf{r}} + \frac{\mu}{r^3}\mathbf{r}\right)\cdot(\mathbf{r}\times\dot{\mathbf{r}}) = br\sigma_1,$$

is the first integral of Equation (5.5). It can also be written in the easily verified form

$$\mathbf{r}\cdot(\dot{\mathbf{r}}\times\ddot{\mathbf{r}}) = b\boldsymbol{\sigma}\cdot\mathbf{r} \tag{5.6}$$

If we successively carry out scalar multiplication of (5.5) by $\dot{\mathbf{r}}$ and $\ddot{\mathbf{r}}$, we obtain the two relationships

$$\mathbf{r}\cdot(\dot{\mathbf{r}}\times\dddot{\mathbf{r}}) = b\boldsymbol{\sigma}\cdot\dot{\mathbf{r}} \tag{5.7}$$

$$\mathbf{r}\cdot(\ddot{\mathbf{r}}\times\dddot{\mathbf{r}}) = b\boldsymbol{\sigma}\cdot\left(\ddot{\mathbf{r}} + \frac{2\mu}{r^3}\mathbf{r}\right) \tag{5.8}$$

where the last is obtained by using (5.6); Equation (5.7), being a direct consequence of (5.6), need not be discussed. The problem has reduced to discussion of the system of three scalar equations (5.4), (5.6), and (5.8). These equations determine the covariant components of the vector $\dddot{\mathbf{r}}$ in the vector basis $\mathbf{r}\times\dot{\mathbf{r}}$, $\mathbf{r}\times\ddot{\mathbf{r}}$, $\dot{\mathbf{r}}$, by means of which $\mathbf{r}$ can be expanded in terms of the vectors of the relative basis:

$$\frac{1}{b\Delta}(\mathbf{r}\times\ddot{\mathbf{r}})\times\dot{\mathbf{r}}, \quad \frac{1}{b\Delta}\dot{\mathbf{r}}\times(\mathbf{r}\times\dot{\mathbf{r}}), \quad \frac{1}{b\Delta}(\mathbf{r}\times\dot{\mathbf{r}})\times(\mathbf{r}\times\ddot{\mathbf{r}}),$$

where, by (5.6),

$$b\Delta = \mathbf{r}\cdot\dot{\mathbf{r}}(\mathbf{r}\times\dot{\mathbf{r}})\cdot\ddot{\mathbf{r}} = \mathbf{r}\cdot\dot{\mathbf{r}}\,b\boldsymbol{\sigma}\cdot\mathbf{r}. \tag{5.9}$$

We are led to the expression

$$\dddot{\mathbf{r}}\Delta = \boldsymbol{\sigma}\cdot\mathbf{r}\Big\{\mathbf{r}\Big[\dot{\mathbf{r}}\cdot\dot{\mathbf{r}}\frac{\mu}{r^3} + \frac{1}{2}\Big(\ddot{\mathbf{r}}\cdot\ddot{\mathbf{r}} - \frac{\mu^2}{r^4}\Big) - h + \frac{3\mu}{r^5}(\mathbf{r}\cdot\dot{\mathbf{r}})^2\Big] - \dot{\mathbf{r}}\,\mathbf{r}\cdot\dot{\mathbf{r}}\frac{2\mu}{r^3}\Big\} + \boldsymbol{\sigma}\cdot\dot{\mathbf{r}}\,\dot{\mathbf{r}}\times(\ddot{\mathbf{r}}\times\mathbf{r}) + \boldsymbol{\sigma}\cdot\ddot{\mathbf{r}}\,\dot{\mathbf{r}}\times(\mathbf{r}\times\dot{\mathbf{r}}), \tag{5.10}$$

which holds when $\mathbf{r}\cdot\dot{\mathbf{r}} \neq 0$ and $b \neq 0$ -- spherical and plane motions are eliminated. It is a system of three third-order differential equations, where (5.6) is its first integral. The system contains three independent constants -- the constant unit vector $\boldsymbol{\sigma}$ and h, and the first integral contains one more constant, b.

The law for control of the acceleration due to thrust is, of course, given by the expression

$$\underline{w} = \ddot{\underline{r}} + \frac{\mu}{r^3}\underline{r} \qquad (5.11)$$

5.2 Plane motion. In motion in plane orbits, the vectors $\underline{r}$, $\dot{\underline{r}}$, $\ddot{\underline{r}}$, and $\dddot{\underline{r}}$ are coplaner and it is easy to use (5.6) - (5.8) to conclude that $b = \tilde{b}\,\underline{n}$, $\tilde{b} = \pm b$. Equation (5.5) can be written in the form

$$\dddot{\underline{r}} \times \underline{r} = (\ddot{\underline{r}} + 2\frac{\mu}{r^3}\underline{r}) \times \dot{\underline{r}} + \tilde{b}\,\underline{n}$$

If we carry out vector multiplication of both sides by $\dot{\underline{r}}$ and substitute for $\dddot{\underline{r}} \cdot \underline{r}$ by means of (5.4), we are led to the following representation of the vector $\dddot{\underline{r}}$:

$$\begin{aligned}\dddot{\underline{r}}\,\underline{r}\cdot\dot{\underline{r}} &= \underline{r}\Big[\tfrac{1}{2}(\ddot{\underline{r}}\cdot\ddot{\underline{r}} - \tfrac{\mu^2}{r^4}) + \tfrac{\mu}{r^3}\dot{\underline{r}}\cdot\dot{r} + \\ &+ \tfrac{3\mu}{r^5}(\underline{r}\cdot\dot{r})^2 - h\Big] - \dot{\underline{r}}\,\tfrac{2\mu}{r^3}\underline{r}\cdot\dot{\underline{r}} - \\ &- \dot{\underline{r}} \times (\dot{\underline{r}} \times \ddot{\underline{r}}) + b\,\dot{\underline{r}} \times \underline{n}.\end{aligned} \qquad (5.12)$$

We have obtained a system of two third-order differential equations that contains two constants, $\tilde{b}$ and h. In the special case of plane motion $\tilde{b} = 0$.

5.3 Spherical motion. In accordance with what we said in 4.1, differential equation (5.1) must be supplemented by the term ν r which is colinear with r. But the presence of this term does not change the form of the integrals (5.4) and (5.5). Together, they constitute the differential equations of the problem.

When dimensionless variables (4.10) are introduced, these equations take the form

$$\begin{aligned}&-(\underline{e}_r''\cdot\underline{e}_r')' + \tfrac{3}{2}\underline{e}_r''\cdot\underline{e}_r'' + \underline{e}_r''\cdot\underline{e}_r = \tfrac{1}{2} \\ &(\underline{e}_r''\times\underline{e}_r)' - 2(\underline{e}_r''+\underline{e}_r)\times\underline{e}_r' = \underline{\sigma}\delta, \quad (\delta = \text{const.})\end{aligned} \qquad (5.13)$$

where we have allowed for the fact that $\underline{e}_r\cdot\underline{e}_r' = 0$ and $\underline{e}_r'\cdot\underline{e}_r' = -\underline{e}_r\cdot\underline{e}_r''$ and we have assumed that τ_1 is not preassigned (h = 0).

Here we must add the kinematic relationships

$$\underline{e}_r' = \nu\underline{e}_\varphi, \quad \underline{e}_\varphi' = -\nu\underline{e}_r + \frac{\rho}{\nu}\underline{n}, \quad n' = -\frac{\rho}{\nu}\underline{e}_\varphi, \qquad (5.14)$$

where ν is the velocity and ρ is the projection of the acceleration due to thrust on the normal to the instantaneous orbital plane, represented in dimensionless form. These quantities are treated as new variables.

We have

$$\underline{e}_r'' = \nu'\underline{e}_\varphi - \nu^2\underline{e}_r + \rho\underline{n} \qquad (5.15)$$

and after substitution into (5.13) we obtain the four relationships

$$\rho\nu = \sigma_1\delta, \qquad (5.16)$$

$$\frac{\nu'}{\nu}\rho + \rho' = \sigma_2\delta, \qquad (5.17)$$

$$\frac{1}{3\nu} - \frac{4}{3}\nu - \frac{1}{3}\left(\nu'' + \frac{\nu'^2}{\nu}\right) + \nu^3 = \sigma_3\delta, \qquad (5.18)$$

$$\rho^2 = \frac{1}{3} + \frac{2}{3}\nu^2 + \frac{2}{3}\nu\nu'' - \frac{1}{3}\nu'^2 - \nu^4, \qquad (5.19)$$

where (5.19) has been used in the notation of Equation (5.18). Only two of these four relationships are independent. Indeed, because $\underline{\sigma}$ is a constant unit vector,

$$\begin{aligned}&\sigma_1' = \nu\sigma_2, \quad \sigma_2' = \nu\sigma_1 + \frac{\rho}{\nu}\sigma_3, \\ &\sigma_3' = -\frac{\rho}{\nu}\sigma_2, \quad \sigma_1^2 + \sigma_2^2 + \sigma_3^2 = 1\end{aligned} \qquad (5.20)$$

It is immediately clear that (5.17) is a consequence of (5.16). But this same relationship (5.17) can be obtained by differentiation of (5.18) and accounting for (5.19). It remains to subject the right sides of expressions (5.16) - (5.18) to the second relationship of (5.20). This leads to the differential equation

$$\rho'' - \rho'^2 + \frac{\nu'}{\nu}(\rho^2)' = 2\frac{\rho^2}{\nu^2}(1 - 2\rho^2 - 2\nu^4), \qquad (5.21)$$

which must be considered together with (5.19). A first integral of the system of differential equations (5.19), (5.21) is known -- it is a consequence of the second part of (5.20) expressed in terms of the left sides of (5.16) - (5.18).

Five parameters -- four constants of integration and τ_1 -- must be set in order to arrange for the quantities ν^o, ν', and the Euler angles, Ω', i', and u', to take given values. This establishes the possibility of stating the general boundary-value problem for motion along sperical orbits.

The law for control of thrust is determined by (5.11) and (5.15):

$$\underline{w} = \frac{\mu}{r^2}(\underline{e}_r'' + \underline{e}_r) = \frac{\mu}{r^2}\left[(1-\nu^2)\underline{e}_r + \nu'\underline{e}_\varphi + \rho\,\underline{n}\right]$$

The control region in which the thrust remains directed along the normal to the instantaneous orbital plane is not among the optimal regimes, because $\nu = 1$, $\rho = \pm 1$ is not a particular solution of systems (5.19), (5.21). These differential equations can be satisfied with ν^2 and ρ^2 respectively equal to 1/4 and 7/16; but in this regime with a preassigned constant velocity ν = 1/2, the choice of τ_1 makes it possible to assign only one of the three Euler angles Ω', i', and u'.

5.4 Concerning application of the Ritz method. The solution of the problem of controlling thrust with limited power can be obtained without the method we used above, reduction to the Mayer-Bolza problem. Indeed, returning to the initial equations (1.2), (1.3), (1.4), and (1.12), we have

$$w^2 = |\ddot{\underline{r}} + \frac{\mu}{r^3}\underline{r}|^2 = \frac{c^2 q^2}{m^2} = 2N\frac{q}{m^2} = 2N\left(\frac{1}{m}\right); \qquad (5.23)$$

from which it follows that

$$\tilde{I} = \frac{1}{m_1} - \frac{1}{m_0} = \frac{1}{2}\int_0^{t_1}\frac{1}{N}|\ddot{\underline{r}} + \frac{\mu}{r^3}\underline{r}|^2 dt \qquad (5.24)$$

We should note that the acceleration w is determined by two independent functions of time: N(t) and q(t). It is possible to choose arbitrarily a law of variation N(t) and then use appropriate variation of the consumption q(t) to generate any preassigned functional relationship w(t). Thus, the integrand in (5.24) contains independent variables. When (1.12) is taken into account, this makes it possible to establish the inequality

$$\tilde{I} \geq I = \frac{1}{2N_{max}} \int_0^{t_1} \left|\ddot{\underline{r}} + \frac{\mu}{r^3}\underline{r}\right|^2 dt = \frac{1}{m^1} - \frac{1}{m^0} \qquad (5.25)$$

The problem of minimum total consumption (maximum final mass m^1) reduces to finding a law of motion r(t) that will minimize integral (5.25). When $t_1 = t_1^*$ is fixed, this is a "simple problem" of variational calculus, while when t_1 is not preassigned, it is a problem with a "moveable boundary."

Under the assumption that the boundary values $\underline{r}^1$, $\dot{\underline{r}}^1$, and t_1 are related by the expression

$$\theta = \theta_1(\underline{r}^1, \dot{\underline{r}}^1) + h(t_1 - t_1^*) = 0 \qquad (5.26)$$

we must, according to the rules of variational calculus, set the variation of

$$\Omega_2 = \theta_1(\underline{r}^1, \dot{\underline{r}}^1) + h(t_1 - t_1^*) + \frac{1}{2}\int_0^{t_1}\left(\ddot{\underline{r}}\cdot\ddot{\underline{r}} + 2\mu\frac{\underline{r}\cdot\ddot{\underline{r}}}{r^3} + \frac{\mu^2}{r^4}\right)dt \qquad (5.27)$$

to zero. If we assume that the upper limit of integration also varies, we find that

$$\delta\Omega_2 = \int_0^{t_1} \underline{L}(\underline{r})\cdot\delta\underline{r}\,dt + \left\{\left(\ddot{\underline{r}} + \frac{\mu}{r^3}\underline{r} + grad_{\dot{\underline{r}}^1}\theta_1\right)\cdot\Delta\dot{\underline{r}}^1 - \left[\left(\ddot{\underline{r}} + \frac{\mu}{r^3}\underline{r}\right)^{\cdot} - grad_{\underline{r}^1}\theta_1\right]\cdot\Delta\underline{r}^1\right\} - \left[(H)_{t_1} - h\right]\delta t_1 = 0 \qquad (5.28)$$

where Δ denotes the total variation at the right end:

$$\Delta\underline{r}^1 = \delta\underline{r}^1 + \dot{\underline{r}}^1\delta t_1, \quad \Delta\dot{\underline{r}}^1 = \delta\dot{\underline{r}}^1 + \ddot{\underline{r}}^1\delta t_1 \qquad (5.29)$$

The left end is assumed to be fixed ($\delta\underline{r}^0 = 0$, $\delta\dot{\underline{r}}^0 = 0$); H in formula (5.28) denotes the left side of (5.4), while $\underline{L}(\underline{r})$ denotes differential operator (5.1).

The condition that the variation $\delta\Omega_2$ vanish leads, of course, to differential equation (5.1) and boundary conditions (3.2).

Reduction of the problem to finding the minimum of integral (5.25) makes it possible to use the Ritz method for determination of the vector r. The last is sought in the class of functions that are continuous together with their derivatives of up to third order, inclusive, that satisfy the boundary conditions of the problem. By selection of functions in this class that are equipped with undetermined constants $C_1, \ldots, C_s$, we are led to the problem of finding the minimum of $I(C_1, \ldots, C_s)$, which leads to finding these constants from a system of s finite equations. When the "geometric boundary conditions" -- the vectors r^1 and $\dot{r}^1$ -- are given and t_1 is fixed, it is easier to choose an appropriate expression for r; this case of boundary conditions is, for construction of an "exact" solution to the Mayer-Bolza problem, least simple, since the values of λ^1 and $\dot{\lambda}^1$ are entirely unknown. On the other hand, "force" boundary conditions (in the terminology of problems on bending of rods) corresponds to the case $\underline{\lambda}^1 = 0$, $\dot{\underline{\lambda}}^1 = 0$ which is the simplest Mayer-Bolza problem, and these boundary conditions make it difficult to find an appropriate expression for r(t). Moreover, it follows from (5.28) that it is possible to neglect verifying whether "force conditions" hold, because they are automatically satisfied when $\delta\underline{r}^1 \neq 0$, $\delta\dot{\underline{r}}^1 \neq 0$ (and $\delta t_1 = 0$) at the point at which I achieves its minimum. Of course, it is doubtful whether it is possible to trust such solutions in practice.

A very complete discussion of optimal problems for the case of limited-power engines may be found in (16), which has an exhaustive bibliography.

SECTION 6. SINGULAR-CONTROL REGIONS (VARIABLE THRUST q = q(t)).

6.1 Statement of the variational problem. In 1.8 we showed that in the case of thrust devices with limited fuel consumption the possibility of achieving a "singular-control region" is not excluded. In this regime the consumption is determined by Equation (1.37), where the statement of the problem requires the presence of inequalities (1.14).

In this regime condition (1.36) is satisfied; as a result, η and the derivative $\dot{\eta}$ of the regime criterion vanish, and, by (1.33), we have $\dot{\lambda} = 0$, $\lambda = const$. In accordance with (1.27), differential equation (1.28) becomes the equation for a unit vector in the direction of the thrust e:

$$\underline{L}(\underline{e}, \underline{r}) = \ddot{\underline{e}} + \frac{\mu}{r^3}\left(\underline{e} - \frac{3}{r^2}\underline{r}\cdot\underline{e}\,\underline{r}\right) = 0 \qquad (6.1)$$

This result may be obtained by direct statement of the variational problem of minimizing the total fuel consumption. By (1.4), we have

$$\frac{cq}{m} = -c\frac{\dot{m}}{m} = \left|\ddot{\underline{r}} + \mu\frac{\underline{r}}{r^3}\right|, \qquad \ln\frac{m^0}{m} = I = \frac{1}{c}\int_0^{t_1}\left|\ddot{\underline{r}} + \frac{\mu}{r^3}\underline{r}\right|dt \qquad (6.2)$$

where

$$\left|\ddot{\underline{r}} + \frac{\mu}{r^3}\underline{r}\right| = \left(\ddot{\underline{r}} + \frac{\mu}{r^3}\underline{r}\right)\cdot\underline{e} \qquad (6.3)$$

As in (5.4), the problem reduces to finding the variation of the functional

$$\Omega_3 = \theta_1(\underline{r}^1, \dot{\underline{r}}^1) + h(t_1 - t_1^*) + \int_0^{t_1}\left|\ddot{\underline{r}} + \frac{\mu}{r^3}\underline{r}\right|dt \qquad (6.4)$$

Assuming that the upper limit t_1 need not be fixed, we find that

$$\delta\Omega_3 = grad_{\underline{r}^1}\theta_1\cdot\Delta\underline{r}^1 + grad_{\dot{\underline{r}}^1}\theta_1\cdot\Delta\dot{\underline{r}}^1 + h\delta t_1 + \left(\ddot{\underline{r}} + \frac{\mu}{r^3}\underline{r}\right)\cdot\underline{e}\Big|_{t_1}\delta t_1 + \frac{1}{2}\int_0^{t_1}\frac{1}{\left|\ddot{\underline{r}} + \frac{\mu}{r^3}\underline{r}\right|}\delta\left[\left(\ddot{\underline{r}} + \frac{\mu}{r^3}\underline{r}\right)\cdot\left(\ddot{\underline{r}} + \frac{\mu}{r^3}\underline{r}\right)\right]dt, \qquad (6.5)$$

where Δ denotes the total variation.

The variation of the bracketed expression can be written in the form

$$\frac{1}{2}\delta\left(\ddot{\underline{r}}+\frac{\mu}{r^3}\underline{r}\right)\cdot\left(\ddot{\underline{r}}+\frac{\mu}{r^3}\underline{r}\right)=\left(\ddot{\underline{r}}+\frac{\mu}{r^3}\underline{r}\right)\cdot\delta\ddot{\underline{r}}+$$
$$+\frac{\mu}{r^3}\left[\left(\ddot{\underline{r}}+\frac{\mu}{r^3}\underline{r}\right)\cdot\delta\underline{r}-\frac{3}{r^2}\left(\ddot{\underline{r}}+\frac{\mu}{r^3}\underline{r}\right)\cdot\underline{r}\,\underline{r}\cdot\delta\underline{r}\right]$$

Now, using (6.3), we can represent the integrand in the form

$$\underline{e}\cdot\delta\ddot{\underline{r}}+\frac{\mu}{r^3}\underline{e}\cdot\delta\underline{r}-\frac{3}{r^2}\underline{e}\cdot\underline{r}\,\underline{r}\cdot\delta\underline{r}=$$
$$=(\underline{e}\cdot\delta\dot{\underline{r}})^{\cdot}-(\dot{\underline{e}}\cdot\delta\underline{r})^{\cdot}+\underline{L}(\underline{e},\underline{r})\cdot\delta\underline{r}$$

If we use (5.20), integration and substitution into (6.5) leads to the following expression for the variation:

$$\delta\Omega_3=\int_0^{t_1}\underline{L}(\underline{e},\underline{r})\cdot\delta\underline{r}\,dt+(\mathrm{grad}_{\underline{r}^1}\theta_1-\dot{\underline{e}}^1)\cdot\Delta\underline{r}^1+$$
$$+(\mathrm{grad}_{\dot{\underline{r}}^1}\theta_1+\underline{e}^1)\cdot\Delta\dot{\underline{r}}^1+(h-H)_{t_1}\delta t_1\,, \qquad (6.6)$$

where

$$H=-\left(\dot{\underline{e}}\cdot\dot{\underline{r}}+\frac{\mu}{r^3}\underline{r}\cdot\underline{e}\right). \qquad (6.7)$$

This is the previously introduced quantity H_λ defined by (2.2) for λ = const and $\eta = 0$. Of course, by setting the variation $\delta\Omega_3$ to zero, we obtain differential equation (6.1) and boundary conditions of the form (3.2) for λ = const.

6.2 <u>The differential equations of the problem.</u> The unknown function -- the consumption q(t) -- has been eliminated from the statement of the problem, and it is determined from formula (1.37) after the solution has been found. The problem consists in determining the vector $\underline{r}$ and the unit vector $\underline{e}$ from Equation (6.1) and the equation

$$\left(\ddot{\underline{r}}+\frac{\mu}{r^3}\underline{r}\right)\times\underline{e}=0, \qquad (6.8)$$

which represents the result of eliminating q(t) from the equations of motion. This system of equations admits the vector and the scalar integrals

$$\dot{\underline{e}}\times\underline{r}-\underline{e}\times\dot{\underline{r}}=\underline{a}=a\underline{\sigma} \qquad (6.9)$$

$$H=-\left(\dot{\underline{e}}\cdot\dot{\underline{r}}+\frac{\mu}{r^3}\underline{r}\cdot\underline{e}\right)=h\,. \qquad (6.10)$$

By (2.20),

$$\dot{\underline{e}}=\underline{\nu}\times\underline{e}\,,\quad \ddot{\underline{e}}=\dot{\underline{\nu}}\times\underline{e}-\nu^2\underline{e} \qquad (6.11)$$

and substitution into (6.1) yields

$$\dot{\underline{\nu}}\times\underline{e}-\nu^2\underline{e}=\frac{\mu}{r^3}3(\underline{e}_r\alpha_1-\underline{e})\quad,\quad(\alpha_1=\underline{e}_r\cdot\underline{e})$$

This gives us

$$\nu^2=\frac{\mu}{r^3}(1-3\alpha_1^2)\,,\qquad \dot{\underline{\nu}}=\frac{3\mu}{r^3}\alpha_1\underline{e}\times\underline{e}_r \qquad (6.12)$$

The relationship $\alpha_1\le\frac{1}{\sqrt{3}}$ mentioned in (17) is a consequence of the first of these equations.

6.21 <u>Plane motion (18, 19).</u> In Equations (3.15) and (3.16) we set $\lambda'/\lambda=0$ and $\vartheta/\lambda=0$. We obtain two equations that can be written in the form

$$-\frac{h}{r}=\frac{\mu}{r^3}\cos\psi+\nu^2\cos\psi-\tilde{\nu}\left(\frac{\dot{r}}{r}\sin\psi+\dot{\psi}\cos\psi\right) \qquad (6.13)$$

$$-\frac{\tilde{a}}{r}=2\tilde{\nu}\cos\psi-\left(\frac{\dot{r}}{r}\sin\psi+\dot{\psi}\cos\psi\right) \qquad (6.14)$$

Here $\tilde{\nu}$ denotes the projection of the vector ν on the normal $\underline{n}$ to the orbital plane, so that

$$\nu=\tilde{\nu}n\,,\quad \tilde{\nu}=\dot{\psi}+\frac{k}{r^2}=\dot{\psi}+\dot{\varphi}\quad ; \qquad (6.15)$$

it follows from (6.12) that

$$\tilde{\nu}^2=\frac{\mu}{r^3}(1-3\cos^2\psi),\quad \dot{\tilde{\nu}}=-\frac{3\mu}{r^3}\sin\psi\cos\psi \qquad (6.16)$$

It follows from Equations (6.13), (6.14), and the first equation of (6.16) that

$$\tilde{a}\tilde{\nu}=h+\frac{3\mu}{r^2}\cos^3\psi \qquad (6.17)$$

$$\left(h+\frac{3\mu}{r^2}\cos^3\psi\right)^2=\mu\frac{\tilde{a}^2}{r^3}(1-3\cos^2\psi)\;; \qquad (6.18)$$

Moreover, by differentiating (6.17) and using the second equation of (6.16), we find that

$$2\frac{\dot{r}}{r}\cos^2\psi+3\dot{\psi}\sin\psi\cos\psi=\frac{\tilde{a}}{r}\sin\psi \qquad (6.19)$$

This equation is solved simultaneously with (6.14). We find that

$$\frac{\dot{r}}{r}=\frac{2\sin\psi}{3-5\cos^2\psi}\left(\frac{\tilde{a}}{r}+3\tilde{\nu}\cos\psi\right), \qquad (6.20)$$

$$\dot{\psi}=\frac{1}{\cos\psi(3-5\cos^2\psi)}\left[\frac{a}{r}(1-3\cos^2\psi)-4\nu\cos^3\psi\right] \qquad (6.21)$$

Equation (6.18) defines r in terms of ψ and the constants $\tilde{a}$ and h. Moreover, the remaining unknowns ν, $\dot{r}$, and $\dot{\psi}$ are expressed in terms of these same quantities ψ, $\tilde{a}$, and h by (6.17), (6.20), and (6.21); if we know ν and $\dot{\psi}$, we can also find $\dot{\varphi}$. It is not difficult to show that substitution of the values found for $\dot{r}$ and $\dot{\psi}$ into the result of differentiation of (6.17) leads to an identity -- we used it to describe the evaluation of (6.16) for $\dot{\tilde{\nu}}$, which can easily be seen because the initial system of three equations -- (6.13), (6.14), and the first equation of (6.16) -- is sufficient for determination of the three unknowns r, ψ, and k.

The results can be represented in explicit form (18) if t_1 is not fixed and h = 0. Then, by (6.18),

$$\frac{\mu}{r}=\frac{\tilde{a}^2}{9}\,\frac{1-3\cos^2\psi}{\cos^6\psi} \qquad (6.22)$$

and, moreover,

$$r\tilde{\nu}=\frac{\tilde{a}}{3}\,\frac{1-3\cos^2\psi}{\cos^3\psi}\,,\quad \dot{r}=2\tilde{a}\,\frac{1-2\cos^2\psi}{3-5\cos^2\psi}\,\frac{\sin\psi}{\cos^2\psi} \qquad (6.23)$$

$$r\dot{\psi} = \frac{\tilde{a}}{3\cos\psi}\frac{1-3\cos^2\psi}{3-5\cos^2\psi}, \quad r\dot{\varphi} = \frac{\tilde{a}}{3\cos^3\psi}\frac{1-3\cos^2\psi}{3-5\cos^2\psi}(3-4\cos^2\psi) \tag{6.24}$$

It follows from (6.23) that

$$\frac{d\varphi}{d\psi} = 4 - \frac{3}{\cos^2\psi}, \quad \varphi = 4\psi - 3\,\mathrm{tg}\,\psi + C_1 \tag{6.25}$$

All of the desired quantities have thus been expressed in terms of ψ -- the angle between the thrust and the radius vector. This same parameter can be used to express the time, if we use (6.22) and (6.24):

$$\frac{\tilde{a}^3}{27\mu}dt = -\frac{3-5\cos^2\psi}{(1-3\cos^2\psi)^2}\cos^7\psi\, d\psi \tag{6.26}$$

Integration introduces one more constant, C_2. Finally, returning to formula (1.37) and using (6.10), we find that

$$\begin{aligned} \frac{q}{m} &= \frac{1}{c}\left[(\underline{v}\cdot\underline{e})^{\cdot} + \frac{2\mu}{r^2}\cos\psi\right] \\ \ln\frac{m_0}{m} &= \frac{1}{c}\,\underline{v}\cdot\underline{e} + 2\int_0^t \mu\frac{\cos\psi}{r^2}dt \end{aligned} \tag{6.27}$$

This determines the mass and, consequently, the fuel consumption q; it is still necessary to check satisfaction of inequalities (1.14).

The solution contains four constants: a, C_1, C_2, and m_0. It is clear from its structure that these constants can be chosen so that initial conditions concerning mass, the real anomaly φ^0 and only two of the three quantities r^0, $\dot{r}^0$, $r^0\dot{\varphi}^0$ are satisfied. At the right end it remains possible to preassign only one of these quantities by selection of an appropriate t_1. If, however, we are dealing with the problem of injection into an extremal arc in a singular-control region, at the transition point it is required (by the Erdmann-Weierstrass conditions) that we ensure continuous conjugacy not only with respect to the "coordinates" r, φ, $\dot{r}$, and $r\dot{\varphi}$, but also with respect to the controls ψ and $\dot{\psi}$, which are not eliminated in the case of strictly constrained specialization of the initial and boundary conditions. It seems that we are compelled to evaluate "singular control" as a mathematical oddity (or monstrosity) and we can eliminate any thought of bringing it to practical realization.

REFERENCES

(1) Leitmann, G., On a Class of Variational Problems in Rocket Flight, "Journal of Aero-space Science 26," N 9 (1959).

(2) Miele, A., General Variational Theory of Flight Paths of Rocketpowered Aircraft, Missiles and Satellite Carriers, "Astronautica Acta 4" N 4, 1958.

(3) Lawden, D. F., Interplanetary Rocket Trajectories, "Advances in Space Science I," N I, 1959.

(4) Lawden, D. F., Optimal Programming of Rocket Thrust Direction, "Astronautica Acta I," N I, 1953.

(5) Leitmann, G., Minimum Transfer Time for a Power-Limited Rocket, "Journal of Applied Mechanics, 6I - APM - 6, 1961.

(6) Bliss, G. A., Lectures on the Calculus of Variations, University of Chicago Press, Chicago, 1946.

(7) Troitskii, V. A., On Variational Problems for Optimizing Control Processes, "PMM," Volume 26, No. 1, 1962.

(8) Berkovitz, L. D., Variational Methods in Problems of Control and Programming, "Journal of Mathematical Analysis and Application," 3, pp. 145-187, 1961.

(9) Tarasov, E. V., Optimal Flight Regime for Space Flight, "Iadatel'stvo Oborongiz," Moscow, 1963 (in Russian).

(10) Dahlard, L., Application of Pontryagin's Maximum Principle in Determining the Optimum Control of a Variable Mass-Vehicle, "Progr. Astronautics and Rocketry," 8, pp. 21-29, Academic Press, 1962.

(11) Isaev, V. K. and Sonin, V. V., On One Nonlinear Optimal-Control Problem, Avtomatika i Telemekhanika," Volume 23, No. 9, 1962.

(12) Lur'e, A. I., Free Fall of a Material Point in the Cabin of a Satellite, "PMM," Volume 26, No. 1, 1962.

(13) Illarionov, V.F., and Shkadov, L. M., "Turning of the Satellite's Orbital Plane (Circular Orbit)," Prikl. Mat. Mekh. 26, No. 1, 1962.

(14) Lass, H., Solloway, C., Motion of a Satellite under the Influence of a Constant Normal Thrust, "ARS-Journal," 32, N I, 1962.

(15) Gus'kov, Yu. P., A Method for Controlling Rotation of the Plane of a Circular Satellite-Orbit, "PMM," Vol. 27, No. 3, 1963.

(16) Grozdovskii, G. L., Ivanov, Yu. P., and Tokarev, V. V., The Mechanics of Low-Thrust Space Flight, "Inzhenernyi zhurnal," Vol. 3, No. 3, 1963; ibid, Vol. 3, No. 4, 1963; ibid, Vol.4, No. 1, 1964.

(17) Fried, B. D., Trajectory Optimisation for Powered Flight in Two or Three Dimensions, "Space Technology," Chap. IV, Wileq, New York, 1959.

(18) Lawden, D. F., Optimal Powered Arcs in an Inverse Square Law Field, "ARS-Journal," 31, N. 4, 1961.

(19) Lawden, D. F., Optimal Intermediate - Thrust Arcs in a Gravitational Field, "Astronautica-Acta," 8, N 2-3, 1962.

PILOTED GUIDANCE AND CONTROL OF THE SATURN V LAUNCH VEHICLE

by Robert W. Gunderson
Supervisory Research Engineer
NASA-Marshall Space Flight Center
Huntsville, Alabama

Gordon H. Hardy
Research Scientist and Pilot
NASA-Ames Research Center
Moffett Field, California

ABSTRACT

The characteristics of man as a part of a vehicle control loop are studied within the context established by the potential roles of the astronaut in guidance and control of the Saturn V launch vehicle. Emphasis is placed on astronaut participation during the atmospheric phase of the flight profile.

INTRODUCTION

Primarily as a result of the unmanned mission assignments of booster systems during the past decade, the guidance and control of current launch vehicles is usually considered in terms of fully automatic systems. However, with the advent of manned launch vehicle systems such as Saturn V, an increased interest has been shown in the question of how, if at all, the astronaut should participate in the guidance and control functions. Advocates of increased astronaut participation refer to the unique capabilities of the human controller as a potential means of substantially increasing system reliability and flexibility. Critics of astronaut participation reply that the addition of the human controller would only serve to unnecessarily compound and confuse an already complex problem. Some serious study of the question has been accomplished for vehicles of configuration and mission profile different from the Saturn V, most notably the studies of references 1 and 2. These results do seem to at least permit the general conclusion that the problem of guidance and control of large launch vehicle systems is sufficiently within the capabilities of human controller performance and control technology to warrant further consideration.

This paper examines manual control during the atmospheric phase of the Saturn V trajectory. The results and conclusions are based upon simulation and analytical studies conducted jointly by the Marshall Space Flight Center and the Ames Research Center.

SATURN V CONFIGURATION AND FLIGHT PROFILE

The basic Saturn V configuration and flight profile for the lunar orbital rendezvous mission are shown in Figure 1. Consisting of three stages and the Apollo spacecraft, the overall vehicle length is approximately 111 m (364 ft), the maximum diameter is 10 m (396 in.), and the fully fueled weight is 2.7×10^6 kg (6,600,000 lb). The first, or SI-C, stage is powered by five F-1 engines, providing a total liftoff thrust of 33.3×10^6 N (7,500,000 lb). The four outboard engines are swiveled and provide the control force during powered flight. First stage burnout occurs at an altitude of about 61 km (200,000 ft) and, after jettison of the SI-C stage, powered flight is resumed upon ignition of the five J-2 engines of the S-II stage. Again, powered flight control is accomplished through swiveling of the four outboard engines, each engine providing approximately 889,000 N (200,000 lb) thrust. After separation of the S-II stage at an altitude of approximately 183 km (600,000 ft), the single J-2 engine of the S-IVB stage is ignited and maintained to injection into the parking orbit. A second burn of the S-IVB engine then injects the Apollo spacecraft into a translunar trajectory from the parking orbit.

SATURN V GUIDANCE AND CONTROL

Examination of the flight profile shows that the first stage of flight is terminated outside the sensible atmosphere. During atmospheric flight, the guidance and control problem is dominated by the requirement of assuring the structural integrity of a large, flexible, liquid-fueled, and aerodynamically unstable vehicle in the face of possibly severe wind disturbances. Outside the sensible atmosphere, the dominant problem becomes one of assuring precise injection into orbit, within a minimal fuel-expenditure constraint. Thus, the control problem dominates first stage guidance and control requirements, while the guidance requirement dominates during the upper stages of flight.

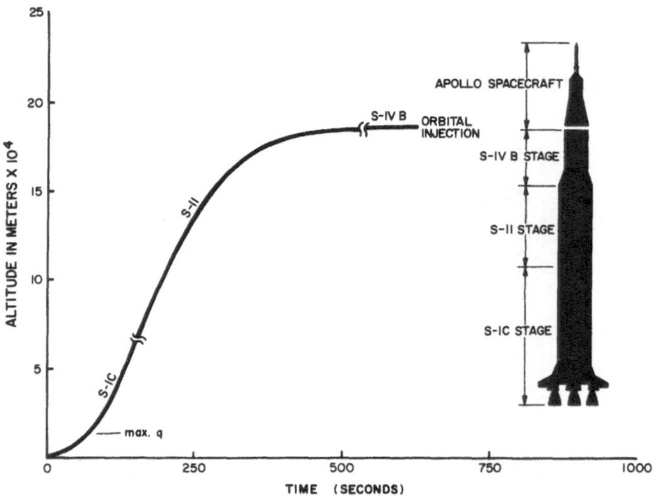

FIGURE 1. SATURN V FLIGHT PROFILE AND BASIC VEHICLE CONFIGURATION.

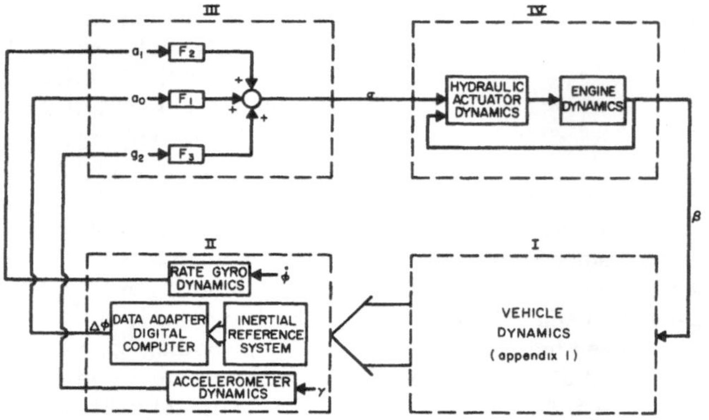

FIGURE 2. CONVENTIONAL CONTROL SYSTEM CONFIGURATION.

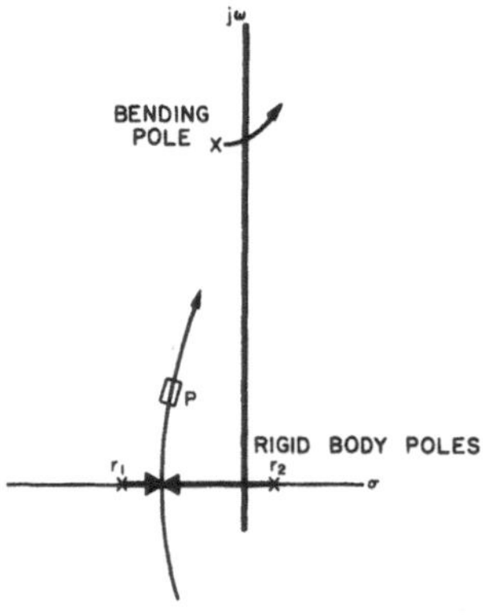

FIGURE 3. SIMPLIFIED ROOT LOCUS–CONVENTIONAL CONTROL WITHOUT FLEXIBLE BODY STABILIZATION NETWORKS.

One approach to alleviate the structural loading problem associated with atmospheric flight is to considerably reduce first stage trajectory constraints. By not constraining the vehicle to a given trajectory, the control system can be permitted to exercise greater flexibility in reducing the effect of severe wind disturbances. If this approach is followed, however, comparatively large deviations from an unperturbed trajectory must be expected. When the guidance loop is closed outside of the sensible atmosphere, the scheme must be capable of determining a new fuel-optimal flight path. In the case of Saturn V, first stage guidance is limited to a precomputed, time-dependent, roll and pitch maneuver; i.e., "open-loop" guidance. Shortly after launch, the vehicle is first commanded to perform the preprogramed roll maneuver to alignment with the desired flight azimuth. Following alignment, the tilt program is initiated by a pitch command to the control system. While the tilt program is active throughout first stage flight, no attempt is made to constrain the vehicle to the trajectory defined by execution of the tilt program under unperturbed or assumed wind conditions. The principal concern of this paper is with the atmospheric phase of flight; however, some mention should be made of the expected closed-loop guidance scheme. This scheme, the iterative guidance mode, relies on the adequacy of a simplified mathematical model to permit a closed form solution to the calculus of variation problem arising from the attempt to determine optimum thrust direction. An explicit equation for the thrust direction is obtained in the form

$$X_p = \arctan (A + Bt),$$

where X_p is the direction for minimum propellant consumption. The constants A and B are determined by the required cutoff velocity, cutoff position, initial values of the state variables, vehicle acceleration, and engine specific impulse. Values for A and B are computed continuously during flight. Considerable flexibility is gained through this scheme because (1) the same set of command equations is applicable to almost all orbital missions, (2) the command equations are not dependent upon number of high thrust stages, (3) a small number of calculated presettings are involved, and (4) the accuracy and propellant economy are excellent. The scheme, developed by the Marshall Space Flight Center, has been thoroughly investigated in both simulated flight studies and in actual flight tests with excellent results. A considerably more comprehensive and detailed discussion can be found in reference 3.

"CONVENTIONAL" CONTROL

This section discusses a fully automatic control scheme for the first stage of Saturn V. This discussion provides a basis for determining the implications of astronaut participation. Again, a more complete discussion can be found in reference 3.

The open-loop first stage guidance discussed in the preceding section reduces guidance and control during atmospheric flight to three somewhat contradictory functions: (1) provide attitude stabilization to the reference established by the open-loop guidance program, (2) reduce the structural loading effects of severe wind disturbances, and (3) minimize dispersion from the assumed nominal trajectory to an extent consistent with the structural loading criteria and system simplicity. The basic form of the conventional control system is shown in the simplified block diagram of Figure 2. As indicated, the closed-loop system can be represented as four major functional elements. Note that lead time required for hardware development will generally limit the ability of the control engineer to impose requirements on the characteristics of blocks II and IV, except during the early phases of system development. Subsequent changes in vehicle configuration and mission assignments, which could lead to substantial variation in the vehicle dynamics, cannot always be matched with compensating variation in the configurations of these blocks. In essence then, the control engineer's problem is to provide satisfactory accomplishment of the control functions through synthesis and implementation of the control law represented by block III.

The basic form of the conventional control law, as shown in Figure 2, is a simple attitude-error ($\Delta\phi$) plus attitude-rate ($\dot{\phi}$) feedback, with the gains a_0 and a_1 essentially established by the response requirements of an assumed rigid vehicle. The uncontrolled vehicle dynamics ($a_0 = a_1 = 0$) are dominated by the divergent aerodynamic instability resulting from the location of vehicle center of pressure forward of the center of gravity. If a nominal value of rate gain (a_1) is assumed and the rate loop closed, then a simplified representation of system behavior can be given as shown in Figure 3. It has been assumed that linearization of the equations of motion describing blocks I, II, and IV is legitimate as well as the assumption that the time variation of parameters is sufficiently small to permit the equations of motion to be considered as autonomous at a particular time of flight. If the attitude gain (a_0) is considered as an open-loop gain, the locus, originating at the poles corresponding to vehicle instability, will travel roughly as shown. For conventional control, the gains a_0 and a_1 will be chosen such that the closed-loop position of roots r_1 and r_2 represents a desired rigid body control frequency and damping. Unfortunately, the value of gain required to position the rigid body mode at a desired location (P) is generally sufficient to also cause the loci, originating at the open-loop position of flexible body modes, to cross the imaginary axis. Thus values of gain required to permit desired rigid body response will generally result in unstable flexible body dynamics. It then becomes necessary to add the stabilization filters f_1 and f_2. In addition to

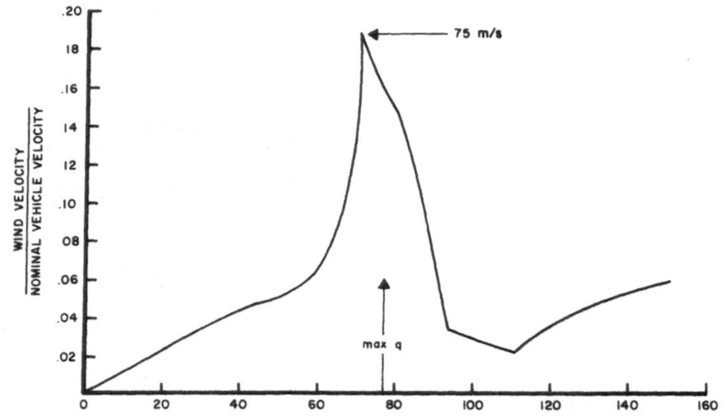

FIGURE 4. ASSUMED WIND DISTURBANCE.

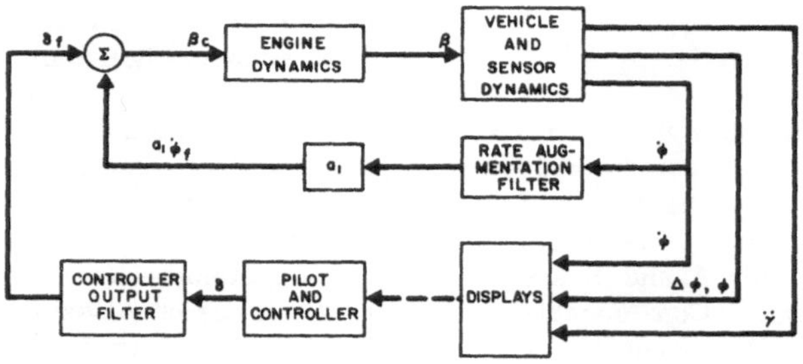

FIGURE 5. BASIC MANUAL CONTROL SYSTEM CONFIGURATION.

adequately stabilizing flexible body dynamics, these filters must not have any deleterious effect on the rigid body response. This problem is intensified in the case of Saturn V by the relatively close proximity of the first bending mode natural frequency to the desired control frequency. The filters F_1 and F_2 are not generally required to stabilize the fuel sloshing dynamics; sufficient damping can be obtained through tank baffling to stabilize all but extremely small slosh amplitudes. A design goal of minimization of slosh amplitudes is imposed, however, to minimize the contribution of the slosh mass to structural loading.

The adherence to a control law based entirely on the attitude stabilization requirement could, in the presence of disturbing winds, result in excessive structural loads. However, the structural loading criterion is generally not sufficiently stringent to justify the employment of a load-minimum control law, with the attendant trajectory deviation penalties. Instead, a compromise control law is usually favored which, in addition to reducing wind-induced structural loads, has a tendency to reduce lateral drift from the nominal trajectory. This approach, called drift-minimum control, necessitates the closing of an additional loop, with lateral acceleration ($\ddot{\gamma}$) as sensed by body-fixed accelerometers serving as the additional control variable. The gains a_0, a_1, and g_2 shown in Figure 2 are determined by first assuming rigid body, quasi-steady-state conditions ($\ddot{\phi} = \ddot{z} = 0$) and then imposing a drift minimum condition ($\ddot{z} = 0$). These assumptions lead to a set of three linear equations that define the minimum drift condition, approximate control frequency, and damping, in terms of the variables a_0, a_1, and g_2. A unique solution is obtained through specifying the desired control frequency and damping. The load reducing property of the drift-minimum philosophy results from the tendency of the vehicle to rotate into the wind flow as the means of reducing drift. Hence, the drift-minimum condition implies an induced wind-load condition which, while not a load minimizing condition, does have a load reducing effect.

If the resulting loads are still in excess of design conditions, the drift-minimum policy can be departed from; by weighting the g_2 gain with respect to the attitude error gain, a greater load reduction effect can be achieved. However, the $\ddot{\gamma}$ loop furnishes a relatively high gain feedback path for the flexible body dynamics, resulting in the necessity of the stabilization filter F_3. If the structural criteria permit, reduction of the gain g_2 from its drift-minimum value results in somewhat less severe filter requirements. Thus, the accelerometer loop gain is generally varied from its drift-minimum value depending on the structural loading design criteria. In addition, for reasons of relative system simplicity, the accelerometer loop is not closed until some time after launch, when wind effects become important, and is again opened after the vehicle has passed through the region of high dynamic pressure. While the preliminary choice of gains may be determined by the drift-minimum conditions, these values are usually somewhat modified during the developmental process.

The final major requirement placed on control system performance is that which introduces the adaptive control aspect; i.e., the flight control system must be capable of satisfactory performance over sufficiently large deviations from the predicted flight environment. Tolerances must be imposed on such factors as center-of-pressure location, center-of-gravity location, and others, even if it is assumed that the vehicle will experience no perturbation from the nominal trajectory. In fact, the absence of closed-loop guidance and the disturbing winds will result in trajectory deviations and even greater variations from predicted nominal values will result. In addition, tolerances must be imposed on the description of hardware performance characteristics, predicted flexible body dynamics, and predicted fuel sloshing dynamics. As a result, any control system configuration capable of satisfactory performance must be inherently an adaptive system. In the case of conventional control, the control gains a_0, a_1, and g_2 and filters F_1, F_2, and F_3 must be predetermined to permit satisfactory performance over the expected potential range of control parameters. Time variation of gains are preprogramed to permit desirable values for particular times of flight; however, the conventional system in no way senses the parameter environment and "self-adjusts" gains. Such a control system has been classified as passive-adaptive control by Aseltine and has the obvious advantage of being relatively simple in terms of component and scheme complexity as well as the less apparent advantage of being relatively tractable to analytical study. The more salient disadvantages include the consequences of unexpected variations in control parameters and the dependence of selected gains and filter configurations on specific vehicle characteristics and mission assignments.

ASTRONAUT PARTICIPATION

Clearly the presence of the astronaut in the control loop will have no effect on the uncompensated vehicle dynamics nor influence the requirement for the hydraulic-actuator engine system. Conceivably, the astronaut could establish a visual attitude reference, at least for short time periods, if the capsule configuration permitted. At best, however, such a capability would probably be of only emergency situation value. It will therefore be assumed that the structure of blocks I, II, and IV remain unchanged from that corresponding to the case of conventional control, with the exception that the open-loop guidance command $X_{p,r}$ will be considered a part of the control input if desired.

FIGURE 6. DISPLAY PANEL.

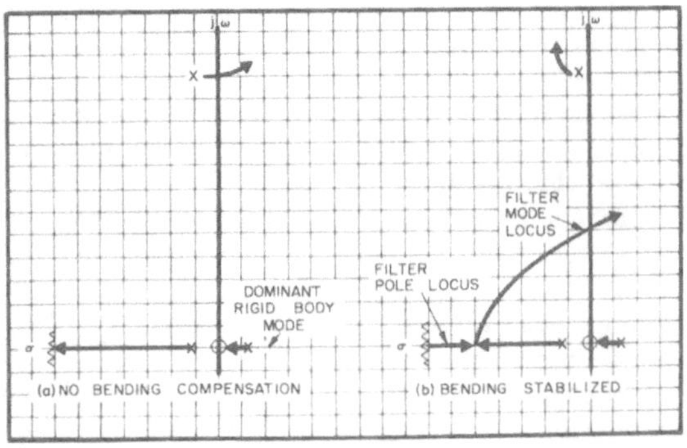

FIGURE 7. SIMPLIFIED ROOT LOCUS-MANUAL CONFIGURATION

It will be further assumed that the basic control requirements remain the same for the piloted system as for the fully automatic system; i.e., the system must be capable of satisfactorily accomplishing:

a. Open-loop guidance

b. Stabilization to the attitude reference

c. Reduction of wind-loading effects

d. Reduction of lateral drift, without compromising the effectiveness of a., b., and c. or undue increase in system complexity.

First, the control system configuration required for satisfactory performance of these functions will be established. Finally, the question of whether the inherent adaptive characteristics of the human controller can provide an effective means of compensating for variations of control parameters will be discussed.

SIMULATION STUDIES

The simulation results presented here were obtained principally from Saturn V piloted guidance and control simulation studies conducted at the Ames Research Center. Both fixed-base simulation and moving-base simulation were accomplished, with the latter performed using the Ames Research Center's five-degree-of-freedom human centrifuge to provide the inflight acceleration environment of the Apollo capsule. The equations of motion used in this study to describe the uncompensated vehicle dynamics were a perturbation set with respect to a reference frame moving along an assumed nominal trajectory. After simplification, these equations reduce to the linear, time-varying set (shown in Appendix I) and permit study of astronaut control action involving five degrees of control freedom (3 rotational and 2 translational). The first two modes of flexible body dynamics were included in the simulation as were the propellant sloshing dynamics of the first stage fuel and LOX tanks. Control system hardware was simulated by programing the differential equations corresponding to complex curve fits of experimentally obtained transfer functions. Major nonlinearities such as the hydraulic system velocity saturation characteristic and engine position limiting characteristics were also included. The primary disturbance source was assumed as a horizontal wind, with no restriction on direction. Figure 4 illustrates the disturbance as sensed by the astronaut as an angle of attack caused by wind. The corresponding maximum wind velocity is approximately 75 m/s in the sensible atmosphere, while the maximum value of wind shear occurs near the point of maximum dynamic pressure. Low amplitude high-frequency gusts and turbulence were initially included in the wind profile but were found to be neglectable effects. The direction of the wind was chosen in a random sequence from run to run to help eliminate pilot adaptation to particular wind profile and direction. Particular wind velocities and wind buildups were in accordance with specified Saturn V design winds.

PERFORMANCE MEASURES

Performance measures were specified for the simulation as a means of determining the degree to which the manual control system satisfied design constraints. The principal measure used was the body bending moment occurring at a critical location on the vehicle. The moment was continuously computed according to the expression

$$M_x = \frac{\partial M_x}{\partial \alpha}\alpha + \frac{\partial M_x}{\partial \beta}\beta + \sum_{i=1}^{n} \frac{\partial M_x}{\partial \eta_i}\ddot{\eta}_i .$$

Data relating slosh mass accelerations to body bending moment were not available at the time of simulation; however, the slosh mass accelerations were monitored and later computation indicated a relatively small contribution. A second performance measure was obtained by monitoring the distance and velocity dispersions normal to the nominal trajectory at the time of first-stage cutoff, while a third measure used was the numerical Cooper Pilot Opinion Rating (reference 4). This rating related the pilot's subjective opinion of system controllability to the assigned task. Additional measures were obtained by monitoring the amplitudes of flexible body and propellant sloshing accelerations as well as attitude and attitude rate amplitudes.

MANUAL CONTROL SYSTEM CONFIGURATION

Figure 5 is a block diagram representation of the basic control system configuration established during the simulation studies. In particular, the system of Figure 5 was determined to be the minimal configuration necessary for astronaut accomplishment of the assigned control tasks. The displays are further illustrated by Figure 6. Here, the two large instruments are standard aircraft all-attitude indicators, with the upper indicator serving as the primary flight instrument except during the region of high dynamic pressure. Pitch, yaw, and roll vehicle attitudes were displayed on the sphere of this instrument, with attitude error in pitch and yaw displayed on the flight director needles. Pitch and yaw lateral accelerations, as measured by body-fixed accelerometers, were displayed on the cross needles of the lower all-axis indicator with pitch, yaw, and roll attitude error displayed by the sphere of the instrument. Attitude rates, as sensed by body-fixed rate gyros, were displayed on the three dc meters at the upper right of the display

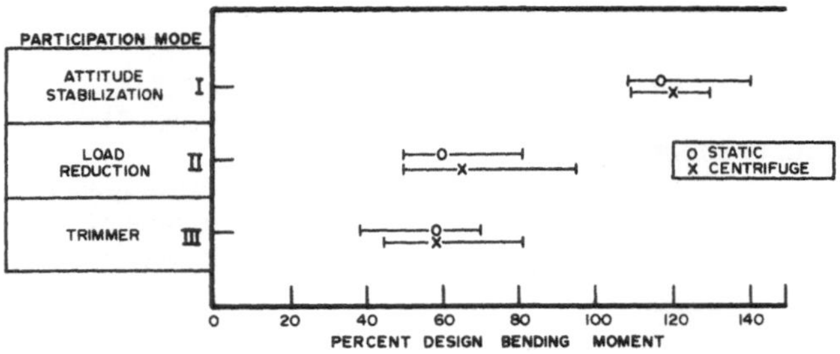

NOTES: 1. CONTINUOUS TIME OF FLIGHT SIMULATION
2. ELASTIC BODY WITH SLOSHING
3. DESIGN DAMPING AND CONTROL POWER

FIGURE 8. MAXIMUM BENDING MOMENT.

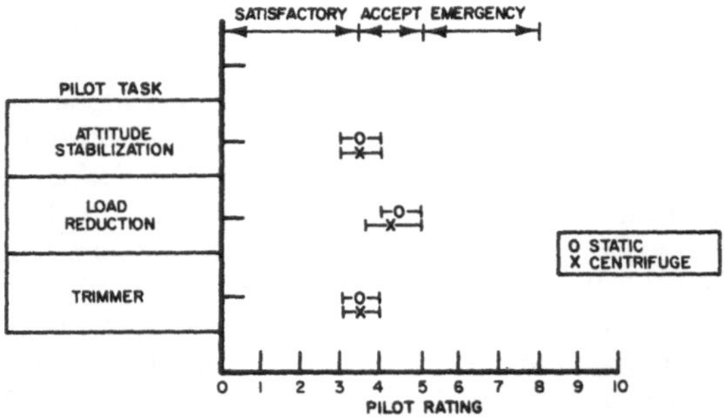

FIGURE 9. PILOT RATINGS.

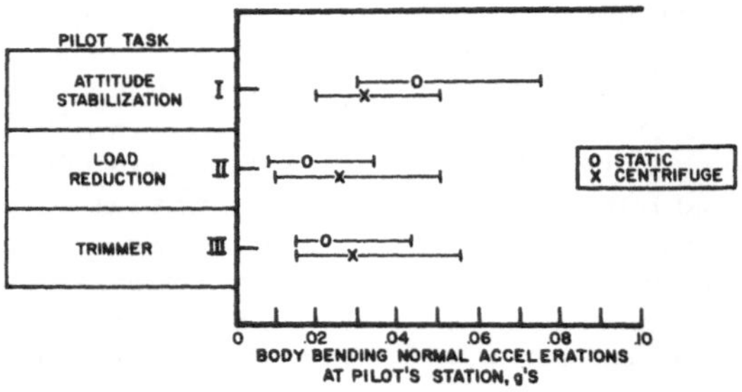

NOTES: 1. CONTINUOUS TIME OF FLIGHT SIMULATION
2. ELASTIC BODY WITH SLOSHING
3. DESIGN DAMPING AND CONTROL POWER

FIGURE 10. ACCELERATION AT THE CREW STATION RESULTING FROM FLEXIBLE BODY DYNAMICS.

panel. Time of flight was displayed on the elapsed-time clock at the left of the all-axis attitude indicators. A three-axis sidearm controller was used during the study, representative of the controller configuration to be used in the Apollo capsule.

Inspection of Figure 5 indicates that the primary design problem is to establish satisfactory values for scaling of displayed variables, controller sensitivity, rate gain, and adequate rate and controller output filter configurations. Analogous to the design of fully automatic systems, the display, controller, and rate feedback gains were first determined on the basis of assumed rigid body dynamics, with values chosen to permit satisfactory performance of task schedule I, shown in Table 1. Values finally determined through parametric simulation study are as follows.

$\ddot{\gamma}_d$ max $\sim 1/8$ g

$\dot{\phi}_d$ max $\sim 4°/s$

$\delta_{a_o} = 3°$/full stick deflection.

$\Delta\phi_d$ max $\sim 10°$

$a_1 \sim .87$ 1/s

δ_{a_o} roll = .3°/full stick deflection.

Note that rate augmentation was required for satisfactory performance, even under rigid body assumptions. Figure 7 is a simplified root locus plot of an aerodynamically unstable vehicle showing system response as seen from the point of pilot insertion. A closed rate augmentation loop is assumed, but no stabilization filter. The two poles on the real axis represent the dominant rigid body response to control inputs, while the flexible body response is represented by the complex pole near the imaginary axis. The zero at the origin is introduced by the rate augmentation loop and, as the rate gain is increased, the closed loop roots will move along the indicated loci. In particular, the locus originating at the bending pole will generally cross rapidly into the right half plane, and the system will be unstable at bending mode frequencies. As was the case for fully automatic systems, it is necessary to insert a flexible body stabilization network into the rate loop. Figure 7 (b) illustrates the general effect of adding such compensation. Notice particulary that while the locus of the divergent root is generally unchanged, an additional locus has resulted and the dominant vehicle response is now largely determined by a combination of the divergent mode and the newly introduced filter mode. Thus, caution must be exercised in providing flexible body stabilization in order not to degrade rigid body handling qualities. A typical pitch and yaw channel filter configuration which provided satisfactory stabilization characteristics is given by the transfer function:

$$F(p) = \frac{336}{(p+6)(p+7)(p+8)} .$$

The response mode introduced by this filter was heavily damped and found to be satisfactory for nominal parameter variations.

In general, even though the bending may be stabilized, the closed-loop bending modes will remain relatively lightly damped. It then remains to assure that:

a. The component of bending moment caused by bending acceleration terms remains small.

b. Flexible body motions do not obscure rigid body control information at the display.

c. Flexible body motion cues at the capsule do not degrade astronaut performance.

In the case of the manual control system, the major disturbance to the flexible body dynamics is attributable to the astronaut's engine swivel command signals. Consequently, a "smoothing" filter was placed at the output of the controller, as shown in Figure 5, in an effort to reduce the bending frequency content of the command signal. Only second order controller-output filters were investigated, with the configuration described by

$$F(p) = \frac{(2.7)^2}{p^2 + 2.7p + (2.7)^2}$$

resulting in the most satisfactory performance. With this controller configuration, no advantage could be gained by adding filtering between sensors and the displays.

SYSTEM PERFORMANCE FOR EXPECTED FLIGHT CONDITIONS

Figures 8, 9, and 10 summarize typical results obtained during evaluation of astronaut control capabilities under the assumption of nominal flight conditions. Results are shown for the three modes of astronaut participation summarized in task schedules of Table 1. System gains, filter configurations, and display scaling remained constant for each of the various modes. In all cases, the roll and pitch maneuvers were accomplished by presenting the command signal as an attitude error command. With respect to mode III, a typical automatic attitude-plus-attitude-rate control system was assumed to satisfy the attitude stabilization requirement. The primary responsibility given to the astronaut was to "close" the accelerometer loop during the time of high dynamic pressure and, by attempting to null the displayed accelerometer signals, to alleviate the structural loading resulting from high winds.

The data shown were taken from a total of 18 test

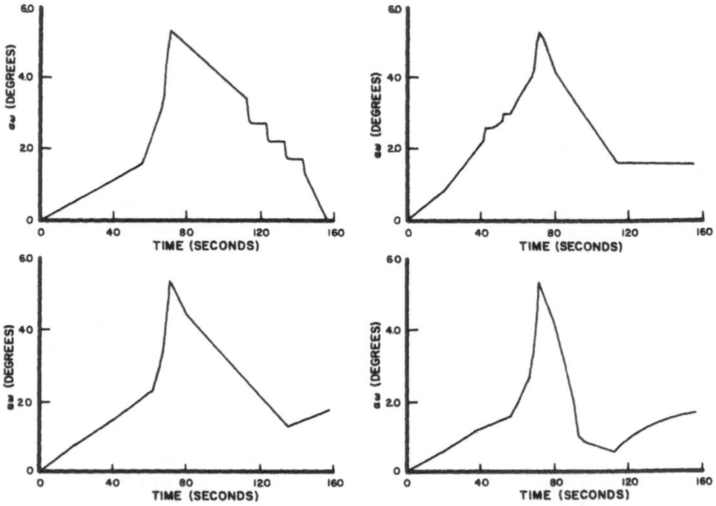

FIGURE 11. WIND PROFILE SET.

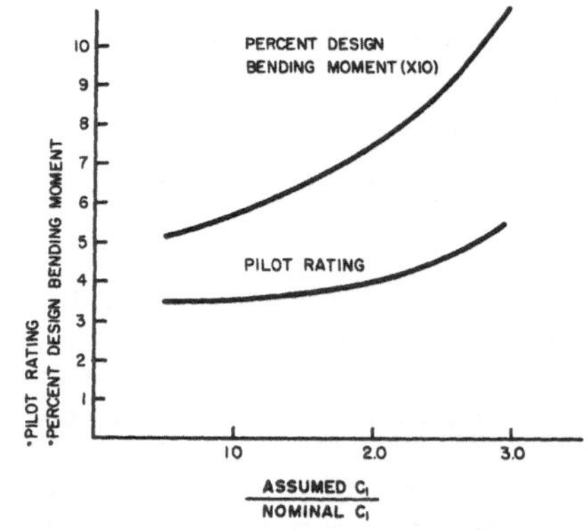

FIGURE 12. AERODYNAMIC MOMENT COEFFICIENT VARIATION AT HIGH q.

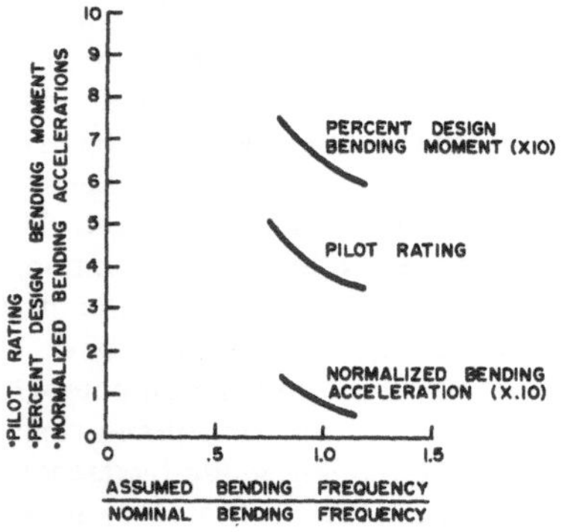

FIGURE 13. FIRST BENDING MODE NATURAL FREQUENCY VARIATION.

flights for each participation mode; nine on fixed-base simulation and nine on the centrifuge simulation. Runs were equally divided between three skilled test pilots, each pilot accomplishing three fixed-base and three centrifuge flights per participation mode. The data spread is described by the indicated greatest and least values together with the averaged value.

The percent of design bending moment appearing as the ordinate of Figure 8 is the maximum rms value of pitch and yaw bending moments occurring during each flight. As expected, the attitude stabilization mode resulted in bending moments somewhat in excess of design values; however, the average value is consistent with that obtained using a fully automatic system with attitude stabilization control law only. The bending moments resulting from the load reduction and trimmer modes are substantially below design limits, with the slightly smaller moments of the trimmer mode probably reflecting the reduced task loading. After a test pilot had completed his three fixed-base or centrifuge flights, he was requested to rate the participation mode according to the Cooper rating system. These results are shown in Figure 9. All modes were rated as either satisfactory or acceptable, with the higher ratings of the load-reduction mode again probably reflecting the higher degree of task loading. The final figure records the maximum values of acceleration resulting from flexible body dynamics as sensed at the pilot station. During all test flights, the pilots were instructed to attempt to input "smooth" control commands. It is interesting to note that in the load reduction and trimmer modes the pilots primary display during the time of greatest control activity (high q) was the accelerometer display, where the bending motion was much more noticeable than on the attitude display used for the attitude stabilization mode. Apparently, the "smoothness" cue thus provided resulted in less severe motions for the load reduction mode even though it was judged more difficult to accomplish than the attitude stabilization mode.

VARIATIONS FROM THE NOMINAL

As discussed in a preceding section, it is not sufficient to base the design of a booster control system, or to draw conclusions on performance capabilities, solely on the basis of the predicted nominal flight environment. In this section, typical results will be presented which should indicate the capability of the manual booster control system to adapt to non-nominal conditions.

Of the potential deviations from predicted flight conditions, one of the most important to consider is the wind environment. Unfortunately, it is also one of the more difficult to treat, since it implies an attempt to account for the statistical character of the wind environment by repeated simulated flights through a large number of profiles. Even then, there is a question of wind representation; e.g., the adequacy of wind representations based upon analysis of radiosonde balloon measurements. The presence of unexpected wind conditions could result, however, in at least one of two effects. First, vehicle response to unexpected severe winds could result in excessive structural bending moments. Second, severe winds could result in sufficient deviation from the nominal trajectory to substantially alter the values of the assumed nominal control parameters. In an effort to gain an insight into the ability of the human controller to compensate for these effects, simulated flights were conducted for the set of wind profiles shown in Figure 11. The control system configuration remained the same as the configuration previously described for all wind cases. The only variation in system characteristics permitted was the control behavior of the pilot. Each wind was varied with respect to direction and with respect to the time of occurrence of maximum wind velocity, with wind directions limited to 45°, 135°, 225°, and 315° relative to vehicle heading. The pilot participation mode assumed was the three-axis task of mode I, Table 1.

The results obtained during these studies were consistent with those already presented for the nominal wind profile (Fig. 4). In particular, variation of the wind profiles with respect to time of occurrence of maximum velocities resulted in greatest bending moments when the maximum velocity occurred during the time of maximum dynamic pressure. As expected, maximum winds occurring sufficiently before or after the high q region appeared as relatively less severe disturbances to the pilot, resulting in improved performance for these cases. Dependence of satisfactory control response on wind direction could not be detected from the results of this study. The quartering wind directions were selected because of the effect of presenting a simultaneous but nonsymmetrical disturbance in the pitch and yaw control axes. While the direction of the wind was randomly selected from the four choices, results obtained remained consistent with those previously presented. Again, note that while the results indicate at least a degree of adaptivity to a changing wind environment, they are also heavily dependent upon the form of the assumed winds and upon the number of simulated flights.

Figures 12 and 13 are given as representative results obtained during studies conducted to determine the adaptability of the human controller to variations from nominal control parameters. Particularly since the control system configuration relied only upon rate augmentation, the variation of system performance with aerodynamic stability c_1 was considered significant as a major contributor to vehicle response. The bending moment results shown in Figure 12 were those recorded at the time of high q for participation mode I

TABLE I. PARTICIPATION MODES.

	0 TO 3 S	3 TO 18 S	18 TO 60 S	60 TO 95 S	95 TO CUT-OFF
I. ATTITUDE PLUS LOAD CONTROL	3-AXIS ATTITUDE CONTROL	ATTITUDE CONTROL: ROLL MANEUVER	ATTITUDE CONTROL: PITCH MANEUVER	LOAD CONTROL: ROLL ATTITUDE CONTROL	ATTITUDE CONTROL: PITCH MANEUVER: NULL ATTITUDE RATES 3 S BEFORE CUT-OFF
II ATTITUDE CONTROL	3-AXIS ATTITUDE CONTROL THROUGHOUT FLIGHT ROLL AND PITCH MANEUVER COMMAND PRESENTED AS ATTITUDE ERRORS. NULL ATTITUDE RATES 3 S BEFORE CUT-OFF				
III TRIMMER MODE	MINOR ATTITUDE CORRECTIONS ONLY			LOAD CONTROL:	MINOR ATTITUDE CORRECTIONS ONLY

TABLE 2. MALFUNCTION STUDY RESULTS.

	TIME OF FAILURE		
FAILURE MODE	$0<t<60$	$60<t<105$	$105<t<150$
ONE ENGINE OUT	O O O O O O O O O	O O O O X_1 O O O O	O O O O O O O O O
TWO ENGINES OUT			O O O O O O O O O
ONE ENGINE HARD OVER	O O O O O X_1 O O O	X_1 X_2 $X_{1,2}$ O X_2 $X_{1,2}$ X_1 $X_{1,2}$ $X_{1,2}$	O O O O O O O O O
RATE GYRO OUT	X_1 O X_1 O X_1 O O X_1 X_1	O X_2 $X_{1,2}$ O X_2 $X_{1,2}$ O X_2 $X_{1,2}$	O O O O O O O O O
$\Delta\phi = 0$	O O O O O O O O O	O O O O O O O O O	O O O O O O O O O
TRIMMER MODE RATE OUT	X_1 X_1 X_1 X_1 X_1 X_1 O X_1 O	O $X_{1,2}$ $X_{1,2}$ O $X_{1,2}$ X_2 O $X_{1,2}$ $X_{1,2}$	O O O O O O O O O

PERFORMANCE CRITERIA EXCEEDED:

X_1 - $M/M_D > 100\%$
X_2 - PILOT RATING > 6.5

and with the wind of Figure 4 varied randomly with respect to the four quartering wind directions. The ratio of actual to assumed c_1 was held constant throughout the trajectory. It can be seen that the human controller is apparently able to adapt to substantial variations from the nominal and still perform at least within the design bending moment. It should be noted that the essential effect of variations in c_1 is to cause a deviation in the closed-loop rigid body frequency. The variations presented resulted in deviations within the range permitting effective control response by the human controller. However, the first bending mode natural frequency is in the neighborhood of 1 Hz, which is approaching the boundary of effective human control response. Figure 13 shows results obtained as the first bending mode natural frequency was varied from its nominal. Of particular interest is the ability of the human controller to perform satisfactorily as the natural frequency is lowered. Here the filter configurations remained constant, so that the first bending mode approached instability as the natural frequency decreased. It can be seen that performance definitely decreased as the bending response became more oscillatory. Further, in contrast to the case of c_1 variation, the pilots were unable to compensate for (i.e., stabilize) the oscillatory instability. These results were consistent throughout the studies conducted on parameter variations. Compared to an equivalent fully automatic conventional control system, the piloted system exhibited a somewhat greater tolerance to parameter variations, providing that the variations affected the vehicle response at sufficiently low frequencies (less than approximately .3 Hz). No advantage was indicated for parameter variations having an effect on the higher frequency range.

Much more gross deviations from the nominal control environment can be expected to result from malfunctions of the control or propulsion systems. Results are presented for the following malfunction conditions:

a. One engine out: engine thrust fails in any one of the four outboard engines.

b. One engine "hard-over:" a hydraulic actuator on one engine fails and fully extends, causing the engine to swivel to full deflection.

c. Rate augmentation failure: the rate augmentation is lost in one axis of the nominal manual control system.

d. Rate augmentation failure: the rate feedback of one axis of the trimmer mode control system (see mode III, Table 1) fails.

e. Loss of attitude error command: the pitch and yaw attitude error command signals (nominal manual control system configuration) fail.

f. Two engines out: engine thrust fails in two of the four outboard engines (failure only after time of high q).

The effect of a propulsion system failure or an engine hard-over failure can be readily seen by examining the sketch of Figure 14. Since the four outer engines are swiveled for pitch, yaw, and roll control, the loss of an engine will create a decrease of effective control force and introduce substantial coupling between control axes. In the sketch, the pilot has commanded a pitch control force but, because of the loss of an engine, will also command a roll control torque. If there were an initial roll error, a yaw coupling would also result. In addition to the nominal control task required by the trimmer participation mode, the loss of rate feedback required the pilot to adequately stabilize the defective control axes. Following an attitude error failure (case e), the pilot utilized a nominal pitch attitude program placed on a cutout around the elapsed-time clock. By reading the correct value of pitch attitude as a function of time and comparing with the actual attitude presented by the sphere of the all-axis attitude indicator, the pilot could derive the approximate pitch attitude error. The yaw and roll attitudes could be controlled directly from the sphere indication.

Three pilots, each performing at least nine fixed-base simulated flights for each failure case, participated in these studies. Because of simulation limitations, each pilot knew what the failure mode would be, but not the time of failure or the control axes that would be affected. In all cases except the failure of case d, the participation mode was that of mode I, both before and after the failure.

Results obtained are summarized in Table II. Here the three columns correspond to the general time of failure occurrence. The first column ($0 < t < 60$) corresponds to the time of flight prior to region of high dynamic pressure; the second column ($60 < t < 105$) corresponds to the region of high dynamic pressure; and the third ($105 < t < 150$) corresponds to the time remaining until first stage engine shutdown. The six rows correspond to the six failure conditions. The appearance of a small circle indicates that none of the performance limits listed below the table were violated when the failure (indicated by the row) occurred some time during the indicated time interval. Thus, the circle indicates a successful flight even though performed under malfunction conditions. An x indicates a flight during which at least one of the performance limits was exceeded, with the subscripts indicating the performance limit or limits exceeded. For example, nine simulated flights were successfully completed under the assumption that an engine was lost some time during the time interval of $0 < t < 60$ s. Eight successful simulated flights were completed assuming the same failure to occur some time during flight through the region of high dynamic pressure. However, the design bending moment

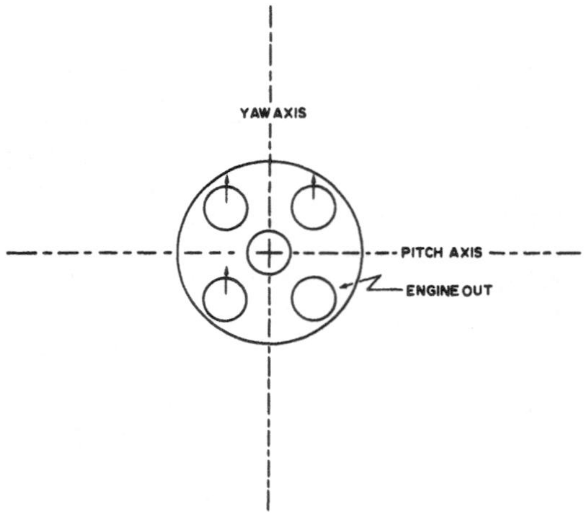

FIGURE 14. ENGINE-OUT EFFECT.

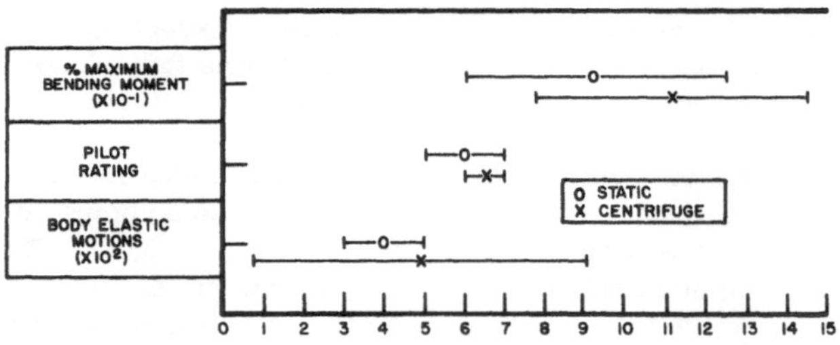

NOTES: 1. CONTINUOUS TIME OF FLIGHT SIMULATION
2. ELASTIC BODY WITH SLOSHING
3. DESIGN DAMPING AND CONTROL POWER

FIGURE 15. SINGLE AXIS CONTROL WITHOUT RATE AUGMENTATION.

was exceeded during one flight. Again none of the performance limits were violated when the engine failed during the final time period, $105 < t < 150$. Thus, out of a total of 27 simulated flights during which an engine was assumed to malfunction, the performance limits were violated only once and that was for a flight with the engine failure occurring during the time of highest dynamic pressure.

In general, all of the malfunction conditions were controllable during the essentially nonaerodynamic flight period following the time of high dynamic pressure. It should also be noted that a malfunction occuring during the first time period generally resulted in exceeding the performance limits not during that time period but during the flight through the more troublesome time of high dynamic pressure. The pilots could generally control satisfactorily during the first time period, even with a malfunction, but would exceed performance limit value upon entering the second time period. The cases involving loss of rate augmentation were particularly critical. The inherent phase lag of the pilot's three-axis sidearm controller combined with the lag of the controller filter and the statically unstable aerodynamics to create a rather undesirable condition, resulting in a pilot tendency to amplify oscillations. Note that even in these cases the pilots were able to control for a sufficient time to at least perform a "prepared" abort.

To supplement these studies and to gain further information on astronaut control capabilities in the case of a loss of rate feedback, an additional study was conducted during which the manual system was responsible for controlling a single control axis only, but without the aid of rate augmentation. Such a control function might be expected if the astronaut were to act as a backup to a fully automatic primary control system. These studies were conducted on both fixed-base and centrifuge simulations. The results shown in Figure 15 indicate that the single axis, no-rate, control case could be considered acceptable for emergency situations. It was believed that the somewhat less desirable results of the centrifuge simulation runs could be attributed to severe track noise, which is inherent to the centrifuge drive system and not present during actual flight. The vibration effect made it difficult to resolve display information as accurately as could be done in fixed-base simulation. This opinion, however, remains to be confirmed.

CONCLUSION

The results of these studies show that it is feasible to assign first stage aerodynamic flight control responsibilities to the astronaut. The problem becomes one of determining the degree of responsibility which will permit the capabilities of the human controller to be exploited but which minimizes the effect of human controller deficiencies. Several degrees of responsibility are conceivable; for example:

a. Replacement of the conventional fully automatic system as the primary control system by the manual configuration.

b. Retention of the conventional fully automatic system as the primary control system with a manual override capability.

c. Retention of the conventional fully automatic system in the form of attitude-plus-attitude-rate feedback with provision for astronaut control of load reduction (Table 1, trimmer mode).

d. Provision of manual system takeover from fully automatic primary system in the event of an emergency situation (i.e., system malfunctions, severe winds, and parameter estimate deficiencies). Responsibility prior to emergency situation could range from performance and system monitoring to active control responsibility of override or trimmer modes.

To determine which, if any, of these alternatives permit the greatest advantage of utilizing astronaut control capabilities, it is first desirable to summarize some relative advantages and disadvantages with respect to conventional control.

Attitude Control.—A comparison of the integrals of the squared attitude error emphasizes the relative coarseness of manual attitude control. However, the manual configuration is easily capable of controlling attitude within the bounds established for acceptable performance.

Structural Loading Control.—The high gain of the body-fixed acceleration feedback loop imposes a severe flexible body stabilization requirement on the conventional control system. To stabilize this loop, it is often necessary to provide such heavy filtering that little, if any, load reduction effect is achieved. Satisfactory load control has been consistently achieved with the manual configuration for the wind profiles assumed.

Trajectory Control.—Stabilization to the nominal trajectory has not been considered as a major control task. To the extent monitored, manual and conventional systems results were comparable.

Flexible Body Dynamics.—The flexible body dynamics stabilization requirement is reduced from the relatively complex, multiloop, stabilization problem of conventional control to compensation of the rate-augmentation loop only in the case of manual control. However, the disturbance of pilot control commands to the lightly damped bending dynamics necessitates an additional filter at the controller output.

Propellant Sloshing Dynamics.—Tank baffling is sufficient for both conventional and manual control to obviate the requirement of artificial stabilization.

Sensitivity to Parameter Variations.—The manual configuration is capable of somewhat greater tolerance to variation of control parameters from the predicted nominal values, providing the parameter variations have effect on response only at sufficiently low frequencies.

Sensitivity to System Malfunctions.—While simulation limitations necessitated a number of assumptions, the studies conducted to date definitely indicate the capability of the manual configuration to compensate for the effects of a fairly large class of guidance and control and propulsion system malfunctions. The advantage of the manual configuration here is the elimination of complex malfunction detection and identification schemes, as well as gain and network switching schemes required for comparable compensation by fully automatic means.

Scheme Flexibility.—One of the recognized attributes of a manual control system is its relative flexibility. Thus, without change of the basic configuration, it is possible for the manual configuration to perform an attitude control function, a load control function, or combine the attitude and load control functions as the requirement may be.

Complexity.—While the structure of the manual configuration is similar to that of the conventional configuration, the addition of the displays and controller results in greater component complexity.

Performance Predictability.—The most obvious disadvantage of manual control, especially when compared to conventional techniques, is the difficulty in predicting, within a given degree of confidence, the probability of successful system operation. Only by the greatest of coincidences are the results of any two piloted flights the same, even though the conditions of each flight are identical. Variation in pilot performance (from flight to flight) is certainly not the only cause of deviation from expected system performance, but must be considered as a definite problem area of manual control.

Design Procedures.—In the case of conventional control, the design technique is one of applying analysis and synthesis techniques of linear control theory and finally verifying the design through simulation studies, including as much actual hardware in the simulation as practicable. For example, the method of D-decomposition is used as a convenient technique to study the effects of parameter variation on system stability behavior. These results are in turn verified through simulation studies. However, the design procedure for manual control is largely empirical and depends heavily on simulation studies; not only as a means of verifying a design, but also as a means of arriving at a system configuration. The lack of analytical design techniques not only results in costly simulation study programs, but also is a major contributor to the problem of performance predictability cited previously.

On the basis of these results, it seems reasonable to conclude that astronaut control capabilities would be best utilized in the role of manual backup to a primary conventional system. In this role, the principal responsibility of the astronaut would be to maintain control of the vehicle in the event of an emergency situation not controllable by the primary system. Since the automatic abort system would remain operative, failure of the astronaut to regain control after an emergency situation would result only in the same automatic abort that would have occurred even if control responsibility had not been assigned. Thus, if the astronaut were to be utilized in such a manner he would, at least ideally, assume control of the vehicle only in the event of an emergency situation which normally (i.e., in the absence of astronaut backup) would result in automatic abort of the mission. Assuming it possible to assure that the astronaut did not assume control unnecessarily, the use of the astronaut could only result in an increase in the reliability of control.

Extensive further study of the backup role is currently being conducted by personnel at the Marshall Space Flight Center and Ames Research Center of NASA. Typical objectives of these studies are to:

a. Determine the degree of astronaut "involvement" in control prior to an emergency situation which is necessary to permit effective takeover.

b. Investigate the causes and consequences as well as the means of avoiding unnecessary control takeover.

Depending on the results of this investigation, a final recommendation will be presented on the utilization of astronaut control capabilities during the atmospheric phase of Saturn V flight.

APPENDIX

Equations of Motion.—To facilitate the discussion in this paper, a rather simplified set of equations of motion are included. Major assumptions leading to the set of equations include:

a. The symmetric configuration of Saturn V permits inertia product and aerodynamic coupling effects to be neglected.

b. Deviations from an assumed nominal trajectory are sufficiently small to permit the adequacy of perturbation equations derived with respect to the nominal.

c. The magnitude of perturbation equation variables and time derivatives are sufficiently small to permit linearization and small-angle approximations.

d. The adequacy of lumped aerodynamic effects.

e. The adequacy of the theoretical models describing the effects of flexible body and propellant sloshing phenomena.

Assuming the validity of these assumptions, the equations describing the pitch plane dynamic behavior of the vehicle can be written as follows.

Moment Equation

$$\ddot{\phi} = -c_1\alpha - c_2\beta + \ddot{\phi}_s + \ddot{\phi}_e$$

Normal Force Equation

$$\ddot{Z} = a_{1a}\Delta\phi + a_{2a}\beta + a_{3a}\alpha + \ddot{Z}_s + \ddot{Z}_e$$

Equation of the ith Bending Mode

$$\ddot{\eta}_i = -b_{1i}\dot{\eta} - b_{2i}\eta_i + b_{3i}\beta + b_{4i}\dot{\beta} + \ddot{\eta}_{is}$$

Propellant Slosh Equation of the jth Tank

$$\ddot{\xi}_j = -c_{1j}\dot{\xi}_j - c_{2j}\xi_j + c_{3j}\ddot{\phi} - \ddot{Z} - \ddot{\xi}_b$$

Angle of Attack Equation

$$\alpha_\omega = \tan^{-1}\frac{V_w}{V_m}$$

$$\alpha = \alpha_\omega + \Delta\phi - K\dot{Z}$$

Acceleration Sensed by Control Accelerometer

$$\ddot{\gamma} = a_{2a}\beta + a_{3a}\alpha + a_{4a}\ddot{\phi} + \ddot{\gamma}_b + \ddot{\gamma}_s$$

NOMENCLATURE

$\dot{x}$ — $\frac{dx}{dt}$

ϕ — vehicle attitude

$\Delta\phi$ — vehicle attitude error

$\ddot{\phi}_s$ — contribution of propellant sloshing dynamics to moment equation

$\ddot{\phi}_e$ — contribution of engine dynamics to the moment equation

$\ddot{Z}$ — acceleration normal to vehicle velocity vector

$\ddot{Z}_s$ — contribution of propellant slosh dynamics to normal acceleration

$\ddot{Z}_e$ — contribution of engine dynamics to normal acceleration

$\ddot{\eta}_i$ — generalized acceleration of the ith bending mode

$\ddot{\eta}_{is}$ — contribution of propellant slosh dynamics to the ith bending mode equation

$\ddot{\xi}_j$ — acceleration of the lumped propellant mass of the jth tank

$\ddot{\xi}_b$ — contribution of bending to the jth propellant slosh equation

α — angle of attack

α_ω — angle of attack caused by wind

V_w — velocity of assumed wind

V_m — velocity of the vehicle in still air

f — subscript denoting filter output

d — subscript denoting the full-scale value of a display variable

δ_{a_o} — control power: degrees of engine deflection for maximum controller output

$\frac{\partial M_x}{\partial\alpha}, \frac{\partial M_x}{\partial\beta}, \frac{\partial M_x}{\partial\ddot{\eta}_i}$ — coefficients of bending moment equation, dependent upon time of flight and vehicle station

REFERENCES

1. Holleman, E. C., Armstrong, N. A., and Andrews, W. H., "Utilization of the Pilot in the Launch and Injection of a Multi-Stage Orbital Vehicle," presented at the IAS 28th Annual Meeting, New York, N. Y., January 26, 1960.
2. Muckler, F. A., and Obermayer, R. W., "The Use of Man in Booster Guidance and Control," presented at the AIAA Guidance and Control Conference, August 1963.
3. Haeussermann, W., and Duncan, R. C., "Status of Guidance and Control Methods, Instrumentation, and Techniques as Applied in the Apollo Project," presented at the Lecture Series on Orbit Optimization and Advanced Guidance Instrumentation, Advisory Group for Aeronautical Research and Development, North Atlantic Treaty Organization, Duesseldorf, Germany, October 21-22, 1964.
4. Cooper, G. E., "Understanding and Interpreting Pilot Opinion," Aero. Engr. Rev., March 1957.

DISCUSSION

R. W. Gunderson and G. W. Hardy

"Piloted Guidance and Control of the Saturn V Launch Vehicle".

Q. What is the maximum time of operation of the inertial guidance system during the flight?

A. The boost phase only -- about the first two minutes; into a 100 nautical mile orbit.

Q. How do you assure that winds don't cause excessive deviation from the nominal trajectory?

A. First stage guidance is open loop. Winds cause deviations which are within the capability of the second-stage closed loop guidance system to correct.

Q. When might the astronaut take over?

A. Typically during a failure, such as an actuator loss. Also, the "adaptive" astronaut can compensate for changes in loop gain.

Q. Are there other modes in which the astronaut could take over?

A. There may be deficiencies in design, such as failure to predict elastic behavior which may require astronaut intervention.

Q. What are the sensors that allow astronauts to know they must react?

A. An accelerometer located at the center of percussion which provides angle of attack. The astronaut's control signal is filtered and sent to the rudder.

Q. Has there been simulation with a human operator?

A. Yes, but the mathematical representation of the human can be used if care is taken.

Q. What filter is used with the operator in a simulation?

A. It has poles at -6, -7, and -8.

EXPERIENCE WITH THE APPLICATIONS OF RADIO COMMAND GUIDANCE TO SATELLITE LAUNCH VEHICLES

by Dr. George H. Myers
Member of Technical Staff
Bell Telephone Laboratories, Incorporated
Whippany, New Jersey

ABSTRACT

The results of five years of experience in guiding satellites into orbit by means of radio command guidance will be described. Some flight results from the DELTA series, which includes the TIROS weather satellites; the TELSTAR, RELAY, and SYNCOM communications satellites; and various other satellites and probes will be presented and discussed.

The guidance system uses a single precision monopulse radar to measure missile position and transmits corrective orders and sequencing commands to the missile, using the beam as a communications channel. The missile-borne equipment acts only as a transponder and communications channel, which gives it a considerable weight advantage over inertial systems.

As a result of this experience, suitable methods for adapting the control system to handle faults in input data and in the communications path to the missile, have been determined. In addition, efficient methods for controlling radar look-angles; for limiting missile bending moments during aerodynamic phases; and reducing range safety dispersions have been developed. These techniques will be described, as well as the general methods used in applying a single guidance system and guidance computer program to such a diverse spectrum of satellite types.

INTRODUCTION

The Radio Command Guidance System described in this paper, designed by Bell Telephone Laboratories and built and operated by the Western Electric Company, has guided into orbit all satellites launched by the DELTA Satellite Launcher, starting in 1960 and continuing to the present time. Table 1 lists all of the missions, their purposes, and the orbits achieved, which in all cases were very close to the desired ones. Inspection of this table shows that the range of missions and orbits has been very large, ranging from space probes, through elliptical orbits for communications satellites, to the essentially circular orbits required for the TIROS Weather Satellites. Indeed, all of the communications satellites placed in orbit by the United States have been guided into orbit by this guidance system. This includes the active Telstar, Relay, and Syncom, as well as the passive Echo balloons. Also, all of the TIROS Weather Satellites have been launched into orbit in this manner. Basically the same set of guidance equations has been used for all these missions, which have performed with a very high reliability record. In the four years that this guidance system has been used for satellites, it has been found that the major control system problems are not the theoretical ones usually discussed in text books, but are of a considerably more practical nature. It turns out that so many constraints on the control system, which are not often considered, are required that the optimization turns out to be considerably different from many of the cases considered in the literature. An extra complication is that the control system actually is of quite high order. Fortunately, the problem is sufficiently constrained so that this high order has not proved to be any particular problem. In this paper, after a brief description of the operation of the Guidance System, both from a hardware and analytic standpoint, the problems encountered in actual operation, both operational and technical, will be discussed. Following this, the philosophy used in solving these problems will be presented.

GUIDANCE SYSTEM DESCRIPTION

The guidance system is illustrated in Figure 1. The missile-borne equipment, housed in the second-stage of the DELTA vehicle, serves as a radar beacon to provide return pulses to the tracking radar and as the receiving portion of the command data link between the ground and the missile. The tracking radar functions as the transmitting portion of the data link and also serves as a sensor to determine the slant range, azimuth angle, and elevation angle of the missile during its flight. The guidance computer shown in the figure was designed and manufactured by the Univac Division of the Sperry Rand Corporation.

The three-stage DELTA missile, designed by the Douglas Aircraft Company, consists of two liquid propellant stages and a solid propellant third stage. The powered flight portion of a

TABLE I

DELTA PROGRAM TO 1/31/64

Delta No.	Launch Date	Satellite Name	Satellite Function	Orbit Achieved Apogee(N.M.)/Perigee(N.M.)/ Inc.(Deg)
1	5/13/60	ECHO	Passive reflector for communications experiment. 100 ft. diameter aluminum coated mylar sphere. Payload weight equals 150 lbs.	Orbit not achieved
2	8/12/60	ECHO I	Same as above	907/812/47.2
3	11/23/60	TIROS II	Meteorological satellite. Two TV cameras for photographic cloud coverage and infrared radiometers: Payload weight equals 280 lbs.	392/337/48.5
4	3/25/61	P-14 EXPLORER X	Magnetometer measurements on strength and direction of magnetic fields in unexplored regions to 100,000 N.M. Weight equals 280 lbs.	157,000/92/33
5	7/12/61	TIROS III	Meteorological satellite. Weight equals 280 lbs.	443/397/47.9
6	8/15/61	S-3 EXPLORER XII	Energetic particle and radiation detectors to map Van Allen radiation belts. Weight equals 84 lbs.	41,723/169/33.1
7	2/8/62	TIROS IV(D)	Meteorological satellite. Weight equals 280 lbs.	452/386/48.3
8	3/17/62	S-16 OSO-I	Orbiting Solar Observator for solar measurements. Weight equals 446 lbs.	311/301/32.9
9	4/26/62	UK-1/S51 ARIEL I	United Kingdom satellite for studying the composition of the ionosphere and the effect of solar radiation on ionosphere. Weight equals 128 lbs.	646/218/53.8
10	6/19/62	TIROS V(E)	Meteorological satellite. Weight equals 280 lbs.	524/318/58.1
11	7/10/62	TELSTAR I	Active communication satellite for multi-channel communications and television transmission. Weight equals 170 lbs.	3041/515/44.8
12	9/18/62	TIROS VI(F)	Meteorological satellite. Weight equals 280 lbs.	383/369/33.1
13	10/2/62	S-3A EXPLORER XIV	Measurements on solar particles, cosmic radiation, and solar winds. Weight equals 90 lbs.	52,969/152/33
14	10/27/62	S-3B EXPLORER XV	Electron, Proton and Ion Detectors. Weight equals 98 lbs.	9514/170/18.0
15	12/13/62	RELAY I	Active Communications Satellite. Weight equals 170 lbs.	4014/713/47.5

Delta No.	Launch Date	Satellite Name	Satellite Function	Orbit Achieved Apogee(N.M.)/Perigee(N.M.)/ Inc.(Deg)
16	2/14/63	SYNCOM I	Synchronous orbit active communications satellite. Weight equals 150 lbs.	19,839/18,571/33.0
17	4/2/63	S6 EXPLORER XVII	A satellite designed to measure structure of the atmosphere in terms of density, composition and electron temperatures. Weight equals 405 lbs.	483/130/57.8
18	5/7/63	TELSTAR II	Active communications satellite. Weight equals 175 lbs.	5832/526/42.7
19	6/19/63	TIROS VII(G)	Meteorological satellite. Weight equals 300 lbs.	351/335/58.2
20	7/26/63	SYNCOM II	Synchronous orbit active communications satellite. Weight equals 147 lbs.	19,486/19,368/33.2
21	11/26/63	IMP-I EXPLORER XVIII	Interplanetary monitoring platform for solar flare measurements. Weight equals 138 lbs.	106,713/104/33.2
22	12/21/63	TIROS VIII(H)	Meteorological satellite. Weight equals 260 lbs.	407/378/58.5
23	1/21/64	RELAY II	Active communications satellite. Weight equals 185 lbs.	4000/1127/46.3

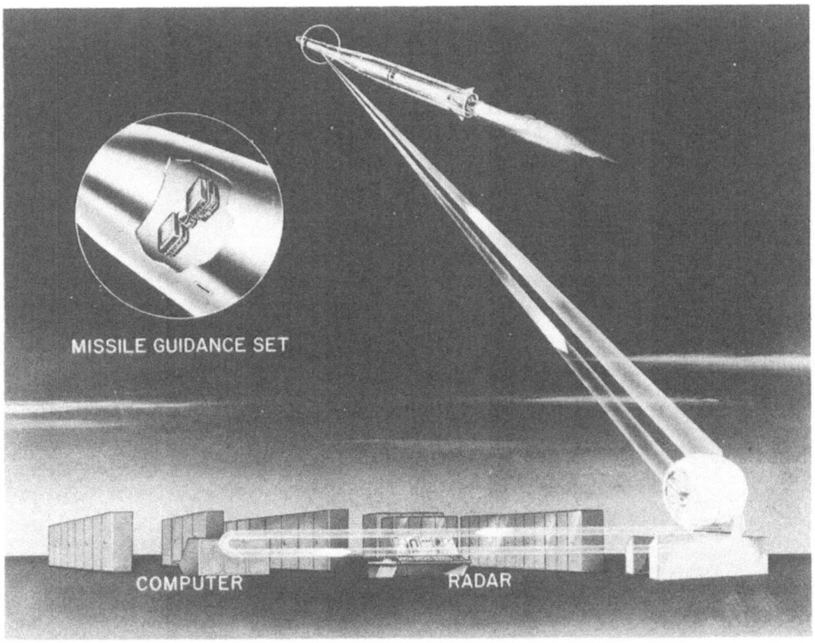

Figure 1.

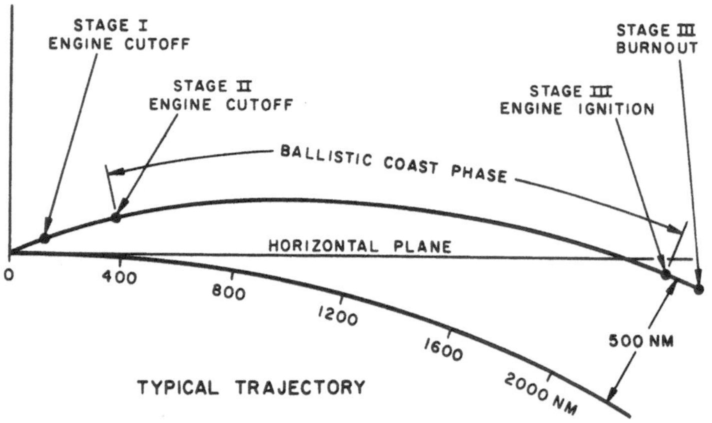

Figure 2.

typical satellite trajectory is shown in Figure 2. The guidance system transmits corrective pitch and yaw steering commands during first- and second-stage powered flight. Second-stage engine cutoff is ordered by the guidance system when the position and velocity of the missile are such that the addition of the third-stage velocity impulse at the end of the ballistic coast phase would yield the desired orbit. The unguided third-stage is spin-stabilized to maintain attitude control. For most satellite trajectories, the third-stage velocity impulse is added at the apogee of the transfer ellipse, but this is not always true. The length of the ballistic coast phase is determined by a timer in the missile which is set on the ground. It is not under the control of the guidance system in any way.

The steering and cutoff commands during the first and second stage ascent are calculated in the guidance computer, using the radar tracking data of the missile's position as the basic input information. The computer is programmed with a set of guidance equations that process the radar data and compute the desired commands to the missile. The missile-borne equipment acts only as a transponder and communications terminal, which gives it a considerable weight advantage over inertial guidance systems. Almost all satellites have been guided into orbit by radio systems, probably because of the weight and flexibility advantages of such systems.

It should be noted that no Doppler information is available from the radar; all velocities and accelerations must be determined by smoothing position data.

A block diagram of the Pitch Steering System is shown in Figure 3. A similar diagram could be drawn for yaw steering.

The computer, operating on the radar data, obtains its measure of the vehicle position in the Z direction, Z_c. After steering has started, the measured trajectory variables are compared with a reference trajectory and corrective pitch turning rates, $\hat{\dot{\theta}}$, are sent to the missile. The pitch rate programmed in the missile, $\dot{\theta}_p$, and the ordered turning rates are the inputs to the autopilot's reference integrating gyro. The function of the autopilot is to align the direction of the acceleration vector, θ, with the desired attitude, as indicated by the gyro output, $\hat{\theta}$. The coordinate system used is shown in Figure 4, where the missile roll axis points almost along the Y axis.

If the missile's roll axis is aligned with the Y axis of Figure 4, a small change in pitch attitude, θ, multiplied by the thrust acceleration, a, is the change in Z-direction acceleration, Z. Two integrations, represented by their Laplace notation in Figure 3, then give the vehicle position, Z.

In Figure 5, the programmed pitch rate and the reference trajectory it describes have been removed from the steering loop, and all of the missile dispersions are combined in a single attitude error source, E_θ. With the reference trajectory removed, the variables in the diagram represent dispersions about the reference values. The quantity to be minimized by steering may be a linear combination of some of the above variables.

Figure 6 shows the effects of θ and $\dot{Z}$ on the total insertion velocity vector of the DELTA missile with its unguided third stage (assuming θ is small) to be

$$E = \dot{Z} + V_{III}\theta \tag{1}$$

The 0 subscript in Figure 6 indicates reference values, and V_{III} is the magnitude of the velocity increment of the third stage. In complex frequency notation:

$$E(s) = \frac{a}{s}\left(1 + \frac{s}{\omega_2}\right)\theta(s) \ , \qquad \omega_2 = \frac{a}{V_{III}} \tag{2}$$

The block diagram of Figure 5 is drawn as if the control system were a continuous one, while of course it is actually a sampled data system. As a matter of fact, it is actually a multirate one. It turned out that so much time was required to perform the computations, that there was insufficient time between samples to do all the necessary calculations. Instead of slowing down the sampling rate, the sampling rate was maintained and presmoothing used to cut down the noise bandwidth. Of course, the basic signal bandwidth is so low that any reasonable smoothing will not affect it.

The optimization problem is to determine the transfer function of Y_L shown in Figure 5 which minimizes the integral over all frequencies of the power spectrum of the error signal. As stated, this problem has been solved many times. However, when the practical necessities are taken into account, the numerous constraints give an optimum which is not often discussed in the literature.

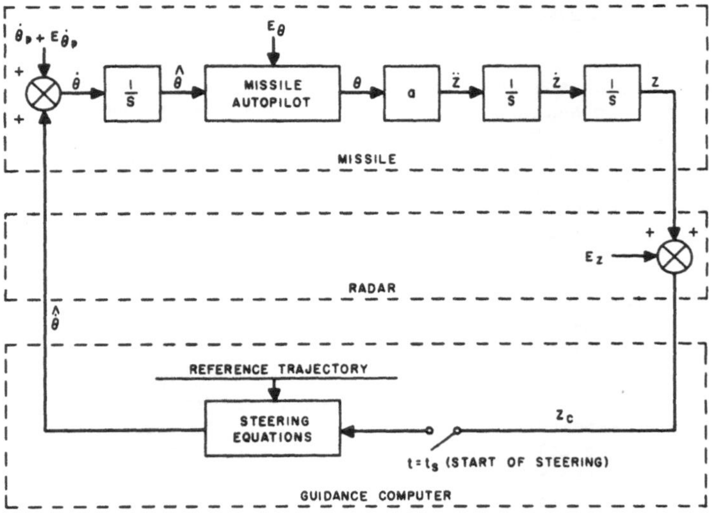

a IS THE MISSLE THRUST ACCELERATION

$\dot{\theta}_p$ IS THE PROGRAMMED TURNING RATE

$E_{\dot{\theta}_p}$ IS THE ERROR IN THE PROGRAMMED RATE

E_θ IS THE ERROR IN THE THRUST VECTOR DIRECTION

E_Z IS THE RADAR ERROR IN MEASURING THE POSITION Z

Z_C IS THE COMPUTERS MEASUREMENT OF Z

$\hat{\dot{\theta}}$ IS THE GUIDANCE ORDERED TURNING RATE

PITCH STEERING SYSTEM

Figure 3.

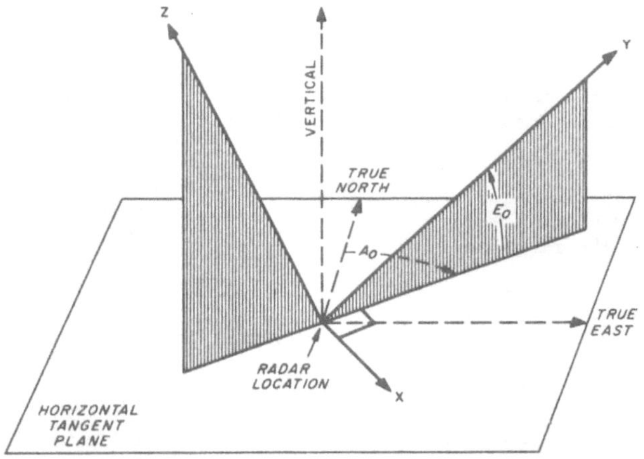

NOTES:

1. X AXIS IS IN THE HORIZONTAL PLANE.
2. Y AXIS IS AT AZIMUTH ANGLE A_O FROM TRUE NORTH AND AT ELEVATION ANGLE E_O ABOVE HORIZONTAL PLANE.
3. Z AXIS IS PERPENDICULAR TO X AND Y AXES, COMPLETING THE RIGHT-HAND CARTESIAN COORDINATE SYSTEM.

BTL COMPUTATIONAL COORDINATE SYSTEM

Figure 4.

III. OPERATIONAL AND TECHNICAL PROBLEMS

The types of problems encountered in guiding satellites can be classified as either operational or technical. Among the operational problems that are encountered are the large variety of types of missions that must be guided, and the relatively short lead times between the receipt of a reference trajectory and the scheduled launch time. In addition, there are the last-minute weight and trajectory changes desired by the payload builders, and the problem of the operator confidence. He must be given sufficient experience and indications so that he may know that everything is working properly.

Besides these operational problems, there are also numerous technical constraints. For a radio guidance system, perhaps the most important of these is maintaining a proper look-angle (the angle between the radar line-of-sight and the missile roll axis). The antenna gain is directly related to this angle and in all cases if the angle gets outside of its allowed range, there is the possibility of low signal level, either from interference patterns between the antennas, or from attenuation caused by the flame. While the look-angle requirements will be different for different missiles and antennas, such a problem will probably always exist because the missile itself is a large, conducting surface. In addition to this guidance constraint, there are several missile constraints. For example, when guiding within the atmosphere bending moments in the missile induced by guidance commands shall not be excessive. In addition, of course, fuel consumption caused by guidance cannot be too great. However, it is not fair to charge to the guidance system excess fuel required just because a missile is non-nominal (for example, because it is launched on the wrong azimuth) and therefore needs extra fuel to steer it back to the proper trajectory. In addition, precautions must be observed so that loss of radar data or loss of steering orders do not cause catastrophic results. And, of course, there is the ever-present problem of range safety.

IV. PROBLEM SOLUTION

These problems are attacked in two basic steps. The first is to satisfy all technical requirements with a reference preflight trajectory. The next is to compute steering orders with the constraint that if no more orders are sent, a nominal attitude profile will be followed for the rest of the flight, and also the correct terminal conditions will be reached (consistent with present knowledge). The fact that a nominal attitude profile is followed determines the order of the system (that is, the number of integrations in the system). Also, it requires that the reference pitch program be instrumented "on-board". The constraint also implies that only three terminal conditions at insertion into orbit can be controlled. Many optimum guidance schemes can satisfy more than three conditions, but in general such a method requires precomputation of a control sequence for the remainder of the flight. In addition, if more than three terminal conditions are to be satisfied it turns out that the flight regime is such that it is very difficult to maintain the proper look-angle constraints. Of course, when we say that the orders will produce the correct terminal conditions with no further corrections, we mean that they will only do so after the missile has responded to the orders, which means after the initial transient is over. It can be seen that a constraint of this sort, with missiles which are within tolerances, can easily be made to satisfy the look angle, structural, fuel and range safety problems. In addition, the effects of data loss can be minimized in this way, because the missile is already on a correct trajectory and additional data presumably only serves to modify the steering. As a matter of fact, it has been found that such a technique is quite insensitive to loss of data for reasonably short periods, the major effect being a slightly larger transient when tracking is recovered. The technique is almost completely insensitive to loss of isolated pieces of data, both because there is so much redundancy in the input, and because of these facts just discussed. A widely erroneous piece of data, caused by a bad bit, could produce some problems, but such data is weeded out by comparing input data with predicted gates in a standard manner. Any points which pass through these gates, even if they are in error, are heavily filtered and do not cause much harm. The problem of the very non-nominal missile is still present, and it would be desirable to guide such missiles into as good an orbit as is possible, subject to the technical constraints. This is accomplished by nonlinear limiting on both the integrated orders (that is, the total ordered attitude) and also the instantaneous orders transmitted to the missile. In addition, finite-memory polynomial smoothing is of considerable help in the case of bad data points, because such smoothing insures that data taken before a certain time has no effect on the guidance.

The philosophy described above for solving the technical constraints actually helps simplify the operational problems. Because the missile is to follow the reference trajectory as closely as possible, both the reference trajectory and the ballistic functions can be inserted in the guidance computer as polynominal approximations (including Taylor Series expansions of some of the functions about the reference trajectory). Thus, a trajectory for a particular mission can be

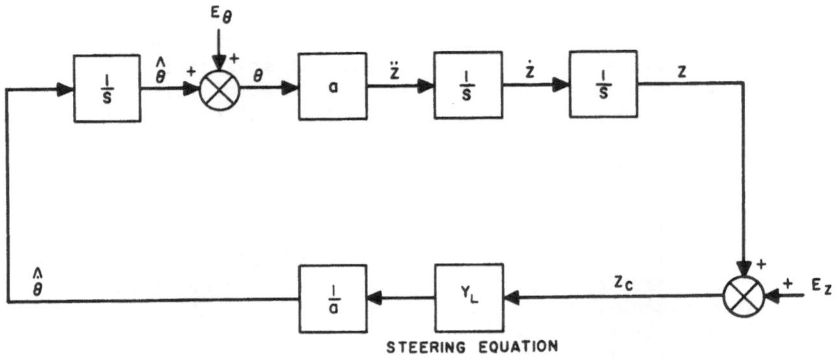

SIMPLIFIED STEERING LOOP
Figure 5.

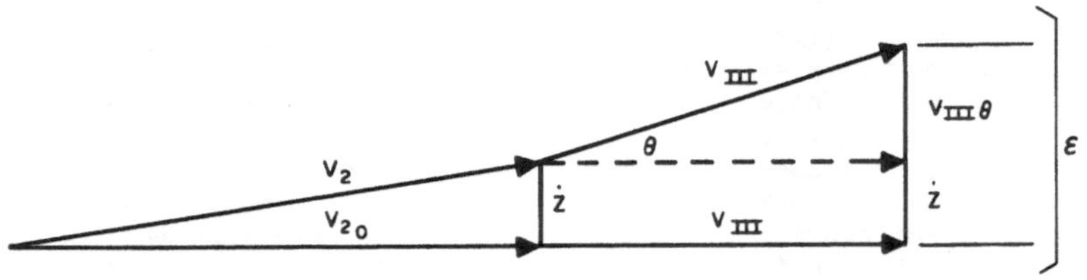

VELOCITY ERRORS
Figure 6.

represented completely by a set of constants giving both the nominal mission profile, and the action to be taken in the case of non-nominalities. In this guidance system and computer, such constants are punched on a roll of Mylar tape. This tape allows detailed checking of the constants in preflight operations and also allows automatic preparation of the punched tape by a digital computer, thus minimizing the human operations involved and therefore increasing reliability. While it is physically possible to insert constants and corrections into the computer memory by hand, in general these have not been allowed in the interest of reliability. All such corrections or changes to trajectories must be made on the punched tapes, which then can be checked. One exception to this rule has been made to permit the ever-present last minute changes in payload weight and desired orbits. It is possible to insert a few constants into the program by means of switches on the console which stay at their preset position. These are used to make the necessary last minute modifications. However, the range of correction of these handset constants is deliberately made very small, so that even if somebody sets them completely at random, no serious results will ensue. In addition, the settings of these constants can be printed out on a typewriter, thus giving a permanent record of what was set in.

Operator confidence is an important factor, even though there is practically nothing for the operator to do during a guided flight. However, in spite of this fact, it has been found highly desirable to indicate the status of the mission. This is done in two ways. An analog simulator is used before the flight to simulate the mission, and the operator can conduct simulated flights so that he may learn just what should be happening at each instant during the actual flight. In addition, numerous status lamps are activated by the guidance computer at critical periods to indicate, for example, the start of guidance; termination of powered stages; discrete signals such as cutoff transmitted to the missile; and such malfunctions as premature low acceleration, which can be caused either by a premature shutdown, or by steering which is too far off. In addition, the steering signals are presented to the operator by means of lights.

V. CONCLUSION

This paper has presented some of the operational constraints required in guiding satellites, and the way they have been handled for the DELTA series of missiles. However, it seems that such constraints and restrictions are probably both desirable and necessary for most cases of radio-guided missiles. While formal control theory gives many clues on how to properly optimize a system, many practical constraints must be considered.

VI. REFERENCES

(1) Myers, G. H. and Thompson, T. H., "Guidance of TIROS I", ARS Journal, vol. 31, pp. 636-640, May, 1961.

(2) Evans, M. J., Myers, G. H., and Timko, J. W., "Command Guidance of Telstar Launch Vehicle, BSTJ, vol. 42, pp. 2153-2168, Sept. 7, 1963.

DISCUSSION

G. H. Myers

"Experience with the Applications of Radio Command Guidance to Satellite Launch Vehicles".

Q. Is the trajectory computed during flight?

A. A pre-computed trajectory is used as the basis for guidance commands.

Q. How do you provide protection from random disturbances?

A. Subsequent sets of measurements are predicted from measured data; if the next measurements deviate by more than a preset amount they are rejected. No special protection at the output is provided.

Q. What maximum acceleration was set in the design?

A. Levels were set which were large enough to take into account all expected levels.

Q. What injection accuracies were achieved?

A. The injection errors are almost entirely due to those introduced by the unguided, timer controlled, 3rd stage.

Optimal Correction Program for Motion in a Gravitational Field Under Restricted Power

V. V. Tokarev

We wish to consider the problem of delivering maximum payload with a restricted-power propulsion system (low-thrust powerplant), taking into account correction of the trajectory. It is required to select a powerplant operational program and the proper relation between the weight of the power source and fuel store so as to ensure the maximum payload that can be delivered, on the average, with complete reliance.

For the ideally controlled propulsion system the problem reduces to one of minimizing the mathematical expectation associated with the increment in the integral of the reactive acceleration squared. We have to take into account the errors in realizing the optimal reactive acceleration program, coordinate and velocity measurement errors, and deviations from the initial conditions. We regard as already known the probabilistic characteristics of these stochastic variables (mathematical expectation, correlation function, variance). Restrictions are imposed on the variance of the final deviations of the coordinates and velocity.

The powerplant functions continuously. On the basis of measurement data taken at certain instants in time (correction times), optimal correction of the reactive acceleration program is carried out. From the condition of minimal mathematical expectation of the increment in the functional of the problem, we decide upon the optimum distribution and optimum number of correction times [1]

1. Realization Errors in the Reactive Acceleration Program and Deviations from the Computed Trajectory

The problem of maximizing the payload G_N for an ideal restricted-power system, as we know [2], is broken down into its dynamic and weight aspects. In the former it is required to minimize the integral

$$J = \int_0^T a^2 dt$$

of the square of the reactive acceleration a (thrust divided by the instantaneous mass) over the time T. In the second aspect of the problem the known value of J is used to optimize the weight of the powerplant G_N and weight of the fuel supply G_M:

$$G_N = \sqrt{\Phi} - \Phi, \quad G_M = \sqrt{\Phi} \quad \left(\Phi = \frac{\alpha}{2g} J\right)$$

so as to ensure maximum payload

$$G_N = (1 - \sqrt{\Phi})^2$$

(all weights are normalized to the initial value, α, is the specific gravity of the powerplant and is a proportionality coefficient between the weight of the powerplant and programmed power).

The extremal law for $a_*(t)$ found by solving the variational problem of minimizing J will be realized with certain error:

$$\delta \vec{a} = \vec{a}(t) - \vec{a}_*^{(0)}(t) \qquad (1.1)$$

This produces a deviation in the actual trajectory $\vec{r}(t)$ from the computed trajectory:

$$\delta_a \dot{\vec{r}}(t) = \dot{\vec{r}}(t) - \dot{\vec{r}}_*^{(0)}(t), \qquad (1.2)$$
$$\delta_a \vec{r}(t) = \vec{r}(t) + \vec{r}_*^{(0)}(t)$$

Assuming these deviations to be small, we can write down for them a linearized system of equations

$$\delta_a \ddot{\vec{r}} = \delta \vec{a} + (\partial \vec{R}/\partial \vec{r})_{\vec{r}=\vec{r}_*^{(0)}} \delta_a \vec{r} \qquad (1.3)$$

where $\vec{R} = \vec{R}(r, t)$ is the acceleration vector due to gravitational forces.

In order to obtain analytical results, the second term on the right-hand side of Equation (1.3) is neglected:

$$\delta_a \ddot{\vec{r}} = \delta \vec{a} \qquad (1.4)$$

For this purpose the initial unperturbed motion is regarded as uniform motion simulating interorbital transfer in the field of the sun, i.e., motion in a force-free field between two rest points separated by a distance L during a period T:

$$a_*^{(0)}(t) = \frac{6L}{T^2}\left(1 - 2\frac{t}{T}\right), \quad r_*^{(0)}(t) = L\frac{t^2}{T^2}\left(3 - 2\frac{t}{T}\right) \qquad (1.5)$$

To find the average deviations $\delta_a \dot{\vec{r}}, \delta_a \vec{r}$ caused by the error $\delta \vec{a}$ we need to know the probabilistic characteristics of the latter: the mathematical expectation and correlation function. Let these characteristics be given, for example:

$$M[\delta a(t)] \equiv 0 \qquad (1.6)$$
$$K[\delta a(t), \delta a(t')] = \sigma_a^2 A(t) A(t') e^{-|t-t'|/\delta t}$$

Here M and K are the symbols denoting mathematical expectation and correlation function, σ_a and δt may be interpreted as constants characterizing the accuracy and response time of the control system executing the program of acceleration due to thrust, $A(t)$ is a known nonstochastic function determined by the control mode: $A(t) = a(t)$ corresponds to control based on the relative deviation; $A(t) =$ const. corresponds to control based on the absolute deviation [it is further assumed that $A(t) \equiv L/T^2$].

Let us calculate the mathematical expectation and variance of the deviations $\delta_a \dot{r}$ and $\delta_a r$. The mathematical expectation and integration operators are permutable, so that

$$M[\delta_a \dot{r}(t)] = M\int_0^t \delta a \, dt = \int_0^t M(\delta a) \, dt = 0 \qquad (1.7)$$
$$M[\delta_a r(t)] = M\int_0^t dt \int_0^t \delta a \, dt = \int_0^t dt \int_0^t M(\delta a) \, dt = 0$$

The variances of $\delta_a \dot{r}$ and $\delta_a r$ will have nonzero values. Here we again make use of the permutability of the expectation and integration operators:

$$D[\delta_a \dot{r}(t)] = M[\delta_a \dot{r}(t) \delta_a \dot{r}(t)] = M\left[\left(\int_0^t \delta a(\xi) d\xi\right)\left(\int_0^t \delta a(\zeta) d\zeta\right)\right] =$$
$$= M\int_0^t d\xi \int_0^t [\delta a(\xi) \delta a(\zeta)] d\zeta = \int_0^t d\xi \int_0^t M[\delta a(\xi) \delta a(\zeta)] d\zeta \qquad (1.8)$$

The last integrand is, by definition, the correlation function (1.6), so that

$$D[\delta_a \dot{r}(t)] = \left(\frac{L}{T^2}\right)\sigma_a^2 \int_0^t d\xi \int_0^t e^{-|\xi-\zeta|/\delta t} d\zeta \qquad (1.9)$$

Calculating this integral and following an anlogous procedure for $\delta_a r$, we obtain a final expression for the variances of the stochastic variables (1.2) due to the error δa of the type (1.6):

$$D[\delta_a v(\theta)] = 2\sigma_a^2 \tau \left[\theta - \tau(1 - e^{-\theta/\tau})\right] \quad \left(v = \frac{2T}{L}, \theta = \frac{t}{T}, \tau = \frac{\delta t}{T}\right)$$
$$D[\delta_a \ell(\theta)] = \frac{2}{3}\sigma_a^2 \tau \left\{\theta^2 - 3\tau\left[\frac{1}{2}\theta^2 + \tau\theta e^{-\theta/\tau} - \tau^2(1 - e^{-\theta/\tau})\right]\right\} \quad \left(\ell = \frac{r}{L}\right) \qquad (1.10)$$

which is obtained by addition of the variances $D(\delta_a v)$ and $D(\delta_0 v)$, $D(\delta_a \ell)$ and $D(\delta_0 \ell)$, since the deviations δ_a and δ_0 are independent, and $D(\delta_a + \delta_0) = D(\delta_a) + D(\delta_0)$. Terms of the order τ^2 and higher are neglected here, because in the physical sense $\delta t \ll T$.

Deviations from the analytical trajectory are caused not only by the error but also by discrepancy in the realization of the initial conditions:

$$\delta_0 \dot{\vec{r}} = \dot{\vec{r}}(0) - \dot{\vec{r}}_0$$
$$\delta_0 \vec{r} = \vec{r}(0) - \vec{r}_0 \qquad (1.11)$$

The errors $\delta_0 \dot{\vec{r}}$ and $\delta_0 \vec{r}$ are assumed to be independent stochastic vectors, whose components are uncorrelated and are distributed about zero with a known standard deviation; for the uniform motion under consideration it is assumed that

$$M(\delta_0 v) = M(\delta_0 \ell) = 0 \;, \quad D(\delta_0 v) = \sigma_{0v}^2 \;, \quad D(\delta_0 \ell) = \sigma_{0\ell}^2 \qquad (1.12)$$

If toward the end of the motion the expected deviations of the actual from the computed trajectory exceed the prescribed admissible deviations $\delta \dot{r}_{MAX}$ and δr_{MAX}, then it will be necessary to correct the trajectory during the motion. In the given example, this situation amounts to violation of the system of inequalities

$$2\sigma_a^2\tau + \sigma_{0v}^2 \leq \delta v_{MAX}^2$$
$$\frac{2}{3}\sigma_a^2\tau + \sigma_{0v}^2 + \sigma_{0\ell}^2 \leq \delta \ell_{MAX}^2 \qquad (1.13)$$

2. Optimal Correction Program

We will presume that the inequalities (1.13) are not satisfied, i.e., during the motion it becomes necessary to correct the trajectory. For correction to be executed it is necessary to know the actual values of the coordinates and velocity. The accuracy of single measurements of these variables cannot be accepted as satisfactory, whence it is assumed that measurements are performed on a regular basis at sufficiently frequent intervals, and the results of the measurements performed between the times t_{i-1} and t_i ($i = 1, 2, \cdots, \mu$) are translated to the time t_i and averaged. However, a certain error will still persist:

$$\delta_n \dot{\vec{r}}_i = \dot{\vec{\rho}}(t_i) - \dot{\vec{r}}(t_i)$$
$$\delta_n \vec{r}_i = \vec{\rho}(t_i) - r(t_i) \qquad (2.1)$$

where $\vec{\rho}(t_i)$ and $\dot{\vec{\rho}}(t_i)$ are the radius vector and velocity obtained as the result of translation and averaging of the measurements. In the considered example it is supposed that

$$M(\delta_n v_i) = M(\delta_n l_i) = 0, \quad D(\delta_n v_i) = (\sigma_{nv}^{(i)})^2, \quad D(\delta_n l_i) = (\sigma_{nl}^{(i)})^2 \qquad (2.2)$$

and that the errors $\delta_n v_i$ and $\delta_n l_i$ $(i = 1, \ldots, \mu)$ are uncorrelated.

The standard deviations of the measurements $\sigma_{nv}^{(i)}$ and $\sigma_{nl}^{(i)}$ will depend on the length of the averaging interval $(t_i - t_{i-1})$. As this interval is increased they will decrease at first due to the elimination of stochastic errors, but then they will increase because in normalizing to the more remote time t_i the error δa injected by the powerplant will begin to be felt. As a result of averaging the measurements at the times t_i $(i = 1, \ldots, \mu)$, the deviations of the actual trajectory from the $(i-1)$th computed trajectory $r_*^{(i-1)}(t)$ will be unknown with a certain degree of error:

$$\Delta \dot{\vec{r}}_i = \dot{\vec{\rho}}(t_i) - \dot{\vec{r}}_*^{(i-1)}(t_i), \quad \Delta \vec{r}_i = \vec{\rho}(t_i) - \vec{r}_*^{(i-1)}(t_i) \qquad (2.3)$$

The corrections to the thrust acceleration program $\vec{a}_*(t)$ are chosen from the condition of optimality "in the large"; the optimum correction

$$\Delta \vec{a}_*^{(i)}(t) = \vec{a}_*^{(i)}(t) - \vec{a}_*^{(i-1)}(t) \qquad (2.4)$$

computed at the instant t_i (ith correction time) must ensure minimization of the functional

$$J_i = \int_{t_i}^{T} a^2 dt \qquad (2.5)$$

during motion between the points $\{\vec{\rho}(t_i), \dot{\vec{\rho}}(t_i)\}$ and $\{\vec{r}_f, \dot{\vec{r}}_f\}$ (predetermined end point) in phase space during the time $\Delta t_i = T - t_i$.

Thus it is demanded of the correction $\Delta \vec{a}_*^{(i)}$ that it reduce the deviations $\Delta \dot{\vec{r}}_i$ and $\Delta \vec{r}_i$ to zero at the end of the trajectory with minimum power consumption. It is not required to come out on the computed trajectory at the end of the motion.

It is assumed that errors are not introduced in computation of the correction $\Delta \vec{a}_*^{(i)}$. For uniform motion in a force-free field the correction to the acceleration program is determined by the following relation

$$\Delta a_*^{(i)}(\theta) = \frac{6L}{T^2} \frac{1}{1-\theta_i} \left[\left(2 \frac{\Delta l_i}{1-\theta_i} + \Delta v_i\right) \frac{\theta - \theta_i}{1-\theta_i} - \left(\frac{\Delta l_i}{1-\theta_i} + \frac{2}{3} \Delta v_i\right) \right] \qquad (2.6)$$

In the time interval between instants t_i and t_{i+1} must be executed the program $\vec{a}_*^{(i)}(t)$, which at the time t_{i+1} is changed over to the new optimal program $\vec{a}_*^{(i+1)}(t)$, and so forth.

In the case of motion with errors and corrections the expression for the functional J can be written in the following form:

$$J = \sum_{i=0}^{\mu} \int_{t_i}^{t_{i+1}} (\vec{a}_*^{(i)} + \delta \vec{a})^2 dt, \quad (t_0 = 0, t_{\mu+1} = T) \qquad (2.7)$$

In a force-free field the equations for the optimal corrections are linear, so that

$$a_*^{(i)}(t) = a_*^{(0)}(t) + \Delta a_*^{(1)}(t) + \cdots + \Delta a_*^{(i)}(t) \qquad (2.8)$$

which permits the functional J to be transformed as follows:

$$\Delta J = J - J_0 = T\Big\{ 2\Big[\sum_{i=1}^{\mu} \int_{\theta_i}^{1} a_*^{(0)} \Delta a_*^{(i)} d\theta + \int_0^1 a_*^{(0)} \delta a \, d\theta \Big] + 2\Big[\sum_{i=3}^{\mu} \int_{\theta_i}^{1} \Delta a_*^{(i)} \Big(\sum_{j=1}^{i-2} \Delta a_*^{(j)} \Big) d\theta + \sum_{i=1}^{\mu} \int_{\theta_i}^{1} \Delta a_*^{(i)} \delta a \, d\theta \Big] + \Big[2 \sum_{i=2}^{\mu} \int_{\theta_i}^{1} \Delta a_*^{(i)} \Delta a_*^{(i-1)} d\theta + \sum_{i=1}^{\mu} \int_{\theta_i}^{1} (\Delta a_*^{(i)})^2 d\theta + \int_0^1 (\delta a)^2 d\theta \Big] \Big\} \qquad (2.9)$$

where

$$J_0 = T\int_0^1 (a_*^{(0)})^2 d\theta = \frac{12L^2}{T^3} \qquad (2.10)$$

Assuming that $\Delta J/J \ll 1$ and stopping with linear terms relative to $\Delta J/J_0$, we obtain an approximate expression for the payload for motion with corrections:

$$G_n \approx (1-\sqrt{\phi_0})^2 - \Delta\phi\left(\frac{1}{\sqrt{\phi_0}} - 1\right) \qquad (2.11)$$
$$\left(\Delta\phi = \frac{\alpha}{2g}\Delta J,\ \phi_0 = \frac{\alpha}{2g} J_0\right)$$

For minimization of the averaged quantity G_n it is necessary to determine the distribution of correction times θ_i $(i=1,\ldots,M)$ and their M such that, with the stipulation of guaranteeing a prescribed accuracy (on the average) at the end of the motion, the mathematical expectation of ΔJ will be a minimum.

To calculate $M(\Delta J)$ it is necessary to know the expectation of the integrands in (2.9). These expressions include the corrections $\Delta a_*^{(i)}$, the errors δa, and their derivatives. The corrections $\Delta a_*^{(i)}$, according to Equation (2.6), depend on the total deviations Δv_i and Δl_i. The latter are comprised of the deviations due to the error δa and of the error incurred in the $(i-1)$th and ith measurements. With the aid of Equations (1.2), (1.4), (2.1), and (2.3) it can be shown that

$$\Delta v_i = \frac{T^2}{L}\int_{\theta_{i-1}}^{\theta_i} \delta a\, d\theta + \delta_n v_i - \delta_n v_{i-1}$$
$$\Delta l_i = \frac{T}{L}\int_{\theta_{i-1}}^{\theta_i} d\theta \int_{\theta_{i-1}}^{\theta} \delta a\, d\theta + \delta_n l_i - \delta_n l_{i-1} - \delta_n v_i(\theta_i - \theta_{i-1}) \qquad (2.12)$$

The expectation of the terms contained in the first brackets of Equation (2.9) is equal to zero, since the integrands of these terms are linear with respect to the stochastic variables Δv_i, Δl_i, and δa, the expectation values of which are equal to zero (1.6), (1.7).

Considering that the stochastic deviations $\delta a(t)$ before and after the instant t_i of changeover to the new program $a_*^{(i)}(t)$ are uncorrelated:

$$K[\delta a(t), \delta a(t')] = 0 \quad \text{for}$$
$$t_{i-1} < t < t_i,\quad t_{j-1} < t' < t_j,\quad i \neq j \qquad (2.13)$$

we obtain the same result for the second bracketed expression in Equation (2.9), since the indices of the factors $\Delta a_*^{(i)}$ differ by more than unity. This means, on the basis of (2.12), that the factors are uncorrelated and their mathematical expectations are each equal to zero.

The first two terms contained in the last bracketed expression in (2.9) are expressed as follows in terms of the deviations Δv_{i-1}, Δv_i, Δl_{i-1}, and Δl_i:

$$\int_{\theta_i}^1 \Delta a_*^{(i)} \Delta a_*^{(i-1)} d\theta = \frac{12L^2}{T^3}\frac{1}{1-\theta_{i-1}}\left[\frac{\Delta l_i \Delta l_{i-1}}{(1-\theta_{i-1})^2} + \frac{\Delta l_i \Delta v_{i-1}}{2(1-\theta_{i-1})} + \left(1 - 2\frac{\theta_i-\theta_{i-1}}{1-\theta_{i-1}}\right)\frac{\Delta v_i \Delta l_{i-1}}{2(1-\theta_{i-1})} + \left(\frac{1}{3} - \frac{1}{2}\frac{\theta_i - \theta_{i-1}}{1-\theta_i}\right)\Delta v_i \Delta v_{i-1}\right]$$
$$\int_{\theta_i}^1 (\Delta a_*^{(i)})^2 d\theta = \frac{12L^2}{T^3}\frac{1}{1-\theta_i}\left[\frac{(\Delta l_i)^2}{(1-\theta_i)^2} + \frac{\Delta l_i \Delta v_i}{1-\theta_i} + \frac{1}{3}(\Delta v_i)^2\right] \qquad (2.14)$$

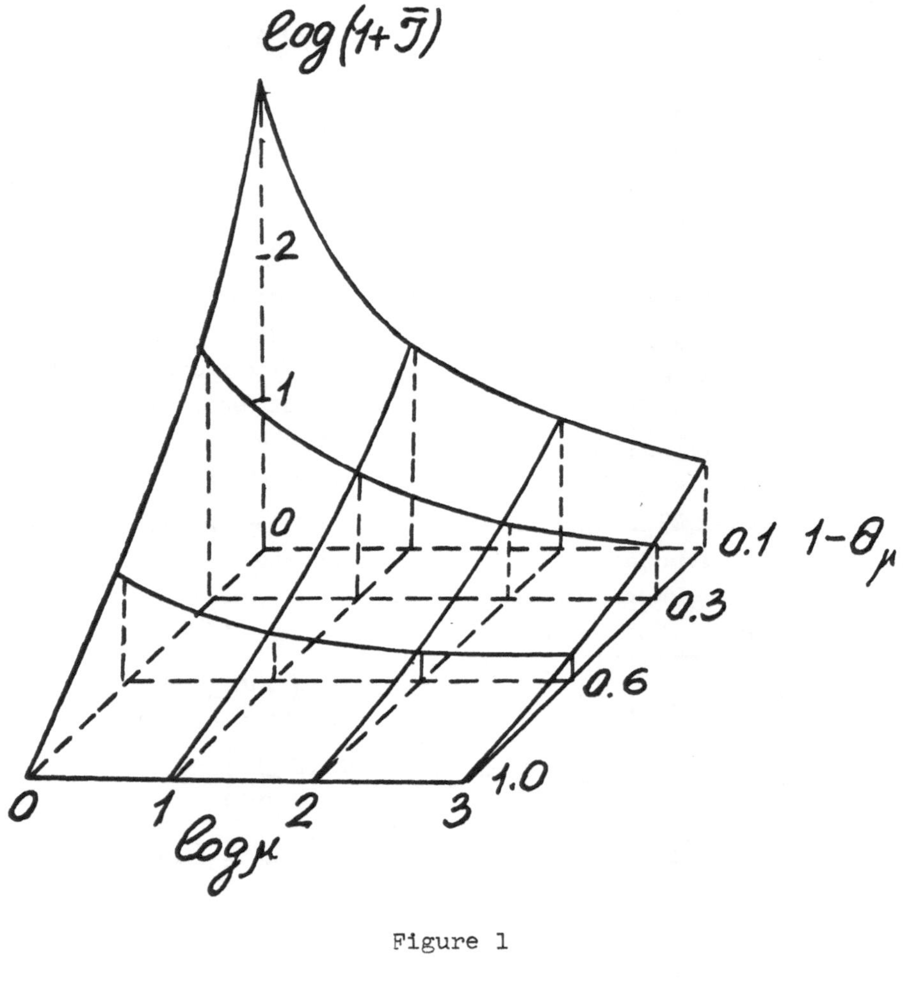

Figure 1

Consequently, the expectation value of the increment in the functional, correct to squared terms in τ, is equal to

$$M(\Delta J) = \frac{12L^2}{T^3}\left\{\sum_{i=1}^{M}\left[\frac{1}{3}\sigma_a^2\tau\left(2\xi_i^3 - 3\xi_i^2 + 2\xi_i - 1\right) + \frac{(\sigma_{nv}^{(i-1)})^2}{1-\theta_i}\xi_i(\xi_i - 1) + \frac{(\sigma_{nv}^{(i-1)})^2 - (\sigma_{nv}^{(i)})^2}{3(1-\theta_i)} + \frac{(\sigma_{nl}^{(i-1)})^2 - (\sigma_{nl}^{(i)})^2}{(1-\theta_i)^3}\right] + \frac{\sigma_a^2}{12} + \frac{2}{3}\frac{(\sigma_{nv}^{(M)})^2}{1-\theta_M} + \frac{(\sigma_{nl}^{(M)})^2}{(1-\theta_M)^3}\right\} \quad (2.15)$$

$$\left(\xi_i = \frac{1-\theta_{i-1}}{1-\theta_i}, \quad \sigma_{nv}^{(0)} = \sigma_{0v}, \quad \sigma_{nl}^{(0)} = \sigma_{nl}\right)$$

At the end point of the motion ($\theta = 1$) the trajectory must fall within the prescribed region of final values of the coordinates and velocity, i.e., the expected deviations from the computed trajectory at the end point must have an upper bound. This condition is spelled out in the form of two inequalities analogous to (1.13):

$$2\sigma_a\tau(1-\theta_M) + (\sigma_{nv}^{(M)})^2 \le (\delta v_{max})^2 \quad (2.16)$$
$$\frac{2}{3}\sigma_a^2\tau(1-\theta_M)^3 + (\sigma_{nv}^{(M)})^2(1-\theta_M) + (\sigma_{nl}^{(M)})^2 \le (\delta l_{max})^2$$

These inequalities impose limitations on θ_M in seeking to minimize $M(\Delta J)$ with respect to θ_i.

If the measurement error is disregarded $[\sigma_{nv}^{(i)} = \sigma_{nl}^{(i)} = 0, \ i = 1, \ldots, M]$ and if the initial conditions are assumed to be exactly realized ($\sigma_{0v} = \sigma_{0l} = 0$), then for the optimum distribution of correction times we obtain a geometric progression:

$$(\xi_i)_{opt} = \left(\frac{1-\theta_{i-1}}{1-\theta_i}\right)_{opt} = (1-\theta_M)^{-\frac{1}{M}} \quad (2.17)$$
$$(i = 1, \ldots, M)$$

The minimum expectation value for the increment in the functional, for a given number of correction times,

$$M(\Delta J)_{min} = \frac{12L^2}{T^3}\sigma_a^2\left(\frac{\tau}{3}\bar{J} + \frac{1}{12}\right) \quad (2.18)$$

$$\left(\bar{J} = M\left[\frac{2}{(1-\theta_M)^{3/M}} - \frac{3}{(1-\theta_M)^{2/M}} + \frac{2}{(1-\theta_M)^{1/M}} - 1\right]\right)$$

decreases monotonically with increasing M, where

$$\lim \bar{J} = -\ln(1-\theta_M) \quad \text{as } M \to \infty$$

[see log ($1 + \bar{J}$) in Figure 1 as a function of M and ($1 - \theta_M$)]. It is also apparent that $\bar{J}$ decreases with increasing ($1-\theta_M$) so that the optimum choice of time for the last correction must be made from the inequalities (2.16) so as to maximize the distance at that instant from the end of the trajectory:

$$(1-\theta_M)_{opt} = \min\left\{\frac{1}{2}\frac{\delta v_{max}^2}{\sigma_a^2\tau}, \left(\frac{3}{2}\frac{\delta l_{max}^2}{\sigma_a^2\tau}\right)^{1/3}\right\} \quad (2.19)$$

In the case of a nonzero constant variance in the measurement error $[\sigma_{nv}^{(i)} = \sigma_v, \sigma_{nl}^{(i)} = \sigma_l, i = 0, 1, \dots \mu]$ the optimum distribution of correction times is no longer expressed by such a simple equation as (2.17) and must be determined numerically. If, however, Equation (2.17) is used anyway, the nature of the dependence of $M[\Delta J]$ on μ and $(1 - \theta_\mu)$ will be preserved.

Conclusion

In this paper the problem of optimal correction is stated such that just before realization of the trajectory the optimum (on the average) weight proportions and optimum reactive acceleration programs are chosen so as to ensure the maximum payload that can be delivered with complete reliance.

It is found that the optimum distribution of correction times in the case of exact measurements obeys a geometric progression, Equation (2.17). The correction times fall closer and closer together toward the end of the trajectory. The instant of the last correction must be as far as possible from the trajectory end point (in guaranteeing impact within a predetermined region of final coordinates and velocity).

The clustering of correction times toward the trajectory end point is fully reasonable. At the beginning of the motion considerable time is left in which to eliminate deviations from the trajectory, and these deviations can be allowed to take on fairly large values with a sufficiently small level of correction Δa (2.16), i.e., the vehicle can move for a longer period of time without correction than near the end of the motion. The same distribution law governing the correction times is also obtained by Lawden [3], who investigates the impulsive correction of ballistic trajectories. In contrast with impulsive correction, however, where the error is introduced solely at the instants the impulses are applied, in the case investigated in the present paper, where error is continuously accumulated (regardless of whether or not a correction is actually made) with the stipulation of exact measurements, there does not exist an optimum number of correction times.

However, if the variance of the measurement errors is a nonzero quantity and depends on the intervals between correction instants, the existence of an optimum number of correction times, for which $M(\Delta J)$ has a minimum, is inferred from Equation (2.15).

Literature Cited:

1. V. V. Tokarev, "Influence of Random Deviations from the Optimum Thrust Program on the Motion in a Graviational Field of a Variable-Mass Body with Constant Power Consumption", Prikl. Matem. i Mekhan., 27, No. 1 (1963) [Appl. Math. Mech., Vol. 27, p. 33].
2. G. L. Grodzovskii, Yu. N. Ivanov, and V. V. Tokarev, "Mechanics of Space Flight with Low Thrust", Inzhenernyi Zhurnal, 3, No. 3 (1963).
3. D. F. Lawden, "Optimal Programm for Correctional Maneuvers", Astronaut. Acta, 6, No. 4 (1960).

DISCUSSION

V. V. Tokarev

"An Optimal Program
of Correction of Motion using
Limited Power in a Gravitation Field".

Q. What quantities are measured periodically?

A. Speed at given times during flight.

Q. By speed, you mean vector velocity?

A. It is according to formula (3).

OPTIMIZATION OF MIDCOURSE VELOCITY CORRECTIONS

by Dr. Robert G. Stern and Dr. James E. Potter

Staff Engineers Experimental Astronomy Laboratory
Massachusetts Institute of Technology
Cambridge, Massachusetts

ABSTRACT

The concept of a six-dimensional state space is used to develop the fundamental equations of linearized midcourse guidance. Both fixed end-point (fixed-time-of-arrival) and variable end-point (variable-time-of-arrival) problems are considered. It is shown that the variable-time-of-arrival problem can be simplified mathematically by the introduction of a special coordinate system, which is called the critical-plane coordinate system.

A method is developed for determining the optimum time at which to apply a single midcourse correction the effect of which is to satisfy a set of position constraints. The correction is "optimum" in the sense that its magnitude is minimized. For a given nominal trajectory, the time of the correction depends on the predicted miss vector at the destination. The method is particularly simple to apply in the case of variable-time-of-arrival guidance; by exploiting the critical-plane coordinate system, a single curve can be prepared prior to the flight to indicate the optimum correction time as a function of a miss parameter which is determined from in-flight navigational measurements.

Multiple-correction strategies are then investigated. A method is developed for determining an optimum schedule of midcourse corrections. The optimum schedule is the one for which the sum of the magnitudes of all corrections is minimized. It is proved that the number of corrections in an optimum schedule is no greater than the number of constraints to be satisfied at the nominal time of arrival at the destination.

In position-constrained variable-time-of-arrival guidance there are only two constraints at the nominal time of arrival; hence there are at most two corrections in the optimum schedule. The optimum two-correction strategy is compared with the optimum single-correction strategy. It is shown that for certain ranges of the miss parameter two corrections can effect a saving in total magnitude of velocity correction, while in other ranges no improvement can be obtained from two corrections. A geometric construction, based on the theory of convex sets, is used to determine the ranges of miss parameter in which two corrections are preferable, and also the times and components of both corrections when they are preferable.

It may be noted that the developments in this paper are deterministic rather than statistical. No consideration is given to the uncertainties of the navigational measurements; it is assumed that a sufficient number of measurements has been made during the flight so that the error in the predicted miss vector at the destination is negligible. The control action taken is determined by the predicted miss vector.

1. INTRODUCTION

Midcourse guidance systems for space vehicles can be separated into two classes – fixed-time-of-arrival (FTA) systems and variable-time-of-arrival (VTA) systems. In the former a set of constraints on the vehicle's state relative to the target body must be satisfied at a specified terminal time; in the latter the same types of constraints must be satisfied, but the actual terminal time at which they are satisfied is permitted to vary slightly from the predetermined nominal terminal time. By relaxing the specification on time of arrival, the VTA system permits more flexibility in the development of guidance laws.

The constraints most commonly specified are the three components of the vehicle's position relative to the target. Battin[1]* Noton, Cutting, and Barnes[2], and McLean, Schmidt, and McGee[3], among others, have shown how linear theory can be applied to compute the components of a single small step change in velocity that will satisfy a set of three position constraints. The theory depends on the availability of a precomputed reference trajectory relative to which the vehicle's actual trajectory can be defined. The components of the required velocity change, usually referred to as the "correction," vary linearly with the predicted deviation of the vehicle's actual position from the desired position at the nominal time of arrival at the destination.

Breakwell[4], [5] attacks the problem of selecting the optimum times at which a series of midcourse corrections should be made. The optimization criterion is that the total fuel expenditure (i.e., the sum of the magnitudes of the velocity steps) be minimized. The analysis is statistical, based on a priori knowledge of the variances of the uncertainties in the vehicle's initial state, in the observations, and in the corrections applied. A more recent statistical study, containing computer results for a number of simulated interplanetary missions, has been made by White, Callas, and Cicolani.[6]

The present paper, like those just cited, deals with optimizing a correction schedule. First, a method is developed for determining the optimum

*Superior numbers refer to similarly-numbered references at the end of this paper.

time at which to apply a single midcourse correction that satisfies all the position constraints. Then optimum multiple-correction strategies are investigated and compared with the optimum single-correction strategy. The strategies developed in this paper, unlike those in the previous papers, are deterministic rather than statistical. The correction schedule depends on the predicted position variation at the nominal time of arrival at the destination; it does not depend on the uncertainties in the measured or controlled quantities.

As a preliminary to the development of the optimization procedure with which this paper is primarily concerned, the next two sections formulate the basic equations of midcourse guidance and describe the critical-plane coordinate system, which simplifies the analysis of VTA guidance.

2. MIDCOURSE GUIDANCE THEORY

The position and velocity of a space vehicle on its trajectory can be represented by the six-component state vector $\underline{x}$.

$$\underline{x} = \begin{bmatrix} \underline{r} \\ \underline{v} \end{bmatrix} \tag{2.1}$$

$\underline{r}$ and $\underline{v}$ are the position and velocity vectors, respectively, of the vehicle with respect to the origin of the coordinate system. The equations of motion of the vehicle have the form

$$\dot{\underline{x}} = \underline{f}(\underline{x}, t) + \overset{*}{M} \sum_{k=1}^{p} \underline{c}_k \, \delta(t - t_k) \tag{2.2}$$

$\underline{f}$ is the rate of change of the state vector with time in the absence of any control action. $\underline{c}_k$ is a corrective velocity change resulting from an acceleration impulse applied at time t_k. $\delta(t - t_k)$ is the Dirac delta function at t_k. There are p impulses during the flight. $\overset{*}{M}$ is a 6-by-3 compatibility matrix, relating the six-dimensional $\underline{x}$ to the three-dimensional $\underline{c}_k$

$$\overset{*}{M} = \begin{bmatrix} \overset{*}{0}_3 \\ \overset{*}{I}_3 \end{bmatrix} \tag{2.3}$$

$\overset{*}{0}_3$ and $\overset{*}{I}_3$ are the 3-by-3 zero matrix and the 3-by-3 identity matrix, respectively.

The solution of Eq. (2.2) with ideal initial conditions and no corrective impulses constitutes the nominal, or reference, trajectory. This solution is precomputed numerically and stored for use during the flight. Because the actual initial conditions are not ideal, inflight corrections are required; the corrective acceleration at time t_k produces a step change $\underline{c}_k$ in the velocity at t_k. To first order in the variation $\delta\underline{x}_I$ of the initial state from its nominal value and to first order in the velocity corrections $\underline{c}_k$, the variation $\delta\underline{x}_D$ in the state vector at the nominal time of arrival at the destination is given by

$$\delta\underline{x}_D = \overset{*}{\Phi}_{DI}\, \delta\underline{x}_I + \sum_{k=1}^{p} \overset{*}{\Phi}_{Dk} \overset{*}{M} \underline{c}_k$$

$$= \delta\underline{x}_D^- + \sum_{k=1}^{p} \overset{*}{\Phi}_{Dk} \overset{*}{M} \underline{c}_k \tag{2.4}$$

$\delta\underline{x}_D^-$ is the state vector variation that would exist at the nominal time of arrival if no corrections were applied. $\overset{*}{\Phi}_{ji}$ is the 6-by-6 state transition matrix.

$$\overset{*}{\Phi}_{ji} = \left[\frac{\partial \underline{x}_j}{\partial \underline{x}_i} \right] \tag{2.5}$$

For fixed t_i and variable t_j the elements of $\overset{*}{\Phi}_{ji}$ can be determined by numerical integration, as shown in Ref. (1). It is convenient for analytic work to partition $\overset{*}{\Phi}_{ji}$ into four 3-by-3 submatrices.

$$\overset{*}{\Phi}_{ji} = \left[\begin{array}{c|c} \overset{*}{A}_{ji} & \overset{*}{B}_{ji} \\ \hline \overset{*}{C}_{ji} & \overset{*}{D}_{ji} \end{array} \right] \tag{2.6}$$

Let the number of constraints to be satisfied at the destination be m. In general, the constraints are functions of the position and velocity (i.e., of the state) of the vehicle relative to the target at the actual time of arrival. In FTA guidance the actual arrival time t_A is the same as the nominal arrival time t_D; in VTA guidance the two arrival times are usually not the same. The relative state of the vehicle at t_A is designated $\underline{x}_{RA}$.

$$\underline{x}_{RA} = \underline{x}_A - \underline{x}_{TA} \tag{2.7}$$

$\underline{x}_A$ and $\underline{x}_{TA}$ are, respectively, the state of the vehicle and the state of the target at t_A. The constraint equations may be written in the form

$$\gamma_k(\underline{x}_{RA}) = 0 \ ; \ k = 1, \ldots\ldots, m \tag{2.8}$$

Since $\underline{x}_{RA}$ is a six-component vector, the maximum number of linearly independent constraints is six. The vector form of (2.8) is

$$\underline{\gamma} = \underline{0}_m \tag{2.9}$$

where $\underline{0}_m$ is the m-component zero vector.

If the vehicle is on the reference trajectory, all the constraints are satisfied at t_D. On the actual trajectory, satisfying a particular constraint γ_k requires that, to first order,

$$\delta\gamma_k = \underline{h}_k^T (\underline{x}_{RA} - \underline{x}_{RD}) = 0 \tag{2.10}$$

where $\underline{h}_k$ is a six-component column vector of partial derivatives.

$$\underline{h}_k = \left[\frac{\partial \gamma_k}{\partial \underline{x}} \;\middle|\; t = t_D \right] \tag{2.11}$$

$\underline{x}_{RD}$ is the relative state at t_D. Note that $\delta \underline{x}_D$ is the state variation of the actual trajectory from the reference trajectory at the nominal time of arrival, while $(\underline{x}_{RA} - \underline{x}_{RD})$ is the difference between the relative state of the actual trajectory at the actual time of arrival and the relative state of the reference trajectory at the nominal time of arrival. The difference in relative states can be determined as follows:

$$\begin{aligned} \underline{x}_{RA} &= (\underline{x}_D + \delta\underline{x}_D + \underline{f}_D\, \delta t_D) \\ &\quad - (\underline{x}_{TD} + \underline{f}_{TD}\, \delta t_D) \\ &= \underline{x}_{RD} + \delta\underline{x}_D + \underline{f}_{RD}\, \delta t_D \end{aligned} \tag{2.12}$$

$$\underline{x}_{RA} - \underline{x}_{RD} = \delta\underline{x}_D + \underline{f}_{RD}\, \delta t_D \tag{2.13}$$

δt_D is the change in time of arrival. $\underline{f}_{RD}$ is the time rate of change of the relative state at t_D.

$$\delta t_D = t_A - t_D \tag{2.14}$$

$$\underline{f}_{RD} = \underline{f}_D - \underline{f}_{TD} \tag{2.15}$$

The variational form of the i-th constraint equation can now be written as

$$\begin{aligned} \delta\gamma_i &= 0 = \underline{h}_i^T\, (\delta\underline{x}_D + \underline{f}_{RD}\, \delta t_D) \\ &= \underline{h}_i^T \left[\delta\underline{x}_D^- + \sum_{k=1}^{p} \overset{*}{\Phi}_{Dk}\, \overset{*}{M} \underline{c}_k + \underline{f}_{RD}\, \delta t_D \right] \end{aligned} \tag{2.16}$$

For FTA guidance δt_D is equal to zero. All m constraints can be combined into a single vector equation.

$$\delta \underline{\gamma}_{FTA} = \overset{*}{H}\, \delta\underline{x}_D^- + \overset{*}{H} \sum_{k=1}^{p} \overset{*}{\Phi}_{Dk}\, \overset{*}{M} \underline{c}_k = \underline{0}_m \tag{2.17}$$

where the m-by-6 matrix $\overset{*}{H}$ is given by

$$\overset{*}{H} = \begin{bmatrix} \underline{h}_1^T \\ \cdot \\ \cdot \\ \cdot \\ \underline{h}_m^T \end{bmatrix} \tag{2.18}$$

In VTA guidance one of the constrants can be used as a "stopping" condition, that is, as a means of determining δt_D. This is true because δt_D is not an independent variable but is actually dependent on $\delta \underline{x}_D$. The constraint used for this determination must be one for which the scalar product of $\underline{h}_k$ and $\underline{f}_{RD}$ is not equal to zero. Suppose that γ_m meets this requirement. Then, from (2.16),

$$\delta t_D = \frac{-\underline{h}_m^T \left[\delta\underline{x}_D^- + \sum_{k=1}^{p} \overset{*}{\Phi}_{Dk}\, \overset{*}{M} \underline{c}_k \right]}{\underline{h}_m^T\, \underline{f}_{RD}} \tag{2.19}$$

The variational equation of the i-th constraint (where $i \neq m$) becomes

$$\delta\gamma_i = \underline{h}_i^T \left[\overset{*}{I}_6 - \frac{\underline{f}_{RD}\, \underline{h}_m^T}{\underline{h}_m^T\, \underline{f}_{RD}} \right] \left[\delta\underline{x}_D^- + \sum_{k=1}^{p} \overset{*}{\Phi}_{Dk}\, \overset{*}{M} \underline{c}_k \right] = 0 \tag{2.20}$$

There are (m - 1) independent relations of this type that must be satisfied. These are combined in a vector equation as follows:

$$\delta \underline{\gamma}_{VTA} = \overset{*}{L}\, \delta\underline{x}_D^- + \overset{*}{L} \sum_{k=1}^{p} \overset{*}{\Phi}_{Dk}\, \overset{*}{M} \underline{c}_k = \underline{0}_{m-1} \tag{2.21}$$

where

$$\overset{*}{L} = \begin{bmatrix} \underline{h}_1^T \\ \cdot \\ \cdot \\ \cdot \\ \underline{h}_{m-1}^T \end{bmatrix} \left[\overset{*}{I}_6 - \frac{\underline{f}_{RD}\, \underline{h}_m^T}{\underline{f}_{RD}^T\, \underline{h}_m} \right] \tag{2.22}$$

Equation (2.21) indicates that m constraints at the variable end point t_A in VTA guidance are equivalent to only (m - 1) constraints at the nominal end point t_D.

Constraint equations (2.17) and (2.21) may be generalized in the single equation

$$\sum_{k=1}^{p} \overset{*}{P}_k\, \underline{c}_k + \underline{\rho} = \underline{0}_n \tag{2.23}$$

where

$$(\overset{*}{P}_k)_{FTA} = \overset{*}{H} \overset{*}{\Phi}_{Dk} \overset{*}{M} \tag{2.24}$$

$$(\overset{*}{P}_k)_{VTA} = \overset{*}{L} \overset{*}{\Phi}_{Dk} \overset{*}{M} \tag{2.25}$$

$$(\underline{\rho})_{FTA} = H \delta \underline{x}_D^- \tag{2.26}$$

$$(\underline{\rho})_{VTA} = \overset{*}{L}\delta\underline{x}_D^- \tag{2.27}$$

n is the number of constraints at t_D. The n-dimensional vector $\underline{\rho}$ is known as the miss vector.

The preceding discussion applies to arbitrary constraints, six or less in number, at the time of arrival at the destination. In many guidance systems the only quantities constrained are the components of the vehicle's final position relative to the target. For such systems Eqs. (2.17) and (2.21) can be simplified considerably, as shown below.

When final position relative to the target is completely specified and there are no other constraints, the basic vector equation of the constraints is

$$\underline{\gamma} = \underline{r}_{RA} = \underline{0}_3 \tag{2.28}$$

For FTA guidance

$$\delta\underline{\gamma}_{FTA} = \delta\underline{r}_D \tag{2.29}$$

$$\overset{*}{H} = \left[\overset{*}{I}_3 \;\vdots\; \overset{*}{0}_3\right] \tag{2.30}$$

$$(\overset{*}{P}_k)_{FTA} = \overset{*}{B}_{Dk} \tag{2.31}$$

so that the final guidance equations are given by

$$\delta\underline{\gamma}_{FTA} = \sum_{k=1}^{p} \overset{*}{B}_{Dk}\underline{c}_k + \delta\underline{r}_D^- = \underline{0}_3 \tag{2.32}$$

For VTA guidance, $\underline{h}_m^T$ can be taken as

$$\underline{h}_m^T = [0 \quad 0 \quad 1 \quad 0 \quad 0 \quad 0]$$

Then

$$\overset{*}{L} = \begin{bmatrix} 1 & 0 & 0 & 0 & 0 & 0 \\ 0 & 1 & 0 & 0 & 0 & 0 \end{bmatrix} \left[\overset{*}{I}_6 - \frac{\underline{f}_{RD}\,\underline{h}_m^T}{\underline{f}_{RD}^T\,\underline{h}_m}\right]$$

$$= \begin{bmatrix} 1 & 0 & -\dfrac{v_{RDx}}{v_{RDz}} & 0 & 0 & 0 \\ 0 & 1 & -\dfrac{v_{RDy}}{v_{RDz}} & 0 & 0 & 0 \end{bmatrix} \tag{2.33}$$

$$(\overset{*}{P}_k)_{VTA} = \begin{bmatrix} 1 & 0 & -\dfrac{v_{RDx}}{v_{RDz}} \\ 0 & 1 & -\dfrac{v_{RDy}}{v_{RDz}} \end{bmatrix} \overset{*}{B}_{Dk} \tag{2.34}$$

where v_{RDx}, v_{RDy}, v_{RDz} are the components of the relative velocity vector $\underline{v}_{RD}$. With these relations substituted into (2.21), the VTA guidance relations reduce to.

$$\delta\underline{\gamma}_{VTA} = \overset{*}{L}' \sum_{k=1}^{p} \overset{*}{B}_{Dk}\underline{c}_k + \overset{*}{L}'\delta\underline{r}_D^- = \underline{0}_2 \tag{2.35}$$

where $\overset{*}{L}'$ is the "reduced" $\overset{*}{L}$ matrix.

$$\overset{*}{L}' = \begin{bmatrix} 1 & 0 & -\dfrac{v_{RDx}}{v_{RDz}} \\ 0 & 1 & -\dfrac{v_{RDy}}{v_{RDz}} \end{bmatrix} \tag{2.36}$$

The miss vector is simply $\delta\underline{r}_D^-$ for position-constrained FTA guidance while for position-constrained VTA guidance it is the two-dimensional vector ($\overset{*}{L}'\delta\underline{r}_D^-$).

3. CRITICAL-PLANE COORDINATE SYSTEM

A special rotating coordinate system, called the critical-plane coordinate system, is developed in this section for the purpose of simplifying the analysis of VTA guidance.

If $\overset{*}{B}_{Dk}$ is nonsingular, the vector $\underline{w}_k$ is defined by

$$\underline{w}_k = \overset{*}{B}_{Dk}^{-1}\,\underline{v}_{RD} \tag{3.1}$$

Consider the effect of premultiplying $\underline{v}_{RD}$ by $\overset{*}{L}'$ and of premultiplying $\underline{w}_k$ by ($\overset{*}{L}'\overset{*}{B}_{Dk}$).

$$\overset{*}{L}'\,\underline{v}_{RD} = \underline{0}_2 \tag{3.2}$$

$$(\overset{*}{L}'\,\overset{*}{B}_{Dk})\,\underline{w}_k = \overset{*}{L}'\,\overset{*}{B}_{Dk}\,\overset{*}{B}_{Dk}^{-1}\,\underline{v}_{RD} = \underline{0}_2 \tag{3.3}$$

If the predicted position variation $\delta\underline{r}_D^-$ is parallel to $\underline{v}_{RD}$, $\overset{*}{L}'\delta\underline{r}_D^-$ in Eq. (2.30) is equal to $\underline{0}_2$, and no corrections are required; thus the component of any given $\delta\underline{r}_D^-$ that is parallel to $\underline{v}_{RD}$ has no effect on the VTA constraints. Similarly, the component of the velocity correction $\underline{c}_k$ that is parallel to $\underline{w}_k$ has no effect on the constraints. For every correction time t_k there is some direction in which a small velocity step produces a position variation at t_D that is parallel to $\underline{v}_{RD}$ and hence has no effect on the constraints.

The physical interpretation of these observations is that the VTA constraints are satisfied if the vehicle's actual position at the nominal time of arrival is on the line through the nominal target point parallel to $\underline{v}_{RD}$. Position-constrained

FTA guidance is guidance to a specified point in position space at a specified time; position-constrained VTA guidance is guidance to a specified line in position space at the same specified time. As indicated by Eq. (2.29),

$$(\delta \underline{r}_D)_{FTA} = \underline{0}_3 \tag{3.4}$$

while for VTA,

$$(\delta \underline{r}_D)_{VTA} = -\underline{v}_{RD}\ \delta t_D \tag{3.5}$$

The direction of $\underline{v}_{RD}$ is designated the <u>non-critical direction</u> at t_D, and the direction of $\underline{w}_k$ is designated the <u>noncritical direction</u> at t_k. The noncritical direction varies with t_k because the elements of $\overset{*}{B}_{Dk}$ are functions of t_k. As t_k approaches t_D, $\overset{*}{B}_{Dk}$ approaches singularity; $\underline{w}_k$ approaches infinity in magnitude, and its direction approaches that of $\underline{v}_{RD}$. Thus, stipulating that $\underline{v}_{RD}$ defines the noncritical direction at t_D is consistent with stipulating that $\underline{w}_k$ defines the noncritical direction at t_k. $\underline{w}_k$ is the <u>noncritical vector</u>. The plane perpendicular to $\underline{w}_k$ is called <u>the critical plane</u>; only the components of velocity correction in this plane can affect the constraints.

The concept of the critical plane and the noncritical direction can be exploited by the formulation of a coordinate system in which two of the axes lie in the critical plane and the third axis is in the noncritical direction. This system is a rotating system due to the dependence of $\underline{w}_k$ on t_k. It will be referred to as the <u>critical-plane coordinate system</u>. The subscript W will be used to indicate that a vector or matrix is expressed in the critical-plane system.

Figure 1 illustrates the relation between the critical-plane system axes, labeled $\xi\ \eta\ \zeta$, and the axes x y z of the reference coordinate system. The ξ and η axes are in the critical plane, the ξ axis lying along the line of intersection of the critical plane with the reference plane; the ζ axis is in the noncritical direction, and the η axis is so directed that $\xi\ \eta\ \zeta$ form a right-handed orthogonal triad. Angles θ and ϕ serve to align $\xi\ \eta\ \zeta$ relative to x y z. The range of θ is 0° to 360°; the range of ϕ is 0° to 180°.

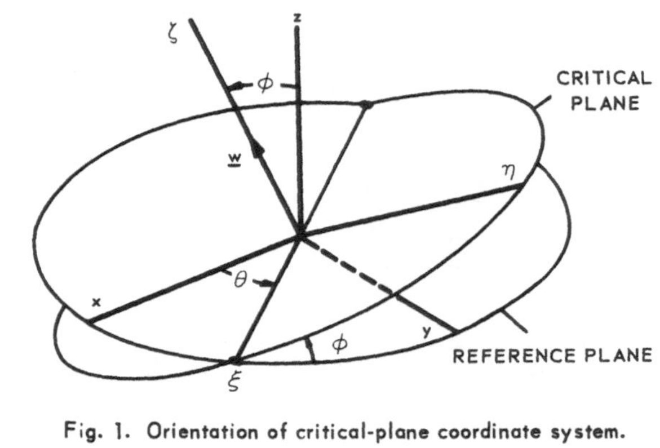

Fig. 1. Orientation of critical-plane coordinate system.

The vector $\underline{r}$ is transformed from the reference coordinate system to the critical-plane coordinate system by the orthogonal transformation matrix $\overset{*}{X}$.

$$\underline{r}_W = \begin{bmatrix} \xi \\ \eta \\ \zeta \end{bmatrix} = \overset{*}{X}\,\underline{r} = \overset{*}{X}\begin{bmatrix} x \\ y \\ z \end{bmatrix} \tag{3.6}$$

$$\overset{*}{X} = \begin{bmatrix} \cos\theta & \sin\theta & 0 \\ -\sin\theta\ \cos\phi & \cos\theta\ \cos\phi & \sin\phi \\ \sin\theta\ \sin\phi & -\cos\theta\ \sin\phi & \cos\phi \end{bmatrix} \tag{3.7}$$

From the definition of an orthogonal matrix,

$$\overset{*}{X}^{-1} = \overset{*}{X}^T \tag{3.8}$$

The transformations for $\underline{v}_{RD}$ and $\underline{w}_k$ are

$$\underline{v}_{RD} = \begin{bmatrix} v_{RDx} \\ v_{RDy} \\ v_{RDz} \end{bmatrix} = \overset{*}{X}_D^T(\underline{v}_{RD})_W = v_{RD}\,\overset{*}{X}_D^T\begin{bmatrix} 0 \\ 0 \\ 1 \end{bmatrix}$$

$$= v_{RD}\begin{bmatrix} \sin\theta_D\ \sin\phi_D \\ -\cos\theta_D\ \sin\phi_D \\ \cos\phi_D \end{bmatrix} \tag{3.9}$$

$$\underline{w}_k = \begin{bmatrix} w_{kx} \\ w_{ky} \\ w_{kz} \end{bmatrix} = \overset{*}{X}_k^T\ (\underline{w}_k)_W = w\,\overset{*}{X}_k^T\begin{bmatrix} 0 \\ 0 \\ 1 \end{bmatrix}$$

$$= w_k\begin{bmatrix} \sin\theta_k\ \sin\phi_k \\ -\cos\theta_k\ \sin\phi_k \\ \cos\phi_k \end{bmatrix} \tag{3.10}$$

v_{RD} and w_k, without the underlining, are the magnitudes of $\underline{v}_{RD}$ and $\underline{w}_k$, respectively. From these equations the orientation angles can be found in terms of the components of $\underline{v}_{RD}$ and $\underline{w}_k$ in the reference coordinate system.

$$\cos \phi_D = \frac{v_{RDz}}{v_{RD}} \tag{3.11}$$

$$\tan \theta_D = \frac{v_{RDx}}{v_{RDy}} \tag{3.12}$$

$$\cos \phi_k = \frac{w_{kz}}{w_k} \tag{3.13}$$

$$\tan \theta_k = \frac{w_{kx}}{w_{ky}} \tag{3.14}$$

The vector relationship of Eq. (2.30) can be further simplified if it is expressed in the critical-plane coordinate system. The matrices involved are $\overset{*}{L}'$ and $\overset{*}{B}_{Dk}$. From (2.31),

$$(\overset{*}{L}')_W = \begin{bmatrix} 1 & 0 & 0 \\ 0 & 1 & 0 \end{bmatrix} \tag{3.15}$$

$(\overset{*}{B}_{Dk})_W$ can be obtained from $\overset{*}{B}_{Dk}$, the elements of which are computed by numerical integration, by a matrix transformation which can be derived from Eqs. (2.4), (2.5), and (2.6). If $\delta \underline{r}_i$ is zero and there are no velocity corrections between t_i and t_j, then

$$\delta \underline{r}_j = \overset{*}{B}_{ji} \, \delta \underline{v}_i \tag{3.16}$$

$$\overset{*}{X}_j^T (\delta \underline{r}_j)_W = \overset{*}{B}_{ji} \overset{*}{X}_i^T (\delta \underline{v}_i)_W \tag{3.17}$$

$$(\delta \underline{r}_j)_W = \left(\overset{*}{X}_j \overset{*}{B}_{ji} \overset{*}{X}_i^T\right)(\delta \underline{v}_i)_W \tag{3.18}$$

Thus,

$$(\overset{*}{B}_{ji})_W = \overset{*}{X}_j \overset{*}{B}_{ji} \overset{*}{X}_i^T \tag{3.19}$$

Correspondingly,

$$(\overset{*}{B}_{Dk})_W = \overset{*}{X}_D \overset{*}{B}_{Dk} \overset{*}{X}_k^T \tag{3.20}$$

The elements in the third column of $\overset{*}{X}_D \overset{*}{B}_{Dk} \overset{*}{X}_k^T$ can be derived from (3.1).

$$(\overset{*}{B}_{Dk})_W (\underline{w}_k)_W = (\underline{v}_{RD})_W \tag{3.21}$$

$$\left(\overset{*}{X}_D \overset{*}{B}_{Dk} \overset{*}{X}_k^T\right) \begin{bmatrix} 0 \\ 0 \\ w_k \end{bmatrix} = \begin{bmatrix} 0 \\ 0 \\ v_{RD} \end{bmatrix} \tag{3.22}$$

For this equation to be valid for non-zero w_k, it is necessary that the elements in the third column of the triple scalar product be zero, zero, and v_{RD}/w_k. Consequently,

$$(\overset{*}{L}')_W (\overset{*}{B}_{Dk})_W (\underline{c}_k)_W = \left[\begin{array}{c|c} \overset{*}{Q}_k & \begin{matrix} 0 \\ 0 \end{matrix} \end{array} \right] (\underline{c}_k)_W$$

$$= \overset{*}{Q}_k \underline{a}_k \tag{3.23}$$

where $\overset{*}{Q}_k$ is the 2-by-2 matrix consisting of the elements in the first two rows and the first two columns of $\overset{*}{X}_D \overset{*}{B}_{Dk} \overset{*}{X}_k^T$, and $\underline{a}_k$ is the two-dimensional vector consisting of the components of $\underline{c}_k$ along the ξ_k and η_k axes in the critical plane. Then the vector equation of the constraints is

$$\sum_{k=1}^{p} \overset{*}{Q}_k \underline{a}_k + \underline{\rho} = \underline{0}_2 \tag{3.24}$$

Here miss vector $\underline{\rho}$ is the projection of of $\delta \underline{r}_D^-$ in the critical plane.

Eq. (3.24) brings out the fundamental advantage of the critical-plane concept. With this concept the constraints of the VTA guidance problem, which is physically a three-dimensional problem, can be expressed mathematically in a rotating two-dimensional coordinate system. The concept is exploited in the following sections in devising optimum midcourse correction strategies.

4. SINGLE-CORRECTION STRATEGY

If the number of constraints at t_D is equal to or less than three, a single velocity correction normally is sufficient to satisfy these constraints. In this section a method is developed for determining the time at which to apply a single correction such that the magnitude of the correction is a minimum.

In position-constrained FTA guidance, for which the number of constraints at t_D is three, the single correction $\underline{c}_{FTA}$, applied at time t_C, that satisfies the constraints, is, according to Eq. (2.32),

$$\underline{c}_{FTA} = - \overset{*}{B}_{DC}^{-1} \, \delta \underline{r}_D^- \tag{4.1}$$

This equation can be solved as long as t_c is such that $\overset{*}{B}_{DC}$ is not singular.

In position-constrained VTA guidance, for which there are only two constraints at t_D, the single correction $\underline{c}_{VTA}$ satisfies the constraints if $\underline{a}$, its projection in the critical plane corresponding to t_C, is given by

$$\underline{a} = -\overset{*}{Q}_C^{-1} \underline{\rho} \tag{4.2}$$

Time t_C is chosen such that $\overset{*}{Q}_C$ is not singular. The third component of $\underline{c}_{VTA}$, the one in the non-critical direction at t_C, does not affect the VTA constraints; obviously the magnitude of $\underline{c}_{VTA}$ is minimized if the third component is zero, so that

$$\underline{c}_{VTA} = \begin{bmatrix} \underline{a} \\ 0 \end{bmatrix} \qquad (4.3)$$

For either FTA or VTA the magnitude of the correction depends on t_C and the corresponding miss vector. The object now is to determine the t_C which minimizes the magnitude of $\underline{c}$ when $\delta \underline{r}_D^-$ or $\underline{\rho}$ is specified. Because it is simpler to analyze, the VTA case will be considered first.

Miss vector $\underline{\rho}$ can be expressed in target-centered polar coordinates ρ and ψ in the critical plane. ρ is the magnitude of the vector $\underline{\rho}$; ψ is the angle between $\underline{\rho}$ and the ξ_D axis.

$$\underline{\rho} = \begin{bmatrix} \delta\xi_D^- \\ \delta\eta_D^- \end{bmatrix} = \rho \begin{bmatrix} \cos\psi \\ \sin\psi \end{bmatrix} \qquad (4.4)$$

$$\psi = \tan^{-1}\left(\frac{\delta\eta_D^-}{\delta\xi_D^-}\right) \qquad (4.5)$$

The square of the magnitude of $\underline{c}_{VTA}$ is

$$c_{VTA}^2 = \underline{a}^T \underline{a} = \underline{\rho}^T (\overset{*}{Q}_C^{-1})^T \overset{*}{Q}_C^{-1} \underline{\rho} \qquad (4.6)$$

Then

$$c_{VTA} = \rho \left\{ [\cos\psi \quad \sin\psi] \cdot (\overset{*}{Q}_C^{-1})^T \overset{*}{Q}_C^{-1} \begin{bmatrix} \cos\psi \\ \sin\psi \end{bmatrix} \right\}^{1/2} \qquad (4.7)$$

The magnitude of $\underline{c}_{VTA}$ is a linear function of the magnitude of $\underline{\rho}$ but varies nonlinearly with ψ and t_C. The only property of $\underline{\rho}$ that affects the optimum time of correction is the angle ψ.

The optimum single-correction strategy is predetermined (before the flight) by computing and plotting a one-parameter family of curves, the parameter being ψ. The abscissa is the time of correction (or, equivalently, one of the anomalies), and the ordinate is the miss correctable per unit of velocity correction, i.e., the ratio of ρ to c_{VTA}. For each value of ψ there is generally a value of t_C at which the miss correctable is a maximum; this is the optimum t_C for that ψ. A cross-plot of optimum t_C versus ψ is the basis for selecting the time of correction to be used when some ρ has been inferred from measurements made during the flight.

For FTA guidance, $\delta \underline{r}_D^-$ is expressed in terms of target-centered spherical coordinates. The magnitude of $\underline{c}_{FTA}$ varies linearly with the magnitude of $\delta \underline{r}_D^-$ but nonlinearly with the two angle coordinates. The optimum single correction strategy involves a two-parameter family of curves, the parameters being the two angles. Thus, the procedure is considerably more laborious than the one-parameter procedure outlined for VTA. However, the basic strategy is the same; curves of miss correctable versus t_C are plotted for fixed values of the two angles, and the values of t_C corresponding to the maxima of the individual curves are cross-plotted as functions of the two angles to indicate the optimum correction time for any given $\delta \underline{r}_D^-$.

The procedure is illustrated for VTA guidance in Figs. 2 and 3. The reference trajectory is an outbound (perifocus to apofocus) Hohmann transfer with eccentricity of 0.95. The trajectory of the target is assumed to lie in the plane of the reference trajectory, and the relative velocity of the vehicle with respect to the target at t_D is assumed to be parallel to the velocity of the vehicle relative to the attractive focus. Obviously, this is a highly simplified case, the reference trajectory being both two-dimensional and two-body; nevertheless the guidance problem is still three-dimensional, because out-of-plane and in-plane components of the miss vector are both taken into consideration. The simplicity of the case reduces the amount of numerical computation, yet adequately illustrates the fundamental principles of single-correction strategy.

Physically in this trajectory the noncritical direction at t_D is parallel to the minor axis of the reference ellipse. The critical plane at t_D is perpendicular to the reference trajectory plane, and the line of nodes between the two planes is the major axis of the reference ellipse. Thus, when $\psi = 0°$, $\underline{\rho}$ is parallel to the major axis; when $\psi = 90°$, $\underline{\rho}$ is perpendicular to the reference plane.

The range of ψ is 0° to 360°. Since an increase of 180° in ψ between one miss vector and another changes the sign of the required correction but not its magnitude, the miss correctable, which is a ratio of magnitudes, is not affected, and hence the computations can, in general, be confined to the range $\psi = 0°$ to $\psi = 180°$. In this simple special case, there is also symmetry about $\psi = 90°$; that is, the miss correctable at $(90° + \psi)$ is equal to the miss correctable at $(90° - \psi)$ for the same t_C, so that the actual computations cover only the range 0° to 90° in ψ.

Fig. 2 is a plot of normalized miss correctable versus E_C, the eccentric anomaly at the time of correction. The range of E_C for an outbound Hohmann transfer is 0° to 180°. The normalized miss correctable is the miss correctable at the given ψ and E_C divided by the miss correctable at $\psi = 0°$, $E_C = 0°$. Curves are drawn for $\psi = 0°$, 5°, 15°, 30°, 60°, and 90°. The dotted curve in the figure is the locus

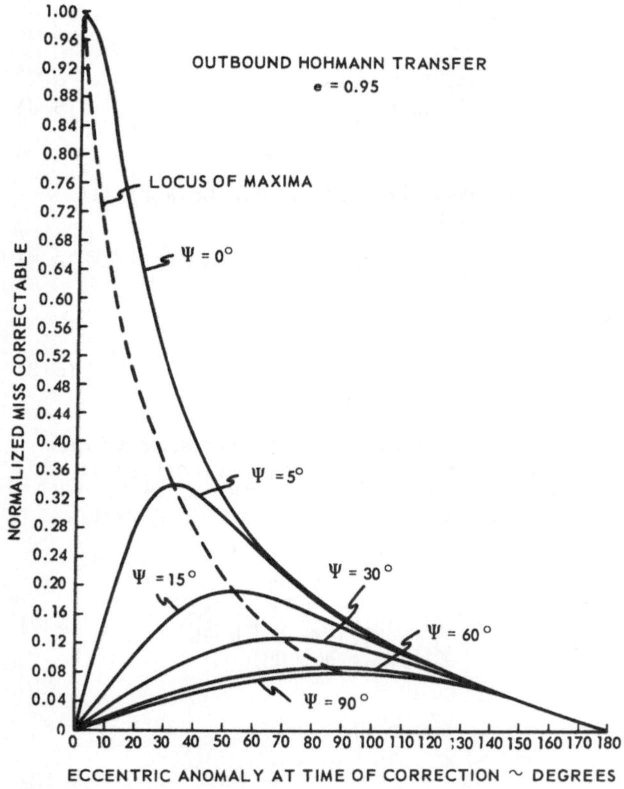

Fig. 2. Normalized miss correctable versus eccentric anomaly at time of correction.

of the points of maximum miss correctable for each ψ.

Fig. 3 is a re-plot of the dotted curve of Fig. 2. Eccentric anomaly at optimum correction time and maximum normalized miss correctable are plotted as functions of ψ. The optimum correction time curve is the only one that is needed to select the time of correction in an optimum single-correction VTA strategy.

5. MULTIPLE-CORRECTION STRATEGIES

Section 4 presents a method of selecting the optimum time for a single midcourse correction which completely nullifies the miss vector at the destination. However, no consideration has yet been given to the problem of multiple velocity corrections, i.e., of determining whether several partial corrections, made at different times, may result in a total magnitude of velocity change that is smaller than the magnitude of the single correction already described.

It was shown above that the miss at the nominal time of arrival at the destination can be described by an n-dimensional vector where n is the number of terminal constraints for FTA guidance and one less than the number of terminal constraints for VTA guidance. In this section, it will be shown that for a given set of injection errors there is an optimum velocity correction

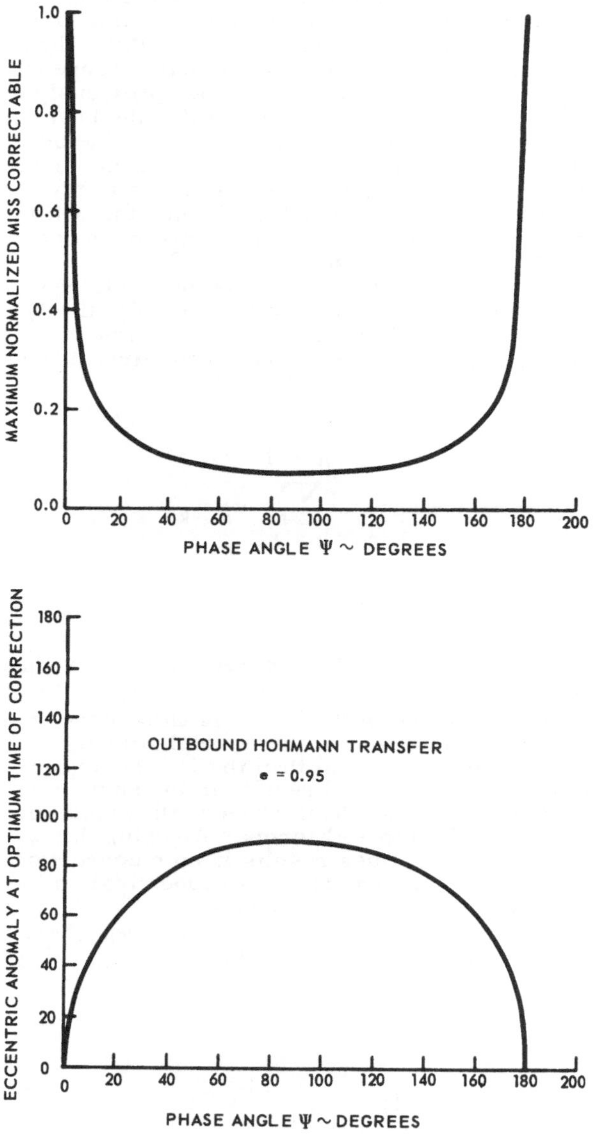

Fig 3. Maximum miss correctable and optimum eccentric anomaly at time of correction versus phase angle Ψ.

schedule consisting of at most n velocity corrections which completely nulls out the miss due to injection errors.

The following relation between injection errors and velocity corrections was obtained in Section 2:

$$\sum_{k=1}^{p} \overset{*}{P}_k \underline{c}_k + \underline{\rho} = \underline{0}_n \tag{5.1}$$

where $\underline{\rho}$ is the n dimensional vector representing the miss that would occur without velocity corrections, $\underline{c}_k$ represents the kth velocity correction applied at time t_k and $\overset{*}{P}_k$ is the influence matrix relating the kth velocity correction to the miss.

It will be shown below that any velocity correction schedule containing more than n corrections can be reduced to an n correction schedule without increasing the total velocity change required. This result was obtained previously by Neustadt.[7] It then remains to find the best correction schedule from among all possible schedules containing n or fewer corrections. This is always possible in principle, and a fairly simple graphical method for finding the optimum schedule in the case when n is two will be described in the next section.

In reducing the p correction schedule of Eq. (5.1) to an n correction schedule, only (n + 1) corrections will be considered at a time. Therefore, let $\underline{\rho}_a$ denote the effect of the first (n + 1) velocity corrections on the miss.

$$\underline{\rho}_a = - \sum_{k=1}^{n+1} \overset{*}{P}_k \underline{c}_k \tag{5.2}$$

The effect of the (n + 1) corrections is to reduce the miss from $\underline{\rho}$ to $(\underline{\rho} - \underline{\rho}_a)$.

It will be shown that there is an n correction schedule which uses no more total velocity change (probably less) than the (n + 1) corrections of Eq. (5.2) and results in the same corrective effect ρ_a. This reduces the summation in Eq. (5.1) to (p - 1) terms. Applying this reduction (p - n) times results in an n correction schedule which is at least as economical as the original p correction schedule.

To carry out the reduction of the Eq. (5-2) schedule to n corrections, let

$$\underline{u}_k = \frac{1}{c_k} \overset{*}{P}_k \underline{c}_k \tag{5.3}$$

Then

$$\underline{\rho}_a = - \sum_{k=1}^{n+1} c_k \underline{u}_k \tag{5.4}$$

Since $\underline{u}_k$ is an n-dimensional vector, the vectors $\underline{u}_1, \ldots, \underline{u}_{n+1}$ are linearly dependent, and there are scalars $\lambda_1, \ldots, \lambda_{n+1}$ such that

$$\sum_{k=1}^{n+1} \lambda_k \underline{u}_k = \underline{0}_n \tag{5.5}$$

Let

$$\lambda = \sum_{k=1}^{n+1} \lambda_k \tag{5.6}$$

If λ is negative, the signs of all the λ_k's are changed so that

$$\lambda \geq 0 \tag{5.7}$$

Now, let

$$a_k = \frac{\lambda_k}{c_k} \tag{5.8}$$

and choose r so that a_r is the maximum of the a_k's. Then $a_r \geq a_k$ for $k = 1, \ldots, n+1$ and $a_r > 0$ since $\lambda \geq 0$. Multiplying Eq. (5.5) by $1/a_r$ and adding the result to Eq. (5.4) yields

$$\underline{\rho}_a = - \sum_{k=1}^{n+1} \mu_k \underline{u}_k \tag{5.9}$$

with

$$\mu_k = c_k - \frac{\lambda_k}{a_r} = \frac{c_k}{a_r} \left\{ a_r - a_k \right\} \tag{5.10}$$

By the last equality above $\mu_k \geq 0$ and $\mu_r = 0$. Finally, by Eqs. (5.3) and (5.9) it follows that

$$\underline{\rho}_a = - \sum_{k=1}^{n+1} \overset{*}{P}_k \underline{c}_k' \tag{5.11}$$

with

$$\underline{c}_k' = \frac{\mu_k}{c_k} \underline{c}_k \tag{5.12}$$

Eq. (5.11) represents an n correction schedule since $\underline{c}_r'$ is zero. The total velocity change required by the new correction schedule is

$$c' = \sum_{k=1}^{n+1} c_k' = \sum_{k=1}^{n+1} \mu_k$$

$$= \sum_{k=1}^{n+1} c_k - \frac{\lambda}{a_r} \tag{5.13}$$

Since $\lambda \geq 0$, c' is less than or equal to $c_1 + \ldots + c_{n+1}$, and thus the new correction schedule is at least as economical as the old schedule.

6. MULTIPLE-CORRECTION STRATEGY FOR POSITION-CONSTRAINED VTA GUIDANCE

The multiple correction case for position-constrained VTA guidance is studied in this section. It was shown in the preceding section that the optimum velocity correction schedule for position-constrained VTA guidance need never contain more than two corrections (in some cases it consists of only one correction). This section outlines a geometric construction for determining the optimum set of velocity corrections needed to null out the miss due to given injection errors.

The analysis in this section makes extensive use of the equation of a line segment joining two given vectors in the target critical plane. Thus, in Fig. 4, the vectors $\underline{a}$ and $\underline{b}$ define a line segment $\overline{AB}$. If the head of the vector $\underline{h}$ lies on this line segment and λ denotes the ratio of $\overline{HB}$ to $\overline{AB}$, then

$$\underline{h} = \underline{b} + \lambda(\underline{a} - \underline{b}) = \lambda\underline{a} + (1-\lambda)\underline{b} \tag{6.1}$$

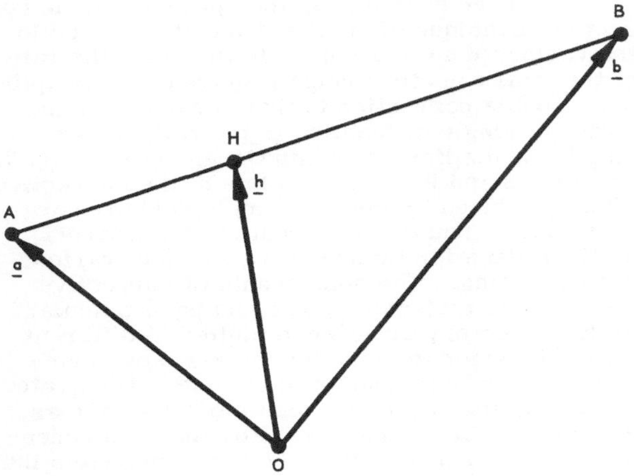

Fig. 4. Vector relation in the critical plane.

As the point H moves along the line segment, λ ranges from zero to one. Thus the equation of the line segment $\overline{AB}$ is

$$\underline{h} = \lambda\underline{a} + (1-\lambda)\underline{b}, \quad 0 \leq \lambda \leq 1 \tag{6.2}$$

Suppose that the miss vector $\underline{\rho}$ due to injection errors is given and it is desired to find the optimum velocity correction schedule which nulls out this miss. Since the use of two corrections may result in a saving in total velocity change, the total corrective effect will be written as the sum of two components $\underline{\rho}_1$ and $\underline{\rho}_2$, to be produced by separate velocity corrections. Thus, with

$$\underline{\rho}_k = \overset{*}{P}_k\underline{c} = c_k\underline{u}_k \;;\; k = 1, 2 \tag{6.3}$$

Eq. (5.1) becomes

$$\underline{\rho}_1 + \underline{\rho}_2 = -\underline{\rho} \tag{6.4}$$

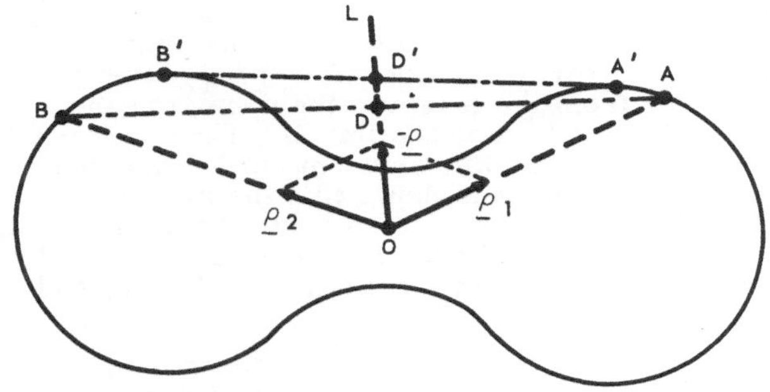

Fig. 5. Construction for optimum double velocity correction.

Fig. 5 illustrates this vector sum. The solid curve in the figure is a polar plot of maximum miss correctable as a function of ψ; that is, it represents the same type of information as that shown in the upper half of Fig. 3.

In order to obtain the minimum total velocity change, corrective effects $\underline{\rho}_1$ and $\underline{\rho}_2$ should each be obtained by use of the single-correction strategy of Section 4; i.e., the correction which produces $\underline{\rho}_k$ is made at the time for which the miss correctable is maximum for the ψ corresponding to $\underline{\rho}_k$. If A and B are the points where lines drawn from the origin along vectors $\underline{\rho}_1$ and $\underline{\rho}_2$ intersect the solid curve, then, since the solid curve is the locus of the maximum corrective effect of a unit velocity change and since $\underline{u}_k$ is defined in Eq. (5.3) as the corrective effect of a unit velocity change, $\underline{u}_1$ and $\underline{u}_2$ are the vectors from the origin to A and B, respectively. The magnitude of the correction that produces $\underline{\rho}_k$ is simply the ratio of the magnitude of $\underline{\rho}_k$ to the magnitude of $\underline{u}_k$.

$$c_k = \frac{\rho_k}{u_k} \;;\; k = 1, 2 \tag{6.5}$$

Fig. 5 illustrates this ratio; the equation follows directly from Eq. (6.3). The sum of the magnitudes of the two constituents of the multiple correction is

$$c_{12} = c_1 + c_2 \tag{6.6}$$

Since ρ is given, minimizing c_{12} is the same as maximizing the ratio ρ/c_{12}. This ratio represents the miss correctable with unit total velocity change when the required corrective effect $-\underline{\rho}$ is divided into components parallel to $\underline{\rho}_1$ and $\underline{\rho}_2$.

$$\frac{-\underline{\rho}}{c_{12}} = \frac{\underline{\rho}_1 + \underline{\rho}_2}{c_{12}}$$

$$= \lambda\underline{u}_1 + (1-\lambda)\underline{u}_2 \tag{6.7}$$

where

$$\lambda = \frac{c_1}{c_1 + c_2} \tag{6.8}$$

Thus $0 < \lambda < 1$ and the head of the vector $-\underline{\rho}/c_{12}$ must lie on the line segment $\overline{AB}$. However $-\underline{\rho}/c_{12}$ also lies on the line L drawn from the origin along the vector $-\underline{\rho}$, and therefore $\underline{\rho}/c_{12}$ must be the vector from the origin to the point D at the intersection of $\overline{AB}$ and the line L. Then ρ/c_{12} is the length of the line segment $\overline{OD}$. More generally, this argument shows that the line segment $\overline{AB}$ is the locus of miss vectors which can be corrected with unit total velocity change after being split up into components parallel to $\underline{u}_1$ and $\underline{u}_2$.

To maximize the ratio ρ/c_{12}, the points A and B must be moved around the solid curve until the intersection of $\overline{AB}$ and L is farthest from the origin. This occurs when A and B are positioned so that the line segment $\overline{AB}$ meets the solid curve at points of tangency, as the line segment $\overline{A'B'}$ does in Fig. 5.

If the line L intersects the solid curve at a point where the solid curve is convex as in Fig. 6 and one tries to represent $-\underline{\rho}$ as the sum of two partial corrections, one finds that the intersection of the line segment $\overline{AB}$ with L is always closer to the origin than the intersection of L with the solid curve and no double correction is as economical as the best single correction.

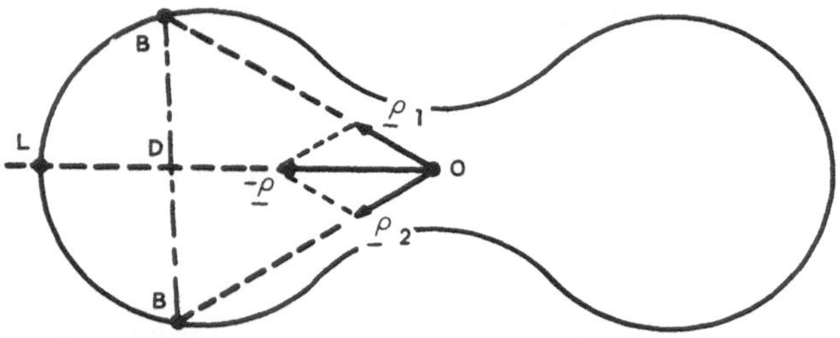

Fig. 6. Example of a vector $-\underline{\rho}$ for which a single correction is optimum.

Finally, the locus S of vectors $-\underline{\rho}$ which are obtained with an optimum single or double velocity correction with unit total velocity change is made up of the convex parts of the curve for optimum single corrections plus the tangent line segments as illustrated in Fig. 7.

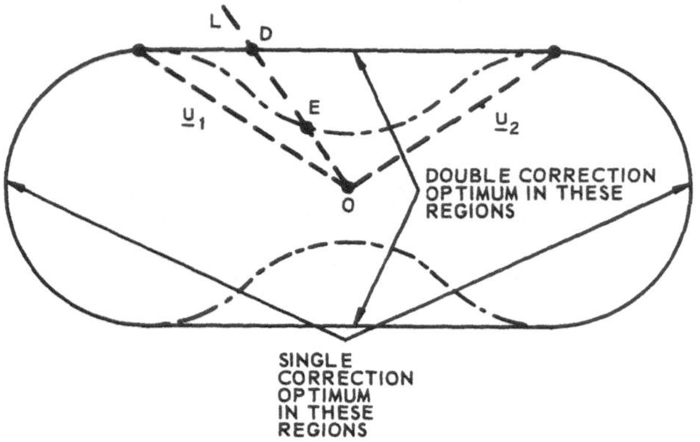

Fig. 7. Plot of corrective effect obtained with optimum unit double or single velocity correction.

The curve S is the basis for calculating optimum velocity correction schedules. Suppose that it is desired to find the optimum correction schedule which produces the corrective effect $-\underline{\rho}$.

If the line L from the origin along the vector $-\underline{\rho}$ intersects the curve S along a curved portion, $-\underline{\rho}$ should be obtained by means of the best single correction. However if the line L from the origin along the vector $-\underline{\rho}$ intersects the curve S along a straight line portion, $-\underline{\rho}$ should be split up into components parallel to the vectors $\underline{u}_1$ and $\underline{u}_2$ from the origin to the ends of the straight line portion. This operation involves solving a pair of simultaneous linear equations. Thus, scalars α and β must be found such that

$$-\underline{\rho} = \alpha \underline{u}_1 + \beta \underline{u}_2 \tag{6.7}$$

where $-\underline{\rho}$, $\underline{u}_1$ and $\underline{u}_2$ are given. Writing this vector equation out in components results in two scalar equations which may be solved for α and β provided $\underline{u}_1$ and $\underline{u}_2$ are not collinear. $\underline{u}_1$ and $\underline{u}_2$ cannot be collinear since they extend from the origin to opposite ends of a line segment. Once α and β have been found, the optimum single correction technique of Section 4 may be applied to the vectors $\alpha\underline{u}_1$ and $\beta\underline{u}_2$. In this case the ratio of the total velocity change required for the optimum double correction to that required for the optimum single correction is the ratio of the lengths of the line segments $\overline{OE}$ and $\overline{OD}$ in Fig. 7.

Figs. 8 and 9 are polar plots of the corrective effect produced by the best single velocity correction as a function of direction in the target critical plane for Hohmann transfers of various eccentricities. The components of corrective effect in the trajectory plane and perpendicular to the trajectory plane were plotted to different scales in order to make the curves more symmetrical. For the outbound transfers illustrated in Fig. 8, the curves become more concave as the eccentricity increases. For the 0.95 eccentricity outbound transfer, the curve indicates that, in the directions which favor a double correction the most, the best single correction uses about fifty percent more total velocity change than the best double correction. The curves for inbound Hohmann transfers in Fig. 9 are always convex, indicating that single corrections are always better than double corrections on inbound transfers of this type.

In more precise terms the locus of $-\underline{\rho}$ vectors which are obtained with an optimum single or double velocity correction with total velocity change less than or equal to unity is the convex hull of the set of $-\underline{\rho}$ vectors which may be obtained with optimum single corrections with total velocity change less than or equal to unity. A convex set is a set having the property that any line segment joining two points in the set lies entirely within the set. The convex hull of a set is the smallest convex set containing the original set. The operation of drawing in tangent line segments described above thus corresponds to constructing the convex hull of the set of $-\underline{\rho}$ vectors which may be obtained with optimum single corrections with total velocity change less than or equal to unity.

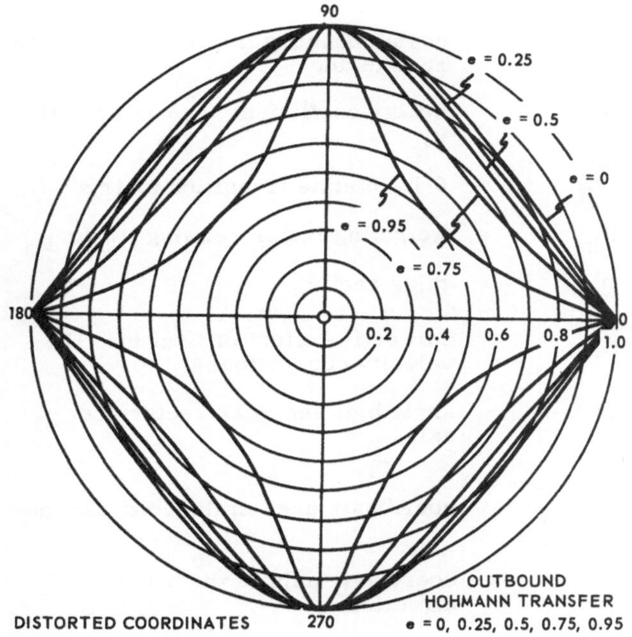

Fig. 8. Polar plot of normalized miss correctable versus phase angle Ψ.

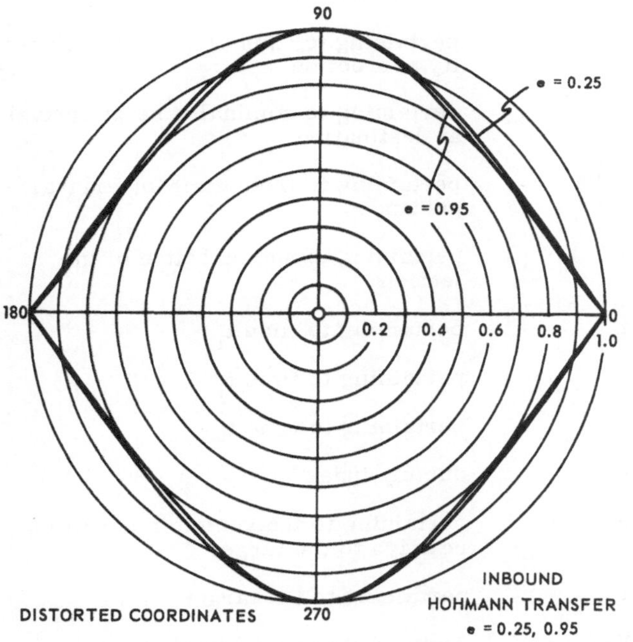

Fig. 9. Polar plot of normalized miss correctable versus phase angle Ψ.

NOMENCLATURE

General

An asterisk over a capital letter indicates a matrix.

An underlined lower-case letter indicates a column vector. A vector symbol without the underlining indicates the magnitude of the vector.

A single dot over a vector symbol indicates the first derivative with respect to time in an inertially non-rotating coordinate system.

Square brackets around a partial derivative involving vector quantities indicates a column vector or matrix consisting of the partial derivatives of the components of the vectors.

English Symbols

a_k	ratio of λ_k to the magnitude of $\underline{c}_k$
a_r	maximum of the a_k's
$\overset{*}{A}_{ji}$, $\overset{*}{B}_{ji}$, $\overset{*}{C}_{ji}$, $\overset{*}{D}_{ji}$	submatrices of the state transition matrix
$\underline{c}$	single velocity correction which satisfies the constraints
$\underline{c}_k$	k-th of a series of velocity corrections which, taken together, satisfy the constraints
c'	sum of magnitudes of $\underline{c}_k'$ vectors
$\underline{c}_k'$	k-th velocity correction in revised schedule
c_{12}	sum of magnitudes of velocity corrections in a two-correction schedule.
$\underline{f}$	rate of change of the state vector with time in the absence of any control action
FTA	fixed-time-of-arrival
$\underline{h}_k$	six-dimensional vector of the partial derivatives of constraint γ_k with respect to the components of $\delta\underline{x}_D$.
$\overset{*}{H}$	m-by-6 matrix relating the constraints to $\delta\underline{x}_D$ in FTA guidance
$\overset{*}{I}_k$	k-by-k identity matrix
$\overset{*}{L}$	(m - 1)-by-6 matrix relating the constraints to $\delta\underline{x}_D$ in VTA guidance
$\overset{*}{L}'$	2-by-3 matrix relating position constraints to $\delta\underline{r}_D$ in VTA guidance
m	number of contraints at actual time of arrival
$\overset{*}{M}$	6-by-3 compatibility matrix
n	number of constraints at nominal time of arrival
$\underline{0}_k$	k-component zero vector

$\overset{*}{0}_k$ — k-by-k zero matrix

p — number of velocity corrections

$\overset{*}{P}_k$ — influence matrix relating the k-th velocity correction to the miss vector

$\overset{*}{Q}_k$ — 2-by-2 matrix indicating the effect on the target miss vector of a velocity correction at t_k

$\underline{r}$ — position vector

S — locus of $\underline{\rho}$ vectors obtained from optimum strategy with unit total velocity change

t — time

$\underline{u}_k$ — corrective effect of a unit velocity correction in the direction of $\underline{c}_k$

$\underline{v}$ — velocity vector

v_{RD} — magnitude of $\underline{v}_{RD}$

VTA — variable-time-of-arrival

$\underline{w}_k$ — noncritical vector at time t_k

w_k — magnitude of w_k

$\underline{x}$ — state vector

$\overset{*}{X}$ — transformation matrix from reference coordinate system to critical-plane coordinate system

x, y, z — axes of reference coordinate system

Greek Symbols

$\underline{\alpha}_k$ — component of velocity correction in the critical plane

α — weighting factor relating $\underline{u}_1$ to $-\underline{\rho}$

β — weighting factor relating $\underline{u}_2$ to $-\underline{\rho}$

γ_k — k-th constraint

$\underline{\gamma}$ — constraint vector

δ — first variation

δt_D — change in time of arrival at destination

$\delta(t - t_k)$ — Dirac delta function at time t_k

λ — sum of $(n + 1)$ λ_k's

λ — scalar quantity limited to the range zero to one

λ_k — scalar coefficient associated with $\underline{u}_k$

$\overset{*}{\Phi}_{ji}$ — 6-by-6 state transition matrix

μ_k — weighting factor relating $\underline{u}_k$ to $\underline{\rho}_a$

$\underline{\rho}$ — miss vector

$\underline{\rho}_a$ — corrective effect of first $(n + 1)$ velocity corrections

ψ — angle between miss vector and ξ_D axis

ξ, η — coordinate axes in the critical plane

ζ — coordinate axis in the noncritical direction

Subscripts

A — pertaining to actual time of arrival at destination

C — pertaining to time of a single velocity correction

D — pertaining to nominal time of arrival at destination

FTA — pertaining to fixed-time-of-arrival guidance

I — pertaining to nominal time of injection

i — pertaining to time t_i

j — pertaining to time t_j

k — pertaining to time t_k

k — dummy index

R — pertaining to the vehicle's condition relative to the target

T — pertaining to the target

VTA — pertaining to variable-time-of-arrival guidance

W — pertaining to critical-plane coordinate system

x, y, z — pertaining to vector components along the axes of the reference coordinate system

ξ, η, ζ	pertaining to vector components along the axes of the critical-plane coordinate system

Superscripts

T	transpose of a vector or matrix
–	pertaining to conditions that would exist if no corrections were applied
-1	inverse of a matrix

ACKNOWLEDGEMENT

The authors wish to express their appreciation to the United States National Aeronautics and Space Agency, which supported the preparation and presentation of this report under NASA Research Grant NsG 254-62. They also wish to thank Mrs. Linda P. Abrahamson and Mr. Berl P. Winston, who programmed the computations and prepared the figures used in the numerical example.

Acknowledgment is made of the MIT Computation Center for the work done on its computer under Problem Number M3008.

The publication of this report does not constitute approval by the National Aeronautics and Space Administration of the findings or the conclusions contained therein. It is published only for the exchange and stimulation of ideas.

REFERENCES

(1) Battin, R. H., Astronautical Guidance, McGraw-Hill, Inc., New York, 1964.

(2) Noton, A. R. M., Cutting, E., and Barnes, F. L., "Analysis of Radio-Command Mid-Course Guidance," JPL Technical Report No. 32-28, Sept. 1960.

(3) McLean, J. D., Schmidt, S. F., and McGee, L. A., "Optimal Filtering and Linear Prediction Applied to a Midcourse Navigation System for the Circumlunar Mission" NASA Technical Note D-1208, 1962.

(4) Breakwell, J. V., "Fuel Requirements for Crude Interplanetary Guidance," Advances in the Astronautical Sciences, Vol. 5, Plenum Press, Inc., New York, 1960, pp. 53 – 65.

(5) Breakwell, J. V., "The Spacing of Corrective Thrusts in Interplanetary Navigation," Advances in the Astronautical Sciences, Vol. 7, Plenum Press, Inc., New York, 1961, pp. 219 – 235.

(6) White, J. S., Callas, G. P., and Cicolani, L. S., "Application of Statistical Filter Theory to the Interplanetary Navigation and Guidance Problem," NASA TN D-2697, 1965.

(7) Neustadt, L. W., "Optimization, a Moment Problem, and Nonlinear Programming," Journal of the Society for Industrial and Applied Mathematics, Series A, Control, Vol. 2, No. 1, Philadelphia, 1964, pp. 33-53.

DISCUSSION

R. G. Stern and J. E. Potter

"Optimization of
Midcourse Velocity Corrections".

Comment by Stern

The number of corrections being equal to the number of constraints has probably been known, but has not been published before.

OPTIMIZATION OF SPACECRAFT FLIGHT CONTROL

by

A. K. Platonov, A. A. Dashkov, and V. N. Kubasov

The problem of in-flight control of spacecraft is of interest today in connection with the development of interplanetary communications. This problem also features some degree of independent significance from the standpoint of automatic control theory, since the latter theory contains a fairly complete and at the same time quite simple model of finite-state control processes, enabling us to study and understand many of the aspects of processes of this type.

This paper investigates the properties of an optimized spacecraft control process in the case where complete information is available on the craft's motion and where control is exerted in order to sustain predetermined values of certain functionals defined on the trajectory. Functionals pertinent here may be trajectory parameters characterizing the conditions for encountering a planet or rendezvous with another spacecraft in outer space, or parameters defining the position in space of the craft at a fixed instant of time.

We consider a space $\bar{b}$ of characteristic trajectory parameters. In this space there exists a region in which the conditions imposed on the flight trajectory are satisfied. A certain known point corresponds to the uncontrolled trajectory in this space. The purpose of the control is to bring the starting point into the region referred to.

Acceleration of the spacecraft brought about by the propulsion plant will serve as a control input. The specific features of this type of control are determined by the nature of the dependence of the parameters $\bar{b}$ on the control acceleration vector. This dependence is a nonlinear one in the general case. In some cases it is to be hoped that the problem will lend itself to linearization given a judicious choice of characteristic parameters. The study of the problem in a linear framework would make it possible to ascertain the fundamental regularities of the process for a whole class of possible controls.

We cite here an example of assignment of linear parameters. In lunar and planetary flights one of the important sources of nonlinearity are the gravitational fields of the target bodies. This nonlinearity can be eliminated if the osculating characteristics of the planetocentric motion at infinity are considered as the control parameters; these are the relative velocity vector at infinity $V_{rel\,\infty}$ and the target range $\bar{b}$ (Figure 1). The variable b is the distance at which the craft would be seen to proceed from the center of the planet were there no attraction. This variable is determined by the values of the coordinates of the planetocentric radius vector ρ and by the components of the relative velocity V_{rel} near the planet as follows:

$$b = \frac{|\bar{c}|}{|V_{rel\,\infty}|}$$

$$\bar{c} = \bar{\rho} \times V_{rel}$$

$$|V_{rel\,\infty}| = \sqrt{|\bar{V}_{rel}|^2 - \frac{2\mu}{|\bar{\rho}|}}$$

$$\mu = fM$$

Here f is the gravitational constant, and M is the mass of the planet.

We may consider the components of the vector $\bar{b}$ as an example of the parameters characterizing the approach to a planet, ascribing to this vector the direction in the plane of the planetocentric trajectory orthogonal to the osculating relative velocity vector at infinity $\bar{V}_{rel\,\infty}$. In this convention, the vector $\bar{b}$ will be parallel to the plane of the diagram, which is by definition orthogonal to the vector $\bar{V}_{rel\,\infty}$. The projections of the vector $\bar{b}$ onto any axes in this plane are two convenient trajectory parameters. These parameters ensure maximum linearity in their relationship to the components of the control acceleration input, and take into account all perturbations acting on the craft. The vector $V_{rel\,\infty}$ and the vector $\bar{b}$ are defined as follows:

$$\bar{V}_{rel\,\infty} = |V_{rel\,\infty}| \left[\sin\gamma \frac{\bar{f}}{|\bar{f}|} + \cos\gamma \frac{\bar{c} \times \bar{f}}{|\bar{c} \times \bar{f}|} \right]$$

$$\bar{b} = b \left[\cos\gamma \frac{\bar{f}}{|\bar{f}|} - \sin\gamma \frac{\bar{c} \times \bar{f}}{|\bar{c} \times \bar{f}|} \right],$$

where $\bar{f}$ is the Laplace vector,

$$\bar{f} = \bar{V}_{rel} \times \bar{c} - \mu \frac{\bar{\rho}}{|\bar{\rho}|}$$

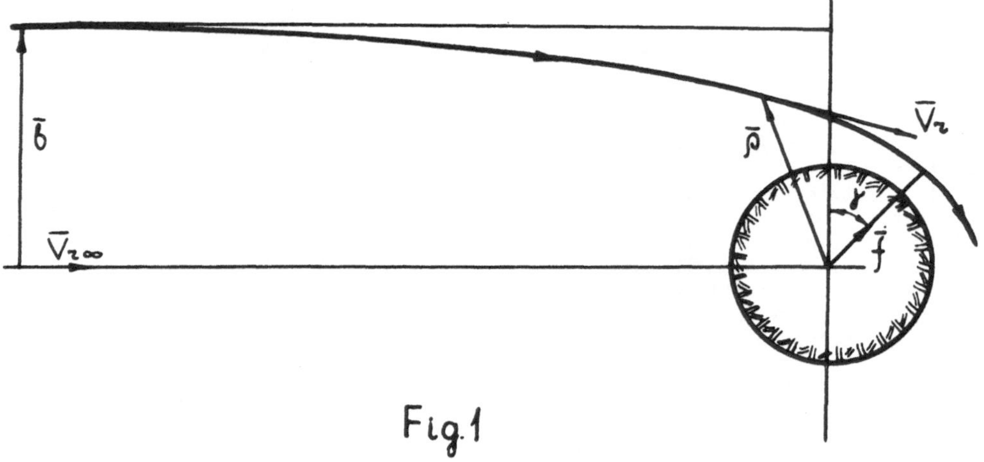

Fig.1

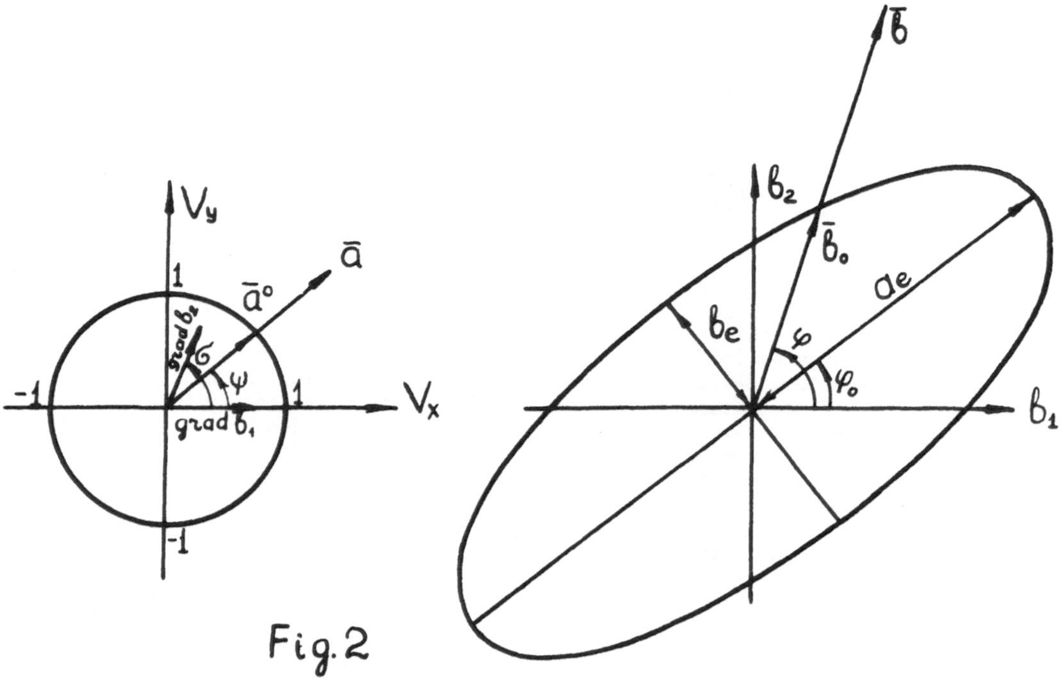

Fig.2

$$\sin \gamma = \frac{a}{\sqrt{a^2+b^2}}$$

$$\cos \gamma = \frac{b}{\sqrt{a^2+b^2}}$$

$$a = \frac{\mu}{|V_{rel\,\infty}|}$$

In our further discussion, we shall assume that the characteristic parameters $b_1, b_2, \ldots, b_n$ are assigned in the general case such that the problem may be treated in a linear approximation.

Within a linear formulation, the relationship between the control parameters b and the control acceleration $\bar{a}$ may be stated as:

$$\bar{b} = \int_0^T N \bar{a}\, dt$$

The matrix $N(t)$ consists of derivatives of the components of the vector $\bar{b}$ with respect to the components of the velocity vector $\bar{V}$ at time t.

$$N(t) = \left\| \begin{matrix} \frac{\partial b_1}{\partial V_x} & \frac{\partial b_1}{\partial V_y} & \frac{\partial b_1}{\partial V_z} \\ \frac{\partial b_2}{\partial V_x} & \frac{\partial b_2}{\partial V_y} & \frac{\partial b_2}{\partial V_z} \\ \frac{\partial b_n}{\partial V_x} & \frac{\partial b_n}{\partial V_y} & \frac{\partial b_n}{\partial V_z} \end{matrix} \right\|$$

As a result of the control in effect over a time interval from 0 to T, integrals (1) must assume specified values. The power used up in effecting the control is defined single-valuedly by the integral

$$V = \int_0^T a\, dt$$

The acceleration a is subject in the general case to the constraint:

$$0 \le a \le a_{max}$$

The problem of minimized power losses during controlled flight may be formulated as follows:

> the vector function $\bar{a}(t)$ satisfying constraint (2) and the isoperimetric condition (1) must be satisfied in such a way that the functional V will be minimized.

In order to solve the problem we investigate the variation δH of the auxiliary functional H which coincides with V when the isoperimetric condition (1) is satisfied.

$$H = \int_0^T \left[a + \bar{\lambda}^* (\dot{\bar{b}} - N \bar{a}^\circ a) \right] dt$$

$$\delta H = \int_0^T \left\{ \left[1 - \bar{\lambda}^* N \bar{a}^\circ \right] \delta a - a \bar{\lambda}^* N \delta \bar{a}^\circ \right\} dt$$

Here $\bar{\lambda}$ is a constant undetermined vector. The asterisk denotes the transpose, and $\bar{a}^\circ$ is the unit control acceleration vector. Denoting

$$\bar{v}(t) = N^* \bar{\lambda} = v \cdot \bar{v}^\circ$$

we find that the control input must be applied at those points of the trajectory at which

$$[1 - v] < 0$$

and must be maximized, i.e.

$$a = a_{max}$$

The direction in which the control acceleration acts is then defined by the formula

$$\bar{a}^\circ = \bar{v}^\circ$$

The resulting solution $\bar{a}(t)$ depends on the value assigned to the constant vector $\bar{\lambda}$. We suppose that it is possible to select a vector $\bar{\lambda}$ such that the isoperimetric conditions (1) is fulfilled. Then $\bar{a}(t)$ will be the solution required, capable of realizing a constrained minimum of the functional V.

Investigation of this solution reveals the following. At each point on the trajectory, depending on the dimension of the vector $\bar{b}$, there exists an optimal control space (3) either identical with the three-dimensional velocity vector space at that point or else possessing a smaller dimension. In a particular case it may be an optimal control plane or an optimal control curve. The control input vector $\bar{a}(t)$ must belong to the optimal control space at every instant of time.

The effectiveness of the control at a given point on the trajectory may be characterized by the influence of a set of unit velocity bursts of equal sign on the parameters $\bar{b}$. In the case where the direction of

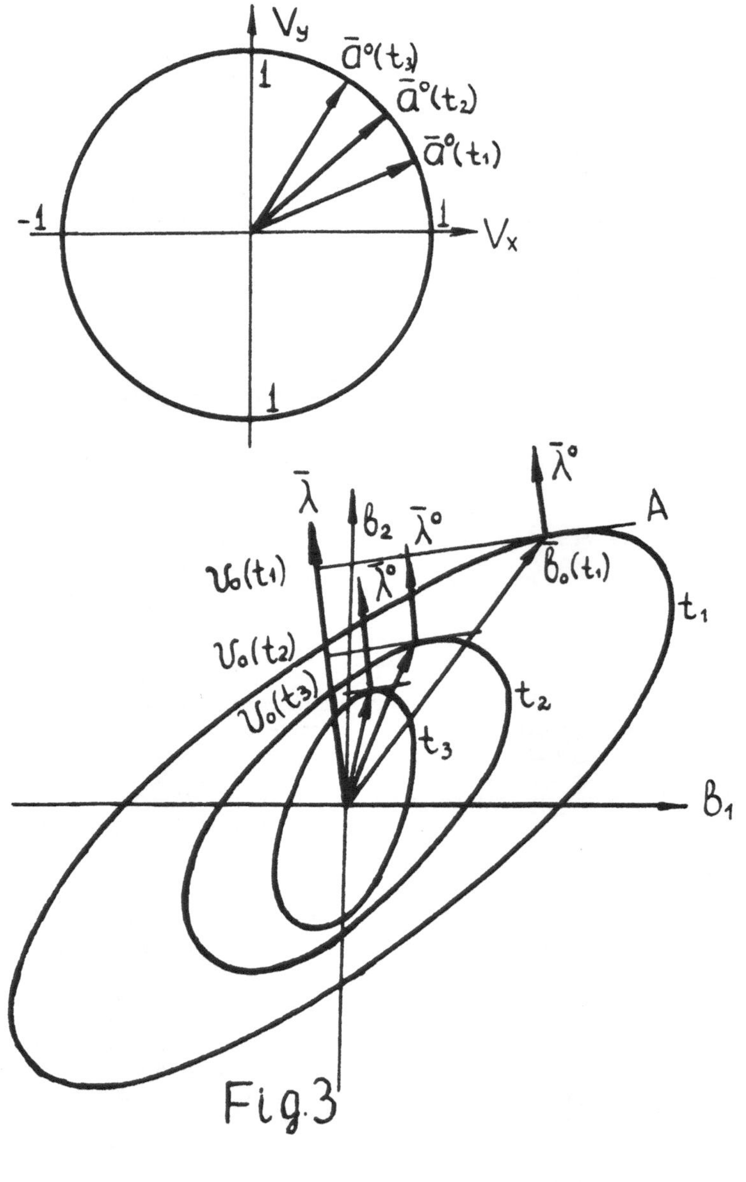

Vy
Vx
1
-1
$\bar{a}^o(t_3)$
$\bar{a}^o(t_2)$
$\bar{a}^o(t_1)$
$\bar{\lambda}$
$\bar{\lambda}^o$
A
b_2
b_1
$v_0(t_1)$
$v_0(t_2)$
$v_0(t_3)$
$\bar{b}_0(t_1)$
t_1
t_2
t_3

Fig.3

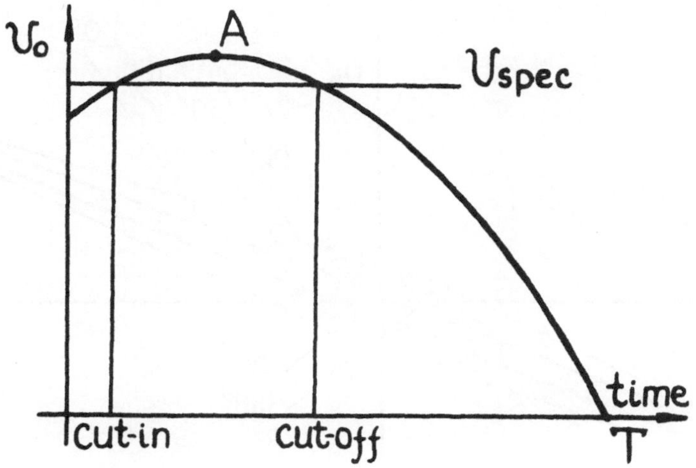

v_o
A
v_{spec}
time
cut-in
cut-off
T

Fig.4

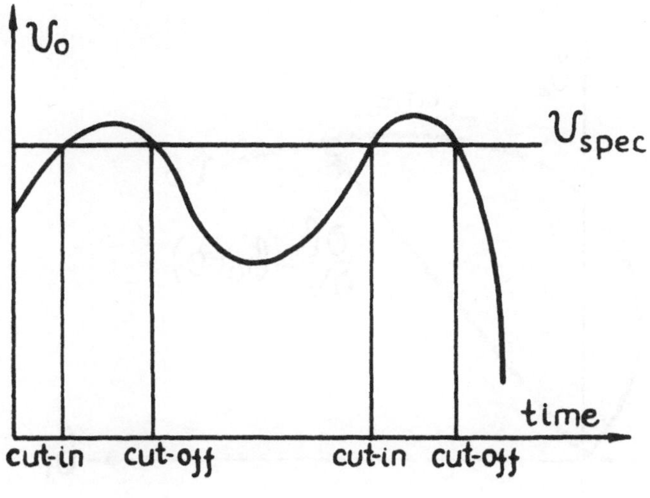

v_o
v_{spec}
time
cut-in
cut-off
cut-in
cut-off

Fig.5

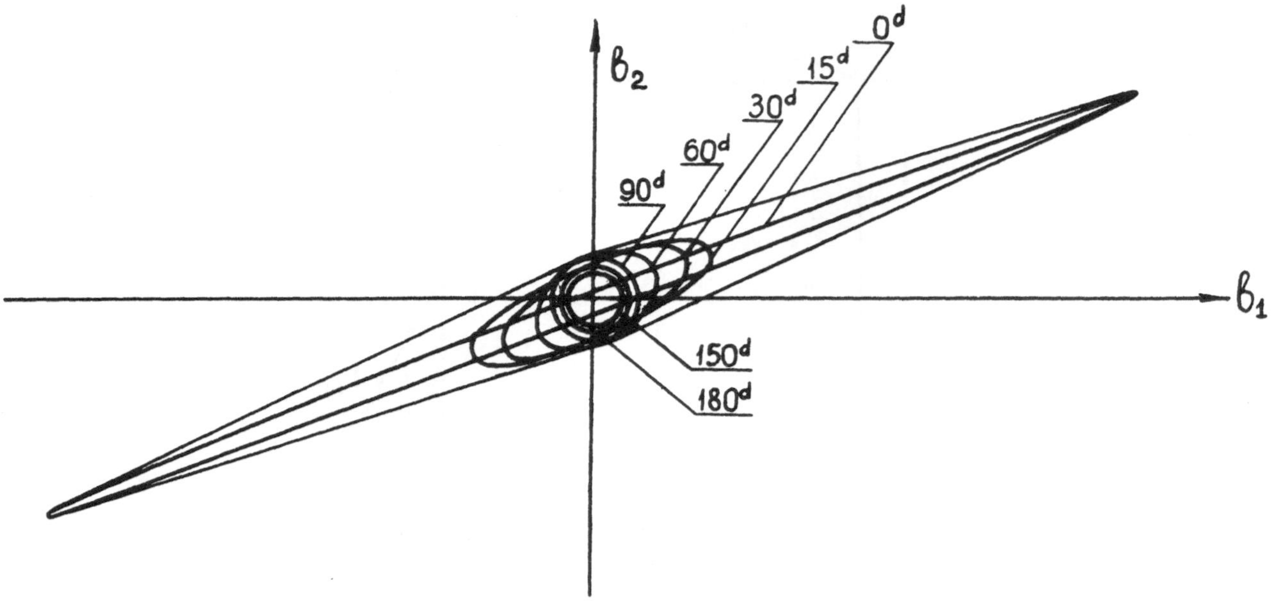

Fig 6

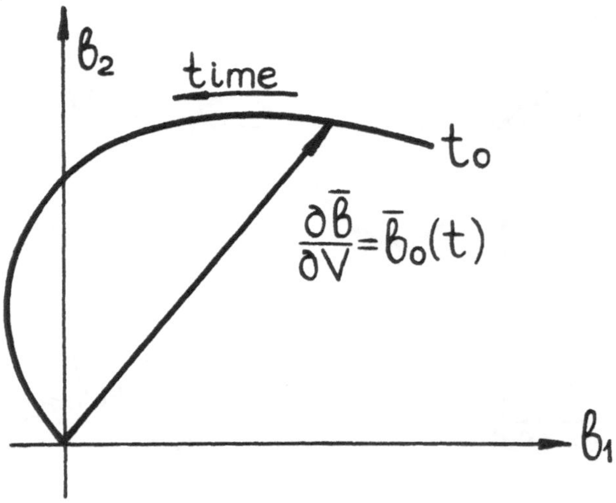

Fig.7

the control acceleration is arbitrary, the unit sphere may be the set in question. To the unit sphere

$$\bar{a}^{o*}\bar{a}^{o} = 1$$

in the optimal control space corresponds, by virtue of the linearity of the transformation, the ellipsoid

$$\bar{b}^{*}N^{-1*}N^{-1}\bar{b} = 1$$

in the space of control parameters $\bar{b}$. This ellipsoid characterizes the influence of the unit control vectors at the point in question on the trajectory (figure 2).

Since the vector $N\,\delta\bar{a}^{o}$ in the second term of the formula for the variation δH is a vector tangent to the ellipsoid of influence in the $\bar{b}$ space, the condition that the second term of the variation δH vanish means that the vector $\bar{\lambda}$ is directed normal to the surface of the ellipsoid. Hence, the direction of the control input $\bar{a}^{o}(t)$ must point to that point on the ellipsoid having the maximum projection onto the constant vector $\bar{\lambda}$ in the control parameter space $\bar{b}$. The control is then carried out at those points on the trajectory at which the projection of the influence ellipsoid is maximized,

$$v_0 = \bar{\lambda}^{o*} N \bar{a}^{o}$$

and exceeds a certain specified value (figure 3, figure 4):

$$v_{spec} = \frac{1}{\lambda}$$

If a_{max} is sufficiently small, the characteristics of the influence ellipsoid will manage to change markedly over the control time. In that case fulfillment of the requirements formulated earlier will ensure maximum effectiveness in the direction $\bar{\lambda}^{o}$ decided upon.

There exists on the trajectory a point A at which the effectiveness v_0 in the indicated direction $\bar{\lambda}^{o}$ will be maximized. This point corresponds to the maximum projection of the influence ellipsoids of all points of the trajectory onto the selected direction $\bar{\lambda}^{o}$. Hence the conclusion that the pulsed nature of the control is feasible holds precisely at that point. Actually, for a finite value $\bar{a}_{max}$ the control will occur in some neighborhood of the point A. As $\bar{a}_{max}$ is increased, this neighborhood will dwindle and the control will occur at points on the trajectory with greater effectiveness. The minimized functional V is therefore obtained as $a_{max} \to \infty$, that is in the limit of a pure impulse acceleration.

If the function $v(t)$ possesses internal minima over the extent of the flight time, then there exist deviations in $\bar{b}$ from the initial trajectory such that optimal correction of these deviations will call for spacing of the control input at different points on the trajectory. This much is inferred from the condition for switching on the control engine when $v_0 > v_{spec}$, which may be satisfied in this case over several segments of the trajectory (figure 5).

Sequential shifting along the most effective directions in the space of control parameters $\bar{b}$ is effected in such a way in this type of control that the total shift will turn out equal to that specified in advance.

Hence, the solution of the problem of optimal flight control contains within it cases of repeated engine starts even in the absence of control errors.

In the light of the optimal control principles formulated above, the influence ellipsoids corresponding to the times the engine is switched off and on again in sequence have a common tangent whose equation is

$$\bar{b}^{*}\bar{\lambda}^{o} = v_{spec}$$

The radius vectors $\bar{b}_{off}$ and $\bar{b}_{on}$ of the points of contact with the influence ellipsoids on that straight line indicate the input directions in the control acceleration space $\bar{b}$ at the times of motor switch-off and subsequent switch-on. As a_{max} is increased, the engine-on time is reduced and this tangent becomes transformed in the limit to the tangent to the envelope of the entire set of influence ellipsoids over the flight time interval.

Hence, at both high and low values of a_{max} the control input may prove optimal at several points in cases where the set of influence ellipsoids is not everywhere convex over the flight time interval. In those cases there exist directions $\bar{\lambda}_o$ such that the influence function $v_0(t,\bar{\lambda})$ related to them exhibits the shape seen in figure 5.

A similar shape of the function $v_0(t)$ may be related to the degeneracy of the matrix N(t) at some point on the trajectory. In that case we will observe a nonuniformity in the directions in $\bar{b}$ space from the standpoint of effectiveness of control. One example of such a situation may be seen in a set of influence ellipsoids in the plane of the diagram described above and appearing in figure 6, corresponding to a flight to the planet Mars. In this case, the principal influence exerted on the trajectory at the beginning

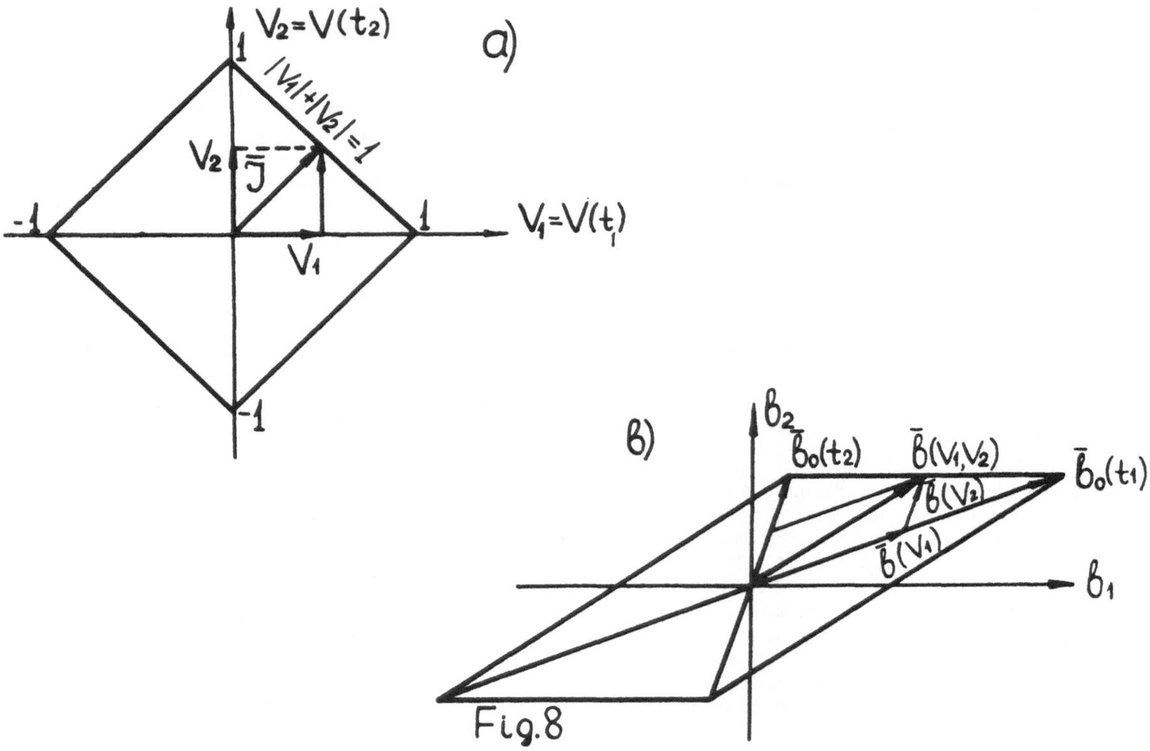
a)
$V_2 = V(t_2)$
$|V_1| + |V_2| = 1$
V_2
$\bar{J}$
V_1
$V_1 = V(t_1)$
1
-1
1
-1
b)
b_2
$\bar{b}_0(t_2)$
$\bar{b}(V_1, V_2)$
$\bar{b}_0(t_1)$
$\bar{b}(V_2)$
$\bar{b}(V_1)$
b_1

Fig.8

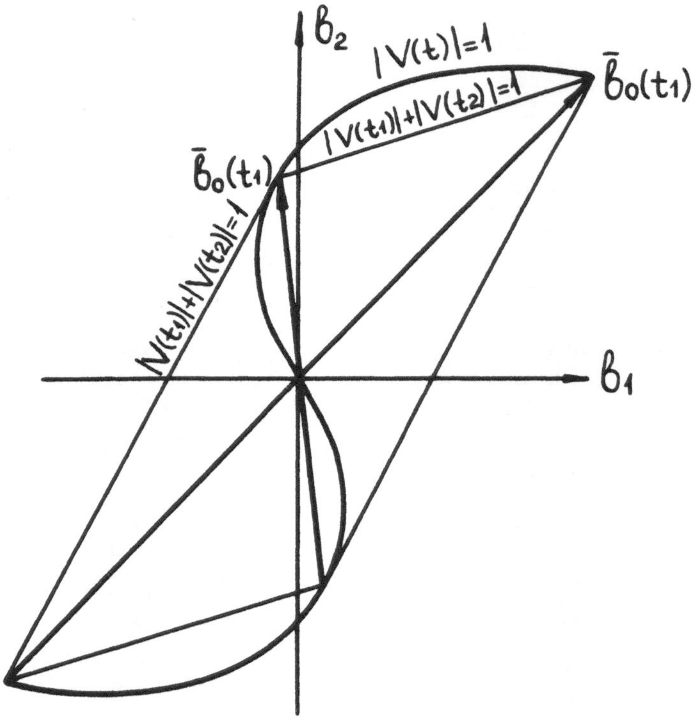
b_2
$|V(t)| = 1$
$\bar{b}_0(t_1)$
$|V(t_1)| + |V(t_2)| = 1$
$\bar{b}_0(t_1)$
$|V(t_1)| + |V(t_2)| = 1$
b_1

Fig.9

of the flight is that exerted by a change in the energy of geocentric motion, while the matrix N is close to degenerate. To correct most of the deviations in the plane of the diagram, we find that the optimum response with respect to power savings is a program calling for switching the engines on twice during the trip, once on the first day of flight and once again on the ninth. Each time the disturbance affecting the trajectory is applied in the direction of maximum effectiveness so that the required outcome will be brought about as the end result after the second disturbance input is applied.

Consider the case of impulsive control in greater detail. From the given set of influence ellipsoids of a single control input we can obtain the influence figure of the optimal unit control input disturbance, that is a disturbance such that the functional $V=\sum V_i$ will be unity. For this purpose, the rectifying plane must be rolled on the given, not everywhere convex, set of influence ellipsoids for a single impulsive control. The points of contact of the rectifying plane and the original set of influence ellipsoids determine the possible times and directions of application of the optimal velocity control impulses. The rectified segments of the resulting convex figure correspond to intermittent starting of the engine, while the segments belonging to the original set of influence ellipsoids correspond to single engine starts.

The influence figure obtained thereby is the region of deviations in the $\bar{\ell}$ space which deviations may be eliminated by the total unit input disturbance. It can be shown that no more than n control impulses are needed to change n parameters in the absence of errors in the correction system.

These results do not depend on the form of the original set of influence figures. In particular, this set may correspond to those control impulses whose direction is fixed in space in one way or another. In that case, as well as in the case where the number of parameters controlled exceeds the number of independent control disturbances at each point on the trajectory, the use of the control technique described above may prove necessary independently of any concepts involving minimization of the functional V.

An example of a situation of this sort may be found in the correction of a flight trajectory with the aid of a one-parameter control disturbance. For simplicity we assume the vector $\bar{\ell}$ to be two-dimensional, i.e. the trajectory is assumed to have only two characteristic parameters. In that case the set of influence ellipses may be depicted in the plane $\ell_1 \ell_2$ of these parameters. If there is only one free parameter for the control V(t) at each point on the trajectory, then the corresponding influence ellipse in the plane of parameters $\ell_1 \ell_2$ will degenerate into a straight-line segment (figure 7). The length and slope of this line segment are determined by the partial derivatives of the control parameters $\ell_1 \ell_2$ with respect to the control disturbance input V(t). The set of influence figures for all points on the trajectory is mapped in the ℓ_1, ℓ_2 plane with the aid of a hodograph of the vector $\bar{\ell}_0(t)$ with the components $\frac{\partial \ell_1}{\partial V}$, $\frac{\partial \ell_2}{\partial V}$.

The influence figure for the control at two fixed points t_1, t_2 on the trajectory constitutes a parallelogram spanned by the vectors $\pm\bar{\ell}_0(t_1)$, $\pm\bar{\ell}_0(t_2)$. Actually, in the case of two engine starts in flight the analog of the unit sphere in the space V_1V_2, i.e. the region of impulses of like sign, is the figure depicted in figure 8a. Its image in $\ell_1\ell_2$ space is the parallelogram seen in figure 8b.

The optimal influence figure seen in figure 9 is obtained by rolling the rectifying line on the hodograph of $\pm\bar{\ell}_0(t)$. This figure determines a different control tactic depending on the direction of the deviation vector $\bar{\ell}$. In the given case either one start or two starts of the engine during the trip will be optimal. The point for one engine start depends on the direction of the corrected deviation $\bar{\ell}$. The optimal instants of time for two starts of the engine in flight are fixed and correspond to the points of contact of the hodograph of $\bar{\ell}_0(t)$ and the rectifying line.

We may conveniently take the heliocentric radius vector as the direction of the one-parameter disturbance on the interplanetary trajectory. The control may be carried out then by applying a disturbance input in the direction to the sun or in the direction away from the sun, using a single-axis orientation. Let $\ell_1 \ell_2$ be the components of the deviation vector in the plane of the diagram with the coordinate axes ξ, η.

In that case two control disturbances are required to eliminate the deviations in the plane of the diagram. In the linear case, clearly, the components of the deviations may be represented as:

$$\ell_1 = \frac{\partial \xi}{\partial \dot{r}}(t_1)V_1 + \frac{\partial \xi}{\partial \dot{r}}(t_2)V_2$$

$$\ell_2 = \frac{\partial \eta}{\partial \dot{r}}(t_1)V_1 + \frac{\partial \eta}{\partial \dot{r}}(t_2)V_2$$

where V_1 and V_2 are the control disturbance inputs at points t_1 and t_2 respectively $\partial\xi/\partial\dot{r}\,(t)$, $\partial\eta/\partial\dot{r}\,(t)$ are the corresponding velocity derivatives along the direction of the heliocentric radius vector.

This system has a solution if its discriminant does not vanish, and this is usually the case on interplanetary trajectories.

Since the control input lies in the plane of the heliocentric trajectory in this case, the plane may not be changed. As a result of this type of control, therefore, the time of approach to the planet coincides with the time the planet crosses the nodal line of its orbit and the plane of the heliocentric trajectory. The number of control parameters may not exceed four independently of the number of times the control engine is switched on in flight.

Literature Cited

1. Okhotsimskii, D. E., Notes on the theory of rocket motion, Prikladnaya Matematika i Mekhanika, 10, No. 2 (1946).

2. Lawden, D. F. and Long, R. S., The theory of correctional maneuvers in interplanetary space, Astronaut, acta 1960.

3. Rosenblum, A., Control systems with integral restraints and quality criterion determined by the state of the system at a predetermined time. Proceedings of the 1st International IFAC Congress, Moscow, 1961.

4. Lorell, J., Velocity increments required to reduce target miss on coasting trajectories, Advances Astronaut Science, Vol. 6, 1961.

Specific Systems

GUIDANCE AND CONTROL OF THE MARINER PLANETARY SPACECRAFT*

by John R. Scull
Chief, Guidance and Control Division
Jet Propulsion Laboratory
California Institute of Technology
Pasadena, California

ABSTRACT

The unmanned exploration of the Moon and nearby planets imposes very significant requirements on the guidance and control system of both the booster rocket and the spacecraft. The booster rocket must be launched into a narrow, moving corridor with great precision within a limited period of time. After separation, the spacecraft must be capable of providing corrections to the trajectory and controlling the orientation of solar panels, antennas, and scientific instruments.

The guidance and control systems of the Mariner planetary spacecraft are described. Some of the techniques developed to provide redundant operation during the long interplanetary flight are discussed. Flight performance of the Mariner guidance and control systems is reported.

INTRODUCTION

The challenge of planetary exploration requires continuous research and development to achieve mission goals. Larger, multistage launching vehicles are being built to provide more thrust; spacecraft and scientific instruments for measuring conditions in outer space have become more complex and have been further refined to achieve greater accuracy and reliability. Successful completion of proposed planetary investigations requires the solution of many difficult problems in guidance and control.

The permissible launching period for a planetary spacecraft is limited to approximately one month every two years. During this period, a maximum of two hours per day is considered a suitable "window" for launching the spacecraft.

The spacecraft must be launched from a platform moving around the Sun at 66,000 miles per hour toward a destination planet moving 10,000 or 15,000 miles per hour faster or slower than the Earth and located hundreds of millions of miles away! It must pass near this target with a degree of accuracy much higher than that of the best military artillery; yet, if the spacecraft is unsterilized, it must have a very low probability of impacting on the destination planet so as to avoid surface contamination. This interception requires that the spacecraft pass within the confines of a narrow zone, only a few thousand miles on a side, to enable the scientific instruments to make measurements of the planet.

The components of the control systems must operate successfully over an extremely long lifetime in a very hostile environment. The length of the Venus trajectory, 180 million miles over a 109-day period, is short when compared to that of Mars, 325 million miles over an eight-month period.

Along the way, the orientation system must direct high-gain antennas back toward the Earth to communicate scientific and engineering measurements. The solar panels must be pointed toward the Sun in order to generate power required for spacecraft operation. Scientific instruments must be properly oriented toward the planet during the encounter sequence to obtain measurements required for each experiment aboard the spacecraft. The accuracy of orientation must be relatively high to provide directional control of a midcourse rocket motor used for trajectory corrections.

The foregoing accomplishments must be made while the spacecraft is traveling through outer space, where it is exposed to vacuum conditions greater than those found in the best radio tubes, while being bombarded with charged particles, cosmic dust, and meteorites. Owing to the change in solar constant over the trajectory, the variation in radiant incoming heat is nearly 2 to 1. The major scientific payoff occurs at the end of the mission when a few hours of data are obtained during the time of closest approach to the planet. On some missions, successful operation beyond this time is required in order to

*This paper presents the results of one phase of research carried out at the Jet Propulsion Laboratory, California Institute of Technology, under Contract No. NAS 7-100, sponsored by the National Aeronautics and Space Administration.

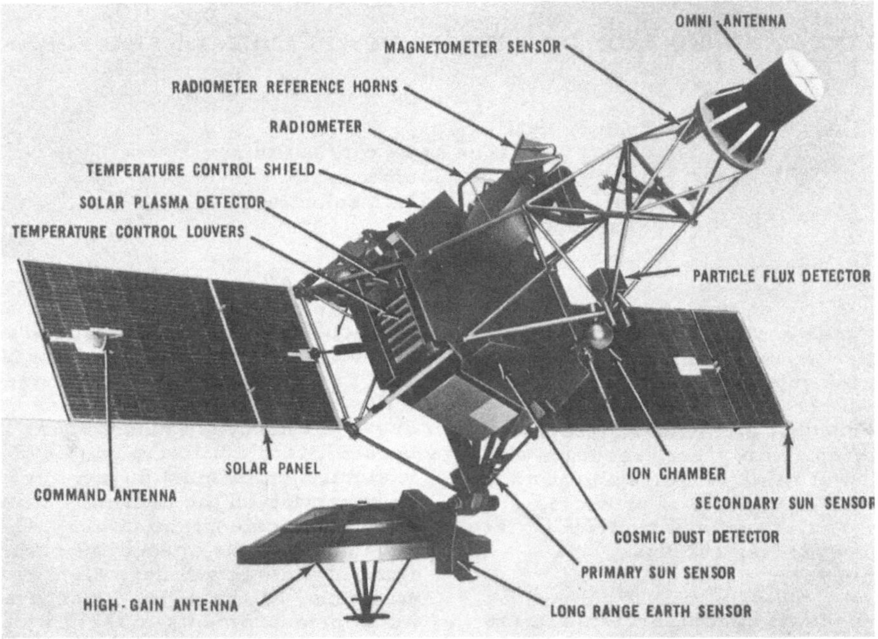

Fig. I. *MARINER* II

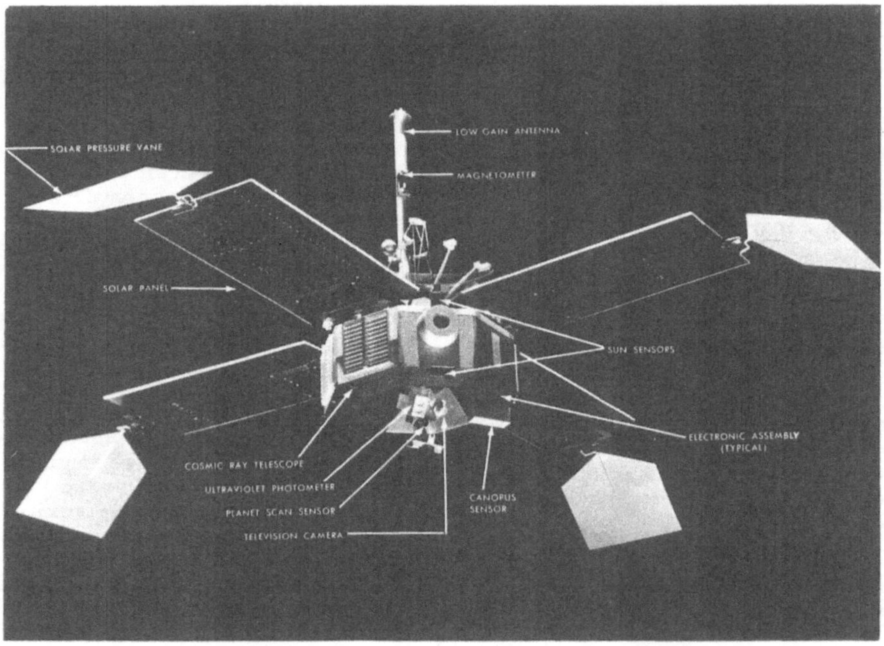

Fig. 2. *MARINER* IV

transmit data, recorded during planetary encounter, back to the Earth at a low rate.

LAUNCH VEHICLE CHARACTERISTICS

The Mariner planetary spacecraft were launched on their voyages by the Atlas-Agena booster. The first-stage rocket, developed by Convair-Astronautics Division, was first flown in 1957. The booster is one which has been modified by the National Aeronautics and Space Administration to be used as a space launch vehicle. The Atlas is steered by gimballing the rocket engines, rather than by using conventional control fins. Attitude control provided by a programmed autopilot is overriden by radio commands from a radar system and computer located on the ground.

The upper stage of the booster rocket, the Agena, is produced by Lockheed Missiles and Space Division and was first flown in 1959. The Agena has been used as an upper stage on both the Thor and Atlas boosters. It is steered by gimballed rocket nozzles and controlled during coast flight by small gas-jet thrustors. Rocket motor thrust is terminated at the proper times on a signal derived from an on-board velocity meter. Attitude of the Agena stage is determined by gyroscopes and infrared horizon scanners.

The spacecraft and Agena stage are boosted into a parking orbit of approximately 100 nautical miles; the vehicle then coasts to the appropriate injection point. A coast period, ranging from a few minutes up to a maximum of about 20 minutes duration, is required. At injection, the Agena rocket motor is again fired to launch the spacecraft into its interplanetary trajectory. The accuracy of this type of injection guidance system for Venus and Mars missions is shown in Table 1. The table also shows the amount of midcourse velocity correction required to compensate for injection errors and the total accuracy obtained.

SPACECRAFT DESCRIPTION

Mariner II

The Mariner II spacecraft (Fig. 1) launched toward Venus in 1962 weighed 447 lb and was 11 ft, 11 in. high and 16 ft, 6 in. wide. Its mission was to conduct scientific measurements during the long, interplanetary flight and to pass within a zone 8,000 to 40,000 miles from the surface of Venus in such an orientation that the terminator could be scanned by a radiometer.

The power system consisted of two solar panels, totaling 27 sq ft, and a rechargeable battery. Power consumed during the mission varied from 148 to 222 watts.

Scientific and engineering data were transmitted to receiving stations on Earth by an omnidirectional antenna and a high-gain steerable antenna. Commands from the ground were received through two command antennas and a command receiver. Orientation in three degrees of freedom was provided by the attitude control system. Cadmium-sulfide sensors directed the solar panels toward the Sun with an accuracy of 1 deg. The roll orientation of the spacecraft was sensed by an Earth detector containing a photomultiplier and a vibrating-reed chopper. Actuating devices for the long cruising flight consisted of small 0.01-lb-thrust gas jets fed from a source of pressurized nitrogen.

The spacecraft was tracked by NASA's Deep Space Net (DSN) for several days after it was launched, and deviations from the proper trajectory were determined at a central computing station. The velocity differential required to correct the trajectory, thus enabling the spacecraft to pass near the planet, was determined. Roll and pitch angles through which the spacecraft must turn and delta velocity required were transmitted to the spacecraft by radio command.

The midcourse correction system consisted of a 50-lb-thrust hydrazine monopropellent rocket motor capable of imparting a 200-ft-per-sec velocity increment. During the thrusting period of the midcourse maneuver, the force from the gas jets was too small to correct for thrust milalignment torques. Small, postage-stamp-size jet vanes, located in the rocket exhaust, were used to provide directional control.

The scientific instruments included in the Mariner II spacecraft consisted of a magnetometer, solar plasma probe, cosmic dust detector, charged particle and radiation detector, plus two radiometers for measuring planetary surface temperature at encounter.

Mariner IV

The Mariner IV spacecraft is illustrated in Fig. 2. It was launched toward Mars in 1964, weighed 575 lb, stood 10 ft high, and was 22 ft, 7 in. wide. It was designed to carry scientific instruments for investigating the interplanetary region between Earth and Mars.

During planetary encounter, a series of television pictures of the surface will be taken and recorded on magnetic tape. These pictures will be played back to Earth at reduced bandwidth following closest approach. As the spacecraft passes behind the planet, the refraction and attenuation of the radio signal during occultation will be used to measure the properties of the Mars atmosphere.

The power for the spacecraft is provided by four solar panels, having a total area of 70 sq ft, and a rechargeable storage battery. Engineering and scientific measurements are telemetered back to Earth via redundant 10-watt radio transmitters. One transmitter is a cavity amplifier; the other contains a traveling-wave tube.

Telemetry is transmitted over an omnidirectional antenna on the first third of the flight. In the latter portion of the trajectory, commands and telemetry are handled by a high-gain directional antenna. Unlike the Mariner II design, the high-gain antenna of the Mariner-Mars spacecraft is not steerable but is rigidly attached to the spacecraft structure. This mechanization is possible because of the unique trajectory characteristics of the mission which allow a fixed-antenna orientation.

The attitude control system in the pitch and yaw axes is similar to that of the earlier Mariner, and solar orientation is sensed by cadmium-sulfide detectors. The reference for the roll attitude is the star Canopus, which is the second brightest star in the heavens and is located at nearly the south ecliptic pole. Canopus rather than Earth was selected as a roll reference because Earth passes between the spacecraft and the Sun during part of the interplanetary trajectory. This causes the Earth to appear as a narrow crescent, as well as the Sun to be within the field of view of the Earth sensor.

The Canopus star tracker uses an electrostatic image dissector as its detector, thus requiring no moving parts. Nitrogen gas jets and jet vane actuators, similar to those of the Mariner Venus design, are employed. Compensation for an unbalance in solar radiation pressure is provided by moveable paddles located on the tips of the solar panels.

The midcourse correction employed by Mariner IV was somewhat like that of the Venus probe but contained an added provision for multiple restarts which allowed any additional vernier trajectory correction required later in the mission to be made.

SEQUENCE OF EVENTS

The nominal flight sequence for the Mariner-Mars spacecraft begins just prior to launch when the inhibit to the central timer is removed, permitting the timer to begin operating. Many of the components of the guidance and control system are turned on and operated during the boost period in order to telemeter their performance while in a high-acceleration environment.

Five minutes after liftoff, when the booster reaches an altitude where the atmospheric pressure is sufficiently low, the nose shroud is separated and the high-voltage power supplies of the transmitter and cruise science instruments are turned on.

Separation of the spacecraft and booster occurs after burnout of the Agena stage, and about 1 min later solar panels and solar vanes are deployed.

The attitude control system is activated and the Sun acquired approximately one hour after liftoff.

During the next 15 hr, the spacecraft is rolled about the Sun line for calibration of the magnetometers. At the conclusion of this period, the Canopus sensor is turned on and a roll search initiated.

After Canopus is acquired, the gyroscopes used for providing control during the search and acquisition modes are turned off.

The midcourse maneuver sequence is initiated several days following launch. Times for pitch- and yaw-commanded turns are transmitted from the Deep Space Net, along with motor burn duration. An additional ground command begins the maneuver sequence. The gyroscopes are again turned on and allowed to warm up for about an hour, at which time the spacecraft is placed on inertial control and executes the predetermined angles in pitch and roll. The midcourse motor is ignited and burns for the prescribed time.

About an hour after ignition of the midcourse rocket motor, the timer initiates an automatic reacquisition of the Sun and Canopus. Upon completion of this maneuver, the logic inhibit of the solar vane system is removed and the vanes begin their adaptive compensation for unbalanced solar radiation pressure.

Approximately one month after launch the data rate of the telemetry system is reduced from 33.3 to 8.3 bits per sec.

The nominal angle for the field of view of the Canopus sensor is adjusted four times during the coasting trajectory. Canopus lies approximately 15 deg away from the south ecliptic pole; thus, in the trajectory around the Sun, the angle between the Sun and Canopus may vary as much as 30 deg.

Three months after launch the telemetry signal is switched from the omnidirectional antenna to the fixed high-gain antenna.

Nine hours prior to encounter, the science instruments for the encounter sequence are activated. The television system begins taking pictures 45 min prior to closest approach. All other encounter science instruments continue to operate for approximately 6 hr after the 22 television pictures have been recorded on magnetic tape. Ten hours after closest approach the tape recorder plays back the television pictures at a reduced data rate. The mission will probably conclude approximately 20 to 40 days after encounter, when the combined effects of increased range and fixed antenna pointing error cause the communications system to exceed its threshold.

RELIABILITY ENHANCEMENT

Extremely long operating times required of the Mariner spacecraft caused redesigns of the guidance and control system used previously on Ranger. Changes were made primarily to obtain longer life and redundant operation in the event of subsystem failure.

First to be eliminated for the cruise portion of the flight were the gyroscopes which provide rate damping to the attitude control system. Since the spin motor bearings of most gyroscopes have a life expectancy of from 1,000 to 2,000 hr, they are among the lowest-reliability components in the spacecraft. However, they are necessary for initial acquisition, controlled maneuvers, and inertial stabilization in case of Sun- or Canopus-sensor failure.

A synthetic rate-damping signal is generated by integration of the voltage applied to the solenoid valve of the gas jets. Since the actuation of the gas jets causes a fixed angular acceleration, the integral of this signal is proportional to the rate differential applied. This "derived rate" damping system was used on both Mariner II and Mariner IV. An attitude control system using derived rate damping is much less susceptible to noise which causes double gas jet pulses than one employing gyro damping. Any disturbance in the spacecraft attitude resulting in Canopus leaving the field of view of the star sensor will automatically turn on the gyros and cause the system to go through a reacquisition sequence.

The solar radiation pressure control vanes added to the Mariner IV spacecraft are designed to reduce the unbalance torque acting on the pitch and yaw axes to nearly zero. Each time a gas jet is actuated, the solar radiation pressure vanes are stepped differentially $1/100^{th}$ deg, thereby compensating for any unbalance in external torques over a period of several days. An unbalance of 70 dyne-cm, caused by solar radiation pressure, biased the attitude control system on Mariner II to one side of the limit cycle. In the Mariner-Mars spacecraft these torques are reduced to below 10 dyne-cm. In addition, the vanes produce a restoring torque which would cause the spacecraft to return to Sun orientation in the absence of gas-jet operation. The adaptive balance action of the control vanes is limited to ±20 deg from their nominal position to minimize the torque unbalance which would result if a vane actuator failed. The actuator and its associated electronics are designed to consume power only when actually in motion. The primary benefit from the solar vanes is a reduction in consumption of control jet gas and resultant minimization in the number of times the solenoid valves must be actuated. This should markedly improve the reliability of the gas valves and reduce the probability of leakage caused by wear. The solar pressure balancing system is capable of compensating for an initial unbalance torque of up to 80 dyne-cm.

The total amount of pressurized nitrogen for the control jets is not reduced from that required without the balancing operation of the solar vanes. The amount of gas carried is sufficient to complete the mission, even if one solar vane should fail in either of its extreme positions. Solar vanes improve the overall reliability of Sun-line attitude control by (1) reducing the requirements placed on the gas-jet system and (2) providing a redundant restoring torque in case of gas valve failures.

The gas system is further redundant in that each of the two valves providing a torque couple operates from a separate gas reservoir and regulator. If a gas valve fails in a closed position, the remaining valve providing torque around that axis will still continue to operate. Conversely, if a valve fails in an open position, the pressure from that tank would go to zero. Half the gas in the remaining tank would be dissipated in providing a torque to counter that from the leaking valve. Since the amount of gas carried in either tank is more than sufficient for the entire mission, the remaining gas in the second system would be capable of providing control for the duration of the mission.

Logic is built into the guidance and control and other systems to provide automatic mode switching should non-catastrophic failure occur. Automatic reacquisition of Canopus in the event of loss of track was described in a preceding section. Automatic switchover from one transmitter to a spare occurs if proper operation is not sensed at the end of a prescribed time period. The main power regulator is so designed that, if the output voltage varies from the required limits, automatic switchover to an additional unit used during maneuvers will occur.

As an alternative to mode switching controlled by internal logic, the operating characteristics of the spacecraft may be changed by overriding ground commands. The spacecraft can be commanded to reject the star being tracked by the Canopus sensor and go through a roll search until a new star of the predetermined brightness falls within its gates. The center of the field-of-view of the Canopus tracker may be varied relative to the Sun by ground command as well as by the internal sequencer. Internal logic, which causes the roll attitude control system to go into an automatic roll search when certain Canopus brightness bounds are exceeded, may be removed by ground command. The spacecraft may be commanded to go into an inertially controlled mode if an attitude sensor fails, or to assume a new attitude by sending incremental angle commands, as well as holding a preset attitude.

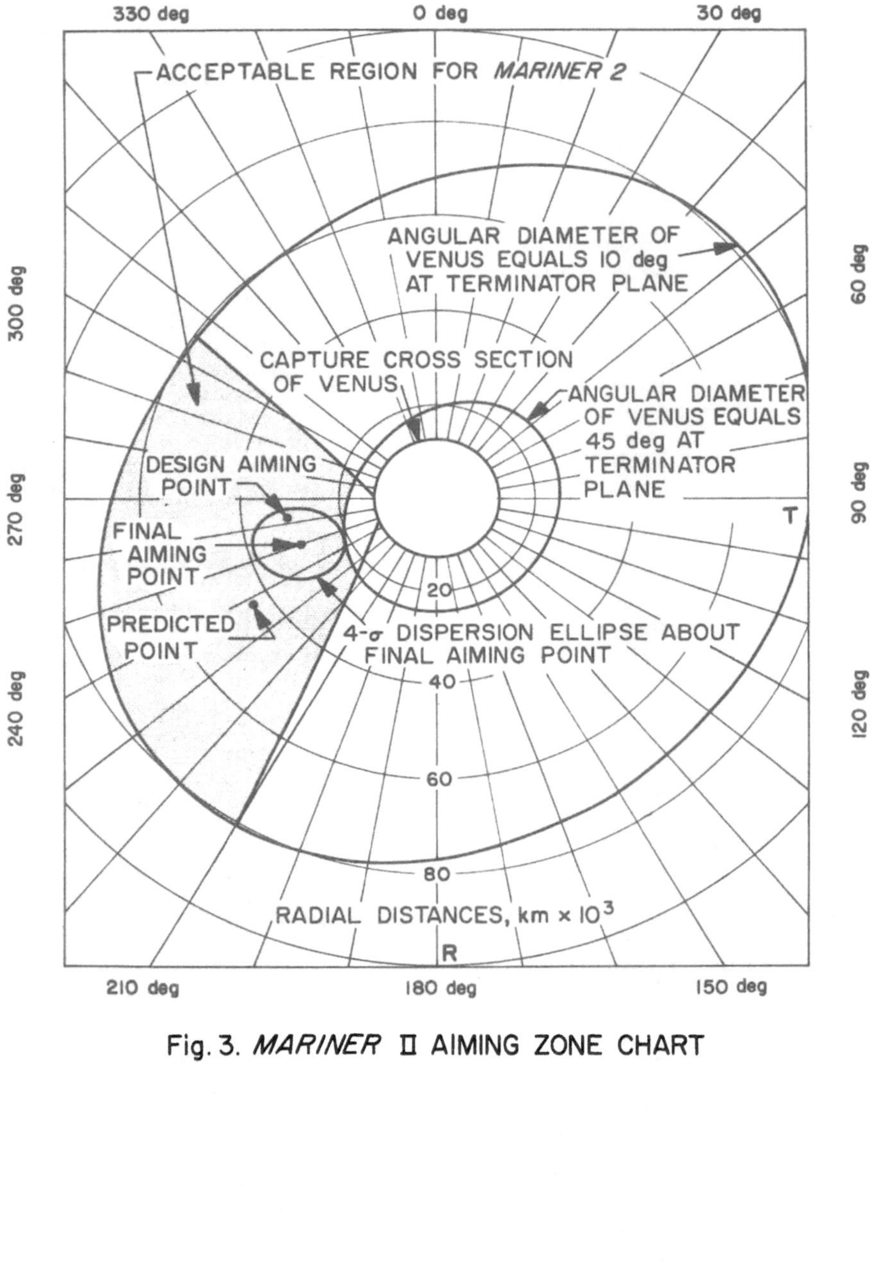

Fig. 3. *MARINER* II AIMING ZONE CHART

Determination of system failures requires that a large proportion of the total telemetry channels be assigned for failure detection. Within the 20 attitude control channels and 22 channels for power system measurements, failure detection takes a higher priority than performance evaluation. Fault isolation down to the subsystem level may be determined on the ground to assist in making command override decisions.

Another major factor in reliability enhancement is spacecraft qualification testing, which occurs at all levels, from overall spacecraft systems testing to that of initial component selection.

Many of the electronic components used to fabricate Mariner subsystems are subjected to rigid screening and aging processes prior to assembly. Component parts are selected from a small group of qualified types and go through further screening tests to eliminate units evidencing significant behavior variations. Component characteristics are measured before and after a temperature aging process, and components showing variations beyond a predetermined level are rejected. Individual serial numbers are assigned to each electronic component part and an IBM record is kept showing its screening history. Failures occurring either at the component or subsystem level can be correlated with the performance of similar components taken from the same batch and selected by the same process. Thus systematic failures are detected and all components from a suspect batch are rejected.

Early in the program, prototype units of all subsystems are assembled for a Proof Test Model (PTM). The PTM is a complete spacecraft used to investigate any subsystem interface problems and evaluate overall system characteristics. Component or subsystem modifications determined from PTM evaluation may then be introduced into the flight spacecraft units during their fabrication process. Changes made to subsystem units subsequent to the beginning of PTM assembly must be checked out on this model before modifications are introduced into flight units.

A type-approval (TA) unit is used to determine possible subsystem weaknesses prior to acceptance testing of the flight units. One unit, identical to each flight subsystem, is exposed to environments significantly above those expected during flight. Vibration levels 50 percent higher than those expected during the boost period are typical of type-approval testing. Other tests performed include high- and low-temperature testing in a vacuum and testing in environments of humidity, shock, static acceleration, rf interference, drop testing and handling.

Prior to delivery for assembly into the spacecraft, each flight subsystem must undergo flight acceptance (FA) tests. These tests are similar to those described in type-approval testing but are made at environmental levels expected during actual flight.

After the flight spacecraft is assembled, subsystem and system functional tests are performed. When the spacecraft is considered completely flightworthy, it undergoes a series of mission tests which attempt to duplicate, at the entire spacecraft system level, a sequence of operations expected in actual flight, which include operation of all major elements of the spacecraft in a vacuum and simulated solar radiation environment. The entire spacecraft must successfully pass a series of mission tests without a major failure before it is considered acceptable for launch.

A unit similar to the flight module is operated in a vacuum and variable temperature environment over a prolonged operating time, equal to that expected during the mission, in order to determine any time-dependent characteristics of the subsystems. From time to time during these life tests critical parameters are measured and variations from previous test results are determined.

Thus it is seen that the reliability program necessary for long-duration missions is not only one of qualification testing, but, in reality, is a way of life for spacecraft personnel requiring meticulous attention to detail throughout the entire program, from initial component selection through mission testing.

FLIGHT TEST RESULTS

Mariner II

The Mariner II spacecraft was launched on its way toward Venus on August 27, 1962. Shortly after separation, Sun acquisition occurred in a normal manner. Seven days after launch, when the Earth was far enough away so that its intensity was not too great for the dynamic range of the Earth sensor, the central computer and sequencer (CC&S) initiated Earth acquisition. After the spacecraft stopped its roll motion and the Earth was apparently acquired, the telemetered output of the brightness reading from the Earth sensor was considerably lower than it should have been if it were locked properly on the Earth. The actual intensity monitored was that which might have been expected if the spacecraft were locked on the Moon. The strength of the radio signal transmitted from the high-gain directional antenna, however, indicated that the roll axis of the spacecraft was properly directed toward the Earth.

The midcourse maneuver was delayed several days from the time originally planned pending verification of the proper Earth lock. On September 4, 1962, it was established that the

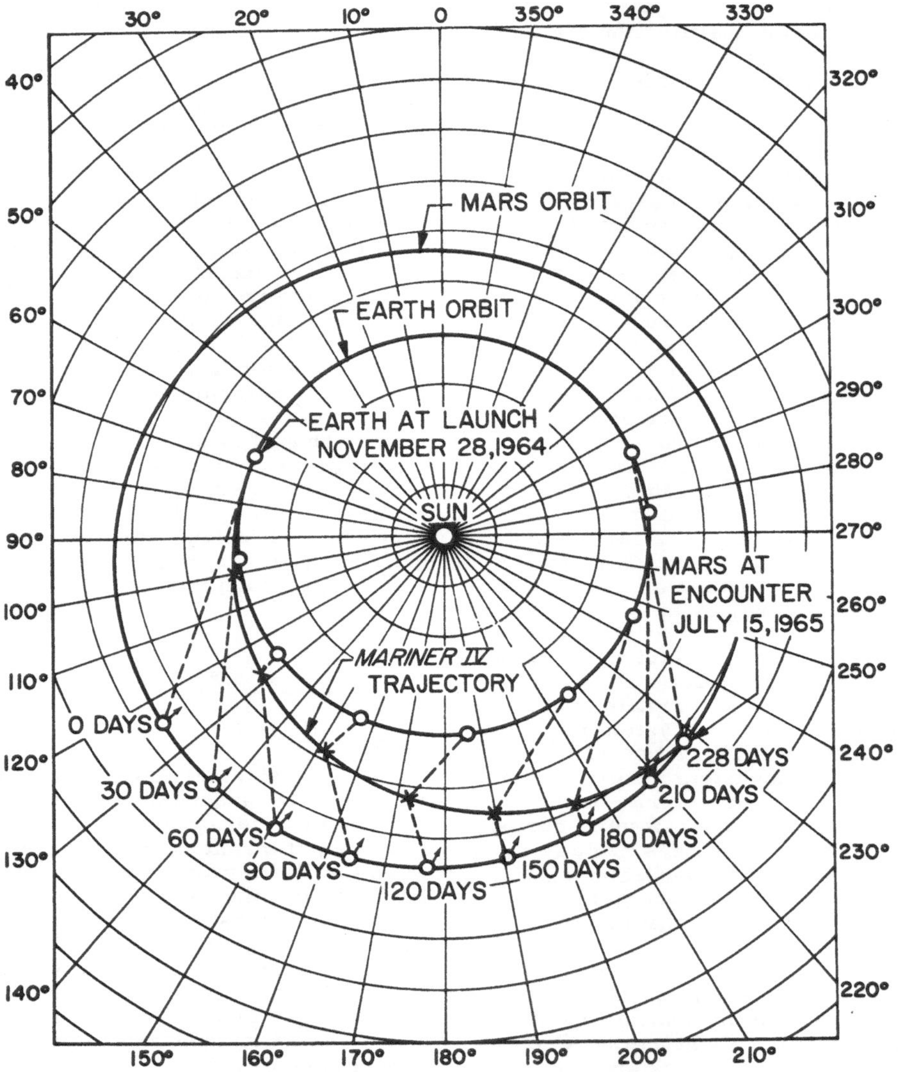

Fig. 4. *MARINER* IV TRAJECTORY

spacecraft was properly oriented toward the Earth, not the Moon, and the midcourse sequence was commanded. The midcourse rocket motor ignited and burned for 27.8 sec. Thrust was terminated when the velocity measured by an accelerometer was equal to the value commanded from the ground. The midcourse maneuver decreased the relative speed of the spacecraft with respect to the Earth by 47 ft per sec and resulted in a correction of the Venus miss distance from 233,000 to 21,598 miles. The shaded area of Fig. 3 shows the acceptable aiming zone for the Mariner II mission. The final aiming point shown was the nominal point commanded by the midcourse maneuver. The actual point of flyby is shown as the predicted point on the figure. Although the dispersion of the midcourse maneuver was larger than intended because of velocity shutoff errors, the point of closest approach fell well within the acceptable region for the mission. The Mariner II midcourse maneuver set a record for trajectory correction when the spacecraft was 1,492,000 miles away.

After Sun and Earth reacquisition following midcourse maneuver, the brightness telemetry measurement from the Earth sensor was still considerably below the anticipated amount. As the intensity was approaching the threshold value, a transient, apparently caused by a meteorite hit, caused the logic of the attitude control system to turn on the gyros and initiate an Earth acquisition sequence. Before the gyros were fully up to speed, the Earth brightness jumped up to the proper value and remained at a normal level throughout the remainder of the mission. The exact cause for this anomaly has never been determined. It has been suggested that an internal reflection in the optical system of the Earth sensor was being sensed in the early part of the mission and, as the Earth brightness decreased because of the increasing distance, this "flare" reached a sufficiently low value and the proper object was tracked.

The temperature rise of the Mariner spacecraft as it approached the Sun was much more rapid than predicted. By December 12, six of the temperature sensors exceeded their upper limits and the equipment was exposed to an environment far beyond the tested levels. Several of the subsystems, including the solar panels and the internal sequencer, suffered partial failures due to the high temperatures. The encounter sequence for scanning the planet was designed to be initiated by the internal sequencer, but 3 days prior to closest approach there was reason to believe that it would not function. A redundant sequence initiation provided by an Earth command was designed into the spacecraft. The command was transmitted from the Goldstone Tracking Station on the 109th day of travel and, 36 million miles away, the Mariner II responded and began its encounter sequence.

Twenty days after the spacecraft had passed Venus, the telemetered brightness of the Earth had reached its design limit. When the Woomera tracking station made a normal search for the spacecraft signal it could no longer be found. On succeeding days other DSN stations also searched in vain. Nevertheless, Mariner II set two spacecraft records: for longer continuous operation (129 days) and for traveling a greater distance than any previous spacecraft (53.9 million miles from the Earth).

During the 129 days of operation the gas jets of the attitude control system used only 2 of the 4 lb of pressurized nitrogen gas. The limiting factors which terminated the mission were (1) temperature extremes beyond the design limits and (2) increasing Earth-spacecraft distances, which caused the Earth sensor and communications to reach their threshold.

Mariner IV

The Mariner IV spacecraft was launched toward Mars on November 28, 1964, on a trajectory shown in Fig. 4. Sun acquisition was normal; however, some difficulty occurred in the acquisition of the roll reference star, Canopus. Since the absolute brightness of Canopus had not been measured from above the Earth's atmosphere, narrow intensity limits could not be used. In roll searching for Canopus, the spacecraft locked on several stars of almost comparable brightness located at the same angle away from the Sun. A provision was built into the system to allow an override and subsequent roll search initiation based on Earth command. Canopus was acquired, as indicated by the proper telemetered brightness of both Canopus and a simple Earth detector. It was then determined to proceed with the midcourse maneuver.

The maneuver sequence was commanded December 4, 1964. However, the initial starting transient of the gyroscopes caused the spacecraft to lose roll reference. Canopus had to be reacquired and the midcourse maneuver was postponed until the next day while the situation was analyzed. The second attempt at the midcourse maneuver was made December 5, 1964, and this time was completely successful. The commanded velocity increment of 51 ft per sec reduced the miss distance from 158,679 to 5,729 miles from the surface. The accuracy of the maneuver is described in Table 1. Figure 5 shows the acceptable target region about Mars, the aiming point, and the closest approach points determined from the first four trajectory computations.

Following the midcourse maneuver, the Sun and Canopus were reacquired. However, the Canopus tracker caused a loss of roll lock, and automatic reacquisition was required several times. It was determined, from analysis of the

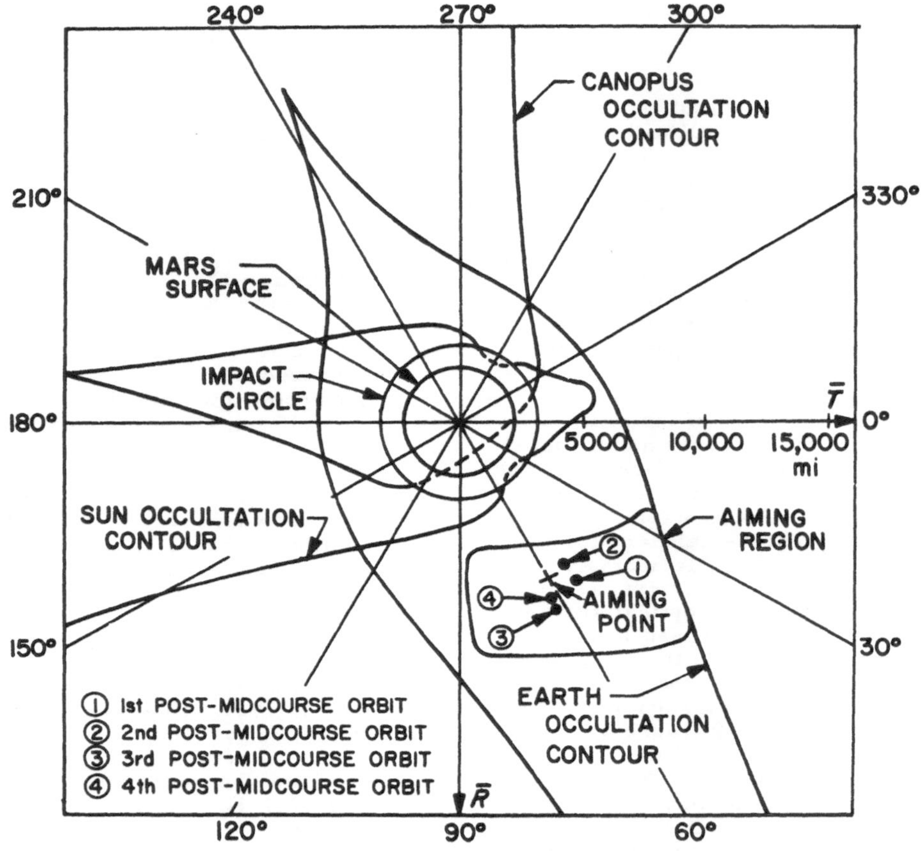

Fig. 5. *MARINER* IV AIMING ZONE AND CONSTRAINT REGION

telemetry records, that bright objects were coming within the field of view of the Canopus tracker and causing the intensity and error channels to initiate logic for a reacquisition sequence. One of these bright particles occurred shortly after a space science instrument telemetered a meteorite impact on the spacecraft.

It was concluded that the bright objects causing Canopus tracker difficulties were probably dust particles dislodged from the spacecraft following micrometeorite impact. An Earth command was sent to the spacecraft which removed the intensity and angular error logic of the roll control system so these bright particles would not cause automatic roll search. Since transmission of this command, no inadvertent roll searches have occurred, although several bright particles have passed in and out of the field of view of the tracker.

The rate of gas consumption of the Mariner IV nitrogen gas supply is 0.03 lb per day, and the remaining life of the gas system is approximately 4 years.

Solar vanes have reduced the initial radiation pressure unbalance from 30 dyne-cm to below an average of 5 dyne-cm. On one day, December 15, 1964, the radiation pressure unbalance acting on the yaw axis of the spacecraft was so well balanced that a 9-hr interval occurred between successive firings of the gas jets. It has not been possible to completely balance the torques acting on the spacecraft so that the gas jets are never actuated because the torques appear to have a random value of ±5 dyne-cm and reversals of torque have been observed over as short a period as 1 to 2 hr. This effect was not observed in the Mariner II spacecraft because of the presence of a high biasing torque.

V. CONCLUSIONS

The Mariner systems were specially designed to assure reliable operation of extremely long mission durations. These special considerations include:

1. Simplicity of design.
2. Reduction in the number of moving parts.
3. Automatic redundancy features which provide backup in the event of failure.
4. A large number of telemetry measurements to evaluate system and subsystem behavior.
5. Ground command.
6. Override capability for most backup functions.
7. A thorough testing and qualification program prior to sending the spacecraft on its long voyage into space.

The resultant spacecraft, based on these considerations, have shown the ability to accomplish difficult mission tasks even in the presence of partial failures of components and in a hostile environment.

Table 1. Midcourse guidance performance

Target	(1) Miss due to injection guidance km	(2) Orbit determination accuracy from tracking accuracy km	(3) Midcourse maneuver to correct (1) m/sec	(4) Error due to midcourse maneuver km	(5) Total accuracy (RSS of 2 and 4) km
Typical values, assuming:	Accuracy of maneuver: pointing, 0.5 deg; magnitude, 1.0% Tracking accuracy: 2×10^{-3} rad, 0.15 m/sec				
Moon	6,000	10	40	64	65
Mars	500,000	2,500	20	5,400	6,000
Venus	300,000	1,000	20	2,700	2,900
Accuracies obtained during flight test					
Moon (RA-VII)	2,060	15	30	25	29
Mars (M-IV)	255,000	1,080	17	2,360	2,600
Venus (M-II)	374,000	1,500	31	23,400	23,600

DISCUSSION

J. R. Scull

"Guidance and Control of the Mariner Planetary Spacecraft".

Q. What are the parameters of the trajectory which are measured?

A. Tracking antenna angles and doppler.

Q. Is range measured?

A. Only during the early part of flight.

Q. Are angular rates measured?

A. No.

Q. What is the angular measurement accuracy?

A. Approximately 0.01 degrees.

Q. What is the interaction between the solar pressure compensation system and the main attitude control system?

A. The main system provides immediate compensation; the solar pressure system is long-term and may require several days to react.

Q. Are there redundant gas systems?

A. The supplies and jets are completely redundant. An open or closed failure of any single valve can be overcome.

THE ORBITING GEOPHYSICAL OBSERVATORY ATTITUDE CONTROL SYSTEM

By Mr. E. P. Blackburn
Senior Staff Member
TRW Space Technology Laboratories
Redondo Beach, California

ABSTRACT

The Orbiting Geophysical Observatory is a general purpose satellite designed for scientific exploration of the earth and its near-space environment. It is capable of carrying a variety of scientific experiments in a wide range of earth orbits. Two basic orbits are planned: a highly elliptical, low-inclination orbit, and a more nearly circular polar orbit. Experiments can be oriented toward the earth or sun or along a line normal to the vehicle-earth line and contained in the orbital plane. The attitude control system is capable of meeting these orientation requirements simultaneously by controlling the attitude of the vehicle and the orientation of appendages hinged to the vehicle.

In this paper, the attitude control system is described, then the design considerations leading to its present configuration are given. The attitude control requirements are described as they relate to the experiments, the vehicle and the various orbits. Based upon these requirements, the synthesis process is reviewed showing the interaction between the vehicle and the attitude control control system design. The impact of the space environment on the control system design is described as are the design features which were incorporated to enhance the reliability of the system.

The experimental verification of the design is the subject of a separate section of the paper. This section summarizes the environmental testing, integrated control system testing, air bearing testing and the integration of the control system into the spacecraft.

INTRODUCTION

The Orbiting Geophysical Observatory is a fully-oriented scientific satellite designed for use in earth orbit. It is a general purpose vehicle capable of accommodating a large number of experiments in varying combinations for each specific launch. The design orbital lifetime is one year.

Two basic orbits will be flown: A highly eccentric, low-inclination orbit reaching an apogee of 80,000 nautical miles and a more nearly circular, polar orbit. The eccentric orbit will provide measurements outside the influence of the earth's magnetic field and will provide two crossings of the earth's radiation belts per orbit. The polar orbit will provide aurora and magnetic field measurements in the vicinity of the magnetic poles. By passing near the poles, the sensitivity of cosmic ray measurements is also increased. The two orbits are:

	Eccentric	Polar
Inclination, degrees	31	82 to 90
Perigee, n. miles	150	140
Apogee, n. miles	80,000	500

The OGO attitude control system is capable of pointing experiments toward (or away from) the earth and sun and with respect to the orbit plane. This attitude control capability is used:

1) Where a vector measurement is to be made, such as an earth's field measurement.

2) Where a streaming phenomena, such as solar plasma, is to be detected and the experiment has a preferred orientation relative to the flow.

3) Where a component of the spacecraft velocity may influence the effective temperature of a molecule or atom being detected and this component must be known to reduce the data gathered by the observatory.

4) When an experiment is made in cooperation with ground stations (such as Lawrence's electron density experiment) and the orientation relative to the station must be known.

5) Where an experiment must be protected from solar radiation.

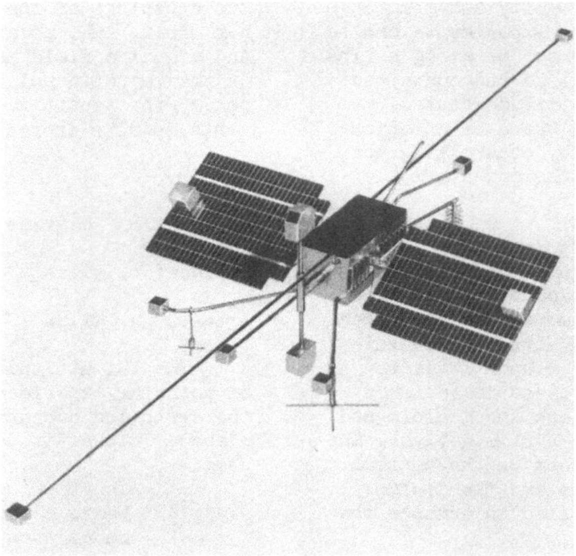

Fig. 1. Vehicle Configuration

Examples of each type of experiment were carried on OGO-A. Appendix A is a table of those OGO-A experiments which required either preferred or known (or both) spacecraft attitude.

Based upon considerations such as these, NASA specified the vehicle orientation requirements, their allowable variations and their allowable rates. The salient features of this specification are given in Appendix B.

In examining these specifications, it should be remembered that the OGO is only one of a family of NASA Observatory satellites (1). The Orbiting Astronomical Observatory will provide an extremely stable, accurate platform for astronomical measurements, and the Orbiting Solar Observatory serves an accurate vehicle for solar measurements. The solar measurement capability will be extended by the Advanced Orbiting Solar Observatory.

Because these other vehicles are or will be available, the OGO is not called upon for extremely accurate pointing at distant bodies. For instance, OGO is capable of relatively coarse pointing at the sun ($\pm$ 5 degrees), hence can be used for measuring streaming phenomena from the sun. However, if a specific point on the sun, such as a sunspot, is to be measured, the experiment would be assigned to the Advanced Orbiting Solar Observatory.

With this background, the design of the OGO attitude control system can be described. In Section B, the vehicle is discussed as it interfaces with the control system. Section C contains a description of the control system. The reasons for the selection of this particular configuration are described in Section D. Ground testing of the system is discussed in Section E, and the results of the first vehicle flight are given in the final section.

B. VEHICLE CONFIGURATION

The external configuration of the vehicle is shown in Fig. 1. The vehicle subsystems and the earth-oriented experiments are contained in the central box structure. Solar Power is collected by the solar cell arrays which are hinged along a line connecting them and passing through the box structure. The large booms, which are deployed after launch, support those experiments which must be isolated from the magnetic field generated by the vehicle.

(1) Superior numbers refer to similarly-numbered references at the end of this paper.

The experiment pointing requirements are met as follows:

- Earth-oriented experiments are located within the central box. The attitude control system maintains one face of the box toward the earth (the face pointing down in the figure).
- Solar-oriented experiments are housed in the two rectangular compartments located at either end of the solar array. The solar array is maintained toward the sun by controlling the yaw attitude of the vehicle (about the vehicle-earth line) and by controlling the orientation of the array relative to the vehicle about the array hinge line.
- Experiments oriented in the orbit plane are contained in the two boxes shown connected by a vertical shaft at the front of the vehicle (Fig. 1). The shaft angle is controlled to place the face of each package normal to the orbit plane. These packages are called the Orbit Plane Experiment Packages (OPEP).

The experiment orientation requirements are met by controlling the three vehicle rotational degrees of freedom and the solar array and OPEP hinge angles. In all, five degrees of freedom are used.

The requirement for isolating some experiments from the vehicle's magnetic field could have been satisfied with only one boom. However, a single boom would have resulted in large aerodynamic torques about the yaw axis. It was found that the weight of a second boom was less than the additional control system weight which would have been required to overcome the aerodynamic torques.

The experiment orientation requirements had a primary influence on the vehicle configuration. The requirement to orient experiments toward the sun led to the use of a fully oriented solar array for supplying vehicle power. This array utilizes the same sensors and control circuitry which are required for orienting the experiments. The experiment orientation requirements are also exploited in the design of the vehicle thermal control system. The orientation geometry is such that the sun and earth are located in a plane normal to the solar array drive shaft, and the two sides of the box structure adjacent to the solar panels are not exposed to direct sun or earth radiation during normal operation of the

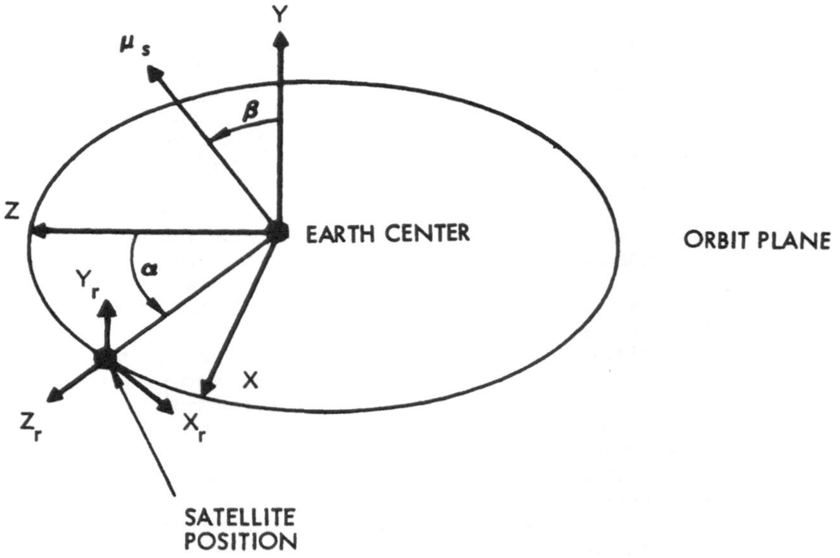

Fig. 2. Orbit Coordinate System

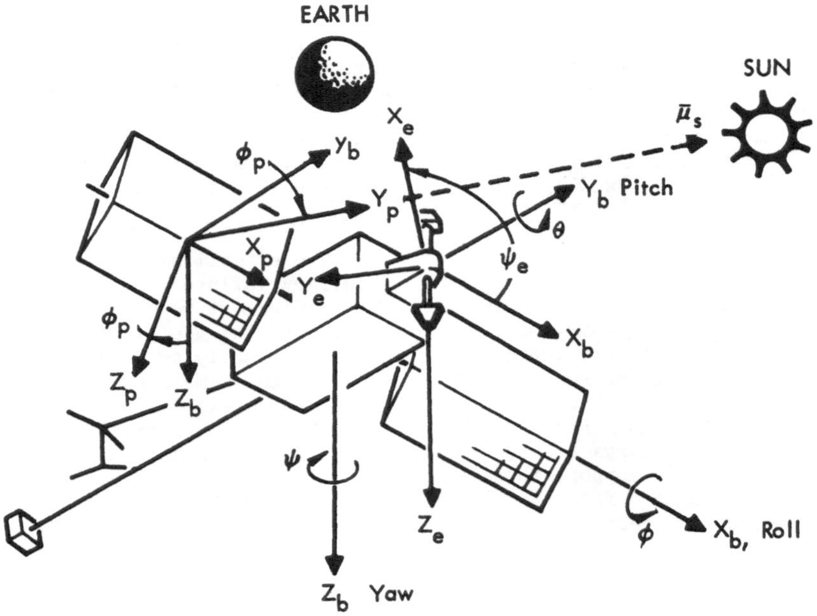

Fig. 3. Vehicle Coordinate System

vehicle. These two sides are used as the primary radiating surfaces for rejecting heat from the vehicle. Their emissivity is controlled by a system of movable louvers whose positions are controlled by passive bi-metal actuators. This arrangement results in the accurate control of vehicle temperature. The requirement to orient experiments toward the earth also resulted in savings in the communication system since narrower beam antennas could be used.

The power, thermal, and communication systems were designed to operate within the attitude accuracies required by the experiments. However, the power and thermal systems do impose one additional constraint on the attitude control system. Following separation of the spacecraft from the booster, the control system is required to assume a sun-oriented attitude before the vehicle batteries are depleted and before the internal temperature of the vehicle exceeds its limits. These same constraints apply when the vehicle is emerging from the eclipse condition.

C. ATTITUDE CONTROL SYSTEM DESCRIPTION

1. Geometry

Before proceeding to a description of the attitude control system, the geometry of the orientation requirements and of the vehicle will be given (2). The kinematics are complicated somewhat by the fact that the solar panel has only one degree of freedom with respect to the vehicle body.

The coordinate system defined in Fig. 2 will be used to describe the nominal vehicle orientation. The X, Y, Z triad is inertially fixed with the Y axis along the orbital angular velocity vector. The Z axis is oriented such that a unit vector, μ_s, along the earth-sun line will be in the Y-Z plane. The sun angle from the Y axis is defined as β.

A second triad X_r, Y_r, Z_r is defined by a rotation α about the Y axis. The angle α is defined as the orbital clock angle of the vehicle, hence the Z_r axis will be along the earth-vehicle line. The vehicle body angles will be defined with respect to this rotating coordinate system.

Fig. 3 defines the vehicle body axis system. In the case of $\beta=0$, the body axis system is ideally controlled to be coincident with the X_r, Y_r, Z_r axis system.

The nominal pitch and roll orientation (about the Y_b and X_b axes respectively) is defined as that attitude where the Z_b and Z_r axes are coincident; e.g., when one face of the vehicle is earth oriented.

The nominal yaw control (about the Z_b axis) is defined such that the Y_b-Z_b plane contains μ_s. If this constraint is met, a single rotation of the solar array (about the X_b axis) will bring its face normal to the sun line. The geometry of the array and the vehicle yaw attitude is shown in Fig. 4 as a function of the sun angle β.

The array geometry shown in Fig. 4 is of particular interest when the sun lies in the orbit plane ($\beta=90^\circ$). For this condition, the vehicle is called upon to yaw through 180° each time it crosses the earth-sun line. This is called the "noon turn."

Two control approaches are available for this condition. The first possibility is that of holding the yaw angle at either plus or minus 90 degrees throughout the orbit. In this case, the array would continue to rotate the same direction at the orbital rate. The second possibility, and the method chosen, is that shown in Fig. 4. A noon turn is commanded, and the array rotation is reversed. This method was chosen because its use limits the array rotation to plus or minus 90 degrees. Had the other method been chosen, slip rings would have been required for all electrical connections between the array and vehicle body. With the limited array rotation, hard wiring can be used and the wiring allowed to wrap up and unwrap as the array rotates back and forth.

The nominal orientation of the OPEP package is achieved when the X_e vector (Fig. 3) is contained in the X_r-Y_r plane. The nominal OPEP angle relative to the body is simply the negative of the vehicle yaw angle.

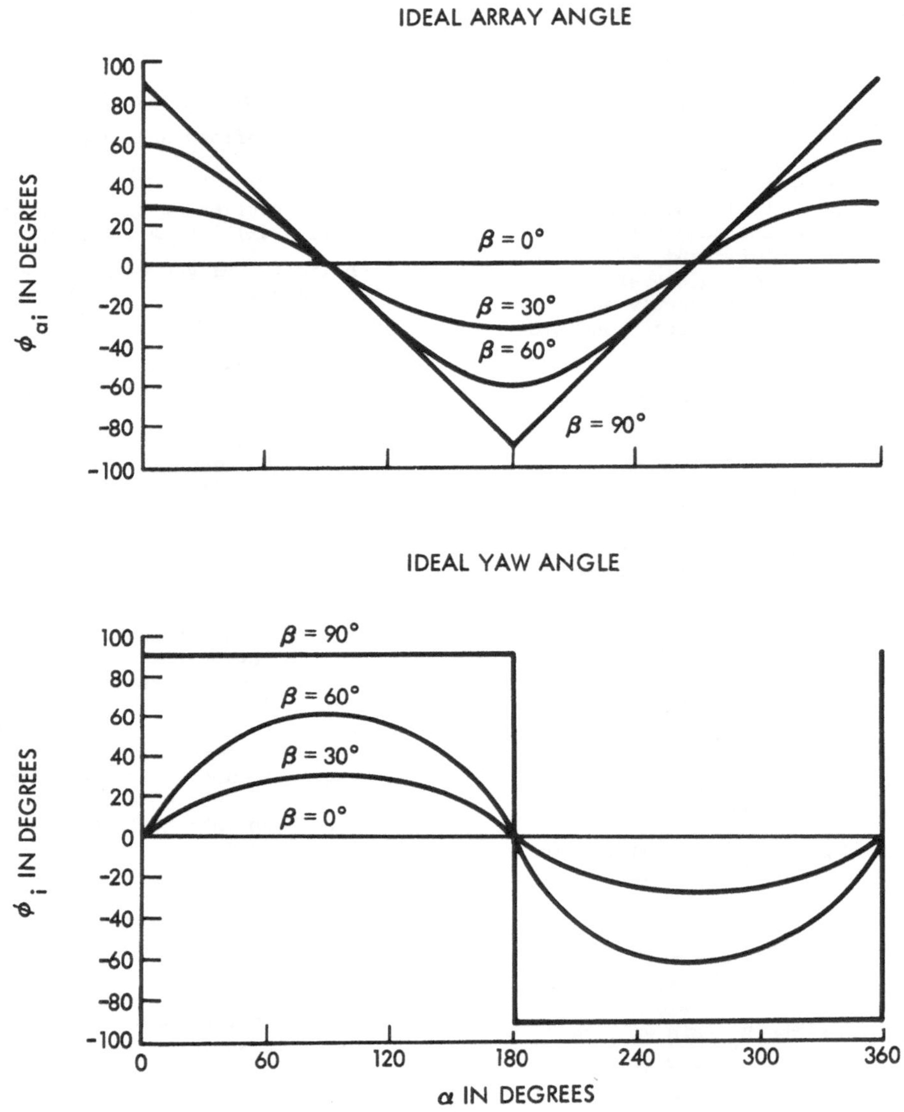

NOTE: PRINCIPAL ANGLES ONLY ARE SHOWN

Fig. 4. Nominal Yaw and Array Angles

2. Control System Configuration

The pitch and roll body control channels receive error signals from an infrared earth sensor which scans the thermal discontinuity between earth and space. This horizon scanner provides body attitude error signals in two axes.

The vehicle yaw axes is referenced to the sun by a sun sensor mounted on the solar array. This sensor is a two-axis device with the second axis used to control the array angle with respect to the body. These two sensing axes will be referred to as the "yaw" and "array" sun sensor axes.

The OPEP orientation is controlled by a rate gyro operating in a gyro compassing mode. The package orientation relative to the body is driven such that the gyro output is nulled. The nulled condition implies that the orbital rate vector is normal to the gyro input axis, hence the OPEP is oriented in the orbital plane.

A block diagram of the attitude control system is shown in Fig. 5. The system sequencing is shown in the normal (N) position. The second position (A) corresponds to the acquisition mode of operation where the vehicle is being maneuvered to bring the sun and earth within the sensor fields of view. The OPEP control system is not shown in the figure.

In the normal mode, the array axis of the sun sensor is connected to a drive motor which positions the array relative to the body. The feedback in this loop is implicit since the sun sensor is mounted on the movable array. The array angle relative to the body is sensed for telemetry purposes, but this angular transducer is not used in the normal mode.

The horizon scanner is connected to the pitch and roll control channels which are identical. Reaction wheels driven by a.c. motors supply the body torques. Pneumatic reaction jets are operated in parallel with the wheels to remove stored angular momentum, desaturating the reaction wheels. As shown by the block diagram, the wheels and jets are both operated in an on-off mode with deadband. The reaction jet deadband is set well outside the wheel deadband. Once the wheel is saturated (operating at its design speed where its torque output is effectively zero), the attitude error will increase and the appropriate pneumatic jet will fire. This addition of angular momentum causes the wheel to slow down as it returns the body to a small attitude error.

The reaction wheels perform two functions. They store the periodic components of the external disturbance torques and they provide a vehicle maneuver capability, in each case without requiring the expenditure of reaction control gas. The major maneuvers are:

a) The noon turn, described earlier.

b) Post eclipse reacquisition, where the vehicle emerges from the sun eclipse with two body axes oriented toward the earth and must reacquire the sun in the yaw axis.

c) Maintaining the body rate equal to the orbital rate in order to keep the $-Z_b$ axis oriented toward the earth. This maneuver is most demanding for the highly eccentric orbit.

The yaw control channel is similar to the pitch and roll channels except that it is referenced to the yaw axis of the sun sensor. Also, in the normal mode, the yaw pneumatic jets are not used. They are disabled to prevent the explosion of gas during the post eclipse sun reacquisition maneuver.

During the course of an orbit, the vehicle attitude will vary while the total angular momentum vector remains relatively stationary in inertial space. As a result, the momentum stored in the reaction wheels will be handed from wheel to wheel as the vehicle turns.* It is this effect which

*If the vehicle-sun line lies along the orbital rate vector, the pitch reaction wheel spin axis will be space fixed and its angular momentum will not be transferred to another wheel. However, the yaw wheel spin axis is directed toward the earth, and its momentum will be transferred to other wheels twice per orbit.

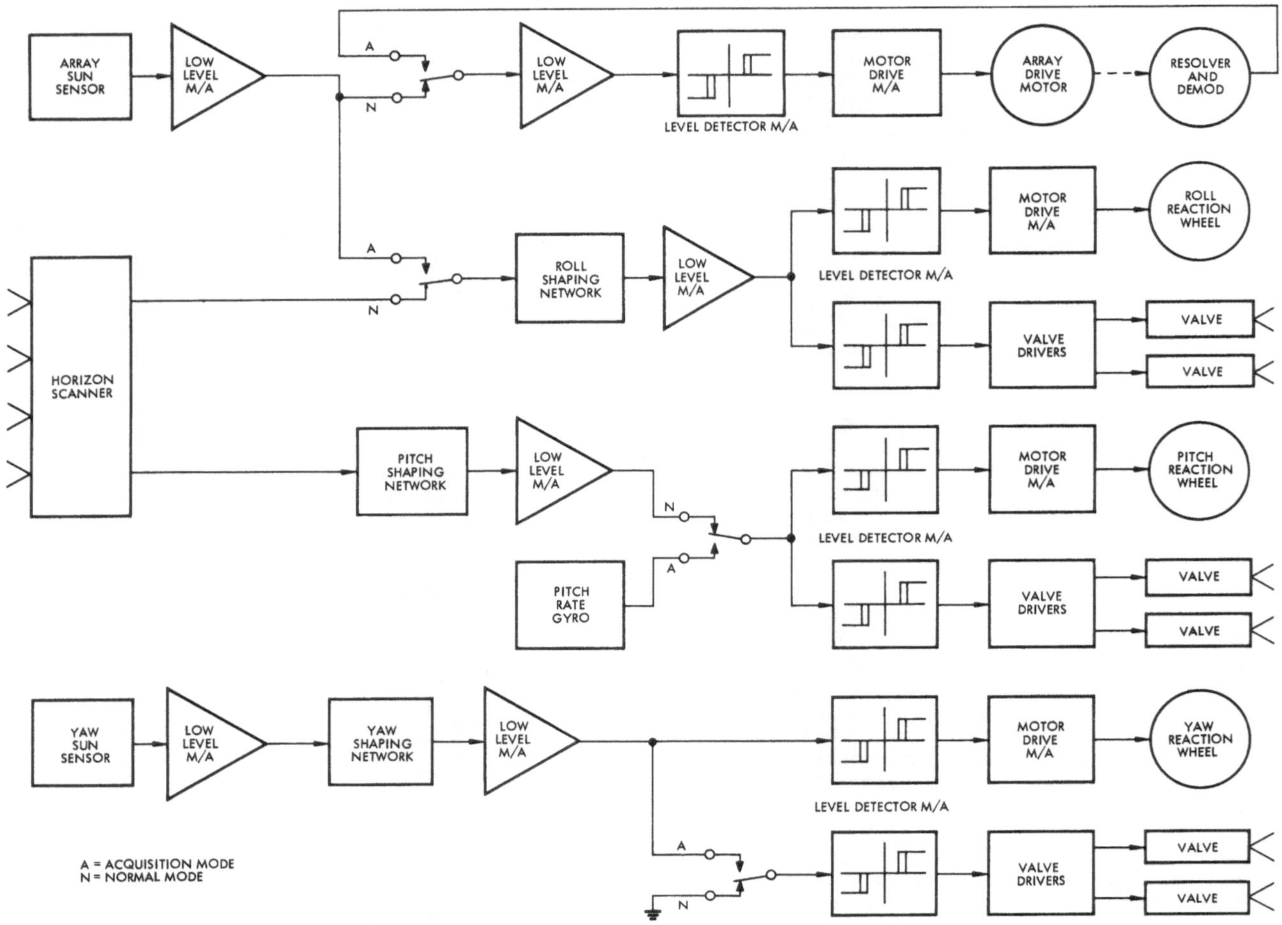

Fig. 5. Control System Block Diagram

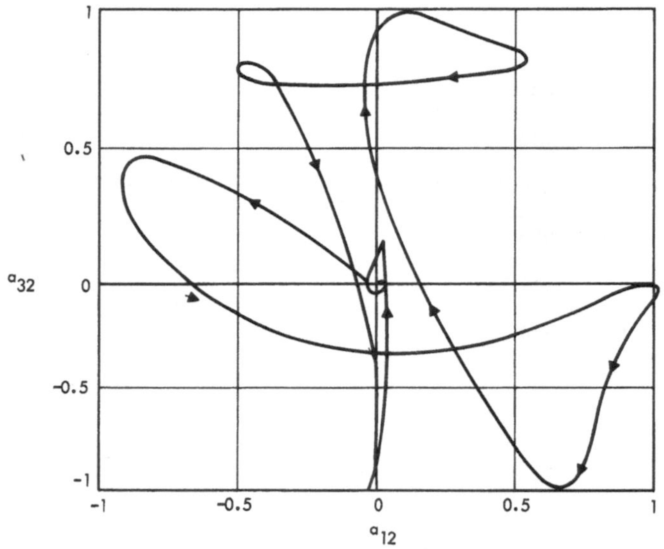

Fig. 6. Sun Acquisition Maneuver

permits disabling the yaw reaction jets. If excess momentum builds up in the yaw wheel, it will be transferred to the pitch or roll axes and removed by the reaction jets operating about these axes.

The sun sensor is mounted on the solar array and the yaw axis error signal is a function of the array angle with respect to the body. For small angles,

$$\epsilon_y = \psi \cos \phi - \theta \sin \phi$$

where

ϵ_y = sensed yaw axis error

ψ = body yaw axis error (Figure 3)

θ = body pitch axis error (Figure 3)

ϕ = array angle with respect to body.

As the solar array angle approaches 90 degrees, the yaw gain approaches zero. However, this indeterminant point coincides with the noon condition where yaw control is not required for keeping the solar array normal to the sun line. As the vehicle leaves the noon condition, the yaw error signal will inherently change polarity and command a noon turn. No special sequencing is required.

Since the noon turn is preceded by a period of zero yaw gain, the direction of the turn cannot be predicted with certainty. This does not effect the solar array drive since it will reverse its direction and unwrap for either clockwise or counter-clockwise noon turns. The OPEP drive is sensitive to the direction of the turn, however, and will "wrap up" if successive turns are made in such a manner that they are cumulative. As a result, limit switches are provided which sense overtravel of the OPEP drive and return it to a favorable condition. In all other respects, the OPEP drive operates without special logic. Its speed of response is greater than that of the vehicle and it follows the noon turn maneuver without the use of special logic.

In the acquisition mode, the solar array is caged to the vehicle body and the array axis of the sun sensor is connected to the body roll control channel. The yaw axis pneumatic jets are enabled for this mode. The roll and yaw body axes are then referenced to the sun with each channel capable of gas jet operation. If the sun is outside the field of view of the sun sensor, a course sensor with a full spherical field of view is automatically sequenced in. The sun acquisition is accomplished without rate gyros in the roll and yaw axes.

The pitch axis is connected to a biased rate gyro during the acquisition phase, and a vehicle rotation of 0.49 degrees/second is commanded. This manuever continues until the horizon scanner senses the earth, at which time the normal mode is entered.

D. SYSTEM SELECTION

The requirements placed upon the control system are moderate in terms of accuracy and maneuvering. In fact, the system operates as a regulator rather than a controller throughout most of its useful life. As a result, the dominant consideration in designing the system was reliability rather than performance. This design goal was achieved by exploiting the allowable system tolerances to simplify the airborne system and to make it relatively immune to the space environment. In this section the control system elements will be described in the context of this design philosophy.

1. Control Signal Processing

Early implementation studies showed that significant savings in the number of electronic components could be effected if on-off rather than proportional signal processing were used. As a result, the reaction wheels were driven by magnetic amplifiers operating in a switching mode. This configuration results in higher limit cycle attitude errors and rates than would be achieved by a proportional system, but the predicted errors and rates were shown to be within the allowable band, and the hardware simplification was adopted.

As discussed earlier, the pneumatic jets are operated in parallel with the reaction wheels, and a significant attitude error must be incurred before they will turn on and desaturate a wheel. Had accuracy been the dominate consideration, proportional wheel control would have been used with the gas jet pulses initiated by tachometers measuring the actual wheel speed. Again, it was found that

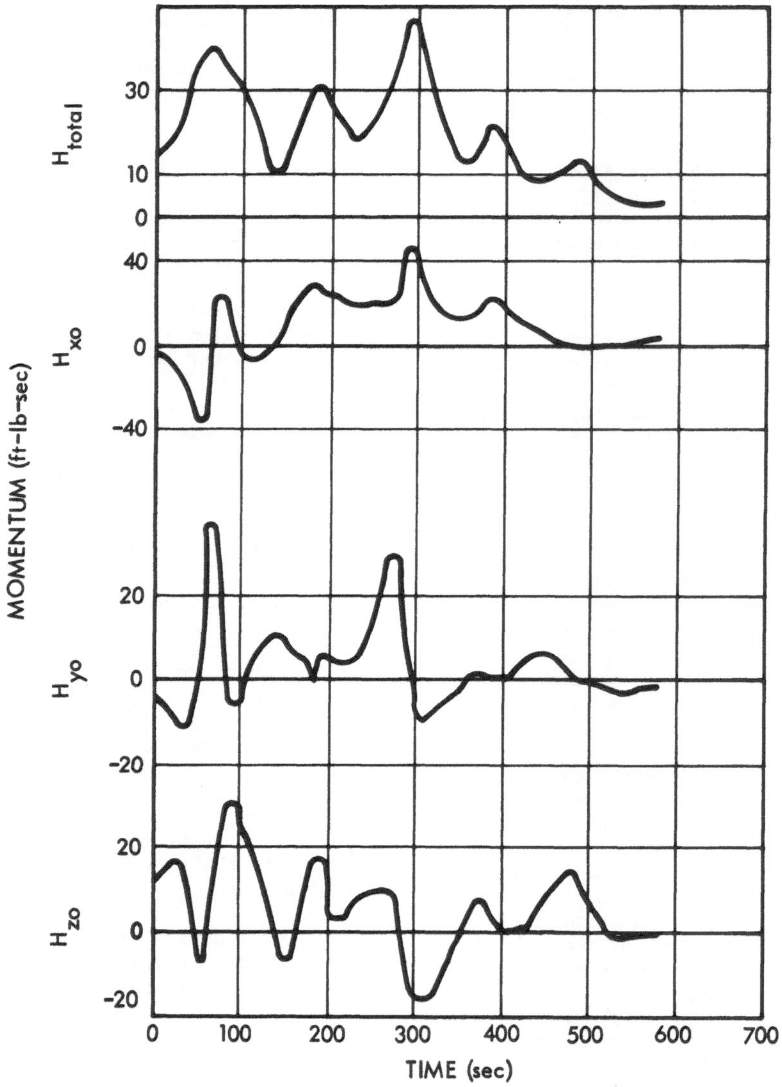

Fig. 7. Body Angular Momentum During Sun Acquisition

the simpler system was capable of meeting the control requirements and this design was adopted.*

In the normal mode of operation, lead-lag compensation is used rather than rate gyro compensation. This method has obvious life and reliability advantages, but leads to rather inefficient maneuvering during the initial sun acquisition maneuver. The sun sensor has full spherical coverage. However, it is not a truly proportional device in that it saturates and gives a constant output for large error angles. In the saturated regions, the lead-lag compensation is not effective, and the vehicle rates can build up.

This effect is shown by Figs. 6 and 7. Fig. 6 is a plot of the sun vector trace in the roll-yaw plane where a_{12} is the roll component of the vector, a_{32} the yaw. Initially, the back of the vehicle is faced toward the sun. The trace ends with the proper sun orientation. Fig. 7 shows the total vehicle angular momentum during the maneuver and the components of angular momentum along the three control axes.

The maneuver shown is a "worst case" maneuver with adverse initial attitude and attitude rate conditions. Note that the total vehicle momentum actually increases before the desired final conditions are reached. This is wasteful of reaction control gas, but the maximum requirement for initial acquisition is only one pound of gas. This penalty is quite acceptable when balanced against the system simplification realized by omitting rate gyros.

In the pitch axis, a rate gyro is used during the earth acquisition phase. The earth is acquired by pitching at a constant rate about the sun line until the earth intercepts the horizon scanner field of view.

*Tachometers are used for telemetry and checkout purposes, but they are not required for proper system operation.

This can require up to half an orbit, and a rate sensor is required. However, the rate gyro is turned off after acquisition, and the pitch axis reverts to passive compensation thereafter.

2. Sun Sensor

A two-stage sun sensor is used with the coarse stage providing full spherical coverage and the fine stage providing more accurate coverage in a 17 degree cone about null.

The coarse stage consists of two groups of silicone solar cells mounted at the outer end of each solar array. The cells are mounted on the faces of rectangular boxes and their outputs are summed in resistor networks. The accuracy of the coarse sensor is primarily dependent upon the thermal stability of the solar cells, and a heat sink is incorporated for each set of cells.

The fine sensor utilizes a single two-axis detector called the "Radiation Tracking Transducer." This detector is used in conjunction with pin-hole optics and receives a single spot of sunlight on its face. Current is induced through the detector in the vicinity of the light bundle in much the same manner as in a solar cell. This current diffuses back through the detector in its dark regions setting up a lateral voltage field across the detector face. The field is measured about the periphery of the detector and provides a voltage proportional to the deviation of the image from the center of the detector. The output is reasonably linear over the sensor's 17 degree field of view. Accuracy is on the order of 0.1 degrees.

3. Horizon Scanner

The horizon scanner consists of four separate detectors for sensing the edge of the earth at a point in each quadrant of the apparent disc. Each head measures the angle between the yaw axis of the spacecraft and the line of sight to the edge of the earth. The difference in angular measurement between opposite heads constitutes the attitude error in one axis.

Fig. 8. Horizon Scanner

Fig. 9. Wabble Gear Drive

Each head utilizes a thermistor bolometer detector. Scanning is effected with a flexure-mounted mirror driven by a coil. This device, called a "positor", is used in order to avoid bearings in the mirror drive. The detectors and positors are mounted in two packages. These packages are shown in Figure 8.

In operation, each positor scans over its total travel until the earth's edge is located by the detector logic. At this point, the head locks onto the edge and the positor scans across a narrow angle (1.6 degrees) centered on the horizon. The scan frequency is 13 cycles per second.

The scanning motion is achieved by operating the detector heads in a limit cycle about the horizon. The radiation gradient at the horizon is used to command a change in the direction of the positor motion. This change occurs only after a time delay, and a stable limit cycle is achieved. The angular measurement is made by averaging the positor position.

The scanner can be operated over a range of apparent earth sizes from 5 to 150 degrees. Its accuracy is 0.25 degrees exclusive of uncertainties in the horizon. The system is redundant in that three of the four heads will supply the required attitude information in two axes. Logic has been included which permits operation using any three heads.

4. Reaction Wheels

The three reaction wheels and their motors are packaged in pressurized containers and run at a maximum speed of 1250 revolutions per minute. Alternating current induction motors are used, both to avoid brushes and to reduce the external magnetic field generated by the motors.

5. Pneumatic System

Regulated Argon gas is used for momentum removal, operating through six, 0.05 pound thrust, gas jets. The jets are arranged in two sets of three and located on two booms which increase the lever arm of the jets about the vehicle center of gravity. A total of 640 pound-seconds of impulse is available for one year's operation.

6. Solar Array and OPEP Drive

The solar array is positioned using an error signal from the sun sensor while the OPEP drive is referenced to an electrically caged, single-degree-of-freedom gyro. Two gyros are used for redundancy. The respective drive systems of the two appendages are identical in other respects. Both use on-off signal processing. Dynamic compensation is not used since the parasitic damping of each drive system is sufficient to stabilize its motion.

The torque source for each drive is a 0.6 inch-ounce motor driving through a gear reduction of 24,000:1. A conventional gear train provides a 240:1 reduction and is connected in series with a "wabble" drive (see Figure 9). The wabble drive consists of two conical face gears with pitch line contact at one point on each gear. The input gear is then caused to nutate and the contact line in effect rotates about the gears. The two gears have different numbers of teeth and output motion is produced. The gear ratio of the wabble drive is 100:1.

This drive was selected because it was ideally suited to the space environment. The conventional gear train is completely sealed by the bellows shown in Figure 9 and operates in a grease vapor environment. The output gear operates slowly, has a large contact area, and is lubricated with carbon impregnated in the metal.

E. TESTING

The stabilization and control system was subjected to a series of tests designed to demonstrate the functional and environmental capabilities of the system. This section covers engineering evaluation tests using motion simulators, qualification testing, calibration and system tests and vehicle tests. In many cases, a control system test is only one element of a larger vehicle test. While this has been noted, the scope of the paper has been restricted to the control system portions of these tests.

1. Engineering Simulation Tests

In order to gather dynamic data on the operation of the system, a series of engineering tests were run under conditions which simulated the low-torque space environment.

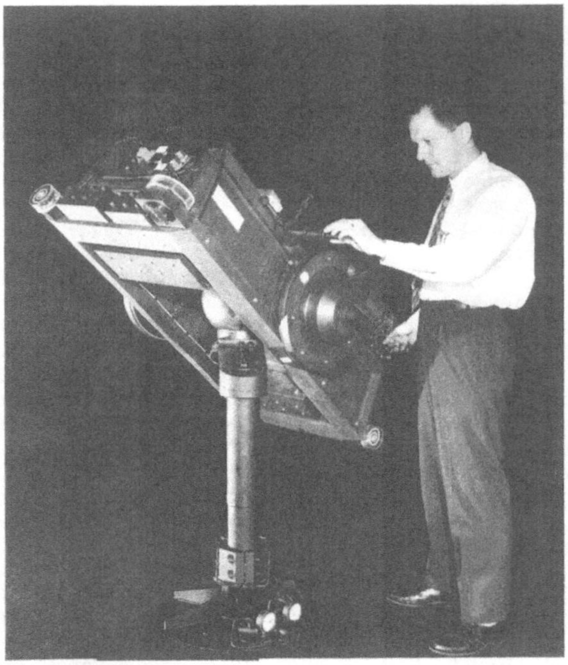

Fig. 10. Air Bearing Motion Simulator

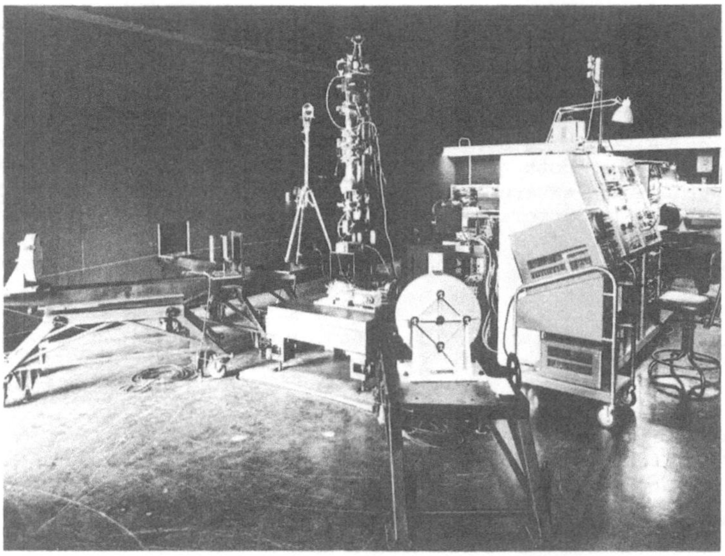

Fig. 11. Integrated System Test Facility

Two simulators were used in these tests: a single-axis torsion-wire suspended table and a three-axis air bearing supported table. The single-axis simulator, while less useful than the three-axis, was available much earlier in the program, and, as a result, had a comparable impact on the system design.

Neither of these two simulators were designed to test for vehicle disturbance torques. In fact, the parasitic torques associated with the simulators exceeded the expected external vehicle torques. Their usefulness, then, was primarily in testing for large angle maneuvers and for system sequencing.

a. Single-Axis Simulator

The single-axis simulator consisted of a circular platform suspended from its center by a "torsion wire" (3). The control system was mounted on the platform and used to control its angular position about the axis defined by the torsion wire. This "torsion wire," designed to minimize the spring restoring torque on the platform, consisted of several highly stressed wires, rectangular in cross-section and separated from each other by a few thousandths of an inch to prevent their rubbing and adding damping to the system. The use of several flat wires resulted in lower restoring torques than would have been achieved with a single wire. The length of the suspension wires was approximately 10 feet.

Spring torques were further reduced through the use of a servo system which slaved the upper end of the suspension wires to the rotational motion of the platform. The interaction of the servo and suspension wires can be seen from a simplified transfer function for the restoring torque:

$$T = \theta \left[\frac{s}{s + 1/\tau} \right] K$$

where

T = restoring torque

θ = platform rotational angle

K = torsional spring rate of the suspension

τ = equivalent time constant of the servo

s = Laplace Operator

From the transfer function, it is apparent that the torsion wires were used to minimize the effect of the servo following error.

This simulator was used for checking the initial sun acquisition maneuver in one axis. It also provided gross data on the limit cycle behavior of the system. As a result of these tests, the mathematical model of the control system was refined, and minor gain changes were effected in the design.

b. Three-Axis Simulator

The three-axis simulator consisted of a rigid stainless steel table supported from below by a spherical air bearing (4). The table was capable of free rotation about its yaw axis and ±92 degrees about the pitch and roll axes. The control system was mounted to the table and controlled its attitude with respect to simulated sun and earth sources. Fig. 10 shows this simulation facility.

The table weighed 1500 pounds, was supported on a 10-inch spherical bearing, and exhibited 1/10 the inertia of the actual spacecraft. Because of the reduced inertia, the control reaction wheels and gas jets were scaled to 1/10 the torque and momentum capability of the flight units. The solar array drive was also scaled down. In other respects the control system simulated was identical to the flight system.

Disturbance torques were minimized by careful grinding of the air bearing (supplied by U.S. Beryllium) which was spherical to within 10 micro inches. The simulator was kept free of external signal and power leads. A battery pack on the table supplied control system power and a 6-channel command receiver and 8-channel FM/FM telemetry transmitter were mounted on the table to handle information flow.

The earth simulators, used to excite the horizon scanner, were simply heated metal discs. Three different sizes were used in order to simulate different scheduled orbit altitudes. The largest, 84 inches in diameter, was permanently attached to a wall of the simulation room. The other two, 16 and 35 inches in diameter, were portable and could be placed at different positions depending upon the test to be run.

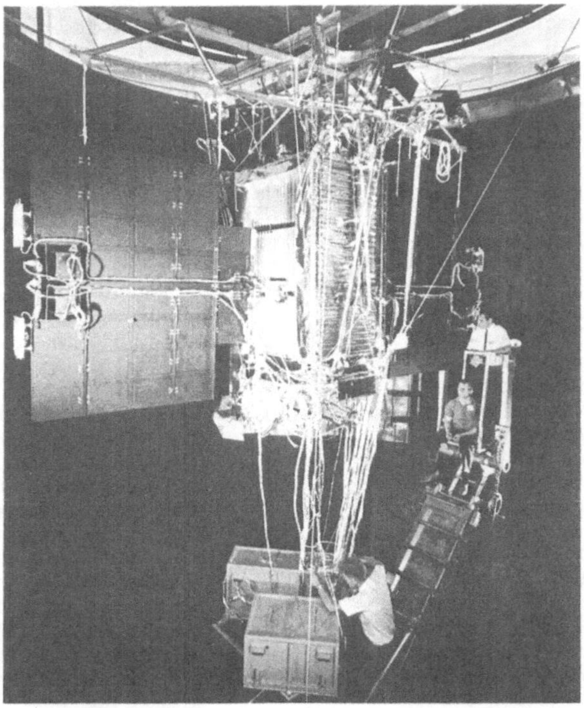

Fig. 12. Preparation for 12-Day Environmental Test

A 500 watt incandescent bulb was used as a simulated sun source. Because its output was considerably below the sun's flux density as seen from earth, a double convex lens was fitted to the sun sensor to raise the energy level at the detector. The source was movable through an arc below the simulator to reproduce the apparent motion of the sun due to the vehicle's orbital rate.

The simulator was used to test the control system in the noon turn maneuver, in the initial acquisition maneuver (sun and earth), and in the post eclipse reacquisition maneuver (sun only).

2. Qualification Testing

Each component of the stabilization and control system was qualified for a humidity environment, temperature storage, thermal-vacuum operation and for vibration and shock. These tests were performed only once for each prototype assembly. The actual flight units were subjected to environmental testing at reduced levels to demonstrate their flight readiness.

The vibration experienced by a given assembly is a function of the vibration imposed upon the spacecraft by the booster and the location of the component in the spacecraft. The booster environment was specified by NASA using earlier booster flight data. The vibration transmitted to a given component was determined analytically, then by test. The experimental data was obtained by exciting a structural test vehicle with the simulated booster vibration environment and measuring the resulting environment at various points within the spacecraft. This measured vibration level, with a suitable safety factor, was used during qualification of the system elements.

3. Calibration and System Tests

Prior to integration into the spacecraft, critical control system components were calibrated and the system components were integrated and tested as an operating system. A large part of this work could have been accomplished during the process of integrating the system into the vehicle. However, by performing these tests in the laboratory, the spacecraft assembly operation was shortened considerably and an overall program schedule advantage was realized.

a. Sun Sensor Calibration

The sun sensors were designed to be virtually independent of the vehicle thermal environment. Each of the two sensors was heavily insulated from energy radiated from the vehicle and low conductive paths were used in attaching the sensors to the structure. As a result, the sun sensors could be calibrated as units prior to installation on the solar panels.

Each sensor was calibrated for null and scale factor using a carbon arc sun simulator. The sensors were also calibrated thermally to insure the stability of the null and scale factor calibration.

The sensors are maintained within acceptable temperature limits by reradiating solar energy from the back of the sensor assembly. In addition the sensors contain heat sinks to maintain their temperature during eclipse periods. This passive thermal control system was calibrated by adjusting the emissivity of the back surface of the sensor through the addition or deletion of mylar tape to the radiating surface. Thermal measurements were taken under vacuum conditions, again using a carbon arc sun simulator.

b. Horizon Sensor Calibration

The horizon scanner was checked for calibration using the test fixture shown in the foreground and center of Figure 11. Since the sensor operates on the gradient between the earth and cold space, it was expedient to simulate only the edges of the earth. This was done using a heated plate with a cold plate masking a portion of it to simulate the earth-space thermal interface. The radiation from this simulator is collimated through an off-axis parabolic mirror. As shown by Figure 11, two hot plates are used for a single scanner head to simulate both edges of the earth. The scanner head is mounted on a rotary table, and calibration is accomplished by rotating the head through known angles with respect to the simulated earth sources. The collimating mirrors can also be adjusted to simulate changes in altitude.

c. Pneumatic System Calibration

The thrust and specific impulse of the reaction jet system was measured under vacuum conditions. This problem was complicated by the low thrust levels to be measured (0.05 pound thrust jets are used), and a special thrust measuring fixture utilizing a spring balance and an eddy current damper was required.

Each jet was measured to ascertain its thrust level, and the flight units were selectively assembled for matched thrust. The impulse stored in the lines between the jet solenoid valves and the jets was measured and the specific impulse of the system was determined. These latter two tests were design verification tests rather than flight hardware proofing tests and were performed only once.

d. Integrated Control System Tests

Prior to assembling the control system components into the spacecraft, they were connected and operated as a complete system. The test area used for these checks is shown in Figure 11. The solar array drive is mounted vertically as shown in the center of the picture. The array drive and one horizon sensor head are mounted on a turntable which simulates the vehicle pitch motion. The other horizon scanner head is mounted on a similar fixture at the far end of the test area. The test console is shown on the right and portions of the control electronics can be seen behind the console.

The horizon scanner heads were excited using their calibration fixtures. Incandescent sources were used for the sun sensors, and the orbit plane experiment package was excited using earth's rate, and by torquing the gyros. The reaction jets were checked for polarity by visual observation of a ribbon tied in front of each nozzle, and the reaction wheels were attached on free turntables to permit polarity tests by visual observation. The entire test was controlled from a central panel.

Using this test setup, the system was checked for compatibility of components, end-to-end polarities and scale factors and for the proper sequencing of the equipment. Systems passing these tests were approved for integration into the spacecraft.

4. Vehicle Tests

During the process of assembling the control system into the vehicle, the system was again checked for component-to-component compatibility, in this case using the actual flight wiring harnesses. Vehicle compatibility was also checked including electromagnetic interference testing.

Once the control system was installed in the vehicle, the telemetry and command systems became the prime means of communicating with it, both during vehicle tests and during launch checkout. As a result, telemetry calibration was particularly important for the OGO vehicle.

Telemetry calibration and end-to-end polarity tests were made by exciting the sun sensors and the horizon scanner with portable source simulators and by torquing the OPEP gyros. The pitch rate gyro polarity could not be verified conveniently at this time. However, it was checked when the spacecraft was suspended by its handling sling. The vehicle was physically rotated and the nozzle exhaust observed along with the reaction wheel and reaction jet solenoid telemetry.

Following the installation and test of all subsystems, the spacecraft integrity was verified by flight readiness environmental testing. The entire vehicle was subjected to vibration testing with linear vibration imposed along each of the three axes and torsional vibration imposed about the launch thrust axis.

The vehicle was also subjected to a 12-day thermal vacuum environment with all subsystems operating. In this test, the sun sensors and horizon scanner were removed from the vehicle and attached to source simulators. They remained within the vacuum chamber, however, and were connected electrically to the vehicle. A cold-wall chamber was used for this test with carbon-arc simulation of the solar radiation. The chamber is also capable of simulating the eclipse condition. Figure 12 shows the first flight vehicle being installed in the vacuum chamber for test.

Because of the stringent magnetic field requirements imposed by the experiments, it was necessary to verify the radiation level of the vehicle. This test was performed in a Helmholtz coil facility with all subsystems operating.

F. DESCRIPTION OF FIRST OGO LAUNCH

The first OGO spacecraft was launched from Cape Kennedy on September 4, 1964. A successful orbit was achieved and the sun acquisition maneuver was successfully completed. However, a boom failed to deploy properly from its stowed position and blocked the horizon scanner field of view.

The horizon sensor locked onto the undeployed boom and indicated an attitude error. In response to this error, the reaction wheels saturated and the pneumatic jets were turned on. This was the correct response of the attitude control system, but since its action could not affect the apparent attitude error, all of the control gas was expended, and the vehicle was spun up.

Since sun acquisition had been completed, the spin vector was oriented approximately toward the sun as were the solar panels. However, the vehicle was not spinning about its axis of maximum moment of inertia and it quickly assumed a new, minimum angular energy orientation.

In this orientation, the solar panels were no longer normal to the sun and the battery charging level fell dangerously low. In response to this situation, the solar array drive was commanded on and automatically turned the array toward the maximum power position. The drive has been turned on periodically since then to correct for the apparent motion of the sun with respect to the spin vector.

Given adequate power, the spacecraft assumed operation as a spin stabilized vehicle. All of the experiments operated, and 80 percent of the experimental data was of scientific value even though the vehicle was spinning.

APPENDIX A, PARTIAL EXPERIMENT LIST, OGO-A

The experiments listed in Table A-1 are those flown on OGO-A where the experimenters desired either a known or controlled vehicle attitude. The desired orientation is given in the left column. Those experiments which did not have a preferred orientation are not listed.

The OGO-A contained a scanning package which was attached to the orbit plane experiment package (OPEP). This package (noted "scanned OPEP") provided a scanning motion in and out of the orbit plane.

APPENDIX B, ATTITUDE REQUIREMENTS

The following attitude and attitude rate requirements were used as the basis for the OGO control system design (5).

1. Earth Pointing

   Maximum attitude error = 2 degrees
   Maximum attitude rate = .001 radians/second
   (exclusive of orbital rate)

2. Sun Pointing

   Maximum attitude error = 5 degrees
   Maximum attitude rate = .03 radians/second

3. OPEP Orientation

   Maximum attitude error = 5 degrees
   Maximum attitude rate = .03 radians/second

   (These requirements apply to the polar orbits. In the case of the highly eccentric orbits, they need be met only when the orbital rate is equal to or greater than that for the polar orbit).

4. Acquisition Time

   Maximum initial acquisition time = 10 min.
   Maximum post-eclipse reacquisition time = 10 min.

Table A-1 (1 of 2)

ORIENTATION	EXPERIMENTER	AGENCY	TYPE OF DETECTOR
Solar	Anderson	University of California at Berkeley	Solar Cosmic Ray
Solar/Scanned OPEP	Wolfe	Ames Research Center	Solar Plasma (proton)
1) Solar (proton) 2) Earth (electron)	Bridge	Massachusetts Institute of Technology	Solar Plasma (proton & electron)
Solar	Cline	National Aeronautics and Space Administration Goddard Space Flight Center	Solar Positrons
Scanned OPEP	Davis	Goddard Space Flight Center	Cosmic Ray
Relative to Earth	Van Allen	State University of Iowa	Trapped Radiation
Relative to Earth	Smith	Jet Propulsion Laboratory	Search Coil Magnetometer
Relative to Earth	Heppner	Goddard Space Flight Center	Rubidium Field Magnetometer
Knowledge of Spacecraft Wake Orientation	Sagalyn	USAF Cambridge Research Center	Spherical Ion Probe
OPEP	Whipple	Goddard Space Flight Center	Planar Ion Probe

Table A-1 (2 of 2)

ORIENTATION	EXPERIMENTER	AGENCY	TYPE OF DETECTOR
Earth Oriented and Low Rates	Lawrence	National Bureau of Standards	Electron Density by EM Radiation Properties
OPEP	Taylor	Goddard Space Flight Center	Bennett Mass Spectroscope
Earth Oriented	Alexander	Goddard Space Flight Center	Micrometerite Detector
Earth Oriented	Helliwell	Stanford Research Institute	VLF Detector
Solar (but known orientation required in other axis)	Haddock	University of Michigan	Radio Astronomy
Anti-Solar (Protection from Solar Radiation)	Mange	Naval Research Laboratory	Lyman-Alpha Detector
Solar and Anti-Solar (Viewing of Earth Counter-Glow)	Wolff	Goddard Space Flight Center	Gegenschein Photometer

ACKNOWLEDGEMENTS

The OGO spacecraft was developed by TRW/Space Technology Laboratories under the direction of the National Aeronautics and Space Administration, Goddard Space Flight Center, Greenbelt, Maryland, and represents the efforts and contributions of many individuals from these two organizations. In particular, the author wishes to acknowledge the prior work of Mr. D. D. Otten and the helpful suggestions of Mr. M. Robinson and Dr. E. I. Ergin.

REFERENCES

(1) NASA, Goddard Space Flight Center, Greenbelt, Maryland, The Observatory Generation of Satellites, March, 1963.

(2) D. D. Otten, "Design of the Attitude Control System for the Orbiting Geophysical Observatory," presented at the 14th International Astronautical Congress, Paris, France, September, 1963.

(3) N.H. Beachley, et al., "Testing OGO's Attitude Controls," Control Engineering, October, 1964.

(4) N.H. Beachley, "OGO Attitude Control System Studies on the Air-Bearing Simulator," 2313-6011-KU-000, Space Technology Laboratories, Redondo Beach, California, August, 1963.

(5) NASA, Goddard Space Flight Center, Greenbelt, Maryland, Specifications for the Orbiting Geophysical Observatory, SO-S49/50-1, Revision 1, August, 1962.

DISCUSSION

E. P. Blackburn

"The Orbiting Geophysical Observatory Attitude Control System".

Q. Why do you not use a tachometer to unload the wheel momentum, but instead tolerate the error which will accumulate before the gas jets act?

A. The reliability increase due to omission of the tachometer is considered to be more important than the error. Also, we need the jets in the initial acquisition mode to remove rates caused by booster separation; the wheels have insufficient momentum for this.

Q. Will you please explain the "shift" of spin axis?

A. The vehicle initially was spun up about the axis of minimum moment of inertia. As energy was dissipated internally, coning began, and the final spin axis became that of maximum moment of inertia.

Q. What was the spin speed?

A. For Syncom II, 120 rpm; for Syncom III, 175 rpm; for Early Bird, 147 was nominal, 153 was achieved. The speed can vary between 120 rpm and 200 rpm. At the upper limit centrifugal effects become limiting. One millisecond corresponds to about one degree, and the jet pulses must operate consistent with this speed.

Q. How is nutation damped?

A. Two types of oscillation are present during precession. One is at the spin frequency and its harmonics; the other is free body nutation. The total amplitude does not exceed 1/4 degrees. Nevertheless a damper consisting of a tube about 1/3 full of mercury is used. The tube is mounted away from the axis and acts by dissipating energy in sloshing. Its time constant is between 60 and 90 seconds - we are not able to measure it more accurately.

Q. Is pseudo-rate information used in roll and yaw during acquisition?

A. A lead-lag network is used and provides very light damping. The system is not optimum for gas usage, but is perhaps best for reliability.

Q. Are roll and yaw acquisitions performed simultaneously?

A. Yes.

Q. Are position gyros used?

A. Only during sun acquisition. They are then turned off and not used again. The satellite contains no position gyros. The gyroscopic effect of the reaction wheels is small.

CONTROL OF THE SYNCOM COMMUNICATION SATELLITE

By: Mr. Donald D. Williams
Chief Scientist
Communications Satellite Laboratory
Hughes Aircraft Company
El Segundo, California

ABSTRACT

The spin-stabilized Syncom communication satellite is provided with a simple but versatile reaction control system for precession of the spin axis and correction of the orbit. The required control is achieved by use of two jets which may be pulsed on during a controlled sector of each spin revolution or operated continuously over many spin revolutions. The design requirements, configuration of the control system, and its operational use are described, and the in-orbit maneuvers of Syncom 2, the first synchronous satellite, and Syncom 3, the first stationary satellite, are summarized.

DESIGN REQUIREMENTS

The basic objective of the Syncom program was the demonstration of a communication satellite in synchronous orbit. Becuase of its proven performance and moderate cost, the Thor-Delta was selected by the NASA as the launching vehicle. In order to obtain a nominally circular final orbit using this vehicle, it is necessary to add a rocket engine to the spacecraft for final injection into the synchronous orbit. The three-stage Delta vehicle injects the spacecraft near the perigee of a transfer orbit, with an altitude of a few hundred kilometers and a velocity of about ten kilometers per second. The injection sequence approximates a Hohmann transfer, with the spacecraft engine providing a velocity increment of about 1.6 kilometers per second at apogee of the transfer orbit. Other requirements related to the orbit included the ability to control the period and if possible the eccentricity and inclination within the limits of impulse available from the control system. In regard to the communications function, it was necessary to provide adequate prime power from solar cells as well as a practical maximum of directivity of the vehicle's communication transmission antenna. Two redundant transponders were specified for reliability, the up-link frequency of approximately 7 gc and the down-link frequency of about 1.8 gc being dictated by planned use of existing ground equipment for which these frequencies had been allocated. For compatibility with the NASA Minitrack network, two 136 mc telemetry transmitters were specified; to simplify the design of the transponder and to eliminate problems associated with the narrow beamwidth of the ground antenna at 7 gc, the use of redundant command receivers completely separate from the transponders in the allocated 148 mc band was indicated. The performance of the vehicle restricted the mass to less than about 70 kg including the apogee engine.

SELECTION OF SPIN STABILIZATION

It is clear from the above requirements that a random attitude cannot be tolerated and, especially when the overturning moments (thrust eccentricity) of the apogee engine are considered, the stabilization must be quite "stiff". The use of continuous active control by means of momentum wheels and reaction jets was out of the question because of weight considerations; while stabilization by gravity gradient would give restoring torques many orders of magnitude too low even if the necessary hardware were developed. The selection of spin stabilization was quite natural especially in view of its other advantages, including the ease of providing a mild and uniform thermal environment throughout the vehicle as well as the presence of a centrifugal force field ("artificial gravity") for expulsion of liquid propellant used in the control system. The use of spin stabilization was facilitated because the third stage solid engine of the vehicle is spin-stabilized and imparts its spin to the satellite payload; had this not been the case it would have been necessary to provide spin-up means for the payload.

With spin stabilization, the only useful antenna directivity readily achievable is to limit the beamwidth measured in a plane containing the spin axis, with symmetry around this axis. In order for such directivity to be useful at all times, it is clear that the axis should be oriented essentially perpendicular to the plane of the orbit. Unless geographic coverage within the area of visibility is to be limited, the beamwidth should be no less than the angle subtended by the Earth at the orbital altitude, or about 17.5 degrees at the circular synchronous altitude of roughly 36000 km. This permits a gain of about 8 db; however, at 1.8 gc the physical length allowed for the antenna permitted only about 6 db to be actually realized.

If the spin axis is normal to the plane of the orbit, the angle between the direction of the Sun and the axis varies during the year over the range $90^\circ \pm \alpha$, where α is the angle between the plane of the orbit and the ecliptic. The initial Syncom spacecraft were designed for an orbit inclined at 33 degrees to the equator; by suitable choice of launch time the angle α could be made no more than 23.5 degrees, the value for the ultimate objective of an equatorial orbit. This modest variation of the incidence angle of the Sun's rays permits efficient operation of a cylindrical array of solar cells around the axis, and also has a favorable influence on the thermal design of the vehicle.

CONTROL SYSTEM REQUIREMENTS

Since the direction of the spin axis must be initially in or nearly in the plane of the orbit because of the required direction of thrust of the rocket stages, it is evident that means must be provided for reorienting the axis during the mission. Furthermore, once the axis is normal to the plane of the orbit, the energy (hence period) can be effectively controlled only by thrust perpendicular to the spin axis.

CONTROL SYSTEM CONCEPT

The spin axis may be precessed by the application of torque normal to this axis; such a torque is readily provided by a propulsion jet on the craft, but to be effective this torque must have a time-average component in the spatial direction in which we wish the axis to move. Likewise, acceleration perpendicular to the spin axis may be produced by a jet, but must have a time-average component in the spatial direction of the desired net acceleration.

It is evident that if the control jets are operated over some fixed portion of each spin cycle relative to an external reference for spin phase angle, a net effect of precession or acceleration may be produced; and that the direction of the precession or acceleration may be controlled by selecting the sector over which the jets are turned on. This concept is illustrated in Figure 1. The torquing jet thrusts parallel to the axis and may also be operated continuously over a number of spin cycles to produce a net acceleration along the axis.

It is evident that in the case of precession the dynamics of the system must be taken into account, since although we have indicated that we may change the direction of the angular momentum vector it is not immediately obvious that in the process we do not cause the geometric axis to nutate about the angular momentum vector with ever increasing amplitude. The latter in fact does not occur even without means for damping the nutation, as was first demonstrated experimentally by the Hughes Aircraft Company early in 1960. The geometric axis (actually, principal axis of inertia) has during precession a "wobbling" motion superimposed on its uniform motion; but the amplitude of the "wobble" is of the order of the precession per revolution; with practical parameters, this is a fraction of a degree and constitutes no problem whatsoever. That this should be the case may be understood by examination of the Euler equations; for a body with spin angular velocity ω_z, spin moment of inertia I_z and equal transverse moments of inertia I_x, the response to torques defined in body axes is that of an undamped system with resonant frequency $|\Omega|$, where

$$\Omega = (1 - \frac{I_z}{I_x})\,\omega_z$$

Since the torque due to pulsing has components at angular frequency ω_z and harmonics thereof, and since for long-term stability $I_z/I_x > 1$ but cannot exceed 2, it is apparent that the resonance is not excited.

ATTITUDE DETERMINATION

The application of the concept requires that a reference be provided for determination of the instantaneous angle of the vehicle about the spin axis. A device using simple slit optics and solar cells readily provides a well-defined pulse centered when the Sun is in the plane of the slits, or the solar sensor "beam". One such device with the plane of the "beam" parallel to the spin axis provides the needed phase information. A second device whose beam plane is not parallel to the axis may be used to measure the angle (ϕ) between the spin axis and Sun line, since the difference in spin phase of the two pulses is a function of ϕ. In practice, the beam planes intersect along a line perpendicular to the spin axis (Figure 2); the angle I is 35° in the Syncom vehicles.

Since solar sensing leaves the direction of the axis undetermined with regard to orientation around the Sun line, the attitude of the axis cannot be determined immediately without additional data. This difficulty is overcome by observing the orientation of the plane of polarization (electric vector) of the signal from the transponder, which is parallel to the spin axis. Over several weeks, however, the orbital motion of the Earth produces sufficient change in the direction of the Sun to permit the attitude to be practically determined from solar aspect (ϕ) data alone.

CONTROL SYSTEM SELECTION

Two types of control propulsion systems have been used on Syncom vehicles, cold gas (nitrogen) and monopropellant (hydrogen peroxide). Alternatives considered include hydrazine monopropellant and hypergolic bipropellant; these were rejected for the initial system because of known developmental difficulties.

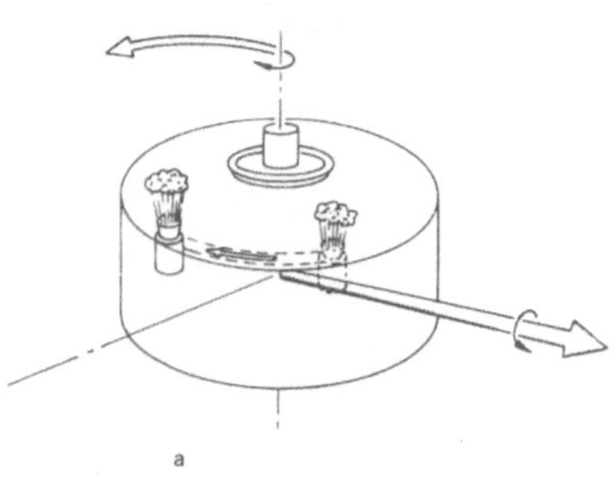

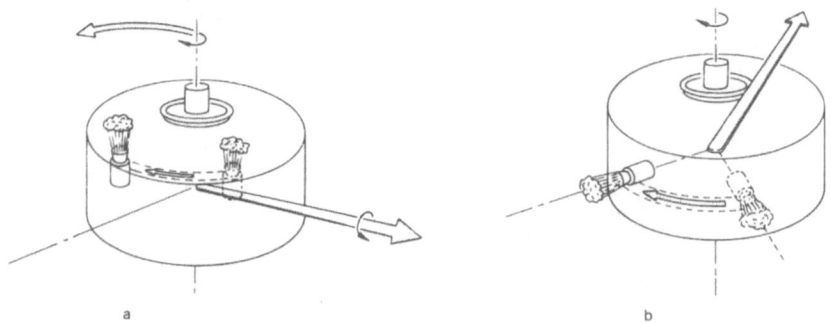

Fig. 1. Pulsed jet control: (a) Orientation control. (b) Velocity control.

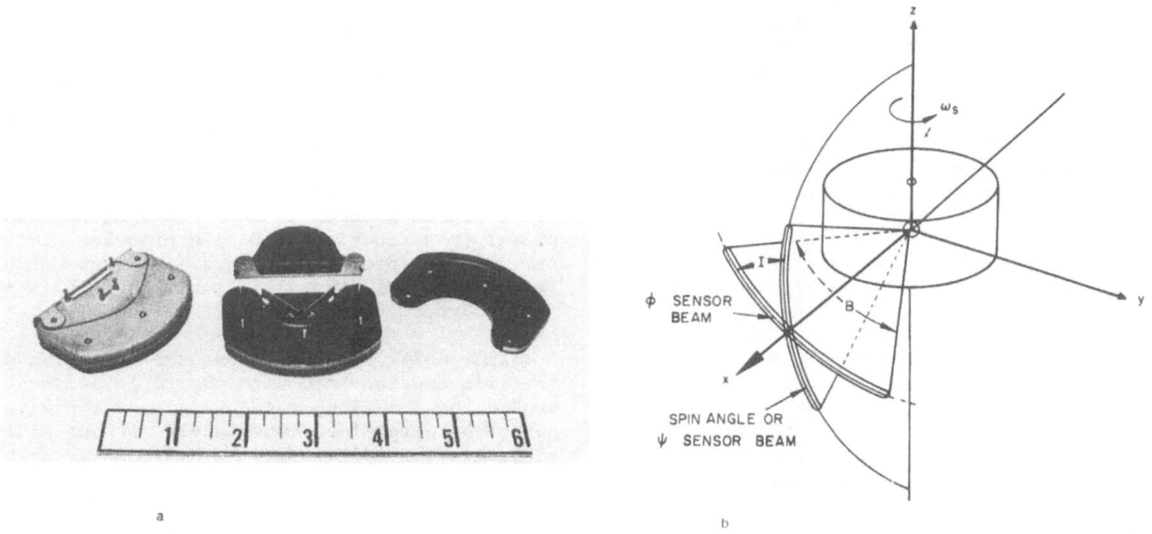

Fig. 2. (a) Solar sensors. (b) V-beam formed by the sensors.

The initial design philosophy was to incorporate a cold gas system and an independent hydrogen peroxide system in each vehicle. Both systems had the same basic geometry of Figure 1, but it was intended to utilize cold gas, with pulse rise and decay times of less than 10 milliseconds and specific impulse of about 70 seconds, for maneuvers requiring pulsed operation; while hydrogen peroxide was to be used mainly for gross corrections using the axial engine to produce continuous thrust along the axis, prior to reorientation. The hydrogen peroxide system provided superior performance (specific impulse over 150 seconds, with further a much more favorable ratio of propellant to tankage weight) in this application; however, because of low confidence in the reproducibility of the pulses, with rise and fall times of the order of 30 milliseconds, the hydrogen peroxide system was considered as a back-up to nitrogen for pulsed maneuvers although the pulsed specific impulse was about 100 seconds. At typical spin speed, one millisecond corresponds to 1° of spin.

Syncom I, launched February 14, 1963, was accordingly provided with a nitrogen gas system charged with about 1.1 kg of gas at a pressure of 250 atmospheres. Tankage was titanium and had a weight of 1.9 kg; the nozzles developed about 0.36 kg of thrust at initial pressure. No pressure regulation was used. A hydrogen peroxide system was also incorporated.

Just prior to burnout of the solid-propellant apogee engine, all signals from Syncom I terminated abruptly. An extensive investigation strongly suggested that the failure was due to explosion of one of the gas tanks.

Accordingly, the pressure in the nitrogen system in Syncom 2, launched July 26, 1963, was reduced to 170 atmospheres with no change in tank volume, giving about 0.7 kg of gas. Syncom 2 has been highly successful and the nitrogen system performed quite satisfactorily when called upon; however, it leaked gas at an unexpectedly high rate and the thrust due to leakage caused a moment about the roll axis which gave a substantial drop in the spin speed.

In the case of Syncom 3, launched August 19, 1964, it was desired to make orbit plane changes at transfer orbit perigee and apogee, leading to a final equatorial orbit. Because of these plane changes, the dispersion of the energy after apogee boost required an increased control velocity capability. Experience with the hydrogen peroxide system performance in pulsed mode in Syncom 2 had been excellent; accordingly, the nitrogen system was removed and a second hydrogen peroxide system substituted. Total fuel was 4.4 kg, which if used entirely in continuous axial jet mode would give as much as 180 m/sec velocity increment. Operation of both of these systems in Syncom 3, including a critical reorientation maneuver prior to apogee boost, has been highly satisfactory.

HYDROGEN PEROXIDE HARDWARE

A Syncom hydrogen peroxide system is shown in Figure 3; Figure 4 shows a complete spacecraft with a solar cell panel removed to reveal the interior.

The spherical hydrogen peroxide tanks are made of essentially pure aluminum to minimize catalysis of the fuel. About 60 percent of the volume of the tanks is occupied initially with 90 percent hydrogen peroxide and the remainder with nitrogen at about 15 atmospheres. The tanks contain no bladder or diaphragm, and there is no external source of pressurizing gas and no pressure regulation. The tanks are manifolded by a tube connected to each tank at the position farthest from the spin axis; spin provides a pressure head at these outlets and as fuel is depleted the tank pressure falls off (blow-down principle). The engines employ solenoid valves, silver screen catalyst to induce decomposition, and nozzles with an expansion ratio of 17:1. Thrust at initial pressure is about 1.3 kg, and would drop to about 0.5 kg as fuel is depleted were it not for evolution of oxygen due to decomposition of fuel. Although no pressure rise is observed in Syncom 2, the pressure increases in the Syncom 3 systems at a rate on the order of 0.1 atmosphere per day. This gradual decomposition gives a useful life of several years; excessive pressure is vented by a pressure relief valve above about 18 atmospheres. The Syncom hydrogen peroxide system was designed and manufactured by the Walter Kidde Company. Problems encountered in the design of such a system are primarily those of material selection in valves to achieve both compatibility (low catalysis) and proper mechanical function; in addition, scrupulous cleanliness is mandatory.

The systems are installed and aligned by Hughes; the alignment is of cardinal importance since the spin speed may change by as much as 25 rpm even with the allowable misalignment of 0.2 degree. Since dynamic unbalance can contribute an effective misalignment, the vehicle is balanced to about 10 kg cm^2. Jet alignment is performed by optical means using mandrels in the nozzles, the adequacy of this method having been verified by tests made on Syncom 1 on an air bearing prior to launch.

GROUND CONTROL EQUIPMENT

The ground control equipment provides for sending the necessary commands to the spacecraft, and for measuring the solar aspect angle ϕ. The requirements are essentially the capability to send command pulses in real time with the proper time delay (relative to the primary solar sensor pulse) and of the correct duration (nominally 60° of spin), the pulses being synchronized with the spin; and to measure the spin angle between the pair of solar sensor pulses to obtain ϕ. In the Syncom system, these functions are both performed using a rotating mechanical contactor

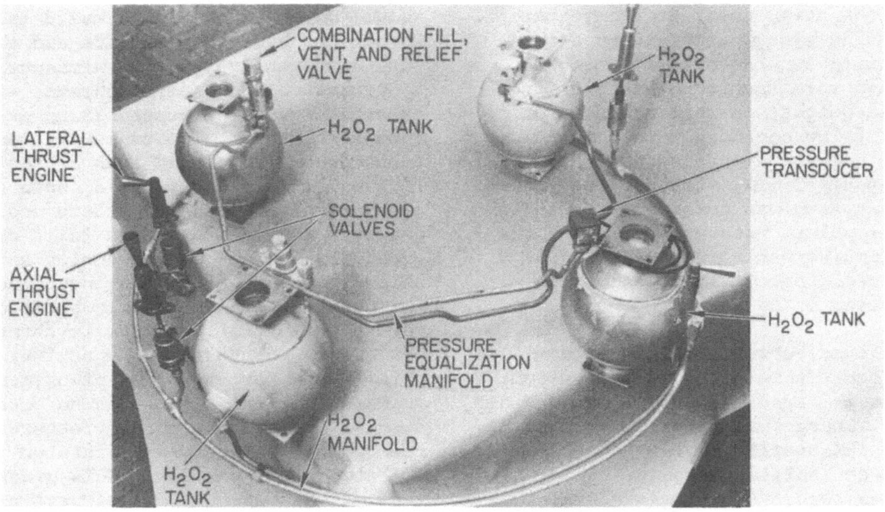

Fig. 3. Hydrogen peroxide systems.

Fig. 4. Syncom spacecraft.

device which is driven synchronously with the spin at a fixed phase relation, the lock being maintained by an electronic control loop which causes the drum to "track" the primary solar sensor pulses (called ψ pulses). The device, known as the synchronous controller drum unit, is shown in Figure 5. The dial on the left positions the command contacts around the periphery of the drum, and is so arranged that its reading is equal to the number of degrees of spin from the time of reception of the ψ pulse to the start of the command pulse. The dial setting is computed taking account of the desired direction of thrust or torque relative to the direction of the Sun; the geometry of the control jets relative to the solar sensor beam plane; the pulse duration; and the two-way propagation delay. The effect of the latter can be calibrated out directly, since it is possible to send pulses to the vehicle without actuating the jets, these pulses being transponded through special provisions made in the telemetry system. If the command dial is rotated until the leading edges of the received pulses coincide with the ψ sensor pulses, the resulting dial setting is simply added to the value calculated for zero propagation time. This process is known as "delay calibration". In practice this procedure is used only as a check on values calculated from predicted slant range to the vehicle, since there are different circuit delays in the command and delay calibration paths.

The dial on the right moves contacts which generate a cursor, which is displayed together with the pulse from the second solar sensor (ψ_2 pulse), the dial being positioned so that the two coincide in time. The spin angle between the two pulses (known as ψ_2) is read from the inner graduations. The nonlinear outer graduations give ϕ directly.

The command system itself operates on the principle of first selecting a command (by means of a train of logic pulses), then executing by means of pulse of a special execute tone. The latter is controlled by the contactor. In both pulsed and continuous operations (in the latter the contactor being by-passed) the total time of pulsing is controlled by an operator using a push button and stop watch. Syncom ground control station construction and operation is performed by the Hughes Aircraft Company.

The Syncom vehicles have additionally a self-contained mode of pulsed operation known as the quadrant mode. The primary sensor plus three others define four quadrants. By suitable logic circuitry it is possible to pulse the jets on during the time when the Sun line is in a selected quadrant. This mode has been successfully used in Syncom 2, but is not generally employed because only four discrete directions of thrust or torque are available.

CONTROL ANALYSES

Only a brief summary of the calculations entering into orbit control maneuvers can be given here. For large maneuvers, inclined orbits, and high eccentricities such as occur in the initial phase shortly after launch, the orbit transfer and geometry problems involved in producing a desired orbit change are solved on a digital computer. For small maneuvers with the orbit nearly circular, a linearized theory may be used; and for Syncom 3 a number of maneuvers have been calculated entirely by hand; this of course is quite possible, but tedious and subject to error, for any maneuver. In the design of maneuvers, the effect of future orbit perturbations must be anticipated and taken into account.

In reorientation, the direction of the Sun at the time of maneuver may be regarded as fixed during the operation. Also, the pulse phasing is not changed during transmission of a train of command pulses. Therefore, during precession the mean motion of the spin axis makes a fixed angle with the plane of spin axis and Sun line. Therefore, on the unit sphere the path of the spin axis is a rhumb line (loxodrome) with the direction of the Sun line as the pole. Since the spin axis is never far from normal to the Sun line, the rhumb line has an arc length negligibly greater than that of a great circle and the efficiency is very high (greater than 99 percent).

The geometric analysis required is performed in advance by means of a digital computer; however, during a precession of considerable magnitude it is desirable to be able to stop, evaluate the results, and if necessary continue with perhaps a corrected phase or revised duration of pulsing. Therefore, before execution of the maneuver a plot is prepared. The plot is on the Mercator projection with the direction of the Sun at the pole. The rhumb line is plotted at an angle to the meridian equal to the spin phase angle of the torque. From the output of the computer a series of lines are plotted through points on the nominal path which represent approximations to the great circles of constant polarization angle. The "latitude" corresponds to $90^\circ - \phi$; when polarization and ϕ are given, the analyst may immediately plot the attitude, and measure the phase angle and angle to be precessed through to reach the objective.

The calibration of the control system is of considerable importance since accurate calibration data decrease the number of maneuvers needed to attain a precise orbit or attitude.

Calibration data may be divided into two parts, phase calibration and magnitude calibration. Phase calibration is performed by calculation from measured pressure - time data for pulses from the engines. Prior to launch of Syncom I,

Fig. 5. Synchronous controller drum unit.

the results were verified by limited testing on a spin table mounted on knife-edge supports. Magnitude calibration data give the required duration of pulsing to produce a given effect, as a function of pressure; and also the weight of fuel consumed. The data are derived from chamber pressure and flow rate measurements. Because of varying specific impulse with number of pulses and pressure, the magnitude data have been a greater source of difficulty than the phase calibration. In practice, the prelaunch calibration data are adjusted based on in-flight performance. In reorientation maneuvers, this may be done immediately from the plot. For orbit control maneuvers, the orbit must be determined from tracking data (a function performed by the NASA Goddard Space Flight Center) and the transfer evaluated. Essentially, control is open-loop insofar as the automatic equipment is concerned, except for the drum servo loop which tracks the spin. The attitude and orbit control loop is closed by the analyst, who achieves the desired result by successive corrections, if necessary. All calculations, with the exception of orbit determination, are the responsibility of the Hughes Aircraft Company.

IN-ORBIT PERFORMANCE

As of the date of writing, 27 maneuvers have been performed by Syncom 2 and Syncom 3. All maneuvers have been successful, although a few have had exasperating calibration errors.

Five continuous axial jet maneuvers have been executed with a total velocity increment of about 100 m/sec, the largest single maneuver being about 30 m/sec. Both energy and plane changes have been made.

Six reorientations have been performed, the largest being about 80° in a single pulse train, and the smallest only 4°. The total angle precessed through is about 230° for two spacecraft.

16 pulsed lateral velocity corrections have been made, ranging from about 15 to 0.1 m/sec and totalling about 50 m/sec.

On the basis of this experience, the control system concept is thoroughly proven, and experienced personnel can derive and execute maneuvers on a routine basis with high confidence.

LATER DEVELOPMENTS

The first commercial communication satellite, the HS-303 "Early Bird" vehicle to be launched initially in March, 1965, is designed and built for the Communications Satellite Corporation by Hughes. It is very similar to Syncom 3 in most respects except for the transponder, with which the command system is now integrated. The control system is virtually identical, although for reasons of economy the contactor in the ground equipment has been replaced by an electronic device.

The HS-304 is a larger and more advanced configuration proposed for the Communications Satellite Corporation. In addition to a hydrogen peroxide system for large initial corrections, it is proposed to incorporate a system which electrolyzes water (or possibly hydrogen peroxide) to provide hydrogen and oxygen for the control rocket engines. This system provides a high specific impulse for counteracting perturbations over extended periods of time; the limited velocity increment available in a single correction make it somewhat unattractive for correcting initial errors. The HS-304 also incorporates an electronically phased antenna array which effectively "despins" the antenna beam and permits about 10 db additional gain.

The use of the basic Syncom configuration is by no means limited to communications.

The Applications Technology Satellite (ATS) program will extend its usefulness to meteorological observation and environmental measurements, as well as test a mechanically despun antenna. One version of ATS is a synchronous satellite stabilized by gravity gradient; however, during apogee injection and until the orbit and attitude are perfected for gravity gradient boom deployment, the vehicle is spin-stabilized and controlled in the manner of Syncom. The Syncom control system has in fact been suggested for such diverse applications as a lunar probe, a planetary probe; and even a lunar soft lander, where it may be remarked that model tests show that spinning a main body about non-spinning legs produces a spectacular improvement on the ability to remain upright on touchdown.

DISCUSSION

D. D. Williams

"Control of the Syncom Communication Satellite".

Q. What is the lifetime?

A. On Syncom II we have demonstrated an 18-month life. The fuel is now exhausted. We feel that with additional fuel this could be extended to three years. The catalyst bank is not the limiting factor. We have intentionally induced nutation to observe the action of the damper.

USE OF SINGLE-AXIS ORIENTATION FOR ARTIFICIAL EARTH SATELLITES

R. F. Appazov, V. P. Legostaev, B. V. Raushenbakh

In order for a spaceship orbiting the earth to effect reentry, it must be given a certain momentum to transfer it to a descent orbit. It is important that the direction of the imparted momentum be variable over a certain range. For example, the effect of transfer to the descent orbit may be achieved either by deceleration (horizontal impulse) or by rotating the satellite velocity vector about the normal to the trajectory. In the general case, an optimum direction for applying this momentum to minimize the amount of fuel required for the maneuver may always be found.

If we pose the problem of simplifying the guidance system of the satellite to the maximum, and directing the thrust of the rocket engine in the way required, we can make use of the sun as celestial checkpoint. Tracking the sun makes it possible to utilize the simplest optical tracking device and to rely on single-axis orientation. Of course this method also has its disadvantages – the engine may be started only over the sunlit half of the earth, and the satellite – sun direction must not make a large angle with the orbit plane. These circumstances impose certain restrictions on the orbiting conditions.

The launch time may be decided upon on a day-by-day or month-by-month basis such that the decelerating impulse given in that direction will bring about a landing of the spaceship in a predetermined region. Nevertheless, this system is entirely acceptable for all important cases of practical interest.

The deceleration velocities actually used in practice range from 100 to 200 m/sec in magnitude; because it is impossible to direct the impulse in the direction of the range gradient (with respect to velocity), the launch ranges will depend on the time of year. For Vostok-I type orbits, this is illustrated in Table I listing velocity derivatives of range and orientation angle of the thrust vector for various times of the year.

Here L is the range over the earth's surface from the point where the decelerating power plant is switched on to the landing point.

L'_{ν} is the deceleration velocity derivative of the launch range.

$L'_{\gamma 1}$ is the orientation angle derivative of the launch range in the orbit plane.

$L'_{\gamma 2}$ is the orientation angle derivation of the launch range in the plane perpendicular to the velocity vector.

Clearly, from the table, the requirements imposed on orientation precision and on speed to keep deviations within several tens of kilometers are not overly stringent.

For orbits tilted 65°, such as used in launching objects of the Vostok class, winter flights are more suited than summer flights from this viewpoint.

The use of a simple orientation technique will be justified only in the case where instrumental solution is also simple.

DESCRIPTION OF THE SYSTEM

We now introduce a right-hand coordinate from OXYZ (Fig. 1) fixed in the principal inertial axes of the vehicle. Let the x-axis coincide with the direction of the reaction thrust of the engine designed to effect a transfer to the descent orbit.

Now turn to the layout of the guidance system (Fig. 2). Clearly, the system consists of a photoelectric sensor, three two-degrees-of-freedom gyroscopes responding to the several projections of the satellite's angular velocities ω (ω_x, ω_y, ω_z), a logic device and several actuators (retrorockets). The photoelectric sensor (sun sensor) responds only to rotations of the vehicle (relative to the sunward direction) about the y- and z-axes, since a rotation about the x-axis will have no effect on the direction of the reaction force of the primary rocket engine.

Control of rotation about the x-axis is the simplest approach in the light of the foregoing. It reduces to a damping of the angular velocity ω_x of the angular velocity. The extreme simplicity of this mode of stabilization enables us to forbear further discussion of it in this paper. In what follows, therefore, our assumption is that $\omega_x = 0$. Rotations about the z- and y-axes may be viewed then arbitrarily as rotations in pitch (ϑ) and in yaw (ψ), where $\omega_y = \dot{\psi}$ and $\omega_z = \dot{\vartheta}$. We further stipulate that the angles $\vartheta = 0$ and $\psi = 0$ correspond to the x-axis directed sunward.

Stabilization in pitch and in yaw is achieved by differencing signals from the photoelectric sun tracker and the angular velocity sensors supplying information through pitch and yaw channels in the logic circuit.

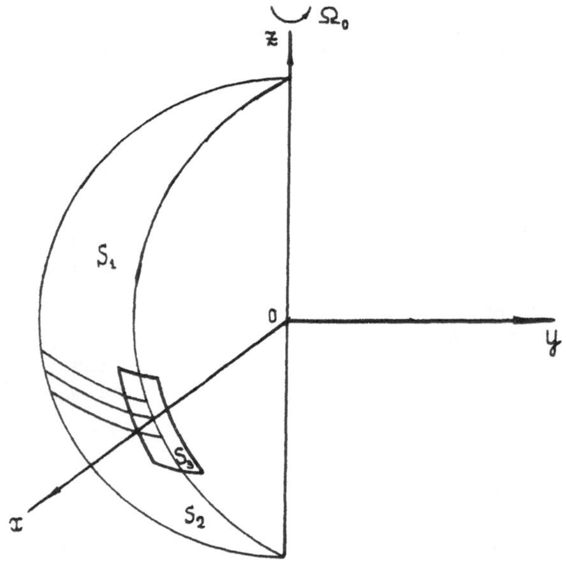

Fig. 1

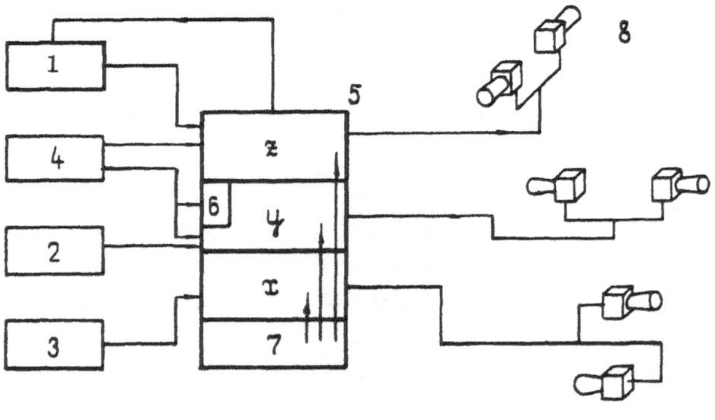

1. Pitch channel rate sensor
2. Yaw channel rate sensor
3. Roll channel rate sensor
4. Photoelectric solar sensor
5. Amplifier-transducer logic unit
6. Time delay
7. Pulse generator
8. Actuators

Fig. 2

We now consider some components of the guidance system in greater detail. The primary functions of the photoelectric sun tracker in the guidance system is to track the sun, which may present any arbitrary initial position relative to the satellite, and to lock on the sun subsequently with a predetermined precision about the values $\vartheta = \psi = 0$.

The sun tracker executes these functions, while mounted in fixed position on the satellite frame, with the aid of three independent cells whose fields of vision, as well as their arrangement on the satellite, may be seen in Fig. 1.

The fields S_1 and S_2 overlapping near the x-axis have the function of seeking and locking onto the sun on the x-axis. Field S_3 is an auxiliary field. The total angular extent of fields S_1 and S_2 in the XOZ plane is 180° (slightly greater in practice). This makes it possible to "find" the sun in a single revolution of the satellite about its axis via rotation in pitch.

The zone where fields S_1 and S_3 overlap provides a neutral zone for yaw oscillations of the satellite.

The auxiliary field S_3 provides information on the correct orientation of the satellite when angles ϑ and ψ are close to zero (i.e., with the x-axis directed sunward to the required precision), and field S_3 is illuminated by the sun.

We realize from this brief description that the photoelectric sensor gives only three on-off type signals indicating whether or not the corresponding field of vision S is sunlit.

A two-degrees-of-freedom gyroscope processing these signals into a function of the appropriate angular velocity component is installed to damp out angular motions of the satellite on each stabilization axis. The output response of these instruments is in on-off form, with a dead band in the rotation angle of the gyroscope outer gimbal ring.

The angular velocity sensor mounted on the pitch axis works in two modes: in the first the dead band is symmetric about the angular velocity value $\dot{\vartheta} = 0$, in the second the dead band is shifted by an amount $\dot{\vartheta} = \Omega_0$. Accordingly, control signals are generated in the first mode when the angular velocity $\dot{\vartheta}$ exceeds the instrument sensitivity; in the second mode control signals provide information that the pitch angular velocity differs by some predetermined magnitude Ω_0 from the amount exceeding the instrument sensitivity.

The angular velocity sensor mounted on the yaw axis functions only in the first of these modes.

In considering the operational logic, and for engineering calculations, we may represent these gyroscopic instruments together with their output devices in the form of a relay element with dead band and lag (Fig. 3 and Fig. 4).

As for the actuators, the retrorockets used yield a constant thrust when switched on. They can be either shut off or switched on, and from the viewpoint of automatic control theory must also be viewed as relay, or discontinuous on-off, control elements. A more detailed examination reveals that the lag in retrorocket cutoff must be taken into account in analyzing the dynamics of the satellite, i.e., the time from the appearance of the electrical signal commanding the cutoff of the retrorocket and the actual drop of the retrothrust to zero. This time is made up of the time required to close the appropriate valves and the time required to eject the volumes of gas enclosed between these valves and the nozzle tip. Since the time required to empty out a certain semiclosed volume in a vacuum is theoretically equal to infinity, in the calculations we have to work with an effective gas ejection time arrived at by dividing the momentum of the gas escaping completely after the valves have been closed by the nominal retrorocket thrust.

In the logic device of the orientation system, together with the components comparing the control signals arriving from the pick-offs, there is a pulse generator shaping signals of constant frequency. If the signal coming from the logic unit to switch on the retroengine is supplied directly, then the retroengine will work continuously and throughout the entire time this signal is operative (continuous mode). But if this signal is fed to the retroengine via the pulse generator, then the engine will be switched on and off periodically, during the entire time this signal is operative. Since the switch-on time is less than the pause, e.g. less by n times, within the time involved, this pulsed mode of operating the retroengine corresponds dynamically to the operation of the engine at some lower effective thrust (n + 1) times less than the nominal thrust. The system of actuators is thereby capable of operating in three modes: a high control torque (with the engines operating continuously), a low effective control torque (in pulsed operation of the retroengine), and zero torque (with retroengine shut off).

Summarizing the above, we may state that this orientation system is an on-off pulsed automatic control system. It is reasonable to attempt a mathematical analysis of the performance of this system after a brief description of its operational logic.

OPERATIONAL LOGIC OF THE SYSTEM

After a certain time has elapsed with the retroengine switched on, the single-axis sun-tracking system is switched on.

If the sun is not found within the field of vision of the photoelectric sun sensor, the signals arriving from the angular velocity sensors via the amplifying the transducing logic circuitry to the actuators will damp the angular velocities of the satellite about the x-axis and y-axis, and its regular rotation in pitch at angular vel-

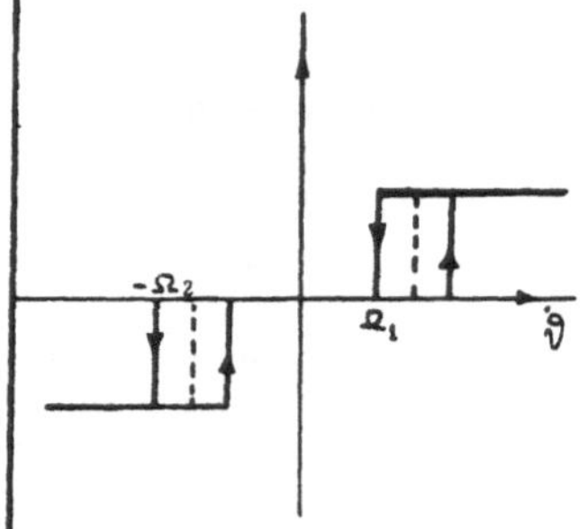

Fig. 3

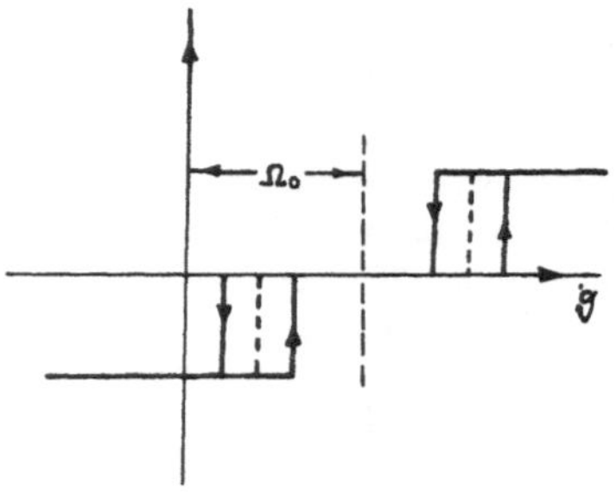

Fig. 4

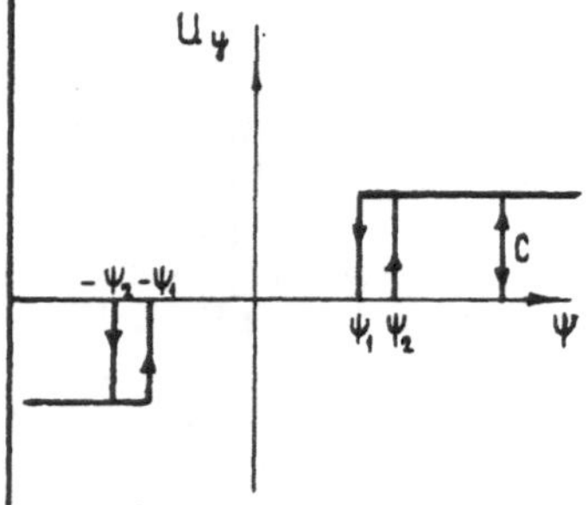

Fig. 5

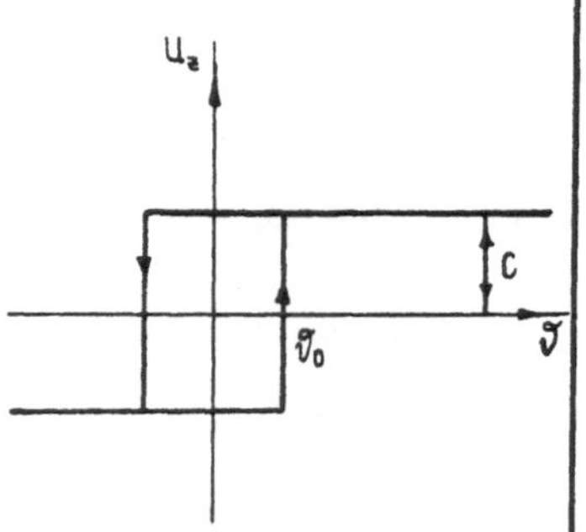

Fig. 6

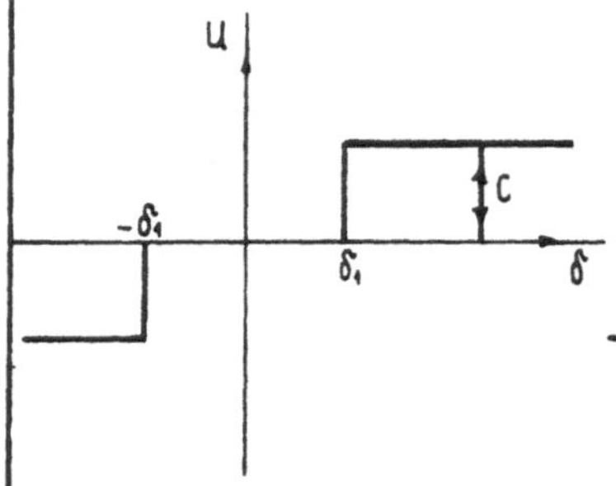

Fig. 7

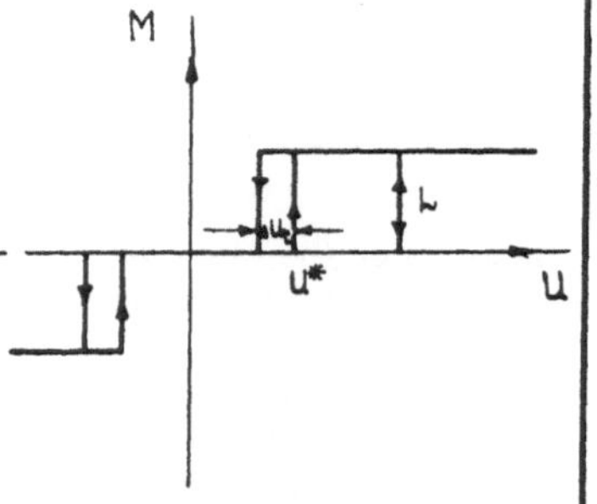

Fig. 8

ocity $\Omega_0 > 0$, since the pitch angular velocity sensor functions when the sun is being tracked in the second operational mode.

On achieving satellite angular velocities $\dot{\psi} = 0$, $\omega_x = 0$, $\dot{\vartheta} = \Omega_0$, (accurate to the dead band), the actuators cease to function, and the satellite will be rotating by inertia. Once the sun is acquired in the field of vision S_1 or S_2, the tracking system will operate as follows.

a) In pitch.

The angular velocity of the satellite rotation in pitch will commence to slow down, so that the retroengines of the orientation system will be in continuous operation with the object of achieving rapid deceleration. The reason for this deceleration is that after the sun has illuminated fields S_1 or S_2, the constant shift of the neutral zone of the angular velocity sensor by the amount Ω_0 will be eliminated and that sensor will be converted to the first operational mode: the usual null-seeking operation.

As the absolute magnitude of the pitch angular velocity decreases from the value $\dot{\vartheta} = \Omega_0$ to the closest boundary of the dead band (which we designate $+ \Omega_1$, see Fig. 3), the system will commence operating in the pulsed mode, retaining the previous sign on the torque. After $\dot{\vartheta}$ has diminished to a magnitude corresponding to the opposite boundary of the dead band (which we designate as $- \Omega_2$), the control signal to the actuators will be eliminated. The satellite will continue to rotate by inertia at an angular velocity close to $- \Omega_2$, and eventually the sun will escape from the fields of vision S_1 and S_2 of the sensors, which will "lose" the sun. The orientation system will then resume its search for the sun, a bias Ω_0 will be imposed on the angular velocity sensor, and since the satellite will not have time to effect a pronounced turn away from the sun because of the low value $\dot{\vartheta} = \Omega_2$, the sun will again fall into the field of vision S_1 or S_2. Then, during the short switch-on time (and we should not overlook the fact that the search for the sun is carried on with the retroengines operated in pulsed fashion), the angular velocity of satellite rotation will be close in absolute magnitude to angular velocity at which the sun was "lost," i.e., close in absolute magnitude to Ω_2. And since the bias Ω_0 imposed on the sensor will be removed after the field S_1 or S_2 is again illuminated by the sun, the orientation system will revert to the previous operational mode, except that the sign of the torque acting on the satellite will change and the satellite angular velocity will attain the value $\dot{\vartheta} = - \Omega_2$. After some time has elapsed this will lead to a very brief "loss" of the sun and the entire process will be recapitulated. This is the gist of the oscillations in pitch characterized by angular velocities close to the angular velocity sensor sensitivity. Clearly, as seen in Fig. 1, the oscillations will be executed at the boundaries of the fields of vision S_1 and S_2 lying in the XOZ plane, i.e., at ϑ values close to zero degrees.

b) In yaw.

Correct satellite orientation requires that the angle $\psi = 0$. To achieve this goal, the orientation system begins to yaw the satellite in the negative direction of angles ψ (in a right-handed coordinate system) when the sun falls, say, into the field S_1. The actuators controlling satellite yaw operate in the following mode in this case. If the angular velocity of rotation exceeds the neutral of the angular velocity sensor and is of positive sign, the actuators will operate in a continuous mode. In all remaining cases the system of actuators will operate in the pulsed mode.

In order to achieve more rapid rotation of the satellite in yaw a time lag is introduced into the system in order to conserve the yaw control torque for a certain time (when the sun escapes from the field of vision of the solar sensor because of self-sustained oscillations in pitch). This lag must be slightly greater than the period of the pitch oscillations. This lag is elminated when the sun appears in field S_3.

When fields S_1 and S_2 are illuminated by the sun simultaneously, the system operates in the same manner as in the absence of sunlight, i.e., yaw control goes over into the damping mode. When the sun falls into the field S_2, the actuators operate in the continuous mode if the rotation rate exceeds the threshold of the neutral zone and is of negative sign. The actuators operate in the pulsed mode in all remaining cases.

Table II provides a clearcut picture of the system operational logic in the pitch and yaw channels.

Our notation here is:

for S + if the sensor is illuminated by the sun*;
- − if the sensor is not illuminated by the sun;

for the angular velocity sensors (in the first mode)
- + if the angular velocity is positive and greater than the dead band;
- − if the angular velocity is negative and greater in absolute value than the dead band;
- 0 if the angular velocity is within the dead band;

for the angular velocity sensors (in the second mode):
- + if the angular velocity is greater than the null bias Ω_0 (with dead band taken into account);
- − if the angular velocity is less than the bias Ω_0 (with dead band taken into account
- 0 if the angular velocity is equal to the bias Ω_0 (with dead band taken into account);

for retroengine torques:
- + if the system is operating in the continuous mode, producing a positive (negative) torque; if the system is operating in the pulsed mode, producing a positive (negative) torque;
- 0 if the cluster of retroengines is cut off.

EQUATIONS OF MOTION AND SYSTEM PERFORMANCE ANALYSIS

The logic circuitry of the system described above and the high sensitivity of the angular rate sensors make it

* The sensor is considered sunlit during the time sunlight is falling on it plus the lag time.

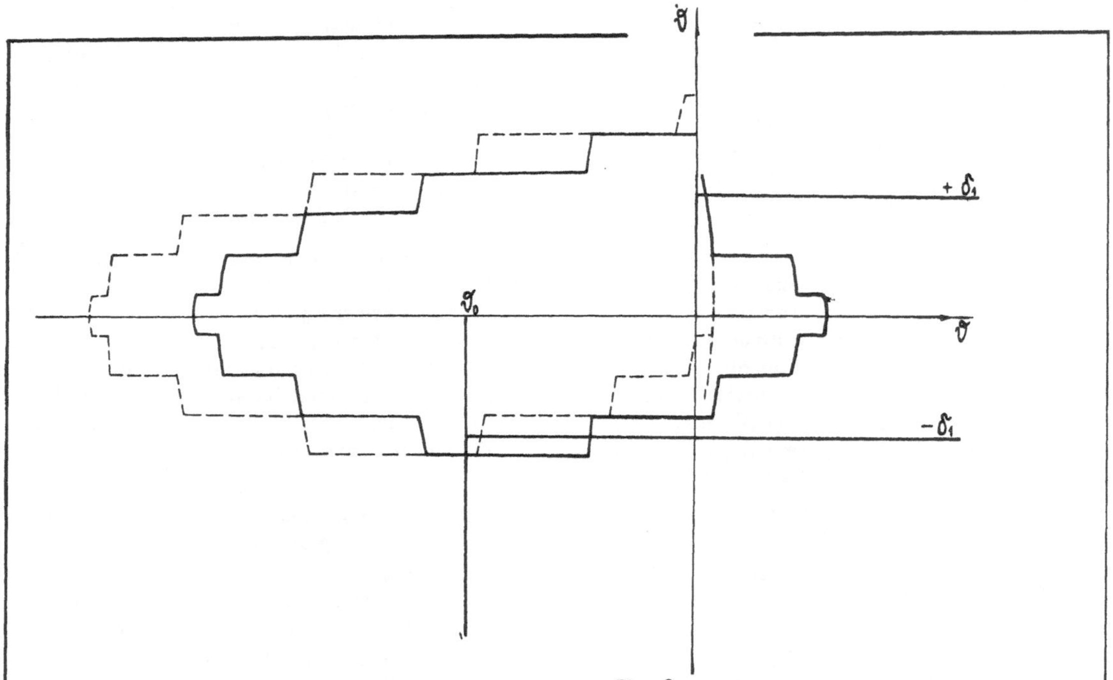

$\dot{\vartheta}$
$+\delta_1$
ϑ_0
ϑ
$-\delta_1$

Fig. 9.

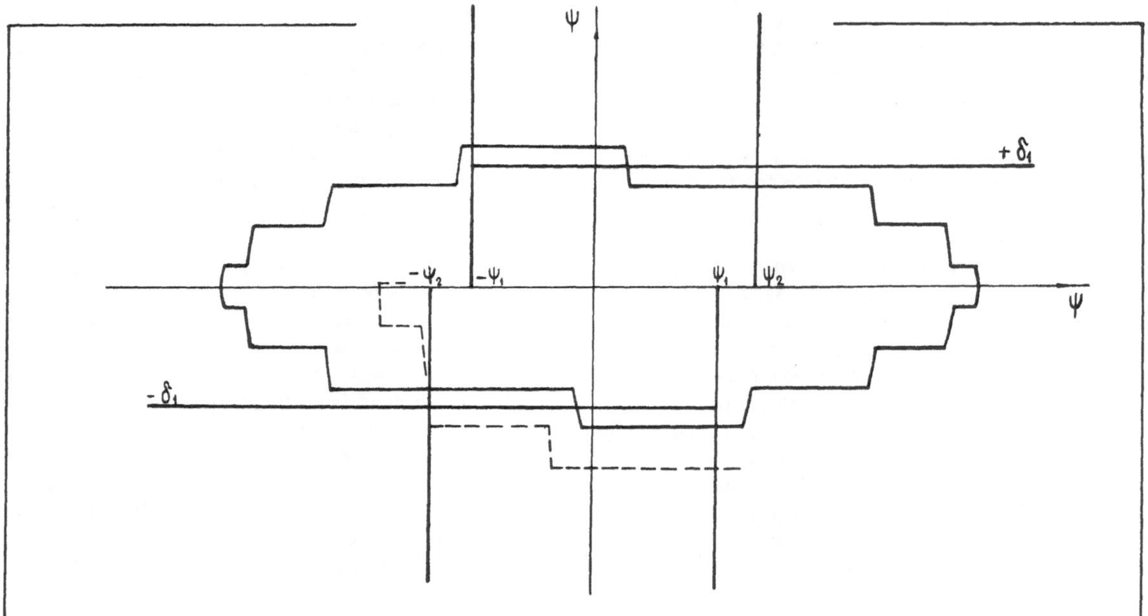

ψ
$+\delta_1$
$-\psi_2$
$-\psi_1$
ψ_1
ψ_2
ψ
$-\delta_1$

Fig. 10.

TABLE I

Characteristic deceleration speeds	Winter			Summer			Spring, Fall		
	100 m/sec	150 m/sec	200 m/sec	100 m/sec	150 m/sec	200 m/sec	100 m/sec	150 m/sec	200 m/sec
L (km)	10000	8500	7400	8000	6000	5000	8700	7000	6000
$L'_{\nu}\left(\frac{\text{km}}{\text{m/sec}}\right)$	-100	-71	-45	-58	-32	-21	-65	-43	-29
L'_{γ_1} (km/deg)	140	199	237	14	24	26	52	70	73
L'_{γ_2} (km/deg)	117	128	133	15	11	9	46	43	37

TABLE II

S_1	+	+	+	0	0	0	+	+	+	+	+	+	0	0	0	—	+	+	0	0	0	0	0	0
S_2	0	0	0	+	+	+	+	+	+	0	0	0	+	+	+	+	+	+	0	0	0	0	0	0
S_3	0	0	0	0	0	0	0	0	0	+	+	+	+	+	+	+	+	+	+	+	+	0	0	0
$\dot{\vartheta}$ Angular velocity sensor (in the first mode)	-	0	+	-	0	+	-	0	+	-	0	+	-	0	+	-	0	+						
$\dot{\vartheta}$ Angular velocity sensor in the search mode (in the second mode)																			+	0	-	+	0	-
$\dot{\psi}$ Angular velocity sensor	+	0	-	-	0	+	+	0	-	+	0	-	+	0	-	+	0	-	+	0	-	+	0	-
Pitch control torque	1 +	$\frac{1}{1+n}$ +	0	1 +	$\frac{1}{1+n}$ +	0	1 +	$\frac{1}{1+n}$ +	0	1 +	$\frac{1}{1+n}$ +	0	1 +	$\frac{1}{1+n}$ +	0	1 +	$\frac{1}{1+n}$ +	0	$\frac{1}{1+n}$ -	0	$\frac{1}{1+n}$ +	$\frac{1}{1+n}$ -	0	$\frac{1}{1+n}$ -
Yaw control torque	1 -	$\frac{1}{1+n}$ -	$\frac{1}{1+n}$ -	1 +	$\frac{1}{1+n}$ +	$\frac{1}{1+n}$ +	$\frac{1}{1+n}$ -	0	$\frac{1}{1+n}$ +	$\frac{1}{1+n}$ -	0	$\frac{1}{1+n}$ +	$\frac{1}{1+n}$ -	0	$\frac{1}{1+n}$ +	$\frac{1}{1+n}$ -	0	$\frac{1}{1+n}$ +	$\frac{1}{1+n}$ -	0	$\frac{1}{1+n}$ +	$\frac{1}{1+n}$ -	0	$\frac{1}{1+n}$ +

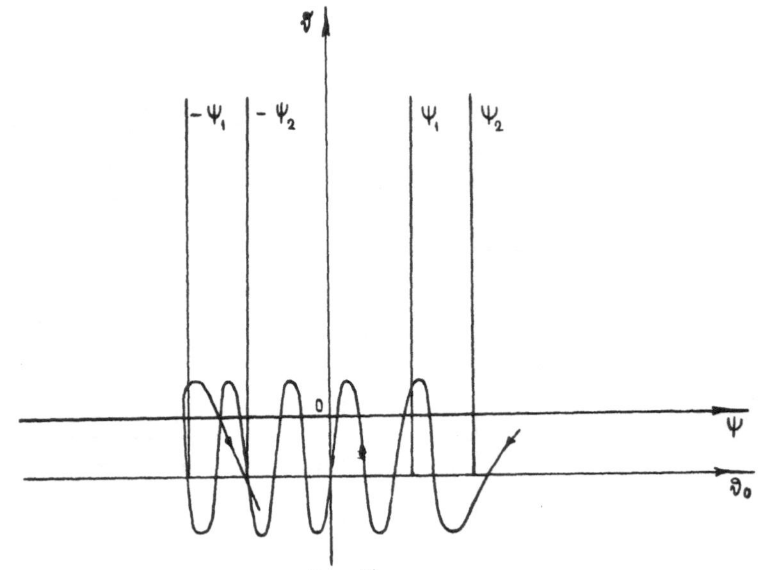

Fig. 11.

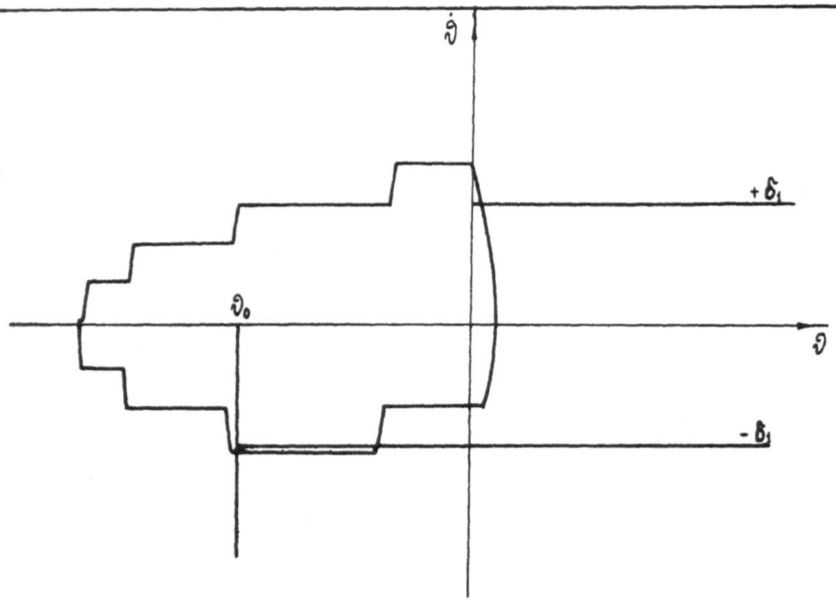

Fig. 12.

possible to view the satellite dynamics in terms of the pitch and yaw channels separately in all cases of practical importance (excluding the stage of damping of the initial perturbations).

a) We here denote as A and B the principal moments of inertia of the satellite about the y-axis and z-axis respectively. M_y and M_z respectively denote the control moments about those axes. The satellite equations of motion about the center of mass in the pitch and yaw channels may now be presented in the form

$$\begin{aligned} A\ddot{\psi} &= M_y \\ B\ddot{\theta} &= M_z \end{aligned} \tag{1}$$

b) The output response of the photosensor is an on-off pulse type with a characteristic hysteresis loop. The equations of the nonlinear control unit may be represented in the form in the yaw plane (Fig. 5)

$$\begin{aligned}
U_y &= 0 \begin{cases} \text{when } -\psi_1 < \psi < \psi_2 \;\&\; \dot{\psi} > 0 \\ -\psi_2 < \psi < \psi_1 \;\&\; \dot{\psi} < 0 \end{cases} \\
U_y &= C \begin{cases} \text{when } \psi > \psi_2 \;\&\; \dot{\psi} > 0 \\ \psi > \psi_1 \;\&\; \dot{\psi} < 0 \end{cases} \\
U_y &= -C \begin{cases} \text{when } \psi < -\psi_1 \;\&\; \dot{\psi} > 0 \\ \psi < -\psi_2 \;\&\; \dot{\psi} < 0 \end{cases}
\end{aligned} \tag{2}$$

The presence of the sun in fields S_1 and S_2 simultaneously defines the neutral zone; in the pitch plane (Fig. 6)

$$\begin{aligned} U_z &= C \,\mathrm{sgn}\,(\theta - \theta_0) \text{ when } \dot{\theta} > 0 \\ U_z &= C \,\mathrm{sgn}\,(\theta + \theta_0) \text{ when } \dot{\theta} < 0 \end{aligned}$$

Here the absence of the sun in the instrument field of view is seen as a negative signal.

c) The differential equation of the rotation of the outer gimbal of the two-degrees-of-freedom gyro may be stated in simplified form as: in the yaw plane

$$\ddot{\delta}_y + 2n\dot{\delta}_y + k^2\delta_y + a\,\mathrm{sgn}\,\dot{\delta}_y + b\,\mathrm{sgn}\,\delta_y = m\dot{\psi} \tag{3}$$

in the pitch plane

$$\ddot{\delta}_z + 2n\dot{\delta}_z + K^2\delta_z + a\,\mathrm{sgn}\,\dot{\delta}_z + b\,\mathrm{sgn}\,\delta_z = m(\dot{\vartheta} + \varkappa\Omega_0)$$

where δ is the rotation angle of the outer gimbal ring, and $\varkappa = 0$ when $1 \cdot \mathrm{sgn}\,(\vartheta + \vartheta_0)$ 0 at $\dot{\vartheta} < 0$
$\varkappa = 1$ when $1 \cdot \mathrm{sgn}\,(\vartheta - \vartheta_0)$ 0 at $\dot{\vartheta} > 0$

The output response of the angular rate sensor is of on-off type with a dead band for the rotation angle of the outer gimbal ring (Fig. 7)

$$\begin{aligned} u &= 0 \text{ when } -\delta_1 < \delta < +\delta_1 \\ u &= \mathrm{sgn}\,\delta \text{ when } |\delta| > \delta_1 \end{aligned} \tag{4}$$

d) The retrothrust steering engines may be represented as an on-off operated control unit with a cut-off time lag (Fig. 8):

$$\begin{aligned}
M &= L \begin{cases} \text{when } u > u^* \text{ if } \dot{u} > 0 \\ u > u^* - |\Delta u_\tau| \text{ if } \dot{u} < 0 \end{cases} \\
M &= 0 \begin{cases} \text{when } -(u^* - |\Delta u_\tau|) < u < u^* \text{ if } \dot{u} > 0 \\ u^* - |\Delta u_\tau| > u > -u^* \text{ if } \dot{u} < 0 \end{cases} \\
M &= -L \begin{cases} \text{when } u < -u^* \text{ if } \dot{u} < 0 \\ u < -(u^* - |\Delta u_\tau|) \text{ if } \dot{u} > 0 \end{cases}
\end{aligned} \tag{5}$$

where $\Delta U_\tau = U(t^* + \tau) - U^*$ when $u\dot{u} < 0$, τ is the delay time.

The performance of the retrothrust steering engines may be represented with ease as a function of the system operational logic by making use of equations (2), (4), and (5):

$$M_y = 0 \text{ when } |\delta_y| < \delta_{1y} \begin{cases} \text{when} \\ -\psi_1 < \psi < \psi_2 \;\&\; \dot{\psi} > 0 \\ -\psi_2 < \psi < \psi_1 \;\&\; \dot{\psi} < 0 \\ \text{if } \theta + \theta_0 > 0 \text{ when } \dot{\theta} < 0 \\ \text{or } \theta - \theta_0 > 0 \text{ when } \dot{\theta} > 0 \end{cases}$$

$$M_y = -\frac{L_y}{1+n} \,\mathrm{sgn}\,\delta_y \text{ at } |\delta_y| > \delta_{1y} \begin{cases} \text{for any } \psi, \text{ if} \\ \theta - \theta_0 < 0 \text{ at } \dot{\theta} > 0 \\ \theta + \theta_0 < 0 \text{ at } \dot{\theta} < 0 \end{cases}$$

$$\left.\begin{aligned}
&M_y = 0 \text{ when } |\delta_y| < \delta_{1y} \\
&M_y = -L_y \text{ when } \delta_y > \delta_{1y} - \Delta\delta(\tau) \;\&\; \dot{\delta} < 0 \\
&\qquad\qquad \delta_y > \delta_{1y} \;\&\; \dot{\delta} > 0 \\
&M_y = -\frac{L_y}{1+n} \text{ when } \delta_y < \delta_{1y} \;\&\; \dot{\delta} > 0 \\
&\qquad\qquad \delta_y < \delta_{1y} - \Delta\delta(\tau) \;\&\; \dot{\delta} < 0
\end{aligned}\right\} \begin{cases} \theta + \theta_0 > 0 \;\&\; \dot{\theta} < 0 \\ \theta - \theta_0 > 0 \;\&\; \dot{\theta} > 0 \\ \psi > \psi_2 \;\&\; \dot{\psi} > 0 \\ \psi > \psi_1 \;\&\; \dot{\psi} < 0 \end{cases} \tag{6}$$

$$\left.\begin{aligned}
&M_y = L_y \text{ when } \delta_y < -\delta_{1y} + \Delta\delta(\tau);\; \dot{\delta} > 0 \\
&\qquad\qquad \delta_y < -\delta_{1y} \qquad \dot{\delta} < 0 \\
&M_y = \frac{L}{1+n} \text{ when } \delta_y > -\delta_{1y} \qquad \dot{\delta} < 0 \\
&\qquad\qquad \delta_y > \delta_{1y} + \Delta\delta(\tau) \qquad \dot{\delta} > 0
\end{aligned}\right\} \begin{cases} \theta + \theta_0 > 0 \quad \dot{\theta} < 0 \\ \theta - \theta_0 > 0 \quad \dot{\theta} > 0 \\ \psi < -\psi_2 \quad \dot{\psi} < 0 \\ \psi < -\psi_1 \quad \dot{\psi} > 0 \end{cases}$$

$$\left.\begin{aligned}
&M_z = 0 \text{ when } |\delta_z| < \delta_{1z} \\
&M_z = \frac{L_z}{1+n} \,\mathrm{sgn}\,\delta_z \text{ when } |\delta_z| > \delta_{z1}
\end{aligned}\right\} \begin{cases} \theta - \theta_0 < 0 \;\&\; \dot{\theta} > 0 \\ \theta + \theta_0 < 0 \;\&\; \dot{\theta} < 0 \end{cases} \tag{7}$$

$$\left.\begin{aligned}
&M_z = -L_z \text{ at } \delta_z > \delta_{1z} - \Delta\delta(\tau);\; \dot{\delta} < 0 \\
&\qquad\qquad \delta_z > \delta_{1z} \qquad \dot{\delta} > 0 \\
&M_z = -\frac{L}{1+n};\; -\delta_{1z} < \delta_z < \delta_{1z} - \Delta\delta(\tau) \;\&\; \dot{\delta} < 0 \\
&\qquad\qquad -\delta_{1z} < \delta_z < \delta_{1z} \;\&\; \dot{\delta} > 0 \\
&M_z = 0 \qquad \delta_z < -\delta_{1z}
\end{aligned}\right\} \begin{cases} \theta + \theta_0 > 0 \;\&\; \dot{\theta} < 0 \\ \theta - \theta_0 > 0 \;\&\; \dot{\theta} > 0 \end{cases}$$

The switching conditions are stated in the assumption that the period between pulses T is appreciably longer than the time lag of the rate sensors and actuators.

Systems (1), (3), and (6), (7) comprise a complete system of equations making possible an analysis of the dynamics of the satellite steering in the pitch and yaw channels.

As an example, we have in Fig. 9 a phase portrait of the pitch channel at various numerical values of the system parameters. A characteristic feature of the cycle of self-sustained oscillations about the pitch axis is the cycle asymmetry, a consequence of the system asymmetry. Here, as in the subsequent diagrams, the broken curve denotes presumed departures from the steady-state cycle.

System oscillations in the yaw plane depend on large measure on the system parameters and particularly on the nature of the pitch oscillations. The reason is that the retroengines operate in a double generator fashion in the yaw channel — one mode is specified by the pulse generator, the other by self-sustained pitch oscillations.

Figure 10 shows a cycle of self-sustained yaw oscillations with parameters selected when the self-sustained pitch oscillations exert virtually no effect at all on the yaw.

The phase portrait of the system in the plane of the variables $\vartheta \psi$ (Fig. 11) clearly reflects the trajectory traversed by the sun in the field of vision of the "solar sensor."

Tests of the system confirmed the correctness of the results extracted from theoretical calculations, and their high reliability.

Figure 12 depicts (as an example) the phase portrait of the system in the pitch plane, based on the processing of telemetered in-flight data from Vostok-I, piloted by Yu. A. Gagarin, and received... one minute prior to firing of the retrorocket set.

DISCUSSION

R. F. Appazov,
V. P. Legostaev and B. V. Raouschenbakh

"The Use of
Single-Axis Orientation
for Artificial Earth Satellites".

Q. What are the sensors 1, 2, and 3?

A. Photoresistors.

Q. What is the overlap of the 2 sun sensors?

A. About 1/2 degree.

Q. Are retro rockets fired as a function of time or of sun-earth angle?

A. The sun must be in S_3, the firing is determined by time.

Q. Is the system used in all Vostoks?

A. Yes.

Q. What is the system accuracy?

A. 1 to 1 1/2 degrees.

Q. Is control required during the firing of the braking rocket?

A. The system described in this paper is used only prior to retro firing.

FINE SOLAR ALIGNMENT SYSTEM FOR OPERATION IN STABILISED VEHICLES

W.S. Black and D.B. Shenton,
Culham Laboratory,
U.K.A.E.A.,
Abingdon, Berkshire,
England.

ABSTRACT

Experiments being undertaken by Culham Laboratory, U.K.A.E.A. are carried in the attitude controlled heads of Skylark sounding rockets which point the whole payload at the sun to better than 1 degree. This paper **reports** on the fine alignment system within the experiment which maintains the position of a solar image to better than 1 arc minute. The system comprises a platform carrying a small imaging mirror which directs sunlight on to a split field sensor whose outputs, suitably processed and amplified energise the platform actuators to maintain the required position. Mounted on the same platform is the main imaging component of the experiment which thus maintains a solar image in fine alignment on the equipment.

INTRODUCTION

The research programme of the Culham Laboratory of the United Kingdom Atomic Energy Authority is mainly concerned with plasma physics and nuclear fusion which involves the study of high temperature plasmas in the laboratory. It also includes studies of some of those naturally occurring regions of plasma which are widely distributed throughout the universe. One of these regions is the sun. Here there is considerable interest in the general behaviour of the solar corona and chromosphere, particularly the physical processes by which the corona is heated to a temperature greatly in excess of both the solar surface or photosphere and the chromosphere or zone immediately above the photosphere. Spectroscopic observations of the radiation from these regions give important experimental data needed in these studies. Experience with high temperature laboratory plasmas already has shown the most informative spectral region lies below 3000 Å which requires the use of vacuum ultraviolet spectroscopic instrumentation. It further requires that this must be carried to altitudes where atmospheric absorption of the ultraviolet radiation is no longer significant, thus involving the use of space research. techniques.

On this basis a series of experiments are being undertaken by Culham Laboratory which form part of the British National Space Research Programme. These are designed to use the Skylark high altitude sounding rocket. All the experiments require to be aligned on the sun with varying degrees of accuracy determined by the particular experimental requirements. The Attitude Control Unit[1] is capable of aligning the payload head of a Skylark rocket with the solar vector to better than one degree and also controls the position in roll about the solar vector.

The aim of one of the Culham Laboratory experiments is to set the edge of a solar image on the entrance slit of a spectrograph so as to obtain spatial resolution in the chromospheric and coronal spectra and to prevent the very intense photospheric radiation of the main solar disc from entering the instrument directly. For this experiment the position of the image must be maintained with an accuracy of 1 arc minute or better during any recording interval which may extend over a large part of the effective rocket flight. This requirement has been met by the development of a fine solar alignment system which moves the solar imaging optical component of the experiment so as to compensate for any movement of the head with respect to the solar vector.

GENERAL DESIGN

The design provides for a platform to carry the main imaging component. This is normally a mirror whose reflecting properties vary to suit particular aspects of the experiment but may sometimes be a diffraction grating to give wavelength dispersion. The platform is arranged to tilt about two mutually perpendicular axes over angles greater than the maximum specified pointing errors of the Attitude Control Unit. Ideally a solar image position error detector should be placed at the spectrograph entrance slit but this is precluded by the possibility of a wavelength dispersed image and the complication introduced by varying the reflecting properties of the mirror. Consequently an auxiliary mirror is provided on the same platform whose sole purpose is to form an image on a simple split field error detector situated near the spectrograph entrance slit.

"Superior numbers refer to similarly-numbered references at the end of this paper".

The amplified and modulated error detector output currents drive motors in the mirror platform actuator. The mirror, its platform and suspension are on the ambient pressure side of a bulkhead and the motors, gears, cams and electronics are in a compartment maintained at atmospheric pressure. Under certain conditions of large pointing errors which may arise during acquisition it is possible for a solar image from the main mirror to fall on the error detector and the image from the auxiliary mirror to miss it altogether. A Target Eye is used to prevent the system attempting to lock on to the wrong image. Provision is made for monitoring signals to check performance on the ground and in flight. The power supply available within the rocket is an unregulated battery from which the system inverters and stabilisers are run. A block diagram of the system is shown in Fig. 1.

POSITION SENSORS

Target Eye The general construction is shown in Fig. 2.

A simple lens of 0.84 inches focal length is used to produce a solar image at the focal plane where an aperture is provided whose size is arranged to be about half a degree less than the actual field of view required, which is 3 degrees square in this case. A silicon solar cell whose peak response occurs at 8500 Å is placed behind the aperture to detect the amount of light passing through giving the characteristic in Fig. 3. A red glass filter of passband 7000 Å to 11000 Å placed in front of the cell maintains a more constant solar response from ground level to upper atmosphere conditions and reduces the otherwise significant effects of atmospheric scattering particularly below 5000 Å.

Error Detector

The static split field detector comprises four triangular silicon solar cells, covered by filters, placed in a square array giving a field of about 4 degrees. A solar image falling upon them is divided according to the magnitude and direction of the misalignment from the centre of the array. The reference axes about which the errors are measured can be arranged in two ways as shown in Fig. 4. In the first arrangement axes P and Y are used and the individual cell outputs are fed into auxiliary addition and subtraction stages which resolve the error into two components anywhere within the field. Axes (P+Y) and (P-Y) are used in the second arrangement which is that adopted for the Alignment System. Here diagonally opposite pairs of cells are connected in parallel opposition which permits resolution of errors into two components without the use of auxiliary circuits but over the central portion of the field only and with lower sensitivity. The outer part of the field is used only during acquisition when it is not necessary for control to be exercised in both axes simultaneously. The characteristic curves, Fig. 5, show the response to deviation of the solar image from the reference axis and illustrate the variation in detector sensitivity over the field for both configurations. The sensitivity of the arrangement adopted is very nearly linear and uniform for any misalignments from the centre up to values safely in excess of the required operating accuracy when tracking.

The four cells are selected for matched performance before assembly and to take advantage of the dependence of temperature coefficient on load a low output impedance is provided. The detector sensitivity is about 12 mA per degree and adjustable by varying the nominal f/60 aperture of the imaging mirror. The assembly is held in an adjustable mounting close to the spectrograph entrance slit as shown in Fig. 6.

ELECTRONICS

The appropriate error detector load impedance is obtained by using the input transistor emitter circuit of the error detector differential amplifier. Basically this amplifier is an impedance converter in which the error signal current produces a voltage proportional to the error at the detector. The amplifier output is 3 volts per mA of error current or 36 volts per degree error. For a pointing accuracy of 1 arc minute or 600 mV referred to the output the control of temperature drift in the amplifier is not a severe problem. A satisfactory level is achieved using a dual NPN silicon planar unit enclosed in one envelope as the input stage to reduce the effect of differential heating, the main source of drift. Typical values obtained correspond to less than 1 arc second per degree C.

Examination of the system transfer functions shows that a compromise is necessary to achieve satisfactory response to both steady state tracking and transients. Series compensation is adopted comprising passive networks for phase advance and integration. These are designed for a velocity lag of 1.2 arc seconds at a tracking speed of 1 degree per second and overshoot not exceeding 15% in response to a non saturating step function. Gain and phase margins at the gain crossover frequency are adequate to ensure stability under laboratory, field testing and flight conditions. Allowance is made for the unavoidable variation in error detector sensitivity experienced in the field due to varying atmospheric absorption. The mirror platform actuator drives are standard 2 phase 400 c/s inducation type servo motor. The error detector differential amplifier output is modulated at this frequency using a silicon PNP transistor selected for low leakage current and wide range of operating temperature. The transistor is operated in the inverted condition to reduce the

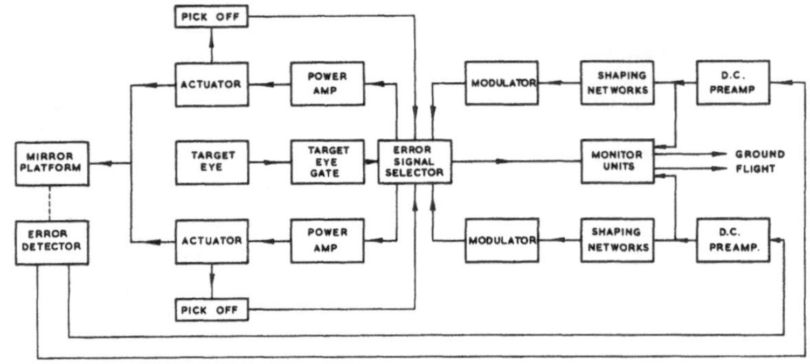

FIG.1. BLOCK DIAGRAM.

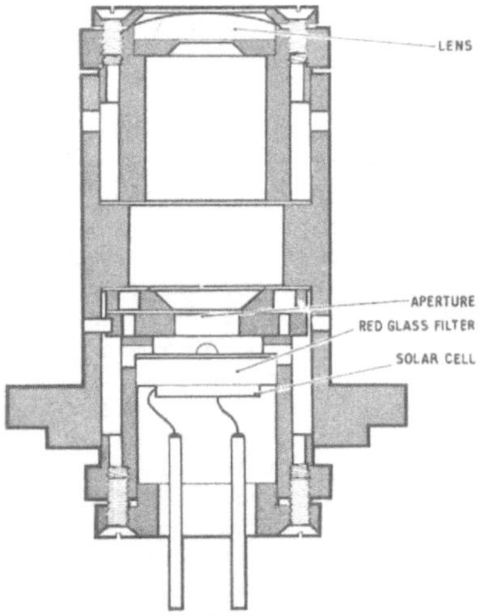

Fig. No 2 TARGET EYE

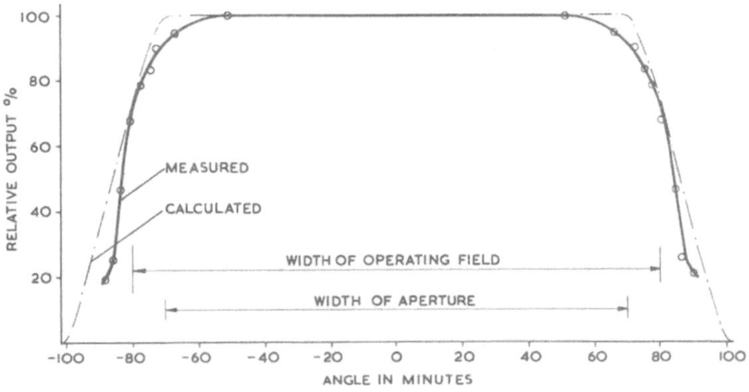

FIG.3 TARGET EYE CHARACTERISTIC

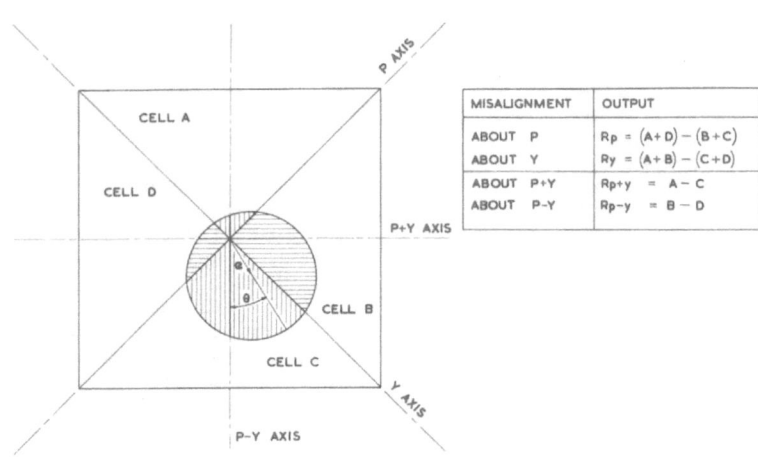

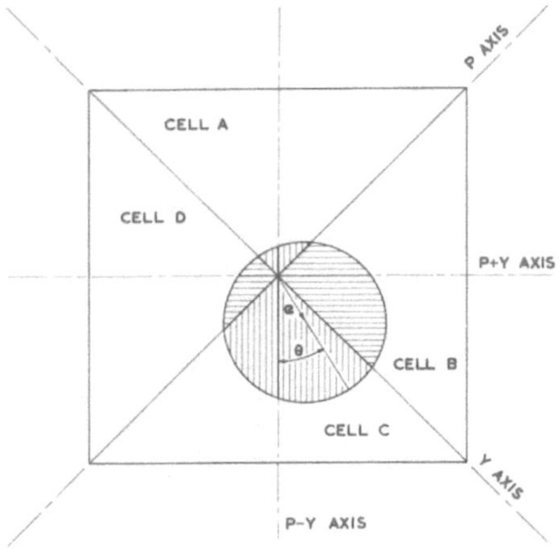

MISALIGNMENT	OUTPUT
ABOUT P	Rp = (A+D) − (B+C)
ABOUT Y	Ry = (A+B) − (C+D)
ABOUT P+Y	Rp+y = A − C
ABOUT P-Y	Rp-y = B − D

Fig. 4 ERROR DETECTOR ARRAY

effect of offset voltage which is thus limited to an error equivalent of 0.5 arc seconds. A square wave formed from a 400 c/s supply reference voltage is applied between base and collector to prevent drifting of the modulator waveform. The power amplifier stages following the modulator provide a 90 degree phase shift to eliminate the need for large phase shift capacitors. The motor reference winding is excited directly from the main power supply, a 400 c/s solid state inverter.

When the target eye is energised, the target eye gate opens and the servo loop is controlled by the error detector. If the sun is outside the target eye field the gate is closed and the servo loop controlled by the signal derived from a pick off on the mirror platform actuator which is thus centred. The target eye gate circuit incorporates an adjustment of the operating level to give fine control of the field width.

Monitoring facilities are provided for ground checkout and in flight telemetry of important functions of the associated experimental equipment in addition to the fine alignment system.

MIRROR PLATFORM AND ACTUATOR

The arrangement of the mirror platform and its actuator is shown in Fig. 7. Ideally the two axes about which the mirror platform rotates should pass through the poles of both mirrors to avoid lateral image movement due to relative movement of the poles but this is clearly not possible with two mirrors mounted on a single platform. One such suspension axis is approximated for the (P+Y) axis which is parallel to the spectrograph slit as greatest accuracy is required for image movement across the slit. To meet local space limitations the other axis (P-Y) is below the first axis and in a plane perpendicular to it passing through the pole of the main mirror. Both axes are also arranged to be perpendicular to the principal reflected ray of the main mirror in its nominal central position. The main and auxiliary mirror principal reflected rays are parallel and the mirror focal lengths are equal. This system minimises the errors due to lateral image movement of the main mirror image with respect to the auxiliary mirror image. These errors are a function of the angle of incidence of the principal rays, the angle through which the platform is required to rotate about both axes away from the nominal central position, the focal length of the mirrors, the distance between their poles and the separation distance of the two axes. The greatest errors are suffered when deviation from the nominal centre position occurs in both axes simultaneously but this does not exceed 2 arc seconds about (P+Y) and 12 arc seconds about (P-Y) for the maximum excursions specified for the Attitude Control Unit.

An arm projecting from the rear face of the mirror platform passes through a flexible bellows vacuum seal in the mounting flange and into the actuator unit. The mirror platform suspension is carried on one side of the flange and the actuator unit on the other so that the latter is in the pressurised body of the rocket. The arm is spring loaded to maintain contact with two cams whose mutually perpendicular axes are in the same planes as the mirror platform suspension axes. Angular movement of the arm in one axis by rotation of one cam results in the arm sliding along the other cam without angular movement. The cam shaft is rotated by a precision gear which engages with a standard motor driven gear reduction head mounted on a swinging plate to give backlash control by adjustment of the depth of gear engagement. Preloaded angular contact ball bearings are used to eliminate cam shaft and mirror platform suspension slackness. The overall reduction ratio is 10^4 so that the system inertia tends to be dominated by the motor and the friction by the first few passes of the gear reduction. Mounted on the end of the cam shafts are A.C. angular displacement pick offs used to sense cam shaft position. Subsidiary microswitch operating cams are also carried on the shafts to prevent the effective polarity reversal caused by cam rotation through more than $\pm$ 90° from its correct central position.

PERFORMANCE

Testing of the fine alignment system in the laboratory requires the use of a solar simulator It is necessary to simulate the radiation intensity at the error detector only allowing the use of a weaker source and a large collecting area of the imaging mirror which is conveniently achieved by using an aluminised main mirror to produce the image on the error detector. The radiation source is a filament projector lamp which illuminates a circular aperture simulating the solar disc. The aperture lies in the focal plane of the projection lens so producing an approximately parallel beam.

Before flight the experimental assembly, Fig. 8 is operated in full sunlight, Fig. 9. The position of the error detector is then adjusted until the solar image at the spectrograph slit observed with a low power microscope is in the required position. Measurement of the mirror platform angular position when the Attitude Control Unit is zeroed is made at this time by means of the calibrated cam shaft pick offs. Records obtained during the flight of the first stabilised Skylark vehicle from Woomera on August 17th 1964[(2)] show that the fine alignment system completed acquisition of the solar image within one second of the sun appearing in the target eye field, and retained

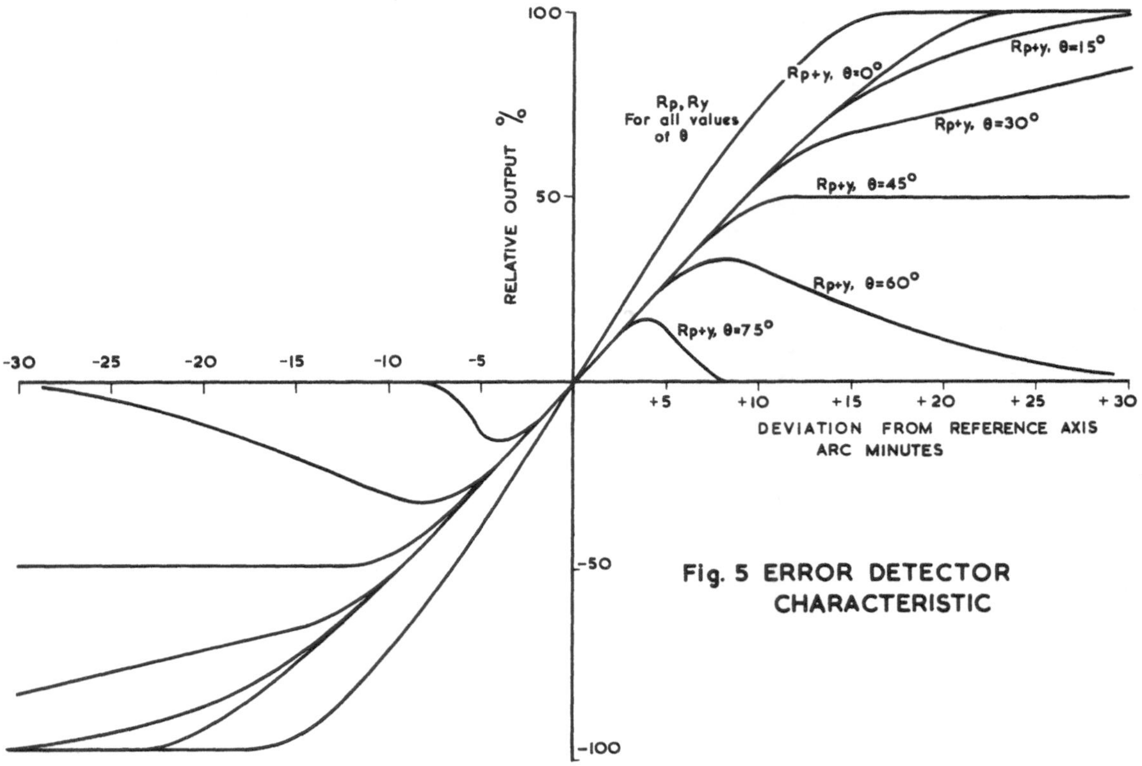

Fig. 5 ERROR DETECTOR CHARACTERISTIC

Fig. 6. Error detector and spectrograph entrance slit.

control over a flight period of about four minutes, with errors not exceeding 12 arc seconds peak to peak. A typical part of the flight telemetry record is shown in Fig. 10 in which it is seen that errors do not exceed 8 arc seconds peak to peak in either axis, and the RMS value does not exceed 2 and 3 arc seconds for the (P+Y) and (P-Y) axes respectively. The mirror platform position signal indicates the movements of the whole rocket head about the two lateral axes, these movements being 6 and 8 arc minutes respectively.

Fig. 7. Mirror platform and actuator.

CONCLUSIONS

The general aim of aligning a solar image to better than 1 arc minute has been achieved with this design using mainly conventional components to limit the amount of development work. However, the ultimate performance is limited by backlash and friction inherent in the gear trains and cams used and there is little room available for significant improvement. An alternative system which avoids these problems has been under development in which the platform is carried in flexure pivots and actuated by windings carried directly on the platform which are in the field of a permanent magnet.

ACKNOWLEDGEMENTS

Many persons have contributed to the development of this alignment system and the flight programmes. We especially wish to acknowledge the encouragement and support for the whole experimental programme of Dr. J.B. Adams, Director; Dr. R. Wilson, Spectroscopy Division Head; and the assistance of Mr. J.G. Firth and other members of the Natural Plasma Group, all of Culham Laboratory, U.K.A.E.A. Our acknowledgements are also due to Elliott Brothers, Frimley, for the Attitude Control Unit; the Royal Aircraft Establishment, Farnborough and the Weapons Research Establishment, Australia for the Skylark rockets SL 301 and 302 and their successful flights.

Fig. 8. Experimental assembly and rocket nose cone.

REFERENCES

(1) Cope, P.E.G., J. British Interplanetary Soc., Vol. 19, No. 7, p.285-291, 1964.

(2) Wilson, R., Annales D'Astrophys., Vol. 27, p. 769, 1964.

Fig. 9. Pre-flight operation check.

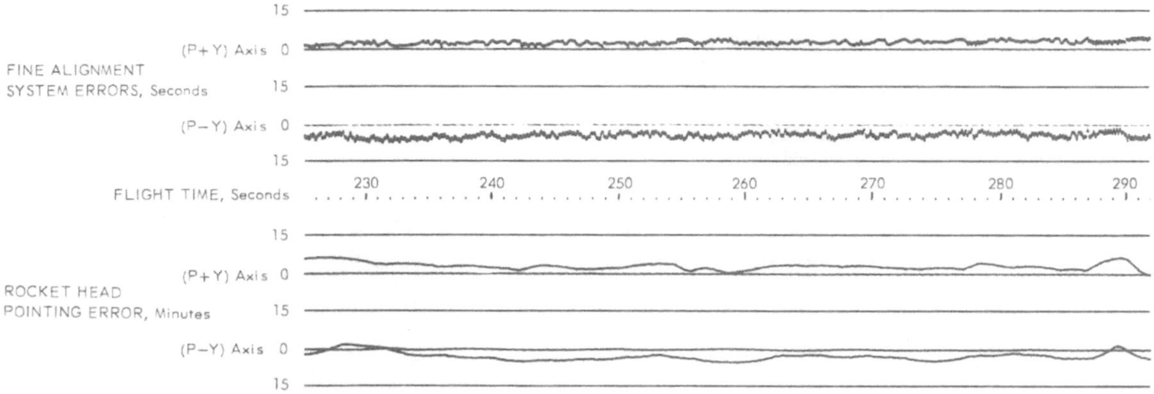

Fig. 10 FLIGHT TELEMETRY RECORD

DISCUSSION

D. B. Shenton and W. S. Black

"Fine Solar Alignment System for Operation in Stabilized Vehicles".

Q. How is the rocket head controlled?

A. The control system consists of coarse and fine solar sensors which actuate nitrogen jets for pitch and yaw. Roll is controlled by a magnetometer and nitrogen jets. The head which weighs 450 pounds, can be aligned to within 20 arc-seconds of the sun -- this has been demonstrated on three flights.

Attitude Control

ON ONE CLASS OF SYSTEMS FOR CONTROLLING THE ORIENTATION OF ARTIFICIAL EARTH SATELLITES

by V. A. Bodner, K. V. Alekseev, and G. G. Bebenin

ABSTRACT

Combination systems of control are used for orienting satellites designed for protracted flight times. The best control for satellites travelling in relatively low orbits is provided by control systems in which the load on a flywheel is imposed by a magnetic torquer. Investigation of such systems (1, 2, 3, 7) has explained the most characteristic properties of practical application of such systems, their design, and their construction.

Many problems of design and computation of combination systems with magnetic drives for the case of three-dimensional orientation of satellites, however, require further investigation. The greatest difficulties in such investigations are due to the difficulty of combining the processes of controlling angular motion and loading of flywheels in a general dynamic pattern. Another important problem is synthesis of flywheel systems characterized by control-channel crosstalk.

The present paper is devoted to carrying investigation of these problems to the point required for indicating possible methods for solving them. Since the article is written from the theoretical viewpoint, the satellites under discussion are not assigned concrete purposes. The requirements imposed on the system, however, are quite specific:

1) Minimum mean-square error in satellite orientation;

2) Minimum energy consumption for torquing flywheels.

The same assumptions as adopted in (3) lie at the base of the investigation.

1. THE EQUATION OF MOTION

We introduce two right coordinate systems with a common origin at the center of mass of the satellite: an orbital system $Ox_0y_0z_0$ and a system, $Oxyz$, rigidly tied to the satellite. We direct the Oy_0 axis along the vertical in the direction away from the earth, and the Oy axis perpendicular to the orbital plane along the vector $\bar{\nu}_0$ of the orbital angular velocity of the satellite. The axes of both systems coincide when the satellite is oriented in the assigned position. In the general case, the orientation of the system $Oxyz$ relative to the axes of the orbital system $Ox_0y_0z_0$ is given by the Euler angles ε_x, ε_y, and ε_z, which, following aviation terminology, we will henceforth call the bank, yaw, and pitch angles, respectively.

The transformation from the $Oxyz$ system to the $Ox_0y_0z_0$ system is given by the matrix expression

$$[x_0, y_0, z_0]^T = D[x, y, z]^T, \tag{1.1}$$

where, for small deviations of the satellite from the oriented position,

$$D = \begin{Vmatrix} 1 & -\varepsilon_z & \varepsilon_y \\ \varepsilon_z & 1 & -\varepsilon_x \\ -\varepsilon_y & \varepsilon_x & 1 \end{Vmatrix} \tag{1.2}$$

(here and henceforth the superscript T denotes transposition).

Our hypothetical satellite is oriented by means of three flywheels. Denoting the resultant angular momentum of the flywheels by $\bar{H}$, we have

$$\bar{H} = H_x i + H_y j + H_z k, \tag{1.3}$$

where H_x, H_y, and H_z are the angular momenta of the gyroscopes along the axes of the associated coordinate system, while i, j, and k are unit vectors.

In order to guarantee a given satellite orientation, the angular momentum vectors of the flywheels are changed in accordance with control rules adopted for this purpose.

The differential equation for the angular motion of the satellite in the $Oxyz$ coordinate system is of the form

$$\frac{d}{dt}(\bar{\bar{I}}\cdot\bar{\omega}+\bar{H}) + \bar{\omega}\times(\bar{\bar{I}}\cdot\bar{\omega}+\bar{H}) = \bar{M}, \tag{1.4}$$

where $\bar{\bar{I}} = \|J_x, J_y, J_z\|$ is a diagonal matrix composed of the moments of inertia of the satellite,

$\bar{\omega} = |\omega_x, \omega_y, \omega_z|$ is the vector of the absolute angular velocity of the satellite,

$\bar{M}$ is the principal vector of the external moment.

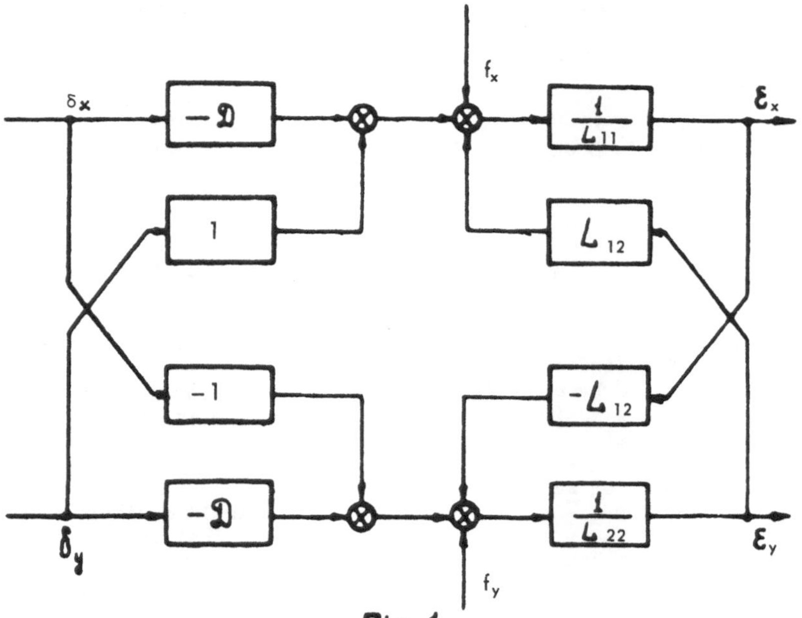

f_x
δ_x
$-\mathcal{D}$
$\frac{1}{L_{11}}$
ε_x
1
L_{12}
-1
$-L_{12}$
$-\mathcal{D}$
$\frac{1}{L_{22}}$
δ_y
ε_y
f_y

Fig. 1

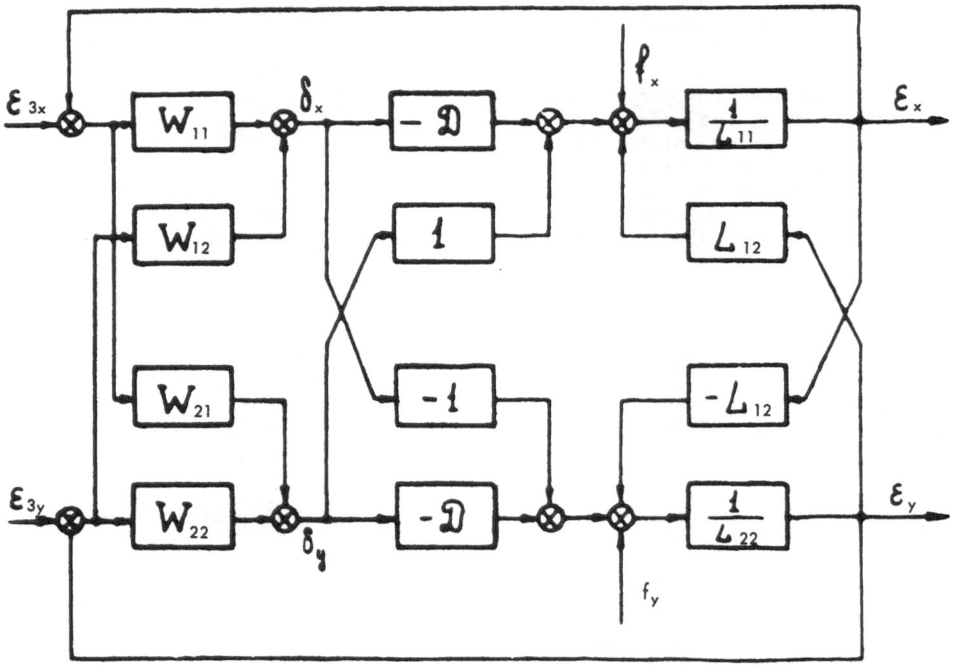

ε_{3x}
W_{11}
δ_x
$-\mathcal{D}$
f_x
$\frac{1}{L_{11}}$
ε_x
W_{12}
1
L_{12}
W_{21}
-1
$-L_{12}$
ε_{3y}
W_{22}
$-\mathcal{D}$
$\frac{1}{L_{22}}$
ε_y
δ_y
f_y

Fig. 2

The external moment is created by gravity, the electric and magnetic fields of the earth, pressure due to solar radiation, forces due to aerodynamic deceleration, etc. From this moment we eliminate the gravitational moment, whose components along the axes of the system Oxyz are known (6). We will assume that the remaining part of the moment, which we will call the perturbing moment $\bar{M}_b$, is a stationary random function of time.

Then, if we write the vector for the angular velocity of the satellite

$$\bar{\omega} = \bar{\omega}_0 + D^T \nu_0$$

where $\bar{\omega} = [\dot{\varepsilon}_x, \dot{\varepsilon}_y, \dot{\varepsilon}_z]$ is the vector of the relative angular velocity, we can write Equation (1.4) in scalar form in terms of the projections on the axes of the associated coordinate system:

$$\begin{aligned} &J_x \ddot{\varepsilon}_x + [4\nu_0^2(J_z - J_y) + H_z \nu_0]\varepsilon_x - [\nu_0(J_x + J_y - J_z) - H_z]\dot{\varepsilon}_y - \\ &\quad - H_y \dot{\varepsilon}_z = -\frac{dH_x}{dt} + H_y \nu_0 + M_{bx} \\ &J_y \ddot{\varepsilon}_y + [\nu_0^2(J_z - J_x) + H_z \nu_0]\varepsilon_y + [\nu_0(J_x + J_y - J_z) - H_z]\dot{\varepsilon}_x + \\ &\quad + H_x \dot{\varepsilon}_z = -\frac{dH_y}{dt} - H_x \nu_0 + M_{by} \\ &J_z \ddot{\varepsilon}_z + [3\nu_0^2(J_x - J_y)]\varepsilon_z - (H_y \dot{\varepsilon}_x - H_x \nu_0 \varepsilon_x) - (H_x \dot{\varepsilon}_y - H_y \nu_0 \varepsilon_y) = \\ &\quad = -\frac{dH_z}{dt} + M_{bz} \end{aligned} \qquad (1.5)$$

where M_{bx}, M_{by}, and M_{bz} are the components of the vector of the perturbing moment.

It follows from the system of equations that we have obtained that crosstalk exists in the control channel both between the controlled and control coordinates. This circumstance greatly complicates analysis and synthesis of control systems. These equations can be simplified, however, if we load each flywheel not when it achieves its limiting angular momentum, but in proportion to the inclination of this last. The best results will be obtained when the loading is such that the constant components of the angular momenta for the bank and yaw channels are equal to zero and the variable components (the average deviations from the constant values) are so small that terms of the form $H_y \varepsilon_z$, $H_x \dot{\varepsilon}_z$, $H_x \nu_0 \varepsilon_y$, etc., in Equations (1.5) can be neglected.

As far as the pitch channel is concerned, it is desirable to load the flywheel relative to some constant value of H_{zo} that is not equal to zero in order to increase system rigidity and decrease the crosstalk with the bank and yaw channels (see Equations (1.5)).

The deviation of H_z from this value is also assumed to be small. The terms $H_y \nu_0$ and $H_x \nu_0$ in the right sides of Equations (1.5) are abandoned because, in comparison with them, the rates of change of the angular momenta H_x and H_y may be of the same order.

The above considerations can be used as the physical reasoning justifying linearization of Equations (1.5) and, at the same time, provide the technical requirements for loading of the flywheels. Assuming that the magnetic torquer realizes these requirements, we rewrite Equations (1.5) in the form

$$\begin{aligned} L_{11}(D)\varepsilon_x + L_{12}(D)\varepsilon_y &= -D\delta_x + \delta_y + f_x \\ -L_{12}(D)\varepsilon_x + L_{22}(D)\varepsilon_y &= -\delta_x - D\delta_y + f_y \\ L_{33}(D)\varepsilon_z &= -D\delta_z + f_z \end{aligned} \qquad (1.6)$$

where

$$L_{11}(D) = D^2 + 4(\sigma + 1) + z_0, \quad L_{33}(D) = \beta D^2 + 3(1-\alpha),$$

$$L_{12}(D) = (\sigma + z_0)D,$$

$$L_{22}(D) = \alpha D^2 + \sigma + \alpha + z_0,$$

$$\alpha = \frac{J_y}{J_x}, \quad \beta = \frac{J_z}{J_x}, \quad \sigma = \beta - \alpha - 1, \quad D = \frac{1}{\nu_0}\frac{d}{dt},$$

$$\delta_i = \frac{H_i}{J_x \nu_0}, \quad f_i = \frac{M_{bi}}{J_x \nu_0} \quad (i = x, y, z) \quad z_0 = \frac{H_{zo}}{J_x \nu_0}$$

Analysis of Equations (1.6) permits us to conclude that the dynamic properties of a satellite with flywheels are equivalent, with respect to the bank and yaw channels, to the dynamics of a two-channel system with crossed feedforward and feedback. (Figure 1).

2. SYNTHESIS OF THE CONTROL LOOP FOR A SYSTEM WITH FLYWHEELS

Choosing the parameters for the pitch-control system, as consideration of the last Equation (1.6) shows, is the usual problem of automatic control theory. We will therefore omit its solution here, and turn to choosing structures and parameters for the bank and yaw systems. The gyroscopic relationship between the parameters of the angular movements of a satellite and the angular momenta of the flywheels determine the physical characteristics of the control for these channels. This relationship greatly complicates analytic solution of the problem of system synthesis and, as we will show below, permits us to obtain concrete results only in special cases. In general, however, additional computer-aided investigations are necessary.

We take a flywheel-control law of the form (4)

$$\begin{aligned} \delta_x &= W_{11}(\varepsilon_{xz} - \varepsilon_x) + W_{12}(\varepsilon_{yz} - \varepsilon_y) \\ \delta_y &= W_{21}(\varepsilon_{xz} - \varepsilon_x) + W_{22}(\varepsilon_{yz} - \varepsilon_y) \end{aligned} \qquad (2.1)$$

where W_{ik} $(i, k = 1, 2)$ is the transfer function of the control part of the system, which is to be determined, and

ε_{xz}, ε_{yz} are given values of the angular deviations of the satellite with respect to bank and yaw.

In general, the signals ε_{xz} and ε_{yz} include useful components ε_{xm} and ε_{ym} and noise ε_{xn} and ε_{yn}, i.e.,

$$\begin{aligned} \varepsilon_{xz} &= \varepsilon_{xm} + \varepsilon_{xn} \\ \varepsilon_{yz} &= \varepsilon_{ym} + \varepsilon_{yn} \end{aligned} \qquad (2.2)$$

Figure 2 shows a block diagram for a system using control (2.1). We choose W_{ik} $(i,k=1,2)$ by requiring compensation of the crossed feedforward and feedback in the control channels. It is easy to show that when

$$-W_{12}D + W_{22} = L_{12},\qquad W_{21}D + W_{11} = L_{12} \tag{2.3}$$

the left sides of Equations (1.6), which describe the angular movements of the satellite, become independent:

$$\begin{aligned}
[L_{11} + W_{11}D - W_{21}]\varepsilon_x &= -(W_{11}D - W_{21})\varepsilon_{xz} - \\
&\quad -(W_{12}D - W_{22})\varepsilon_{yz} + f_x, \\
[L_{22} + W_{22}D + W_{12}]\varepsilon_y &= -(W_{21}D + W_{11})\varepsilon_{xz} - \\
&\quad -(W_{22}D + W_{12})\varepsilon_{xz} + f_y.
\end{aligned} \tag{2.4}$$

Using Equations (2.3) to find expressions for W_{11} and W_{12} and substituting these values into (2.4), we obtain the following characteristic equations for the system:

for the bank channel:

$$D^2(1+\sigma+z_0-W_{21}) + (4\sigma+4+z_0-W_{21}) = 0$$

for the yaw channel:

$$(D^2+1)(\sigma+\alpha+z_0+W_{12}) = 0$$

The two purely imaginary roots of the last equation indicate that conditions for neutral stability exist independently of the choice of W_{12}. Thus, even though the left sides of Equations (1.5) are "decoupled" in the case of the control we have chosen, it is impossible to control an angular movement of a satellite relative to yaw.

We will supplement controls (2.1) by terms that will make it possible to eliminate the crossed feedforward and feedback.

The following controls satisfy this requirement (5):

$$\begin{aligned}
\delta_x &= W_{11}(\varepsilon_{xz}-\varepsilon_x) + W_{12}(\varepsilon_{yz}-\varepsilon_y) + \frac{\delta_y}{D}, \\
\delta_y &= W_{21}(\varepsilon_{xz}-\varepsilon_x) + W_{22}(\varepsilon_{yz}-\varepsilon_y) - \frac{\delta_x}{D}
\end{aligned} \tag{2.5}$$

After substitution into (1.5), assuming that

$$W_{12} = -\frac{L_{12}}{D},\qquad W_{21} = \frac{L_{12}}{D} \tag{2.6}$$

we find that

$$\begin{aligned}
(L_{11}+W_{11})\varepsilon_x &= W_{11}\varepsilon_{xz} - \frac{L_{12}}{D}\varepsilon_{yz} + f_x, \\
(L_{22}+W_{22})\varepsilon_y &= \frac{L_{12}}{D}\varepsilon_{xz} - W_{22}\varepsilon_{zy} + f_y
\end{aligned} \tag{2.7}$$

Setting

$$W_{11} = a_{11} + a_{10}\frac{1}{D}$$
$$W_{22} = a_{22} + a_{20}\frac{1}{D}$$

and $\varepsilon_{xz} = \varepsilon_{yz} = 0$ (stabilization mode), we can rewrite (2.7) in the form

$$\begin{aligned}
[D^2 + a_{11}D + (4\sigma+z_0+4+a_{10})]\varepsilon_x &= f_x \\
[\alpha D^2 + a_{22}D + (\sigma+\alpha+z_0+a_{20})]\varepsilon_y &= f_y
\end{aligned} \tag{2.8}$$

The parameters a_{10}, a_{11}, a_{20}, and a_{22} are chosen by means of ordinary methods. At first sight, control (2.5) corresponds to the demands we have presented to the control system. They ensure both independent control in the different channels and the proper transient process.

Analysis of the possibility of practically realizing this control shows that the control moments are related by the following equations to the parameters of the angular movement of the satellite:

$$D^2\delta_x + \delta_x = [W_{11}D^2 + W_{21}D]\varepsilon_x + [W_{12}D^2 + W_{22}D]\varepsilon_y$$
$$D^2\delta_y + \delta_y = [W_{21}D^2 - W_{11}D]\varepsilon_x + [W_{22}D^2 - W_{12}D]\varepsilon_y$$

If we now use Equations (2.8) to eliminate ε_x and ε_y, we obtain a relationship between the control moments and the perturbing moments acting on the satellite. Analysis of the equations thus obtained shows that in the presence of certain types of influences (unit step, pulse, etc.), the system operates satisfactorily. When the satellite is subject to perturbing moments in the form of stationary random functions and certain types of periodic disturbances with frequency equal to or a multiple of the rotation frequency, the angular momenta of the flywheels increase.

We will now consider a more general approach to system synthesis, this time eliminating the requirements toward control-channel autonomy. In order to do so, we write Equations (1.5) in matrix form:

$$L\varepsilon = p\delta + f, \tag{2.9}$$

where

$$L = \begin{Vmatrix} L_{11} & L_{12} \\ -L_{12} & L_{22} \end{Vmatrix},\qquad p = \begin{Vmatrix} -D & 1 \\ -1 & -D \end{Vmatrix},$$

$$\varepsilon = \begin{Vmatrix} \varepsilon_x \\ \varepsilon_y \end{Vmatrix},\qquad \delta = \begin{Vmatrix} \delta_x \\ \delta_y \end{Vmatrix},\qquad f = \begin{Vmatrix} f_x \\ f_y \end{Vmatrix}$$

and

$$\delta = W(\varepsilon_z - \varepsilon) \tag{2.10}$$

where

$$W = \begin{Vmatrix} W_{11} & W_{12} \\ W_{21} & W_{22} \end{Vmatrix},\quad \varepsilon_z = \begin{Vmatrix} \varepsilon_{xz} \\ \varepsilon_{yz} \end{Vmatrix} = \begin{Vmatrix} \varepsilon_{xm} \\ \varepsilon_{ym} \end{Vmatrix} + \begin{Vmatrix} \varepsilon_{xn} \\ \varepsilon_{yn} \end{Vmatrix}$$

Equation (2.9) immediately implies that

$$\varepsilon = yp\delta + yf$$

where

$$y = \frac{L^*}{\Delta(D)}$$

L^* is the adjoint of $L(D)$, and

$\Delta(D)$ is the determinant of $L(D)$.

It is characteristic that the matrix transfer functions of the satellite for the control (CM) and perturbing (PM) moments are different. Figure 3 shows the schematic diagram for this system. It follows from Figure 3 that

$$\varepsilon = y p W(\varepsilon_z - \varepsilon) + y f$$

or

$$\varepsilon = (E + y p W)^{-1} y p W \varepsilon_z + y (E + y p W)^{-1} f \quad (2.12)$$

where E is the identity matrix.

We will now find the system error,

$$\Delta\varepsilon = \varepsilon_m - \varepsilon \quad (2.13)$$

where $\varepsilon_m = \begin{Vmatrix} \varepsilon_{xm} \\ \varepsilon_{ym} \end{Vmatrix}$ or, when (2.12) is taken into account,

$$\Delta\varepsilon = \varepsilon_m - [E + y p W]^{-1} y p W \varepsilon_m - [E + y p W]^{-1} y p W \varepsilon_n - y (E + y p W)^{-1} f,$$

where $\varepsilon_n = \begin{Vmatrix} \varepsilon_{xn} \\ \varepsilon_{yn} \end{Vmatrix}$

Since $E - [E + y p W]^{-1} y p W = [E + y p W]^{-1}$, we have

$$\Delta\varepsilon = [E + y p W]^{-1} \varepsilon_m^* - [E + y p W]^{-1} y p W \varepsilon_n, \quad (2.14)$$

where

$$\varepsilon_m^* = \varepsilon_m - y f$$

We will call an input signal equal to $\varepsilon_z^* = \varepsilon_m^* + \varepsilon_n$ normalized.

A schematic diagram for the system is shown in Figure 4. Although this diagram has been obtained by formal transformation from the conditions for its equivalent circuit, which is shown in Figure 3, there are fundamental differences between them. While the signal components ε_m and ε_n of the normalized input signal will nonetheless be correlated. If we let

$$\phi(D) = [E + y p W]^{-1} y p W \quad (2.15)$$

denote the closed-system matrix transfer function, expression (2.14) takes the form

$$\Delta\varepsilon = [E - \phi(D)] \varepsilon_m^* - \phi(D) \varepsilon_n \quad (2.16)$$

It then follows that the reproduction error is equal to the sum of two errors: the dynamic error $[E - \phi(D)] \varepsilon_m^*$ and the error $\phi(D) \varepsilon_n$ due to noise. The dynamic error decreases as the response of the system becomes faster, and reaches zero as $\phi \to E$. In this case, however, the noise increases. If $\phi \to 0$, the noise approaches zero, but the dynamic error increases.

We will assume that the expression for $\phi(D)$ that insures minimum mean-square-error and reproduction of the given conditions for the maneuver is defined (below we will discuss possible methods for determining it). Then the problem of synthesis consists in finding the matrix transfer function of the correcting device W(D), for which we can obtain an expression from (2.15):

$$W(D) = p^{-1} y^{-1} \phi(D) (E - \phi(D)]. \quad (2.17)$$

As a result, in order to determine W(D) it is necessary to know dynamic properties of the satellite as an object of control ($y = k^*/\Delta(D)$), the characteristics of the crosstalk between the controlled moments (p), and the properties of the closed-loop system (ϕ (D)). It is clear that ϕ(D) depends on the properties of the signals ε_m^* and ε_n applied to the system inputs.

In order to determine ϕ(D), the control system shown in Figure 4 is given with some filter with a matrix transfer function ϕ^*(D) such that

$$\varepsilon^* = \phi^*(D) \varepsilon_z^*$$

where the ϕ_{ij}^* $(i, j = 1, 2)$ are selected with allowance for the requirements toward system accuracy.

The error introduced by the filter in reproduction of the useful signal is

$$\Delta\varepsilon^* = (E - \phi^*(D) \varepsilon_m^* - \phi^*(D) \varepsilon_n \quad (2.18)$$

Assumed that ϕ^*(D) is selected so that minimum root-mean-square error $\sqrt{\overline{\varepsilon_\rho^{*2}}}$ $(\rho = 1, 2)$ is obtained.

When the condition

$$\phi(D) = \phi^*(D)$$

is satisfied, expressions (2.18) and (2.16) are identically equal. Depending on the perturbations in the system, different forms of the matrix ϕ^*(D) can appear. In particular (11),

$$\phi^*(D) = \begin{Vmatrix} \frac{K_{11}}{D + K_{11}} & 0 \\ 0 & \frac{K_{22}}{D + K_{22}} \end{Vmatrix}$$

The values of K_{11} and K_{22} are determined from the conditions requiring that the errors

$$\Delta\varepsilon_x = \varepsilon_{xm} - \varepsilon_x$$

$$\Delta\varepsilon_y = \varepsilon_{ym} - \varepsilon_y$$

in the channels of the system do not exceed given values.

We will now give one possible way of realizing a correcting device W(D) for the case in which the satellite is subject to random perturbations (9, 10). In order to do so, we use (2.17) to substitute an expression for W(D) into control (2.10):

$$\delta = p^{-1} y^{-1} \phi(D) [E - \phi(D)]^{-1} e(t) \quad (2.20)$$

where $e(t) = \varepsilon_z - \varepsilon$

It now follows that

$$y p \delta = \phi(D) [E - \phi(D)]^{-1} e(t) \quad (2.21)$$

We now determine the right side of Equation (2.21), for which we write

$$\zeta(t) = [E - \phi(D)]^{-1} e(t)$$

or, in integral form,

$$\zeta(t) - \int_0^t K(t - \tau) \zeta(t) d\tau = e(t) \quad (2.22)$$

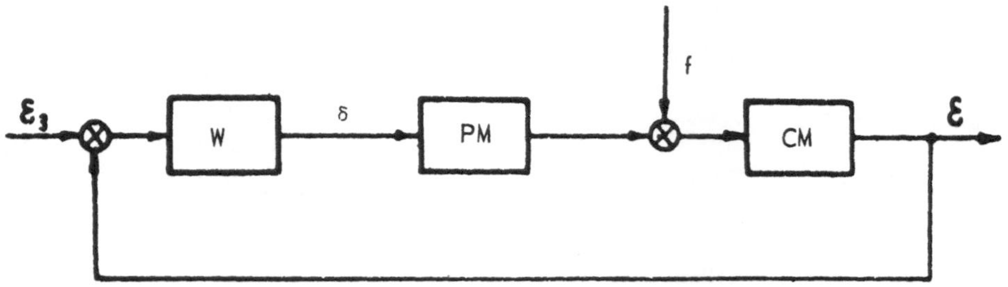

Fig. 3

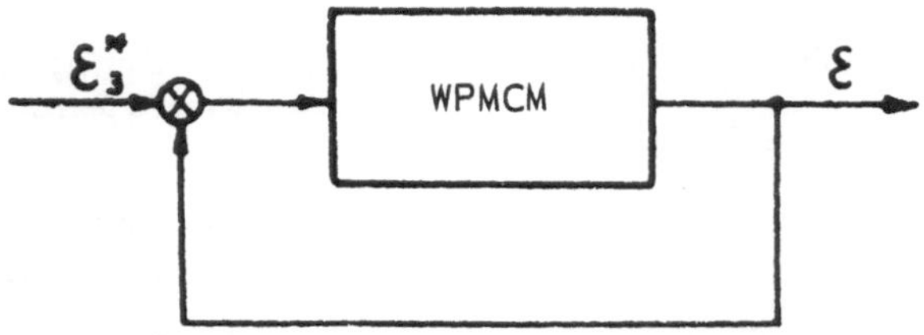

Fig. 4

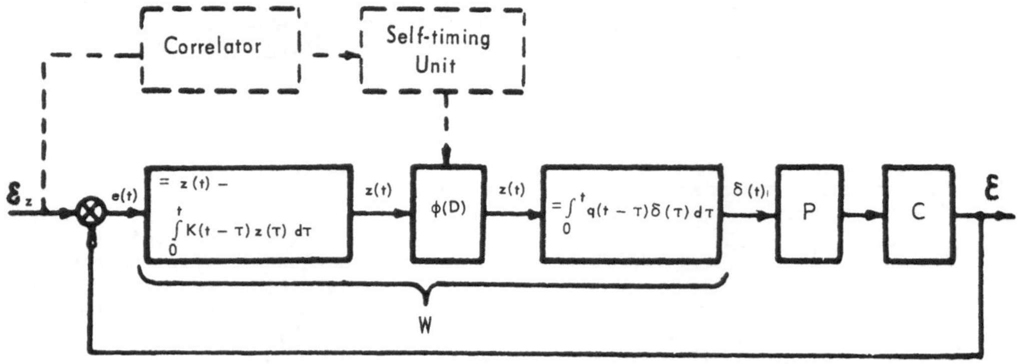

Fig. 5

where $K(t-\tau)$ is a weighting function corresponding to the transfer function $\Phi(D)$.

A computer is used to solve Equation (2.22) for $\zeta(t)$ ($e(t)$ is known). After this, $\zeta(t)$ is applied to the input of a filter with transfer function $\Phi(D)$; at the output we obtain

$$\xi(t) = \Phi(D)\zeta(t)$$

Returning to Equation (2.21), we find that

$$yp\delta = \xi(t)$$

or, in integral form,

$$\int_0^t q(t-\tau)\delta(\tau)d\tau = \xi(t) \qquad (2.24)$$

where $q(t-\tau)$ is a weighting function corresponding to the transfer function of the object of control (CM).

Solution of Equation (2.24) by a second computer (see Figure 5) makes it possible to determine the control moments that must be applied to the satellite in the different control channels.

It is not necessary to explain the procedure for finding an optimal filter $\Phi^*(D)$. We will note that evaluation of $\Phi^*(D)$ must precede correction of the dynamic properties of the satellite such as artificial damping of angular oscillation. This last can be accomplished either by means of a high-speed gyroscope or by the damping properties of a closed loop on the satellite (3).

When the form of the input signals to the filter changes, $\Phi^*(D)$ must change, and even the structure of the filter may require modification, i.e., the filter must be self-adaptive. Such a filter must be used because the system is subject to the influence of perturbations whose spectral composition is difficult to predict. Realization of an optimal filter under such conditions may be accomplished by means of a self-tuning circuit (8).

An optimal filter for the practically important case of an input signal having the form of a low-frequency process and white noise of constant intensity consists of inertial sections. A system (Figure 5) with input-signal correlation can be effectively used for tuning such a filter.

3. SYNTHESIS OF CONTROL LOOPS FOR SYSTEMS WITH MAGNETIC TORQUERS

The physical properties of a magnetic device for torquing in orientation systems are determined by the following factors: 1) variation in the position of the induction vector of the earth's magnetic field in space as the satellite travels along an orbit; 2) crosstalk of moments applied by magnetic torquers to the satellite through control channels.

When a magnetic torquer is used as the fundamental control device in a system, the indicated properties do not permit three-dimensional orientation of a satellite (3). Difficulty in pitch orientation appears when a continuous method of magnetically torquing control devices is used. When a pulse method is used, the possibility of generating inadmissibly large orientation errors is not eliminated, since application of control pulses to the satellite depends not on the state of the system, but on the position of the satellite in an orbit.

Nonetheless, while application of magnetic torquers alone does not answer the requirements of three-dimensional orientation of a satellite, in combination with other devices it provides a rather effective torquing device that eliminates "saturation" of flywheels in all control channels. Two methods of loading flywheels by means of magnetic torquers are possible: continuous and pulse. In the first method the current in the magnetic-torquer coils is controlled by special computers whose inputs receive information about the angular velocities of the flywheels and the direction of their rotation, and information about the components of the magnetic induction of the earth's field along the axes of the related coordinate system. Avoiding the engineering problems of constructing such a system, we will note only that this method is characterized by the additional requirement of energy for artificially maintaining the vectors $\bar{H}$ and $\bar{B}$ perpendicular to each other.

In the pulse method of loading, the torquers are switched on at completely determined positions of the satellite in an orbit. Computation of the required impulse for the loading moment is a relatively simple operation. The necessity of predicting variations in the components of the magnetic-induction vector on the axes along which they are measured greatly complicates actual construction of the requisite computers. The main advantage of this method is its optimality in the sense of energy requirements. We will now use the requirement toward system control to consider a pulse method in more detail.

The approximate expressions for the components of the magnetic-induction vector of the earth's field on the axes of the related coordinate system are of the form (7)

$$B_x = E\cos(u-x)\sin\xi_m$$
$$B_y = -2E\sin(u-x)\sin\xi_m$$
$$B_z = E\cos\xi_m$$

where E is the length of the vector of the earth's magnetic field,

ξ_m is the variable inclination of the orbit relative to the geomagnetic equator,

x is the angle between the ascending node relative to the earth's equator and the ascending node relative to the geomagnetic equator;

$u = \nu_0 t$ is the argument of latitude.

Independent application of momentum pulses is applied by torquers at times when either Bx or By is equal to zero.

Assume, for example, that during a time interval τ_x

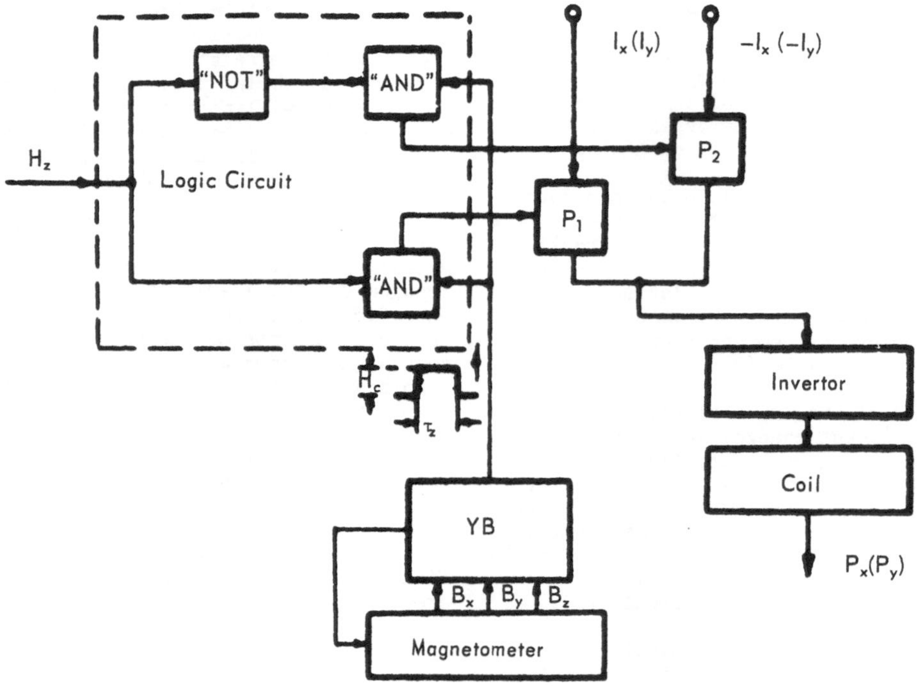

"NOT"
"AND"
$I_x (I_y)$
$-I_x (-I_y)$
H_z
Logic Circuit
P_2
P_1
"AND"
H_c
τ_z
Invertor
Coil
YB
$P_x(P_y)$
B_x
B_y
B_z
Magnetometer

Fig. 6

$$-\Delta B_x \le B_x \le \Delta B_x$$

It follows from the expressions for the moment due to the torquer that when $P_x = 0$

$$M_x = P_y B_z - P_z B_y$$
$$M_y = P_z B_x$$
$$M_z = P_y B_x$$

When the times at which the current through the torquer are symmetric relative to the time at which $B_x = 0$ (P_y and P_z remain constant), the torque pulses satisfy, when the smallness of ΔB_x is taken into account, the equation

$$\Delta m_y = \int_{-\frac{\tau_x}{2}}^{\frac{\tau_x}{2}} \Delta m_y \, dt = 0$$

while

$$\Delta m_z = \int_{-\tau_x/2}^{\tau_x/2} \Delta m_z \, dt = 0,$$

$$\Delta m_x = \int_{-\tau_x/2}^{\tau_x/2} M_x \, dt \neq 0$$

Similar relationships also hold for the other channels.

When the times at which the currents through the coils are symmetrically distributed with respect to the times at which B_x or B_y are equal to zero, the loading-torque pulses are given by the formulas (7)

$$m_x = P_y B_z \tau_x (1 + 2 \operatorname{tg} \zeta_m),$$
$$m_y = P_x B_z \tau_y (1 + \operatorname{tg} \zeta_m),$$
$$m_z = 4 P_x B_x \tau_z (1 - \cos \zeta_m) \sin \zeta_m,$$

where P_x and P_y are the magnetic moments of the coils along the Ox and Oy axes and B_x, B_y, and B_z are the components of the magnetic induction, while τ_x, τ_y, and τ_z are the times at which the current is switched in the corresponding coil.

Thus, if we know the required torque impulse, it is possible to find the times τ_x, τ_y, and τ_z to switch the currents. In this case P_x and P_y must be selected by means of the condition requiring that the time between switching currents in each channel is no greater than 10 minutes. Otherwise there is crosstalk between the loading moments in the control channels. The requisite torque impulse is determined by starting with the limiting admissible angular momentum of the flywheels. Assuming that, in the worst case, the torquer applies only one pulse of loading torque during a rotation in each channel, "saturation" of a flywheel should occur after no more than two circuits of the satellite in its orbit. As a result, system design should be preceded by a careful estimate of the maximum possible perturbing moments. The maximal magnitude of the constant components of these moments is used to select the limiting admissible angular momentum of the flywheels. But in this case, an increase in the loading-torque impulse equal to the accumulated angular momentum of the flywheel prior to the point at which the torquer is switched on is inadequate. It is desirable to choose such an increase by beginning not only from the instantaneous value of H, but by allowing for the sign and rate of its change. However, the presence of a periodoc component in the external perturbing moment requires extraction of the slowly-varying component of the kinetic moment $\bar{H}_c$. The rate of change of the component $\bar{H}_c$ with weight equal to the time of rotation of the satellite between torque pulses and its sign must be taken into account for planning torque applications.

We will now analyze the conditions for selection of the duration of pulses for the case of the bank channel. In order to do so, we consider Equations (3.2). They are satisfied only when the magnetic-induction component B_x changes in accordance with (3.1). In reality, because of various anomalies and and external influences (for example, variation in solar activity), the earth's magnetic field differs from the field of an ideal dipole, which is used for purposes of investigation. Thus, the torque pulse m_x must be applied asymmetrically relative to the zero value of B_x. We represent this last in the form

$$B_x = B_{xng} + B_{xb}$$

where B_{xng} is the induction component defined by the corresponding expression in (3.1)

B_{xb} is the perturbing induction due to a dipole.

Then the right sides of (3.2) take the form

$$\Delta m_y = \int_{-\tau_x/2}^{\tau_x/2} \Delta m_y \, dt = P_z \left[\int_{-\tau_x/2}^{\tau_x/2} B_{xng} \, dt + \int_{-\tau_x/2}^{\tau_x/2} B_{xb} \, dt \right], \qquad (3.3)$$
$$\Delta m_z = \int_{-\tau_x/2}^{\tau_x/2} \Delta m_z \, dt = -P_y \left[\int_{-\tau_x/2}^{\tau_x/2} B_{xng} \, dt + \int_{-\tau_x/2}^{\tau_x/2} B_{xb} \, dt \right]$$

The interval τ_x is selected by means of the conditions required for the first integral in (3.3) to be equal to zero, which, for small B_x corresponds to the interval

$$-\Delta B_x \le B_x \le \Delta B_x$$

The second integral can be expressed in terms of the average value of B_{xb} in the interval τ_x,

$$\int_{-\tau_x/2}^{\tau_x/2} B_{xb} \, dt = B_{x_0} \tau_x$$

Since the interval τ_x is comparatively small, B_{x_0} can be taken equal to the perturbing magnetic field at the point corresponding to $B_{x_0} = 0$. If we denote this value of B_{xb} and B_{xb0}, we can write expression (3.3) in the form

$$\Delta m_y = P_z B_{xb0} \tau_x \qquad (3.4)$$
$$\Delta m_z = -P_y B_{xb0} \tau_x$$

The magnitude of the perturbing induction of the earth's magnetic field is a random function of time and the geomagnetic coordinates (geomagnetic latitude and longitude).

We will now determine the mathematical expec-

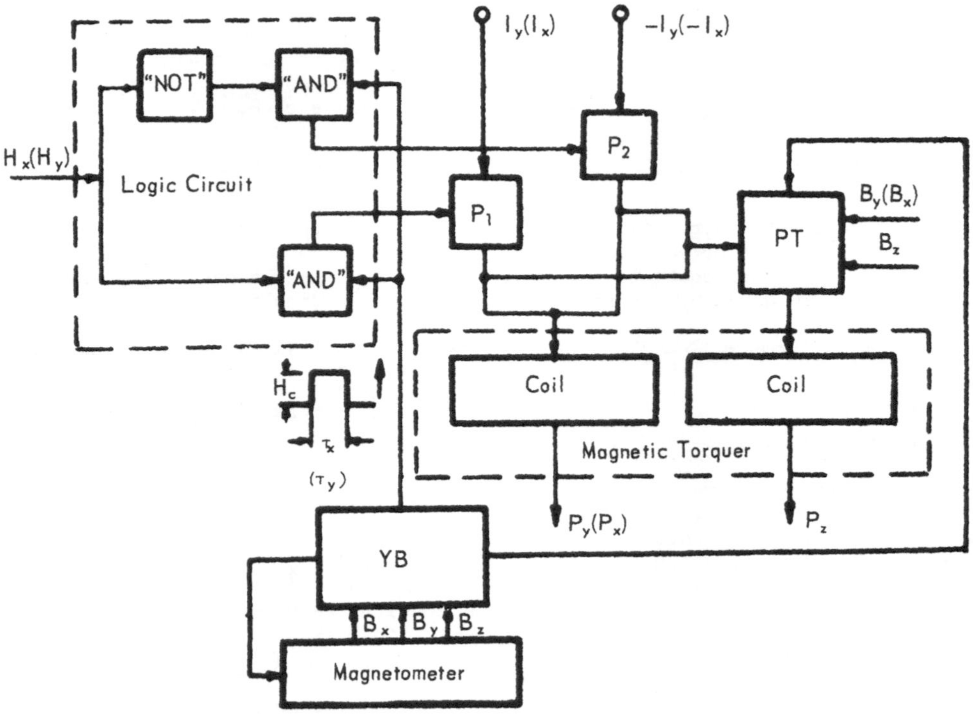
$I_y(I_x)$
$-I_y(-I_x)$
"NOT"
"AND"
"AND"
$H_x(H_y)$
Logic Circuit
P_2
P_1
PT
$B_y(B_x)$
B_z
Coil
Coil
H_c
τ_x
(τ_y)
Magnetic Torquer
$P_y(P_x)$
P_z
YB
B_x
B_y
B_z
Magnetometer

Fig.7

tations $M(\Delta m_x)$ and $M(\Delta m_y)$ for the torque impulses for a flight time t_n, for which we introduce the density of B_{xb},

$$f_{B_x}(\alpha_M, \delta_M t),$$

where α_M and δ_M are the geomagnetic longitude and latitude, respectively. During the flight time of the satellite through each circuit of geomagnetic latitude and longitude, the corresponding loading points take different discrete values. When the flight is sufficiently long, the variation in the geomagnetic longitude covers the entire range

$$0 \leq \alpha_M \leq 2\pi$$

while the latitude variation covers the range

$$-\left|\frac{\pi}{2} + (\varphi_M + i)\right| \leq \delta_M \leq \left|\frac{\pi}{2} - (\varphi_M + i)\right|$$

On the basis of Equations (3.4), we can write the mathematical expectations for the pulses in the following manner:

$$M(\Delta m_y) = P_z \tau_x M(B_{xb}),$$
$$M(\Delta m_z) = P_y \tau_x M(B_{xb}), \tag{3.5}$$

where

$$M(B_{xb}) = \frac{1}{4\pi\left|\pi/2 - (\varphi_M + i)\right|} \int_0^{2\pi} d\alpha_M \int_{-|\pi/2 - (\varphi_M+i)|}^{|\frac{\pi}{2} - (\varphi_M+i)|} d\delta_M \int_0^{t_n} B_{xb} f_{B_x}(\alpha_M, \delta_M t)\, dt$$

The mathematical expectation for the residual impulses should be at least an order of magnitude less than the critical value of the angular momentum of the flywheel, i.e.,

$$M(\Delta m_y) \leq 0.1\, H_k$$
$$M(\Delta m_z) \leq 0.1\, H_k$$

Then, after substitution into expressions (3.5), we obtain the conditions

$$\tau_x \leq \frac{H_k}{P_z M(B_{xb})}$$
$$\tau_x \leq \frac{H_k}{P_y M(B_{xb})} \tag{3.6}$$

for limiting the length of the pulses m_x. It is clear that τ_x must be selected by using the stronger of inequalities (3.6).

A similar technique can be used to obtain conditions constraining τ_y and τ_z. When these constraints and the required torque are taken into account, the necessary values of current for the magnetic torquer are selected.

The magnitudes of the residual impulses and the corresponding constraints on their duration can be reduced by statistically processing information from magnetometers on Bx, By, and Bz with subsequent extrapolation of their variations. Here the points at which torque is applied can be chosen and symmetrically distributed by solving integral equations of the form

$$\int_{t_H}^{t_H + \tau_x} B_x\, dt = 0$$

where t_H is the time at which a pulse begins and B_x is the value of the induction along the Ox axis as determined by extrapolation of its measured values.

The analysis that we have conducted can be used to construct a system for loading flywheels. The corresponding functional system for the bank (yaw*) channel is shown in Figure 6. A control computer (CC) uses data from magnetometers on the magnitudes of the components of the magnetic-induction vector along the Ox, Oy, and Oz axes to determine the point at which the magnetic torquer is switched on to apply torque to a flywheel on the z Ox(Oy) axis and to select a pulse of appropriate duration τ_x (τ_y) (with amplitude equal to the instantaneous value of the angular momentum $\bar{H}$ of the flywheel with allowance for the rate of variation $\dot{H}_c$ in the scale adopted for representation of the physical values and their electrical analogies). This impulse is applied in the form of a pedestal to a logic device. At the same time, the voltage representing the magnitude and sign of the angular momentum H_x (H_y) is applied to this device. When H_x (H_y) is positive, the lower "AND" gate fires, switching on relay R_1, which supplies a current +Iy (+Ix) to a torquer on the Oy(Ox) axis. When Hx (Hy) is negative, the upper "AND" gate** fires, switching on relay R_2. Then a current -Iy (-Ix) flows through the coil. Instead of causing current to flow directly through the coil on the Oz axis, firing of one of the indicated relays causes current to flow through the coil on the Oz axis, firing of one of the indicated relays causes current to flow through a current regulator (CR), which provides the necessary relationship between the currents Iz and Iy (Ix). In order to do so, a signal dependent on By (Bx) and Bz is applied to it. In the interest of economizing electrical energy, which is very important in satellite work, the regulator is switched on immediately before application of the pedestal pulse to the logic circuit by means of a signal from the control computer. Signals from the control computer also switch off the magnetometers while the pedestal pulse is being applied.

---

*In the circuit diagram and the text all material in parentheses indicates the course channel, since the bank and pitch circuits are analogous.

**A "NOT" gate provides voltage inversion and firing of the "AND" gate when Hx (Hy) is negative.

The block diagram shown in Figure 7 for loading on the Oz axis is, in principle, analogous to the circuit discussed above. The difference in this circuit consists only in the fact that it includes an inverter for changing the sign of the current flowing through the coil along the Ox(Oy) axis beginning at time $\tau_z/2$ after switching on. This is necessary so that, as was shown in (7), the sign of the moment Mz will remain constant over τ_z. Here notation with or without parentheses is used depending on when the circuits are switched on. Thus, if switching corresponds to times at which By = 0, then notation without parentheses must be used.

REFERENCES

(1) McElvain, R. J., Satellite Angular Momentum Removal Utilizing the Earth's Magnetic Field, New York, London Academic Press, 1964.

(2) Renuie, R. G., A Magnetic Unloading System for an Ultan Stabbe Unmanned Spacecraft, IEEE Transactions on Aerospace, N2, 1964.

(3) Bodner, V. A., Alekseev, K. B., and Bebenin, G. G., On Application of Magnetic Torquers for Three-Dimensional Stabilization of Artificial Earth Satellites, "Inzhernenyi zhrnal, Vol. 4, No. 4, 1964.

(4) Stuart, W. H., Satellite Attitude Stabilization by Means of Flywheels, "Aerospace Engineering," Vol. 20, No. 9, 1961.

(5) Alekseev, K. B., and Bebenin, G. G., Toward a Thoery for Synthesis of Stabilization Systems for Artificial Earth Satellites with Flywheels, Trudy MVGU im. Bumana (in press).

(6) Alekseev, K. B., and Bebenin, G. G., Control of Cosmic Flight Apparatus, Mashinostroenie, 1964 (in Russian).

(7) Bodner, V. A., Alekseev, K. B., and Bebenin, G. G., Use of Magnetic Torquers for Loading Flywheels.

(8) Bodner, V. A., The Theory of Automatic Flight Control, Izd. Nauka, 1964 (in Russian).

(9) Roitenberg, L. Ya., A Gyroscopic Servosystem for Operation in the Presence of Multidimensional Random Noise, "PMM," XXVI, 2.

(10) Pugachev, V. S., The Theory of Random Functions, Fizmatgiz, 1962 (in Russian).

(11) Leondes, C. T., Roberson, R. E., Aoxi, M., Analysis and Synthesis of a Particular Class of Satellite Attitude-Control Systems, "Aerospace Science," Vol. 29, No. 12, 1962.

DISCUSSION

V. A. Bodner,
K. B. Alekseev and G. G. Bebenin

"On One Class of
Satellite Attitude Control".

Q. Did you take into account aerodynamic moments?

A. No, but there should be no great difficulty in considering these.

INVESTIGATION OF EARTH SATELLITES WITH MAGNETIC ATTITUDE STABILIZATION

by F.Mesch, G.Schweizer and K.Stopfkuchen
Control Division,
DORNIER-WERKE G.m.b.H.,
Friedrichshafen, Germany

ABSTRACT

The dynamic equations for the motion of satellites with passive and semi-passive attitude stabilization are derived. As not only circular but also elliptic orbits are considered, differential equations with periodic coefficients are obtained. It is shown that the equations are considerably simplified if the independent variable "time" is replaced by the true anomaly Ω . Magnetic phenomena are discussed for passive and semi-passive attitude control. Methods are explained for the analysis of the resulting time-varying linear systems.

I. INTRODUCTION

The passive and semi-passive attitude control of artificial satellites is getting more and more interest. Such control systems are simple, reliable and relatively light-weight. Satellites with attitude control have significant advantages over satellites rotating unpredictably along their orbit. Such advantages are:

a) better use of solar cells
b) better control of thermal balance
c) better transmission of signals by directed antennas
d) better ability to measure magnetic and gravitational fields and radiation belts.

Passive stabilization by means of the gravity field of the earth, the remaining aerodynamic moments and the radiation pressure of the sun, have been investigated for example by [1, 2, 3, 4, 5]. Recently detailed work has been done on the use of the magnetic phenomena for attitude stabilization [6]. The magnetic field of the earth can be used for stabilization either for active control by coils, or by permanent magnetic rods for passive control. Permeable rods in addition, allow for damping of oscillations by using the hysteresis energy losses. The latter effect is of most importance because satellites have almost no natural damping.

In the design stage of an artificial satellite, all possible phenomena for passive control should be considered and used, even if one of them, e.g. magnetic control, is to be predominant. Then best control behaviour is achieved and least control moments are required.

In this paper therefore, the equations of motion for the rotation of the satellite about the three principal axes of inertia are derived in general form. No circular orbits are assumed, as usually done, but any elliptic orbit is permitted. Further, no limitations are required concerning the reference coordinate system to which the angular displacements of the satellite are referred. The reference system may undergo any rotation along the orbit.

Studying elliptic orbits, periodic input signals and parametric excitations of the system are inevitable. This means periodically varying coefficients in the differential equations of motion and time-varying moments that influence the satellite. These time-variable equations of motion considerably increase the difficulties of analytical treatment. Therefore in this paper a general method is presented which allows the investigation of such systems. The coupling of roll and yaw axis motions is taken into account. The pitch axis is treated separately, because in almost all cases pitch motion is independent. However, there are no principal difficulties if the motion of the pitch axis is coupled too with roll and yaw.

The dynamic equations for the investigation of various attitude control schemes are derived. Arbitrary numerical values can be inserted to compare different satellite designs. Then two main problems of attitude control can be treated:

a) stability of motion along an orbit
b) stationary oscillations about all three axes due to parametric and periodic excitations along the orbit.

Principally, also transients after large disturbances can be calculated. However, with the equations and methods of this paper one cannot investigate the satellite phase from the separation of the launching vehicle and the de-spin until the attitude control becomes effective. These problems are highly nonlinear. Up to this time, the authors are not aware of any method for solving this problem generally.

Numbers in square brackets refer to similarly-numbered references at the end of this paper.

II. ARTIFICIAL EARTH SATELLITES AS CONTROLLED ELEMENTS

1. Assumptions

The trajectory of the satellite is assumed to be an elliptic orbit. This assumption is justified since the earth's gravity field may be approximated by a central field. Disturbances of the gravity field resulting from the earth's oblateness and inhomogenous structure are neglected. Furthermore, we assume that disturbances of the trajectory do not influence the satellite's attitude and that attitude changes do not influence the trajectory. It becomes thus possible to separate the rotational and translational motions.

Aerodynamic, magnetic and gravitational torques are the predominating moments acting on a satellite near to the earth. Torques due to solar radiation pressure may be neglected here. When determining the aerodynamic torques we assume that the earth's atmosphere does not rotate relative to the earth. The magnetic torques are calculated by approximating the earth's magnetic field by a dipole field. The gravity torques result from the central field directed to the center of the earth.

2. Coordinates used

The choice of suitable sets of coordinates is very important in the investigation of satellite dynamics. The various sets of coordinates used in our investigations may be defined as follows (Fig. 1):

a) Space-fixed System

The origin of the space-fixed system of coordinates is located at the earth's center. The ξ and ζ axes lie in the orbital plane such that the ζ axis is directed towards the apogee and the ξ axis perpendicular to it. If we assume a right-handed system, the η axis is directed downwards (Fig. 1).

b) Horizon-fixed System

The origin of the horizon-fixed system is located at the satellite's center of gravity. The x' axis is rotated about the y' axis with respect to the ξ axis of the space-fixed system by the true anomaly Ω . The z' axis points to the earth's center. The y' axis is perpendicular to the orbital plane so that the x'-y'-z' axes form a right-handed system as well.

c) Reference System

The origin of the reference system is also located at the satellite's center of gravity. The system is rotated about the y' axis by the angle $f(\Omega)$. The reference system may coincide with the horizon-fixed system in the case of a satellite facing the earth during the entire orbit. The reference system is designated x"-y"-z".

d) Body-fixed System

The body-fixed system is tilted with respect to the reference system by the attitude angles φ, ϑ, ψ. Its coordinates are designated x-y-z. It is generally chosen to coincide with the principal axes of the body.

e) Earth-fixed System

The Z-axis points to the South Pole. The X and Y axes lie in the equatorial plane, the X axis intersects the meridian of Greenwich. The earth-fixed axes form a right-handed system.

3. Dynamic Equations of Motion

The Euler equations describe the motions about the three axes of the satellite:

$$\begin{aligned} L &= I_x \cdot \dot{\Omega}_x + (I_z - I_y) \cdot \Omega_y \cdot \Omega_z \\ M &= I_y \cdot \dot{\Omega}_y + (I_x - I_z) \cdot \Omega_z \cdot \Omega_x \\ N &= I_z \cdot \dot{\Omega}_z + (I_y - I_x) \cdot \Omega_x \cdot \Omega_y \end{aligned} \tag{1}$$

where Ω_x, Ω_y and Ω_z are the components of angular velocity. Angular velocities are related to the attitude angles with respect to the space-fixed system as follows: (see App. I)

$$\begin{aligned} \Omega_x &= \dot{\varphi} \cos\psi + \dot{\theta} \cos\varphi \cdot \sin\psi \\ \Omega_y &= \dot{\theta} \cos\varphi \cdot \cos\psi - \dot{\varphi} \sin\psi \\ \Omega_z &= \dot{\psi} - \dot{\theta} \cdot \sin\varphi \end{aligned} \tag{2}$$

The angle θ is defined by $\theta = \vartheta - f(\Omega)$, (3)

where ϑ is the angular displacement of the body-fixed system with respect to the reference system; $f(\Omega)$ denotes the rotation of the reference system with respect to the space-fixed system. If, for example, the satellite is oriented along the gravity field, $f(\Omega)$ equals the true anomaly Ω .

The angular displacements φ, ϑ, ψ of Eq.(2) and (3) are small for attitude-controlled satellites. Then Eq.(2) yields to a first order approximation:

$$\begin{aligned} \Omega_x &= \dot{\varphi} - [\dot{f}(\Omega)] \cdot \psi \\ \Omega_y &= \dot{\vartheta} - \dot{f}(\Omega) \\ \Omega_z &= \dot{\psi} + [\dot{f}(\Omega)] \cdot \varphi \end{aligned} \tag{4}$$

Eq.(4) may be differentiated once more with respect to the time:

$$\begin{aligned} \dot{\Omega}_x &= \ddot{\varphi} - [\dot{f}(\Omega)]\, \dot{\psi} - [\ddot{f}(\Omega)] \cdot \psi \\ \dot{\Omega}_y &= \ddot{\vartheta} - \ddot{f}(\Omega) \\ \dot{\Omega}_z &= \ddot{\psi} + [\dot{f}(\Omega)]\, \dot{\varphi} + [\ddot{f}(\Omega)] \cdot \varphi \end{aligned} \tag{5}$$

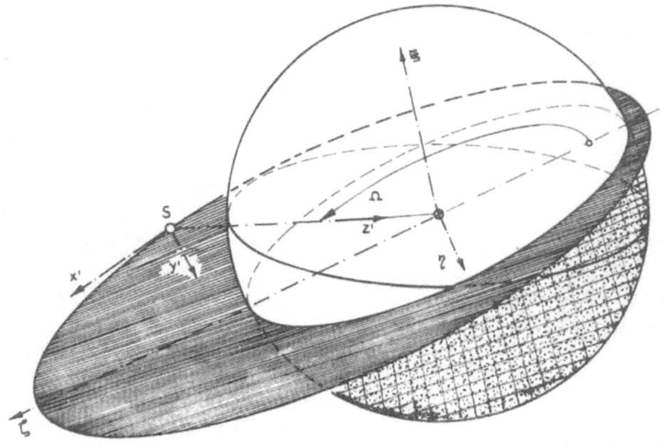

Fig. 1. Space-fixed and horizon-fixed coordinate systems

Substitution of Eq.(4) and (5) into Eq.(1) leads to three differential equations for the angular displacements φ, ϑ, ψ which relate space-fixed and body-fixed coordinates:

$$\frac{L}{I_x} = \ddot{\varphi} - \frac{I_x - I_y + I_z}{I_x} \cdot \dot{f}(\Omega) \cdot \dot{\psi} - [\ddot{f}(\Omega)]\psi + \frac{I_y - I_z}{I_x}[\dot{f}^2(\Omega)]\varphi$$

$$\frac{M}{I_y} = \ddot{\vartheta} - \ddot{f}(\Omega) \tag{6}$$

$$\frac{N}{I_z} = \ddot{\psi} + \frac{I_x - I_y + I_z}{I_z} \dot{f}(\Omega) \cdot \dot{\varphi} + [\ddot{f}(\Omega)] \cdot \varphi - \frac{I_x - I_y}{I_z}[\dot{f}^2(\Omega)]\psi$$

The equations become much simpler and easier to handle if the time t is replaced by the true anomaly Ω as independent variable. Using the relations derived in Appendix I and setting $\varepsilon^2 \ll 1$, we get the set of equations

$$\frac{1}{\dot{\Omega}^2} \cdot \frac{L}{I_x} = \varphi'' - (2\varepsilon \sin\Omega)\varphi' - (1-A) f'(\Omega) \cdot \psi' - [f''(\Omega) - f'(\Omega) 2\varepsilon \sin\Omega]\psi + A f'^2(\Omega) \cdot \varphi$$

$$\frac{1}{\dot{\Omega}^2} \cdot \frac{M}{I_y} = \vartheta'' - (2\varepsilon \sin\Omega)\vartheta' - f''(\Omega) + f'(\Omega) \cdot 2\varepsilon \sin\Omega$$

$$\frac{1}{\dot{\Omega}^2} \cdot \frac{N}{I_z} = \psi'' - (2\varepsilon \sin\Omega)\psi' + (1+B) f'(\Omega)\varphi' + [f''(\Omega) - f'(\Omega) \cdot 2\varepsilon \sin\Omega]\varphi - B f'^2(\Omega) \cdot \psi \tag{7}$$

where
$$A = (I_y - I_z)/I_x$$
$$B = (I_x - I_y)/I_z$$
$$C = (I_z - I_x)/I_y \, .$$

In deriving Eq.(7) we have used the relations

$$\frac{dF}{dt} = \frac{dF}{d\Omega} \cdot \frac{d\Omega}{dt} = F'\dot{\Omega}$$

and

$$\frac{d^2F}{dt^2} = F''\dot{\Omega}^2 + F' \cdot \ddot{\Omega} \tag{8}$$

where $(\dot{})$ denotes a differentiation with respect to t and $(')$ with respect to Ω.

If the satellite faces the earth, $f(\Omega)$ becomes equal to the true anomaly Ω, as mentioned before. In this case Eq.(7) changes to

$$\frac{1}{\dot{\Omega}^2} \cdot \frac{L}{I_x} = \varphi'' - (2\varepsilon \sin\Omega)\varphi' - (1-A)\psi' + 2\varepsilon \sin\Omega\,\psi + A\varphi$$

$$\frac{1}{\dot{\Omega}^2} \cdot \frac{M}{I_y} = \vartheta'' - (2\varepsilon \sin\Omega)\vartheta' + 2\varepsilon \sin\Omega \tag{9}$$

$$\frac{1}{\dot{\Omega}^2} \cdot \frac{N}{I_z} = \psi'' - (2\varepsilon \sin\Omega)\psi' + (1+B)\varphi' - 2\varepsilon \sin\Omega \cdot \varphi - B \cdot \psi$$

If, on the other hand, the satellite is oriented along a magnetic field line and if the orbital plane coincides with the plane of the magnetic field lines, further if the magnetic field is approximated by a dipole field, we get the following equation by using Appendix IV:

$$f(\Omega) = \operatorname{arc\,tg} \frac{\sin 2\Omega}{1/3 + \cos 2\Omega} - 90^\circ . \tag{10}$$

This assumption is true in the case of a polar orbit (90° inclination of the orbital plane to the equatorial plane).

On the left hand of Eq.(7) we find the external torques. We consider damping, gravitational, aerodynamic and magnetic torques. They are derived in the Appendices II to IV.

The gravity torques caused by the gravity field of the earth can be described using the assumptions made in Appendix II:

$$L_M = 3\frac{\mu}{r^3} \cdot (I_z - I_y)\frac{1}{2}\sin 2\varphi \cdot \cos(\theta + \Omega)$$

$$M_M = -3\frac{\mu}{r^3} \cdot (I_x - I_z)\frac{1}{2}\sin[2(\theta + \Omega)] \tag{11}$$

$$N_M = 3\frac{\mu}{r^3} \cdot (I_x - I_y)\frac{1}{2}\sin 2\psi \sin(\theta + \Omega)$$

Substituting Eq.(3) and (I.8) into Eq.(11) and remembering $\varepsilon^2 \ll 1$, we get the normalized gravity torques

$$\frac{1}{\dot{\Omega}^2} \cdot \frac{L_M}{I_x} = -3(1 - \varepsilon\cos\Omega) A \cdot \cos[\Omega - f(\Omega)] \cdot \varphi$$

$$\frac{1}{\dot{\Omega}^2} \cdot \frac{M_M}{I_y} = 3(1 - \varepsilon\cos\Omega) C \cdot \left\{\vartheta \cdot \cos 2[\Omega - f(\Omega)] + \frac{1}{2}\sin 2[\Omega - f(\Omega)]\right\} \tag{12}$$

$$\frac{1}{\dot{\Omega}^2} \cdot \frac{N_M}{I_z} = 3(1 - \varepsilon\cos\Omega) \cdot B \cdot \sin[\Omega - f(\Omega)] \cdot \psi$$

For a satellite facing the earth where $f(\Omega) = \Omega$, this becomes

$$\frac{1}{\dot{\Omega}^2} \cdot \frac{L_M}{I_x} = -3 \cdot (1 - \varepsilon\cos\Omega) \cdot A \cdot \varphi$$

$$\frac{1}{\dot{\Omega}^2} \cdot \frac{M_M}{I_y} = 3 \cdot (1 - \varepsilon\cos\Omega) \cdot C \cdot \vartheta \tag{13}$$

$$\frac{1}{\dot{\Omega}^2} \cdot \frac{N_M}{I_z} = 0$$

From Appendix III, Eq.(III.11) we obtain the normalized aerodynamic torques

$$\frac{1}{\dot{\Omega}^2} \cdot \frac{L_{ae}}{I_x} = 0$$

$$\frac{1}{\dot{\Omega}^2} \cdot \frac{M_{ae}}{I_y} = C(1 - 2\varepsilon\cos\Omega)\left[I_0 + 2\sum_{\nu=1}^{m} I_\nu(C_1)\cos\nu\Omega\right] \cdot \frac{1}{I_y}\sum_{k=1}^{n_1} b_{k\alpha}\sin k\alpha \tag{14}$$

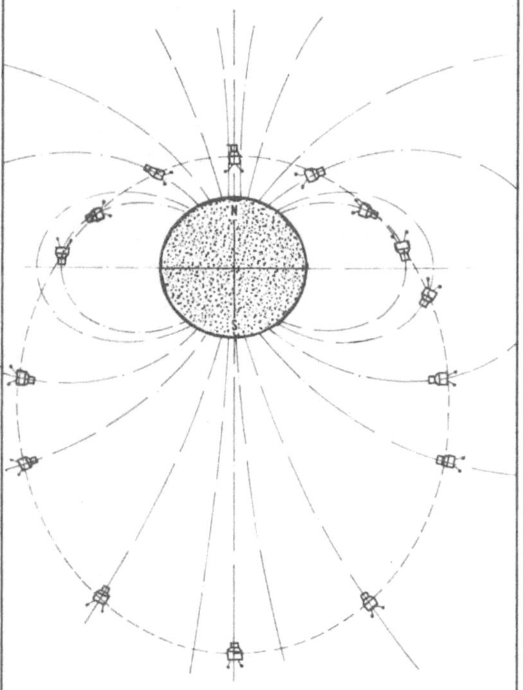

Fig. 2. Satellite with passive stabilization aligning with the magnetic field lines of the earth

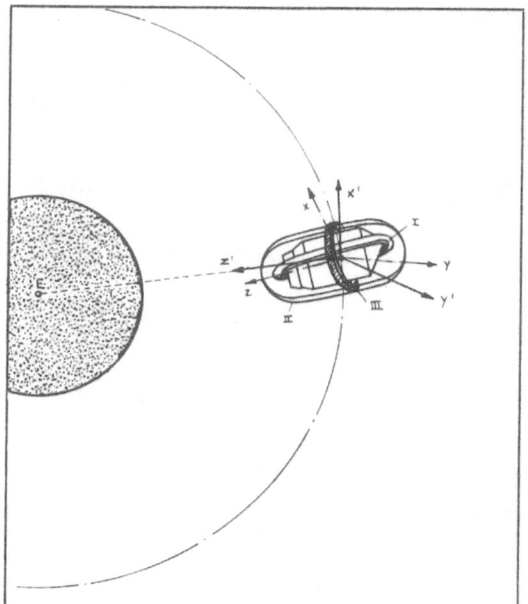

Fig. 3. Active magnetic attitude control by coils

$$\frac{1}{\dot{\Omega}^2} \cdot \frac{N_{ae}}{I_z} = C(1-2\varepsilon\cos\Omega)\left[I_o + 2\sum_{\nu=1}^{m} I_\nu(C_1)\cos\nu\Omega \cdot \right.$$

$$\left. \cdot \beta \frac{1}{I_z} \sum_{i=1}^{n_2} a_{i\alpha} \cos i\alpha \right.$$

In the case of a satellite facing the earth, Eq.(14) is considerably simplified. The static derivatives Eq.(III.7b) and Eq.(III.7c) are constant since there are only small angles of attack. With $f(\Omega) = \Omega$, Eq.(III.15) becomes

$$\alpha = \vartheta - \varepsilon \sin\Omega\,, \tag{15}$$

and using Eq.(III.20), Eq.(14) reduces to

$$\frac{1}{\dot{\Omega}^2} \cdot \frac{L_{ae}}{I_x} = 0$$

$$\frac{1}{\dot{\Omega}^2} \cdot \frac{M_{ae}}{I_y} = C\left[1-2\varepsilon\cos\Omega\right]\left[I_o + 2\sum_{\nu=1}^{m} I_\nu(C_1)\cos\nu\Omega\right] \cdot$$

$$\cdot \frac{C_{M\alpha}}{I_y}\left[\vartheta - \varepsilon\sin\Omega\right] \tag{16}$$

$$\frac{1}{\dot{\Omega}^2} \cdot \frac{N_{ae}}{I_z} = C\left[1-2\varepsilon\cos\Omega\right]\left[I_o + 2\sum_{\nu=1}^{m} I_\nu(C_1)\cos\nu\Omega\right] \cdot$$

$$\cdot \frac{C_{N\beta}}{I_z}\left[C_\beta(1-\varepsilon\cos\Omega)(1-\sin^2\lambda\cos^2\Omega)^{1/2} + \psi\right]$$

III. MAGNETIC CONTROL TORQUES

1. Passive control by means of magnetic rods

Permanent magnetic rods as shown for example in Fig.IV.2, tend to align themselves with the direction of the earth's magnetic field. Thus satellites with passive magnetic stabilization are oriented along the earth's magnetic field. As discussed in Appendix IV, the magnetic field of the earth is approximated here by a dipole field. Thus, if the orbital plane contains the earth's axis, the satellite undergoes two revolutions per orbit (Fig.2), and the function $f(\Omega)$ defining the reference coordinate system is given by Eq.(10). Then we obtain the magnetic torques by virtue of Eq.(IV.2) and (IV.11) of Appendix IV, assuming small perturbations only:

$$L_H = 0$$

$$M_H = -C_H \frac{\sqrt{4-3\sin^2\Omega}}{r^3} \cdot \vartheta \tag{17}$$

$$N_H = -C_H \frac{\sqrt{4-3\sin^2\Omega}}{r^3} \cdot \psi$$

where the constant $C_H = C_E \cdot V \cdot B_{rem}$. Using Eq.(I.8) from Appendix I the normalized torques become, assuming $\varepsilon^2 \ll 1$,

$$\frac{1}{\dot{\Omega}^2} \cdot \frac{L_H}{I_x} = 0$$

$$\frac{1}{\dot{\Omega}^2} \cdot \frac{M_H}{I_y} = -\frac{C_H}{\mu I_y} \cdot (1-\varepsilon\cos\Omega) \cdot \sqrt{4-3\sin^2\Omega} \cdot \vartheta \tag{18}$$

$$\frac{1}{\dot{\Omega}^2} \cdot \frac{N_H}{I_z} = -\frac{C_H}{\mu I_z} \cdot (1-\varepsilon\cos\Omega) \cdot \sqrt{4-3\sin^2\Omega} \cdot \psi$$

Now Eq.(18) together with Eqs.(12) and (14) can be substituted into Eq.(7) and analyzed by the methods shown in the next section. However, satisfactory response is obtained only if additional damping is provided. Damping can be achieved by means of velocity-dependent fluid friction or by means of hysteresis energy losses in permeable rods.

Fluid damping is dealt with in Appendix VI. This paper is mainly concerned with stabilization by magnetic phenomena. Therefore only the average damping torques produced by permeable rods are treated here. As discussed in Appendix IV, we use an average damping factor D derived from the average hysteresis energy losses per period. Provided one has damping torques proportional to velocity as assumed also for Eq. (IV.16), the damping torque produced in roll (analogous in the other axes) is given by

$$L_{DH} = -D_R \cdot \dot{\varphi}(t).$$

Replacing again t by Ω according to Appendix I, the normalized torques become

$$-\frac{1}{\dot{\Omega}^2} \cdot \frac{L_{DH}}{I_x} = \frac{D_R}{I_x(\mu/a^3)^{1/2}}(1-2\varepsilon\cos\Omega)\varphi'(\Omega)$$

$$-\frac{1}{\dot{\Omega}^2} \cdot \frac{M_{DH}}{I_y} = \frac{D_N}{I_y(\mu/a^3)^{1/2}}(1-2\varepsilon\cos\Omega)\vartheta'(\Omega)$$

$$-\frac{1}{\dot{\Omega}^2} \cdot \frac{N_{DH}}{I_z} = \frac{D_G}{I_z(\mu/a^3)^{1/2}}(1-2\varepsilon\cos\Omega)\psi'(\Omega)$$

(19)

2. Active magnetic control

If the magnetic rods are replaced by coils the magnetic vector of the satellite can be rotated independently of the reference system. Thus active magnetic control torques can be provided and used for attitude stabilization.

In general, three coils can be attached to the satellite (Fig.3). Each of these coils generates a magnetic vector normal to the coil plane. The torque of this magnetic dipole acting on the earth's magnetic field is given by [refer to Appendix IV, Eqs.(IV.9) and (IV.18)]:

$$L = -\mu_o\, n\, i\, F \cdot H_E \sin\sigma \tag{20}$$

σ denotes the angle between the magnetic dipole and the direction of the earth's field, n denotes number of windings, F coil area, μ_o magnetic permeability, i current of the coil and H_E magnetic field of the earth.

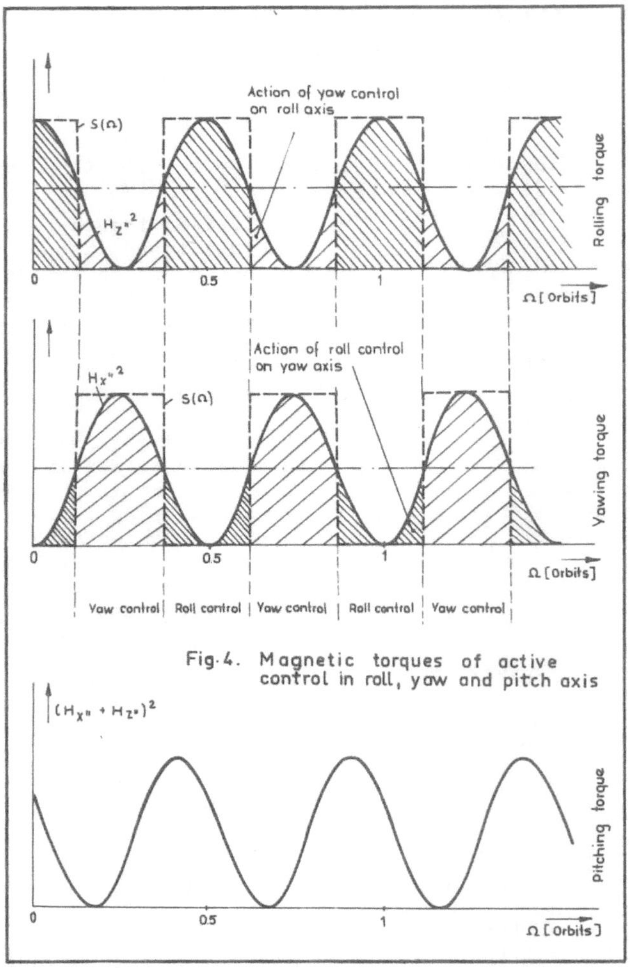

Fig. 4. Magnetic torques of active control in roll, yaw and pitch axis

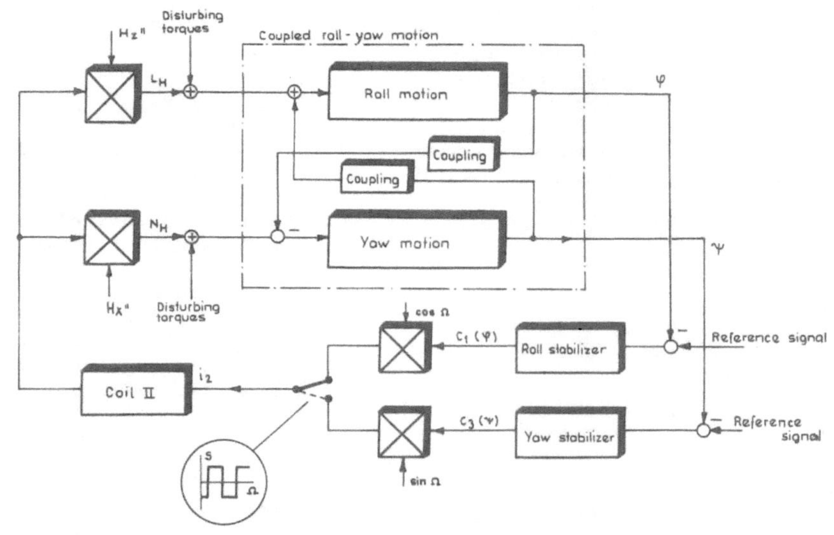

Fig. 5. Block diagram of active magnetic stabilization (roll and yaw)

Now we limit ourselves to a polar orbit, and to $f(\Omega) = \Omega$, that means the satellite is constantly directing the same side to the earth. In the case of general inclination the torques can be computed similarly, but then coupling between all axes results.

The three coils generate magnetic torques when the satellite is displaced from the reference attitude by the angles φ, ϑ, ψ. The input signals of the three coils are designated by the currents i_1, i_2 and i_3. Then the torque components become analogous to Eq.(20) with $C_i = \mu_o \cdot nF$:

$$- \begin{bmatrix} L_H \\ M_H \\ N_H \end{bmatrix} = C_i \cdot \begin{bmatrix} \vec{x} & \vec{y} & \vec{z} \\ [i_1 + \psi i_2 - \vartheta i_3] & [-\psi i_1 + i_2 + \varphi i_3] & [\vartheta i_1 - \varphi i_2 + i_3] \\ H_{x''} & 0 & H_{z''} \end{bmatrix} \quad (21)$$

Assuming small attitude angles, we have neglected higher order terms of the direction cosines. Solving Eq.(21) gives

$$- L_H = C_i \, H_{z''} \cdot \left\{ - \psi i_1 + i_2 + \varphi i_3 \right\}$$

$$- M_H = -C_i \, H_{z''} \cdot \left\{ i_1 + \psi i_2 - \vartheta i_3 \right\} + C_i \, H_{x''} \cdot \left\{ \vartheta i_1 - \varphi i_2 + i_3 \right\}$$

$$- N_H = -C_i \, H_{x''} \cdot \left\{ - \psi i_1 + i_2 + \varphi i_3 \right\} \quad (22)$$

If coils I and III are actuated by signals dependent on ϑ, and coil II by signals dependent on φ and ψ, we obtain

$$- L_H = H_{z''} \cdot \left\{ C_1 (\varphi) + C_3 (\psi) \right\}$$

$$- M_H = -H_{z''} \cdot C_2 (\vartheta) + H_x \cdot C_2 (\vartheta) \quad (23)$$

$$- N_H = -H_{x''} \cdot \left\{ C_1 (\varphi) + C_3 (\psi) \right\}$$

The products $i_1 \cdot \psi$ etc. are small higher order terms and have been neglected. Eq.(23) shows that control moments for rolling and yawing motion are not independent. Pitching torques however are decoupled.

The horizontal and vertical components of the magnetic field H_E are given by Eq.(IV.4) and (IV.5). We obtain

$$- L_H = 2 \, C_E \frac{\cos \Omega}{r^3} \left\{ C_1 (\varphi) + C_3 (\psi) \right\}$$

$$- M_H = C_E \left\{ \frac{\sin \Omega}{r^3} - \frac{2 \cos \Omega}{r^3} \right\} C_2 (\vartheta) \quad (24)$$

$$- N_H = - C_E \frac{\sin \Omega}{r^3} \left\{ C_1 (\varphi) + C_3 (\psi) \right\}$$

It can be seen that the torques vary periodically with average zero, producing negative feedback in one half-period, positive feedback in the next one. This phenomenon is completely untolerable for control purposes. Therefore we square the periodic functions in Eq.(24) and obtain

$$- L_H = \frac{2C_E}{r^3} \cdot \left\{ \frac{1}{2} + \frac{1}{2} \cos 2\Omega \right\} \cdot \left\{ C_1(\varphi) + C_3 (\psi) \right\}$$

$$- M_H = \frac{C_E}{r^3} \cdot \left\{ \frac{5}{2} + \frac{3}{2} \cos 2\Omega - 2 \sin 2\Omega \right\} \cdot C_2(\vartheta) \quad (25)$$

$$- N_H = - \frac{C_E}{r^3} \cdot \left\{ \frac{1}{2} - \frac{1}{2} \cos 2\Omega \right\} \cdot \left\{ C_1(\varphi) + C_3(\psi) \right\}$$

Technically, this can be achieved by measuring the magnetic field components and multiplying them by the output signal from the stabilizer (Fig.5). In this manner one has achieved control torques with no sign changing, consisting of a constant and a periodic term.

As mentioned, rolling and yawing torques are coupled. But there is a phase shift of 180° between the periodic terms of these two torques. This means that one may try to switch coil II alternately between yaw and roll error during one orbit, as shown in Fig.4 and 5. Then rolling and yawing torques become

$$-L_H = \frac{C_E}{r^3} (1 + \cos 2\Omega) \left\{ \left[\frac{1}{2} + \frac{1}{2} \cdot s(\Omega) \right] \cdot C_1(\varphi) + \left[\frac{1}{2} - \frac{1}{2} \cdot s(\Omega) \right] \cdot C_3 (\psi) \right\} \quad (26)$$

$$-N_H = -\frac{C_E}{2r^3} (1 - \cos 2\Omega) \left\{ \left[\frac{1}{2} - \frac{1}{2} s(\Omega) \right] \cdot C_3(\psi) + \left[\frac{1}{2} + \frac{1}{2} s(\Omega) \right] \cdot C_1(\varphi) \right\}$$

Here $s(\Omega)$ denotes the square wave shown in Fig.4; the sqare wave is expressed mathematically by the Fourier series

$$s(\Omega) = \frac{4}{\pi} \cdot \sum_{\nu=1}^{\infty} \frac{(-1)^{\nu+1} \cdot \cos[2\Omega(2\nu-1)]}{2\nu - 1} \quad (26a)$$

Replacing now the radius vector r by Eq.(I.8) we get the normalized torques

$$- \frac{1}{I_x} \frac{L_H}{\dot{\Omega}^2} = \frac{C_E}{\mu I_x} (1 - \varepsilon \cos \Omega)(1 + \cos 2\Omega) \cdot \left\{ \left[\frac{1}{2} + \frac{1}{2} s(\Omega) \right] \cdot C_1(\varphi) + \left[\frac{1}{2} - \frac{1}{2} s(\Omega) \right] \cdot C_3(\psi) \right\}$$

$$- \frac{1}{I_y} \frac{M_H}{\dot{\Omega}^2} = \frac{C_E}{2\mu I_y} (1 - \varepsilon \cos \Omega)(5 + 3 \cos 2\Omega - 4 \sin 2\Omega) \cdot C_2(\vartheta) \quad (27)$$

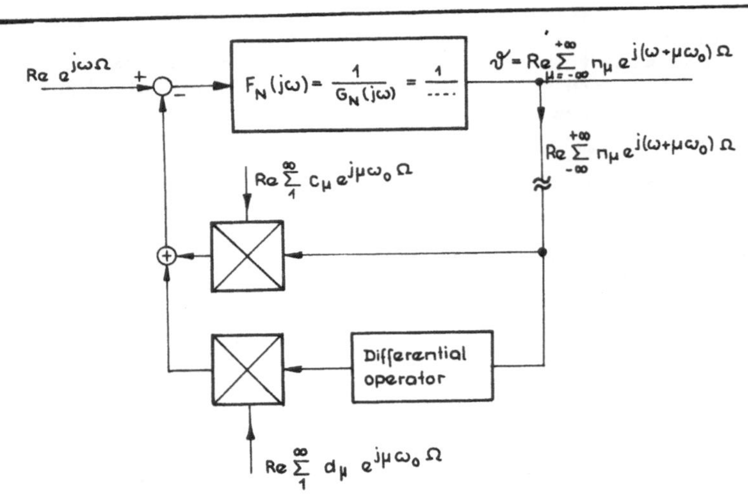

Fig. 6. Block diagram corresponding to Eq. (28), pitch axis.

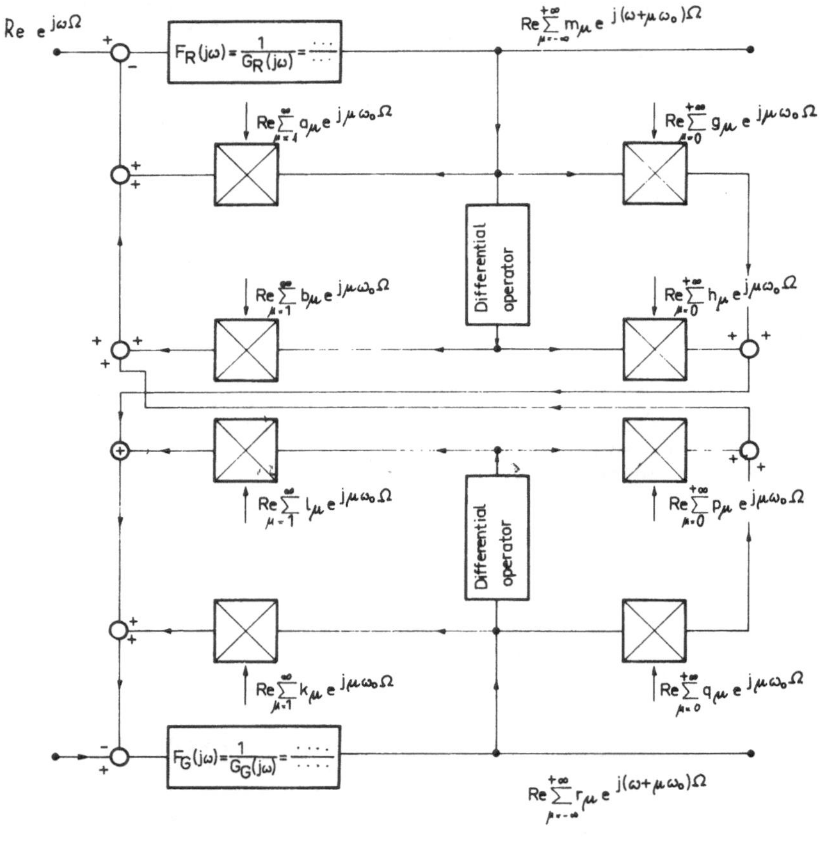

Fig. 7 Block diagram corresponding to Eq. (28), roll and yaw axis

$$-\frac{1}{I_z}\,\frac{N_H}{\dot{\Omega}^2} = -\frac{C_E}{2\mu I_z}\,(1-\varepsilon\cos\Omega)(1-\cos 2\Omega)\cdot\left\{\left[\frac{1}{2}-\frac{1}{2}\,s(\Omega)\right]\cdot C_3(\psi)+\left[\frac{1}{2}+\frac{1}{2}\,s(\Omega)\right]\cdot C_1(\varphi)\right\}$$

Fig.5 shows the complete block diagram of the active magnetic control system. It can be seen that the control system contains large periodical parameters.

We have now derived the complete set of equations for magnetically stabilized satellites as intended, which may be written in abbreviated form:

$$\varphi''(\Omega)+D_R\varphi'(\Omega)+K_R\cdot\varphi(\Omega) = \underbrace{\sum_{\nu=1}^{s} \mathrm{Re}\,\alpha_{R\nu}\, e^{j\nu\Omega}}_{\text{Periodic driving function}} - \underbrace{\left(\sum_{\mu=1}^{\infty} \mathrm{Re}\, a_\mu e^{j\mu\Omega}\right)\varphi(\Omega) - \left(\sum_{\mu=1}^{\infty} \mathrm{Re}\, b_\mu e^{j\mu\Omega}\right)\varphi'(\Omega) - \left(\sum_{\mu=0}^{\infty} \mathrm{Re}\, q_\mu e^{j\mu\Omega}\right)\psi(\Omega) - \left(\sum_{\mu=0}^{\infty} \mathrm{Re}\, p_\mu e^{j\mu\Omega}\right)\psi'(\Omega)}_{\text{Periodic coefficients}}$$

$$\vartheta''(\Omega)+D_N\vartheta'(\Omega)+K_N\vartheta(\Omega) = \underbrace{\sum_{\nu=1}^{s} \mathrm{Re}\,\alpha_{N\nu}\, e^{j\nu\Omega}}_{\text{Periodic driving function}} - \underbrace{\mathrm{Re}\left(\sum_{\mu=1}^{\infty} c_\mu e^{j\mu\Omega}\right)\vartheta(\Omega) - \sum_{\mu=1}^{\infty} \mathrm{Re}\,(d_\mu e^{j\mu\Omega})\,\vartheta'(\Omega)}_{\text{Periodic coefficients}}$$

$$\psi''(\Omega)+D_G\psi'(\Omega)+K_G\psi(\Omega) = \underbrace{\sum_{\nu=1}^{s} \mathrm{Re}\,\alpha_{G\nu}\, e^{j\nu\Omega}}_{\text{Periodic driving function}} - \underbrace{\left(\mathrm{Re}\sum_{\mu=1}^{\infty} k_\mu e^{j\mu\Omega}\right)\psi(\Omega) - \left(\sum_{\mu=1}^{\infty} \mathrm{Re}\, l_\mu e^{j\mu\Omega}\right)\psi'(\Omega) - \left(\mathrm{Re}\sum_{\mu=0}^{\infty} g_\mu e^{j\mu\Omega}\right)\varphi(\Omega) - \left(\sum_{\mu=0}^{\infty} \mathrm{Re}\, h_\mu e^{j\mu\Omega}\right)\varphi'(\Omega)}_{\text{Periodic coefficients}} \qquad (28)$$

The coefficients of the Fourier series follow from Eqs. (7), (12), (14), (18), (19) and (27). The next section contains methods for analyzing such time varying systems.

IV. ANALYSIS OF THE EQUATIONS OF MOTION

The dynamic equations for the motion of the satellite have been derived in the last section. The equations for the roll and yaw axis are coupled second order differential equations with periodic coefficients. The equation for the pitch axis is decoupled. It is a second order system with time-varying parameters.

These equations of motion will now be investigated. In order to do this we have to determine the transfer function of the system. Firstly the motion of the pitch axis will be analyzed, then the motion of the roll and yaw axis.

The equation (28) can be represented by the two block diagrams Fig.6 and 7. In order to determine the transfer function [7, 8] the problem is now to compute the steady state response of the system to the input signal Re $e^{j\omega\Omega}$. The transfer function can then be calculated. It should be noted that the independent variable in the present problem is Ω, the true anomaly. That means if the results of Appendix V are used, one has to replace the variable t by Ω in the corresponding Fourier transforms. In the frequency domain the variable ω relates to Ω and not to t as usual. Since Ω (an angle) is a nondimensional quantity, the variable ω_0 is nondimensional too. Eq.(28) shows furthermore that the functions are periodical with Ω. The "parameter frequency" ω_0 is therefore $\omega_0 = 1$.

It can be shown [7, 9, 10] that the response of the system has the form

$$\vartheta(\Omega) = \mathrm{Re}\sum_{\mu=-\infty}^{+\infty} n_\mu\, e^{j(\omega+\mu\omega_0)\Omega} \qquad (30)$$

if Re $e^{j\omega\Omega}$ is the input signal. The coefficients n_μ are complex functions to be determined now.

For the computation the signal Re $e^{j\omega\Omega}$ and the output signal of Eq.(30) multiplied by the time varying parameters as shown in Fig.6, are assumed to be input signals. The output signal in response to these input signals must be computed and has to be equal to the output of Eq.(30). With the aid of

$$\mathrm{Re}\,(n\cdot C) = \frac{1}{2}\,\mathrm{Re}\,(n\cdot C) + \frac{1}{2}\,\mathrm{Re}\,(n\cdot\bar{C}), \qquad (31)$$

Eq.(32) can be obtained. The dashed parameters in Eq.(31) and (32) stand for complex conjugated variables. The symbol n_μ^+ is the coefficient of the frequency term $(\omega+\mu\omega_0)$, the symbol n_μ^- of the term $(\omega-\mu\omega_0)$.

Eq. (32)

$$
\begin{aligned}
&\mathrm{Re}\sum_{\mu=-\infty}^{+\infty} n_\mu e^{j(\omega+\mu\omega_0)\Omega} = \mathrm{Re}\,\frac{F_N(j\omega)}{2}\cdot\Big\{2+n_1^-\cdot c_1+n_1^+\cdot\bar{c}_1+n_2^-\cdot c_2+n_2^+\cdot\bar{c}_2+\cdots\cdots n_\mu^-\cdot c_\mu+n_\mu^+\cdot\bar{c}_\mu+\cdots\cdots \\
&\qquad +n_1^-[j(\omega-\omega_0)]d_1+n_1^+[j(\omega-\omega_0)]\bar{d}_1+n_2^-[j(\omega-2\omega_0)]d_2+n_2^+[j(\omega+2\omega_0)]\bar{d}_2+\cdots n_\mu^-[j(\omega-\mu\omega_0)]d_\mu+n_\mu^+[j(\omega+\mu\omega_0)]\bar{d}_\mu+\cdots\Big\}e^{j\omega\Omega} \\
&+\frac{\mathrm{Re}[F_N[j(\omega-\omega_0)]]}{2}\cdot\Big\{n_2^-\cdot c_1+n_0\cdot\bar{c}_1+n_3^-\cdot c_2+n_1^+\cdot\bar{c}_2+\cdots\cdots n_{\mu+1}^-\cdot c_\mu+n_{\mu-1}^+\cdot\bar{c}_\mu+\cdots\cdots \\
&\qquad +n_2^-[j(\omega-2\omega_0)]d_1+n_0(j\omega)\bar{d}_1+n_3^-[j(\omega-3\omega_0)]d_2+n_1^+[j(\omega+\omega_0)]\bar{d}_2+\cdots n_{\mu+1}^-[j(\omega-(\mu+1)\omega_0]d_\mu+n_{\mu-1}^+[j(\omega+(\mu-1)\omega_0]\bar{d}_\mu+\cdots\Big\}e^{+j(\omega-\omega_0)\Omega} \\
&+\frac{\mathrm{Re}[F_N[j(\omega+\omega_0)]]}{2}\cdot\Big\{n_0\cdot c_1+n_2^+\cdot\bar{c}_1+n_1^-\cdot c_2+n_3^+\cdot\bar{c}_2+\cdots n_{\mu-1}^-\cdot c_\mu+n_{\mu+1}^+\cdot\bar{c}_\mu+\cdots \\
&\qquad +n_0(j\omega)d_1+n_2^+[j(\omega+2\omega_0)]\bar{d}_1+n_1^-[j(\omega-\omega_0)]d_2+n_3^+[j(\omega+3\omega_0)]\bar{d}_2+\cdots n_{\mu-1}^-[j(\omega-(\mu-1)\omega_0)]d_\mu+n_{\mu+1}^+[j(\omega+(\mu+1)\omega_0]\bar{d}_\mu+\cdots\Big\}e^{j(\omega+\omega_0)\Omega} \\
&+\quad ---\qquad ---\qquad --- \\
&+\quad ---\qquad ---\qquad --- \\
&+\frac{\mathrm{Re}[F_N\cdot[j(\omega-\nu\omega_0)]]}{2}\cdot\Big\{n_{\nu+1}^-\cdot c_1+n_{\nu-1}^-\cdot\bar{c}_1+n_{\nu+2}^-\cdot c_2+n_{\nu-2}^-\cdot\bar{c}_2+\cdots n_{\nu+\mu}^-\cdot c_\mu+n_{\nu-\mu}^-\cdot\bar{c}_\mu+\cdots \\
&\qquad +n_{\nu+1}^-[j(\omega-(\nu+1)\omega_0]d_1+n_{\nu-1}^-[j(\omega-(\nu-1)\omega_0]\bar{d}_1+n_{\nu+2}^-[j(\omega-(\nu+2)\omega_0]d_2+\cdots n_{\nu+\mu}^-[j(\omega-(\nu+\mu)\omega_0)]d_\mu+n_{\nu-\mu}^-[j(\omega-(\nu-\mu)\omega_0]\bar{d}_\mu+\cdots\Big\}\cdot e^{j(\omega-\nu\omega_0)\Omega} \\
&+\frac{\mathrm{Re}[F_N[j(\omega+\nu\omega_0)]]}{2}\cdot\Big\{n_{\nu-1}^+c_1+n_{\nu+1}^+\bar{c}_1+n_{\nu-2}^+c_2+n_{\nu+2}^+\bar{c}_2+\cdots n_{\nu-\mu}^+c_\mu+n_{\nu+\mu}^+\bar{c}_\mu+\cdots \\
&\qquad +n_{\nu-1}^+[j(\omega+(\nu-1)\omega_0)]d_1+n_{\nu+1}^+[j(\omega+(\nu+1)\omega_0]\bar{d}_1+n_{\nu-2}^+[j(\omega+(\nu-2)\omega_0)]d_2+\cdots n_{\nu-\mu}^+[j(\omega+(\nu-\mu)\omega_0)]\cdot d_\mu+n_{\nu+\mu}^+[j(\omega+(\nu+\mu)\omega_0]\bar{d}_\mu+\cdots\Big\}e^{j(\omega+\nu\omega_0)\Omega} \\
&+\quad ---\qquad ---\qquad --- \\
&+\quad ---\qquad ---\qquad ---
\end{aligned}
$$

Eq. (33)

$-G_N[j(\omega-4\omega_0)]$	$\frac{1}{2}[\bar{c}_1+j(\omega-3\omega_0)\bar{d}_1]$	———	———	———	$\frac{1}{2}[\bar{c}_5+j(\omega+\omega_0)\bar{d}_5]$	$\frac{1}{2}[\bar{c}_6+j(\omega+2\omega_0)\bar{d}_6]$	$\frac{1}{2}[\bar{c}_7+j(\omega+3\omega_0)\bar{d}_7]$	$\frac{1}{2}[\bar{c}_8+j(\omega+4\omega_0)\bar{d}_8]$		n_4^-		0
$\frac{1}{2}[c_1+j(\omega-4\omega_0)d_1]$	$-G_N[j(\omega-3\omega_0)]$	$\frac{1}{2}[\bar{c}_1+j(\omega-2\omega_0)\bar{d}_1]$	———	———	———	———	———	———		n_3^-		0
$\frac{1}{2}[c_2+j(\omega-4\omega_0)d_2]$	$\frac{1}{2}[c_1+j(\omega-3\omega_0)d_1]$	$-G_N[j(\omega-2\omega_0)]$	$\frac{1}{2}[\bar{c}_1+j(\omega-\omega_0)\bar{d}_1]$	———	———	———	———	$\frac{1}{2}[\bar{c}_6+j(\omega+4\omega_0)\bar{d}_6]$		n_2^-		0
$\frac{1}{2}[c_3+j(\omega-4\omega_0)d_3]$	$\frac{1}{2}[c_2+j(\omega-3\omega_0)\ {}_2]$	$\frac{1}{2}[c_1+j(\omega-2\omega_0)d_1]$	$-G_N[j(\omega-\omega_0)]$	$\frac{1}{2}[\bar{c}_1+j\omega\bar{d}_1]$	———	———	———	———		n_1^-		0
$\frac{1}{2}[c_4+j(\omega-4\omega_0)d_4]$	———	$\frac{1}{2}[c_2+j(\omega-2\omega_0)d_2]$	$\frac{1}{2}[c_1+j(\omega-\omega_0)d_1]$	$-G_N[j\omega]$	$\frac{1}{2}[\bar{c}_1+j(\omega+\omega_0)\bar{d}_1]$	$\frac{1}{2}[\bar{c}_2+j(\omega+2\omega_0)\bar{d}_2]$	$\frac{1}{2}[\bar{c}_3+j(\omega+3\omega_0)\bar{d}_3]$	$\frac{1}{2}[\bar{c}_4+j(\omega+4\omega_0)\bar{d}_4]$	$\cdot$	n_0	=	1
$\frac{1}{2}[c_5+j(\omega-4\omega_0)d_5]$	———	———	$\frac{1}{2}[c_2+j(\omega-\omega_0)d_2]$	$\frac{1}{2}[c_1+j(\omega)d_1]$	$-G_N[j(\omega+\omega_0)]$	$\frac{1}{2}[\bar{c}_1+j(\omega+2\omega_0)\bar{d}_1]$	$\frac{1}{2}[\bar{c}_2+j(\omega+3\omega_0)\bar{d}_2]$	$\frac{1}{2}[\bar{c}_3+j(\omega+4\omega_0)\bar{d}_3]$		n_1^+		0
$\frac{1}{2}[c_6+j(\omega-4\omega_0)d_6]$	———	———	———	$\frac{1}{2}[c_2+j(\omega)d_2]$	$\frac{1}{2}[c_1+j(\omega+\omega_0)d_1]$	$-G_N[j(\omega+2\omega_0)]$	$\frac{1}{2}[\bar{c}_1+j(\omega+3\omega_0)\bar{d}_1]$	$\frac{1}{2}[\bar{c}_2+j(\omega+4\omega_0)\bar{d}_2]$		n_2^+		0
———	———	———	———	———	$\frac{1}{2}[c_2+j(\omega+\omega_0)d_2]$	$\frac{1}{2}[c_1+j(\omega+2\omega_0)d_1]$	$-G_N[j(\omega+3\omega_0)]$	$\frac{1}{2}[\bar{c}_1+j(\omega+4\omega_0)\bar{d}_1]$		n_3^+		0
$\frac{1}{2}[c_8+j(\omega-4\omega_0)d_8]$	$c_7 \ldots d_7$	$c_6 \ldots d_6$	$c_5 \ldots d_5$	$c_4 \ldots d_4$	$c_3 \ldots d_3$	$c_2 \ldots d_2$	$\frac{1}{2}[c_1+j(\omega+3\omega_0)d_1]$	$-G_N[j(\omega+4\omega_0)]$		n_4^+		0

$$\operatorname{Re}\sum_{\mu=-\infty}^{+\infty} m_\mu e^{j(\omega+\mu\omega_0)\Omega} = \qquad \text{Eq. (36a)}$$

$$\frac{\operatorname{Re} F_R(j\omega)}{2}\Big\{2 + m_1^- a_1 + m_1^+ \bar{a}_1 + m_2^- a_2 + m_2^+ \bar{a}_2 + \ldots\ldots + m_\mu^- a_\mu + m_\mu^+ \bar{a}_\mu + \ldots\ldots$$

$$+ m_1^-[j(\omega-\omega_0)]\, b_1 + m_1^+[j(\omega+\omega_0)]\, \bar{b}_1 + \ldots\ldots m_\mu^-[j(\omega-\mu\omega_0)]\, b_\mu + m_\mu^+[j(\omega+\mu\omega_0)]\, \bar{b}_\mu \ldots$$

$$+ r_0 q_0 + r_1^- q_1 + r_1^+ \bar{q}_1 + r_2^- q_2 + r_2^+ \bar{q}_2 + \ldots\ldots r_\mu^- q_\mu + r_\mu^+ \bar{q}_\mu + \ldots\ldots$$

$$+ r_0 (j\omega) p_0 + r_1^-[j(\omega-\omega_0)]\, p_1 + r_1^+[j(\omega+\omega_0)]\, \bar{p}_1 + \ldots\ldots r_\mu^-[j(\omega-\mu\omega_0)]\, p_\mu + r_\mu^+[j(\omega+\mu\omega_0)]$$

$$\bar{p}_\mu \ldots \Big\} e^{j\omega\Omega}$$

$$+ \frac{\operatorname{Re} F_R\, j(\omega-\omega_0)}{2}\Big\{m_1^- a_1 + m_0 \bar{a}_1 + m_3^- a_2 + m_1^+ \bar{a}_2 + \ldots\ldots m_{\mu+1}^- a_\mu + m_{\mu-1}^+ \bar{a}_\mu + \ldots\ldots$$

$$+ m_2^-[j(\omega-2\omega_0)]\, b_1 + m_0(j\omega)\bar{b}_1 + \ldots\ldots m_{\mu+1}^-[j(\omega-(\mu+1)\omega_0)]\, b_\mu + m_{\mu-1}^+[j(\omega+(\mu+1)\omega_0)]\, \bar{b}_\mu + \ldots\ldots$$

$$+ r_1^- q_0 + r_2^- q_1 + r_0 \bar{q}_1 + r_3^- \bar{q}_2 + \ldots\ldots r_{\mu+1}^- q_\mu + r_{\mu-1}^+ \bar{q}_\mu + \ldots\ldots$$

$$+ r_1^-[j(\omega+\omega_0)]\, p_0 + r_2^-[j(\omega-2\omega_0)]\, p_1 + r_0(j\omega)\bar{p}_1 + \ldots\ldots r_{\mu+1}^-[j(\omega-(\mu+1)\omega_0)]\, p_\mu + r_{\mu-1}^+[j(\omega+(\mu-1)\omega_0)]\, \bar{p}_\mu + \Big\} e^{j(\omega-\omega_0)\Omega}$$

$$+ \frac{\operatorname{Re} F_R\, j(\omega+\omega_0)}{2}\Big\{m_0 a_1 + m_2^+ \bar{a}_1 + m_1^- a_2 + m_3^+ \bar{a}_2 + \ldots\ldots m_{\mu-1}^- a_\mu + m_{\mu+1}^+ \bar{a}_\mu \ldots\ldots$$

$$+ m_0(j\omega)\, b_1 + m_2^+[j(\omega+2\omega_0)]\, \bar{b}_1 + \ldots m_{\mu-1}^-[j(\omega-(\mu-1)\omega_0)]\, b_\mu + m_{\mu+1}^+[j(\omega+(\mu+1)\omega_0)]\, \bar{b}_\mu + \ldots$$

$$+ r_1^+ q_0 + r_0 q_1 + r_2^+ \bar{q}_1 + r_1^- q_2 + r_3^+ \bar{q}_2 + \ldots\ldots r_{\mu-1}^- q_\mu + r_{\mu+1}^+ \bar{q}_\mu + \ldots\ldots$$

$$+ r_1^+[j(\omega+\omega_0)]\, p_0 + r_0(j\omega) p_1 + r_2^+[j(\omega+2\omega_0)]\, \bar{p}_1 + \ldots\ldots r_{\mu-1}^-[j(\omega-(\mu-1)\omega_0)]\, p_\mu + r_{\mu+1}^+[j(\omega+(\mu+1)\omega_0)]\, \bar{p}_\mu \ldots \Big\} e^{j(\omega+\omega_0)\Omega}$$

$$+ \ldots\ldots\ldots\ldots\ldots\ldots\ldots\ldots\ldots\ldots\ldots\ldots$$

$$+ \ldots\ldots\ldots\ldots\ldots\ldots\ldots\ldots\ldots\ldots\ldots\ldots$$

Eq. (36 b)

$$\mathrm{Re}\sum_{\mu=-\infty}^{+\infty} r_\mu \, e^{j(\omega+\mu\omega_o)\Omega} =$$

$$\frac{\mathrm{Re}\,F_R(j\omega)}{2}\Big\{ r_1^- k_1 + r_1^+ \bar{k}_1 + r_2^- k_2 + r_2^+ \bar{k}_2 + \dots r_\mu^- k_\mu + r_\mu^+ \bar{k}_\mu + \dots\dots$$

$$+ r_1^- [j(\omega-\omega_o)] l_1 + r_1^+ [j(\omega+\omega_o)] \bar{l}_1 + \dots r_\mu^- [j(\omega-\mu\omega_o)] l_\mu + r_\mu^+ [j(\omega+\mu\omega_o)] \bar{l}_\mu + \dots$$

$$+ m_o g_o + m_1^- g_1 + m_1^+ \bar{g}_1 + m_2^- g_2 + m_2^+ \bar{g}_2 + \dots m_\mu^- g_\mu + m_\mu^+ \bar{g}_\mu + \dots$$

$$+ m_o(j\omega) h_o + m_1^- [j(\omega-\omega_o)] h_1 + m_1^+ [j(\omega+\omega_o)] \bar{h}_1 + \dots m_\mu^- [j(\omega-\mu\omega_o)] h_\mu + m_\mu^+ [j(\omega+\mu\omega_o)] \bar{h}_\mu + \dots \Big\} e^{j\omega\Omega}$$

$$+ \frac{\mathrm{Re}\,F_R\, j(\omega-\omega_o)}{2}\Big\{ r_2^- k_1 + r_o \bar{k}_1 + r_3^- k_2 + r_1^+ \bar{k}_2 + \dots r_{\mu+1}^- k_\mu + r_{\mu-1}^- \bar{k}_\mu + \dots\dots$$

$$r_2^- [j(\omega-2\omega_o)] l_1 + r_o\, j(\omega) \bar{l} + \dots r_{\mu+1}^- [j(\omega-(\mu+1)\omega_o)] l_\mu + r_{\mu-1}^+ [j(\omega+(\mu-1)\omega_o)] \bar{l}_\mu + \dots\dots$$

$$+ m_1^- g_o + m_2^- g_1 + m_o \bar{g}_1 + m_3^- g_2 + m_1^+ \bar{g}_2 + \dots m_{\mu+1}^- g_\mu + m_{\mu-1}^+ \bar{g}_\mu + \dots$$

$$+ m_1^- [j(\omega+\omega_o)] h_o + m_2^- [j(\omega-2\omega_o)] h_1 + m_o(j\omega) \bar{h}_1 + \dots m_{\mu+1}^- [j(\omega-(\mu+1)\omega_o)] h_\mu + m_{\mu-1}^+ [j(\omega+(\mu-1)\omega_o)] \bar{h}_\mu + \dots \Big\}$$

$$e^{j(\omega-\omega_o)\Omega}$$

$$+ \frac{\mathrm{Re}\,F_R\, j(\omega+\omega_o)}{2}\Big\{ r_o k_1 + r_2^+ \bar{k}_1 + r_1^- k_2 + r_3^+ \bar{k}_2 + \dots r_{\mu-1}^- k_\mu + r_{\mu+1}^+ \bar{k}_\mu + \dots$$

$$+ r_o(j\omega) l_1 + r_2^+ [j(\omega+2\omega_o)] \bar{l}_1 + \dots r_{\mu-1}^- [j(\omega-(\mu-1)\omega_o)] l_\mu + r_{\mu+1}^+ [j(\omega+(\mu+1)\omega_o)] \bar{l}_\mu + \dots$$

$$+ m_1^+ g_o + m_o g_1^+ + m_2^+ \bar{g}_1 + m_1^- g_2 + m_3^+ \bar{g}_2 + \dots m_{\mu-1}^- g_\mu + m_{\mu+1}^+ \bar{g}_\mu + \dots$$

$$+ m_1^+ [j(\omega+\omega_o)] h_o + m_o(j\omega) h_1 + m_2^+ [j(\omega+2\omega_o)] \bar{h}_1 + \dots m_{\mu-1}^- [j(\omega-(\mu-1)\omega_o)] h_\mu + m_{\mu+1}^+ [j(\omega+(\mu+1)\omega_o)] \bar{h}_\mu + \dots \Big\}$$

$$e^{j(\omega+\omega_o)\Omega}$$

$$+ \dots\dots\dots\dots\dots\dots\dots\dots\dots\dots$$

$$+ \dots\dots\dots\dots\dots\dots\dots\dots\dots\dots$$

Eq. (37)

$\frac{1}{2}[q_0+j(\omega-3\omega_0)p_0]$									$\frac{1}{2}[\bar{a}_5+j(\omega+2\omega_0)\bar{b}_5]$		m_3^-		0
$-G_0[j(\omega-3\omega_0)]$									$\frac{1}{2}[\bar{g}_5+j(\omega+2\omega_0)\bar{h}_5]$		r_3^-		0
$\frac{1}{2}[q_1+j(\omega-3\omega_0)p_1]$	$-G_R[j(\omega-2\omega_0)]$								$\frac{1}{2}[\bar{a}_4+j(\omega+2\omega_0)\bar{b}_4]$		m_2^-		0
$\frac{1}{2}[k_1+j(\omega-3\omega_0)l_1]$	$\frac{1}{2}[g_0+j(\omega-2\omega_0)h_0]$	$-G_g[j(\omega-2\omega_0)]$							$\frac{1}{2}[\bar{g}_4+j(\omega+2\omega_0)\bar{h}_4]$		r_2^-		0
$\frac{1}{2}[q_2+j(\omega-3\omega_0)p_2]$	$\frac{1}{2}[a_1+j(\omega-2\omega_0)b_1]$	$\frac{1}{2}[q_1+j(\omega-2\omega_0)p_1]$	$-G_R[j(\omega-\omega_0)]$	$\frac{1}{2}[q_0+j(\omega-\omega_0)p_0]$	$\frac{1}{2}(\bar{a}_1+j\omega\bar{b}_1)$	$\frac{1}{2}(\bar{q}_1+j\omega\bar{p}_1)$	$\frac{1}{2}[\bar{a}_2+j(\omega+\omega_0)\bar{b}_2]$	$\frac{1}{2}[\bar{q}_2+j(\omega+\omega_0)\bar{p}_2]$	$\frac{1}{2}[\bar{a}_3+j(\omega+2\omega_0)\bar{b}_3]$		m_1^-		0
$\frac{1}{2}[k_2+j(\omega-3\omega_0)l_2]$	$\frac{1}{2}[g_1+j(\omega-2\omega_0)h_1]$	$\frac{1}{2}[k_1+j(\omega-2\omega_0)l_1]$	$\frac{1}{2}[g_0+j(\omega-\omega_0)h_0]$	$-G_g[j(\omega-\omega_0)]$	$\frac{1}{2}(\bar{g}_1+j\omega\bar{h}_1)$	$\frac{1}{2}(\bar{k}_1+j\omega\bar{l}_1)$	$\frac{1}{2}[\bar{g}_2+j(\omega+\omega_0)\bar{h}_2]$	$\frac{1}{2}[\bar{k}_2+j(\omega+\omega_0)\bar{l}_2]$	$\frac{1}{2}[\bar{g}_3+j(\omega+2\omega_0)\bar{h}_3]$		r_1^-		0
$\frac{1}{2}[q_3+j(\omega-3\omega_0)p_3]$	$\frac{1}{2}[a_2+j(\omega-2\omega_0)b_2]$	$\frac{1}{2}[q_2+j(\omega-2\omega_0)p_2]$	$\frac{1}{2}[a_1+j(\omega-\omega_0)b_1]$	$\frac{1}{2}[q_1+j(\omega-\omega_0)p_1]$	$-G_R(j\omega)$	$\frac{1}{2}(q_0+j\omega p_0)$	$\frac{1}{2}[\bar{a}_1+j(\omega+\omega_0)\bar{b}_1]$	$\frac{1}{2}[\bar{q}_1+j(\omega+\omega_0)\bar{p}_1]$	$\frac{1}{2}[\bar{a}_2+j(\omega+2\omega_0)\bar{b}_2]$	$\cdot$	m_0	=	1
$\frac{1}{2}[k_3+j(\omega-3\omega_0)l_3]$	$\frac{1}{2}[g_2+j(\omega-2\omega_0)h_2]$	$\frac{1}{2}[k_2+j(\omega-2\omega_0)l_2]$	$\frac{1}{2}[g_1+j(\omega-\omega_0)h_1]$	$\frac{1}{2}[k_1+j(\omega-\omega_0)l_1]$	$\frac{1}{2}(g_0+j\omega h_0)$	$-G_G(j\omega)$	$\frac{1}{2}[\bar{g}_1+j(\omega+\omega_0)h_1]$	$\frac{1}{2}[\bar{k}_1+j(\omega+\omega_0)\bar{l}_1]$	$\frac{1}{2}[\bar{g}_2+j(\omega+2\omega_0)\bar{h}_2]$		r_0		0
$\frac{1}{2}[q_4+j(\omega-3\omega_0)p_4]$	$\frac{1}{2}[a_3+j(\omega-2\omega_0)b_3]$	$\frac{1}{2}[q_3+j(\omega-2\omega_0)p_3]$	$\frac{1}{2}[a_2+j(\omega-\omega_0)b_2]$	$\frac{1}{2}[q_2+j(\omega-\omega_0)p_2]$	$\frac{1}{2}(a_1+j\omega b_1)$	$\frac{1}{2}(q_1+j\omega p_1)$	$-G_R[j(\omega+\omega_0)]$	$\frac{1}{2}[q_0+j(\omega+\omega_0)p_0]$	$\frac{1}{2}[\bar{a}_1+j(\omega+2\omega_0)\bar{b}_1]$		m_1^+		0
$\frac{1}{2}[k_4+j(\omega-3\omega_0)l_4]$	$\frac{1}{2}[g_3+j(\omega-2\omega_0)h_3]$	$\frac{1}{2}[k_3+j(\omega-2\omega_0)l_3]$	$\frac{1}{2}[g_2+j(\omega-\omega_0)h_2]$	$\frac{1}{2}[k_2+j(\omega-\omega_0)l_2]$	$\frac{1}{2}(g_1+j\omega h_1)$	$\frac{1}{2}(k_1+j\omega l_1)$	$\frac{1}{2}[g_0+j(\omega+\omega_0)h_0]$	$-G_g[j(\omega+\omega_0)]$	$\frac{1}{2}[\bar{g}_1+j(\omega+2\omega_0)\bar{h}_1]$		r_1^+		0
$\frac{1}{2}[q_5+j(\omega-3\omega_0)p_5]$								$\frac{1}{2}[q_1+j(\omega+\omega_0)p_1]$	$-G_R[j(\omega+2\omega_0)]$		m_2^+		0
$\frac{1}{2}[k_5+j(\omega-3\omega_0)l_5]$											r_2^+		0

The real parts of Eq.(32) must be equal. If one assumes the imaginary parts to be equal as well, the symbol Re in Eq.(32) can be dropped.

The individual terms of the same frequancy in Eq.(32) must be equal. This assumption leads to a set of algebraic equations which can be used for determining the coefficients n_μ. The set of equations in matrix representation yields Eq.(33). Digital computer programs can be used to solve Eq.(33) with the frequency ω as parameter.

Now the corresponding relations for the roll and yaw axis motion will be derived. Later Eq.(33) will be further analyzed. The motion of the roll and yaw axis can be represented by block diagram 7. Again when determining the transfer function one has to compute the steady state output signal, if the system as shown in Fig.7 is excited by the signal Re $e^{j\omega\Omega}$.

It will be proved that the output signal in the roll axis must have the form

$$\varphi(\Omega) = \text{Re} \sum_{\mu=-\infty}^{+\infty} m_\mu e^{j(\omega+\mu\omega_o)\Omega} \tag{34}$$

and due to the coupling between roll and yaw axis a signal for the yaw angle is obtained:

$$\psi(\Omega) = \text{Re} \sum_{\mu=-\infty}^{+\infty} r_\mu e^{j(\omega+\mu\omega_o)\Omega} \tag{35}$$

In our special case ω_o equals unity.

The complex coefficients m_μ and r_μ are to be determined. The method is essentially the same as in the case of the pitch axis. The signal $e^{j\omega\Omega}$ and the output signals $\varphi(\Omega)$ and $\psi(\Omega)$ multiplied by the corresponding time-varying parameters are assumed to be input signals as shown in Fig.7. Two sets of equations are obtained, one for the roll, another one for the yaw axis (Eq.36a and 36b). The coefficients r_μ determine the coupling from the roll to the yaw axis.

In order to determine the transfer function for the yaw-roll motion one has to excite the yaw input of the system in Fig.7 by the signal Re $e^{j\omega t}$. No specific computations are necessary if one interchanges the symbols R by G, a_μ by k_μ, b_μ by l_μ, g_μ by q_μ and h_μ by p_μ in Eq.(36a) and (36b). If one drops the symbol Re in both sides of (36a) and (36b) one obtains Eq.(37) by assuming all the individual frequency terms to be equal. Eq.(37) can be programmed on digital computers and be solved for r_μ and m_μ. The transfer function is defined as

$$W(\Omega; j\omega) = \frac{\text{steady state output signal}}{\text{input signal } e^{j\omega\Omega}} \tag{38}$$

Furthermore the following individual transfer functions can be defined:

Pitch motion

$$W_N(\Omega; j\omega) = \sum_{\mu=-\infty}^{+\infty} n_\mu e^{j\mu\omega_o\Omega} \tag{39}$$

Roll motion

$$W_R(\Omega; j\omega) = \sum_{\mu=-\infty}^{+\infty} m_\mu e^{j\mu\omega_o\Omega} \tag{40}$$

Roll-yaw coupling

$$W_{RG}(\Omega; j\omega) = \sum_{\mu=-\infty}^{+\infty} r_\mu e^{j\mu\omega_o\Omega} \tag{41}$$

Interchanging the indices yields the corresponding transfer functions for yaw and yaw-roll coupling as explained before:

Yaw motion

$$W_G(\Omega; j\omega) = \sum_{\mu=-\infty}^{+\infty} v_\mu e^{j\mu\omega_o\Omega} \tag{42}$$

Yaw-roll coupling

$$W_{GR}(\Omega; j\Omega) = \sum_{\mu=-\infty}^{+\infty} w_\mu e^{j\mu\omega_o\Omega} \tag{43}$$

The numerical evaluation is done with the aid of general digital computer programs. Thereby the matrix equations are solved for discrete frequencies. It has been proved [11] that the method converges for a limited number of μ. Now the problems of stability analysis and steady state solution with the driving functions of Eq.(28) remain to be solved.

During an orbit the satellite is disturbed by periodic driving functions. These inputs are contained in Eq.(28), namely:

$$\begin{aligned} f_R(\Omega) &= \sum_{\nu=1}^{s} \text{Re}\, \alpha_{R_\nu} e^{j(\nu\Omega)} \\ f_N(\Omega) &= \sum_{\nu=1}^{s} \text{Re}\, \alpha_{N_\nu} e^{j(\nu\Omega)} \\ f_G(\Omega) &= \sum_{\nu=1}^{s} \text{Re}\, \alpha_{G_\nu} e^{j(\nu\Omega)} \end{aligned} \tag{44}$$

The general formula for steady state solutions of linear time-varying systems is given in Appendix V. In our special case Eq.(44) and Eq.(V.9) yield the set of equations

$$\begin{aligned} \varphi(\Omega) &= \text{Re} \sum_{\nu=1}^{s} \alpha_{R_\nu} \cdot W_R(\Omega; j\nu\omega_o)\, e^{j(\nu\omega_o\Omega)} \\ &\quad + \text{Re} \sum_{\nu=1}^{s} \alpha_{G_\nu} \cdot W_{GR}(\Omega; j\nu\omega_o)\, e^{j(\nu\omega_o\Omega)} \\ \vartheta(\Omega) &= \text{Re} \sum_{\nu=1}^{s} \alpha_{N_\nu} \cdot W_N(\Omega; j\nu\omega_o)\, e^{j(\nu\omega_o\Omega)} \end{aligned} \tag{45}$$

Eq. (47)

$$\begin{vmatrix}
\ddots & & & & & & & & & 0 & 0 & 0 \\
& & & & & & & & & & 0 & 0 \\
& & & & & & & & & & & 0 \\
& 0 & 0 & 0 & & & & & & & & \\
& F_N[j(\omega-3\omega_0)] & 0 & 0 & & & & & & & & \\
& 0 & F_N[j(\omega-2\omega_0)] & 0 & & & & & & & & \\
& 0 & 0 & F_N[j(\omega-\omega_0)] & & & & & & & & \\
& & 0 & 0 & F_N(j\omega) & 0 & 0 & 0 & & & & \\
& & & 0 & 0 & F_N[j(\omega+\omega_0)] & 0 & 0 & & & & \\
& & & 0 & 0 & 0 & F_N[j(\omega+2\omega)] & 0 & & & & \\
& & & & 0 & 0 & 0 & F_N[j(\omega+3\omega)] & & & & \\
& & & & 0 & 0 & 0 & 0 & & & & \\
0 & & & & & & & & & & & \\
0 & 0 & & & & & & & & & & \\
0 & 0 & 0 & & & & & & & & & \ddots
\end{vmatrix}$$

Eq. (48)

$$\begin{vmatrix}
\ddots & & & & & & & & & & 0 & 0 & 0 \\
& & & & & & & & & & & 0 & 0 \\
& 0 & 0 & 0 & 0 & & & & & & & & 0 \\
& F_R[j(\omega-2\omega_0)] & 0 & 0 & & & & & & & & & \\
& 0 & F_G[j(\omega-2\omega_0)] & 0 & 0 & & & & & & & & \\
& & 0 & F_R[j(\omega-\omega_0)] & 0 & 0 & & & & & & & \\
& & & 0 & F_G[j(\omega-\omega_0)] & 0 & & & & & & & \\
& & & 0 & 0 & F_R(j\omega) & 0 & 0 & & & & & \\
& & & & 0 & 0 & F_G(j\omega) & 0 & 0 & & & & \\
& & & & & 0 & 0 & F_R[j(\omega+\omega_0)] & 0 & & & & \\
& & & & & & 0 & 0 & F_G[j(\omega+\omega_0)] & & & & \\
& & & & & & & 0 & 0 & F_R[j(\omega+2\omega_0)] & & & \\
0 & & & & & & & & & 0 & & & \\
0 & 0 & & & & & & & & & & & \\
0 & 0 & 0 & & & & & & & & & & \ddots
\end{vmatrix}$$

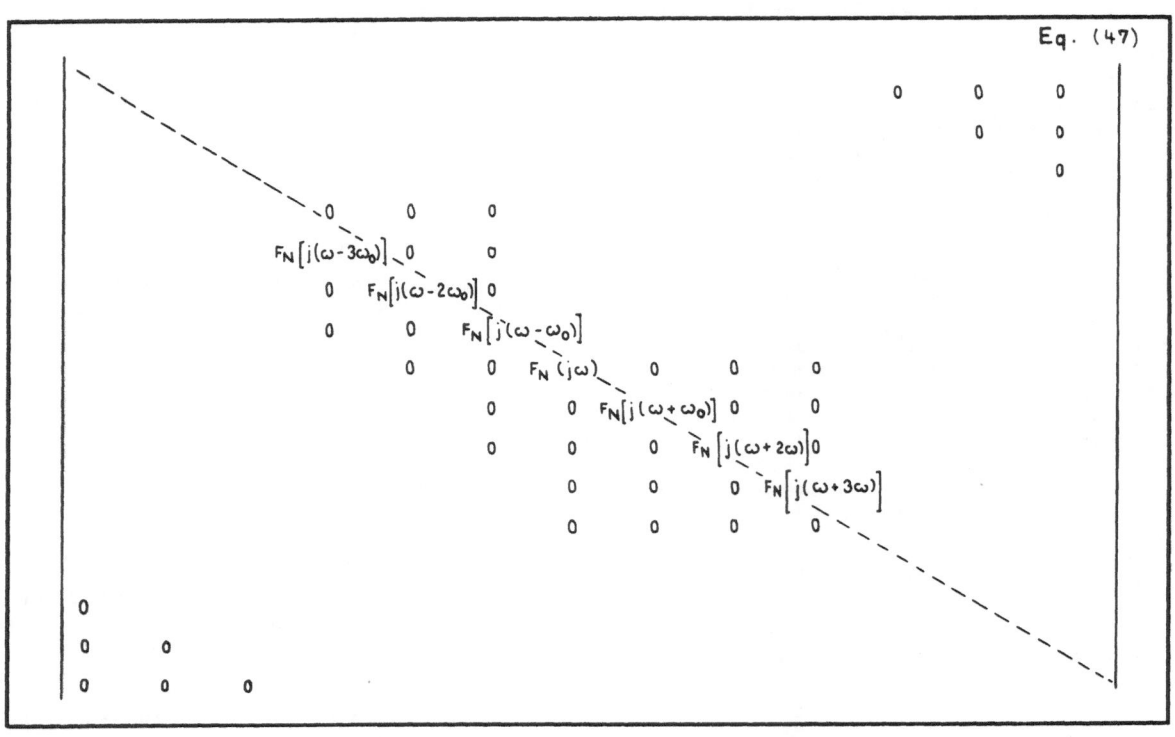

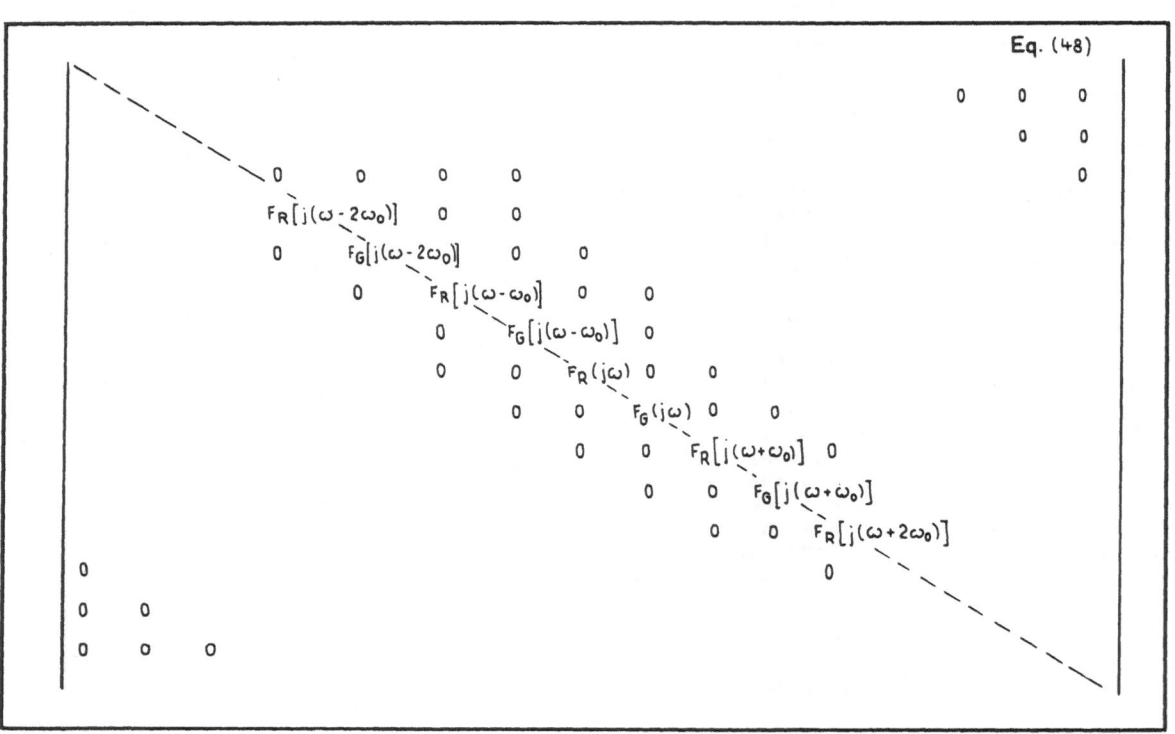

$$\psi(\Omega) = \mathrm{Re}\sum_{\nu=1}^{s} \alpha_{G_\nu} \cdot W_G(\Omega; j\nu\omega_o)\, e^{j(\nu\omega_o\Omega)} + \mathrm{Re}\sum_{\nu=1}^{s} \alpha_{R_\nu} \cdot W_{RG}(\Omega; j\nu\omega_o)\, e^{j(\nu\omega_o\Omega)}$$

The coupling between roll and yaw and vice versa has been taken into account. In our special case ω_o equals unity.

The Nyquist criterion is used for stability analysis. In Appendix V it has been shown that the contour of the function $|F|\cdot|D|$ determines the stability region. The system is stable if the contour does not encircle the origin of the complex plane.

The determinant $|F|$ is given by Eq.(48) and (47), the determinant $|D|$ by Eq.(33) and (37) respectively. $F(j\omega)$ is the transfer function of the fixed parameter part of the differential equation as shown in Eq.(28), without the terms describing the coupling between roll and yaw axis. Thus

$$F_R(j\omega) = \frac{1}{(j\omega)^2 + D_R\, j\omega + K_R} = \frac{1}{G_R}$$

$$F_N(j\omega) = \frac{1}{(j\omega)^2 + D_N\, j\omega + K_N} = \frac{1}{G_N} \tag{46}$$

$$F_G(j\omega) = \frac{1}{(j\omega)^2 + D_G\, j\omega + K_G} = \frac{1}{G_G}$$

The inverse transfer functions are designated by $\frac{1}{F} = G$.

APPENDIX I

1. Definition of Attitude Angles

The orientation of the satellite can be described by the attitude angles φ, θ, ψ (Fig.I.1a). Analogous to the notation used in flight mechanics we denote bank angle by φ, elevation by θ and azimuth by ψ. It should be noted, however, that the definitions used here differ from flight mechanics. The angles are produced by rotations of the body-fixed axes relative to the space-fixed axes in the order θ, φ, ψ. They are positive for clockwise rotation about positive axes.

The relation between body-fixed angular rates and time derivatives of attitude angles follows from Fig.I.1b:

$$\begin{aligned} \Omega_x &= \dot{\varphi}\cos\psi + \dot{\theta}\cos\varphi\sin\psi \\ \Omega_y &= \dot{\theta}\cos\varphi\cos\psi - \dot{\varphi}\sin\psi \\ \Omega_z &= \dot{\psi} - \dot{\theta}\sin\varphi \end{aligned} \tag{I.1}$$

2. Equations of Orbital Motion

According to Kepler's 2nd law the area swept by the radius vector is proportional to time

$$\dot{\Omega} = \frac{c}{r^2}. \tag{I.2}$$

For elliptical orbits the constant c is given by

$$c = \sqrt{\mu a\,(1-\varepsilon^2)}. \tag{I.3}$$

The equation of an ellipse in circular coordinates reads

$$r = \frac{a\,(1-\varepsilon^2)}{1+\varepsilon\cos\Omega}. \tag{I.4}$$

Inserting Eq.(I.3) and (I.4) into (I.2) gives

$$\dot{\Omega} = \frac{(1+\varepsilon\cos\Omega)^2}{(1-\varepsilon^2)^{3/2}} \left(\frac{\mu}{a^3}\right)^{1/2} \tag{I.5}$$

Differentiating Eq.(I.5) once more with respect to t yields

$$\ddot{\Omega} = -\frac{2(1+\varepsilon\cos\Omega)\,\varepsilon\sin\Omega}{(1-\varepsilon^2)^{3/2}} \left(\frac{\mu}{a^3}\right)^{1/2} \dot{\Omega}. \tag{I.6}$$

From Eq.(I.5) and (I.6) follows

$$\frac{\ddot{\Omega}}{\dot{\Omega}^2} = -\frac{2\,\varepsilon\sin\Omega}{1+\varepsilon\cos\Omega}. \tag{I.7}$$

Eq. (I.4) and (I.5) can be expressed as

$$\frac{r^3\dot{\Omega}^2}{\mu} = 1+\varepsilon\cos\Omega. \tag{I.8}$$

The path velocity follows from Eq.(I.4)and(I.5)

$$v = \left(\frac{\mu}{a}\right)^{1/2} \cdot \left[\frac{1+2\,\varepsilon\cos\Omega+\varepsilon^2}{1-\varepsilon^2}\right]^{1/2}. \tag{I.9}$$

The vector of the path velocity and the x'axis form the angle

$$\mathrm{tg}\,\gamma = \frac{\varepsilon\sin\Omega}{1+\varepsilon\cos\Omega}. \tag{I.10}$$

From Eq.(I.5) and (I.9) we obtain

$$\frac{v^2}{\dot{\Omega}^2} = \frac{a^2(1-\varepsilon^2)^2\,(1+2\,\varepsilon\cos\Omega+\varepsilon^2)}{(1+\varepsilon\cos\Omega)^4} \tag{I.11}$$

With Eq.(I.4) and (I.9) we can write

$$\frac{r}{v} = \frac{a^{3/2}\,(1-\varepsilon^2)^{3/2}}{\mu^{1/2}(1+\varepsilon\cos\Omega)(1+2\,\varepsilon\cos\Omega+\varepsilon^2)^{1/2}} \tag{I.12}$$

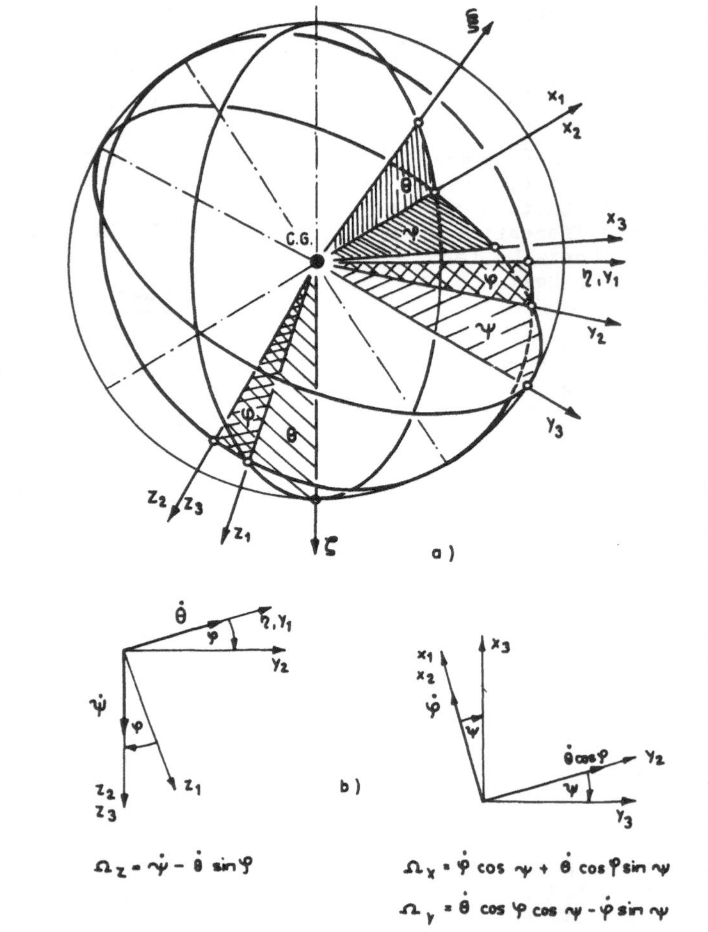

Fig. I.1. Definition of attitude angles. Space-fixed axes ξ, η, ζ; body axes x_3, y_3, z_3.

Fig. II.1. Gravity torque, roll axis

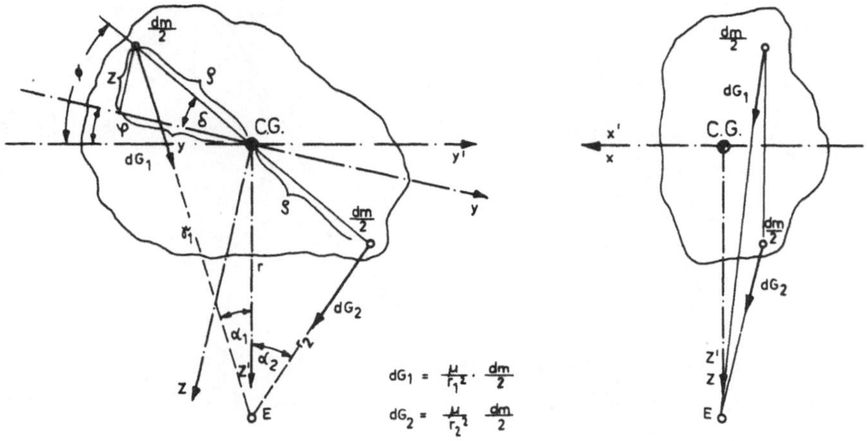

$$dG_1 = \frac{\mu}{r_1^2} \cdot \frac{dm}{2}$$

$$dG_2 = \frac{\mu}{r_2^2} \cdot \frac{dm}{2}$$

$$dL_M = -\mu \frac{dm}{2} \left(\frac{r}{r_1^2} \sin \alpha_1 - \frac{r}{r_2^2} \sin \alpha_2 \right)$$

$$dL_M = -3 \frac{\mu}{r^3} \varrho^2 \sin \phi \cos \phi \, dm$$

$$L_M = 3 \frac{\mu}{r^3} (I_z - I_y) \frac{1}{2} \sin 2\varphi$$

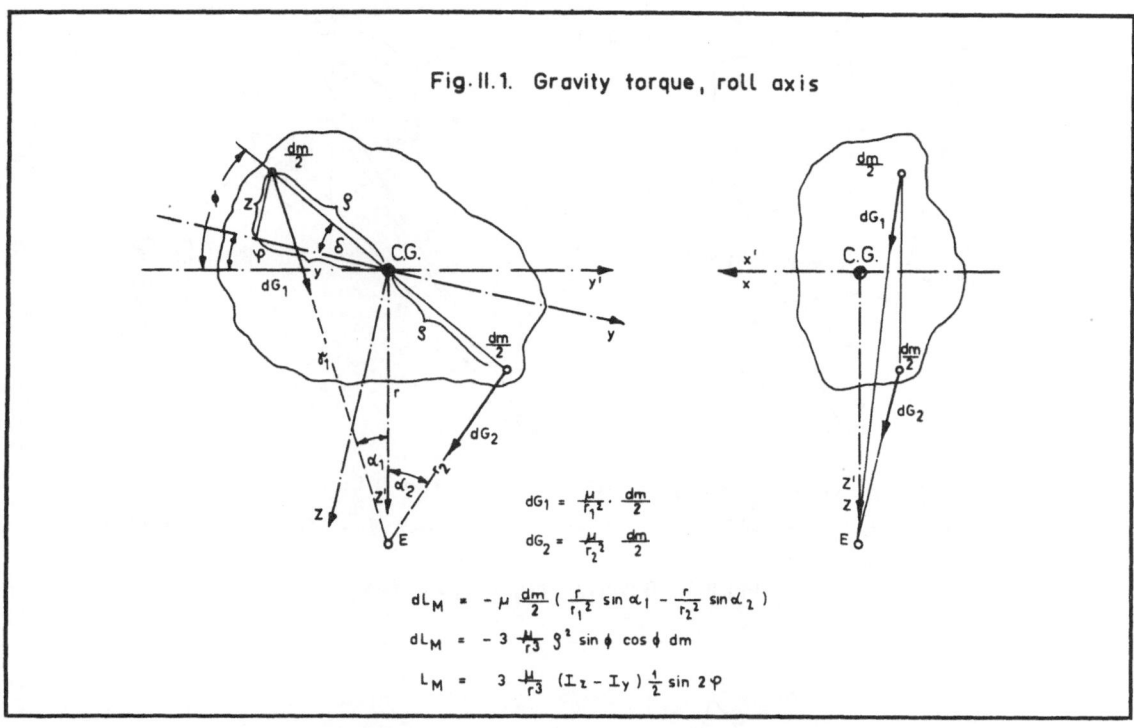

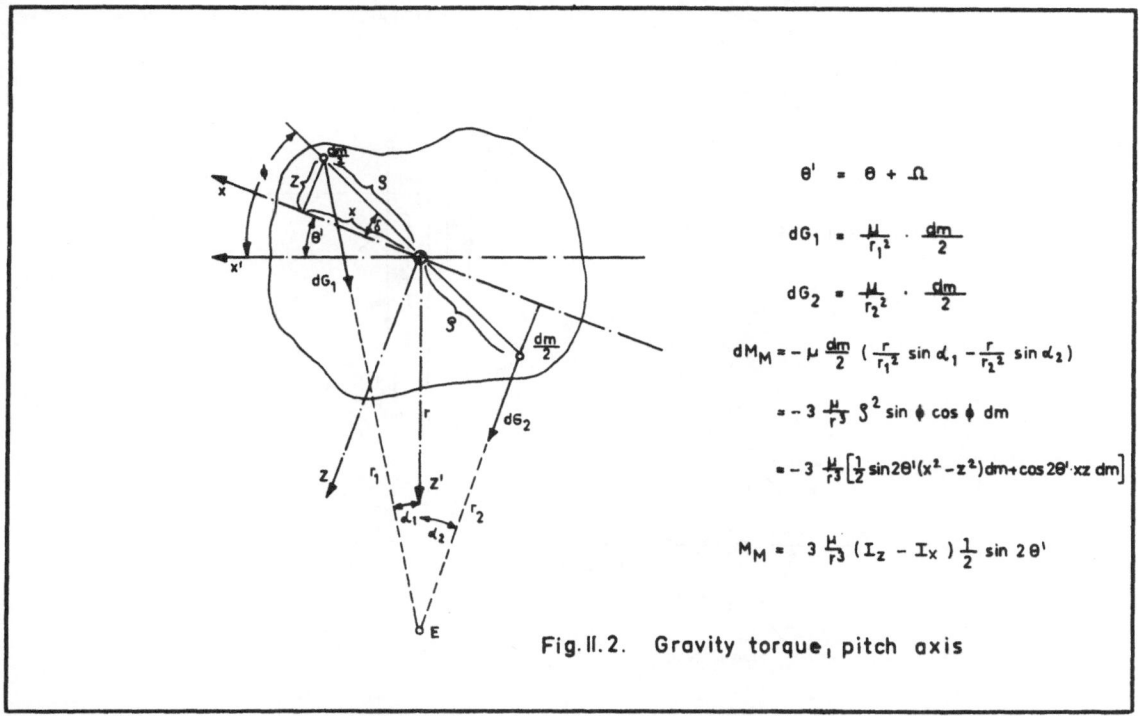

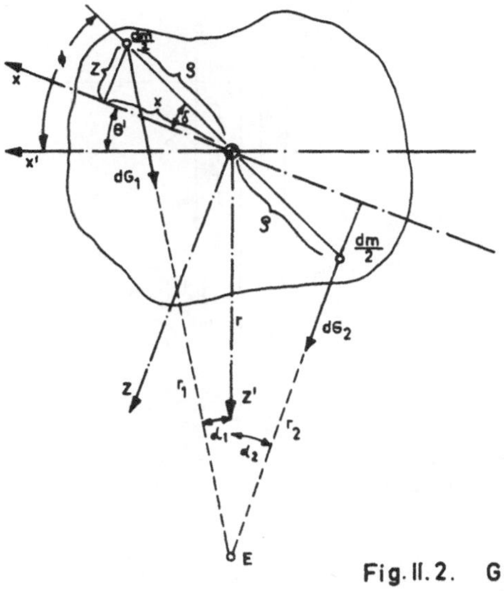

$$\theta' = \theta + \Omega$$

$$dG_1 = \frac{\mu}{r_1^2} \cdot \frac{dm}{2}$$

$$dG_2 = \frac{\mu}{r_2^2} \cdot \frac{dm}{2}$$

$$dM_M = -\mu \frac{dm}{2} \left(\frac{r}{r_1^2} \sin \alpha_1 - \frac{r}{r_2^2} \sin \alpha_2 \right)$$

$$= -3 \frac{\mu}{r^3} \varrho^2 \sin \phi \cos \phi \, dm$$

$$= -3 \frac{\mu}{r^3} \left[\frac{1}{2} \sin 2\theta' (x^2 - z^2) dm + \cos 2\theta' \, xz \, dm \right]$$

$$M_M = 3 \frac{\mu}{r^3} (I_z - I_x) \frac{1}{2} \sin 2\theta'$$

Fig. II.2. Gravity torque, pitch axis

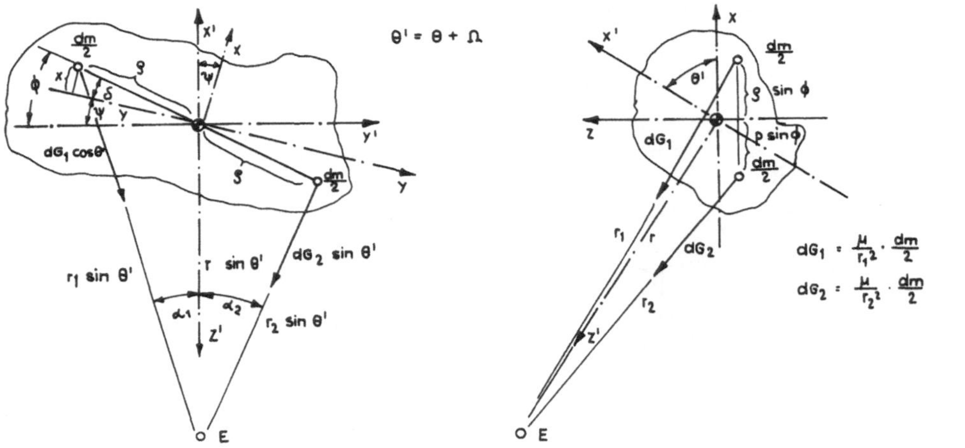

$$dN_M = -\mu \frac{dm}{2} \left(\frac{r}{r_2^2} \sin \alpha_1 - \frac{r}{r_2^2} \sin \alpha_2 \right) \sin^2 \theta'$$

$$r_1 \sin \alpha_1 \sin \theta' = r_2 \sin \theta' = \rho \cos \phi$$

$$r_1 \approx r \left(1 + \frac{\rho}{r} \sin \phi \right)$$

$$r_2 \approx r \left(1 - \frac{\rho}{r} \sin \phi \right)$$

$$N_M = 3 \frac{\mu}{r^3} (I_x - I_y) \frac{1}{2} \sin 2\psi \sin \theta'$$

Fig. II.3. Gravity torque, yaw axis

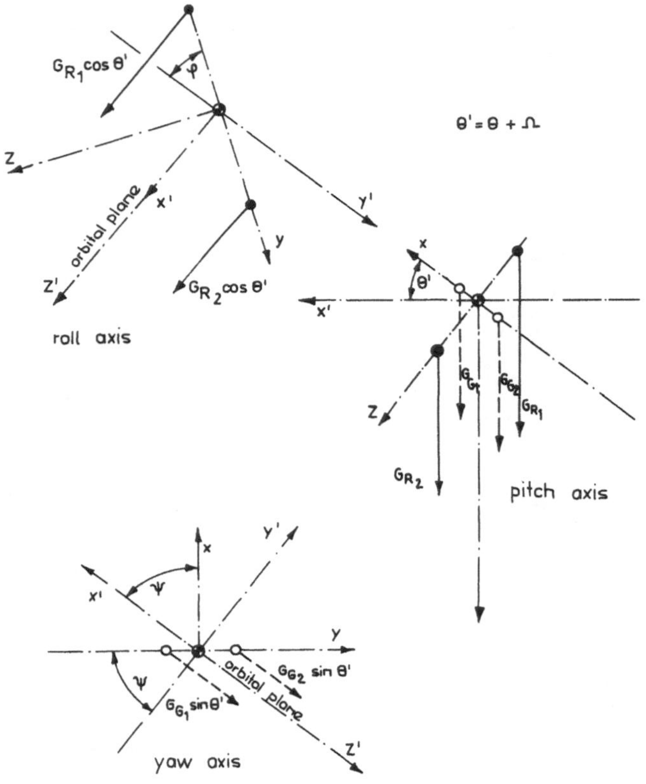

Fig. II.4. Gravity torques for arbitrary reference axes

APPENDIX II: GRAVITY TORQUES

Gravity torques result from the influence of the earth's gravity field upon the satellite. The gravity field here is approximated by a central field. The gravity torques are derived for a body of arbitrary form. The body-fixed coordinate set x, y, z is chosen to coincide with the principal axes so that the products of inertia in the Euler equations vanish.

1. Rolling

Fig. II.1 shows the physical phenomena of the gravity torques about the roll axis (torque about the x-axis). The differential torque is given by

$$d\, L_M = -\mu \frac{dm}{2}\left[\frac{r}{r_1^2} \sin\alpha_1 - \frac{r}{r_2^2} \sin\alpha_2\right]. \quad \text{(II. 1)}$$

With $r_1 \sin\alpha_1 = r_2 \sin\alpha_2 = \varrho \cos\phi$ (II.2)

we get

$$d\, L_M = -\mu \frac{dm}{2}\, r\, \varrho \cos\phi \left[\frac{1}{r_1^3} - \frac{1}{r_2^3}\right]. \quad \text{(II.3)}$$

Setting

$$\begin{aligned} r_1 &\approx r + \varrho \sin\phi = r\left(1 + \frac{\varrho}{r}\sin\phi\right) \\ r_2 &\approx r - \varrho \sin\phi = r\left(1 - \frac{\varrho}{r}\sin\phi\right) \end{aligned} \quad \text{(II.4)}$$

we obtain with $\frac{\varrho}{r} \ll 1$

$$d\, L_M = 3 \frac{\mu}{r^3} \varrho^2 \sin\phi \cos\phi\,. \quad \text{(II.5)}$$

Instead of $\sin\phi \cos\phi$ we can write

$$\sin\phi \cos\phi = \frac{1}{2} \sin 2\phi \frac{y^2 - z^2}{\varrho^2} + \cos 2\varphi \frac{yz}{\varrho^2} \quad \text{(II.6)}$$

Inserting Eq.(II. 6) into (II. 5) we obtain

$$L_M = 3 \frac{\mu}{r^3}\left[\int_k (y^2 - z^2)\, dm\, \frac{1}{2} \sin 2\varphi + \int_k yz\, dm \cos 2\varphi\right] \quad \text{(II.7)}$$

The integrals are the moments of inertia of the body. Because of

$$I_{yz} = \int_k y\, z\, dm = 0 \qquad \text{one gets} \quad \text{(II.8)}$$

$$L_M = 3 \frac{\mu}{r^3} (I_z - I_y) \frac{1}{2} \sin 2\varphi. \quad \text{(II.9)}$$

This equation is valid for the case that x- and x'axes coincide. For a general angle θ the component of rolling moment becomes, according to Fig. II.4,

$$L_M = 3 \frac{\mu}{r^3} (I_z - I_y) \frac{1}{2} \sin 2\varphi \cos(\theta + \Omega) \quad \text{(II.10)}$$

2. Pitching

The situation for the pitch axis (y-axis) can be seen from Fig. II.2. Comparing with Fig. II.1, one sees that the geometry is analogous to the one of the roll axis. Similarly, one gets

$$M_M = 3 \frac{\mu}{r^3} (I_z - I_x) \frac{1}{2} \sin\left[2(\theta + \Omega)\right] \quad \text{(II.11)}$$

3. Yawing

A yawing moment only occurs if x- and x'-axes do not coincide. Fig. II.3 shows how the yawing moment is generated. Similarly to the derivation of rolling and pitching moment, one obtains

$$N_M = 3 \frac{\mu}{r^3} (I_x - I_y) \frac{1}{2} \sin 2\psi \sin(\theta + \Omega) \quad \text{(II.12)}$$

APPENDIX III: AERODYNAMIC TORQUES

Aerodynamic torques are introduced as usually done in flight mechanics:

$$\begin{bmatrix} L_{ae} \\ M_{ae} \\ N_{ae} \end{bmatrix} = \frac{1}{2} \varrho v^2 F\, l \cdot \begin{bmatrix} C_L \\ C_M \\ C_N \end{bmatrix} \quad \text{(III.1)}$$

Reference area F and reference length l depend on the particular configuration. Because in general the elevation θ may change through multiples of π during an orbit, the angle of attack will be of the same order of magnitude. Therefore, the coefficients C_L, C_M and C_N depend not only on the configuration but also on the angle of attack.

1. Air Density

The relation between air density and altitude can be approximated by the exponential law (refer to Fig. III.1)

$$\varrho = \varrho_p\, e^{-k(r - r_p)}\,. \quad \text{(III.2)}$$

Inserting Eq.(I.4) into Eq.(III.2) and taking $\varepsilon^2 \ll 1$ into account, we get

$$\varrho = \varrho_p\, e^{-ka(1 - \varepsilon \cos\Omega) + k\, r_p} \quad \text{(III.3)}$$

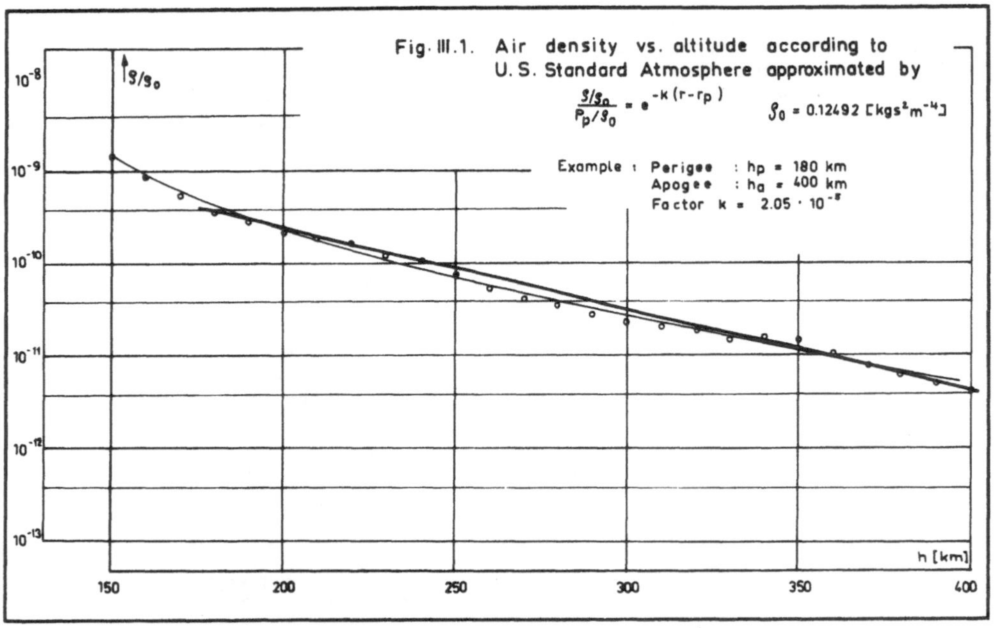

Fig. III.1. Air density vs. altitude according to U.S. Standard Atmosphere approximated by

$$\frac{\rho/\rho_0}{\rho_p/\rho_0} = e^{-k(r-r_p)} \qquad \rho_0 = 0.12492 \; [\text{kgs}^2\text{m}^{-4}]$$

Example : Perigee : h_p = 180 km
Apogee : h_a = 400 km
Factor k = 2.05 · 10^{-5}

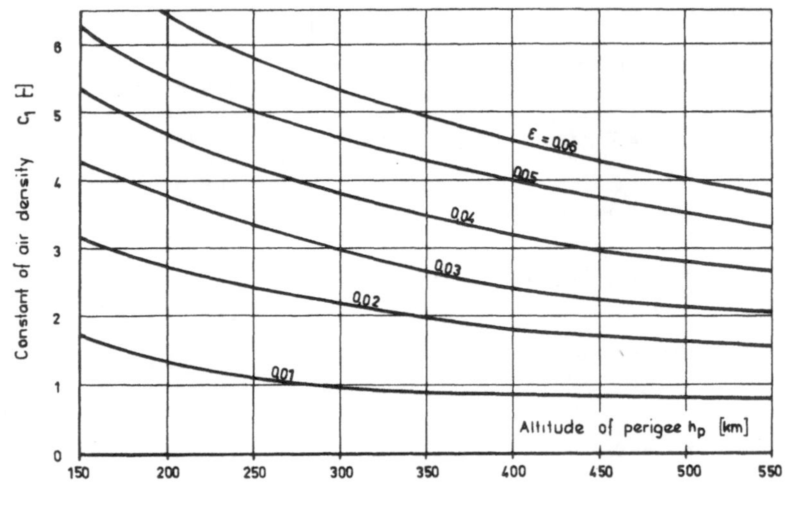

Fig. III.2. Constant C_1 of air density for various eccentricities and perigees

and $$\varrho = \varrho_p C_2 e^{C_1 \cos \Omega} \qquad (III.4)$$

where $$C_1 = k a \varepsilon \qquad (III.4a)$$

$$C_2 = e^{-k a \varepsilon} . \qquad (III.4b)$$

With $$k = \frac{\ln \varrho_p/\varrho_a}{r_a - r_p} \qquad (III.4c)$$

and $$a = \frac{r_p}{1-\varepsilon} \qquad (III.4d)$$

we obtain $$C_1 = \frac{1}{2} \ln \frac{\varrho_p}{\varrho_a} \qquad (III.4e)$$

$$C_2 = \left(\frac{\varrho_a}{\varrho_p}\right)^{1/2} . \qquad (III.4f)$$

Expanding Eq.(III.4) into a Fourier series, we get

$$\varrho = \varrho_p C_2 \left[a_o + \sum_{\nu=1}^{m} a_\nu \cos \nu \Omega\right] \qquad (III.5)$$

where the Fourier coefficients

$$a_o = \frac{1}{2\pi} \int_0^{2\pi} e^{C_1 \cos\Omega} d\Omega \qquad (III.5a)$$

$$a_\nu = \frac{1}{\pi} \int_0^{2\pi} e^{C_1 \cos\Omega} \cos \nu \Omega \, d\Omega \qquad (III.5b)$$

The integrals in Eq.(III.5a) and (III.5b) lead to Bessel functions which are tabulated, for example, in ref. [12]:

$$a_o = I_o (C_1) \qquad (III.5c)$$

$$a_\nu = 2 I_\nu (C_1) . \qquad (III.5d)$$

Eq.(III.5) finally yields

$$\varrho = \varrho_p C_2 \left[I_o (C_1) + 2 \sum_{\nu=1}^{m} I_\nu(C_1) \cos \nu\Omega\right] \quad (III.6)$$

The smaller C_1 the more rapidly the series converges. Values of C_1 for various eccentricities and perigees are shown in Fig. III.2.

2. Coefficients

Near-earth satellites - only these are affected by aerodynamic torques - usually have rotational symmetry. Thus aerodynamic coupling and respective coupling derivatives can be disregarded.

For the coefficients we then can write

$$C_L = \frac{\partial C_L}{\partial \bar{\omega}_x} \bar{\omega}_x \qquad (III.7a)$$

$$C_M = \frac{\partial C_M}{\partial \alpha} \alpha + \frac{\partial C_M}{\partial \bar{\omega}_y} \bar{\omega}_y \qquad (III.7b)$$

$$C_N = \frac{\partial C_N}{\partial \beta} \beta + \frac{\partial C_M}{\partial \bar{\omega}_z} \bar{\omega}_z \qquad (III.7c)$$

where we have introduced non-dimensional angular rates

$$\bar{\omega}_x = \Omega_x \frac{1}{v} \qquad (III.8a)$$

$$\bar{\omega}_y = \Omega_y \frac{1}{v} \qquad (III.8b)$$

$$\bar{\omega}_z = \Omega_z \frac{1}{v} \qquad (III.8c)$$

The damping terms corresponding to the angular rates are small compared with the static terms because the motions about the body axes are relatively slow. Since the main concern of this paper is satellites without spin-stabilization, these damping terms may be neglected.

The static derivatives can be regarded as constant only for small angles of attack; otherwise they depend on the angles of attack. For $C_M(\alpha)$, we may expect a periodic function with non-even symmetry of period 2π. After expansion into a Fourier series we get

$$C_M(\alpha) = \sum_{k=1}^{n_1} b_{k\alpha} \sin k\alpha \qquad (III.9)$$

where the coefficients

$$b_{k\alpha} = \frac{1}{\pi} \int_0^{2\pi} C_M(\alpha) \sin k\alpha \, d\alpha \qquad (III.9a)$$

assuming that $C_M(\alpha)$ is known, perhaps from measurements.

Since the x-z-plane is usually a plane of symmetry for the satellite body we may expect a periodic, even function for $\partial C_N/\partial\beta\,(\alpha)$. A similar Fourier series expansion leads to

$$\frac{\partial C_N}{\partial \beta}(\alpha) = \sum_{i=1}^{n_2} a_{i\alpha} \cos i\alpha \qquad (III.10)$$

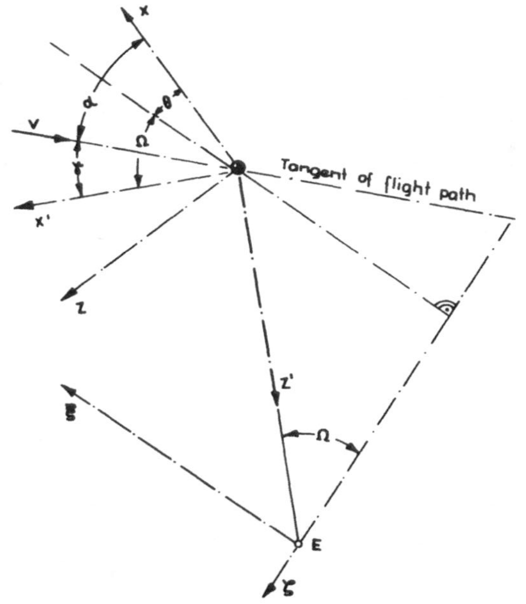

$$\alpha = \theta + \Omega - \gamma$$

$$\gamma \approx \varepsilon \sin \Omega$$

$$\theta = \vartheta - f(\Omega)$$

$$\alpha = \vartheta + \Omega - f(\Omega) - \varepsilon \sin \Omega$$

Fig. III. 3. Definition of angle of attack

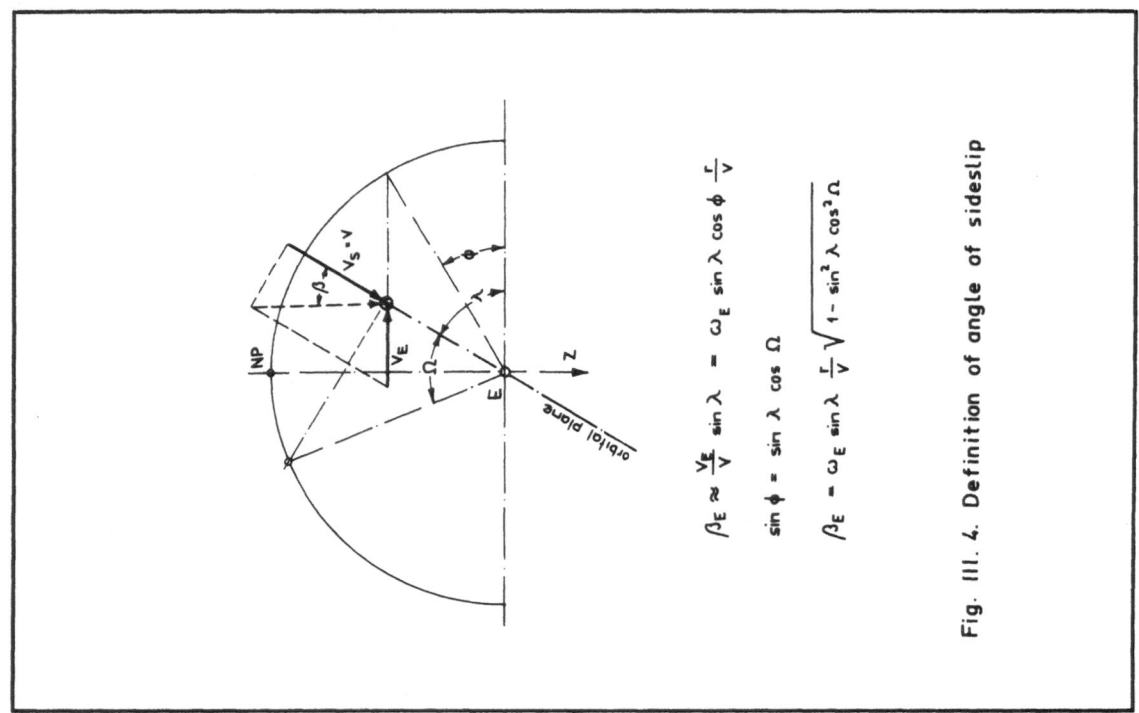

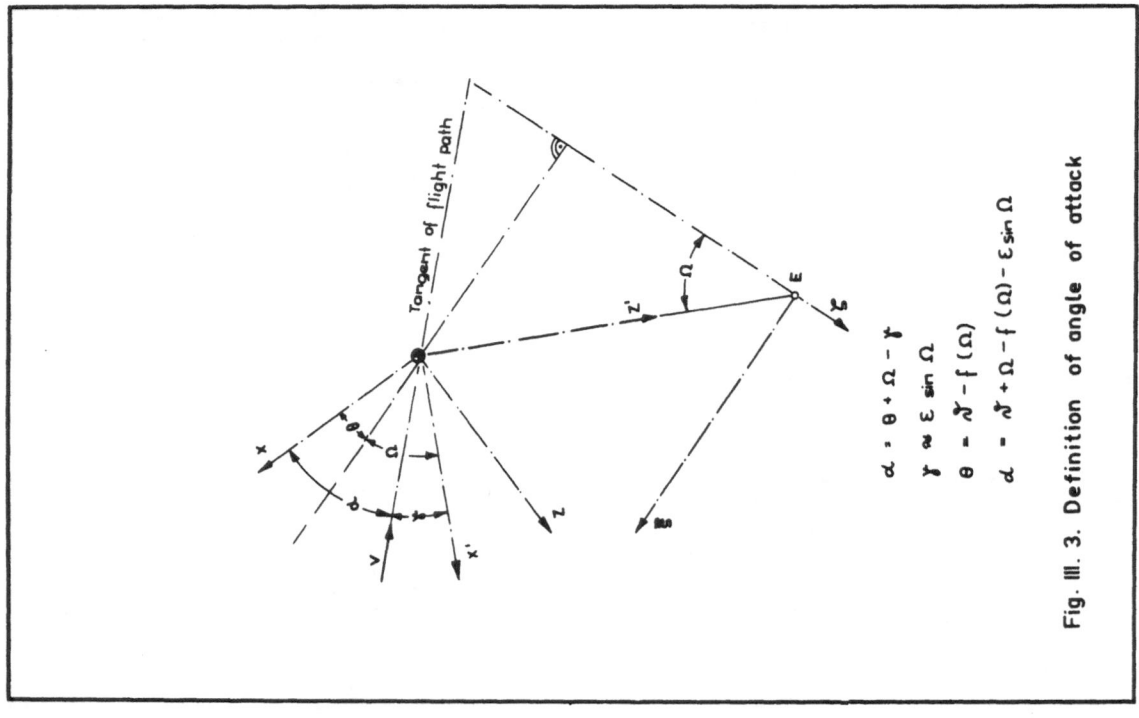

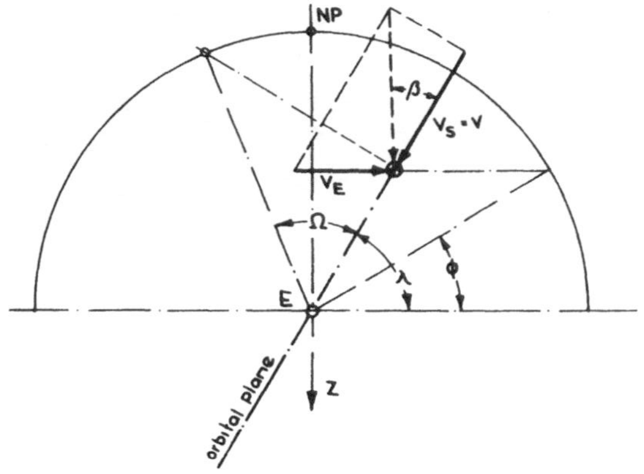

$$\beta_E \approx \frac{V_E}{V} \sin \lambda = \omega_E \sin \lambda \cos \phi \frac{r}{V}$$

$$\sin \phi = \sin \lambda \cos \Omega$$

$$\beta_E = \omega_E \sin \lambda \frac{r}{V} \sqrt{1 - \sin^2 \lambda \cos^2 \Omega}$$

Fig. III. 4. Definition of angle of sideslip

where the coefficients

$$a_{i\alpha} = \frac{1}{\pi}\int_0^{2\pi} \frac{\partial c_N}{\partial \beta}(\alpha)\cos i\alpha \, d\alpha \quad (\text{III}.10a)$$

From Eqs.(III.1), (III.6), (III.9), (III.10) and (I.11), remembering $\varepsilon^2 \ll 1$, one obtains

$$\frac{1}{I_x \dot{\Omega}^2} L_{ae} = 0$$

$$\frac{1}{I_y \dot{\Omega}^2} M_{ae} = C(1-2\varepsilon\cos\Omega)\cdot\left[I_o + 2\sum_{\nu=1}^{m} I_\nu(C_1)\cos\nu\Omega\right]\cdot$$

$$\cdot \frac{1}{I_y}\sum_{k=1}^{n_1} b_{k\alpha} \sin k\alpha$$

$$\frac{1}{I_z \dot{\Omega}^2} N_{ae} = C(1-2\varepsilon\cos\Omega)\cdot\left[I_o + 2\sum_{\nu=1}^{m} I_\nu(C_1)\cos\nu\Omega\right]\cdot$$

$$\cdot \frac{1}{I_z}\sum_{i=1}^{n_2} a_{i\alpha} \cos i\alpha \cdot \beta \quad (\text{III}.11)$$

where the constant $C = \frac{1}{2} Fl\,(\varrho_p \varrho_a)^{1/2}\, a^2$ (III.11a)

depends on the shape of the satellite and of the orbit.

3. Angle of Attack and Angle of Sideslip

Fig.III.3 shows that

$$\alpha = \theta + \Omega - \gamma . \quad (\text{III}.12)$$

Substituting Eq.(3) for θ, one obtains

$$\alpha = \vartheta - f(\Omega) + \Omega - \gamma . \quad (\text{III}.13)$$

From Eq.(I.10) follows, if $\varepsilon^2 \ll 1$,

$$\gamma = \varepsilon \sin\Omega , \quad (\text{III}.14)$$

changing Eq.(III.13) to

$$\alpha = -f(\Omega) + \Omega + \vartheta - \varepsilon\sin\Omega . \quad (\text{III}.15)$$

Yawing of the satellite and rotation of the earth give rise to angles of sideslip. The portion arising from the earth's rotation can be read from Fig.III.4

$$\beta_E = \frac{r\omega_E}{v}\sin\left[1-\sin^2\lambda\cos^2\Omega\right]^{1/2} . \quad (\text{III}.16)$$

Using Eq.(I.4) and remembering $\varepsilon^2 \ll 1$, one obtains

$$\beta_E = \frac{a}{v}(1-\varepsilon\cos\Omega)\omega_E \sin\lambda\left[1-\sin^2\lambda\cos^2\Omega\right]^{1/2} \quad (\text{III}.17)$$

The portion arising from yawing equals the azimuth ψ, so that

$$\beta = \beta_E + \psi \quad (\text{III}.18)$$

Using Eq.(I.12) and inserting it into (III.16) with $\varepsilon^2 \ll 1$ yields

$$\beta = \omega_E\sqrt{\frac{a^3}{\mu}}(1-2\varepsilon\cos\Omega)\sin\lambda\left[1-\sin^2\lambda\cos^2\Omega\right]^{1/2} + \psi \quad (\text{III}.19)$$

From this

$$\beta = C_\beta(1-\varepsilon\cos\Omega)\left[1-\sin^2\lambda\cos^2\Omega\right]^{1/2} + \psi \quad (\text{III}.20)$$

where the constant

$$C_\beta = \omega_E\sqrt{\frac{a^3}{\mu}}\sin\lambda . \quad (\text{III}.20a)$$

APPENDIX IV: MAGNETIC TORQUES

1. The Magnetic Field of the Earth

For the stability analysis in this paper, it is sufficiently accurate to approximate the earth's magnetic field by the field of a magnetic dipole located at the center of the earth and oriented along its axis. The field lines of such a dipole have the form shown in Fig.IV.1 and can be described by the polar-coordinate equation

$$r = r_o \cdot \sin^2\Omega . \quad (\text{IV}.1)$$

The magnitude of the field intensity is given by [6, 13]

$$H_E(r;\Omega) = C_E \cdot \frac{1}{r^3}\cdot\sqrt{4-3\sin^2\Omega} \quad (\text{IV}.2)$$

where Ω = angle between the earth's axis and the radius vector and where $C_E = 8{,}06\cdot 10^{25}$ [Oersted · cm^3].

The angle between H_E and the earth's axis follows from Eq.(IV.1):

$$\delta = \text{arc tg}\,\frac{3\sin 2\Omega}{1+3\cos 2\Omega} . \quad (\text{IV}.3)$$

Also, use is made of the components of H_E in the direction of the radius vector

$$H_r = 2\cdot C_E\,\frac{\cos\Omega}{r^3} \quad (\text{IV}.4)$$

and perpendicular to it, the local "horizontal" component

$$H_\Omega = C_E \cdot \frac{\sin\Omega}{r^3} \quad (\text{IV}.5)$$

Similarly, one obtains the component parallel to the axis of the earth

$$H_Z = C_E\cdot\frac{1}{r^3}\cdot\frac{1}{2}(3+\cos 2\Omega) \quad (\text{IV}.6)$$

and perpendicular to it (in the geographic equator plane)

$$H_X = C_E\cdot\frac{1}{r^3}\cdot\frac{3}{2}\cdot\sin 2\Omega . \quad (\text{IV}.7)$$

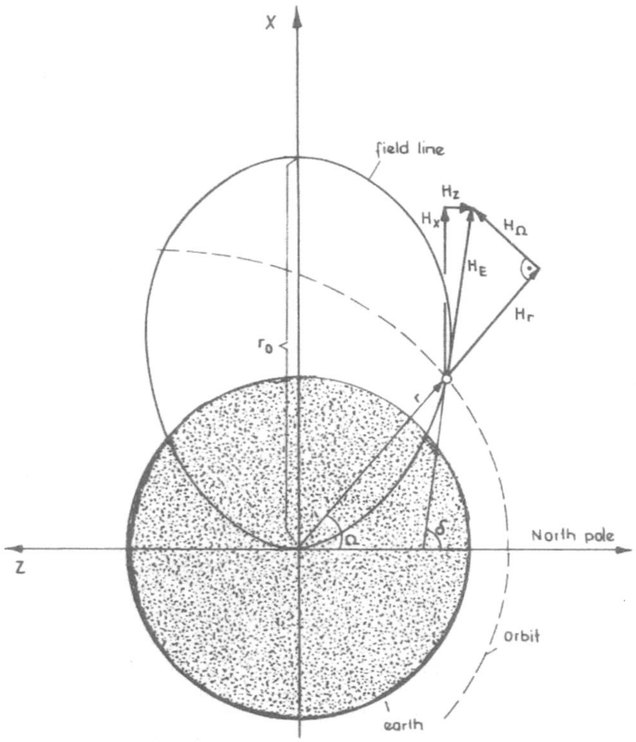

Fig. IV. 1. Components of the earth's magnetic field, and shape of a field line

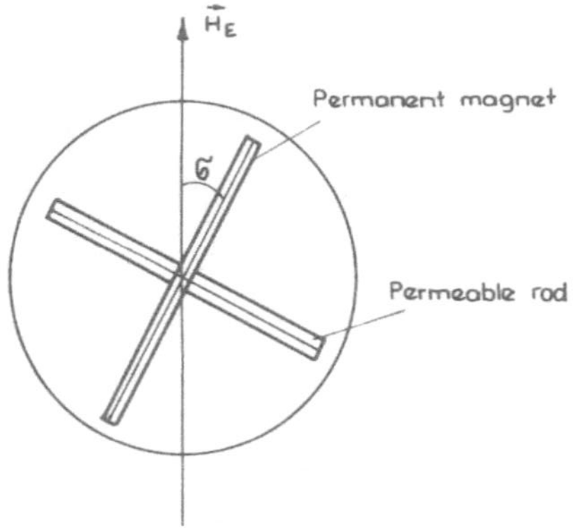

Fig. IV. 2. Permanent magnet and permeable rod for passive stabilization

2. Passive Stabilization

The earth's magnetic field as described, together with bar magnets having the magnetic dipole moment M produce a torque

$$\vec{L} = -\vec{M} \times \vec{H}_E \quad (IV.8)$$

whose magnitude is given by

$$L = -M \cdot H_E \cdot \sin \sigma , \quad (IV.9)$$

where σ denotes the angle between the bar magnet and the vector of the earth's magnetic field; the negative sign shows that the torque tends to decrease the angle σ . Now the magnetic dipole moment of a ferromagnetic rod is proportional to its induction B and its volume V:

$$M = V \cdot B. \quad (IV.10)$$

If the rod consists of a permanent magnet its induction equals approximately the remanent induction B_{rem}; therefore, its torque within the earth's magnetic field is given by

$$L = -H_E \cdot V \cdot B_{rem} \cdot \sin \sigma . \quad (IV.11)$$

This torque of permanent bar magnets may be used for "passive" attitude stabilization. The permanent magnet tends to align itself with the local direction of the earth's magnetic field. If the orbital plane of the satellite coincides with the equator plane of the earth, the attitude will remain approximately constant. In a polar orbit (inclination 90°), however, the satellite will align itself with the direction of the field lines. Because Eq. (IV.3) yields $\delta \approx 2\Omega$ to a first order approximation, the satellite will undergo two revolutions per orbit (Fig.2). However, its angular rate is not constant because of Eq.(IV.3), and if the satellite is to orient itself exactly along the field lines, e.g. for magnetic measurements, angular accelerations are required. Consequently the permanent magnet must be designed not only with regard to disturbing torques but also to these accelerations.

Some numerical values of passively controlled satellites that have been published are shown in Table IV.1, together with corresponding moments of inertia (about an axis perpendicular to the magnet) and with the resulting natural period $T = 2\pi\sqrt{I/M \cdot H_E}$ assuming an average earth's field intensity of H_E = 0,3 Oe.

Table IV.1	Magnetic moment M [dyn.cm/Oe]	Moment of inertia I [g cm²]	Natural period T sec for H_E=0,3 Oe
I N J U N (according to Fischell)	$2,5 \cdot 10^3$	$4,4 \cdot 10^6$	480
T R A N S I T 3B (according to Fischell)	$7 \cdot 10^4$	$1,1 \cdot 10^8$	450
E S R O I	$2,1 \cdot 10^4$	$5 \cdot 10^7$	560

Dimensions

Usually, the field intensity H_E is measured in [Oersted] but induction B in [Gauss = 10^{-8} V · sec/cm³]. Then Eq.(IV.11) reads:

$$L\,[dyn.cm] = -\frac{1}{4\pi} \cdot H_E\,[Oe] \cdot V[cm^3] \cdot B_{rem}\,[Gauss] \cdot \sin \sigma \quad (IV.12)$$

and

$$L\,[m \cdot kp] = \frac{10^{-3}}{9,81} \cdot D\,[dyn. \cdot cm]$$

Note that in the literature on geomagnetics sometimes also the dimension for field intensity is called Gauss instead of Oersted, which is not related to the above-mentioned dimension Gauss of induction.

3. Magnetic Damping

Highly permeable rods of special magnetic materials may be used for damping, using the effect that the oscillation energy is removed by hysteresis energy losses.

A long permeable rod is magnetized, due to the demagnetization effect [14], only along its length. According to Fischell [6], the hysteresis energy losses are at a maximum if the magnetization varies between positive and negative values. Oscillations of the rod by an angle σ therefore are damped maximally if the rod is oriented perpendicular to the field line, that means also perpendicular to the permanent magnet (Fig.IV.2). Then the rod is magnetized by the component $H_\sigma = H_E \sin \sigma$ of the magnetic field. Analogous to the permanent magnet, this magnetization produces a magnetic dipole moment $M = V \cdot B(H_\sigma)$ and a torque $L = -M \cdot H_E \cdot \cos \sigma$ (cosine because the rod is oriented perpendicular to H_E). The torque now is given by

$$L = -V \cdot H_E \cdot B\,(H_\sigma) \cdot \cos \sigma \quad \text{where}$$

$$H_\sigma = H_E \cdot \sin \sigma .$$

Thus the damping torque is a function of the hysteresis characteristic B (H). As this characteristic is a two-valued function, no analytical expression is possible for the exact time function of the damped oscillation. One can only get the hysteresis energy loss per cycle by integrating over the hysteresis characteristic, from which an average damping is derived. In this paper therefore an average damping factor is used.

In section II and III it has been shown that periodic input signals act upon the satellite. When the stationary state is reached, the energy losses of all cycles are equal. The stationary solution has been derived in general form in section IV. In computing the average damping moments, the following procedure is followed: First, a certain damping torque is assumed. Then, the angular displacements φ, ϑ, ψ are

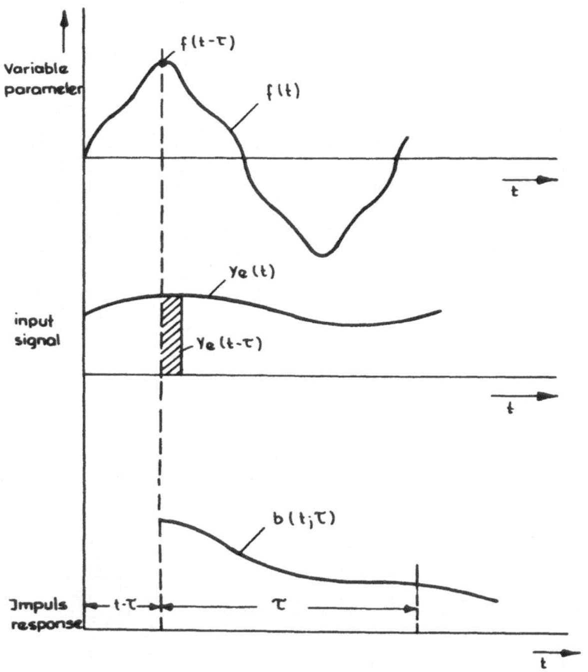

Fig. V. 1. Impulse response, input signal and time - varying parameter

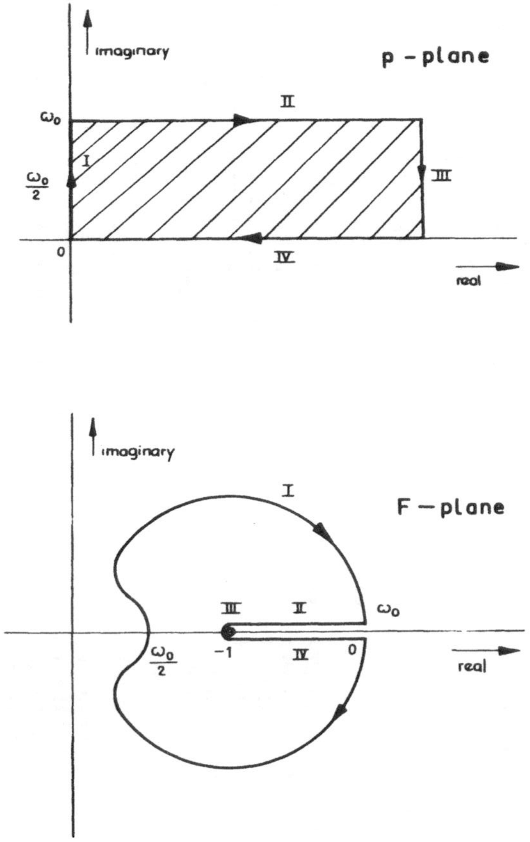

Fig. V.2. Path of integration in p and F plane

computed as functions of Ω, following section IV. For example, the stationary solution of the pitch motion in response to periodic input signals is obtained as

$$\vartheta(\Omega) = \mathrm{Re} \sum_{\nu=1}^{s} \alpha_{N_\nu} W_N(\Omega; j\nu\omega_o) e^{j\nu\omega_o\Omega} \quad (IV.13)$$

The quantities α_N denote the amplitudes of the input signals with frequencies $\nu\omega_o$ (Eq.44).

On the other hand, the hysteresis energy loss per cycle can be determined. Generally

$$E = V \cdot \oint_{1\ \text{period}} H\, dB \quad (IV.14)$$

As mentioned above, for field intensity H the component $H_E \cdot \sin \vartheta$ has to be substituted. Thus one obtains for small ϑ

$$E = \oint_{1\ \text{period}} H_E \cdot \vartheta(\Omega) \cdot d\,B, \quad (IV.15)$$

where B is given by the hysteresis characteristic B(H) and ϑ by Eq.(IV.13).

Now one calculates the energy loss within the system (Fig.6) caused by the damping assumed:

$$E = D \int_{1\ \text{period}} \dot{\Omega}\, \vartheta'^2 (\Omega)\, d\Omega\,. \quad (IV.16)$$

Eq.(IV.13) is substituted for Eq.(IV.16) giving

$$E = D \cdot \left(\frac{\mu}{a^3}\right)^{1/2} \cdot \sum_{\nu=1}^{s} |\alpha_{N\nu}|^2 \sum_{\lambda=-\infty}^{\infty} |n_\lambda(\nu)|^2 \cdot (\nu+\lambda)^2. \quad (IV.17)$$

where D denotes the damping assumed. Eqs. (IV.15 and (IV.17) must be equal. Thus the necessary permeable rods can be determined.

4. Active Stabilization by Magnetic Coils

A plane coil with coil area F, consisting of n windings, carrying current i, has the magnetic dipole moment

$$M = \mu_o \cdot n\, i\, F. \quad (IV.18)$$

The vector of the dipole moment is oriented perpendicular to the coil plane. For the torque again Eq.(IV.8) is valid. In Eq.(IV.18) there is $\mu_o = 4\pi \cdot 10^{-9} \frac{V \cdot S}{A \cdot cm}$. With i [A] and F [$cm^2$] again the dimension of moment is M [V·sec.cm]. With regard to the dimension of torque, refer to Eq.(IV.12).

For the practical design of a coil it is interesting to know its weight and the power consumed. A circular coil has the area $F = R^2 \cdot \pi$ and circumference $2R \cdot \pi$. Thus the total length of the coil wire is $l = n \cdot 2R\pi = 2n\sqrt{\pi \cdot F}$. The resistance is given by $\gamma \cdot \rho \cdot l^2/G$ where G = weight of wire, γ = specific weight and ρ = specific resistance. The power consumed is $N = i^2 \cdot \gamma \rho\, l^2/G$. Substituting this in Eq. (IV.18) one obtains

$$M = \frac{\mu_o}{2} \cdot \sqrt{\frac{N F G}{\pi \gamma \rho}} \quad (IV.19)$$

Thus the torque increases only with the root of power and coil weight. Given a certain power and weight, the torque increases with coil area; consequently, one will choose coil diameters as large as permitted by the form of the satellite.

Furthermore, Eq.(IV.19) shows that the product $\gamma \cdot \rho$ should be made as small as possible. Table IV.2 shows ρ and γ for copper and aluminium.

Table IV.2

	ρ [Ωcm]	γ[p/cm^3]	$\rho \cdot \gamma$[p·Ω/cm^2]
copper	$1,7 \cdot 10^{-6}$	8,9	$15 \cdot 10^{-6}$
aluminium	$3 \cdot 10^{-6}$	2,7	$8 \cdot 10^{-6}$

Consequently, aluminium is far better than copper.

APPENDIX V: METHODS FOR THE ANALYSIS OF SYSTEMS HAVING TIME-VARYING COEFFICIENTS

The theoretical aspects for the analysis of systems having periodical time-varying parameters are discussed here. The method has been theoretically developed in [8] and is discussed in detail in [7] and [15].

A system weighting function as the response to a unit impulse is defined for the linear time-varying system. It is shown in Fig.V.1 that the impulse response depends on the time t related to the time-varying parameter and on the "age" or time τ elapsed since the impulse. The superposition theorem is valid and thus the response to an arbitrary driving function is obtained to be

$$y_a(t) = \int_{-\infty}^{+\infty} b\,(t;\tau) \cdot y_e\,(t-\tau)\, d\tau, \quad (V.1)$$

$b(t;\tau)$ being the system weighting function. The driving function $y_e(t)$ is replaced by its Fourier transform:

$$y_e(t-\tau) = \frac{1}{2\pi} \int_{-\infty}^{\infty} Y_e\,(j\omega)\, e^{j\omega(t-\tau)}\, d\omega \quad (V.2)$$

Substituting Eq.(V.2) into (V.1) gives

$$y_a(t) = \frac{1}{2\pi} \int_{-\infty}^{\infty} b(t;\tau) \left\{ \int_{-\infty}^{+\infty} Y_e(j\omega)\, e^{j\omega(t-\tau)} d\omega \right\} d\tau \quad (V.3)$$

By changing the order of integration one obtains

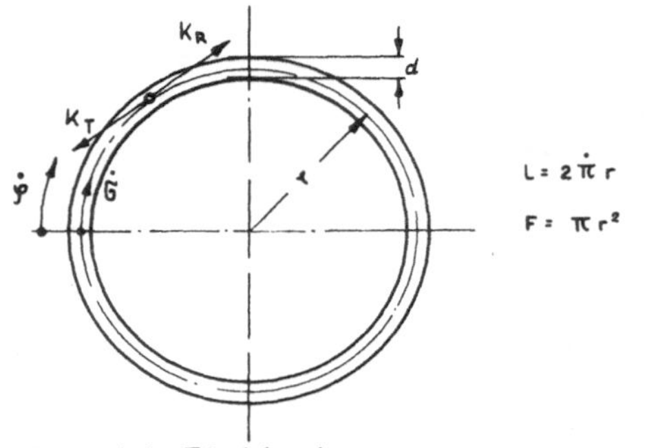

Fig. VI.1 Fluid damper

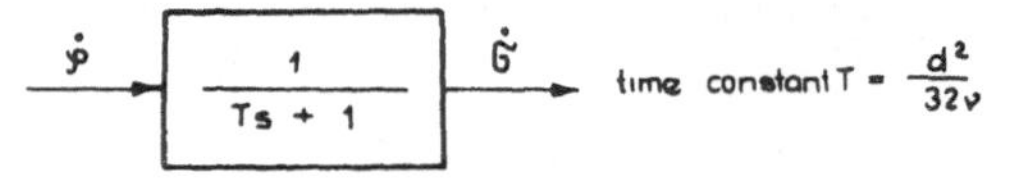

Fig. VI. 2 Signal flow diagram of fluid damper

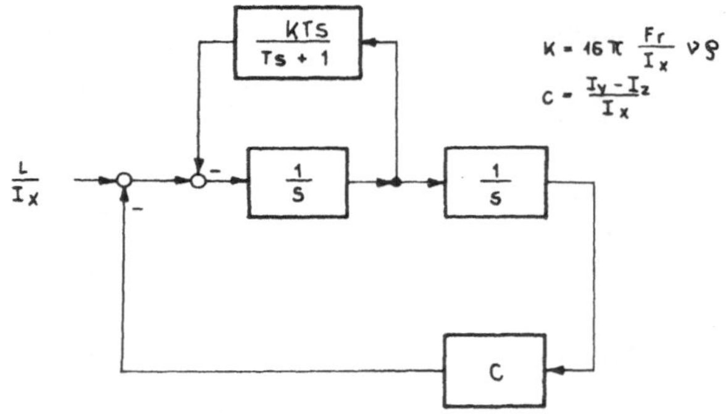

Fig. VI.3 Block diagram of one axis of the satellite with fluid damper

$$y_a(t) = \frac{1}{2\pi} \int_{-\infty}^{+\infty} Y_e(j\omega) \cdot$$

$$\cdot \left\{ \int_{-\infty}^{+\infty} b(t;\tau) e^{-j\omega\tau} d\tau \right\} e^{j\omega t} d\omega. \qquad (V.4)$$

The Fourier transform in the curled brackets is defined as transfer function of the linear time-varying system $W(t;j\omega)$. Only the age variable τ is transformed. $W(t;j\omega)$ must depend on the time t:

$$W(t;j\omega) = \int_{-\infty}^{+\infty} b(t;\tau)\, e^{-j\omega\tau} d\tau. \qquad (V.5)$$

Substituting Eq.(V.5) into Eq.(V.4) yields

$$y_a(t) = \frac{1}{2\pi} \int_{-\infty}^{+\infty} Y_e(j\omega) \cdot W(t;j\omega) e^{j\omega t} d\omega \quad (V.6)$$

Eq.(V.6) corresponds to the well-known relation of the fixed parameter system theory. The output signal is the inverse Fourier transform of the transfer function times the input $Y_e(j\omega)$. Therefore a function

$$Y_a(t;j\omega) = W(t;j\omega) \cdot Y_e(j\omega)$$

is defined. The transfer function $W(t;j\omega)$ is [8].

$$W(t;j\omega) = \frac{\text{stationary output of the system}}{\text{input signal } e^{j\omega t}} \qquad (V.7)$$

In section IV this relation is used to determine the transfer function.

Eqs.(V.5) and (V.7) can be used to solve all problems related to linear time-varying systems. Stability analysis and steady state output signals are primarily important for our satellite problems.

Firstly we investigate the steady state solution of the system having periodic driving functions. The Fourier transform of the driving function is to be

$$Y_e(j\omega) = \pi \cdot \left\{ \sum_{0}^{r} \left\{ (A_\mu - jB_\mu) \cdot \delta(\omega - \mu\omega_0) + \right.\right.$$

$$\left.\left. + (A_\mu + jB_\mu)\, \delta(\omega + \mu\omega_0) \right\} \right\}. \qquad (V.8)$$

Substituting Eq.(V.8) into (V.6) we obtain

$$y_a(t) = \frac{1}{2} \sum_{0}^{r} \left\{ (A_\mu - jB_\mu)\, W(t;j\mu\omega_0) e^{j\mu\omega_0 t} + \right.$$

$$\left. + (A_\mu + jB_\mu) W(t;-j\mu\omega_0) e^{-j\mu\omega_0 t} \right\}$$

$$= \mathrm{Re} \sum_{0}^{r} (A_\mu - jB_\mu) \cdot W(t;j\mu\omega_0)\, e^{j\mu\omega_0 t} \qquad (V.9)$$

which is the solution.

Secondly we make a stability analysis. The matrix representation of Eq.(33) and (37), respectively, is considered. Eq.(33) gives the solution for pitch axis motion:

$$n_\mu = \frac{|DN_\mu|}{|D_N|} \qquad (V.10)$$

$|D|$ is the determinant of the system of Eq.(33) or (37), resp. $|D_\mu|$ is the same determinant, except that the μ-th column has been replaced by the right side of Eq.(33).

All poles of $|D_\mu|$ must be poles of D. Therefore only the zeros of $|D|$ must be considered for stability analysis.

The infinite determinants shown in Eq.(33) and (37), converge because the functions $F[j(\omega+\mu\omega_0)]$ given by Eq.(46) approach zero as $\omega \to \infty$. The numerator of the expansion of the infinite determinant results in a polynomial of infinite degree. It can be seen from Eq. (33) and (37) that the roots of this polynomial are spaced periodically in the p plane, thus if p_0 is a root then $p_0 + j\mu\omega_0$ must be roots also, where μ is an integer. All the independent roots of the numerator of the expanded determinant lie therefore in a strip limited by two lines parallel to the real axis spaced by ω_0. These roots determine the stability of the system because they are the zeros of the determinant.

The Nyquist criterion is used for stability analysis. We consider the change of the angle in the p plane, when taking the contour shown in Fig.V.2. No poles lie on the contour. No singularities will be located within the strip in Fig.V.2 if the net change of the angle around the contour is zero.

The convergence of the problem is better if the numerator and the denominator of Eq.(V.10) are multiplied by the diagonal determinant $|F|$ given by Eq.(47) and (48).

For stability analysis, the $|F|\cdot|D|$ plane is used. The imaginary part of the contour from 0 to ω_0 is mapped in Fig.V.2 as contour I. Hence the polynomial is periodic, the value for $\omega=0$, $\omega=\omega_0$ and $\omega=\omega_0/2$ must be real as Eq.(33) and (37) show. The parallel lines II and IV in the p plane are mapped in the $|F|\cdot|D|$ plane as the real axis, as shown in Fig.V.2. The line III in infinite distance from the imaginary axis in the p plane converges to the point one in the $|F|\cdot|D|$ plane.

The only contribution to the angle is due to the contour I. Therefore, for stability analysis, it is sufficient to investigate whether the frequency plot of the determinant $|F|\cdot|D|$ encircles the origin or not.

APPENDIX VI: FLUID DAMPER

To damp out oscillatory motions of the satellite, the wall friction of a fluid within a closed tube can be used. Fig. VI.1 shows a toroidal tube filled with fluid.

Representative for all axes, damping of the roll axis motion is considered. An angular rate $\dot{\varphi}$ of the satellite causes a motion $\dot{\sigma}$ of the fluid by means of wall friction. The torque produced opposing the satellite motion is proportional to the difference of the velocities between the fluid column and the tube fixed to the frame of the satellite. Laminar flow may be assumed because of the small angular velocities occuring in attitude stabilized satellites; thus the friction force will be [16]

$$K_R = \Delta p_{lam.} \cdot \frac{\pi d^2}{4} = \frac{64}{Re} \cdot \frac{L}{d} \cdot \frac{1}{2} \varrho v^2 (\dot{\varphi} - \dot{\sigma})^2 \frac{\pi d^2}{4} \qquad \text{(VI.1)}$$

Putting Reynold's number $Re = \frac{r(\dot{\varphi} - \dot{\sigma})}{\nu}$, gives

$$K_R = 16 \pi F \nu \varrho (\dot{\varphi} - \dot{\sigma}) \qquad \text{(VI.2)}$$

where ν is the kinematic viscosity, ϱ density.

This friction force is balanced by the inertia force

$$K_T = \varrho \pi F \frac{d^2}{2} \ddot{\sigma} \qquad \text{(VI.3)}$$

giving the differential equation

$$\frac{d^2}{32 \nu} \ddot{\sigma} + \dot{\sigma} = \dot{\varphi} \qquad \text{(VI.4)}$$

The block diagram with the corresponding transfer functions is shown in Fig.VI.2. The damping torque applied to the satellite body follows from Eq.(VI.2):

$$\frac{L_D}{I_x} = - k (\dot{\varphi} - \dot{\sigma}) \qquad \text{(VI.5)}$$

where the constant $K = 16 \pi \frac{F \cdot r}{I_x} \nu \varrho$. The block diagram of the satellite including the fluid damper is shown in Fig. VI.3, yielding the transfer function of the complete system:

$$\frac{\varphi}{L/I_x} = \frac{T s + 1}{T s^3 + (k T + 1) s^2 + c T s + c} \qquad \text{(VI.6)}$$

It can be seen that the differential equation of the roll axis has the damping term $cT\dot{\varphi}$ and a further term $T\ddot{\varphi}$.

If a system design contains a fluid damper as treated here, the additional terms of the differential equations must be considered. The easiest way to do it is in the differential equations with the independent variable t. After this, the true anomaly Ω can be substituted as independent variable.

APPENDIX VII: NOMENCLATURE

a	Major semiaxis of orbit ellipse
C_L; C_M; C_N	Coefficients of aerodynamic torques
D	Damping coefficient; also Symbol for determinant
$f_R(\Omega)$; $f_N(\Omega)$; $f_G(\Omega)$	Periodic driving functions in roll, pitch and yaw axis
h	Altitude of satellite above earth
I_x; I_y; I_z	Principal moments of inertia of the satellite
I_ν	Bessel functions
K_R; K_N; K_G / D_R; D_N; D_G	Abbreviated coefficients for roll, pitch and yaw axis
L	Rolling torque
M	Pitching torque; in Appendix IV: Magnetic dipole moment
N	Yawing torque
r	Radius vector
v	Flight path velocity
Re	Real part
α	Angle of attack
$\alpha_{R\mu}$; $\alpha_{N\mu}$; $\alpha_{G\mu}$	Fourier coefficients of periodic driving functions in roll, pitch and yaw axis
β	Angle of sideslip
γ	Angle of flight path
ε	Numerical eccentricity
θ	Elevation angle
ϑ	Angular displacement of pitch axis
λ	Inclination of orbital plane
$\mu = \sigma m_E$	Gravity parameter $3{,}986 \cdot 10^5$ $km^3 s^{-2}$
ϱ	Air density
σ	Gravity constant $6{,}668 \cdot 10^{-8}$ $cm^3 g^{-1} s^{-2}$
φ	Bank angle, angular displacement of roll axis
ϕ	Geographical latitude
ψ	Azimuth, angular displacement of yaw axis
Ω	True anomaly
Ω_x; Ω_y; Ω_z	Angular velocities about body axes
ω_E	Angular velocity of earth

Suffices

a	Apogee
ae	Aerodynamic
E	Earth
H	Magnetic field of the earth
M	Mechanic
p	Perigee
x; y; z	in direction of body axes
μ, ν	Indices of summation

References

[1] BELETSKIJ, V.V.: "Libration of a Satellite in an Elliptical Orbit" (transl.from Russ.) Artificial Earth Satellites 1961, p.42-53. Plenum Press Inc., New York

[2] GURKO, O.V. and SLABKKIJ, L.I.: "The Use of Forces derived from the Solar Gravitational and Radiation Fields for the Orientation of Cosmic Devices". (transl.from Russ.), Artificial Earth Satellites 1963, p.30-41. Plenum Press Inc., New York

[3] SCHRELLO, D.M.: "Dynamic Stability of Aerodynamically Responsive Satellites". J.of Aerospace Sc. 29(1962), number 10, pp.1145-1155, 1163

[4] De BRA, D.B.: "The Large Attitude Motions and Stability, due to Gravity, of a Sattellite with Passive Damping in an Orbit of Arbitrary Eccentricity about an Oblate Body". Stanford University, SUDEAR 126 (May 1962)

[5] MAGNUS, K.: "Beiträge zur Untersuchung der Drehbewegungen starrer Satelliten auf kreisförmigen Umlaufbahnen". Jahrbuch 1963 der WGL, pp.174-180

[6] FISCHELL, R.E.: "Magnetic and Gravity attitude Stabilization of Earth Satellites". Report CM-996, The Johns Hopkins Univ., Appl. Phys. Lab., May 1961

[7] SCHWEIZER, G.: "Regelkreise mit periodisch sich ändernden Parametern". Regelungstechnik 11(1963), pp.165-169, 196-201

[8] ZADEH, L.A.: "Frequency Analysis of variable Networks", Proc. IRE 38 (1950), pp.291-299

[9] LEONHARD, A.: "Das Verfahren der Beschreibungsfunktion, erweitert für die Untersuchung von parametererregten Schwingungen". Automatic and Remote Control, Proceedings of the 2nd Congr.of IFAC, vol."Theory". Butterworth London and R.Oldenbourg Munich - Vienna, 1965

[10] LAUBER, R.: "Ein neues Verfahren zur Bestimmung der Beschreibungsfunktion aus der Differentialgleichung". Automatic and Remote Control, Proc.of the 2nd Congr. of IFAC, vol."Theory", Butterworth London and R.Oldenbourg Munich - Vienna, 1965

[11] COOLEY, W.W., CLARK, R.N. and BUCKNER, R.: "Stability in Linear Systems having a Time-Variable Parameter". IEEE Trans. on Automatic Control, AC-9 (Oct.1964) pp.426-434

[12] JAHNKE, E. and EMDE, F.: "Funktionentafeln" Teubner-Verlag, Leipzig and Berlin, 1933

[13] JACOBI, W.: "Das Strahlungsfeld im interplanetarischen Raum und seine Einwirkung auf den Menschen im Raumflug". Astronautica Acta, vol. X, pp.150-137

[14] WESTPHAL, W.H.: "Physik", Springer-Verlag Berlin - Göttingen - Heidelberg 1953

[15] SCHWEIZER, G.: "The Analysis of Systems with Periodically Time-Varying Parameters" Proc. Joint Automatic Control Conf., Univ.of Minnesota, Minneapolis/Minn., June 63

[16] KAUFMANN, W.: "Technische Hydro- und Aeromechanik". Springer-Verlag, 1954

DISCUSSION

F. Mesch and G. Schweizer

"Investigation of
Attitude Controlled
Satellites with Magnetic Phenomena".

Q. Is the control loop closed through a ground station?

A. It could be closed either through a ground station or in the satellite. The satellite has not been built.

Q. Did you consider elliptic orbits?

A. Stability is not effected if eccentricity is small. The large disturbances are aerodynamic and magnetic.

Q. Why are only two control loops used?

A. In a polar orbit (considered in this paper) the earth's magnetic field has only two components. For more general orbits a more complicated system would be required.

Q. You have linearized your analysis? Have you studied the validity of this assumption?

A. No, but we think the approximation is good for the small angles to be expected.

Q. What is the experimental accuracy?

A. Unfortunately we don't so far have satellites in Europe.

SPIN CONTROL FOR EARTH SATELLITES

by

Robert E. Fischell
Space Power and Attitude Systems Group Supervisor
The Johns Hopkins University
Applied Physics Laboratory
Silver Spring, Maryland

ABSTRACT

One of the simplest and most frequently used attitude control systems for earth satellites consists of spinning the vehicle about an axis of maximum moment of inertia. Two difficulties that occur with this attitude control system are that magnetic drag will decrease the satellite spin rate and the spin axis will lose its initial orientation due to a variety of perturbing torques. A magnetic torqueing system has been developed at the Applied Physics Laboratory for controlling spin rate and spin axis orientation. The spin rate control system consists of two vector magnetometers mounted perpendicular to the satellite's spin axis whose output is amplified and fed into the coil of an electromagnet. As the satellite spins, this system provides a constant amplitude magnetic dipole moment that is always perpendicular to the component of the earth's magnetic field in the X-Y plane of the satellite. The resulting torque can be used to cause the satellite to spin faster or slower, depending on whether the resulting dipole moment leads or lags the component of the earth's magnetic field in the X-Y plane. Changing from lead to lag can easily be accomplished by a reversing switch operated by radio command; thereby allowing the satellite to have a completely adjustable spin rate. Typical weight for this system is three pounds; typical power consumption, when activated, is less than 2 watts. Control of spin axis orientation is accomplished by precessing the spin axis using a large magnetic dipole mounted along the satellite's Z-axis. This dipole can be magnetized to any state by discharging a condenser through a winding wrapped around a permanent magnet core. The magnet strength and polarity can be adjusted by radio command from a ground station. Typical weight for the spin axis orientation control system is three pounds and there is no steady drain of electrical power. An electronic digital computer has been programmed to calculate the times at which the magnet should be turned on and off to precess the spin axis to any desired orientation.

INTRODUCTION

One of the simplest and most reliable attitude control systems for earth satellites consists of fixing the position of one satellite axis by means of gyroscopic torques resulting from spin about that axis. Among the satellites that have been stabilized in this manner are Vanguard,[1] several Explorer satellites,[2] all Tiros satellites,[3] Alouette,[4] and Syncom,[5] to name just a few. The earliest satellites merely allowed the satellites to spin because it was more convenient than despinning them, and no attempt was made to orient the spin axis after separation from the launch vehicle. However, as space technology developed, two significant improvements have been required for spin controlled satellites: (1) control of spin rate, and (2) control of spin axis orientation.

CONTROL OF SPIN RATE

A satellite spinning in the earth's magnetic field will lose its angular kinetic energy from induced eddy currents in electrical conducting materials and from eddy currents and hysteresis loss in magnetic materials. For example the Vanguard satellite (1958 $\beta 2$) decreased in spin rate from 2.7 rps to 0.55 rps in a period of one year.[6] For this satellite, the principal loss of spin rate was due to eddy currents in its aluminum shell. The satellite 1960-$\eta 1$ despun from 0.785 to less than 0.002 rps in a period of 24 days mostly by magnetic hysteresis loss.[7]

For spin attitude control to be most useful it is often necessary to prevent the decay of spin rate. For some more sophisticated missions, it may be necessary to vary the orbiting satellite's spin rate either up or down at various times during its useful life. Probably the first system employed to alter spin rate was three pairs of spin-up rockets mounted on the outside cylindrical surface of the Tiros I satellite.

A magnetic system for controlling the spin rate of an earth satellite has been designed at the Applied Physics Laboratory for inclusion on two different satellites under the sponsorship of the National Aeronautics and Space Administration, Goddard Space Flight Center: (1) the Atmosphere Explorer satellite (AE-B), and (2) the Direct Measurements Explorer satellite (DME-A). Each of these spacecraft employs a magnetic spin rate control system as indicated in block diagram form in Figure 1. In this system, vector magnetometers are used to sense the magnetic field along the X- and Y- axes of the satellite. The Z-axis will always be taken as the spin axis of

Superior numbers refer to similarly-numbered references at the end of this paper.

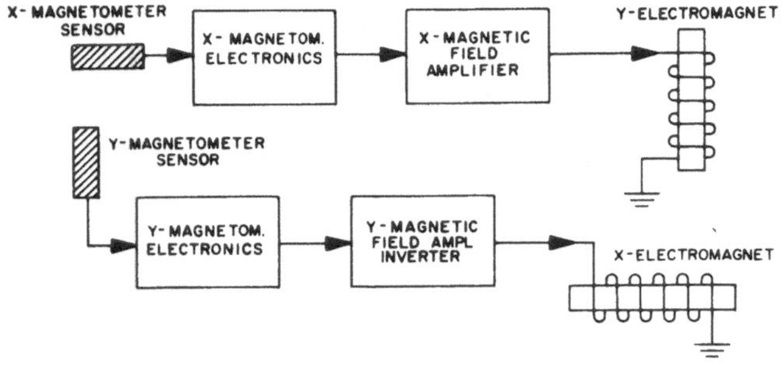

FIGURE 1

BLOCK DIAGRAM OF SPIN CONTROL SYSTEM

Figure 2

DAILY CHANGE OF SPIN RATE AS A FUNCTION OF ALTITUDE

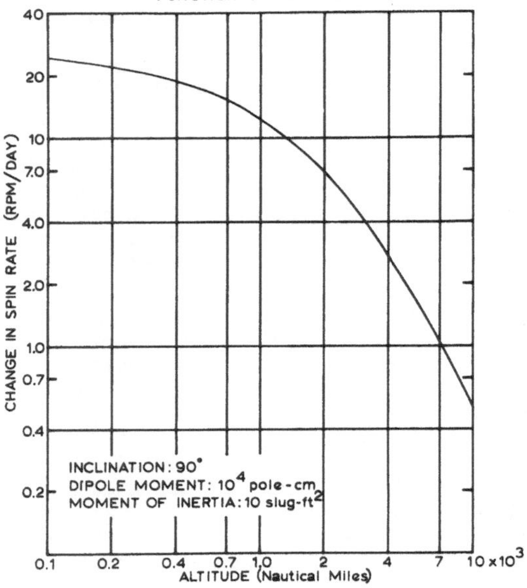

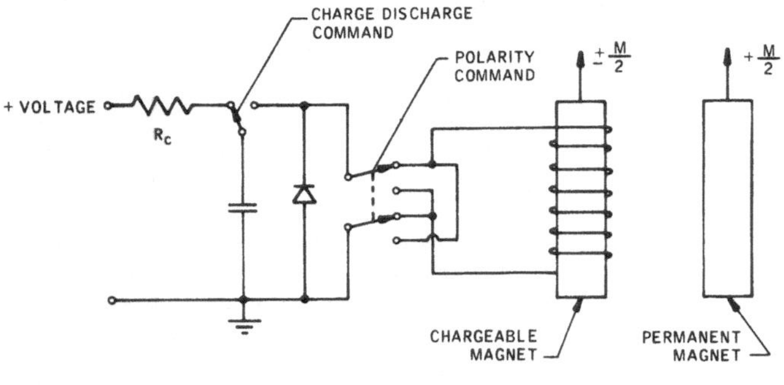

FIGURE 3. CHARGEABLE MAGNET SYSTEM

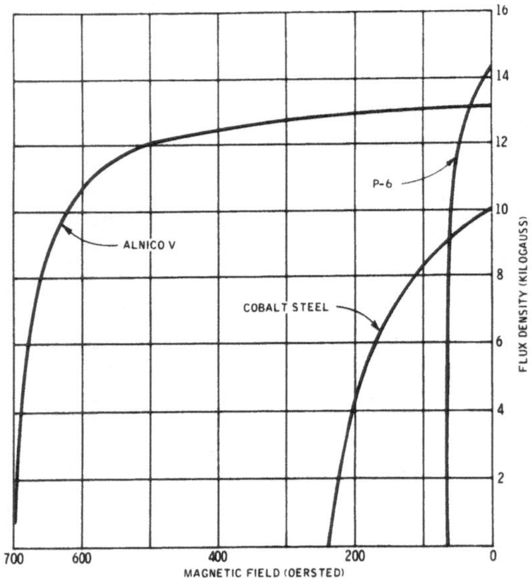

FIGURE 4. DEMAGNETIZATION CURVES FOR ALNICO V, COBALT STEEL AND P-6

the satellite. The output of the X-magnetometer is amplified and fed into a winding of an electromagnet along the satellite's Y-axis. In a similar manner the Y-magnetometer output is amplified, but in this case inverted in sign, and fed into a winding of an electromagnet in the X-direction. If H_{xy} is the projection of the earth's magnetic field in the satellite's X-Y plane, this system of magnetometers and electromagnets will create a magnetic dipole, M_s, in the satellite that is always perpendicular to H_{xy}. The torque then acting on the satellite about its Z-axis is given by

$$\tau_z = M_s H_{xy} \quad \text{(dyne-cm)} \tag{1}$$

where M_s = satellite dipole moment (pole-cm),

and H_{xy} = projection of the earth's magnetic field in the satellite's X-Y plane (oersted).

Depending on which way that satellite is rotating, it will either cause the satellite to increase or decrease its spin rate. If despin results when spin-up is desired, reversing the polarity of the output of both the X and Y amplifiers will accomplish this objective. For either spin-up or despin, the time rate of change of spin rate about the Z-axis will be given by

$$\frac{d\omega_z}{dt} = \tau_z / I_z \quad \text{(radians/sec}^2\text{)} \tag{2}$$

where I_z = spin moment of inertia (gm-cm^2).

ω_z = spin rate about the Z-axis (radians/sec).

When the spin system is in operation there is also a torque acting on the satellite that tends to make it precess. This torque is given by

$$\vec{\tau} = \vec{M}_s \times \vec{H} \quad \text{(dyne-cm)} \tag{3}$$

where $\vec{H}$ = total magnetic field vector (oersted).

If the satellite is spinning rapidly, and if the product $\vec{M}_s \times \vec{H}$ is reasonably small, the induced precession can be kept negligibly small.

If it is desired to keep a constant spin rate, the satellite electronics can include a device to measure the spin rate (viz, by counting zero crossings of one magnetometer) and by comparing this rate to that generated from some accurate timing device such as a tuning fork. The power to the spin-up electronics can be turned on if the rate is too low and kept off when the spin rate exceeds some specified value.

The maximum rate of spin is only limited by the frequency response of the magnetometers and electromagnets. Since this will be on the order of 1,000 cps, it is desirable to provide the satellite with a centrifugal switch for high rate cut-off. This avoids the possibility of the satellite destroying itself from centrifugal force due to an inadvertent command to the spin-up mode. The DME-A spacecraft currently being designed at the Applied Physics Laboratory incorporates a centrifugal switch that turns off spin control power if the satellite spin rate exceeds 25 rps.

Both the AE-B and DME-A satellites require that the satellite spin axis be orthogonal to the orbital plane. Since both these spacecraft are to be placed in near polar orbits, the earth's magnetic field always has a large component in the satellite's X-Y plane. Both these satellites have been designed to provide a spin dipole moment, $M_s = 10^4$ pole-cm. At an average altitude, 400 nautical miles, a spacecraft with $I_z = 10$ slug-ft^2, and $M_s = 10^4$ has the capability of approximately 20 rpm per day change in spin rate.

The DME-A spin control system dissipates a total of 2.0 watts when it operates. To maintain the satellite at the desired spin rate one must add to the spin energy that which is removed by eddy current losses in the satellite's structure and by hysteresis losses in the electromagnets and other magnetic materials in the satellite. Since this is very little energy indeed, it will be possible to operate the satellite spin-up system on a duty cycle of less than 0.1 percent. The average energy dissipation of less than 0.002 watts is completely negligible for this mission.

The total system weight for the DME-A spin control system is 3.8 pounds. This weight includes those parts of the command switching that are used to operate the spin rate control system. The X and Y magnetometers that are used for attitude detection as well as spin rate control are included in the total of 3.8 pounds.

Since the magnetic field intensity decreases inversely as the cube of the distance from the center of a dipole, the earth's magnetic field strength at very high altitudes is drastically reduced. Therefore the capability for spin-up also decreases sharply at higher altitudes. Figure 2 shows the capability for varying spin rate as a function of orbital altitude. This curve is for polar orbiting satellites with the $M_s = 10^4$ pole-cm, $I_z = 10$ slug-ft^2 and with the spin axis oriented perpendicular to the orbital plane.

Compared to using rockets or cold gas jets for maintaining the spin rate of the satellite, the magnetic system described above has the advantage that there is no dissipation of materials and therefore essentially unlimited utilization capability. Also the absence of moving

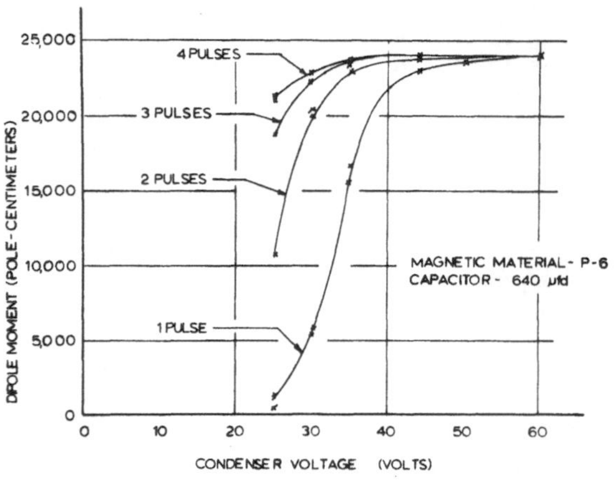

Fig. 5 CHARGEABLE MAGNET CHARACTERISTICS

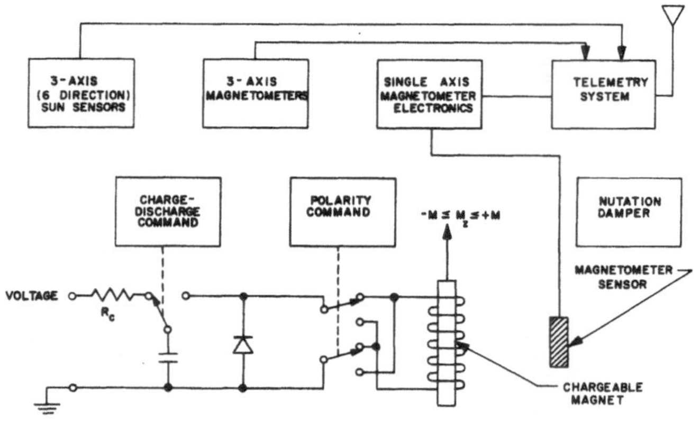

FIGURE 6
BLOCK DIAGRAM OF SPIN AXIS PRECESSION CONTROL SYSTEM

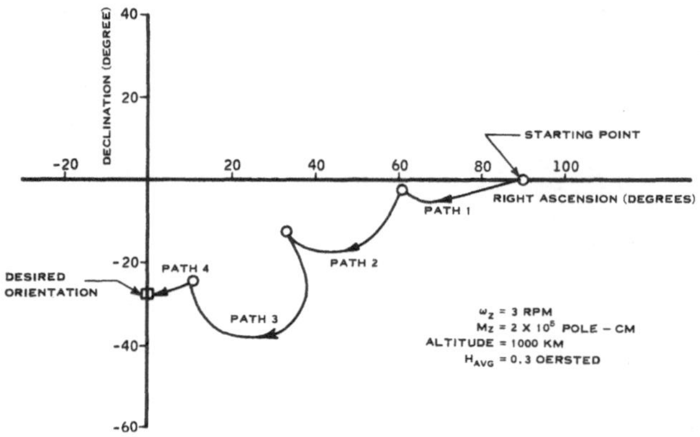

FIGURE 7. SPIN AXIS ATTITUDE MANUEVER FOR A TYPICAL SPACECRAFT

parts should provide improved reliability for extended lifetimes in orbit. Actually either the X- or the Y-magnetometer with its associated amplifier and electromagnet will provide the desired spin rate control, but at half the capability for change in spin rate per day. In this respect the magnetic system provides additional reliability by redundancy. The magnetic system also has the advantage that at any time it can be used to spin-up or despin the satellite to any desired rotation rate and with almost infinitesimal resolution.

CONTROL OF SPIN AXIS ORIENTATION

Many earth satellites such as Tiros, Syncom, AE-B and DME-A require that the spin axis be oriented in some prescribed direction. Tiros was oriented to produce the best meteorological observations; Syncom was oriented perpendicular to its orbital plane to obtain good radio signal strength at the earth's surface. Both AE-B and DME-A are to be oriented with their spin axes perpendicular to their orbital planes so that particle detectors will first be oriented along the satellite's trajectory and then rotated around into the satellite's wake. This provides a most desirable condition for studying particles that are moving slowly relative to the velocity of the satellite.

To provide spin axis orientation Tiros II and later satellites employed a dipole moment generated by electric current flowing in a large coil in the X-Y plane of the satellite.[8] This produced a dipole moment, M_z, which resulted in precession of the spin axis of the satellite. The dipole moment was actuated as required in orbit to achieve the desired orientation of the television cameras.

The Syncom satellite employed pairs of cold gas jets to cause the satellite to have its spin axis oriented perpendicular to the orbital plane.[5]

The AE-B and DME-A satellites being developed at the Applied Physics Laboratory utilize a system of chargeable magnets along the Z-axis of the satellite to produce a magnetic dipole moment which results in precession of the satellite's spin axis so that it can be oriented perpendicular to the orbital plane. Figure 3 illustrates a type of chargeable magnet system that was used for magnetically orienting the satellites 1963-38B and 1963-49B and will be used for orienting the spin axis of the AE-B spacecraft. A source of voltage causes a condenser to be charged and then by radio command discharged through an electrical winding wrapped around a permanent magnet material. When the polarity switch is in the "add" position a dipole moment equal to $+M/2$ is produced in the chargeable magnet. If the polarity switch is reversed and the condenser is charged and then discharged through the windings, a dipole moment $-M/2$ is produced. If the satellite also contains a permanent dipole moment of magnitude $+M/2$ in the Z-direction, the satellite can readily obtain two magnetic states, namely $M_Z = +M$ and $M_Z = 0$. The state $+M$ may be used to precess the satellite's spin axis; the state $M_Z = 0$ is used when no further Z-axis reorientation is desired.

A nutation damper is required to remove angular rates developed about the satellite's X- and Y-axes. For AE-B the nutation damper was designed at Goddard Space Flight Center; the nutation damper for DME-A was developed at the Applied Physics Laboratory.

The selection of materials for the chargeable magnets is based on obtaining a high value for the residual magnetization without requiring too large a value for the energy stored in the condenser. Figure 4 shows the demagnetization curve for three permanent magnetic materials that were considered for this application. Alnico V has a high remanence and a high coercive force so that it retains a large magnetic dipole moment even for comparatively low length to diameter (ℓ/d) ratios. However, the extremely high coercive force makes it difficult to magnetize with a resonable weight of condensers. The permanent magnet material, 40 percent cobalt-steel, requires considerably less condenser energy but has a comparatively low flux density. The material that provided the best system performance was manufactured by the General Electric Company and is termed alloy P-6. This material combines a moderate coercive force with a high flux density. When used with ℓ/d ratios greater than 40, it provides a system for obtaining a large dipole moment for comparatively little weight of condenser and core material.

The dipole moment for these magnets can be enhanced for the same system weight by repeating the charging and discharging cycle of the condenser several times. Figure 5 shows the characteristics for a P-6 magnet that saturated at a dipole moment of 2.4×10^4 pole-cm. For this magnet $\ell/d = 41$, and the total weight of windings plus P-6 core was 1.5 pounds. As can be seen from Figure 5, increasing the condenser voltage or increasing the number of pulse discharges resulted in a greater magnetization of the core material. However, the magnetization always leveled off at the saturation value for this material at this ℓ/d ratio.

This system of obtaining a permanent magnetic dipole in a satellite was used successfully for more than 100 operations over a period of nearly one year for the satellites 1963-38B and 1963-49B. For producing a magnetic dipole moment this system has an advantage over a coil carrying current (such as used on Tiros) in that it requires no steady state power. When large dipole moments are required or when it is not convenient to have a large coil diameter (i.e. coil area greater than 1 m^2), chargeable magnets provide a considerable weight savings as compared to a coil carrying current. A magnetic system has the advantage over a gas jet for exerting torque in that it does not expend material or have moving parts; thereby providing a potentially longer operating life with a greater reliability.

APPLICATION OF MAGNETIC TORQUE FOR SATELLITE SPIN AXIS ORIENTATION

Figure 6 shows a block diagram for a magnetic torque spin axis control system similar to that being used for the DME-A satellite. Instead of using a chargeable and a permanent magnet, this system uses a single chargeable magnet that can be magnetized to any desired level between $-M \leq M_z \leq +M$. The values of +M and -M are obtained as described above. Intermediate values are obtained by charging the condenser for only a short period of time before discharging it. For example, the state $M_z = 0$ can be obtained by the following procedures: (1) charge the magnet to the state $M_z = +M$, (2) command the polarity switch to its negative state, (3) charge the condenser through a large resistor for a short time (typically 10 seconds for the DME-A system), (4) discharge the condenser through the windings of the chargeable magnet. If a state more positive than $M_z = 0$ were desired, the condenser would be charged for less than the time required to achieve zero magnetization; if a magnetic dipole more negative than zero is desired, the charging time is lengthened accordingly. Thus all magnetization states between +M and -M are readily achievable.

In Figure 6 is shown a separate magnetometer that is mounted near the chargeable magnet as a measuring device for determining the magnet's dipole moment. Figure 6 also illustrates the type of attitude detection system that will be employed in the DME-A spacecraft to determine its orientation. This is a requirement for any spin axis orientation system since one must know the satellite's attitude in order to cause it to precess to the desired orientation.

The DME-A system employs a six-direction sun sensing system and a three-axis magnetometer to determine the attitude of the satellite's spin axis. For magnetic spin axis control one must also have a reasonably good knowledge of the satellite's orbit, and one must have a moderately accurate idea of the magnitude and direction of the earth's magnetic field at the position of the satellite.

As previously mentioned, spin axis orientation control a nutation damper so that angular motions other than about the Z-axis will be damped out. Of course it is also necessary that the satellite spin moment of inertia, I_z, be somewhat larger than the other principal moments of inertia, I_x and I_y, so that the satellite will stabilize with its spin about the Z-axis. The DME-A satellite employs a nutation damper consisting of two circular, aluminum tubes each containing a ball that is free to roll inside the tube. These tubes are moved as far away as possible from the satellite's center of mass. The center of curvature of the tubes is located a distance of two inches from the satellite's Z-axis. When the satellite has either $\omega_x \neq 0$ or $\omega_y \neq 0$, the ball in each tube will roll back and forth from its equilibrium (center) position. As the ball moves, six small permanent magnets inside the moving balls cause eddy currents in the aluminum tubes, thereby dissipating the precession energy of the satellite. The magnets in each ball are arranged to have a zero net dipole moment so they will not perturb the spin axis orientation or cause erroneous reading of the magnetometers.

The rate at which the spin axis can be precessed is determined by the strength of the dipole moment in the satellite, the satellite's spin rate and moment of inertia, and the intensity of the earth's magnetic field at the satellite's position. For the DME-A spacecraft, $I_z = 5$ slug-ft^2, $\omega_z =$ 3 rpm and at 1,000 nautical miles altitude near the north magnetic pole, one can achieve a precession rate of approximately ± 5 degrees per minute for $M_z = \pm 10^5$ pole-cm.

Messrs. Frederick F. Mobley and Walter E. Allen, of the Applied Physics Laboratory, have programmed the IBM 7094 computer to determine the best path for moving the spin axis of a satellite to its desired orientation. A typical maneuver is illustrated in Figure 7. The object of this theoretical maneuver was to move the satellite spin axis from a right ascension of +90 degrees and a declination of 0 degrees to right ascension of 0 degrees and a declination of -25 degrees. A further requirement was that all magnet turn-on and turn-off commands would be sent when the satellite was in view of the Applied Physics Laboratory. As can be seen in Figure 7, this required 4 separate attitude control maneuvers.

CONCLUSION

A magnetic spin-up and spin-down system can be provided to maintain a satellite's spin rate at any desired level. This can be accomplished without the dissipation of materials and without the use of moving parts.

A chargeable permanent magnet can be used for precessing the spin axis of a satellite to a desired orientation. This system can provide a large magnetic dipole moment with essentially no dissipation of electrical power and with comparatively little weight. It is readily possible to utilize a digital computer to determine when to turn the dipole moment on and off to cause the spin axis to precess to a desired orientation.

ACKNOWLEDGEMENT

The author wishes to acknowledge the invaluable assistance of Mr. Frederick F. Mobley for data on chargeable magnets and for providing information on the spin axis orientation maneuver described herein.

REFERENCES

1. R. H. Wilson, Jr. "Rotational Decay of Satellite 1960 $\eta 2$ Due to the Magnetic Field of the Earth," Proc. XII, Intern. Astronaut. Congr., Washington, D.C., Academic Press, N.Y., 1963 pp. 368-379.
2. G. Colombo, "The Motion of Satellite 1958-ϵ Around its Center of Mass", Research in Space Sci., Smithsonian Astrophys. Obs. Spec. Rpr. No. 70, July 18, 1961.
3. W. Bandeen and W. P. Manger, "Angular Motion of the Spin Axis of the Tiros I Meterological Satellite Due to Magnetic and Gravitational Torques," J. Geophys. Res., Sept. 1960.
4. J. E. Jackson, "Swept Frequency Topside Sounder S-27," NASA Payload Description N-90005, March 1, 1961.
5. D. D. Williams, "Dynamical Analysis and Design of the Synchronous Communications Satellite," Hughes Aircraft Co., Tech. Memo. 649, May 1960.
6. R. H. Wilson, Jr., "Magnetic Damping of Rotation of the Vanguard I Satellite," Science, Feb. 5, 1960, pp. 355-357.
7. R. E. Fischell, "Magnetic Damping of the Angular Motions of Earth Satellites," ARS Journal, Vol. 31, Sept. 1961, pp. 1210-1217.
8. E. A. Goldberg, "Tiros Pre-flight Testing and Post Launch Evaluation," Journal of Spacecraft and Rockets, July-Aug. 1964, pp. 374-380.

DISCUSSION

R. E. Fischell

"Spin Control for Earth Satellites".

Q. What nutation dampers are used?

A. Two aluminum tubes, each with a ball and magnets. No liquid is used.

Q. What is your magnetometer sensitivity?

A. From 0.5 to 0.025 oersted full scale.

Q. How do you decouple your magnetometers from the field you produce?

A. By orientation and feedback.

Q. Can this be done when more than one magnetometer is used?

A. Yes, by proper location.

Q. How about the effects of electric wiring?

A. If wires are run in pairs, the effects are 3 to 4 orders of magnitude lower than the effects we measure for attitude control. For scientific measurements we use a boom for the instrument.

Q. How large an initial error can be tolerated?

A. There is no limit, since we can move the satellite to reduce any level.

Q. What is initial spin rate?

A. We have an initial spin rate from the booster of from 0.1 to 25 rpm. This eliminates the tumbling which would result if the spin control system were turned on without initial spin.

Q. How many ground stations participate in control?

A. We control only 1 pass out of 100 - therefore one station only is needed.

Q. What accuracy of orientation is achieved?

A. The experimenters require 5 degrees. We can get 1 degree without difficulty.

ATTITUDE CONTROL FOR THE TIROS WEATHER SATELLITES

by Warren P. Manger
Manager, Spacecraft Systems
Astro-Electronics Division
Radio Corporation of America
Princeton, New Jersey

ABSTRACT

The paper describes the evolution of attitude control techniques within the framework of the TIROS weather satellite program, starting in about 1958, which led to a particularly simple stabilization system used in the operational weather satellite system now being implemented for the United States Weather Bureau by NASA and RCA. The stabilization system employs a controlled form of simple spin stabilization, exploiting control torques generated by interaction with the earth's magnetic field, to achieve in effect the same results which would be provided by a completely stabilized earth-oriented spacecraft. Possible future evolution of the basic ideas is also discussed.

INTRODUCTION

The TIROS Operational System (TOS) currently being implemented for the United States Weather Bureau will be one of the first completely operational space systems devoted exclusively to an important peaceful use of outer space; namely, the gathering of weather data on a global basis using satellite-borne sensing equipment.

One of the key problems in the design of any spacecraft is that of providing an attitude-control subsystem which will maintain the proper orientation of the primary sensors or communications subsystems. The problem is especially difficult in the case of the spacecraft for an operational system, where continuous operation over a long period of time, oftentimes several years, is required for reasonable cost effectiveness.

A great deal of work has been done on various approaches to the attitude control problem, and it must be remarked that it often appears that the system designers have been determined to produce hardware equal in complexity to the modern control theory used in the system design; this leads to very interesting and expensive projects, but seems thus far, at any rate, not to have led to very many practical, long-lived, and highly reliable attitude control subsystems.

The TOS spacecraft will employ an extremely simple long-lived stabilization subsystem of proven reliability, achieving quite sophisticated results with great economy of means. It is, therefore, perhaps appropriate to examine in detail the evolution of the stabilization ideas in the TIROS and TOS programs and also to examine the possible implication for future systems.

THE TIROS WEATHER SATELLITE PROGRAM

The TIROS series of world observation satellite (TIROS is an acronym for Television Infra-Red Observation Satellite) is one of the National Aeronautics and Space Administrations most successful programs and has pioneered in use of satellites for gathering data on the world's weather. The program has proved conclusively that meteorological satellites will be one of the outstanding peaceful uses of outer space.

The TIROS program took definite form in 1958. The program was initially conceived as a series of experimental satellites designed to investigate the feasibility of making observations of weather on a worldwide basis from earth satellites. The types of observations considered were two in kind: (1) television images of the earth's cloud cover and (2) the making of infrared maps in a variety of spectral bands. The first launch in the series took place on April 1, 1960. This first satellite performed extremely well and returned results which proved so interesting and useful to the meteorological community that since 1960 there has always been at least one TIROS satellite operating in orbit. There have been 9 launches, all successful, in the series to date. The Weather Bureau is now implementing an operational weather satellite system based on the refined TIROS design, this

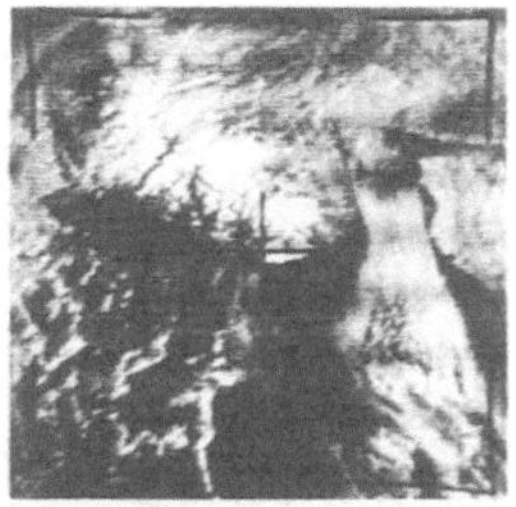

(a) Coast of Norway, TIROS V

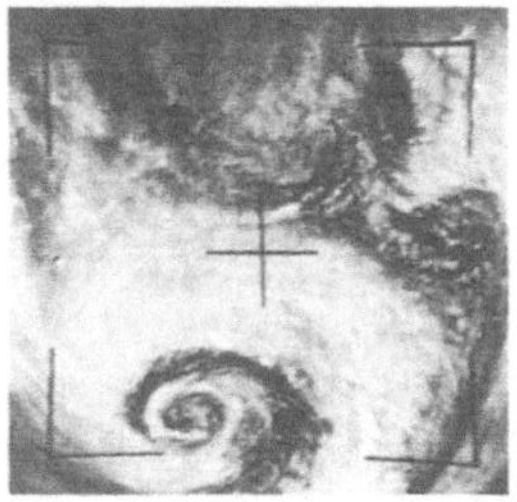

(b) Large-Scale Cloud Organization, TIROS V

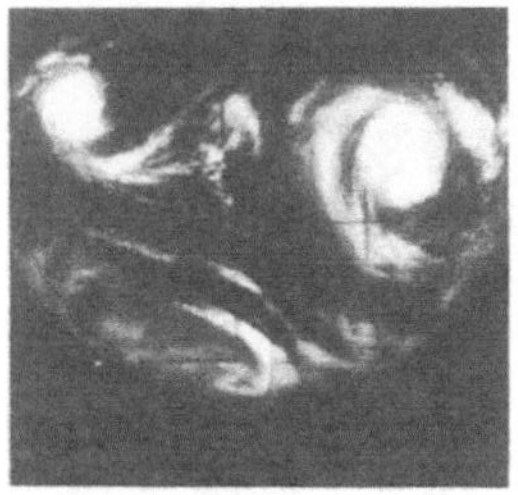

(c) The Earth from 1400 Miles, TIROS IX

Figure 1. Typical TIROS Pictures

Figure 2. TIROS IX in Orbit

systems commonly being referred to as TOS, the TIROS Operational System. Figure 1 shows an example of the typical television pictures received from the TIROS Satellite. Table I is a compilation of the total number of meteorologically useful pictures returned to earth by each of the satellites in the series and also the orbital lifetimes of these satellites. The original design goal for the spacecraft was a three month life in orbit, and it is seen that a number of satellites in the series have lifetimes considerable greater than one year in orbit.

TABLE I

TIROS	Number of Useful Pictures	Useful Lifetime (Days)
I	23,000	89
II	36,100	376
III	35,000	230
IV	32,600	161
V	58,200	321
VI	66,600	389
VII	101,700	653*
VIII	74,100	468*
IX	24,200	70*

*Still operational as of April 1, 1965

The basic TIROS satellite, of the external configuration shown in Figure 2, is a simple pill-box-shaped structure weighing approximately 300 pounds, about 42 inches in diameter, and about 20 inches high. The launch vehicle used in all of the launches to date has been one or another version of the highly reliable Thor-Delta. The planned orbits have generally been circular at about a 450-nautical-mile altitude, with various inclinations from between 50 and 60 degrees up to as high as 100 or 101 degrees for the most recent in the series, TIROS IX. The basic satellite has a solar cell power supply which supplies power to two complete television camera chains and to the infrared radiometers. Each of the camera chains is completely independent, having its own command system, vidicon camera, tape recorder for the storage and remote playback of pictures, and transmitter for returning the pictures to earth. The nature of the television cameras is a key point in the later discussions of stabilization, so it is perhaps worthwhile to explain these in more detail. The basic sensing element is an RCA 1/2 inch vidicon camera tube. The camera is equipped with a shutter operating in such a way that the camera takes a still picture with an exposure time between 1 and 2 milliseconds. The image thus impressed on the face of the vidicon tube is then read out by a scanning electron beam with a frame time of approximately 2 seconds. The resulting video signal is either transmitted directly back to earth, if a ground station happens to be within view, or is fed to the on-board tape recorder for later relay back to earth. The essential point for the discussion here is that the cameras are, in effect, snapshot cameras, so that it is only necessary to have the optical axis of the cameras pointing in a desired direction for a very short period of time; i.e., only during the 1-to-2-millisecond camera exposure. It is also clear that the camera motion during the exposure time must be small enough to avoid blurring. As will be seen later, this happy peculiarity of this particular type of television camera had a strong influence in the development of stabilization systems for the TIROS satellite.

The TIROS program, the satellites themselves, and the scientific results and their meteorological evaluation have been described in great detail in many other publications[1] [2] so that the brief discussion given above will be sufficient for present purposes.

ATTITUDE STABILIZATION FOR TIROS I-VII

A decision was made at the time of the original design of TIROS to employ simple spin stabilization. The limitations of this method of stabilization were recognized, but the conservative choice was made since the basic experimental mission could be carried out with this simple approach. The cameras were mounted in the spacecraft with their optical axes along the spin axis of the satellite, which is the axis of maximum moment of inertia. The satellite was injected into orbit with its spin axis lying in the plane of the orbit and tangent to the orbit at the point of injection, spinning at a nominal rate of about 10 rpm. With this type of stabilization, the spacecraft behaves like a simple ideal gyroscope, maintaining a fixed direction in inertial space, at least in the absence of any disturbing torques. A wide variety of possible disturbing torques were investigated in some detail. In particular, the differential-gravity or gravity-gradient torques acting on the vehicle were computed and found to be large enough to possibly cause appreciable deviation of the spin-axis direction. A careful and detailed integration of the differential-gravity effect, however, showed that it was not large over the design lifetime of 3 months. The interaction between induced eddy currents in the satellite

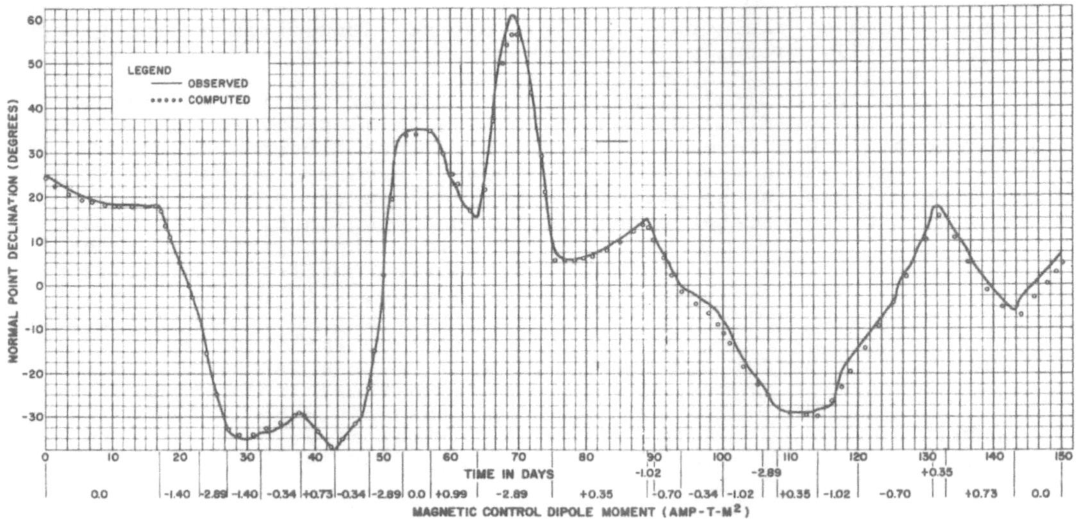

Figure 3. Spin-Axis Declination Versus Time in Days, TIROS VI

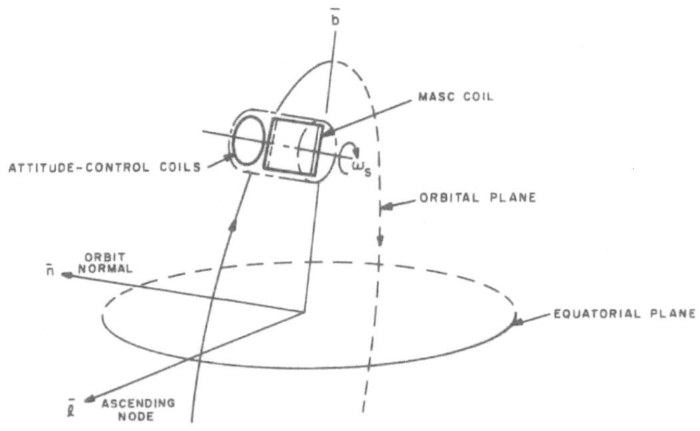

Figure 4. Geometry of Wheel Satellite and Magnetic Torquing Coils

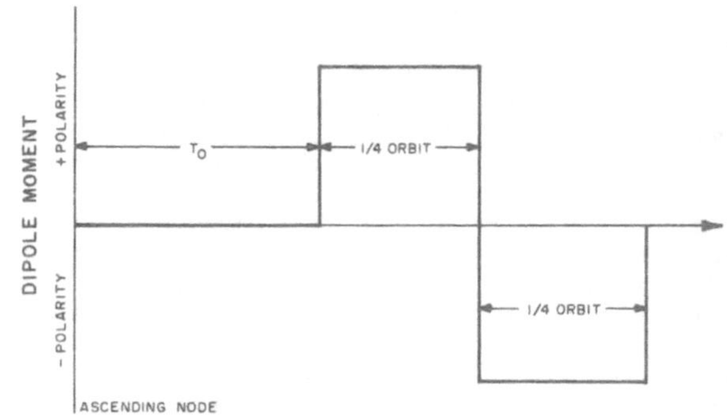

Figure 5. QOMAC Switching Cycle

and the earth's magnetic field were also investigated in great detail, both as regards the decay of spin rate and the torques tending to precess the spin-axis direction, these latter being found to be althogether negligible. The slow spin-rate decay caused by the earth's magnetic field was offset by small solid-propellant spinup rockets incorporated into the satellite design. Thus, to a first approximation, it was believed that indeed the satellite spin axis would maintain a more or less constant direction in inertial space for at least a few months. The geometry of launch and injection were chosen so that, for the first few days or weeks in orbit, the spin axis would point generally downward when the satellite was in the northern hemisphere. With this orientation, the cloud pictures and infrared data obtained could be correlated with ground observations made simultaneously. However, the precession of the orbital plane of about 5 degrees a day changed the geometry of the orbital plane after only approximately two weeks such that the satellite spin axis was no longer in the orbital plane and no useful coverage would be obtained. As time went on and the precession of the orbital plane continued, this condition would correct itself and the initial conditions would be restored again, at which time useful pictures would again be obtained. Thus, the expected results consisted of alternating periods, approximately 2 weeks in length, of good coverage, followed by poor coverage, followed by good coverage, etc. In view of the experimental nature of the program, this deficiency was judged acceptable.

Although it was not part of the basic experiment, it was decided to include in the satellite a means for measuring fairly accurately the direction of the spin axis in space. A simple fixed optical system and infrared-sensitive bolometer were placed in the spacecraft with the optical axis perpendicular to the spin axis so that once each revolution of the satellite, the field of view of this detector (which was about 1 or 2 degrees) swept across the earth. The output of this sensor was telemetered continuously. By observing the time history of the sweeps of this sensor across the earth, it was possible to infer with fair precision the spin-axis direction.

In the first several days following the launch of TIROS I, reduction of this horizon-scanner data showed that, far from maintaining a fixed direction in inertial space, the spin axis was precessing at an appreciable rate; e.g., tens of degrees per day. A detailed study of this data led to a theory and detailed calculations which showed that the observed effect over a period of many weeks could be rather accurately accounted for by hypothesizing that the satellite contained a residual magnetic dipole along its spin axis.[(3)] The approximate strength of the dipole was 1 ampere-turn-meter^2. It was a matter of chance and outstanding good luck that the direction of the dipole moment was such that its interaction with the earth's magnetic field caused the spin axis to precess in a direction which by and large improved the orientation of the cameras throughout the life of TIROS I, so that the aforementioned blank periods of poor coverage never materialized.

With this information in hand, it was obvious that the possibility of using this effect in a controlled fashion should be investigated for incorporation into the subsequent TIROS satellites. Two things became rather obvious almost immediately when means for doing this were studied. In the first place, it would be sufficient to provide the capability to precess the spin axis at a maximum rate of about 25 degrees a day or so. In the second place, so long as the simplest possible system was considered (namely, one in which a constant current was allowed to flow in a coil in the satellite, the coil having its normal aligned parallel to the spin axis) there were severe restrictions on the direction in which the spin axis could be precessed. The theory showed that it would be possible to steer the spin axis in only a few possible directions, determined by the geometrical relationship between the spin axis at any time and the vector average of the earth's magnetic field as seen by the satellite in a complete orbit. Nevertheless, it was found that, even with these restrictions, this simple magnetic torquing scheme would allow sufficient control of the spin-axis direction to achieve useful pictures every day and thus avoid the dead periods caused by the precession of the orbit plane. Such a steering coil was then incorporated into TIROS II and all subsequent vehicles through TIROS VIII. The current in the coil could be selected to be either zero or any of a half dozen other values, either plus or minus, by ground command. The computer programs used to predict the spin axis motion in response to the combined torques of the magnetic torquing coil and differential gravity were gradually improved until extremely accurate prediction and control of the spin axis direction were obtained in all of the subsequent satellites. Fig. 3 is a plot showing the comparison between the predicted and observed values of one of the angles defining spin axis attitude over a period of several months. It is seen that the predictions are extremely accurate, proving rather conclusively that the only significant torques acting on these vehicles at these altitudes are indeed just the magnetic and gravity gradient torques. These results and the computer program used for the predictions have been described in detail elsewhere [4] [5]. The same basic system of spin-axis steering was later incorporated in the Relay communications satellite and in the Telstar satellite. The basic equations used in the computer program by the Bell Telephone Laboratories, according to the above-mentioned references, are substantially the same as those used by RCA in the TIROS program. However, the Bell program is more general, treating the case of an elliptic orbit with rather greater generality than does the RCA program.

As the TIROS program progressed, the usefulness of the meteorological data obtained from the satellites became greater, and of increasing interest and importance. Therefore, it was deemed desirable to continue to use the extremely simple attitude system described above rather than to risk any compromise with mission reliability with a newer type system, even though several modifications of the magnetic scheme had in the meantime been conceived. By the time of the launch of TIROS IX in early 1965, however, some of these new methods were sufficiently well developed to be incorporated in this latest member of the series.

ATTITUDE STABILIZATION FOR TIROS IX AND TOS

The new method of stabilization employed in TIROS IX uses the satellite in the so-called "wheel" configuration. In this mode of operation, the satellite is still spin-stabilized and is launched into orbit in the normal fashion with the spin axis lying in the orbit plane and tangent to the orbit at the point of injection. However, the satellite is immediately precessed, using a magnetic torquing scheme, until the spin axis is perpendicular to the orbital plane, at which time the satellite appears to be rolling around the earth on its orbital track as shown in Fig. 2. Hence, the name "wheel satellite".

The magnetic torquing scheme which is used to turn the satellite spin axis through 90 degrees in order to achieve the wheel mode of operation involves the same sort of coil as used in the earlier spacecraft, having its dipole moment parallel to the spin axis. A slightly more complicated programming of the current in the coil is used, and the coil is effective in maintaining the wheel-mode orientation. It is easily shown that, if the current allowed to flow in the torquing coil is reversed in direction once each quarter orbit (for a circular orbit) and the phase of these current reversals in the orbital cycle can be varied as well, then it is possible to precess the spin axis in any desired direction, thus removing the inherent restrictions of the scheme used in the earlier satellites in the series. In TIROS IX an additional magnetic coil, the plane of which contains the spin axis, is used to generate torques which can be used to control the spin rate of the spacecraft as well. As will be seen later, this spin-rate control results in a considerable simplification of the camera control circuitry.

The tremendous advantage of the wheel mode for a satellite carrying television cameras that obtain pictures of the earth's cloud cover lies in the fact that the cameras can now be placed in the satellite with their optical axes perpendicular to the spin axis, so that once each spin of the satellite the camera is looking straight down at the earth. A signal is provided from an infrared sensor which generates a pulse each time it scans from outer space onto the surface of the earth. This pulse is allowed to shutter the camera at the appropriate times to result in a continuous series of pictures, all taken looking down along the local vertical, still from a basically simple spin-stabilized satellite.

Since the spin-axis steering control involves reversing the coil current every quarter orbit, the scheme is commonly referred to as the Quarter-Orbit Magnetic Attitude Control system, or QOMAC for short. Similarly, since the spin-rate control system depends on interactions with the magnetic field, it is for the sake of brevity referred to as MASC, an acronym for Magnetic Spin Control.

To understand in a little more detail how both these subsystems work, refer to Fig. 4, which shows the overall geometry of the wheel satellite in its orbit with the spin axis perpendicular to the orbit plane, and the geometric location of the various coils used to generate magnetic fields in the satellite which react with the earth's magnetic field.

The cyclic reversals of the spacecraft magnetic dipole polarity in the QOMAC technique effectively produce a variable average direction of the average earth's magnetic field as seen by the spacecraft during the course of any given orbit. It is the proper phasing of these reversals which produce the torques required to precess the satellite in the proper direction to effect an attitude correction.

The detailed theory of this system has been presented in a series of reports.(6)(7)(8) Operationally, the scheme requires the generation of the dipole moment of given polarity for one quarter of an orbit, starting at some preselected time from the ascending node, followed by a negative dipole polarity for the duration of a second quarter orbit. This is defined as one QOMAC cycle and is shown graphically in Fig. 5. The time delay after the ascending node depends on the direction of the attitude correction desired and is determined from the attitude error measurement made as will be described later. The control system in the satellite can be programmed from the ground to accommodate any desired delay time and then to automatically perform the current reversals each quarter orbit with a capability of several orbits programming in storage.

Both high and low torque modes of QOMAC operation are available. The high torque mode is used for correction of gross errors and for rather rapid motions of the spin axis such as are required during the initial wheel orientation maneuver. The satellite nominal spin rate is about 10 rpm, and a dipole moment of about 30 ampere turn-meter^2 provided in the high torque mode will effect a precession rate of approximately 10 degrees per orbit. The low torque mode provides a dipole moment of about 6 ampere turn-meter^2 and causes a precession rate of about 2 degrees per orbit. The coil required for this operation weighs approximately 0.3 pound and consumes just slightly over 1 watt of power when it is being employed. The actual duty cycle is so small that the average power consumed by the torquing coil is essentially negligible. Fig. 6 shows the results of a computer simulation of the initial wheel orientation maneuver requiring a 90-degree change in direction of the spin axis immediately after injection into orbit. It is seen that the turn-around can be accomplished in 12 orbits or so, which amounts to approximately one day. Once the satellite is operating in the wheel mode in a near-polar sun-synchronous orbit, the principal attitude correction required is that caused by the precession of the orbital plane at a rate of approximately 1 degree per day. In order to avoid frequent use of QOMAC cycles to make this correction at essentially a constant rate, provision is made to provide an adjustable constant current to the attitude control coil. If the value of this current is selected appropriately, it causes the spin axis to precess 1 degree a day in very nearly the required corrective direction. This means that QOMAC cycles are required at only very infrequent intervals, on the order of a few times a month, in order to maintain the spin axis perpendicular to the orbit plane. Actual results achieved on TIROS IX prove that this system makes it very simple to maintain the spin axis normal to the orbit plane to within a half a degree.

The foregoing discussion has presumed that it is always known in what direction an attitude correction must be made, implying that a method is always available for knowing the attitude error; that is to say the magnitude and direction of the deviation of the spin-axis direction for the orbit normal. The data for this measurement are provided by telemetering the outputs of a V-head horizon sensor consisting of two infrared bolometers arranged with their optical axes in a "Vee" configuration, as shown in Fig. 7. The plane of the "vee" angle is normal to the spin vector. As the spacecraft spins, the optical axes trace out two conic sections in space. The intersection of these sensor paths with the earth and the corresponding outputs from the sensor are also shown in Fig. 7. When the spin axis in the spacecraft is normal to an earth radius, the outputs of the sensors will be of equal duration as shown in Fig. 7.

Should a roll error (defined as 90 degrees minus the angle between the spin axis and the local vertical) exist, the pulse durations from the two sensors will be unequal as shown in the figure. Should a pure yaw error (i.e., a displacement from the orbit normal about the local vertical) exist at a given point, it will be evidenced as a roll error 90 degrees later in the orbit, and unequal outputs will again be experienced. This shift of the yaw to roll errors due to the inertial rigidity of the spin vector causes the roll and yaw errors to vary sinusoidally, 90 degrees out of phase with one another, as the satellite goes around in orbit. The maximum roll angle must be known in order to determine the number of QOMAC cycles required to correct the error, and the location of its

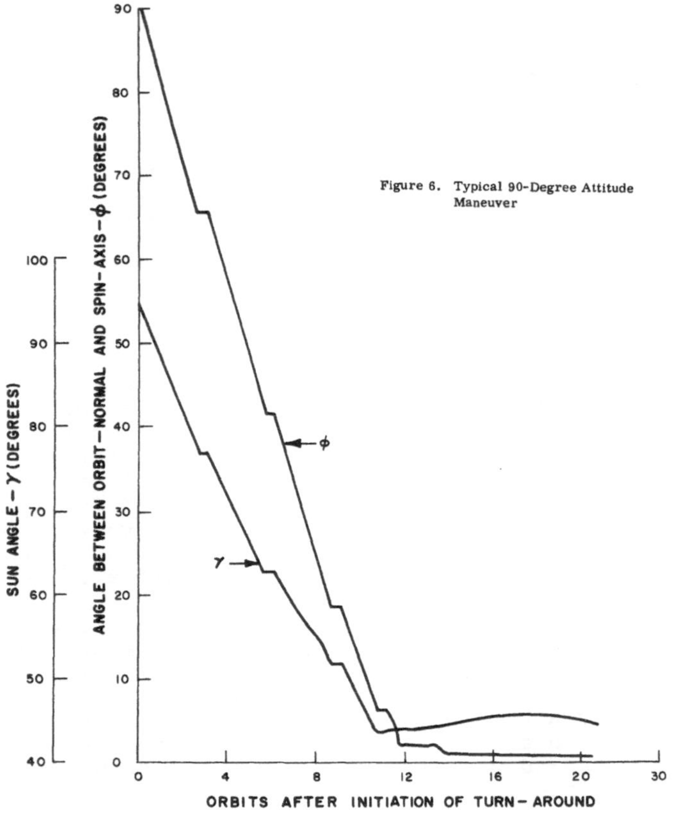

Figure 6. Typical 90-Degree Attitude Maneuver

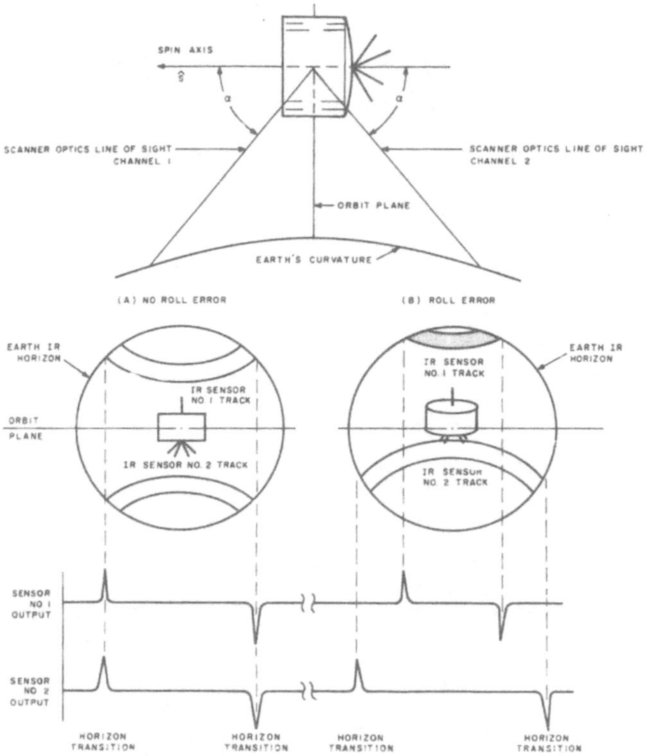

Figure 7. Attitude Sensor Geometry

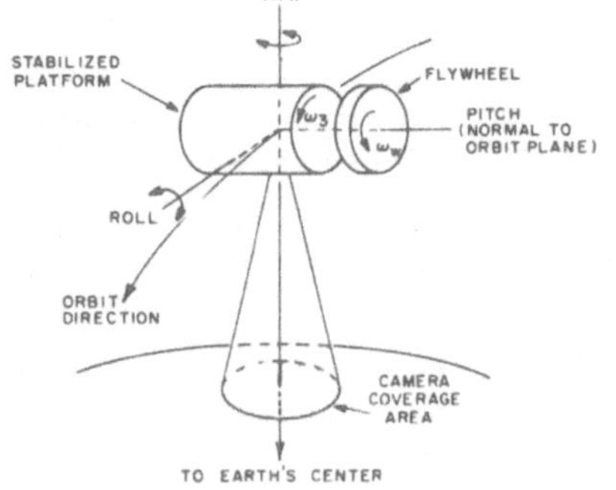

Figure 8. Configuration of "Stabilite" Spacecraft

occurrence in the orbit must be known in order to compute the proper delay time between the ascending node crossing and the start of the QOMAC cycle. The time dependence of the roll and yaw errors permits complete attitude determination from a continuous sequence of instantaneous roll measurements, the accuracy of which is determined by the resolution and reliability of each of the data points. Approximately 10 minutes of roll-error history for a given pass of the satellite over a ground station is required to minimize data reduction problems. A detailed analysis of the possible errors in this method of attitude determination, including such effects as earth temperature variations from one horizon to the other, variations in the horizon temperature gradient, and errors in the data reduction process, shows that the worst case is 0.75 degree and that the probable error is somewhat less than 1/2 degree.

Referring again to Fig. 4, it is seen that the MASC coil is oriented in the spacecraft in such a way that it acts like the armature winding of a motor. Since the earth's magnetic field is uni-directional for any given point in the orbit, it is necessary to reverse the coil current every 180 degrees of spacecraft rotation in order to produce a net torque just as in a d-c motor. The switching of current direction is synchronized by the outputs from the horizon sensors (which also provide signals for synchronizing camera shutters). The angular relationship between the MASC coil and the horizon sensor is selected so that the coil current is commutated about the local vertical. This optimizes the commutation efficiency when the spacecraft is in the vicinity of the earth's magnetic poles, where the field is considerably stronger than at the magnetic equator.

The spin period, as defined by succeeding sky-to-earth horizon pulses from the horizon sensors, must be maintained at the nominal value of approximately 6 seconds with a tolerance of ±25 milliseconds in order to permit simplified camera control circuitry for certain of the cameras used. If a spin correction cycle is defined as the amount of spin-rate correction which can be achieved by activating the MASC coil for the duration of any single pass over a given ground station, it turns out to be simple to size the coil in such a way that three correction cycles per week will maintain the spin period to within a maximum error of about ±8 milliseconds.

The weight and power consumption of the required coil are comparable with those for the QOMAC coil. The maximum daily spin-rate correction, assuming a correction cycle once per orbit, of 1.7 rpm can be achieved with this system. Using this particular method of commutation of the MASC coil, the torques perpendicular to the spin axis causing a change in spin-axis direction are essentially negligible and add no particular burden to the task of maintaining spin-axis direction.

The remaining parameter determining the overall performance of this attitude control system is the accuracy with which the camera can be triggered at the proper phase of rotation of the satellite so that it takes its picture looking straight down along the local vertical. The signal for this is obtained from the infrared horizon scanner. Actually, since the type of camera tube used requires a preparation cycle, it must be commanded into operation at some given point in time prior to the time at which the shutter is triggered. Thus, the situation is a little bit more complex than if it were simply a matter of snapping the shutter at the time the camera is looking straight down, although the basic principle involved is the same. Detailed study of the factors affecting the accuracy with which the horizon scanner pulse can trigger the camera shows that the error is approximately 0.5 degree at most. The basic factor contributing to this error are essentially the same as those contributing to the determination of attitude error in horizon sensors, as discussed previously.

This system has worked very well on TIROS IX. The spin axis is routinely maintained in alignment with the orbit normal to within 0.5 degree. The orbit of this spacecraft was to have been circular, but a booster malfunction resulted in an elliptical orbit having an apogee of about 1500 miles. This wide deviation from the planned altitude causes the cameras to be triggered slightly away from the local vertical, although the usefulness of the pictures is not impaired.

The ideas outlined above describe briefly the evolution from the basic spin stabilization used on TIROS I into a rather simple attitude control system, already proven on TIROS IX, which will be used in the TOS satellites for the operational weather satellite system currently being implemented by NASA and RCA for the United States Weather Bureau. The system achieves an extremely high degree of performance, making it possible to acquire a continuous series of pictures of the earth's cloud cover, all pictures being oriented along the local vertical to the satellite to within an accuracy considerably better than 1 degree, and, in fact, generally as good as 1/2 degree. The system is

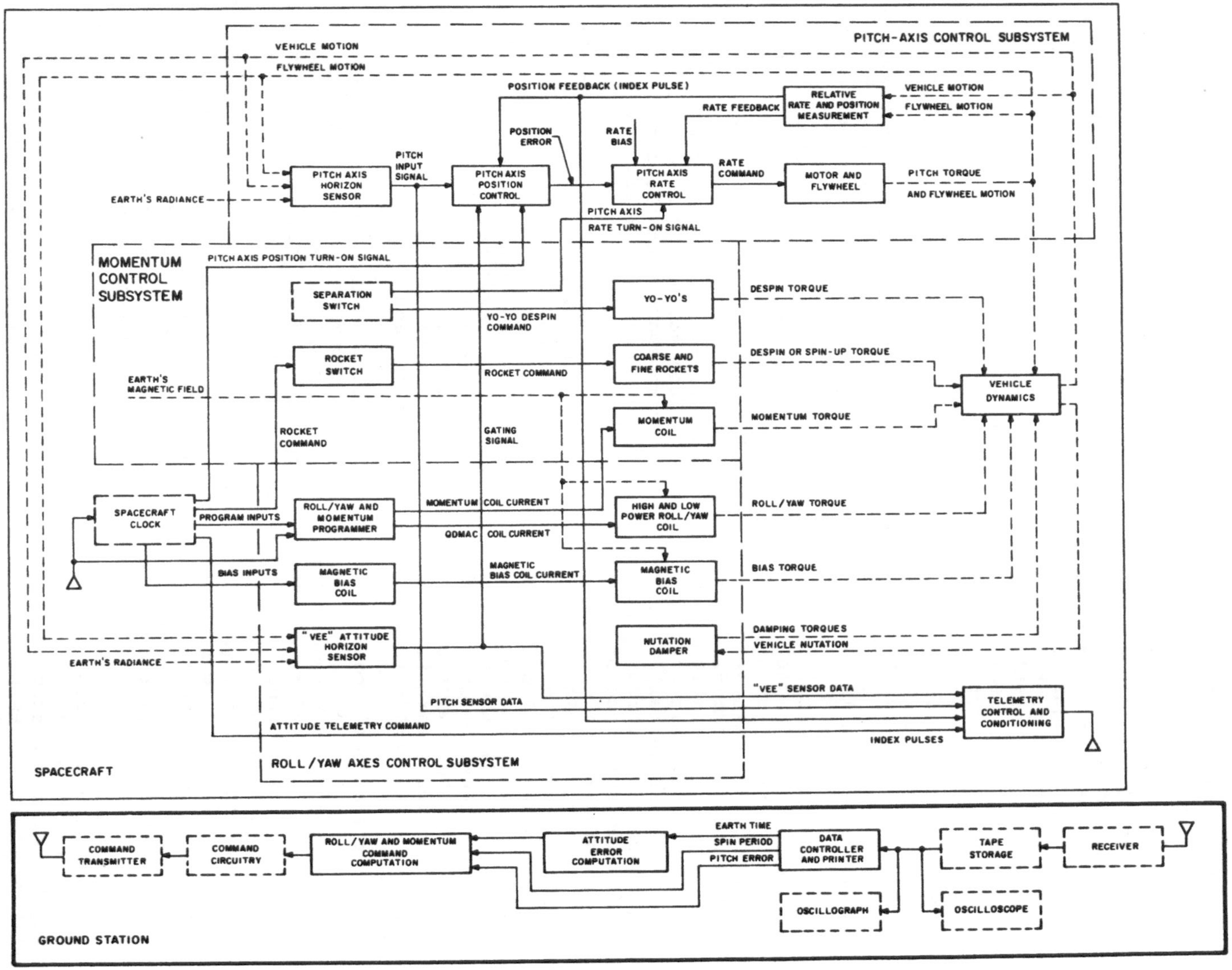

Figure 9. Stabilite System Block Diagram

extremely simple, requiring no mechanical moving parts and a relatively few electronic components. The results achieved are as good as would be obtained from a much more complicated, fully stabilized, earth-oriented vehicle. Such systems typically employ a triad of momentum exchange wheels, together with a three-axis jet torquing system, and are considerably more complicated than the system described here and, therefore, considerably less reliable. On the other hand, the comparison with various gravity-gradient stabilization schemes shows that the magnetic methods suggested here achieve considerably more accurate results than any of the gravity-gradient systems flown to date.

The system is not presented, by any means, as an example of the sophisticated application of modern control technology. It is felt, however, that it is an extremely good example of the achievement of very sophisticated results by extremely simple and reliable means.

In fact, it is even questionable as to whether it is an example of "automatic" control at all. Actually, the angular momentum of the satellite is sufficiently great that the disturbing torques present cannot possibly cause a change in either spin-axis direction or spin rate rapidly enough to require a closed-loop control system altogether within the satellite. The changes in the controlled variables can only take place so slowly that it is entire satisfactory to provide for making the error determinations on the ground and then transmitting the required correction commands back up to the satellite. In accord with the philosophy that complexities belong on the ground rather than in outer space, no serious consideration has been given in the TOS system to making the control systems more automatic. It is clear however that, if the systems application demanded it, both the spin-axis correction and the spin-rate control systems could be made closed-loop within the satellite.

The next section of the report will describe further developments carried out at RCA of the basic ideas discussed thus far. This will lead to the so called "stabilite" concept of three-axis attitude control for spacecraft which has involved considerably more detailed applications of control theory than has the simple TOS system.

STABILITE

The attitude control system described so far as applied in the TIROS IX satellite is ideal for missions satisfying certain constraints. In particular, the on-board instrumentation must be such that it can function adequately from a spin-stabilized satellite. In the case of TIROS IX, the infrared radiation mapping equipment was originally designed to be used from a spin-stabilized spacecraft, and it has already been shown how the particular features of operation of the television cameras permit them to be operated from a spin-stabilized spacecraft so long as they are oriented to point along the local vertical at well defined periodic intervals, at which time the desired pictures will be taken. There are, however, a great many missions which do not satisfy these constraints and for which a completely stabilized, earth-oriented platform is required. Such missions involve more sophisticated infrared mapping measurements and the use of much higher-resolution television cameras, where the spacecraft motions during exposure interval cannot be tolerated. It is only natural then that considerable effort has been expended to see if the ideas developed so far can be extended and applied to a more generally useful stabilization control system.

An examination of the basic ideas behind the TIROS IX attitude control system shows that it depends upon the provision of stored angular momentum to provide gyroscopic stability in the spacecraft, together with magnetic torquing schemes to provide the means to maintain the vector of stored angular momentum perpendicular to the orbit plane and to make it possible to maintain a constant speed of the spinning spacecraft. It happens in the case of TIROS IX that the entire spacecraft is spinning to provide stored angular momentum. The stored angular momentum is fundamentally necessary to the scheme in order to ensure that the existing disturbing torques can cause only very slow changes in desired orientation and spin rate, so that the control loops need not have high response and can be closed through a ground command loop. It is more or less clear from the basic theory of the system, however, that it is not necessary that the entire spacecraft be spinning as in the case of TIROS IX, but that it is sufficient to provide a flywheel comprising a small part of the spacecraft in which the required angular momentum is stored, and that

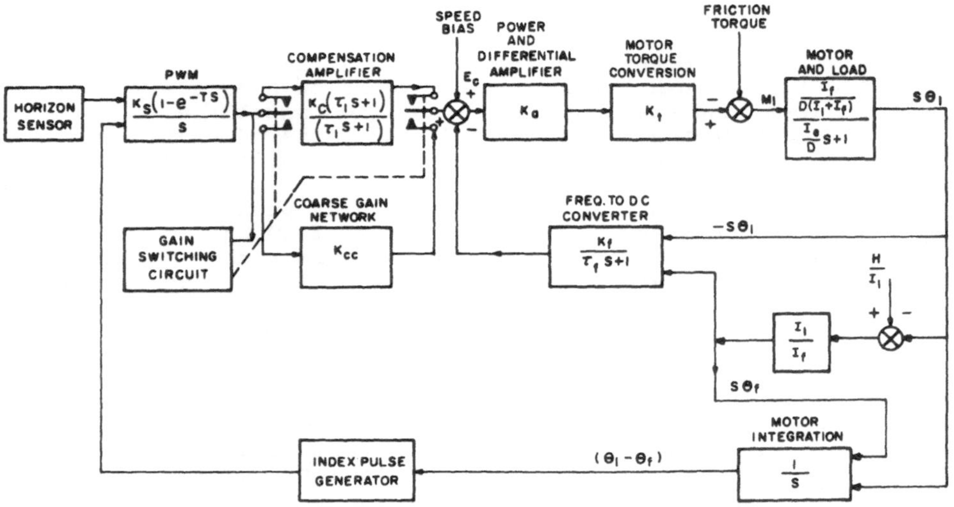

Figure 10. Block Diagram of the Pitch-Axis Stabilization Control Subsystem

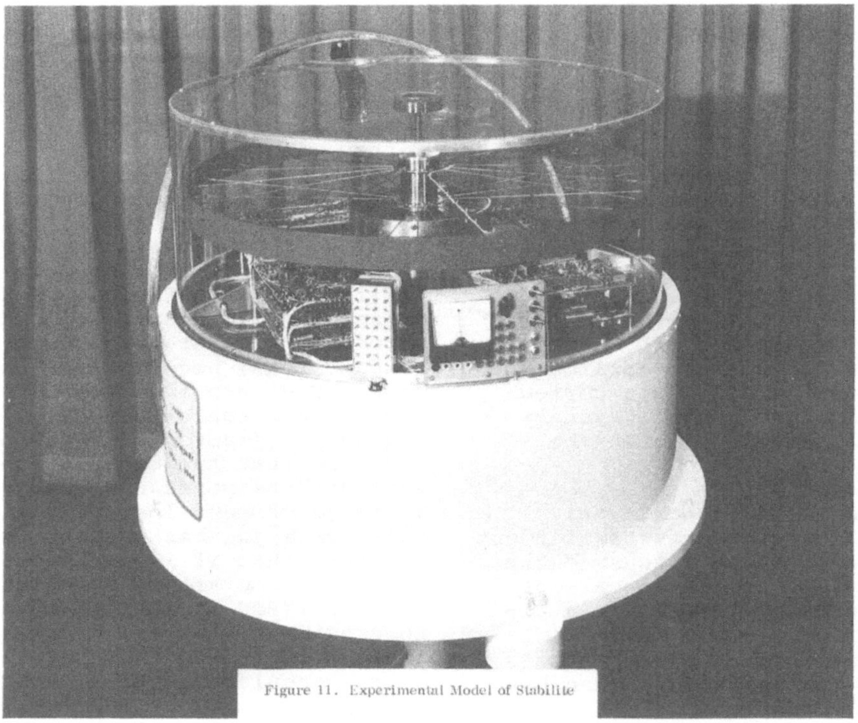

Figure 11. Experimental Model of Stabilite

the remainder of the spacecraft can be earth-oriented. The configuration of the spacecraft then can be as shown as in Fig. 8, which shows the spacecraft in orbit about the earth, the bulk of the spacecraft earth oriented, and the flywheel spinning relative to the spacecraft with its axis along the normal to the orbit plane. The stabilized portion of the spacecraft is then rotating at a rate of one revolution per orbital period. The flywheel is driven relative to the spacecraft by an electric motor at a more or less constant speed of about 100 rpm. The spacecraft spin is maintained precisely at the rate required to achieve earth orientation by exchanging momentum with the spinning flywheel.

The complete system block diagram, Fig. 9, shows that this three-axis attitude control system is made up of three major and nearly independent subsystems which together provide full earth-oriented control of the spacecraft. The first subsystem is a relatively fast closed-loop feedback control system that drives the low-speed flywheel. This subsystem provides two quite different functions. First, it provides the steady-state angular momentum about the pitch axis and, therefore, stabilizes the complete spacecraft about this axis through gyroscopic action. Second, it controls the spacecraft motion about the pitch axis through the exchange of momentum between the flywheel and the remainder of the spacecraft. The complete pitch-axis subsystem consists of simple IR bolometers mounted on the flywheel for the purpose of attitude measurements, much as in the case of TIROS IX; the control electronics; an index generator; and a motor and tachometer. The second subsystem provides control of the direction in inertial space of the angular momentum vector; that is, of precessional motion about the spacecraft roll and yaw axes by the same techniques of quarter-orbit magnetic torquing as described in connection with TIROS IX. This roll and yaw control subsystem also includes a nutation damper that provides damping of nutational motions caused by internal or external disturbances. It should be mentioned that proper design and placement of this nutation damper on the spacecraft make it possible to utilize almost any inertial configuration for the spacecraft; that is, it is not necessary for the axis of stored angular momentum to be the maximum moment of inertia of the entire spacecraft. The theory of this damper and the derivation of the mild constraints on spacecraft configuration when using this system have been discussed in a recent paper.(9)

The third subsystem provides control of the total spacecraft momentum about the vehicle pitch axis. The basic control mode utilizes a magnetic torquing technique to compensate for momentum losses or gains over a long period of time due to external disturbances. The magnetic torquing technique used is in principle very similar to that used on TIROS IX, except that a different type of current switching is required.

All the magnetic torquing techniques, both for the control of the flywheel axis direc- and the control of the net momentum of the vehicle about the pitch axis, utilize ground data processing and ground command to achieve closed-loop operation. The measurement of attitude errors is achieved through the ground processing of data transmitted from the horizon sensors mounted on the flywheel, and both the data and the data processing are substantially the same as in the wheel-mode satellite.

This three-axis attitude control system (which is referred to as the "stabilite") does not use any stored fuel on board in its normal operation; therefore, an extended lifetime can be achieved dependent only on the reliability of the electronic and mechanical components. The system average power consumption is generally considerably less than 10 watts (often less than 5 watts for many systems), and the total system weight is between 5 and 10 percent of the total spacecraft weight. The lower value is generally approached as the spacecraft overall weight and size increase.

Detailed analysis of system performance has shown that pointing accuracies about all three axes of 1.0 degree are easily obtainable; performance better than 0.5 degree is altogether feasible, and it is likely that pointing accuracies of 0.1 degree can actually be obtained. The "jitter" rate (that is the spurious rates of motion of the local-vertical-oriented axes of the spacecraft) can be held to 0.01 degree per second, and it is very likely that an improvement to 0.001 degree per second can be obtained.

From the preceeding discussion, it is clear that the only new element over and above the TIROS IX attitude control system, required to implement the stabilite is the closed-loop control of the flywheel, which basically provides the momentum interchange between the flywheel and the stabilized portion of the spacecraft.

The required momentum transfer is accomplished by using a d-c torque motor to couple the flywheel to the spacecraft.

A horizon scanner, mounted on the flywheel, generates a horizon pulse each time a change from cold space to warm earth is sensed. A variable-reluctance pick-off coil generates an index pulse each time a reference point on the flywheel passes a given reference point on the spacecraft. For a particular orbit altitude, this reference point is geometrically positioned so that the horizon pulse and the index pulse occur simultaneously when the vehicle yaw axis is exactly aligned with the local vertical. Thus, the basic control problem is simply that of exactly synchronizing the two chains of pulses, one from the horizon sensor and the other from the index pulse generator on the flywheel. Since the flywheel typically rotates at about 100 rpm, the pulse repetition frequency is approximately 1.6 pulses per second. This is clearly a sampled data servo.

A block diagram of a typical pitch-axis stabilization control loop is shown in Fig. 10. Its general operation is as follows. The position error, or deviation, of the spacecraft body axis from the local vertical is established by using a pulse-width-modulation (PWM) error detector to measure the time difference between the horizon pulse and the index pulse. The error signal is a constant-amplitude, variable-pulse-width voltage; the width of the error signal is proportional to the angular error. The signal is shaped to an amplitude-modulated voltage by the use of appropriate filtering; this voltage is properly compensated by frequency-sensitive networks and applied to a speed control loop (tachometer loop). The voltages from the fixed speed bias command and the compensated position error voltage are compared, using a differential amplifier, with a signal proportional to the speed in the speed control error detector. A signal proportional to speed is generated by a serial pulse encoder mounted integrally with the drive motors; the pulses are converted to a d-c voltage by means of a frequency to d-c converter and are smoothed using a simple lag circuit. The output of the differential amplifier is fed to a power amplifier which supplies current to the drive motor; the drive motor then applies a torque between the flywheel and the vehicle for proper transfer of momentum and reduction of position error toward a value of zero.

The speed bias command is controlled from the ground. This ground control is required in order to command the appropriate flywheel speed so that essentially all the angular momentum initially in the system at the time of orbital injection, which is not a well controlled variable in general, can be stored in the flywheel. During initial acquisition, prior to closed-loop position control, the flywheel speed relative to the vehicle is increased to the speed commanded by the speed bias voltage. The initiation of subsystem closed-loop position control is ground commanded after the 90-degree QOMAC turn; the vehicle momentum less than or in excess of, that required to rotate at the orbital rate is transferred to or from the vehicle by means of the closed-loop control system. While the control system is performing this function, a low-gain loop is utilized, instead of the normal compensation, in order to compensate for potential satellite-limit cycles during the acquisition mode. The switching between low and high gain is performed as a function of pointing error.

Pitch-axis stabilization servos of this general sort have been analyzed in great detail, and a thorough study of the real characteristics of the various elements of hardware required to implement the system has also been made. Elaborate analog-computer simulations have been carried out, and a complete working model of the stabilite has been assembled and evaluated. Fig. 11 is a photograph of the completed experimental model. The entire unit is mounted on a spherical air bearing at its center of mass to provide an essentially torque-free and force-free suspension. The flywheel used for angular momentum storage is clearly visible. In this simulation, the vector of stored angular momentum is perpendicular to the floor, and the model functions in such a way as to maintain a prescribed horizontal axis of the spacecraft body pointing in a controlled fashion at a collimated light source. All of the performance analyses, computer simulations, and measured experimental results on the laboratory model confirm that the performance figures quoted previously can be obtained.

The stabilite concept, therefore, provides a full three-axis stabilization system for a very general type of spacecraft, of very high degree of performance with respect to both pointing accuracy and minimum spurious rates. It does these things with inherently less complexity than do conventional three-axis systems, basically because in the stabilite, two of the three axes are gyrpscopically stabilized so that the loops can be closed through the ground, since the disturbing torques present cannot cause any changes in orientation to take place except over a relatively long period of time. This means that a relatively more complicated closed-loop control system is required only on the third (pitch) axis. Speaking in very general terms, it would appear that the overall complexity of the system is certainly less than half that of a conventional three-axis system using three momentum exchange wheels and gas jets for wheel desaturation.

CONCLUSION

The evolution of the TIROS weather satellite program has led to very detailed investigation of the class of three-axis spacecraft attitude control systems which feature stored angular momentum normal to the orbital plane. The angular momentum may be provided by spinning the entire spacecraft, as in the case of the TIROS "wheel" satellites, or by spinning only a small portion of the spacecraft. The implementations discussed above also feature controlled interactions with the earth's magnetic field to provide all necessary control torques.

The advantages of this class of systems have been discussed in detail and are believed to be especially significant as applied to operational spacecraft which must achieve long operational lifetimes in orbit.

ACKNOWLEDGMENT

The work described in the paper extends over a period of more than five years, and many persons have made substantial contributions. Special acknowledgment should be made to H. Perkel and his group at the Astro-Electronics Division of RCA.

REFERENCES

(1) "Proceedings of the Symposium on Meteorology from Space," Journal of the British Interplanetary Society, May - June 1964, Vol. 19, No. 9.

(2) "A Quasi Global Presentation of TIROS III Radiation Data," by L. J. Allison, T. I. Gray, Jr., and Guenther Warnecke, NASA SP-53, 1964.

(3) "Angular Motion of the Spin Axis of the TIROS I Meteorological Satellite Due to Magnetic and Gravitational Torques," W. Bandeen and W. P. Manger, Journal of Geophysical Research, September, 1960.

(4) "Magnetic Attitude Control of the TIROS Satellites," E. Hecht and W. P. Manger, American Astronautical Society Paper No. 62-44, March, 1962.

(5) "Attitude Determination and Prediction of of Spin-Stabilized Satellites," L.C. Thomas and J.O. Capellari, Bell System Technical Journal, Vo. 43 pp. 1657-1726, July, Part 2, 1964.

(6) "The TIROS-Wheel Spacecraft with Canted TIROS Cameras, Design Study Report, Contract No. NAS5-3344, prepared by Radio Corporation of America for Goddard Space Flight Center, NASA, Jan. 15, 1964.

(7) "Flywheel Stabilized Magnetically Torqued Attitude Control System for Meteorological Satellites Study Program" Final Report, Contract No. NAS5-3886, prepared by Radio Corporation of America for Goddard Space Flight Center, NASA, Dec. 4, 1964.

(8) "Design Report for the TIROS Operation Satellite (TOS) System," Contract No. NAS5-3173, prepared by Radio Corporation of America for Goddard Space Flight Center, Dec. 30, 1964.

(9) "nutational Stability of an Axisymetric Body Containing a Rotor," V.D. Landon and Brian Stewart, Journal of Spacecraft and Rockets, p. 682, Nov. - Dec. 1964.

THE STATUS OF PASSIVE-GRAVITY-GRADIENT STABILIZATION

by Bruce E. Tinling and
Vernon K. Merrick
Research Scientists
NASA, Ames Research Center
Moffett Field, California

ABSTRACT

In the last few years significant advances have been made in the theory and implementation of passive attitude control systems for earth pointing satellites. In each of the systems that have evolved, the gravity-gradient provides a preferred orientation such that the axis of minimum moment of inertia points toward the center of the earth. Many techniques have been developed for damping attitude motions about this preferred orientation. This is usually accomplished by absorbing the energy of the relative motion between the stabilized body and one or more auxiliary bodies. The various damping techniques differ basically in the ambient field that determines the position of the auxiliary bodies. Systems have been devised in which the auxiliary bodies are oriented by solar, magnetic, aerodynamic, or gravity fields. Hardware for some of these systems is in an advanced state of development.

The choice of a system for passive attitude control will depend upon the orbital altitude and inclination, the anticipated eccentricity, and acceptable steady-state pointing accuracy, whether or not control about the earth-pointing axis is required, the acceptable time to damp transient motions, and the probability of successful deployment of the system. The manner in which the choice of the ambient field utilized to effect the damping enters into the fulfillment of these requirements is discussed to point out the advantages and disadvantages of each system.

To date, passive gravity-gradient stabilized satellites have been flown only at low orbital altitudes. This flight experience is discussed and the performance compared with theoretical predictions. For many applications, it is desirable to place earth-pointing satellites at much higher altitudes, particularly in stationary, or synchronous, orbits. An experimental flight program which could prove the practicability of passive stabilization at this altitude is described.

INTRODUCTION

The fact that the gravity-gradient can cause an earth satellite to point one axis toward the earth has been known since the publication of Lagrange's famous paper on the librations of the moon.[1] This phenomenon can be utilized to orient an artificial earth satellite if a means is provided for damping librations. Many damping systems have been devised. These systems can be divided into three classes; those which require both sensors and power, those which require power, but no sensors, and those which require neither power nor sensors. The first two classes might be considered either semiactive or semipassive. Typical examples are the active damping system which accelerates inertia wheels in response to sensor signals,[2] and the control-moment gyro which requires no sensors.[3] This paper is concerned solely with the completely passive third class, which requires neither sensors nor power.

The first recorded concept of a passive-gravity-gradient control system is believed to be contained in a United States patent filed by Roberson and Breakwell in 1956.[4] The principal activity in the 8 years since this patent declaration has been in theoretical studies leading to practical mechanizations of gravity-gradient control systems. Only within the last 2 years has hardware been developed to the stage of providing solutions to the principal practical problems. As a result we are now on the threshold of developing earth-pointing stabilization systems which promise to be simpler, lighter, cheaper, and more reliable than any previously devised. The recent advances in this technology have stimulated considerable commercial, scientific, and military interest, and plans are being executed to extend the flight experience with these systems to cover the complete range of practical orbits.

The principal difference in the various mechanizations of passive-gravity-gradient control schemes lies in the choice of the ambient field that is used as a reference for the damping system. Systems have been devised that are referenced to either the solar, magnetic, aerodynamic, or gravity fields. The choice of a system for a given application depends upon the altitude, inclination, and anticipated eccentricity of the orbit, the acceptable steady-state pointing accuracy, whether or not control about the earth-pointing axis is required, the acceptable time to damp transient motions, and the probability of

Superior numbers refer to similarly-numbered references at the end of this paper.

successful deployment of the system. The manner in which the choice of ambient field for effecting the damping enters into the fulfillment of mission requirements is discussed to point out the advantages and disadvantages of each system.

The more useful applications of earth oriented satellites are for high altitude orbits, particularly the synchronous orbit. To date, passive systems have been flown only at orbital altitudes of the order of 800 km. This flight experience is discussed and the performance of the passive control systems is compared with theoretical predictions. Means for extending the flight experience to include the synchronous orbit are discussed.

RESTORING TORQUES

Before proceeding to a discussion of the influence of damping technique on the performance of a gravity-gradient control system, it is well to review the state-of-the-art of the technique used to create the necessary restoring torques. It is well known that the magnitude of the gravity-gradient restoring torque varies directly with the difference between the principal moment of inertia about the earth-pointing axis and axes transverse to it and inversely as the cube of the radius of the orbit. If the transverse principal moments of inertia differ, a gyroscopic torque will exist which will orient the satellite about the earth-pointing axis. Furthermore, by choosing the proper ratios of the principal moments of inertia, it is possible to select, within certain bounds, the natural oscillation frequencies of the satellite.[5,6,7]

The satellite structure, as stored in a booster prior to launch, cannot provide the differences in moment of inertia required to limit pointing errors to acceptable levels. Only one technique that has been proposed for creating the required moment of inertia differences has met with any success. This technique was originally proposed by Kamm in a paper published in 1962 which described a practical passive control system.[8] The technique consists in unrolling a prestressed tape, usually manufactured from beryllium-copper, which assumes the shape of an overlapped circular tube. Extending these tubes or rods from the satellite, with or without masses at their tips, affords a method of increasing the moment of inertia by factors as great as 1000 with little expenditure of satellite mass. The technology of manufacturing these elements is considerably older than gravity-gradient technology. They have been manufactured as storable retractable antennas for about 20 years by de Havilland Aircraft Ltd. of Canada.

For orbital altitudes of the order of an earth radius, a single rod of the order of 10 to 20 meters in length with a tip mass of several kilograms will provide sufficient gravity torque to maintain excellent pointing accuracy. However, because the gravity-gradient diminishes as the cube of the orbit radius, balancing of disturbance torques created by solar pressure forces becomes more important with increasing altitude. One method of balancing these forces is to make the satellite symmetrical. Unfortunately, even when deployed symmetrically, gravity-gradient rods will have asymmetries introduced by thermal distortion. The source of this asymmetry is illustrated in Fig. 1. Differential expansion between the illuminated and dark sides of the rod cause it to assume approximately a circular shape. Solar radiation pressure forces will then be greater on the end of the rod which is more nearly normal to the incident rays and a torque about the center of mass will exist. The predominant component of this disturbance varies as even harmonics of the angle ψ with the second harmonic being predominant. The magnitude of the disturbance increases as the cube of the rod length, whereas, since most of the mass is at the rod tips, the restoring torque increases approximately as the square of the rod length. Thus, there is a limit beyond which additional rod length will impair, rather than improve, pointing accuracy.

The situation is alleviated by silver plating the surface of the rods. This serves two functions. The more important function is to reflect most of the incident radiation. The thin coating of silver also improves the thermal conductivity, thereby reducing the thermal gradient. Some research conducted at General Electric Co., sponsored by NASA, has been directed toward obtaining the best possible reflective and conductive properties by forming the rods from a silver alloy. Short rods were formed having suitable structural properties, but manufacturing techniques have not been perfected to provide silver alloy rods of sufficient length and straightness for flight hardware.

A possible alternate solution to the thermal bending problem is to provide an equal and opposite torque by attaching conical reflectors to each rod tip. The size of reflector required is modest, being only slightly larger than the usual tip mass. However, this technique requires that the radius of curvature be known, and that the rod lie in the plane defined by the sun and the position of the undeflected rod. The radius of curvature is calculable, but recent experimental measurements have indicated that the plane of bending deviates as much as 20° from the plane of symmetry. The reason for this is that the tube is asymmetric in cross section (see sketch in Fig. 1) and that the coldest and hottest points are not necessarily diametrically opposed.

At present, no solution to the thermal bending problem is in sight. For rod lengths to 30 meters or less tip masses can provide the required moments of inertia. Limiting the length of the rods is not a serious restriction for orbits up to several earth radii, but the additional tip

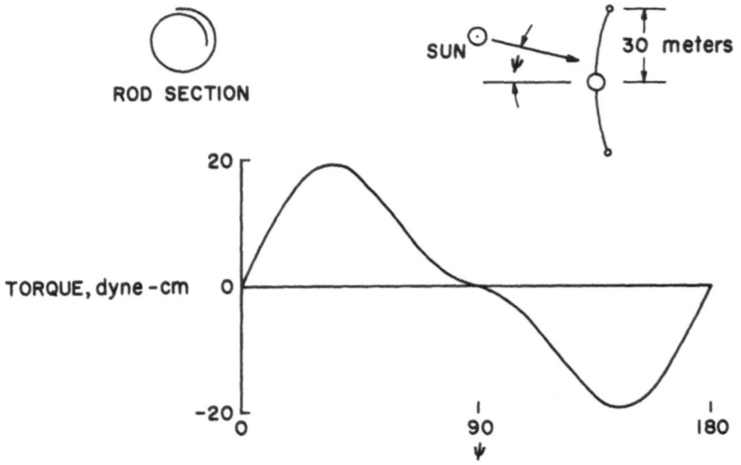

Fig. 1. - Torque from asymmetry induced by solar heating.

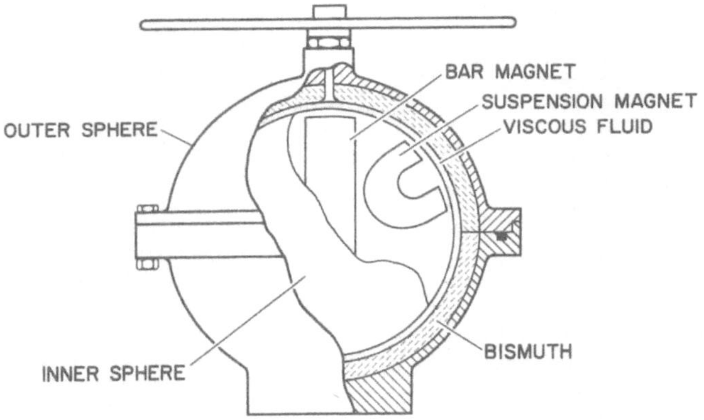

Fig. 2. - The General Electric magnetically oriented viscous damper.

mass increases the control system weight considerably for the synchronous mission.

One other source of error is attributable to the gravity-gradient rods, namely, errors introduced by initial structural curvature. The accepted figure for this error is roughly 1/2°. This error can possibly be reduced for multirod configurations if the curvature is determined on the ground and the rods are paired so that the principal axes of inertia lie in the desired directions.

DAMPING

Techniques dependent on the geomagnetic field

Simple and effective techniques have been devised for damping librational motions by utilizing the geomagnetic field. The ultimate in simplicity is achieved in a system devised by the Applied Physics Laboratories of Johns Hopkins University.[9] The damping device in this instance consists of permeable rods rigidly attached to the satellite structure. As the librational motion causes changes in the direction of the magnetic field imposed on these rods, hysteresis losses occur within the material, thereby dissipating the librational energy. The mass of magnetic material required is small at low altitudes. For instance, the mass of material for a satellite designed for an altitude of 650 km is approximately 200 g. As the altitude is increased, the strength of the geomagnetic field varies inversely as the cube of the orbit radius, and the moment of inertia of the satellite is likely to be required to increase by the same amount to provide sufficient restoring torque. These two factors will cause the required mass of magnetic hysteresis material to increase rapidly with orbit radius.

Another successful damping scheme has been devised in which an auxiliary body, containing a bar magnet, follows the geomagnetic field. The auxiliary body is coupled to the satellite through a spherical viscous damper. This system has been studied by several industrial and governmental organizations within the United States and a successful mechanization has been developed by the General Electric Co. and flown successfully at an altitude of 800 km. The unit weighed approximately 4.5 kg. Units for higher altitudes will be somewhat heavier because the dipole strength must be increased for the auxiliary body to follow the geomagnetic field.

A sketch of the damper unit is shown in Fig. 2. It consists of two concentric spheres separated by a viscous fluid. The internal sphere contains a bar magnet and secondary horseshoe magnets mounted circumferentially within the sphere. The outer sphere consists of two integral concentric shells; an inner shell of bismuth and an outer shell of aluminum alloy. The bismuth shell, in conjunction with the horseshoe magnets in the inner sphere, provides the centering force which prevents contact between the two spheres. The centering force is caused by the repulsive action of a magnetic field upon a diamagnetic material such as bismuth. Therefore, no friction between the two spheres can exist, so any coupling between them is purely viscous.

The viscous fluid between the inner and outer spheres dissipates energy when relative motion takes place between the inner and outer spheres. It also serves to cushion the inner sphere against shock and acceleration during boost. Further mechanical details of this unit as well as planned improvements to secure greater lifetime and reliability can be found in reference 7.

The damper unit can be placed anywhere within the satellite. However, the obvious choice for its location is at the tip of a gravity-gradient rod where it serves the additional function of a tip mass to increase the moment of inertia.

The number of gravity-gradient rods required will depend upon whether two- or three-axis stabilization is required and upon the orbital altitude. For a low altitude satellite, only a single rod will be required to point an axis toward the earth. This is the particular system built by General Electric for a Naval Research Laboratory Satellite (see Fig. 3). As the altitude is increased, the ratio of solar pressure disturbing torque to gravity-gradient restoring torque will increase in proportion to the cube of the orbit radius and the satellite must be made symmetrical by extending rods in opposite directions from the stabilized package. Additional rods must be deployed to provide dissimilar transverse principal moments of inertia if the vehicle is to be oriented about the earth-pointing axis. An additional pair erected along the desired direction of velocity vector will suffice to provide the three dissimilar moments of inertia. However, except for extremely large satellites, the moment of inertia of the gravity-gradient rods contributes nearly 100 percent of the total moment of inertia. Therefore, when only four rods are deployed, the mass distribution will be nearly planar. For this mass distribution, the satellite will have natural frequencies of orbital and twice orbital frequency. Since solar pressure disturbances and disturbance caused by orbital eccentricity also occur at these frequencies, it will usually be necessary to erect a third and shorter pair of rods orthogonal to the others so that the disturbance and natural frequencies differ. A three-axis magnetically damped system will therefore probably require six rods.

The magnetic damper can also be used to advantage for satellites at altitudes of less than 600 km where aerodynamic forces become significant, particularly when a payload must be positioned for retrofiring into the atmosphere.[7] Stabilization in this case is provided by

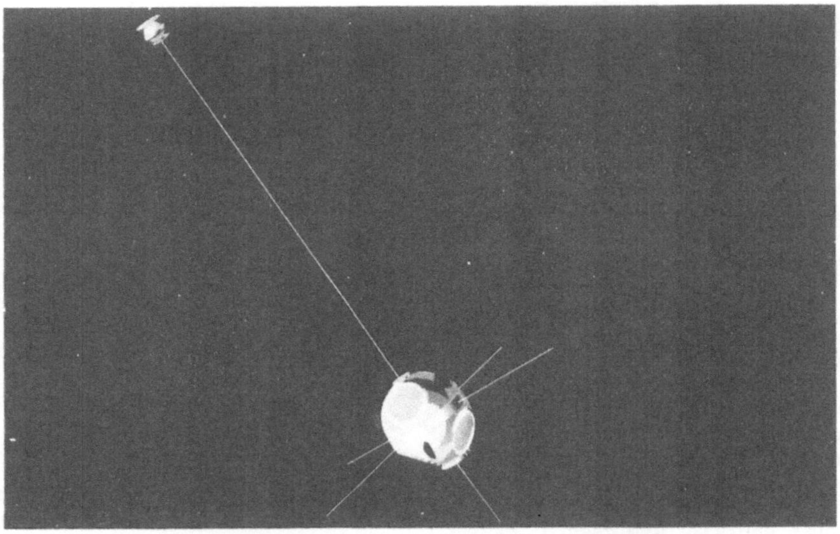

Fig. 3. - Artist's drawing of Naval Research Laboratories Satellite.

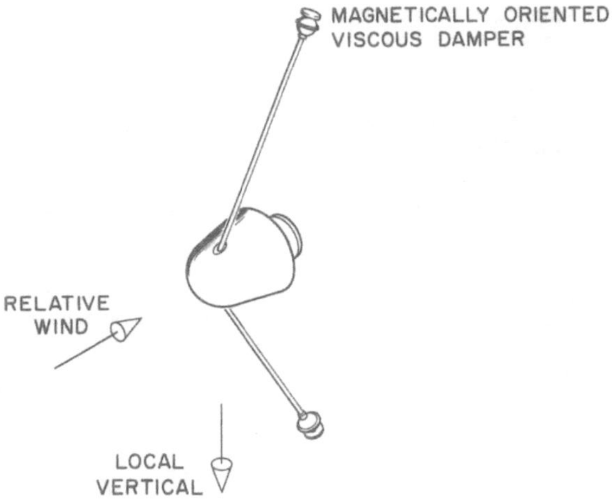

Fig. 4. - Use of the magnetically oriented viscous damper in three-axis stabilization scheme for altitudes where aerodynamic forces are significant.

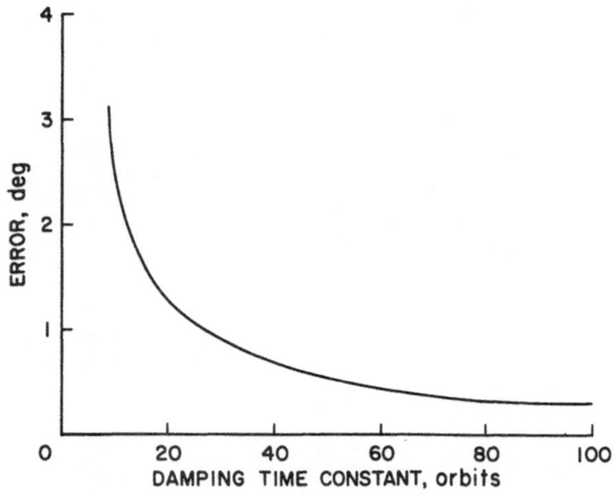

Fig. 5. - Typical relation between time constant and steady-state error in polar orbits for a magnetically oriented viscous damper.

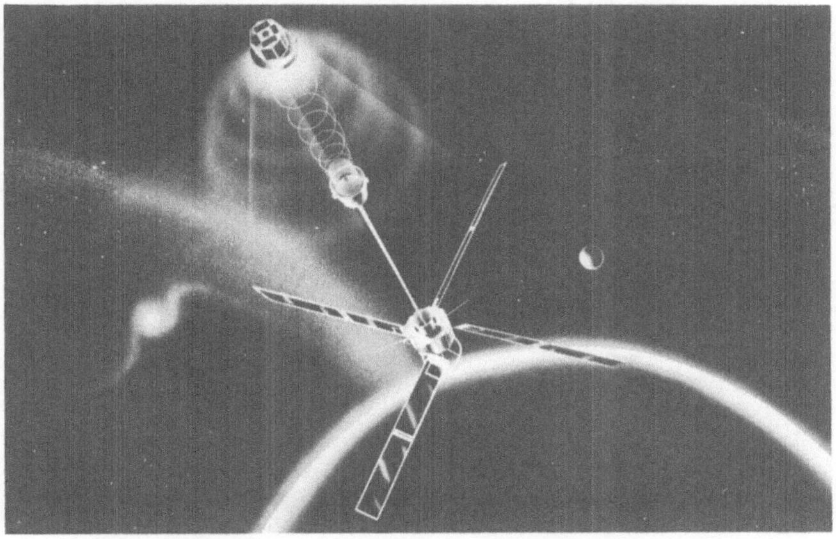

Fig. 6. - Artist's drawing of the Satellite 1963 22A.

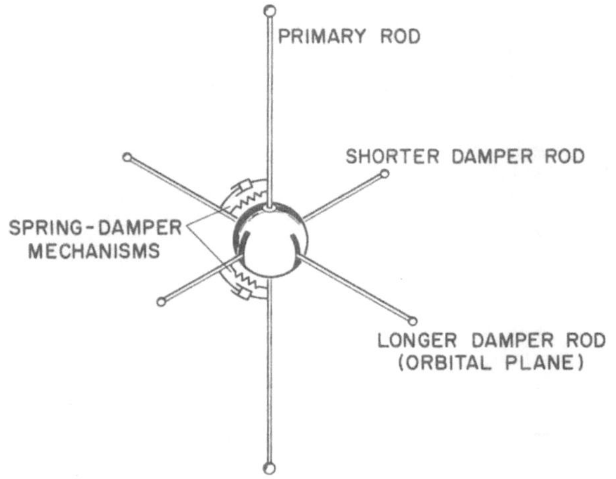

Fig. 7. - Schematic diagram of "Vertistat" configuration.

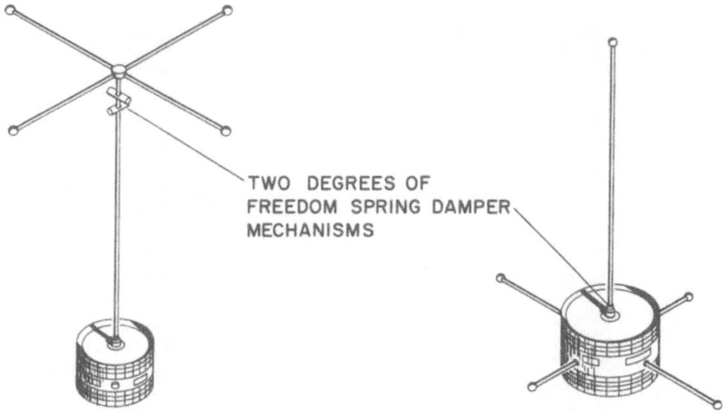

(a) Bell Laboratories system.

(b) Inverted system.

Fig. 8. - Systems relying solely on the gravitational field having a single auxiliary body with two degrees of freedom.

extending two rods about 50 feet in length with the magnetic dampers at their tips. As shown in Fig. 4, the rods are swept back slightly so that the aerodynamic forces provide restoring torques that augment the gravitational torque about the pitch axis and the gyroscopic torque about the local vertical. The angle between the rods must be greater than 90°, otherwise the gravitational and aerodynamic torques will be in opposition and the stable position will be a function of altitude.

All systems that depend upon the geomagnetic field for damping will have a pointing error induced by the damper. The underlying reason is that the magnetic field is nongeocentric so that relative motion exists between the satellite and the direction of the magnetic field vector even for zero attitude error. A part of this motion will be transmitted to the satellite, depending upon the damping level. Results given by Moyer et al. for a system with an oriented auxiliary body are shown in Fig. 5.[7] These results are for polar orbits where the relative motion of the magnetic field vector is the greatest. For a given induced error the decay rate is constant when expressed in orbits and is very nearly independent of altitude. At low altitudes, or high orbital rate, a small damper induced error will not imply an excessive number of days to damp. For example, for a 650 km orbit, an induced error of only 0.5° implies a time constant (time to damp to $1/e$ times the initial amplitude) of the order of 3 days. On the other hand, the same induced error for a 24 hr. orbit implies a time constant on the order of 45 days. To reduce the time to only 10 days implies an induced pointing error on the order of 2°.

The damper induced error is smallest for equatorial orbits and would vanish for equatorial orbits if the geomagnetic and geocentric poles were coincident. It is fortunate that they are not coincident since, in this event, damping of librations within the orbital plane would also vanish. The synchronous equatorial orbit is an interesting limiting case. Here, the satellite is stationary relative to the nominal geomagnetic field so that the induced error does not exist. The magnetic field vector lies between 10° and 20° from the normal to the plane of the orbit, depending upon geomagnetic longitude. There is a component of the field, therefore, that permits damping of the pitching motion within the orbital plane. Damping of the pitching motion within a reasonable period, 20 orbital periods per time constant, for example, implies that the motion in roll about the velocity vector will be critically damped.

The long time constants characteristic of magnetically damped satellites at synchronous altitude are not particularly severe if only the initial transient motion is considered. A 20-day time constant for small angles usually implies stabilization within 50 days which is not a large fraction of a satellite lifetime which may be 5 years or more. However, it is estimated that attitude disturbances as large as 15° caused by micrometeorites will occur about once per year, with smaller disturbances more frequent and larger disturbances less frequent.[10] If this estimation proves to be correct, a considerable fraction of the lifetime of a synchronous, magnetically damped satellite will be spent recovering from micrometeorite hits.

The magnetic field at synchronous altitudes will not be stationary because of the distortion of the magnetic field by the solar activity. Satellite probes have indicated that the solar wind deforms the magnetosphere, compressing it on the sunlit side of the earth and elongating it on the dark side. To a synchronous satellite, this deformation would appear as a diurnal variation in the field direction. A recent model of the distorted field (Mead[11]) indicates this distortion to be small at synchronous altitude. Large nonperiodic variations caused by magnetic storms represent an additional disturbance. Measurements of field strength and direction from Explorer 6 have indicated that the direction of the field can change by as much as 60° and the field strength can be halved during a storm.[12] The occurrence of magnetic storms and the steady-state diurnal variation will cause damper induced errors and, therefore, will influence the selection of damping levels.

Techniques dependent upon the gravitational field

The damping as well as the restoring torque can be made to depend upon the gravitational field. Passive control schemes based upon this concept have an advantage in that there are no damper-induced pointing errors. Two distinct types of damping systems have been devised. Each relies on attitude errors to induce relative motion between the satellite and an auxiliary body.

One of the damping techniques dissipates the energy of the pumping motion of an auxiliary mass attached to the satellite. Theoretical studies of systems of this type have been reported by Newton[13] and Paul.[14] This damping system is effective in damping motions within the orbital plane, but damping of the cross-plane motion vanishes for small attitude motions. Some other technique must therefore be provided to damp the roll motions for a complete system. In a system of this type built by the Applied Physics Laboratories, the cross-plane motion was damped by magnetic hysteresis rods. An artist's conception of this satellite is shown in Fig. 6.[9] The auxiliary body is attached to the end of a gravity-gradient boom by a weak spring. Damping is afforded by mechanical hysteresis within a cadmium coating on the spring. The system is lightly damped, requiring about 50 orbits to reduce the amplitude to $1/e$ times the initial value. It is

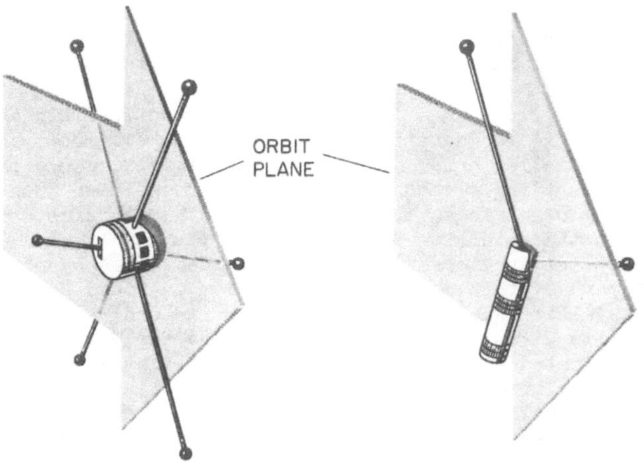

(a) Symmetrical arrangement. (b) Asymmetrical arrangement.

Fig. 9. - Inertially coupled three-axis systems relying solely on the gravitational field having a single auxiliary body with a single degree of freedom.

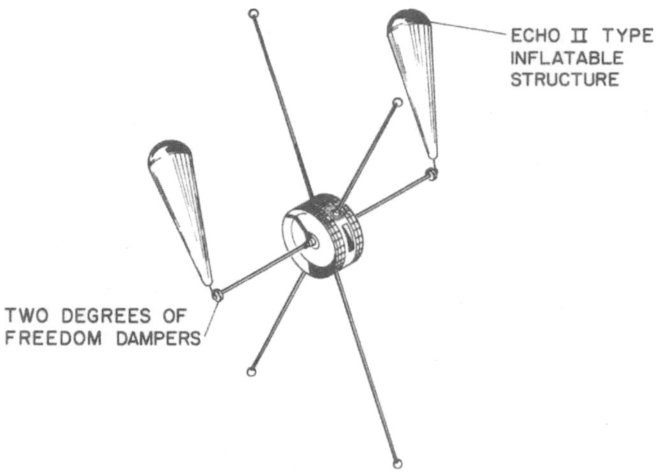

Fig. 10. - Solar-oriented damping system.

postulated by Newton that damping times of the order of 20 orbital periods are possible with this system.

It is possible to improve the damping afforded by pumping motion of a spring-mass system and to provide damping about all axes if the coupling between the torsional and longitudinal modes of the spring is exploited. This coupling exists if the spring wire is noncircular in cross section. The torsional motion is damped by dissipating the energy of the relative motion between the tip mass and an additional body within it. The additional body is suspended in a viscous fluid. A system having these features, known as the Rice/Wilberforce damper, has been developed by Goodyear Aerospace Corp.[15] The damping time constant is about 5 orbital periods.

Many systems have been devised in which attitude errors cause relative rotational motion between the satellite and an auxiliary body. This idea originated with Roberson and Breakwell who applied for a United States Patent in 1956.[4] The first practical mechanization was by Kamm.[8] A sketch of an arrangement of this type is shown in Fig. 7. Kamm suggested the use of the storable tubular rods and recognized the optimum orientation of the auxiliary bodies. The auxiliary body, which in this case is a simple rod, is located in the horizontal plane. In this location, the damper rod is unstable in the gravitational field and must be stabilized with a spring. The reason for placing the damper rod in the horizontal plane can be seen intuitively from the motion of two rods, both alined with the vertical and connected by a viscous damper. If one of the rods is disturbed, the relative motion between them will be damped until they are both oscillating with the same phase and amplitude. This mode of motion will be undamped. The greatest possible damping, therefore, might be expected when the damper rod lies in the horizontal plane. This observation has been rigorously proved by Zajac for simple pitching motion within the orbital plane.[16]

In the Vertistat configuration, two damper rods are required. Each is connected to the stabilized body through a frictionless spring-damper mechanism. The rod with the greater moment of inertia lies in the orbital plane and will damp only motion within that plane. The other damper rod is normal to the orbital plane and, because of gyroscopic coupling, is effective in damping motions about the other two axes. As long as these two rods have different inertias, a preferred orientation about the earth-pointing axis will exist.

A system similar to "Vertistat" has been proposed by Bell Laboratories and others. The Bell System,[10] uses a single auxiliary body consisting of a pair of crossed rods. This body is connected to the satellite through two single-degree-of-freedom spring-damper mechanisms with their axes of rotation at right angles - in effect, a universal joint. A sketch of this arrangement is shown in Fig. 8(a). Another version of this system is possible, if the satellite and damper mass distribution are interchanged (see Fig. 8(b)). The principal difficulty with this inverted arrangement is that the body to be stabilized is the body of least inertia, and, in the presence of a disturbance, will have a greater deviation from the vertical than does the vertical rod.

In the Vertistat configuration, the equilibrium positions of the principal axes of all three bodies lie either within the orbital plane or normal to it. For this situation, the pitch motion within the orbital plane and the roll-yaw motion about the other two axes are independent insofar as the small angle motion is concerned. This simplifies the mathematics of analysis and optimization. However, if two connected bodies are oriented so that at least two of the principal axes are skewed to the orbital plane, it will be found that any attitude motion will cause relative motion of the damper rod. It is possible, therefore, to provide damping about all three axes with a single auxiliary body that has a single degree of freedom.[17] A sketch of a system of this type is shown in Fig. 9(a). The principal axes of the satellite body are either in the plane of the crossed booms or normal to them. The earth-pointing axis bisects the acute angle formed by the booms. The cross products of inertia of the satellite body and the damper rod, relative to a reference frame imbedded in and normal to the orbital plane, are equal and opposite. This is dictated by nature because the principal axes of inertia of the entire satellite must be either normal to or within the orbital plane when the satellite is in equilibrium. For low altitudes, where the gravity-gradient torques are large compared to solar torques, it is possible to simplify the mechanization. This is done, in effect, by cutting the symmetrical configuration in half. The damper rod can be supported from one end as well as from the center when it lies in the horizontal plane since it is in equilibrium regardless of where it is supported. If the satellite body, exclusive of gravity-gradient rods, is reasonably slender, configurations such as shown in Fig. 9(b) are possible. Thus, for this case, three-axis stabilization is possible with the erection of only two gravity-gradient rods, one of which has a single degree of rotational freedom. If the satellite body is a slender cylinder, as shown in Fig. 9(b), its axis will be skewed to both the horizontal and orbital planes.

The damping of the "Vertistat" and all of its derivatives is much better than any of the other systems which have been devised. It is possible for most of the configurations to achieve damping constants (time to reduce amplitude to 1/e) of less than one orbital period for all modes of motion. However, a system designed for the maximum possible damping will usually suffer from reduced restoring torques, greater sensitivity to

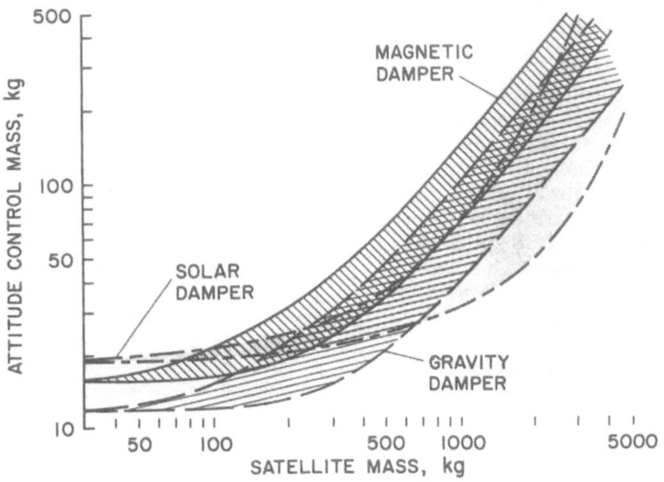

Fig. 11. - The estimated mass of three-axis attitude control systems for synchronous altitude.

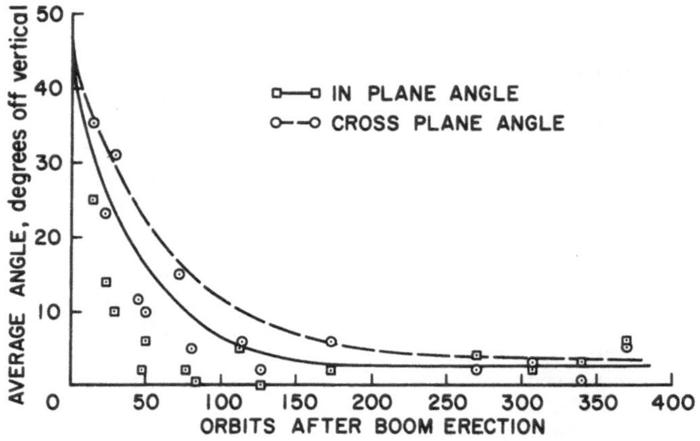

Fig. 12. - Flight test results of Applied Physics Laboratory Satellite 22A.

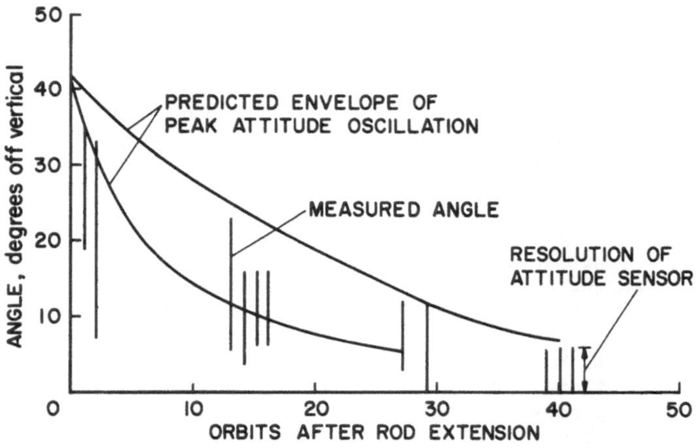

Fig. 13. - Flight test results of Naval Research Laboratories Satellite.

off-design conditions, degradation in the response to the principal oscillatory disturbances torques which occur at once and twice orbital frequency, and degradation of the response to orbital eccentricity. For this reason, a balance between damping time and steady-state pointing accuracy is usually sought in the design. A realistic damping time constant for a practical system, therefore, appears to lie between one and two orbital periods.

The key to a successful mechanization of a "Vertistat" type system is a spring-damper device. There are two devices in an advanced state of development, one at Bell Laboratories[10] and one at General Electric Co.[18] The Bell Laboratories' device relies on a torsional flexure to produce the required spring torques. Damping torque is produced by rotational hysteresis from relative displacement between a magnet, attached to the rotor, and magnetic material attached to the case. The damping torque for this device is amplitude dependent since it depends on magnetic hysteresis. The General Electric device relies on diamagnetic forces to provide a frictionless suspension of the rotor. Damping is achieved by the generation of eddy currents in an aluminum disk as it rotates between poles of permanent magnets. The torsional restraint is obtained from ferromagnetic inclusions in the aluminum disk. These inclusions interact with the magnets to restore the disk to a given rotational position. The inclusions are shaped to give a linear torsional restraint over $\pm 45^\circ$ of rotation. This device differs dynamically from that developed by Bell Laboratories in that the damping is viscous, depending upon angular velocity rather than amplitude.

Technique dependent upon solar field

General Electric Co. has proposed using solar pressure to orient the damping bodies.[18] In this technique, the satellite body is coupled through an eddy current disk to inflatable auxiliary bodies similar in structure to Echo II. The auxiliary bodies are oriented by solar pressure and provide an inertial reference for the damper (see Fig. 10). Since the auxiliary bodies are stationary in inertial space, a predictable steady-state bias from the zero gravity-torque position is introduced which is dependent upon orbital rate (or altitude) and the damping coefficient. Oscillations are induced when the solar reference is lost through eclipse. This effect is less severe as the altitude is increased. In this system, therefore, the damping time associated with a given damper-induced error decreases as the altitude increases. The time constant for errors of the order of 2° is about five days at synchronous altitude.

The dampers are similar in concept to the eddy current damper described for the pure gravity system except that there is no torsional restraint. Two dampers are coupled together at the end of each support rod to provide 360° of freedom about any axis.

The disadvantage of this system is its extreme complexity, requiring six rods, four damper units, and two inflatable structures. A possible future development of this design principle is to place a solar power system within the auxiliary bodies. This would mean further complication since a way must be found to deliver the power through the coupling without incurring either spring or friction torques.

Inflatable structures have also been suggested for use at low altitudes where their orientation will be dictated by aerodynamic forces. The stabilization method appears to hold no advantage over the magnetic technique described earlier.

STABILIZATION SYSTEM WEIGHT

The weight of a passive attitude control system is determined principally by the moment of inertia required to achieve the desired pointing accuracy. For this reason, the weight of the stabilized body influences control system weight only in that larger bodies are likely to have larger disturbance torques.

About 5 kg were attributable to the control system of the passively oriented satellites flown to date.[7,9] This figure can be taken as a current lower limit since it is representative of low-altitude satellites with no stabilization about the earth-pointing axis. The heaviest control systems will be required for three-axis stabilization at synchronous altitude. Estimates of the weight of these systems by Moyer and Foulke are summarized in Fig. 11.[18] For satellites of the weight presently contemplated, the control system weight will be between 12 and 20 kg, depending upon the type of damping. A gravity-oriented damper is shown to be superior to the other types in this respect. However, if stabilization about the earth-pointing axis is not required, there will be little difference between the weights of the magnetic and gravity-oriented damping systems.

FLIGHT PROGRAMS

Two earth-pointing, passively stabilized satellites have been flown by the United States. The excellent results obtained from these flights, which were at altitudes of the order of 600 km, have added increased stimulus to proceed with experiments at altitudes which will lead to successful passive attitude control for synchronous orbits.

The first gravity-gradient-stabilized satellite was conceived and built by the Applied Physics Laboratories of Johns Hopkins University and was launched in mid 1963.[9] As noted earlier, this satellite was damped through pumping librations of an attached spring mass and by magnetic hysteresis rods. The transient motion following deployment of the stabilization system is shown in Fig. 12. The motion damped in about 16 days or about 230 orbits. The final pointing accuracy was from 5° to 10°. Part of this error was caused by thermal bending of gravity-gradient rods. The rods were formed from beryllium-copper and had no reflective coating. When the satellite passed from the earth's shadow into the sunlight, the rapid heating of the rod induced attitude oscillations which had a period of several minutes and an amplitude of 5°. Beryllium-copper has little internal damping and it is reported that these oscillations did not dissipate during an orbital period.

The second gravity-gradient satellite was launched by the Naval Research Laboratories and used the General Electric magnetically oriented damping system (see Fig. 4).[7] The attitude instrumentation for this flight determined when the deviation from the local vertical was within one of several ranges. The smallest of these determined when the error was less than 6°. The librations were damped to within this smallest range after 40 orbits. A precise correlation between measured and predicted damping was not possible because of the limited instrumentation. However, the altitude measurements usually fell within the predicted attitude envelope.

Several future flight experiments are being considered for synchronous and near synchronous altitudes. For a system which uses the General Electric magnetically oriented damper at an altitude of 20,000 to 28,000 km, the deployed satellite must be symmetrical. The goal of this experiment would be to demonstrate an earth-pointing system at near synchronous altitude. Specific design goals are to damp to within $\pm 30^o$ within 20 orbits and to achieve a final pointing accuracy of 4° or better for a circular orbit with additional errors due to eccentricity held to less than 1° per percent eccentricity.

By far the most comprehensive gravity-gradient stabilization experiments are embodied in the Applications Technology Satellite program of the NASA. This program includes satellites stabilized by the inertially coupled system that relies solely upon the gravity-gradient. A deployed satellite similar to that shown in Fig. 9(a) in a 10,000 km orbit could verify the performance predicted by theory. The design may include the ability to vary disturbances by introducing known additional torques through a magnet or possibly a small inertia wheel. Other experiments include varying the length of the rods and the angle between them, viewing the bending of the gravity-gradient rods with a TV camera, and measuring the effectiveness of both the eddy current and magnetic hysteresis dampers.

The design of gravity-gradient-stabilized Applied Technology Satellites launched into a synchronous orbit will be influenced by the extrapolated results of the earlier flight at 10,000 km. The synchronous satellite would determine whether east-west station keeping is compatible with gravity-gradient stabilization. A successful flight of the synchronous satellite will demonstrate that passive gravity-gradient stabilization is feasible for all practical orbits.

REFERENCES

(1) Lagrange, J. L., "Theorie de la Libration de la Lune," Oeuvres, Tome V, p. 1.
(2) Merrick, V. K., Some Control Problems Associated With Earth-Oriented Satellites, NASA TN D-1771, 1963.
(3) Scott, E. D., "Control-Moment Gyro Gravity Stabilization," AIAA paper 63-324.
(4) Roberson, R. E., and Breakwell, J. V., "Satellite Vehicle Structure," United States Patent Office no. 3,031,154, April, 1962.
(5) De Bra, D. B., and Delp, R. H., "Satellite Stability and Natural Frequencies in a Circular Orbit," J. Astron. Sci., Vol. VIII, No. 1, Spring, 1961.
(6) Doolin, B. F., Gravity Torque on an Orbiting Vehicle, NASA TN D-70, 1959.
(7) Moyer, R. G., Katucki, R. J., and Davis, L. K., "A System for Passive Control of Satellites Through Viscous Coupling of Gravity Gradient and Magnetic Fields," AIAA paper 64-659, Aug., 1964.
(8) Kamm, L. J., "Vertistat: An Improved Satellite Orientation Device," J. Am. Rocket Society, No. 32, June, 1962, pp. 911-913.
(9) Fischell, R. E., and Mobley, F. F., "A System for Passive Gravity Gradient Stabilization of Earth Satellites," AIAA paper 63-326.
(10) Paul, B., West, J. W., and Yu, E. W., "A Passive Gravitational Attitude Control System for Satellites," The Bell System Tech. Jour., Sept., 1963, pp. 2195-2238.
(11) Mead, G. D., "Deformation of the Geomagnetic Field by the Solar Wing," J. Geophys. Res., Vol. 69, No. 7, April, 1964, pp. 1181-1195.
(12) Smith, E. J., Sonett, C. P., and Dungey, J. W., "Satellite Observations of the Geomagnetic Field During Storms," J. Geophys. Res., Vol. 69, No. 13, July, 1964, pp. 2669-2688.
(13) Newton, R. R., Damping of a Gravitationally Stabilized Satellite, Applied Physics Laboratories, Johns Hopkins Univ., TR-487, April, 1963.
(14) Paul, B., "Planar Librations of an Extensible Dumbell Satellite," AIAA Journal, Vol. 1, No. 2, Feb., 1963, pp. 411-418.
(15) Buxton, A. C., Campbell, D. E., and Losch, K., Rice/Wilberforce Gravity-Gradient Damping System, IEEE East Coast Conf. on Aerospace and Navigational Electronics, Baltimore, Md., Oct., 1964.

(16) Zajac, E. E., "Damping of a Gravitationally Oriented Two-Body Satellite," J. Am. Rocket Socity, No. 32, 1962, pp. 1871-1875.

(17) Tinling, B. E., and Merrick, V. R., "Exploitation of Inertial Coupling in Passive Gravity-Gradient-Stabilized Satellites," AIAA Journal of Spacecraft, Vol. 1, No. 4, July-Aug., 1964.

(18) Moyer, R. G., and Foulke, H. F., Gravity-Gradient Stabilization of Synchronous Satellites, IEEE 11th Annual East Coast Conf. on Aerospace and Navigational Electronics, Oct., 1964.

DISCUSSION

B. E. Tinling and V. K. Merrick

"The Status of
Passive-Gravity-Gradient Stabilization".

Q. Why is the accuracy in yaw connected to the eccentricity?

A. Accuracies about other axes are also functions of eccentricity, but the largest influence is on the yaw accuracy.

Q. Is it not true that there would be no effect on yaw accuracy if the scheme were symmetric?

A. That is true, but in the case of a symmetric scheme we would require two damper rods.

Q. What is the stiffness of the system?

A. About 20 dyne-cm per degree - not very stiff.

GRAVITY-GRADIENT STABILIZATION OF EARTH SATELLITES

by

Richard B. Kershner and Robert E. Fischell
The Johns Hopkins University
Applied Physics Laboratory
Silver Spring, Maryland

ABSTRACT

The possibility of using the earth's gravity field for vertical stabilization of near-earth satellites has intrigued theoreticians for many years. The very small stabilizing torque available, and the lack of a natural damping mechanism have been recognized as the major problems. The satellite 1963-22A launched in June 1963 was the first orbiting vehicle to successfully achieve passive gravity-gradient attitude stabilization. The initial spin of the satellite was removed by mechanical and magnetic despin devices. An internal electromagnet was then energized to align the satellite along the local magnetic field direction. Twelve hours later when the satellite passed over the north magnetic pole it was vertical with the correct side facing earthward. A 100 foot boom was then extended from the satellite and the magnetic dipole was turned off. Libration damping was provided by the combination of a lossy spring and magnetic hysteresis rods. Within a week after boom deployment the peak angle of oscillation of the satellite was damped below 15 degrees. By June 1964 four other satellites achieved gravity-gradient stabilization using passive damping techniques. The satellites 1963-38B and 1963-49B used the combination of a lossy spring and magnetic rods in a configuration that was virtually identical to the satellite 1963-22A. The satellite 1964-26A achieved gravity stabilization using only 0.6 pound of magnetic hysteresis rods to damp the peak libration angle from 40 degrees to 10 degrees in a period of approximately 4 days. This lightweight, simple and completely passive magnetic damping technique shows great promise for future applications for near-earth orbiting satellites. At satellite altitudes above 5000 miles, vector magnetometers with current amplifiers can be used to enhance the damping capability of the magnetic rods. This semi-passive technique with no moving parts should be suitable for satellites even at synchronous altitude. Thermal bending of the extendible booms has been observed on several satellites. A high frequency, dynamic boom bending has been observed for unplated booms with an amplitude of ±12 degrees. With silver plated booms, the dynamic thermal bending was reduced to less than ±1/4 degree.

INTRODUCTION

In the last several years the advantages of vertical stabilization for spacecraft have become increasingly apparent. Many satellite programs are now approaching the operational phase where high reliability for an extended lifetime in orbit is a requirement. As a result there has been an increasing interest in applying passive gravity-gradient attitude control for several important space missions such as communication and meteorological satellites. The impetus to utilize this method of stabilization has also increased as a result of the successful demonstration in orbit of several gravity-gradient stabilized spacecraft.

As of May 1965 a total of nine satellites have successfully achieved gravity-gradient stabilization. All of these vehicles have been under the sponsorship of the U. S. Navy. Five of these spacecraft were designed at The Johns Hopkins University, Applied Physics Laboratory, Silver Spring, Maryland, and four spacecraft were under the direction of the U. S. Naval Research Laboratory, Washington, D. C.

ORIGIN OF GRAVITY AND CENTRIFUGAL TORQUES

The origin of the torques on an orbiting body are best visualized by considering the gravity and centrifugal forces that act on a simple dumbbell mass distribution. Figure 1 shows the gravity and centrifugal forces that act upon a simple dumbbell that is displaced an angle θ off the vertical in the orbital (pitch) plane. The mass m that is closer to the center of the earth has a gravity force acting upon it that is larger than the gravity force acting on the outermost mass m'. By virtue of being farther from its center of rotation, the centrifugal force F_c' on the outermost mass is greater than the centrifugal force F_c acting on the mass that is closer to the center of rotation.

It is obvious from Figure 1 that the larger gravity force F_g also acts at greater distance from the center of mass of the dumbbell than does the force F_g'. This difference in lever arm tends to accentuate the torque acting on the dumbbell which tends to make the long axis of the body seek the local vertical. The net torque due to gravity forces is given by

$$\tau_g = F_g L - F_g' L' \qquad (1)$$

The larger centrifugal force, on the other hand, has the smaller lever arm and it can be readily shown that,

$$F_c L_c = F_c' L_c' \qquad (2)$$

and therefore there is no torque on a body displaced in the orbital plane due to centrifugal

Superior numbers refer to similarly-numbered references at the end of this paper.

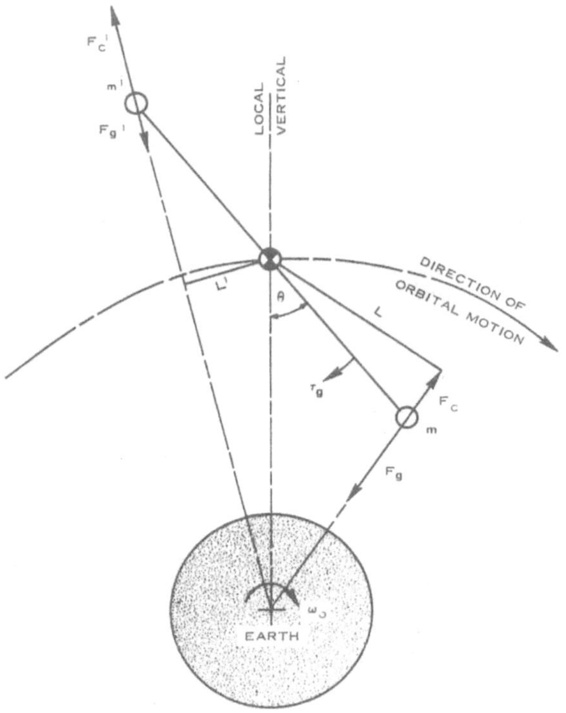

FIGURE 1. GRAVITY AND CENTRIFUGAL FORCES IN THE ORBITAL (PITCH) PLANE.

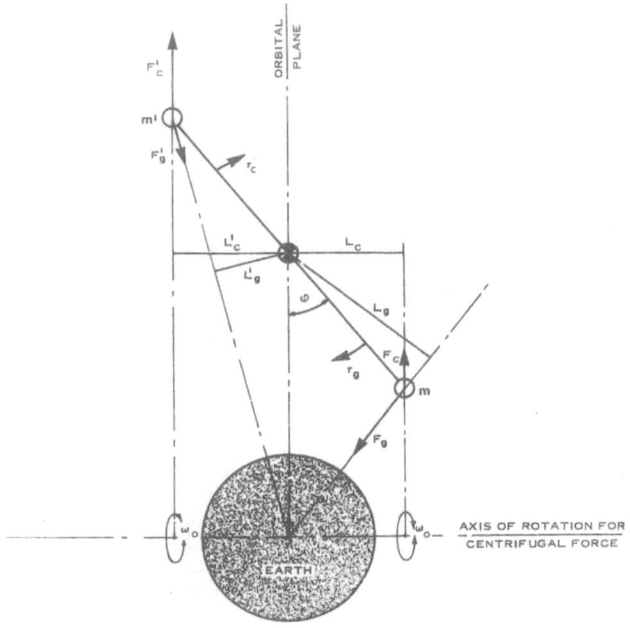

FIGURE 2. GRAVITY AND CENTRIFUGAL FORCES IN THE CROSS-ORBIT (ROLL) PLANE.

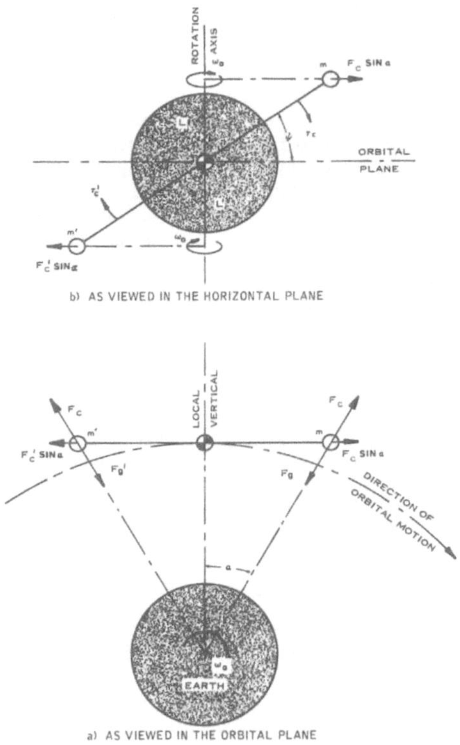

FIGURE 3. GRAVITY AND CENTRIFUGAL FORCES IN THE HORIZONTAL (YAW) PLANE.

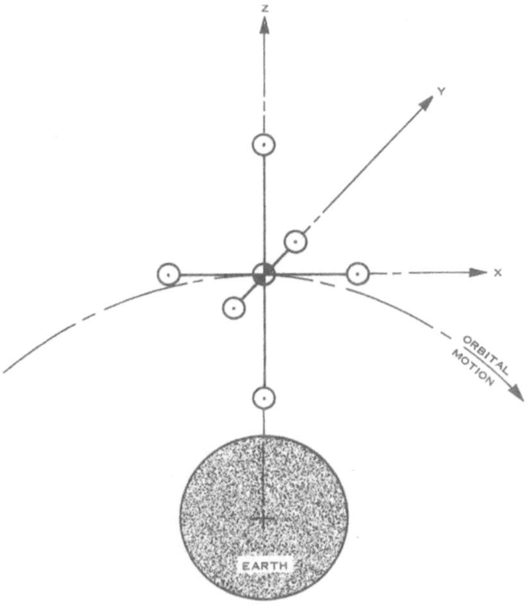

FIGURE 4. STABILIZATION FOR A SATELLITE OF ARBITRARY MOMENT OF INERTIA DISTRIBUTION

forces. It is therefore appropriate that the name "gravity-gradient" torque be applied as the result of forces in the orbital plane because this torque originates entirely from the gravity-gradient field in which the body orbits.

Figure 2 shows the forces that act on a simple dumbbell mass distribution that is displaced an angle ϕ off the vertical in the cross-orbit (roll) plane. The larger gravity force, F_g, has the larger lever arm L_g. The torque acting upon an object displaced in the cross-orbit plane due to gravity-gradient forces is given by

$$\tau_g = F_g L_g - F'_g L'_g \qquad (3)$$

In the roll plane there is a different direction for the centrifugal forces as compared to the gravity forces. Centrifugal forces arise from rotation about an axis rather than originating from a point as is the case with gravity forces. Therefore, as shown in Figure 2, in the roll plane the centrifugal force acting on the dumbbell end-masses have the same lever arms, i.e., $L_c = L'_c$. Therefore there is a net torque due to centrifugal force that tends to restore the dumbbell to the vertical direction, which is given by

$$\tau_c = F_c L_c - F'_c L'_c \qquad (4)$$

The restoring torque in the cross-orbit plane will therefore be larger than in the orbital plane because of the contribution of the centrifugal forces.

Figure 3 illustrates the forces acting upon a horizontal dumbbell whose long axis is displaced by the angle Ψ from the orbital plane. As can be seen in Figure 3, the gravity forces which originate at the mass center of the earth exactly cancel one another, i.e., $F_g = F'_g$. However, from Figure 3b it is obvious that the centrifugal forces acting upon the end-mass m and m' will cause a net torque in the horizontal (yaw) plane that causes the dumbbell to align itself into the orbital plane. The net centrifugal force causing this yaw stabilization is given by

$$\tau_c = LF_c \sin\alpha + L' F'_c \sin\alpha \qquad (5)$$

or

$$\tau_c = 2 L F_c \sin\alpha \qquad (6)$$

where α = half angle at the axis of rotation as shown in Figure 3a.

If we have a body of arbitrary mass distribution as indicated by three sets of orthogonal dumbbells as shown in Figure 4, because of gravity-gradient and centrifugal forces, this body will tend to align itself so that the longest dumbbell will seek the vertical, the next longest dumbbell will stabilize in the orbital plane and the shortest dumbbell will be oriented perpendicular to the orbital plane. If we take the vertical direction as the Z-axis, the direction of forward motion as the X-axis, and with the Y-axis forming an orthogonal set, the torque in the pitch plane can be readily shown to be

$$\tau_p = \frac{3}{2}\omega_o^2 (I_x - I_z) \sin 2\theta \qquad (7)$$

Furthermore, if displaced an angle ϕ off the vertical in the roll plane, the net torque acting upon this body will be given by

$$\tau_r = \frac{4}{2}\omega_o^2 (I_y - I_z) \sin 2\varphi \qquad (8)$$

and if the X dumbbell is displaced an angle Ψ from the orbital plane there will be a yaw stabilization torque due to centrifugal force that is given by

$$\tau_y = \frac{1}{2}\omega_o^2 (I_y - I_x) \sin 2\Psi \qquad (9)$$

where I_x = moment of inertia of the body about the X-axis,

I_y = moment of inertia of the body about the Y-axis,

I_z = moment of inertia of the body about the Z-axis,

and ω_o = orbital angular rate (radians/sec).

The equations listed above are completely adequate for most considerations of passive gravity-gradient stabilization. However, references 1, 2, 3, 4 and 5 are among several which provide a thorough analysis of the gravity-gradient and centrifugal forces giving more precise, but considerably more complex expressions for the restoring torques.

For small angular displacements of θ, φ, and Ψ equations (7), (8) and (9) can be simplified to the form

$$\tau_p = 3\omega_o^2 (I_x - I_z)\,\theta \qquad (10)$$

in the pitch plane;

$$\tau_r = 4\omega_o^2 (I_y - I_z)\,\phi \qquad (11)$$

in the roll plane;

and

$$\tau_y = \omega_o^2 (I_y - I_x)\,\Psi \qquad (12)$$

in the yaw plane. One then obtains for the equation of motion

$$I_y \frac{d^2\theta}{dt^2} + 3\omega_o^2 (I_x - I_z)\,\theta = 0 \qquad (13)$$

$$I_x \frac{d^2\phi}{dt^2} + 4\omega_o^2 (I_y - I_z)\,\phi = 0 \qquad (14)$$

$$I_z \frac{d^2\Psi}{dt^2} + \omega_o^2 (I_y - I_x)\,\Psi = 0 \qquad (15)$$

From these differential equations one can easily obtain the natural periods of libration in the pitch, roll, and yaw planes as follows:

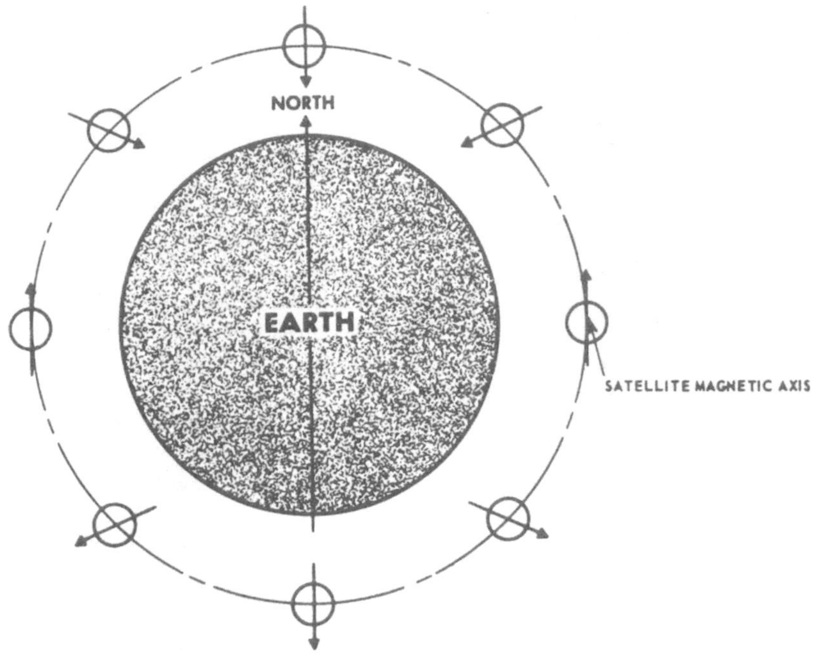

FIGURE 5. MOTION OF A MAGNETICALLY ORIENTED SATELLITE IN A POLAR ORBIT.

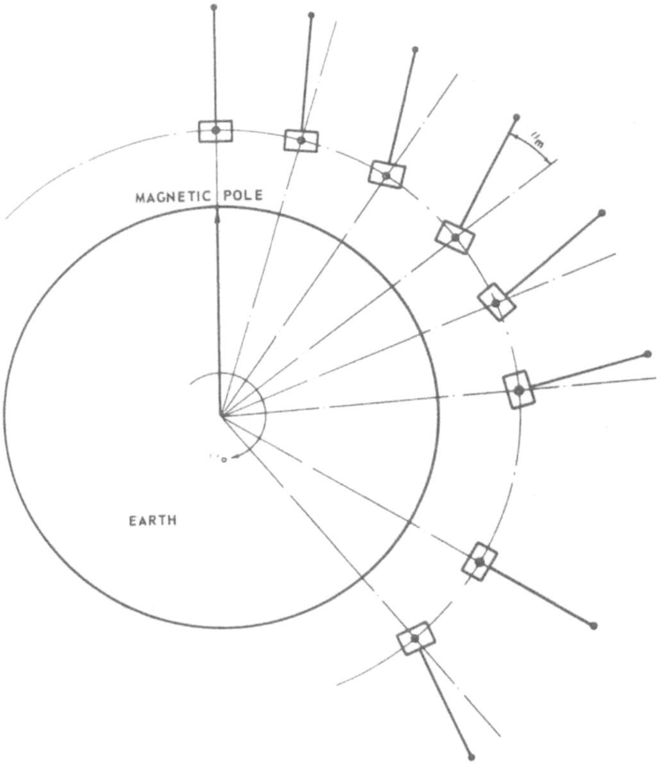

FIGURE 6. INITIAL MOTION AFTER GRAVITY-GRADIENT CAPTURE

$$T_p = T_o \left[3 (I_x - I_z)/I_y\right]^{-1/2} \quad (16)$$

$$T_r = T_o \left[4 (I_y - I_z)/I_x\right]^{-1/2} \quad (17)$$

$$T_y = T_o \left[(I_y - I_x)/I_z\right]^{-1/2} \quad (18)$$

where T_o = orbital period of the satellite.

Substitution of typical numbers that are achievable for satellite moments of inertia readily indicate that the restoring torques are exceedingly small and the natural periods of libration are exceedingly long. Consider a satellite which has the configuration shown in Figure 4, with +Z and -Z boom lengths of 100 feet, +X and -X boom lengths of 40 feet, and +Y and -Y boom lengths of 20 feet, all with end masses of 0.1 slug (3.2 pounds). Then for a satellite orbiting at 600 nautical miles altitude ($\omega_o = 10^{-3}$ radians/sec) we get the following torques:

$$\tau_p = 0.873 \times 10^{-4} \text{ ft-lbs/degree}$$

$$\tau_r = 1.375 \times 10^{-4} \text{ ft-lbs/degree}$$

$$\tau_y = 0.052 \times 10^{-4} \text{ ft-lbs/degree}$$

Although these torques are exceedingly small, when they are applied in the frictionless environment of space, they are able to readily stabilize bodies that are hundreds of feet long. It is however necessary to realize how small these torques really are and to take care in the satellite design to eliminate any other sources of torque that might disturb this real but delicate balance.

The natural frequencies of libration for the "typical" satellite configuration described above are:

$$T_p = 0.63\ T_o = 66.7 \text{ minutes}$$

$$T_r = 0.54\ T_o = 57.2 \text{ minutes}$$

$$T_y = 1.06\ T_o = 112.1 \text{ minutes}$$

The fact that these natural periods of libration are exceedingly long compared to most terrestrial devices makes it difficult to design an effective damper. However, several damping systems have been developed which can effectively cause the satellite to stabilize in the desired manner.

THE PROBLEM OF CAPTURE

For an earth satellite to achieve passive gravity-gradient stabilization, it is necessary to follow certain procedures. Although these procedures will, of course, differ somewhat for various satellite missions, some problems that are common to all will be discussed here.

We will assume for all cases that the boom required to alter the mass distribution of the satellite is extended after the satellite is in orbit. The first thing that must be accomplished is to remove virtually all the spin that may have been imparted to the satellite during the launch procedure. A device that rapidly removes the spin energy of a satellite is the so-called "yo-yo", consisting of two weights attached to cables that are wrapped around the satellite.[6] When the weights are released they spin out from the satellite, causing a tension in the cables that results in a retarding torque on the satellite. This device has been successfully employed on five APL satellites as well as on several Tiros satellites. To guarantee the very low angular rates that are required when erecting a comparatively weak extensible boom, magnetic hysteresis rods can be employed.[7] As these rods rotate in the earth's magnetic field, they remove the spin energy of the satellite because of magnetic hysteresis loss.

The next procedure is to align the satellite vertically with the correct side facing downward. This can be done by energizing an electromagnet that is rigidly attached to the satellite, causing it to follow the earth's magnetic field direction in exactly the same manner as a compass needle. If the direction of the magnet is along the satellite's Z-(symmetry) axis, it can be shown that the satellite will align this axis along the local magnetic field direction.[8] For a polar orbiting satellite, the motion will then be as shown in Figure 5. The magnetic hysteresis rods used to remove the spin energy of the satellite will also damp the oscillations of the satellite about the local magnetic field direction. Since the local magnetic field is vertical at points directly above the earth's magnetic poles, the optimum procedure is to capture the satellite into gravity-gradient attitude stabilization as it passes over one of the poles, having previously stabilized it magnetically with a particular desired face directed toward the earth.

The boom is then erected and the electromagnet turned off by radio command from a ground station. The tumbling rate of the satellite will immediately be reduced by the ratio of the moments of inertia before and after deployment of the boom. For a typical satellite design, the moment of inertia might be increased by a factor of 100, resulting in a decrease in the satellite's tumbling rate to 0.015 revolutions per orbit (rpo).[9] This means that the satellite is essentially stopped in inertial space. To be vertically stabilized the satellite must then achieve an angular rate in inertial space of 1.0 rpo. Immediately after the boom is erected, the satellite continues in its orbital motion with its Z-axis essentially fixed in inertial space. As the satellite moves away from the magnetic pole, as shown in Figure 6, a gravity-gradient torque will act upon it, tending to align the X-axis along the local vertical direction. The angle with the local vertical will continue to increase until

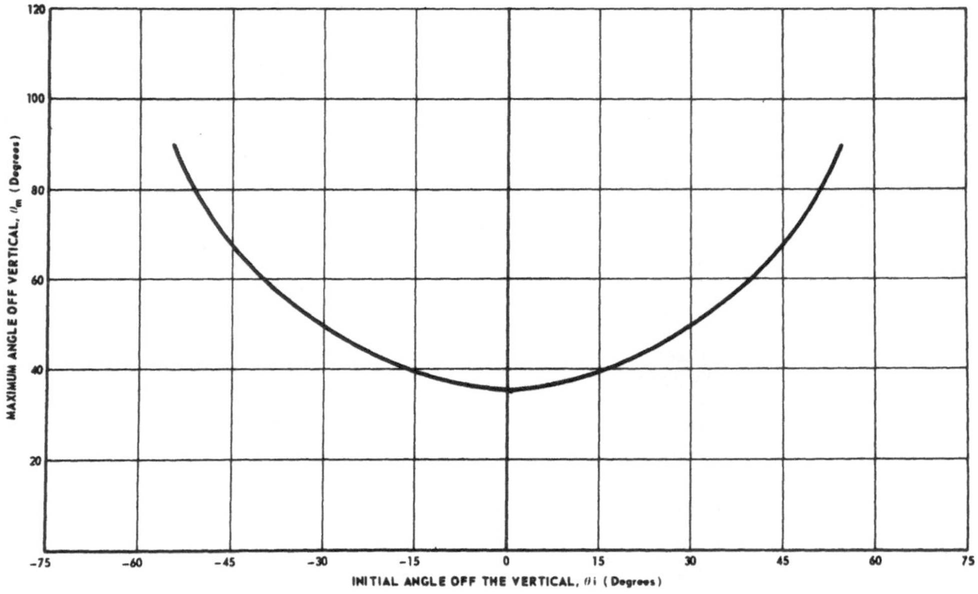

Fig. 7 MAXIMUM ANGLE OFF THE VERTICAL AS A FUNCTION OF INITIAL ANGLE

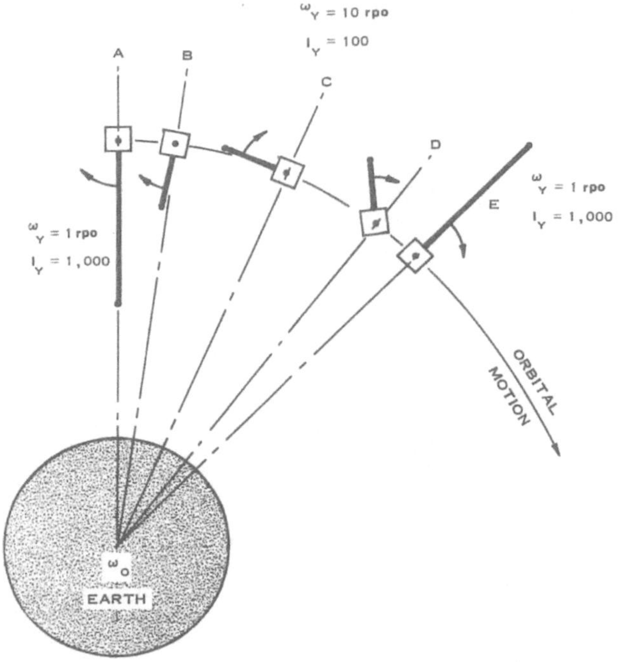

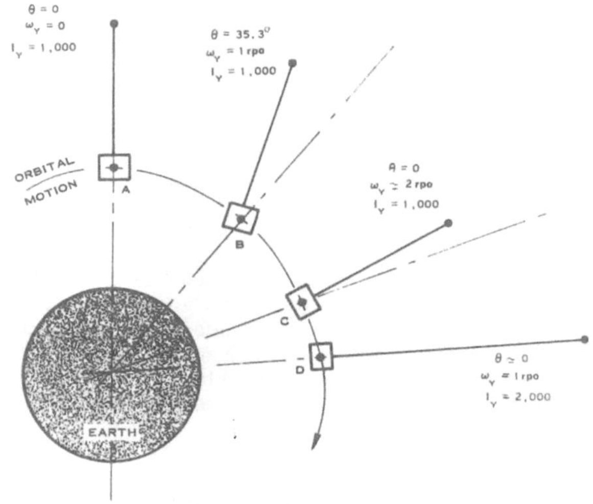

FIGURE 9. CAPTURE SEQUENCE FOR ELIMINATING LIBRATION

FIGURE 8. INVERTING A SATELLITE BY RETRACTING AND EXTENDING A BOOM.

the gravity-gradient torque causes the satellite to develop an angular rate of 1.0 rpo. The satellite angle with the vertical will then decrease as the gravity-gradient torque continues to act, resulting in a planar libration motion for the satellite.

The maximum angle to which the satellite will swing is of great interest. If the maximum angle is less than 90°, capture of the satellite into gravity-gradient attitude stabilization will result. The maximum angle to which the satellite will librate in the pitch plane can be calculated rather simply by equating the angular kinetic energy that the satellite must develop (to achieve an angular rate of 1.0 rpo) with the work done by the gravity-gradient torque as the satellite moves out to that maximum angle.

The satellite kinetic energy when it achieves an angular rate of 1.0 rpo is given by

$$K.E. = \frac{1}{2} I_y \omega_o^2 \tag{19}$$

This work done on the satellite by the gravity-gradient torque is given by

$$W = \int_{\theta_i}^{\theta_m} \tau_p \, d\theta$$

where θ_i = initial angle off the vertical,

and θ_m = the maximum angle with the local vertical direction to which the satellite will swing.

Taking the expression for τ_p from Equation (7) gives

$$W = \frac{3}{2} \omega_o^2 (I_x - I_z) \int_{\theta_i}^{\theta_m} \sin 2\theta d\theta, \tag{20}$$

and therefore

$$W = \frac{3}{4} \omega_o^2 (I_x - I_z) (\cos 2\theta_i - \cos 2\theta_m). \tag{21}$$

Equating (19) and (21) and solving for θ_m yields the result

$$\theta_m = \frac{1}{2} \text{arc cos} \left[\cos 2\theta_i - \frac{2}{3} \left(\frac{I_y}{I_x - I_z}\right)\right]. \tag{22}$$

For $I_x = I_y \gg I_z$ and for $\theta_i = 0$, $\theta_m = 35.36$ degrees which is well below 90 degrees and the satellite will therefore be captured into the vertical stabilization condition. Furthermore, the maximum angle to which the satellite will librate is independent of the orbital period (and therefore independent of the satellite altitude). Also for $I_x = I_y \gg I_z$ the maximum angle is independent of the actual values of I_x, I_y and I_z. When the satellite is not exactly aligned along the local field direction when the boom is erected, the satellite will swing out to a larger angle with respect to the local vertical direction. The result of θ_m as a function of θ_i for $I_x = I_y \gg I_z$ is shown in Figure 7. It is apparent that θ_m increases as the absolute value of θ_i increases; i.e., for positive or negative initial deviation angles off the vertical, the maximum angle θ_m will increase. The limiting angle for capture, θ_m = 90 degrees, occurs at θ_i = 54 degrees.

Another interesting consequence of Equation (22) is that there is an upper limit for the quantity $I_y/(I_x - I_z)$ for which the satellite will capture even for $\theta_i = 0$ (and this value decreases as θ_i increases). For $\theta_i = 0$, the satellite will capture only if

$$\frac{I_y}{I_x - I_z} < 3 \tag{23}$$

If the satellite is provided with a boom that can be extended or retracted to any length by ground command, it is possible to obtain capture with the correct side facing downward regardless of the initial conditions of the satellite relative to the local vertical. In this procedure the boom is extended at any time as long as the satellite's angular rates are low enough so that the boom is not damaged by being extended. The satellite will then librate or possibly turn over depending on the initial conditions of the motion. The libration damper (to be discussed in the next section) would then remove the rotation, if any, and damp the libration down to a small angle. If the satellite is captured with the correct side facing downward, the problem of capture is successfully completed. If, however, the satellite stabilized in an inverted position, it can readily be turned over by retracting and then extending the boom as illustrated in Figure 8. When the satellite is gravity stabilized it is rotating at the rate of 1.0 rpo in inertial space. When the boom is retracted so that the moment of inertia is reduced to some arbitrary value such as 1/10 its original value (as shown in Figure 8) the satellite will acquire an angular rate of 10 rpo. The satellite will therefore rotate rapidly so that the correct side will be facing downward. If the boom is then extended to its pre-retraction length the satellite can be recaptured with the proper angular rate in inertial space (i.e., 1.0 rpo) and with little or no deviation from the vertical. This technique has been successfully accomplished on one of the Naval Research Laboratory satellites.

An interesting suggestion was made by Dr. J. W. Follin of the Applied Physics Laboratory for capture without an initial libration. Let us suppose that we erect the boom when the satellite is vertical with its correct side facing downward as results from magnetic stabilization. One then allows the satellite to swing to its maximum libration angle, θ_m; then the next time it crosses the vertical it will be traveling at 2.0 rpo in inertial space. At this moment,

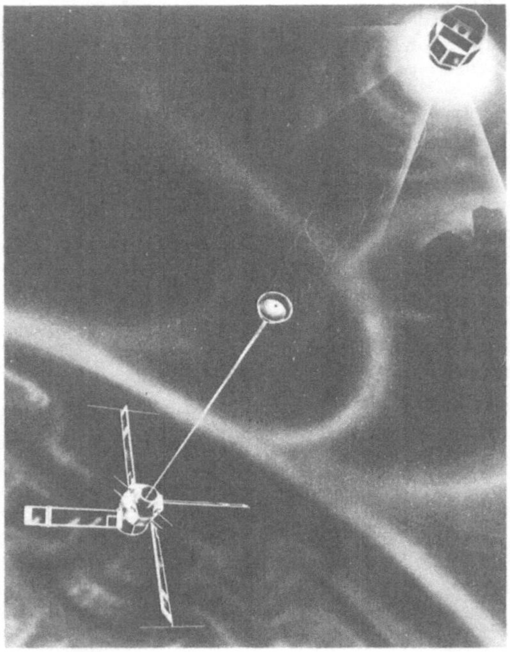

FIGURE 10. ARTIST'S CONCEPT OF THE SATELLITE 1963-22A.

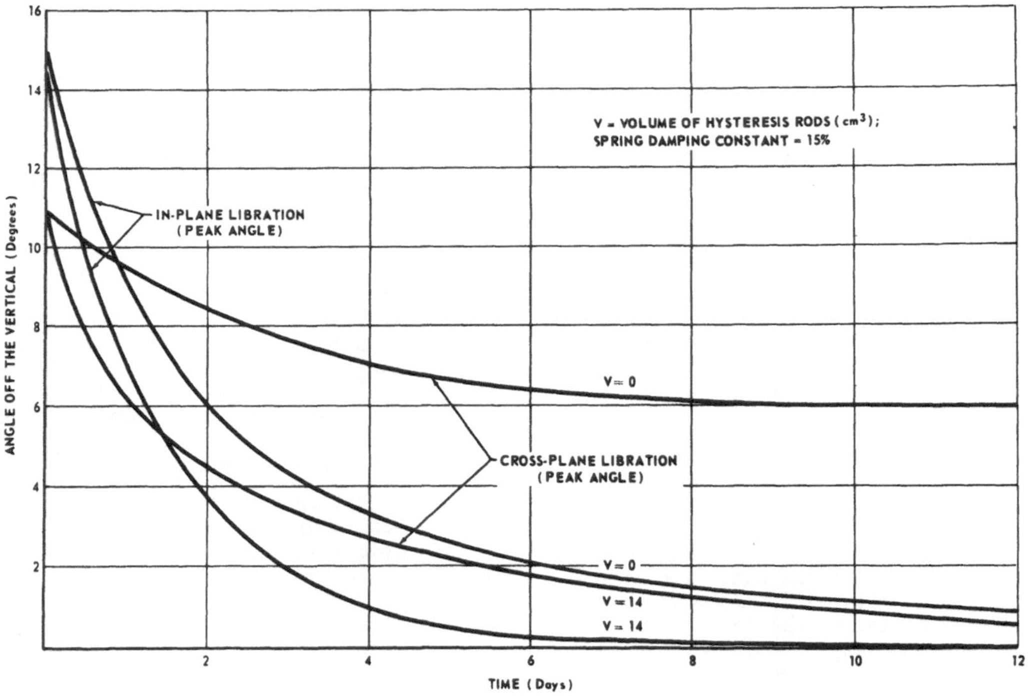

FIGURE 11. THEORETICAL DAMPING OF SATELLITE LIBRATIONS FOR A SATELLITE WITH DAMPING SPRING AND HYSTERESIS

as shown in Figure 9, one extends the boom further to double the moment of inertia of the satellite, or possibly another boom of equal moment of inertia is extended in the -Z direction. Since the angular rate will then be the required value of 1.0 rpo and since the angle off the vertical can be essentially zero, we have captured the satellite into gravity-gradient stabilization without an initial libration. This technique may be desirable for satellites at synchronous altitude where the time to damp the libration to an acceptably small angle might otherwise take many days.

DAMPING SATELLITE LIBRATIONS

After the satellite has been captured so that it does not tumble it is usually necessary to damp the satellite libration about the local vertical direction. In addition to removing the initial libration, effective damping reduces the deviation from the vertical caused by perturbing torques and impulses. Several methods for damping the oscillatory motion of a gravity-gradient stabilized satellite have been suggested and there are undoubtedly many other possibilities. In this section are discussed those damping techniques which have been developed for satellites designed at the Applied Physics Laboratory.

DAMPING SPRING WITH MAGNETIC HYSTERESIS RODS

In June 1963 the satellite bearing the international designation 1963-22A was the first to achieve gravity-gradient stabilization.[10] To obtain the required libration damping this satellite utilized a combination of an ultra-weak spring fastened to the end of a long boom combined with magnetic hysteresis damping rods. In Figure 10 is shown an artist's concept of this satellite with boom and spring extended.

As this type of satellite oscillates about the local vertical, the radial force in the direction of the boom varies because of the difference in gravity-gradient force as a function of angle off the vertical plus an additional force resulting from the libration. This varying force acting on a mass at the end of the spring causes it to move in and out, thus absorbing the libration energy by mechanical hysteresis in the spring. This damping was suggested and analyzed in detail by Dr. R. R. Newton of the Applied Physics Laboratory.[11]

Dr. J. L. Vanderslice of the Applied Physics Laboratory, Dr. R. H. Frick of the Rand Corporation and Dr. B. Paul of the Bell Telephone Laboratories have also analyzed the motion of a boom-and-spring system.[12,13,14] The results of these investigators determined that the spring provides the damping required for gravity-gradient stabilization. The analysis of Dr. Vanderslice has shown that the spring is extremely effective in damping satellite librations in the plane of the orbit and is less effective in damping oscillations perpendicular to the orbital plane. The magnetic rods have some damping effect for motions in the orbital plane and are most effective for damping the cross-plane oscillations. Figure 11 shows the effectiveness of a spring in damping satellite librations for a satellite with and without magnetic damping rods. The effectiveness of the spring and rods for damping all satellite librations is quite apparent from these curves.

Pertinent physical characteristics of satellite 1963-22A, which was the first to achieve gravity stabilization using a passive (spring and magnetic rods) damping system, are as follows:[15]

Number of magnetic rods	4
Magnetic rod dimensions	48 in. x 0.110 in.
Total rod volume	30 cm^3
Magnetic rod material	AEM-4750
Boom length	100 ft.
Damping spring constant	2.14×10^{-6} lb/ft
Zero-force spring length	1.5 in.
Equilibrium spring length	20 ft.
Spring dissipation per cycle	0.50
Weight of spring end mass	3.85 lbs.
Weight of boom end mass	1.5 lbs.

Fabrication of the damping spring, and the design of a simple but reliable means of deploying it in orbit, presented an interesting engineering problem.[16] Satellite 1963-22A employed a helical spring having, under zero force, an equilibrium length of 1.5 in. The helical spring was 7.6 in. in diameter and consisted of 70 turns of 0.008 in. diameter beryllium-copper wire. Since beryllium-copper is an excellent spring material, it does not provide sufficient hysteresis loss. To obtain good damping, a 0.0008 in. thick layer of the mechanically soft material, cadmium, was electrolytically deposited on the outer surface of the beryllium-copper wire. A 0.0002 in. coating of gold was electrolytically deposited over the cadmium to prevent the cadmium from subliming in the hard vacuum of space. When completely fabricated, annealed, and coated, the spring had a constant of 2.14×10^{-6} lb/ft.

To determine effectiveness of the spring under the conditions expected in orbit, a torsional pendulum using the spring material was built and tested in a large vacuum chamber.[17,18] The period of the torsional pendulum was set at 55 minutes to correspond closely to the natural period of libration for the nominal orbit of the 1963-22A satellite. By this method it was determined that the spring had a damping coefficient of 50%; i.e., 1/2 of the maximum energy stored in the spring was dissipated on each oscillation. This compares with an energy loss of less than 1% per cycle for the uncoated beryllium-copper wire.

To prevent tangling or other damage to this ultra-weak spring during handling and launching operations, it was necessary to encapsulate it in a subliming material; the one selected was an

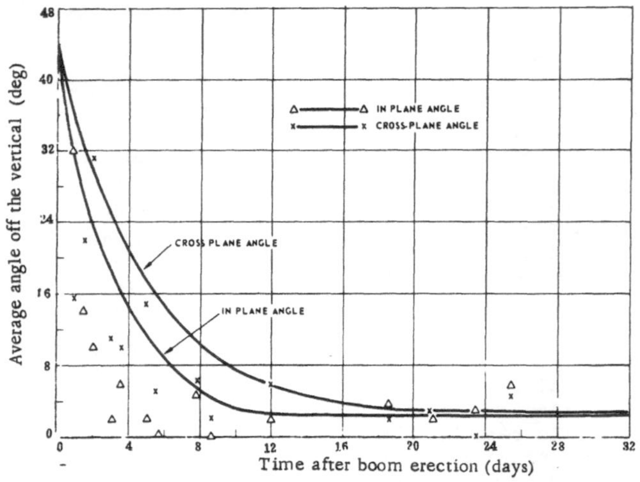

FIGURE 12. DAMPING OF SATELLITE LIBRATION.

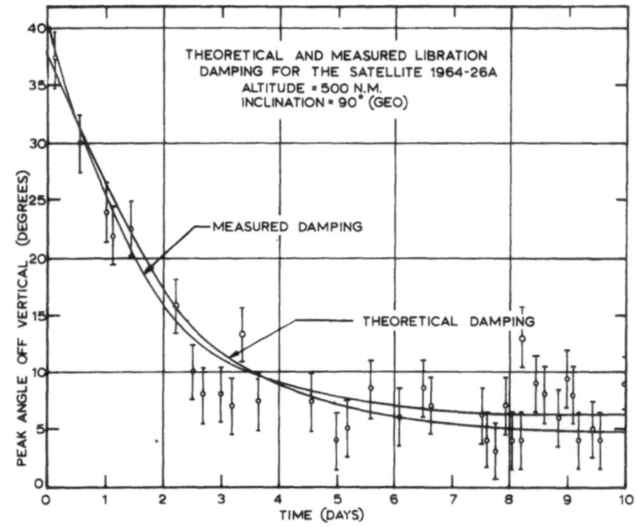

FIGURE 13. DAMPING RESULTS FROM THE SATELLITE 1964-26A.

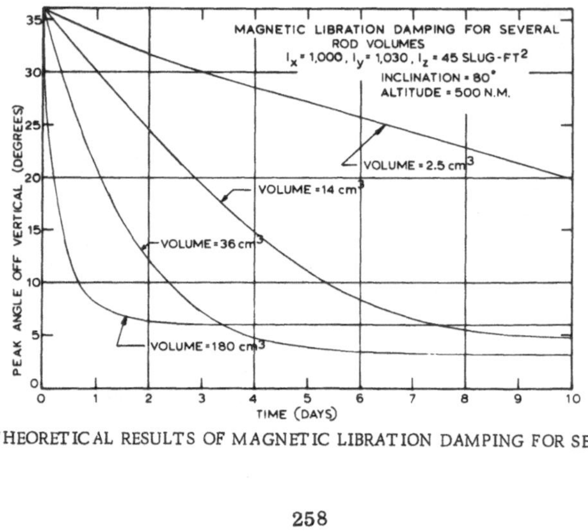

FIGURE 14. THEORETICAL RESULTS OF MAGNETIC LIBRATION DAMPING FOR SEVERAL ROD VOLUMES.

aromatic compound known as biphenyl. Biphenyl was also used to hold the spring end mass securely during extension of the boom. After extension, the biphenyl sublimed away; first the end mass was released, then further sublimation of the biphenyl allowed one coil of the spring to extend at a time. Complete deployment of the spring from the subliming material was accomplished within a period of a day.

The satellite 1963-22A required 2-1/2 days to despin from an initial rate of 1.0 rpm. An additional 1/2 day was required to achieve magnetic stabilization. The 100 foot boom was then successfully erected to its full length. Twelve hours after boom erection the spring started to deploy. After 24 hours it is believed that the spring was completely deployed. As a result of the damping spring and magnetic rods, the angle of the spacecraft relative to the local vertical decreased as shown in Figure 12. The curves shown here represent the average deviation off the vertical as measured at a single telemetry station at the Applied Physics Laboratory. Peak angles off the vertical as high as 10 degrees were observed on some of these passes. Because of the limited telemetry coverage it was never possible to obtain data for a complete libration period of the satellite. A more detailed description of the gravity stabilization system design and performance for the satellite 1963-22A is given in reference 15.

The same technique that was used to gravity stabilize the satellite 1963-22A was later applied to the APL-built satellites 1963-38B and 1963-49B. These satellites were virtually identical to each other in design and external appearance, and were launched into circular, polar orbits at 600 nautical miles altitude.

The satellite 1963-38B was the second satellite to be gravity-gradient stabilized. The initial despin and magnetic stabilization, and boom deployment were completely successful. The satellite was captured into gravity stabilization within 2 days after launch.

A serious problem then developed as the spring deployed from the subliming biphenyl. As this material turned into a gas it acted like a rocket. Although the impulse was only on the order of 10 millionths of a pound, at the end of a 100-foot boom it was sufficient to tumble the satellite.

After tumbling end-over-end for two weeks, the satellite unfortunately captured upside down, with the result of drastically reduced radio signal strength at the surface of the earth.

The satellite 1963-49B was then built with a deflector to reduce the force of the biphenyl subliming sideways. Although this vastly reduced the disturbing effect of the vaporizing biphenyl, there was, nevertheless, just enough force to gently tumble the satellite. This occurred 10 days after the boom was successfully erected. Fortunately, a large magnet was installed in this satellite, just in case it was necessary to turn it over. Although the satellite tried to capture upside down on two occasions, the magnet was energized to push against the earth's magnetic field, and the attitude was corrected. After two weeks when all the biphenyl had sublimed the magnet was activated to capture the satellite along the local vertical with the correct side facing downward.

MAGNETIC HYSTERESIS ROD DAMPING

The fourth APL satellite to achieve gravity stabilization was given the international designation 1964-26A. This satellite was launched into a circular, polar orbit of an altitude of 500 nautical miles. Although this satellite also had a damping spring and magnetic hysteresis rods, the subliming material was changed from biphenyl to benzoic acid so that spring deployment was delayed for 14 days after boom erection.

Magnetic despin from 1.0 rpm was accomplished within one day after injection into orbit. Twelve hours after energizing the electromagnet, the satellite was aligned within 5 degrees of the local direction of the earth's magnetic field. The boom was then successfully deployed to its full 100 foot length when the satellite was near the earth's magnetic north pole. For the next 14 days damping of the satellite librations was accomplished entirely by passive magnetic hysteresis rods.

This satellite employed a total of 4 magnetic hysteresis rods, each 58 inches long, with a diameter of 0.108 inches. The rod material was AEM-4750; a permalloy with a nominal composition 47.5 percent nickel, 52.5 percent iron. In this length-to-diameter ratio, the material had approximately twice the hysteresis loss per rod as compared to the rods used on the satellite 1963-22A. The total rod weight for the satellite 1964-26A was 0.6 pound.

In Figure 13 is plotted the maximum observed libration angles with only magnetic rod damping as seen from telemetry stations at Anchorage, Alaska, and the Applied Physics Laboratory. As can be seen from Figure 13, the initial damping was extremely rapid, leveling off to a highest peak value of approximately 9 degrees by 5 days after boom erection. It should be noted that the maximum angle in a libration cycle will vary considerably from one time to the next. The curve plotted in Figure 13 is the average of the peak angles off the vertical.

Figure 13 also shows a theoretical result for libration damping with magnetic rods as derived by Dr. J. L. Vanderslice of the Applied Physics Laboratory. The agreement between theoretical and experimental results was sufficently good so that other cases could be investigated with a fair degree of confidence in the theoretical results.

The first theoretical investigation was concerned with the effect of rod volume on the

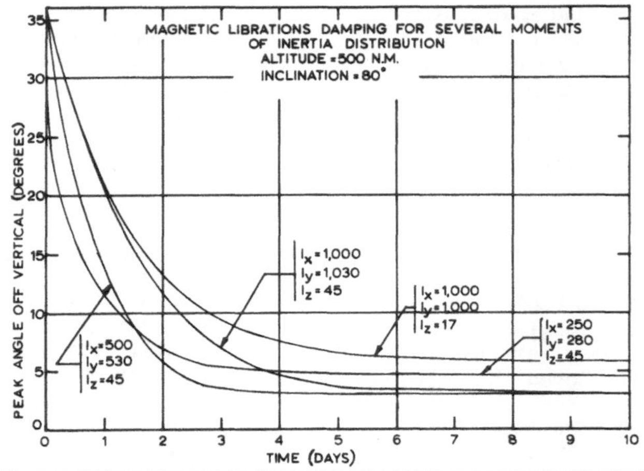

FIGURE 15. MAGNETIC LIBRATION DAMPING FOR SEVERAL MOMENT OF INERTIA RATIOS.

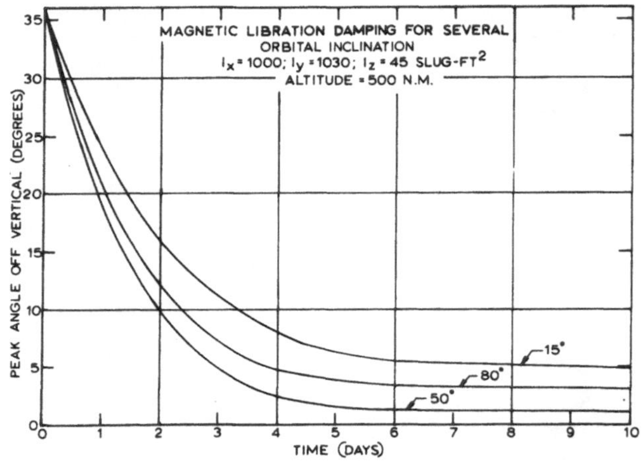

FIGURE 16. MAGNETIC LIBRATION DAMPING AS A FUNCTION OF ORBIT INCLINATI

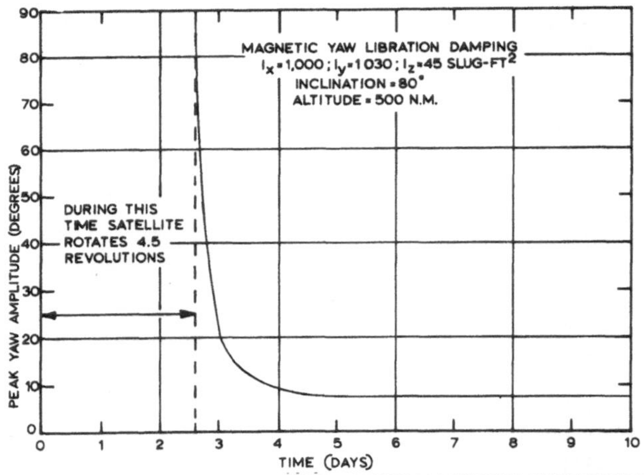

FIGURE 17. YAW LIBRATION DAMPING USING MAGNETIC HYSTERESIS RODS.

damping rate and final stabilization angle. The results for several rod volumes are shown in Figure 14. Increasing rod volume results in a more rapid decay of the initial librations but with a greater magnetic perturbing effect that causes a higher peak deviation off the vertical in the final stabilization condition. One must therefore compromise the desired rate of initial decay with the accuracy of final stabilization.

Another theoretical investigation was concerned with varying the satellite's moments of inertia. The results are shown in Figure 15. From these curves it can be seen that the extremely large ratios of I_x compared to I_z result in less effective gravity-gradient stabilization. It is thought that this may be a result of the perturbing effect of the magnetic rods causing excitation of the natural cross-orbit frequency which is seen from Equation (17) to approach 1/2 the orbital period for $I_x = I_y \gg I_z$.

A third investigation was concerned with the effect of orbital inclination. The results shown in Figure 16 show that the best stabilization is obtained at inclinations near 50 degrees, with the damping rate and final stabilization accuracy both decreasing as one approaches polar or equatorial orbits. This is explained by the fact that magnetic rod pitch damping is less effective in polar orbits and roll damping is ineffective in equatorial orbits, but both modes of damping are effective for middle inclinations.

The capability of magnetic rods to provide damping of the yaw librations for a three axis stabilized spacecraft is shown in Figure 17. For this case, the final yaw stabilization angle was 6 degrees. More accurate yaw stability (with a consequence of lower damping rate) is achievable by decreasing the volume of the magnetic hysteresis rods.

ENHANCED MAGNETIC DAMPING

Since the earth's magnetic field intensity falls off inversely as the cube of the radial distance from the center of the dipole, and since magnetic hysteresis decreases roughly as the cube of the peak magnetizing field, the effectiveness of a fixed volume of magnetic material falls off approximately as the ninth power of radial distance from the center of the earth. Therefore, satellites at very high altitudes would require a very large volume of rods to obtain effective damping. A semi-passive (using electrical power but neither moving parts nor closed-loop sensing) magnetic damping system has been devised to restore this effectiveness of this damping device for higher altitudes. A block diagram of this enhanced magnetic damping system is shown in Figure 18. Magnetometers which have built-in magnetic hysteresis have their output amplified and fed into electromagnets. The result is essentially an amplified hysteresis loop that re-establishes the effectiveness of the magnetic damping system. The required hysteresis could also be put into the system by an electronic device between the magnetometer electronics and the current amplifier.

Figure 19 shows the effectiveness of magnetic rods for libration damping at several orbital altitudes. The effect of enhanced magnetic damping for improved damping at high latitudes is apparent from this illustration. At synchronous, equatorial altitude, where the magnetic field is stationary relative to the local vertical, this system can be made to be quite effective for damping satellite librations.

To decrease the perturbing effects from magnetic torques after the satellite librations have been damped, the gain of the enhanced magnetic damping system could be decreased by radio command to the satellite. In this manner it is possible to achieve rapid initial damping and still obtain accurate final stabilization. When the desired stabilization accuracies are obtained, the enhanced magnetic damping system could probably be turned off, and completely passive magnetic rods could then provide the required low level of damping to prevent long term increase in the peak libration angle.

MAGNETIC HYSTERESIS PLUS EDDY CURRENT DAMPING

Recent studies at the Applied Physics Laboratory indicate that the use of highly permeable magnetic rods, with only moderate hysteresis loss, and surrounded by a sheath of a highly conducting material, can appreciably increase the effectiveness of magnetic rod damping. The increased effectiveness is principally the result of eddy currents in the conducting sheath. The eddy currents provide a psuedo-viscous damper. A damping constant on the order of 20,000 dyne-cm per rad/sec is achievable with long rods.

The fact that rods of the magnetic materials such as mu-metal and 4-79 molybdenum-permalloy have a permeability that is relatively constant, makes it possible to minimize the perturbing effect of the magnetic materials by placing equal volume of rods in the X, Y and Z directions. Since these rods also have comparatively little magnetic remanence, the vector sum of their dipole moments will be approximately along the direction of the external magnetic field. Therefore the perturbing effect due to magnetic torques is drastically reduced.

Figure 20 shows the theoretical results of a combination of X, Y, and Z magnetic rods with eddy current sheaths. The damping constant here was 20,000 dyne-cm per rad/sec, $I_x = I_y = 700$ slug-ft^2, $I_z = 20$ slug-ft^2, i = 50 degrees, perigee = 600 n.m., apogee = 800 n.m. The effectiveness of this type of damper is obivous from this illustration.

By placing electrically conducting sheaths around the electromagnets of the enhanced magnet damping system (Figure 18) and by using an electromagnet in the Z as well as the X and Y directions, one obtains enhanced magnetic hysteresis and eddy current damping.

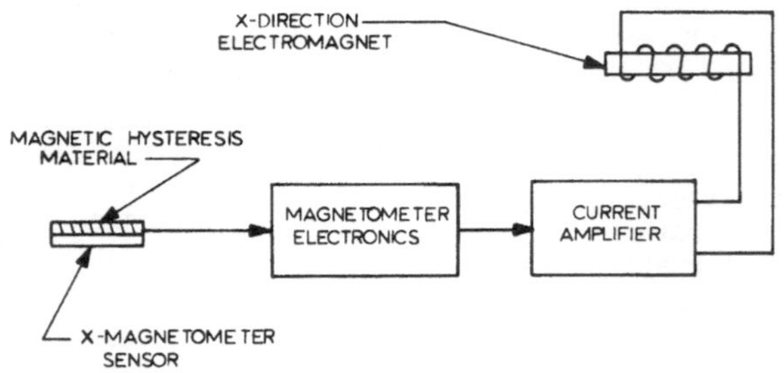

FIGURE 18. BLOCK DIAGRAM --- ENHANCED MAGNETIC DAMPING SYSTEM.

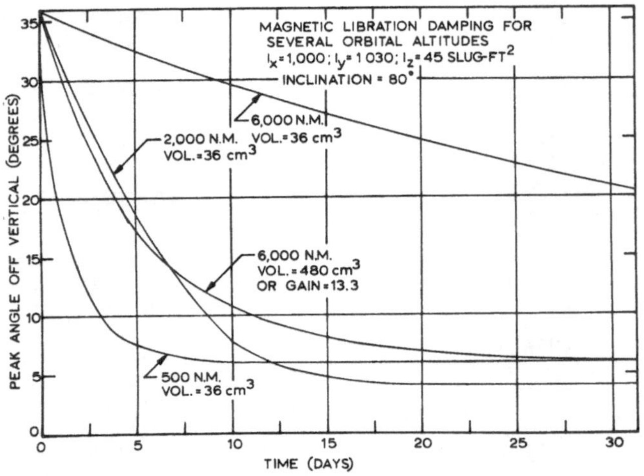

FIGURE 19. MAGNETIC LIBRATION DAMPING FOR SEVERAL ORBITAL ALTITUDES.

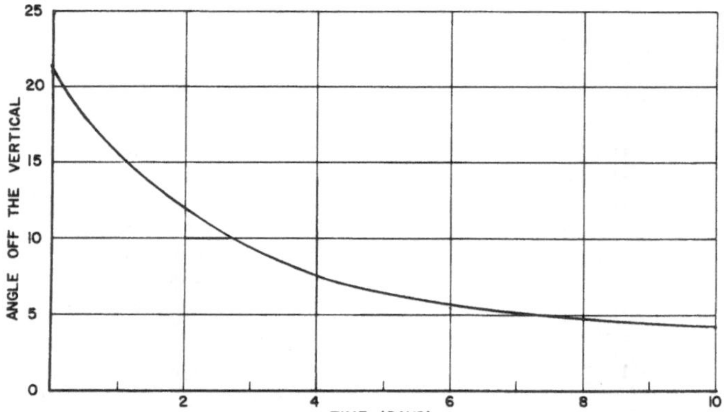

FIGURE 20. THEORETICAL MAGNETIC HYSTERESIS PLUS EDDY CURRENT DAMPING.

EFFECT OF PERTURBING TORQUES

It has been shown that a satellite having $I_x \gg I_z$ with an effective damping mechanism will stabilize vertically along the direction of the earth's gravity gradient. To assure that this is the case for a practical satellite design one must examine all possible torques that tend to perturb the satellite off its vertical position. The principal perturbing torques are (1) magnetic, (2) solar radiation pressure, and (3) aerodynamic. Magnetic interactions are a principal perturbing torque for satellite altitudes below 1000 miles; solar radiation pressure would be the most significant disturbing effect for a satellite in a synchronous orbit; the effect of aerodynamic drag can be significant for satellite altitudes below 500 miles.

The magnetic torque is a result of the interaction of the earth's magnetic field with any permanent or induced dipole moment from permeable material in the satellite. The torque resulting from the interaction with the earth's magnetic field is given by

$$\tau_m = M\,H \sin\,\beta \quad \text{(dyne-cm)} \qquad (24)$$

where M = satellite's magnetic dipole moment (Unit-pole cm),

H = earth's magnetic intensity at the satellite (oersted),

and β = angle between the earth's magnetic field and the magnetic dipole of the satellite.

It has been calculated that the magnetic perturbing torque on the satellite 1963-22A (altitude ≃ 400 nautical miles) results in less than 4.0 degrees displacement of the satellite from the local vertical direction.

Since the earth's magnetic field intensity varies as the inverse cube of the distance from the center of the earth, orbits considerably higher than 400 miles altitude will produce a significantly smaller magnetic perturbing torque.

For a gravity stabilized satellite with an asymmetrical distribution of area about the center of mass there will be a net perturbing torque due to solar radiation pressure. The torque resulting from solar radiation pressure can be readily computed by taking a summation of the solar radiation pressure moments about the center of mass of the spacecraft. The torque resulting from solar radiation pressure is given by

$$\tau_s = \Sigma \left[1 + (C_r)_i\right] A_i\, P_s\, d_i \qquad (25)$$

where $(C_r)_i$ = that fraction of incident photons reflected from the ith surface (one assumes that all others will be absorbed)

A_i = projected area of the ith surface

and P_s = the radiation pressure exerted by the sun on a totally absorbing surface at the mean solar distance (this has a value of 4.8×10^{-5} dyne/cm^2).

d_i = moment arm of the ith surface

For the satellite 1963-22A, this torque causes less than a 0.5 degree displacement from the local vertical direction.

Although the radiation pressure torque is small for the comparatively short boom used for the 1963-22A experiment, the use of much larger booms to stabilize a satellite at higher altitudes would produce appreciable solar radiation pressure torques. This perturbation can be made negligible by designing the spacecraft so that the sum of the solar radiation torques about the center of mass is essentially zero.

Aerodynamic drag can disturb the vertical orientation for satellites at low orbital altitudes. The torque due to aerodynamic drag is given by

$$\tau_a = \frac{1}{2} \sum_i (C_d)_i\, A_i\, \rho\, v^2\, d_i \qquad (26)$$

where $(C_d)_i$ = drag coefficient of the ith surface,

ρ = atmosphere density at satellite altitude

v = orbital velocity.

For the 1963-22A satellite configuration, the aerodynamic torque for an altitude of 400 miles would result in approximately a 1.0 degree deviation off the vertical. At an altitude of 300 miles the displacement off the vertical would be approximately 10 degrees. It is possible to solve the aerodynamic torque problem for satellites at very low altitudes by the use of additional booms that balance the torque about the satellite center of mass, exactly as is the case with the effect of solar radiation pressure at high altitudes.

Another source of perturbing torques are the effect of asymmetrical outgassing or sublimation of materials from the satellite. It is probably advisable to wait at least one day after the satellite is in orbit before erecting the boom so that most of its outgassing is completed. The tumbling of the satellites 1963-38B and 49B were dramatic examples of the effect of the impulse that can be created by subliming materials. Another possible source of thrust that can perturb the satellite attitude is small, satellite-borne rockets, such as those that might be used for station keeping. Also electromagnetic radiation emanating from the satellite can cause a disturbance torque. For most satellites this torque is quite negligibly small.

An additional deviation from vertical stabilization will be observed when the gravity-gradient stabilized satellite is in an eccentric orbit. Dr. Vanderslice of the Applied Physics

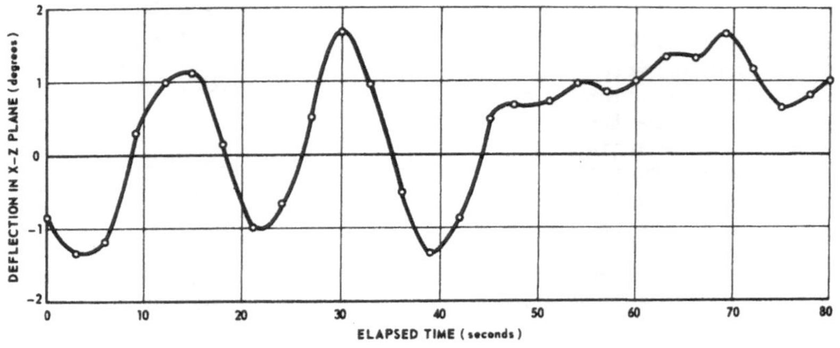

FIGURE 21

DYNAMIC THERMAL BENDING OF THE SATELLITE 1963-22A

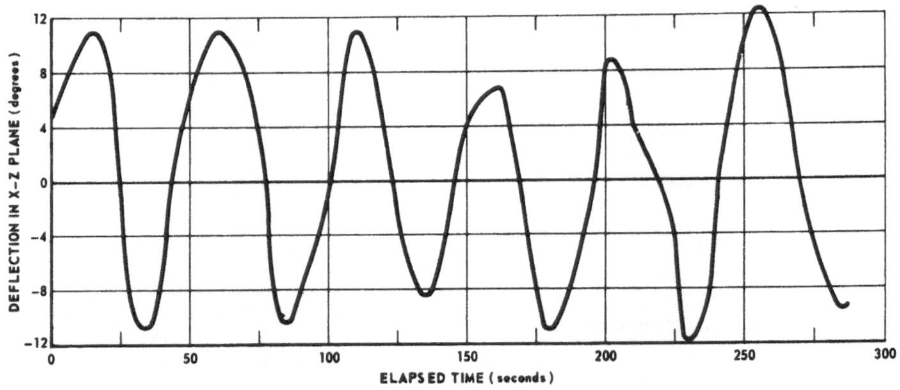

FIGURE 22

DYNAMIC THERMAL BENDING FOR THE SATELLITE 1964-83D

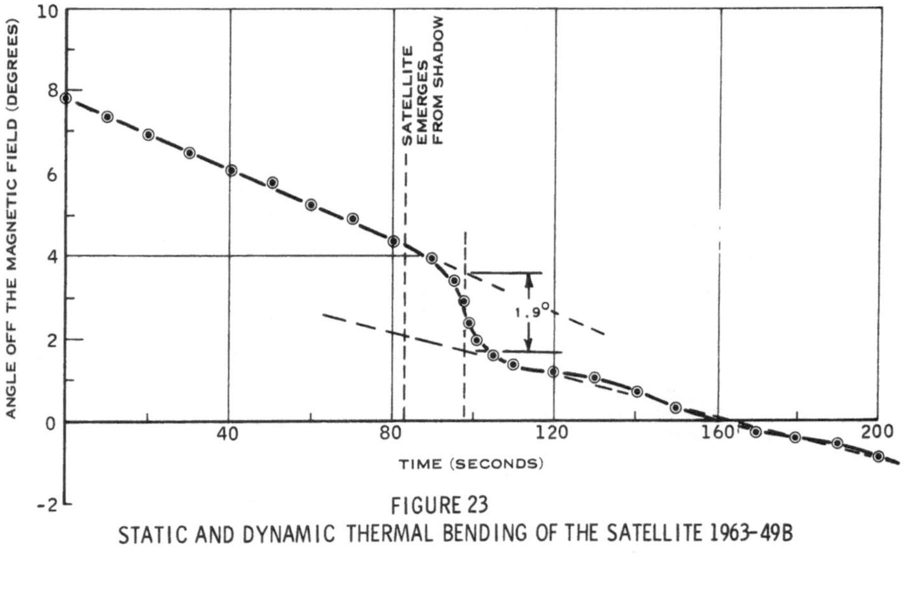

FIGURE 23
STATIC AND DYNAMIC THERMAL BENDING OF THE SATELLITE 1963-49B

Laboratory has shown that, for a well-damped satellite as described above, the maximum deviation off the vertical for non-circular orbits is given approximately by the eccentricity expressed in radians. For example, an eccentricity of 0.01 would cause a maximum deviation off the vertical of 0.57 degree.

THERMAL BENDING OF THE BOOMS

In addition to the perturbing torques described above, there is another effect which causes the final altitude of the satellite to deviate from the local vertical. This effect is caused by the sun shining on one side of the boom so that side expands relative to the dark side, thereby bending the boom. This thermal bending is particularly significant for long boom structures. Thermal bending on the order of 10 degrees is obtained for a beryllium-copper boom, 100 feet long, 1/2 inch in diameter, 2 mils wall thickness, whose longitudinal axis is perpendicular to the sun line.

An unexpected discovery was made in studying the thermal bending of the 100 foot, beryllium-copper boom on the satellite 1963-22A. As the satellite crossed from the earth's shadows into sunlight, there was a snap action of the boom resulting in a high frequency oscillation that has been given the name "dynamic thermal bending." Figure 21 shows the dynamic thermal bending of the satellite 1963-22A. The lack of a repeatable oscillation pattern was typical for this satellite. Peak oscillations as high as 5 degrees were noted on several occasions. The satellite 1964-83D also employed an unplated beryllium-copper boom, 100 feet long. The dynamic thermal bending in the X-Z plane of this satellite is shown in Figure 22. Deviations as large as $\pm$ 12 degrees in both the X-Z and Y-Z planes were observed simultaneously. The fact that the thermal bending was considerably greater for this satellite is probably because there was no mechanical damping spring at the end of the boom as was the case with 1963-22A. For the satellite 1963-22A, the amplitude of the oscillation must have been considerably reduced by energy absorption in the spring connected to the end of the boom.

As a result of studies made at the Applied Physics Laboratory it was determined that silver plating of the exterior surface of the boom would drastically reduce the absorption of sunlight resulting in a significant decrease both of the static and dynamic thermal bending.[19,20] It was also felt that the silver, having a much greater mechanical hysteresis loss as compared to beryllium-copper, would serve to damp the high frequency dynamic thermal bending of the boom. This was verified by in-orbit measurements of the thermal boom bending of the satellite 1963-49B. Figure 23 shows the satellite angle with the magnetic field as it passed near the earth's north magnetic pole going from the earth's shadow into sunlight. A static thermal deflection of slightly less than 2 degrees was observed, and a dynamic thermal bending with a peak amplitude of 0.2 degree was measured immediately upon being fully illuminated by the sun. Seventy minutes later, before the satellite entered the earth's shadow, the dynamic thermal bending was measured to be distinctly less than 0.1 degree. Since the silver plating weighs very little (it is only on one side of the tape and its thickness is only 0.0002 inch) it provides a simple and effective means for reducing both static and dynamic thermal bending.

REFERENCES

1. Roberson, R. E., "Gravitational Torque on a Satellite Vehicle," J. Franklin Institute, Vol. 265, January 1958, pp. 13-22.
2. Klemperer, W. B., and Baker, R. N., "Satellite Librations," Astronautica Acta, Vol. 3, 1957, pp. 16-27.
3. Nidley, R. A., "Gravitational Torque on a Satellite of Arbitrary Shape," ARS Journal, Vol. 30, No. 2, February 1960, pp. 203-204.
4. Roberson, R. E., "Generalized Gravity-Gradient Torques," Chapter 4, Torques and Attitude Sensing in Earth Satellites, Edited by S. F. Singer, Academic Press, N.Y., 1964.
5. Frick, R. M., and Garber, T. B., "General Equations of a Satellite in a Gravitational Field," Rand Corporation Report RM-2527, December 1959.
6. Kershner, R. B., and Newton, R. R., "Attitude Control of Artificial Satellites," Space Astrophysics, Chapter 14, McGraw Hill, 1961.
7. Fischell, R. E., "Magnetic Damping of the Angular Motions of Earth Satellites," ARS Journal, Vol. 31, No. 9, September 1961, pp. 1210-1217.
8. Fischell, R. E., "Passive Magnetic Attitude Control for Earth Satellites," Advances in the Astronautical Sciences, Vol. 11, Western Periodical Company, No. Hollywood, Calif., 1963.
9. Fischell, R. E., "Magnetic and Gravity Attitude Stabilization of Earth Satellites," The Johns Hopkins University, Applied Physics Laboratory, Report CM-996, May 1961.
10. Kershner, R. B., "Gravity-Gradient Stabilization of Satellites," Astronautics and Aerospace Engineering, September 1963, pp. 18-22.
11. Newton, R. R., "Damping of a Gravitationally Stabilized Satellite," The Johns Hopkins University, Applied Physics Laboratory, Report TG-487, April 1963.

12. Vanderslice, J. L., "Dynamic Analysis of Gravity-Gradient Satellite with Passive Damping," The Johns Hopkins University, Applied Physics Laboratory, Report TG-502, June 1963.

13. Frick, R. H., "Equations of Motion of a Passive Satellite Stabilization System," The Rand Corporation, Report RM-4164, June 1964.

14. Paul, B., "Planar Librations of an Extensible Dumbbell Satellite," AIAA Journal, Vol. 1, No. 2, February 1963, pp. 411-418.

15. Fischell, R. E., and Mobley, F. F., "A System for Passive Gravity-Gradient Stabilization of Earth Satellites," Progress in Astronautics and Aeronautics, Edited by R. C. Langford and C. J. Mundo, Academic Press, N.Y., 1964.

16. "Damping Spring for Gravity-Stabilized Satellites," APL Technical Digest, Vol. 2, No. 2, November-December 1962, pp. 20-21.

17. Smola, J. F., Schrantz, P. R., and Tossman, B. E., "The Torsional Damping Capacity of a Thin Wire and Its Application to the Damping of a Librating Satellite," The Johns Hopkins University, Applied Physics Laboratory, Report CM-1039.

18. Tossman, B. E., "The Long Period Damping Capacity of Thin Wires in Torsion," The Johns Hopkins University, Applied Physics Laboratory, Report TG-596, September 1964.

19. Phenix, J. E., "Thermal Bending of Long Thin-Wall Booms Due to Solar Differential Heating in Near-Earth Orbits," The Johns Hopkins University, Applied Physics Laboratory, Report TG-559, March 1964.

20. Kummerer, J. E., "Solar Differential Heating of Thin Wall Booms in Near Earth Orbit," The Johns Hopkins University, Applied Physics Laboratory, Report CF-3073, April 1964.

DISCUSSION

R. B. Kershner and R. E. Fischell

"Gravity Gradient
Stabilization
of Earth Satellites".

Q. How are initial disturbances damped?

A. We have used two types: magnetic hysteresis rods and also an extendable weak spring with mechanical hysteresis. A Cadmium coating on the spring provides a 50% hysteresis loss factor (50% per cycle). The mechanical method has always been used in conjunction with a magnetic damper. We have achieved 5 degrees with both kinds of damper; with magnetic alone we have reached 7 degrees.

Q. Did you consider only axial oscillations of the spring?

A. We considered both, but the damping is most effective in the cross-plane. The magnetic system is least effective in this plane.

Q. Will you please elaborate on the 50% damping factor for Cadmium?

A. It is characteristic of the material itself - the 50% figure applies to the material, not the complete system. In other words, each flexture of such cadmium coated spring dissipates 50% of the energy in the spring itself.

Q. Have you considered "preliminary" damping by means, for example, of gases?

A. We have considered varying the length of the boom for this purpose. If you put the boom out half way, and then the rest of the way you get damping in one oscillation.

SOME PROBLEMS OF OPTIMUM THREE-DIMENSIONAL ATTITUDE CONTROL OF SPACECRAFT

by R. V. Studnev

ABSTRACT

Here a study will be made of the optimum three-dimensional attitude control of a spacecraft. It will be assumed that the stabilization is accomplished by two pairs of reaction jets, one pair of which is fixed in the vehicle so as to produce a torque about the axis of symmetry, and the second pair is mounted in a Cardan suspension and can produce a torque about an arbitrary axis perpendicular to the symmetry axis.

The following problems can be solved:

1. The optimum attitude control of the axis of rotation when this corresponds with the symmetry axis of the spacecraft. Conditions can be found in order that the attitude control can be achieved by one pair of jets fixed in the spacecraft.

2. The optimum attitude control of a spacecraft rotating about a principal axis of inertia which is not a symmetry axis, and

3. The optimum three-dimensional attitude control of an axisymmetric spacecraft in the general case.

INTRODUCTION

One of the urgent problems of the present time is that of the optimum attitude control of a spacecraft by means of reaction jets. Various aspects of this problem have been studied in a series of papers that can be roughly grouped into the following two categories. First, those papers on the optimum attitude control of the motion about a single axis of a spacecraft (5), (8), and, second, optimal corrective controls for a rotating spacecraft, either for minimum time (4), or for minimum fuel consumption (3).

The present paper is devoted to the theoretical investigation of the optimum attitude control of an axisymmetric spacecraft by means of reaction jets under the simplifying assumption that the magnitude of the available controlling torque is so large that the impulsive formulation of the problem can be used. Moreover, as is usual in such studies, it will be assumed that the moments of inertia of the spacecraft are constant (the discharge of fuel (mass carrier) is significantly smaller than the mass of the vehicle).

The investigations will be restricted to small angular errors in the attitude of the spacecraft and will be based on the application of Pontryagin's maximum principle (1, 2) to the above-formulated problems.

1. FORMULATION OF PROBLEMS OF OPTIMUM CONTROL

In this article we will consider problems, which, in general terms, can be formulated as follows. Let the differential equations of motion of the spacecraft be

$$\begin{aligned} \dot{\bar{x}} &= A\bar{x} + B\bar{u} \\ \dot{x}_{n+1} &= \sum_{k=1}^{r} |u_k| \\ \dot{x}_{n+2} &= \lambda_\tau \end{aligned} \tag{1.1}$$

where $\bar{x}$ is the n-dimensional column vector of coordinates in phase space,

$\bar{u}$ is the control vector (r-dimensional) that satisfies the conditions $|u_k| \leq U_k \to \infty$, and

A, B are (n x n) and (n x r) matrices, respectively,

x_{n+1} corresponds to the discharge of fuel (mass carrier),

λ_τ is a constant, and x_{n+2} is a time variable.

It is necessary to find some law of variation of the control vector $\bar{u}(t)$ in order that a functional S assumes a minimum value when system (1.1) changes from an initial state $\bar{x}(o)$ to a terminal state $\bar{x}(T)$, where

$$S = C_{n+1} \cdot x_{n+1}(T) + C_{n+2} x_{n+2}(T) \tag{1.2}$$

We will treat the following three particular cases of this problem:

1. C_{n+1}, C_{n+2}, and value T fixed. This is the problem of minimum fuel consumption.

2. $C_{n+1} = 0$. $C_{n+2} = 1$, and the time T is not fixed but the value of the parameter $x_{n+1}(T)$ is pre-

scribed. This is the problem of minimum settling time for a given fuel consumption.

3. $C_{n+1} = C_{n+2} = 1$, λ_c = const., and the time is not fixed. This is a mixed problem.

We will show that all these problems lead to the same control functions and hence the solutions will be the same when the corresponding boundary values are the same.

This will be proved by making use of Pontryagin's maximum principle (1, 2) and by extracting the part of the Hamiltonian H that contains the control functions:

$$H = \sum_{k=1}^{r} \left[\left(\sum_{i=1}^{r} p_i \cdot b_{ik} \right) u_k + p_{n+1} |u_k| \right] \quad (1.3)$$

where b_{ik} are the elements of matrix B. The canonical equation for p_{n+1} has the form $\dot{p}_{n+1} = 0$, whence it follows that p_{n+1} = const. In the first and third problems, where $C_{n+1} = 1$, we obtain $p_{n+1} = -1$ (see (2)). In all practical versions of the case $C_{n+1} = 0$, the choice of control function is limited by the reverse of fuel and we have $p_{n+1}(T) < 0$. S Since the conjugate variables are only determined to within a constant multiplicative factor, it can be assumed in all the above-mentioned cases that

$$p_{n+1} = -1 \quad (1.4)$$

The requirement that the u_k maximizes H leads to the following control function:

$$u_k = \begin{cases} U_k \cdot \operatorname{sgn}\left(\sum_{i=1}^{r} p_i b_{ik} \right), & \text{when } \left| \sum_{i=1}^{n} p_i b_{ik} \right| > 1 \\ 0, & \text{when } \left| \sum_{i=1}^{n} p_i \cdot b_{ik} \right| < 1 \end{cases} \quad (1.5)$$

The solution of the system of canonical equations can be determined by the boundary conditions for equations (1.1). Then, by virtue of all the boundary conditions (taking x_{n+1} and x_{n+2} into account), it follows from (1.5) that the structure of the control function is preserved and the solutions coincide.

2. OPTIMUM ATTITUDE CONTROL OF A SPACECRAFT ROTATING ABOUT ITS AXIS OF SYMMETRY

Let the spacecraft be rotating about its axis of symmetry (Ox_1) with constant angular velocity ω_x. It will be assumed that the magnitude of the deviation of the Ox_1-axis from the prescribed orientation $\delta = (\alpha_1^2 + \beta_1^2)^{\frac{1}{2}}$ is small. The equations of motion of the spacecraft in the principal central axes of inertia can be written in the form of Euler's equations:

$$\begin{aligned} \dot{\omega}_{x_1} &= \overline{M}_{x_1} \\ \dot{\omega}_{y_1} &= \overline{M}_{y_1} + (1 - \overline{J}_x)\omega_{x_1} \cdot \omega_{z_1} \\ \dot{\omega}_{z_1} &= \overline{M}_{z_1} - (1 - \overline{J}_x)\omega_{x_1}\omega_{y_1} \end{aligned} \quad (2.1)$$

where $\overline{M}_x$, $\overline{M}_y$ and $\overline{M}_z$ are the central torques, $J_y = J_z = J$, and $\overline{J}_x = J_x/J$. We supplement equations (2.1) with kinematic equations determining the orientation of the body-fixed axes $Ox_1y_1z_1$ relative to the space-fixed axes $O\xi\eta\zeta$ (see Figure 1).

$$\begin{aligned} \dot{\gamma} &= \omega_{x_1} \\ \dot{\alpha}_1 &= \omega_{z_1} - \beta_1\omega_{x_1} \\ \dot{\beta}_1 &= \omega_{y_1} + \alpha_1\omega_{x_1} \end{aligned} \quad (2.2)$$

Now we will find the optimum control that changes the orientation of the rotational axis Ox_1 relative to the inertial frame $O\xi\eta\zeta$. As a concrete case we will consider the problem of control with minimum fuel consumption for a fixed time. Since the boundary values of the variational problem are best given in the form of angles and angular velocities of the motion relative to the inertial frame, we will transform the equations of motion (2.1), (2.2) to the nonrotating system of axes $Ox_0y_0z_0$. The origin of these axes is at the center of mass of the spacecraft and the axis Ox_0 coincides with Ox_1, relative to which the axes Oy_1 and Oz_1 rotate with angular velocity ω_x. Moreover, the orientation of axis Ox_0 in the inertial frame $O\xi\eta\zeta$ can be stipulated by the angles α_0 and β_0 as shown in Figure 1.

Making the transformation to the new system of axes, we obtain the equations of motion and the kinematic equations in the form

$$\begin{aligned} \dot{\omega}_{y_0} &= -\overline{J}_x\omega_x\omega_{z_0} + u\sin\psi, \\ \dot{\omega}_{z_0} &= \overline{J}_x\omega_x\omega_{y_0} + u\cos\psi, \\ \dot{\alpha}_0 &= \omega_{z_0}, \quad \dot{\beta}_0 = \omega_{y_0}, \quad \omega_x = \text{const.} \end{aligned} \quad (2.3)$$

The control torque applied to the spacecraft via a Cardan suspension has direction given by the angle ψ appearing in equations (2.3) and magnitude u subject to the condition

$$u \le U_{max} \to \infty \quad (2.4)$$

Equations (2.3) must be supplemented by the equation of fuel consumption, which can be written in the form

$$\dot{g} = u \quad (2.5)$$

The dominant part of the Hamiltonian H containing the controls u and ψ ($\overline{M}_{x_1} = 0$ and ω_x = const.) is

$$H = u\left[(p_1\sin\psi + p_2\cos\psi) - 1 \right] \quad (2.6)$$

In order to derive the equations that guarantee a minimum value of S in (1.2), it is necessary to choose u and ψ so as to maximize H. After some calculations, it is found that the following condition must be satisfied:

$$p_1\cos\psi - p_2\sin\psi = 0 \quad (2.7)$$

i.e.

$$\operatorname{ctg}\psi = \frac{p_2}{p_1} \quad (2.8)$$

and the condition for u is

$$u = \begin{cases} U_{max} & \text{when } p_1\sin\psi + p_2\cos\psi > 1 \\ 0 & \text{when } p_1\sin\psi + p_2\cos\psi < 1 \end{cases} \quad (2.9)$$

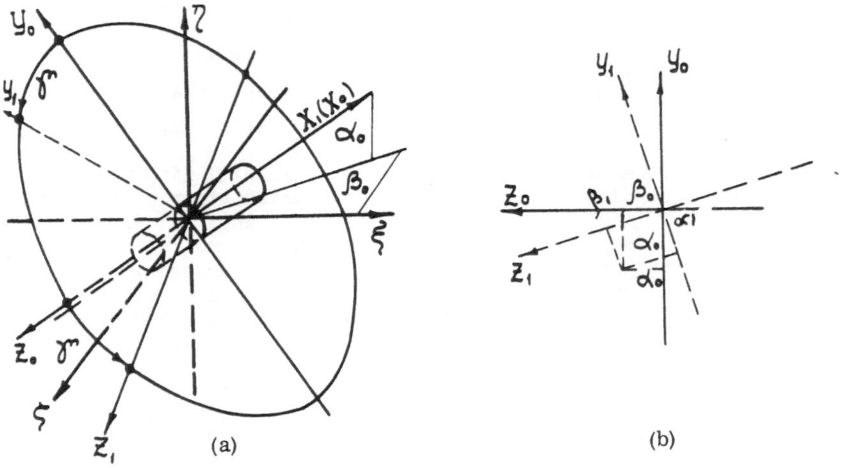

Fig. 1.

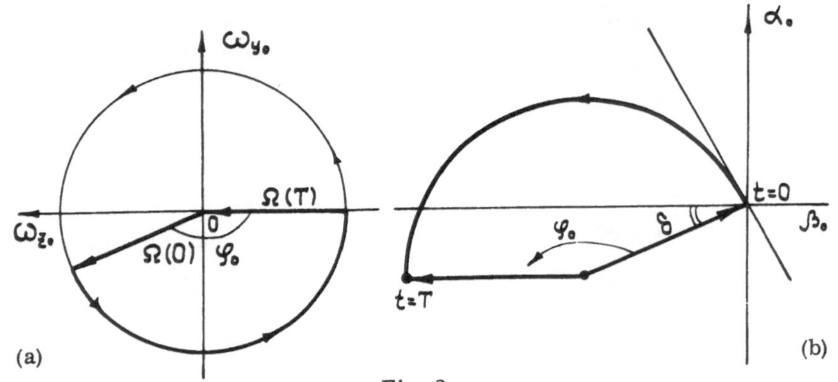

Fig. 2.

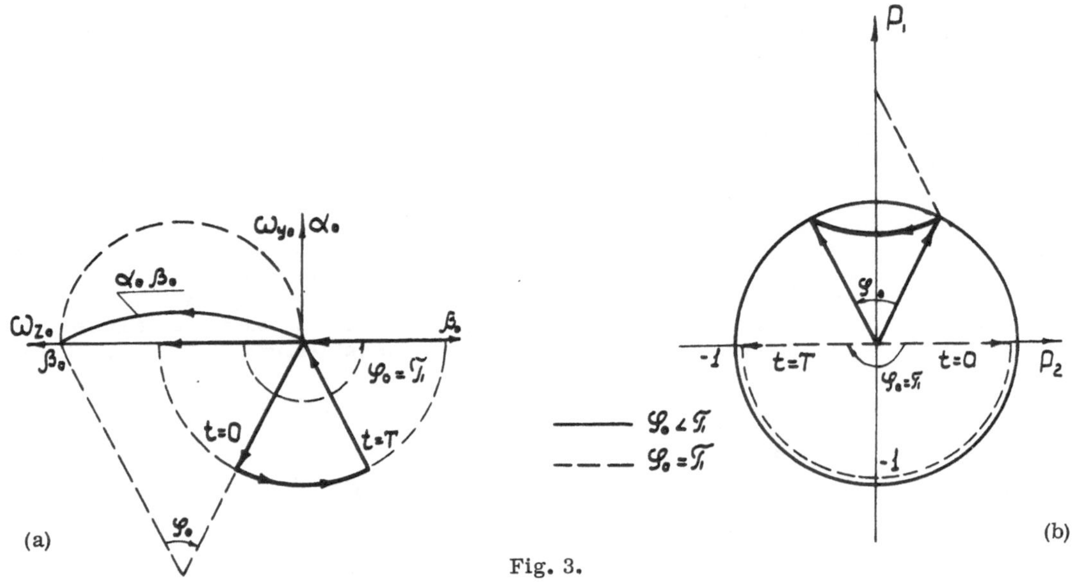

Fig. 3.

Let the functions p and φ be defined as follows

$$p = \sqrt{p_1^2 + p_2^2}$$
$$\sin\varphi = \frac{p_1}{p} \quad ; \quad \cos\varphi = \frac{p_2}{p} \tag{2.10}$$

On account of relations (2.8) and (2.10), conditions (2.9) can be put in the form

$$u = \begin{cases} U_{max} & \text{when} \quad p > 1 \\ 0 & \text{when} \quad p < 1 \end{cases} \tag{2.11}$$

The system of equations for the conjugate variables has the form

$$\begin{aligned} \dot{p}_1 &= -\bar{J}_x \omega_x p_2 - p_y \\ \dot{p}_2 &= \bar{J}_x \omega_x p_1 - p_3 \\ \dot{p}_3 &= 0 \quad ; \quad p_3 = const \\ \dot{p}_y &= 0 \quad ; \quad p_y = const \end{aligned} \tag{2.12}$$

The solution of equations (2.12) can be conveniently analyzed in the phase plane of the conjugate variables (p_1, p_2):

$$(p_1 - p_{10})^2 + (p_2 - p_{20})^2 = C_0^2 \tag{2.13}$$

where

$$p_{10} = \frac{p_{30}}{\bar{J}_x \omega_x} \quad ; \quad p_{20} = -\frac{p_{y0}}{\bar{J}_x \omega_x} \tag{2.14}$$

The great simplification of the present problem brought about by the impulsive formulation consists in the fact that the solution of the equations of motion (2.3) has only to be found for the case u = 0, i.e. the application of an impulsive thrust corresponds to an abrupt change of the angular velocity about the corresponding axis. It is clear that in the impulsive formulation, the p-trajectories can touch the boundary curves but can only intersect them at the beginning and end of the control process.

The solutions of the equations of motion (2.3) in the phase planes (ω_{y_0}, ω_{z_0}) and (α_0, β_0) can be represented in the form

$$\omega_{z_0}^2 + \omega_{y_0}^2 = \omega_{y_0}(0) + \omega_{z_0}(0) = \Omega_0^2(0) \tag{2.15}$$

$$\left(\alpha_0 + \frac{\omega_{y_0}(0)}{\bar{J}_x \omega_x}\right)^2 + \left(\beta_0 + \frac{\omega_{z_0}(0)}{\bar{J}_x \omega_x}\right)^2 = \frac{\Omega_0^2(0)}{(\bar{J}_x \omega_x)^2} \tag{2.16}$$

where $\omega_{y_0}(0)$ and $\omega_{z_0}(0)$ are the components of the impulsive angular velocity $\Omega(0)$ at the beginning of the maneuver (t = 0).

We note that as a consequence of (2.16) the radius-vectors from the center of the circle (2.16) to the initial and final points on the (α_0, β_0) trajectory are parallel, respectively, to the initial and final impulses of angular velocity (see Fig. 2). It is easily proved that the time of motion between two impulses of angular velocity can be determined by the formula:

$$T_1 = \frac{\varphi_0}{\bar{J}_x \omega_x} \tag{2.17}$$

The integrals (2.13), (2.15), and (2.16) enable one to completely solve the problem posed in Section 1.

As an application of the results obtained, we will investigate the problem of turning the axis of rotation of the spacecraft through an angle ($\alpha_0(T) = 0$; $\beta_0(T) = \beta_0^* < 0$) with zero initial velocity and zero terminal velocity at a time $T_1 < \pi/(\bar{J}_x\omega_x)$. It is clear from Figure 3 that the control must consist of two impulsive angular velocities of equal magnitude and having an included angle equal to

$$\varphi_0 = T_1 \bar{J}_x \omega_x \tag{2.18}$$

The p-trajectory corresponding to this maneuver can be seen in Figure 3b. The impulses of angular velocity are applied when the p-trajectory intersects the circle p = I. The solution of the general problem of turning the axis of rotation of a spacecraft through an arbitrary angle ($\alpha_0(T)$, $\beta_0(T)$) in a time T_1 obviously leads to the rotation of the systems of axes for the real and conjugate variables through an angle equal to the angle between the inertial system of coordinates and the plane of the maneuver.

The magnitude of the total impulsive angular velocity $|\Delta\Omega_\Sigma| = |\Omega(0)| + |\Omega(T)|$ can be determined by the formula

$$\Delta\Omega_\Sigma = \frac{|\beta_0^*|\,\bar{J}_x\omega_x}{\sin\left(\frac{\bar{J}_x\omega_x T_1}{2}\right)} \tag{2.19}$$

In the limit, as $\omega_x \to 0$, we obtain a "plane" turn of the spacecraft with a total impulse $\Delta\Omega_\Sigma$ equal to

$$\Delta\Omega_\Sigma = \frac{2|\beta_0^*|}{T_1} \tag{2.20}$$

The limiting case of this type of maneuver is a turn of the spacecraft's axis that is accomplished in time $T_0 = \pi/(\bar{J}_x\omega_x)$. The corresponding phase trajectory is indicated in Figure 3 by the dashed line. The p-trajectory for such a maneuver is a circle that intersects the circle p = 1 at diametral points and has a radius tending to one. When $p_1 < 0$ it lies inside the circle p = 1, and when $p_1 > 0$ it lies outside. Such a turn is the limit of economy

$$\Delta\Omega_\Sigma = |\beta_0^*|\,\bar{J}_x\omega_x \tag{2.21}$$

The optimum maneuver in time $T_1 > \pi/(\bar{J}_x\omega_x) = T_2$ consists in a period of "coasting" $\Delta T = T_1 - T_0$ together with the optimum turning maneuver in time T_0.

3. ATTITUDE CONTROL OF THE ROTATIONAL AXIS OF A SPACECRAFT WITH A SINGLE PAIR OF REACTION JETS

The attitude control system for the rotational axis of a spacecraft is simplified by employing a single pair of reaction jets fixed in the vehicle. Such proposals have been made in several papers (6, 7).

We will study the conditions that must be satis-

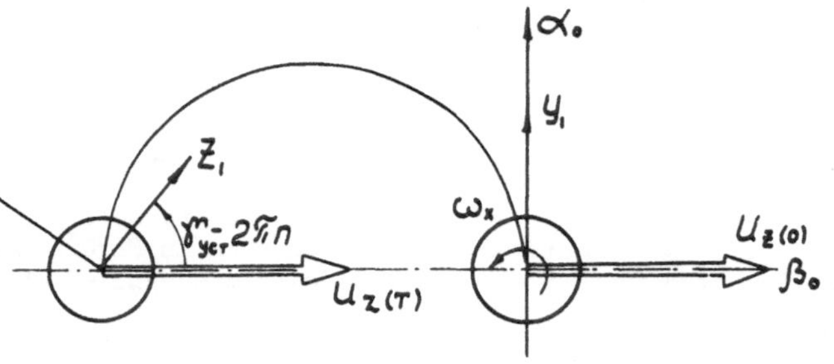
α₀
y₁
y₁
Z₁
ωₓ
β₀

Fig. 4.

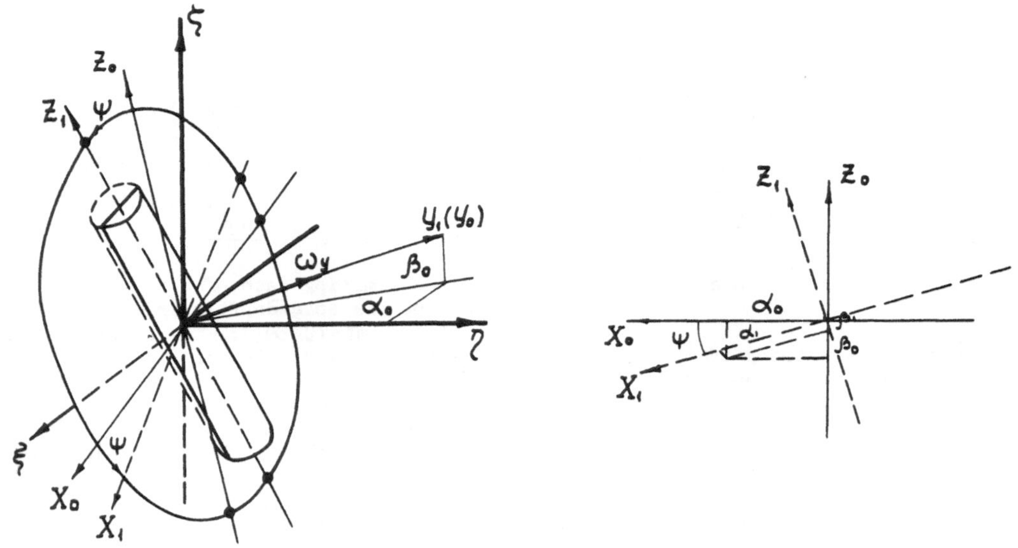
ζ
Z₀
Z₁
ψ
y₁(y₀)
ωy
β₀
α₀
η
ξ
ψ
X₀
X₁
Z₁
Z₀
α₀
X₀
ψ
α₁
β₀
X₁

(a)

(b)

Fig. 5.

fied by the parameters of the spacecraft in order that the compensation of an angular deviation β_o^* can be accomplished with one pair of reaction jets with a minimum consumption of mass carrier. In order to satisfy the above-formulated requirements, it is necessary that after the axis of the spacecraft has crossed the point ($\alpha_o(T_o) = 0$, $\beta_o(T_o) = \beta_o^*$) a finite number of times the angle of spin about the x_1-axis should be exactly 2π (Figure 4).

The conditions for the feasibility of optimal control with a single pair of reactive jets depend on the parameters of the spacecraft and can be written down in the form of formulas for the determination of $\bar{J}_x$:

$$\bar{J}_x = \frac{2k+1}{2n}, \quad k = 0, 1, 2, \ldots, \quad n = 1, 2, 3, \ldots \tag{3.1}$$

where 2k+1 is the number of transits of the x_1-axis of the spacecraft due to the precession during a finite period of the manuever, and

n is the number of revolutions of the spacecraft about the x_1-axis.

The conditions on the choice of the quantity $\bar{J}_x$ can be relaxed, if one drops the requirement of minimum consumption of mass carrier for achieving the attitude correction but maintains the condition that the precession should be eliminated after two control impulses. The latter condition makes it necessary that the difference between the spin angle relative to the x_1-axis and the angle between the corrective impulsive angular velocities should be π plus a multiple of $2\pi n$, i.e.

$$\gamma - \varphi_o = \pi + 2\pi n, \quad n = 0, 1, 2, \ldots \tag{3.2}$$

where γ is the angle of rotation of the spacecraft about the x_1-axis.

It is clear that such a maneuver is feasible when φ_o is close to π. The dependence of the required value φ_o on the number of rotations has the form

$$\varphi_o = \pi \frac{\bar{J}_x}{1 - \bar{J}_x}(2n+1) \tag{3.3}$$

4. OPTIMUM ATTITUDE CONTROL OF SPACECRAFT ABOUT A PRINCIPAL AXIS OF INERTIA THAT IS NOT A SYMMETRY AXIS

We will assume that the rotation takes place about the principal axis of inertia Oy_1, that ω_y = const., and that the deviation from the prescribed direction, given by the η-axis in the inertial reference frame, is small. The motion of the spacecraft can be described by the Euler equations

$$\dot{\omega}_x = u_x, \quad \dot{\omega}_z = -(1 - \bar{J}_x)\omega_y \omega_x + u_z \tag{4.1}$$

where ω_y = const. and the kinematic equations

$$\dot{\beta}_1 = \omega_x - \alpha_1 \omega_y, \quad \dot{\alpha}_1 = \omega_z + \beta_1 \omega_y \tag{4.2}$$

where α_1 and β_1 are similar to the angles introduced in Section 2. They are defined as the angles between the planes passing through the related principal axes of inertia and the η-axis. After transforming the kinematic relations (4.2) to the nonrotating axes (see Section 2), we obtain equations for the angles α_o, β_o that determine the orientation of the body-fixed y_1-axis relative to the axes of the inertial system $-0\xi\eta\zeta$ (Figure 5)

$$\dot{\beta}_o = \omega_x \cos\psi + \omega_z \sin\psi, \quad \dot{\alpha}_o = \omega_z \cos\psi - \omega_x \sin\psi, \quad \dot{\psi} = \omega_y \tag{4.3}$$

It should be noted that u_x and u_z are the control moments produced by the reaction jets fixed in the spacecraft, and ω_x, ω_z are angular velocities measured on the x_1- and z_1-axes, respectively. We will supplement equations (4.1) and (4.3) with the equation for the consumption of mass carrier:

$$\dot{g} = k_x|u_x| + k_z|u_z| \tag{4.4}$$

Now we will form the Hamiltonian H, and write out the canonical equations:

$$H = p_1 u_x - k_x|u_x| + p_2 u_z - k_z|u_z| - p_2(1-\bar{J}_x)\omega_y\omega_x + p_3(\omega_x\cos\psi + \omega_z\sin\psi) + p_4(\omega_z\cos\psi - \omega_x\sin\psi) \tag{4.5}$$

$$\begin{aligned} \dot{p}_1 &= (1-\bar{J}_x)\omega_y p_2 - p_3\cos\psi + p_4\sin\psi \\ \dot{p}_2 &= -(p_3\sin\psi + p_4\cos\psi) \\ \dot{p}_3 &= 0; \quad p_3 = const \\ \dot{p}_4 &= 0; \quad p_4 = const \end{aligned} \tag{4.6}$$

Integration of (4.6) yields the following expressions for $p_1(t)$ and $p_2(t)$:

$$\begin{aligned} p_1 &= c_1 + (1-\bar{J}_x)\omega_y c_2 t - \frac{\bar{J}_x}{\omega_y}(c_3\sin\psi + c_4\cos\psi) \\ p_2 &= c_2 + \frac{1}{\omega_y}(c_3\cos\psi - c_4\sin\psi) \end{aligned} \tag{4.7}$$

From (4.7) it is easily found that the p-trajectory in the phase-plane (p_1, p_2) is generated, as it were, by the translation of the ellipse (see Fig. 6)

$$\left\{\frac{p_1 - [c_1 + (1-\bar{J}_x)c_2\omega_y t]}{\bar{J}_x}\right\}^2 + \{p_2 - c_2\}^2 = c_o^2 \tag{4.8}$$

By maximizing H with u_x and u_y we obtain the conditions that must be satisfied by the control:

$$u_i = \begin{cases} U_{max}\,\mathrm{sgn}\,p_i, & \text{when } |p_i| > k_i \\ 0, & \text{when } |p_i| < k_i \end{cases} \tag{4.9}$$

As an example, we will study the optimum turning of the body-fixed y_1-axis through an angle β_o^* in time $T_o = \pi/\omega_y$ with a minimum consumption of mass carrier and the following boundary conditions:

$$\begin{aligned} &\alpha_o(0) = \omega_x(0) = \omega_z(0) = \beta_o(0) = 0; \quad \omega_y = const > 0 \\ &\alpha_o(T) = \omega_x(T_o) = \omega_z(T_o) = 0; \quad \beta_o(T) = \beta_o^* \end{aligned} \tag{4.10}$$

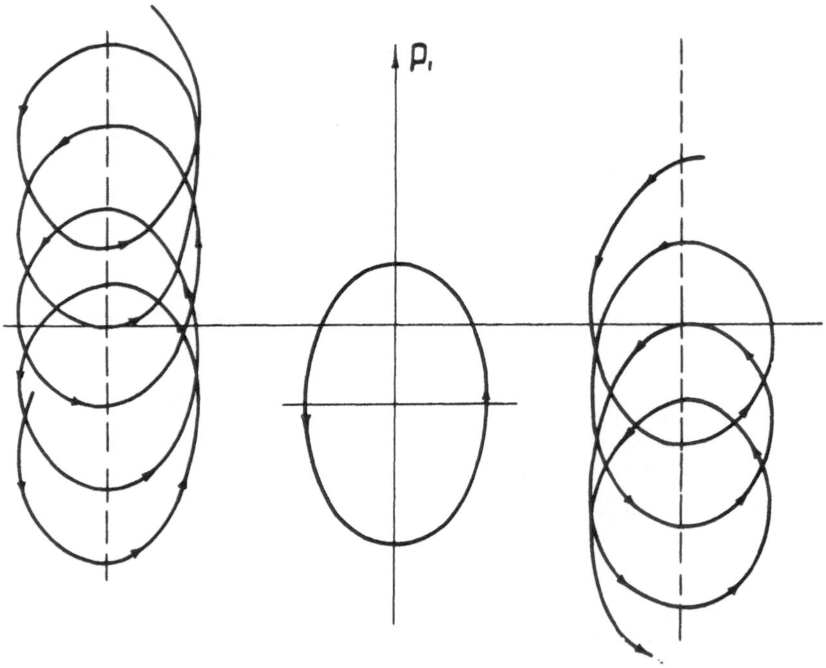

Fig. 6.

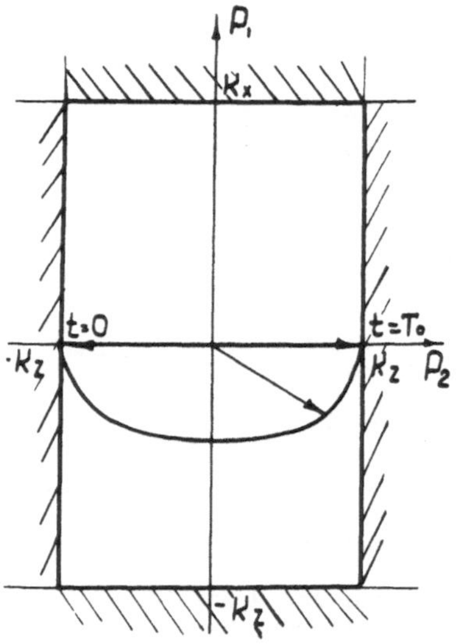

Fig. 7.

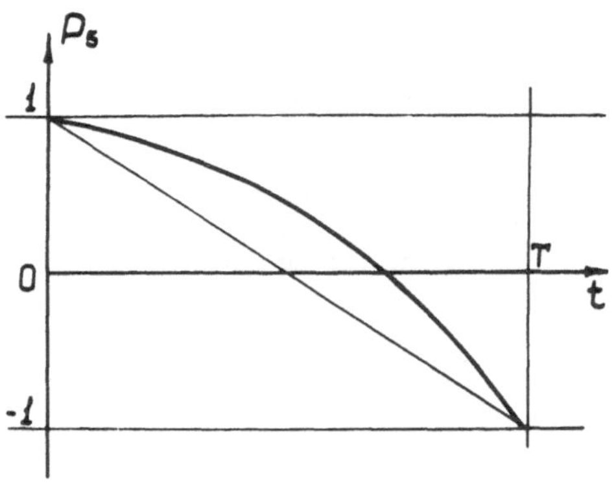

Fig. 8.

From the equations of motion (4.1) and the boundary conditions (4.10) it follows that the number of control impulses u_x is even, which includes zero. The optimum manuever under consideration can be achieved by the u_z control alone (i.e. with $u_x \equiv 0$) which consists of two impulsive angular velocities $\Delta\omega_z$; one at the beginning and one at the end of the maneuver, where

$$\Delta\omega_z = -\frac{\beta_o^* \omega_y}{2} \qquad (4.11)$$

The p-trajectory will be a single ellipse ($C_2 = 0$) which at $t = 0$ and $t = T_o$ is tangential to the boundary lines $|p_2| = k_2$ as shown in Figure 7. During the motion of the spacecraft, the angles α_o, β_o change in accordance with the expressions

$$\alpha_o = \frac{\Delta\omega_z}{\omega_y} \sin\omega_y t$$
$$\beta_o = \frac{\Delta\omega_z}{\omega_y}\left[\cos\omega_y t - 1\right] \qquad (4.12)$$

To carry out a turn in the general case it is necessary to employ control not only with u_z but also with u_x.

5. OPTIMUM THREE-DIMENSIONAL ATTITUDE CONTROL OF AN AXISYMMETRIC SPACECRAFT

In the impulsive formulation for an axisymmetric spacecraft, ω_x is a piecewise constant function and the equations of motion can be taken in the modified form (2.3). We will supplement these with the differential equations for ω_x and γ:

$$\dot\omega_x = u_x$$
$$\dot\gamma = \omega_x \qquad (5.1)$$

and the equation for the consumption of mass carrier

$$\dot g = u + k_x|u_x| \qquad (5.2)$$

In order to find the optimum control we will write down the principal part H which contains the control function:

$$H = u\left[(p_1 \sin\psi + p_2\cos\psi) - 1\right] + p_5 u_x - k_x|u_x| \qquad (5.3)$$

From (5.3) follow the conditions that guarantee optimum control. These conditions are relations (2.8), (2.11), and

$$u_x = \begin{cases} U_x \operatorname{sgn} p_5, & \text{when } |p_5| > k_x \\ 0, & |p_5| < k_x \end{cases} \qquad (5.4)$$

Relations (2.11) and (5.4) indicate that the control is of a limiting nature with rest periods.

The conjugate equations for the variables p_1 to p_4 have been derived earlier (2.14), and here it suffices to write out the equations for p_5 and p_6:

$$\dot p_5 = -\bar J_x \omega_{yo} p_2 + \bar J_x \omega_{zo} p_1 - p_6$$
$$\dot p_6 = 0 \; ; \; p_6 = C_6 \qquad (5.5)$$

In order to simplify the equation for p_5, we multiply the right-hand sides of equations (2.3) by $-p_2$ and p_1, and the right-hand sides of equations (2.12) by ω_{zo} and $-\omega_{yo}$, respectively, and add. We obtain

$$\frac{d}{dt}(\omega_{zo}p_1 - \omega_{yo}p_2) = -C_4\omega_{zo} + C_3\omega_{yo} \qquad (5.6)$$

Integrating (5.6) and substituting the expression in the bracket into the formula for $\dot p_5$, we obtain

$$\dot p_5 = -\bar J_x C_4 \alpha_o + \bar J_x C_3 \beta_o + C_o \qquad (5.7)$$

If $\alpha_o(t)$, $\beta_o(t)$, and C_3, C_4, C_o are known, the problem will reduce to quadratures.

We will examine the problem of the optimum turning of a spacecraft from the initial conditions

$$\gamma(0) = \alpha(0) = \beta_o(0) = \omega_x(0) = \omega_{yo}(0) = \omega_{zo}(0) = 0 \qquad (5.8)$$

thru the angles $\beta_o(T) = \beta_o^*$ and $\gamma(T) = \gamma_o$, i.e. a transition in time T to the conditions

$$\gamma(T) = \gamma_o \; ; \; \beta_o(T) = \beta_o^* ; \; \alpha_o(T) = \omega_x(T) = \omega_{yo}(T) = \omega_{zo}(T) = 0 \qquad (5.9)$$

The optimum manuever can be achieved by two pairs of simultaneous switchings of the controls u_x and u_z, i.e. the spacecraft moves with ω_x having the constant value

$$\omega_x = \frac{\gamma_o}{T} \qquad (5.10)$$

(for the sake of definiteness we will assume that $\gamma_o > 0$, $\beta_o^* < 0$, and $k_x = 1$). The control of a spacecraft rotating with ω_x = const. was treated in Section 2. It remains to prove that the switching of the control u_x at the ends of the time interval (0, T) is not inconsistent with equation (5.7) for p_5 and condition (5.4), i.e. that p_5 satisfies the following conditions

$$p_5(0) = 1 \; ; \; |p_5(t)| < 1 \; ; \; p_5 = -1$$
$$0 < t < T \qquad (5.11)$$

Making use of the results in Section 2, we can write down the following values of C_3 and C_4 for the present problem

$$C_3 = 2\bar J_x \omega_x \cos\frac{\bar J_x \gamma_o}{2}$$
$$C_4 = 0 \qquad (5.12)$$

The expressions for $\alpha_o(t)$ and $\beta_o(t)$ can be put in the following form

$$\beta_o = \frac{\beta_o^*}{2\sin\frac{\bar J_x\gamma_o}{2}}\left[\sin\left(\frac{\bar J_x\gamma_o t}{T} - \frac{\bar J_x\gamma_o}{2}\right) + \sin\frac{\bar J_x\gamma_o}{2}\right]$$
$$\alpha_o = \frac{-\beta_o^*}{2\sin\frac{\bar J_x\gamma_o}{2}}\left[\cos\left(\frac{\bar J_x\gamma_o t}{T} - \frac{\bar J_x\gamma_o}{2}\right) - \cos\frac{\bar J_x\gamma_o}{2}\right] \qquad (5.13)$$

By inserting (5.13) and (5.12) and taking the boundary conditions (5.11) into account after integration, we obtain p_5:

$$p_5 = 1 - \frac{2t}{T} + 2\bar J_x \alpha_o(t)\left[\cos\frac{\bar J_x\gamma_o}{2}\right] \qquad (5.14)$$

The form of the function p_5 is depicted in Figure 8. When $\beta_o^* < 0$, it follows from (5.7) and (5.13) that the maximum value of the derivative $\dot p_5$ occurs at $t = 0$. Whence we obtain a condition in order that the present type of maneuver should be possible:

$$C_o < 0 \qquad (5.14a)$$

The expression for C_o, in the general case where β_o^* and γ_o have arbitrary signs, has the form

$$C_o = -\frac{2}{T} - \frac{(\bar J_x)^2}{T}|\beta_o^*\gamma_o| \cos\frac{\bar J_x\gamma_o}{2} \qquad (5.16)$$

Hence it follows that

$$|\beta_0^* \gamma_0| \cos \frac{\bar{J}_x \gamma_0}{2} < \frac{2}{(\bar{J}_x)^2} \qquad (5.17)$$

Inequality (5.17) is practically always satisfied by virtue of the limitations placed on the magnitude of $|\beta_0^*|$. We note that in the case where

$$\bar{J}_x \gamma_0 > \pi \qquad (5.18)$$

which is similar to the case considered in Section 2, the maneuver can be achieved first with the aid of u_x until the quantity γ attains the value $\pi/\bar{J}_x$, after which control with u_z is commenced.

REFERENCES

(1) Pontryagin, L. S., Boltyanskii, V. G., Gamkrelidze, R. V., and Mishchenko, E. F., The Mathematical Theory of Optimal Processes, English translation by Interscience Publishers, Inc., New York, 1962.

(2) Rozonoer, L. I., Pontryagin's Maximum Principle in the Theory of Optimum Systems, Parts I - III, "Automation and Remote Control," Volume 20, Issues 10-12.

(3) Smol'nikov, B. A., Optimum Braking Regimes of Rotational Motion of a Symmetric Body, "PMM," Volume 28, No. 4, 1964.

(4) Athans, M. and Falb, P. L., Time-Optimal Velocity Control of a Spinning Space Body, IEEE Transactions on Applic. and Ind., No. 67, pp. 206-213, 1963.

(5) Flugse-Lotz, I. and Marbach, H., The Optimal Control of Some Attitude Systems for Different Performance Criteria, "Transaction of the ASME Journal of Basic Engineering," June, 1963.

(6) Reid, H. J. E. and Garner, H. D., Miniguide - A Simplified Attitude Control for Spib-Stabilized Vehicles, AIAA Paper, No. 63-210.

(7) Patapoff, H., Bank Angle Control System for a Spinning Satellite, AIAA Paper No. 63-339, New York, 1963, AIAA Guidance and Control Conference, August 12-14, 1963.

(8) Meditch, J. S., On Minimal-Fuel Satellite Attitude Controls, IEEE Transactions on Applic. and Ind., No. 71, pp. 120-128, 1964.

Asymptotically Stable Rotational Motions of a Satellite

V. A. Sarichev

In this paper we consider the rotational motion of a satellite whose center of mass is following a circular orbit. The satellite shape is a figure of revolution, with A = B, and the axial moment of inertia C. Figure 1 shows the orbital coordinate system Oxyz determined by the radius vector and the transversal and binormal to the orbit, and also the coordinate system $Ox_1y_1z_1$ given by the principal central axes of inertia of the satellite. The orientation of the frame $Ox_1y_1z_1$ with respect to the orbital coordinate system is described by the Euler angles ψ, θ, φ (ψ is the precession angle, θ is the nutation angle, and φ is the proper-motion angle). The satellite's axis of symmetry coincides with the Oz_1 axis, while the Oz axis is directed along the radius vector. (See Figure 1.)

With allowance for gravitational torque, the equations of motion of the satellite take the form:

$$
\begin{aligned}
&A\frac{d}{dt}(\dot\psi\sin^2\theta) - 2A\omega\dot\theta\sin^2\theta\cos\psi - Cr_0\dot\theta\sin\theta - \\
&- A\omega^2\sin^2\theta\sin\psi\cos\psi - Cr_0\omega\sin\theta\sin\psi = 0 \\
&A\ddot\theta + 2A\omega\dot\psi\sin^2\theta\cos\psi - A\dot\psi^2\sin\theta\cos\theta + Cr_0\dot\psi\sin\theta + \\
&+ Cr_0\omega\cos\psi\cos\theta = 0
\end{aligned}
\tag{1}
$$

Here

$$r_0 = \dot\psi\cos\theta + \dot\varphi - \omega\cos\psi\sin\theta = \text{const.}$$

and ω is the angular rate with which the center of mass of the satellite travels in orbit. Equations (1) have a generalized energy integral which may be expressed as:

$$
\begin{aligned}
&\tfrac{1}{2}A(\dot\psi^2\sin^2\theta + \dot\theta^2) + C\omega r_0\cos\psi\sin\theta + \\
&+ \tfrac{1}{2}A\omega^2\cos^2\psi\sin^2\theta + \tfrac{3}{2}\omega^2(C-A)\cos^2\theta = h
\end{aligned}
$$

The stable solutions of the system (1) are determined by the system of algebraic equations

$$
\begin{aligned}
&\sin\theta_0\sin\psi_0\left[\alpha\beta - (\alpha-1)\sin\theta_0\cos\psi_0\right] = 0 \\
&\cos\theta_0\left\{\left[\alpha\beta - (\alpha-1)\sin\theta_0\cos\psi_0\right]\cos\psi_0 - 3(\alpha-1)\sin\theta_0\right\} = 0
\end{aligned}
\tag{4}
$$

where

$$
\begin{aligned}
&\alpha = \frac{C}{A},\ (0 \le \alpha \le 2),\quad \beta = \frac{\dot\varphi}{\omega} = \beta_1 + \sin\theta_0\cos\psi_0, \\
&\beta_1 = \frac{r_0}{\omega},\quad 0 \le \theta_0 \le \frac{\pi}{2}
\end{aligned}
\tag{5}
$$

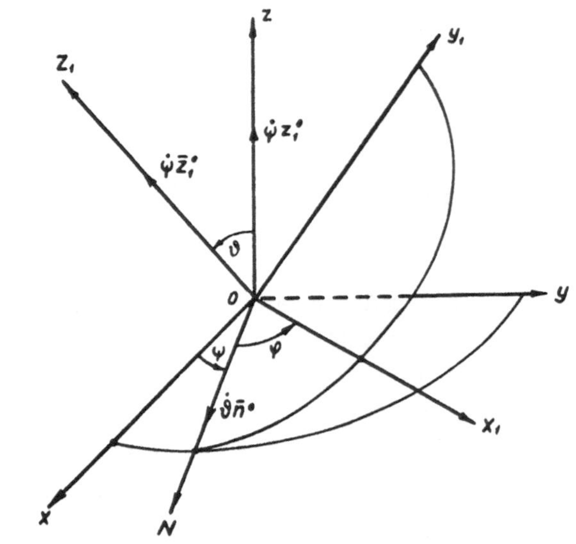

Figure 1. The fundamental coordinate systems.

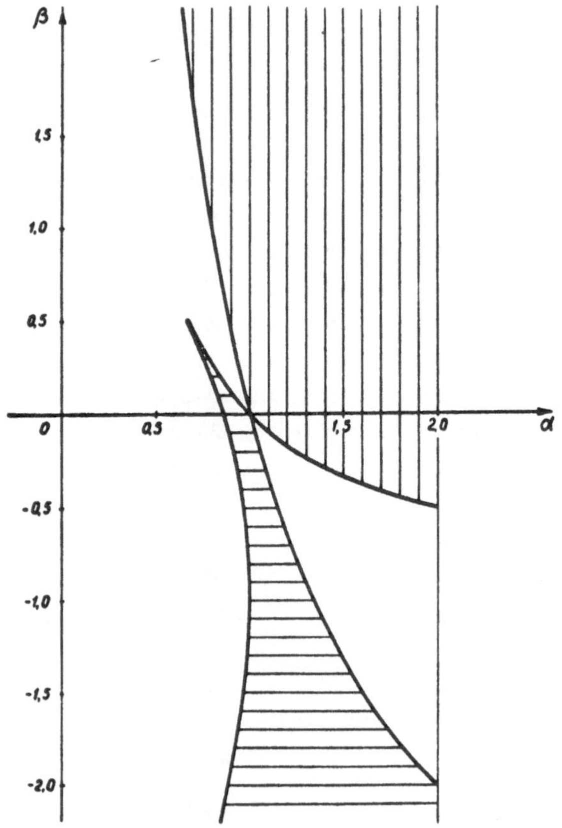

Figure 2. The regions of stability for the solution (6).

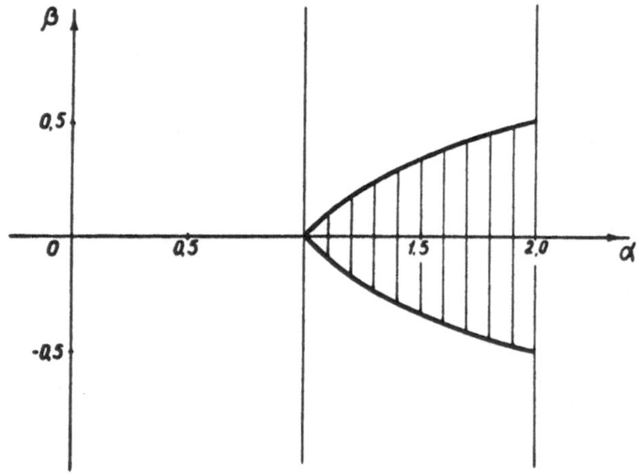

Figure 3. The region of stability for the solution (7).

The system (4) has three solutions:

$$\cos\theta_0 = 0, \quad \sin\psi_0 = 0 \qquad (6)$$

$$\cos\theta_0 = 0, \quad (\alpha - 1)\cos\psi_0 - \alpha\beta = 0 \qquad (7)$$

$$4(\alpha-1)\sin\theta_0 - \alpha\beta\cos\psi_0 = 0, \quad \sin\psi_0 = 0 \qquad (8)$$

In the solutions (6) - (8), the axis of symmetry of the satellite is respectively perpendicular to the orbital plane, or lies within the plane, perpendicular to the radius vector or the velocity vector.

By utilizing the integral (3) and linearizing Equation (1) in the neighborhood of the stable solutions, one readily obtains both necessary and sufficient conditions for the stability of the solutions. Figures 2-4 illustrate the regions in the plane of the parameters α, β where the solutions (6) - (8) are stable. Horizontal shading indicates regions where only the necessary conditions for stability hold; vertical shading represents regions where both necessary and sufficient conditions for stability are fulfilled. (See Figures 2, 3, and 4 for the regions of stability for solutions (6), (7) and (8) respectively.)

The results just described have been obtained by different methods elsewhere ((1-3)).

$$M_{x_1} = -\bar{K}p, \quad M_{y_1} = -\bar{K}q, \quad M_{z_1} = 0 \qquad (9)$$

Here p and q represent the projections of the satellite's absolute angular velocity on the Ox_1 and Oy_1 axes. Corresponding to the guidance moments (9) we have the generalized forces

$$L_\psi = -\bar{K}(\dot{\psi}\sin^2\theta + \omega\sin\theta\cos\theta\cos\psi),$$

$$L_\theta = -\bar{K}(\dot{\theta} + \omega\sin\psi), \qquad (10)$$

$$L_\varphi = 0$$

The moments (9) could be produced by using, for example, two angular-velocity detectors with axes of sensitivity directed along Ox_1 and Oy_1, together with a device to impart moments to the satellite proportional to the signals from the detectors.

This procedure of introducing damping terms is naturally not optimal, as it will tend to change the static solutions. Yet on the other hand there is the advantage that by varying one can correct the position of the satellite's axis of symmetry in the orbital coordinate system.

If we adopt Equation (10), then, the system of algebraic equations governing the static solutions will have the form

$$\sin\theta_0\left[\alpha\beta\sin\psi_0 - (\alpha-1)\sin\theta_0\sin\psi_0\cos\psi_0 - K\cos\theta_0\cos\psi_0\right] = 0$$

$$\alpha\beta\cos\theta_0\cos\psi_0 - (\alpha-1)\sin\theta_0\cos\theta_0\cos^2\psi_0 - 3(\alpha-1)\sin\theta_0\cos\theta_0 + K\sin\psi_0 = 0 \qquad (11)$$

$$K = \frac{\bar{K}}{A\omega}$$

In order to obtain asymptotically stable rotational motions that are stationary in the orbital coordinate system, we propose adding to the gravitational moments acting on the satellite the guidance moments

We shall regard $\sin\theta_0 \neq 0$, since a violation of this condition would correspond to a satellite at rest in the orbital coordinate system, which its axis of symmetry coinciding with the radius vector.

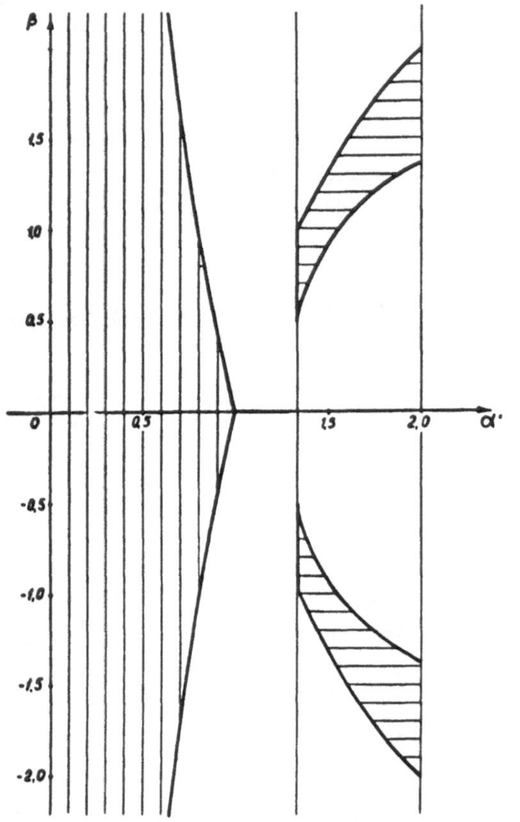

Figure 4. The regions of stability for the solution (8).

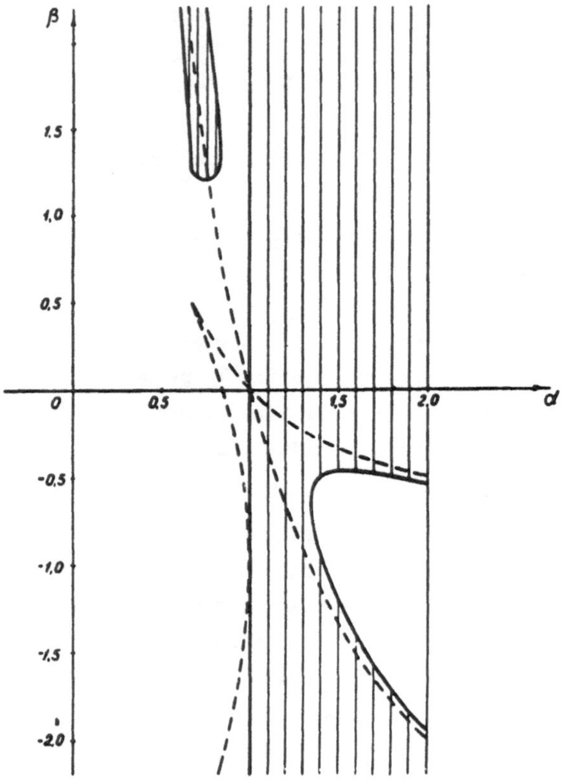

Figure 5. The region of asymptotic stability for the solution (6).

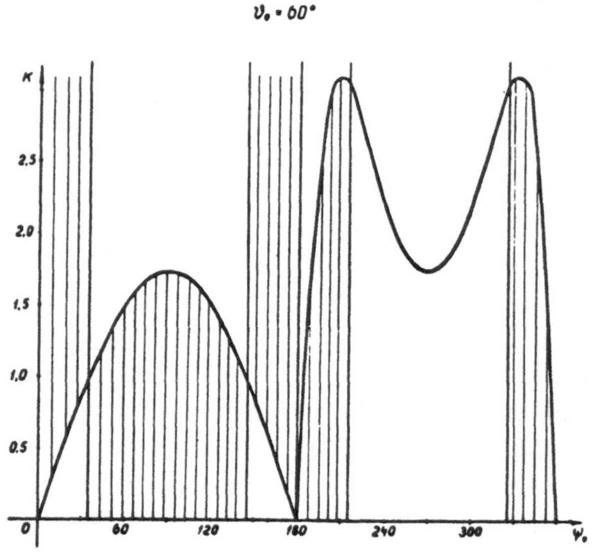

Figure 6. The region of asymptotic stability for the solutions corresponding to the system (11).

The solution (6) satisfies Equation (11). We shall examine the stability of this solution separately. The characteristic equation for the solution (6) becomes

$$\lambda^4 + 2K\lambda^3 + \left[K^2 + (\alpha\beta + \alpha - 1)^2 + 3(\alpha - 1) + 1\right]\lambda^2 + \\ + K(3\alpha - 1)\lambda + K^2 + (\alpha\beta + \alpha - 1)(\alpha\beta + 4\alpha - 4) = 0 \qquad (12)$$

If we use the Lienard and Chipard criteria, we may write the necessary and sufficient conditions for asymptotic stability of the solutions as follows:

$$\begin{aligned} &K > 0 \\ &3\alpha - 1 > 0 \\ &K^2 + (\alpha\beta + \alpha - 1)(\alpha\beta + 4\alpha - 4) > 0 \\ &(\alpha - 1)\left[2K^2 + 2(\alpha\beta + \alpha - 1)^2 - 4(\alpha\beta + \alpha - 1) + 3(\alpha - 1) + 2\right] > 0 \end{aligned} \qquad (13)$$

In the event that $K \geq \frac{3}{2}$, the inequalities (13) will reduce to the condition

$$1 < \alpha \leq 2 \qquad (14)$$

But for small K the region of stability has a more complicated form. As an example, Figure 5 shows the region for the case $K = 0.5$ (the region of stability is indicated by vertical shading). Broken curves bound the region of stability shown in Figure 2 for the solution with $K = 0$. (See Figure 5, "The region of asymptotic stability" for the solution (6)).

Let us now consider the general solution of the system (11). We shall suppose that

$$\sin^2\psi_0 + \cos^2\theta_0 \neq 0$$

It then follows that $\cos\psi_0 \neq 0$. The system (11) constitutes two equations relating the quantities ψ_0, θ_0, α, β, and K. Solving these equations for α and β, we obtain:

$$\alpha = 1 + \frac{\cos^2\theta_0 + \sin^2\psi_0 \sin^2\theta_0}{3\sin\theta_0 \cos\theta_0 \sin\psi_0} K,$$

$$\beta = \frac{\alpha - 1}{\alpha} \cdot \frac{4\cos^2\theta_0 + \sin^2\psi_0 \sin^2\theta_0}{\cos^2\theta_0 + \sin^2\psi_0 \sin^2\theta_0} \sin\theta_0 \cos\psi_0 \qquad (15)$$

The characteristic equation determining the solution $\psi = \psi_0$, $\theta = \theta_0$ that satisfies (11) may be written as

$$\lambda^4 + 2k\lambda^3 + \left(k^2 x \frac{\sin\theta_0 \cos\theta_0}{\sin^2\psi_0} + kxy + 1\right)\lambda^2 + k\left[kxy + (1+\sin^2\theta_0\cos^2\psi_0)\right]\lambda +$$

$$+ kx\frac{\cos^2\theta_0 - \sin^2\psi_0\sin^2\theta_0}{\sin\theta_0\sin^2\psi_0}(k\cos\theta_0 - \sin\theta_0\sin\psi_0) = 0$$

$$x = \frac{\cos^2\theta_0 + \sin^2\psi_0\sin^2\theta_0}{\sin\theta_0\cos\theta_0} \qquad (16)$$

$$y = \frac{1 - 3\cos^2\theta_0}{\sin\psi_0}$$

The parameters α and β in the coefficients of the characteristic equation have been eliminated by means of the relations (15). The necessary and sufficient conditions for asymptotic stability of this solution are:

$$k > 0,$$

$$kxy + (1+\sin^2\theta_0\cos^2\psi_0) > 0,$$

$$(\cos^2\theta_0 - \sin^2\psi_0\sin^2\theta_0)(k\cos\theta_0 - \sin\theta_0\sin\psi_0) > 0, \qquad (17)$$

$$2k^3xy + k^2\frac{\cos^2\theta_0\cos^2\psi_0(5-11\cos^2\theta_0) + \sin^2\psi_0(1+\cos^2\theta_0}{\cos^2\theta_0\sin^2\psi_0} +$$

$$+ 2k\frac{\sin\theta_0\cos 2\psi_0}{\cos\theta_0\sin\psi_0} + (1+\sin^2\theta_0\cos^2\psi_0) > 0$$

An inspection of the stability conditions (17) leads to the following conclusions:

1. If $0 < \theta_0 \leq 30^\circ$, the stability conditions reduce to the inequalities

$$k > 0, \quad 180^\circ < \psi_0 < 360^\circ \qquad (18)$$

2. If $30^\circ < \theta_0 \leq 45^\circ$, the region (18) will be accompanied by some very narrow regions within the interval $0 < \psi_0 < 180^\circ$, in the vicinity of the points $\psi_0 = 0$ and $\psi_0 = 180^\circ$.

3. For $\theta_0 > 45^\circ$, the form of the region of stability becomes considerably more complicated. As an example, Figure 6 shows by vertical shading the region of stability corresponding to the value $\theta_0 = 60^\circ$.

By adopting values for ψ_0, θ_0, and k inside the region of asymptotic stability, the parameters α and β corresponding to these values may be determined by means of the relations (15). The k-dependence of the solutions enables one to make a correction along the orbit for the position of the axis of symmetry of the satellite relative to the orbital coordinate system.

Appendix 1

In an elliptic orbit, the equation of motion of a satellite, excluding the guidance moments (9), will have the form

$$(1+e\cos\nu)^2\psi''\sin\theta - 2e\sin\nu(1+e\cos\nu)\psi'\sin\theta +$$
$$+ 2(1+e\cos\nu)^2\psi'\theta'\cos\theta - 2(1+e\cos\nu)^2\theta'\sin\theta\cos\psi -$$
$$- \alpha\beta_1\theta_1' - (1+e\cos\nu)^2\sin\theta\sin\psi\cos\psi - \alpha\beta_1\sin\psi -$$
$$- 2e\sin\nu(1+e\cos\nu)\cos\theta\cos\psi = 0, \tag{19}$$
$$(1+e\cos\nu)^2\theta'' - 2e\sin\nu(1+e\cos\nu)\theta' + 2(1+e\cos\nu)^2\psi'\sin^2\theta\cos\psi -$$
$$- (1-e\cos\nu)^2(\psi')^2\sin\theta\cos\theta + \alpha\beta_1\psi'\sin\theta +$$
$$+ (1+e\cos\nu)^2\cos^2\psi\sin\theta\cos\theta + 3(1-\alpha)(1+e\cos\nu)\sin\theta\cos\theta +$$
$$+ \alpha\beta_1\cos\psi\cos\theta - 2e\sin\nu(1+e\cos\nu)\sin\psi = 0,$$

Where ν is the true anomaly and e is the eccentricity of the orbit. A prime denotes differentiation with respect to ν.

We seek a particular solution of the system (19) that reduces to a static solution in a circular orbit, expressed as a series in powers of e:

$$\psi = \psi_0 + e\psi_1 + \cdots$$
$$\theta = \theta_0 + e\theta_1 + \cdots \tag{20}$$

The functions ψ_0, θ_0 satisfy the system of equations (4). The solution (6) is also valid for an elliptic orbit, so that in this case all the terms of the development (20) tend to zero, beginning with the second term. For the other two solutions the function ψ_1, θ_1 have the form:

1) for the solution (7),

$$\psi_1 = -2\operatorname{tg}\psi_0\cos\nu$$
$$\theta_1 = 0 \tag{21}$$

2) for the solution (8),

$$\psi_1 = \left\{\frac{\alpha^2\beta^2}{2[\alpha^2\beta^2-8(\alpha-1)^2]} + \frac{2}{2-3\alpha}\right\}\operatorname{ctg}\theta_0\cos\psi_0\sin\nu$$
$$\theta_1 = \frac{\alpha^2\beta^2}{2[\alpha^2\beta^2-8(\alpha-1)^2]}\operatorname{ctg}\theta_0\cos\nu \tag{22}$$

Subsequent terms of the development (20) can also be obtained.

Since $\cos\psi_0 \neq 0$, ψ_1 assumes only finite values for the solution (7). Equations (22) indicate the possibility that a resonance may appear if

$$\alpha^2\beta^2 - 8(\alpha-1)^2 = 0 \tag{23}$$

or if

$$2 - 3\alpha = 0 \tag{24}$$

The curves (23) and (24) lie inside the region of stability of the solution (8), and the quantities ψ_1 and θ_1 assume infinitely great values on these curves.

Appendix 2

We shall demonstrate that for a solid body with unequal moments of inertia moving in a circular orbit, there are no equilbrium positions other than those corresponding to coincidence of the three principle central axes of inertia of the body with the axes of the orbital coordinate system. The equations of motion of a solid body in a circular orbit have the form:

$$
\begin{aligned}
&A\dot{p} + (C-B)qr = 3\,\omega^2 (C-B)\,a_{32} a_{33} && p = \dot{\psi} a_{31} + \dot{\theta}\cos\varphi + \omega a_{21}\\
&B\dot{q} + (A-C)rp = 3\,\omega^2 (A-C)\,a_{33} a_{31} && q = \dot{\psi} a_{32} - \dot{\theta}\sin\varphi + \omega a_{22}\\
&C\dot{r} + (B-A)pq = 3\,\omega^2 (B-A)\,a_{31} a_{32} && r = \dot{\psi} a_{33} + \dot{\varphi} + \omega a_{23}
\end{aligned}
\qquad (25)
$$

$$
\begin{aligned}
&a_{11} = \cos\psi\cos\varphi - \sin\psi\sin\varphi\cos\theta && a_{21} = \sin\psi\cos\varphi + \cos\psi\sin\varphi\cos\theta\\
&a_{12} = -\cos\psi\sin\varphi - \sin\psi\cos\varphi\cos\theta && a_{22} = -\sin\psi\sin\varphi + \cos\psi\cos\varphi\cos\theta\\
&a_{13} = \sin\psi\sin\theta && a_{23} = -\cos\psi\sin\theta
\end{aligned}
$$

$$
\begin{aligned}
a_{31} &= \sin\varphi\sin\theta\\
a_{32} &= \cos\varphi\sin\theta\\
a_{33} &= \cos\theta
\end{aligned}
$$

Taking $\dot{\psi} = \dot{\theta} = \dot{\varphi} = \ddot{\psi} = \ddot{\theta} = \ddot{\varphi} = 0$, we obtain a system of algebraic equations that determines the static solution:

$$
\begin{aligned}
a_{22} a_{33} &= 3\, a_{32} a_{33}\\
a_{23}\, a_{21} &= 3\, a_{33} a_{31}\\
a_{21}\, a_{22} &= 3\, a_{31} a_{32}
\end{aligned}
\qquad (26)
$$

or

$$
\begin{aligned}
&\sin\theta_0 \left[\cos\theta_0 (\cos^2\psi_0 + 3)\cos\varphi_0 - \sin\psi_0\cos\psi_0\sin\varphi_0\right] = 0\\
&\sin\theta_0 \left[\sin\psi_0\cos\psi_0\cos\varphi_0 + \cos\theta_0(\cos^2\psi_0 + 3)\sin\varphi_0\right] = 0\\
&(\sin\psi_0\cos\varphi_0 + \cos\psi_0\sin\varphi_0\cos\theta_0)(-\sin\psi_0\sin\varphi_0 + \cos\psi_0\cos\varphi_0\cos\theta_0 -\\
&\qquad -3\sin\varphi_0\cos\varphi_0\sin^2\theta_0 = 0
\end{aligned}
\qquad (27)
$$

All the solutions of the nonlinear system (27) can in fact be obtained. There are two possible cases for solution of this system:

$$\begin{aligned} &\sin\theta_0 = 0 \\ &(\sin\psi_0\cos\varphi_0 + \cos\psi_0\sin\varphi_0\cos\theta_0)(-\sin\psi_0\sin\varphi_0 + \cos\psi_0\cos\varphi_0\cos\theta_0) - \\ &\quad - 3\sin\varphi_0\cos\varphi_0\sin^2\theta_0 = 0 \quad ; \end{aligned} \qquad (28)$$

$$\begin{aligned} &\cos\theta_0(\cos^2\psi_0 + 3)\cos\varphi_0 - \sin\psi_0\cos\psi_0\sin\varphi_0 = 0 \\ &\sin\psi_0\cos\psi_0\cos\varphi_0 + \cos\theta_0(\cos^2\psi_0 + 3)\sin\varphi_0 = 0 \\ &(\sin\psi_0\cos\varphi_0 + \cos\psi_0\sin\varphi_0\cos\theta_0)(-\sin\psi_0\sin\varphi_0 + \cos\psi_0\cos\varphi_0\cos\theta_0) - \\ &\quad - 3\sin\varphi_0\cos\varphi_0\sin^2\theta_0 = 0 \end{aligned} \qquad (29)$$

The system (28) has the following solutions:

$$\theta_0 = 0, \qquad \sin 2(\psi_0 + \varphi_0) = 0 \qquad (30)$$

$$\theta_0 = \pi, \qquad \sin 2(\psi_0 - \varphi_0) = 0 \qquad (31)$$

Since $\sin\varphi_0$ and $\cos\varphi_0$ cannot vanish simultaneously, the determinant of the first two equations of the system (29) must be equal to zero. Consequently

$$\cos^2\theta_0(\cos^2\psi_0 + 3)^2 + \sin^2\psi_0\cos^2\psi_0 = 0$$

and Equation (29) are equivalent to the system:

$$\begin{aligned} &\cos\theta_0 = 0 \\ &\sin\psi_0\cos\psi_0 = 0 \\ &(\sin\psi_0\cos\varphi_0 + \cos\psi_0\sin\varphi_0\cos\theta_0)(-\sin\psi_0\sin\varphi_0 + \cos\psi_0\cos\varphi_0\cos\theta_0) - \\ &\quad - 3\sin\varphi_0\cos\varphi_0\sin^2\theta_0 = 0 \end{aligned} \qquad (33)$$

The system (33) has the following solutions:

$$\theta_0 = \frac{\pi}{2}, \quad \psi_0 = 0, \quad \sin 2\varphi_0 = 0; \qquad (34)$$

$$\theta_0 = \frac{\pi}{2}, \quad \psi_0 = \pi, \quad \sin 2\varphi_0 = 0; \qquad (35)$$

$$\theta_0 = \frac{\pi}{2}, \quad \psi_0 = \frac{\pi}{2}, \quad \sin 2\varphi_0 = 0; \qquad (36)$$

$$\theta_0 = \frac{\pi}{2}, \quad \psi_0 = \frac{3\pi}{2}, \quad \sin 2\varphi_0 = 0 \qquad (37)$$

All solutions of the system of equations (27) are therefore subsumed by Equations (30) - (31) and (34) - (37). One can easily see from Figure 1 that these solutions correspond to different cases for coincidence of the principal central axes of inertia of the body with the axes of the orbital coordinate system. This result demonstrates that the findings of Michelson ((4-5)) are not correct.

Literature Cited:

1. G. N. Dubshin, Byull. Inst. Teor. Astron., 7 (No. 7), 511-520 (1960).

2. F. L. Chernous'ko, Priklad. Mat. Mekh., 28 (No. 1), 155-157 (1964).

DISCUSSION

V. A. Sarichev

"Asymtotically Stable
Stationary Rotational
Motion of a Satellite".

Q. You show "stability regions" for the undamped system, that is, when the poles are purely imaginary. Wouldn't a non-linear analysis be required to tell if the system is really stable?

A. A non-linear analysis was considered by the authors of the reference quoted in the paper.

FUEL AND ENERGY MINIMIZATION IN THREE DIMENSIONAL ATTITUDE CONTROL OF AN ORBITING SATELLITE

Y. Nishikawa, C. Hayashi, and N. Sannomiya
Department of Electrical Engineering
Kyoto University, Kyoto, Japan

ABSTRACT

This paper deals with fuel and energy minimization problems in the attitude control of a satellite which is in a steady orbit around the earth. The earth-centered triad is taken as a reference frame. The torque producing elements considered include a reaction gas jet and a reaction wheel driven by a d-c motor. The optimal switching policy of the control functions is investigated, and the minimum fuel or energy consumption is determined for various initial states. Influence of a transition time on the amount of fuel or energy consumption is also discussed.

INTRODUCTION

The present paper is concerned with fuel and energy minimization problems in the attitude control of a satellite which is in a steady orbit around the earth. The orbit is either circular or elliptic with a small eccentricity. The satellite is assumed to be a single rigid body. Since the attitude control with respect to the earth is considered, the earth-centered triad is taken as a reference frame of the attitude. In the case of a circular orbit, the gyrodynamical equations of the satellite consist of terms with constant coefficients; while in the case of an elliptic orbit, terms with time-varying coefficients appear in the equations owing to periodical change of the angular velocity of the orbital motion.

The torque producing elements considered in this paper include a reaction gas jet and a reaction wheel driven by a d-c motor. The minimum-fuel control is investigated in a gas-jet system; while the minimum-energy control is discussed in a reaction-wheel system. The maximum principle is used for obtaining the optimal control functions.[1*]

The optimal switching policy of the control functions is investigated, and the minimum fuel or energy consumption is determined for various initial states of angular deviation. Influence of a transition time on the amount of fuel or energy consumption is also discussed.

* Superior numbers refer to similarly-numbered references at the end of this paper.

GYRODYNAMICAL EQUATIONS

The schematic diagram of an attitude control system is shown in Fig. 1.[2]

Figure 2 shows body-fixed axes i, j, k defined relative to reference axes I, J, K by modified Euler angles ϕ, θ, ψ. The I, J, K frame moves with its origin at the gravity center of the satellite, where axis J is directed outward along the local vertical and axis I lies in the direction of the orbital motion. Axis K is perpendicular to the plane of the orbit and directed so as to form I, J, K a right-handed triad. The right-handed triad i, j, k is fixed along the principal axes of the body. The angles ϕ, θ, and ψ are defined as follows:

1. Rotation through angle ϕ about axis K brings I, J, K to $\bar{I}$, $\bar{J}$, K frame (pitching).
2. Rotation through angle θ about axis $\bar{J}$ brings $\bar{I}$, $\bar{J}$, K to i, $\bar{J}$, $\bar{K}$ frame (yawing).
3. Rotation through angle ψ about axis i brings i, $\bar{J}$, $\bar{K}$ to i, j, k frame (rolling).

The following symbols are introduced to describe the motion of the satellite.

A, B, C: principal moments of inertia about axes i, j, k, respectively.
$\mathfrak{h}$: angular momentum of the satellite.[†]
ω : angular velocity of the body axes i, j, k relative to the inertial space; ω_1, ω_2, ω_3 are its i, j, k components, respectively.
$\mathfrak{M}$: external torque applied to the satellite; M_1, M_2, M_3 are its i, j, k components, respectively.

The angular velocity ω is written in terms of the i, j, k components as[‡]

$$\left.\begin{aligned} \omega_1 &= -(\Omega+\phi')\sin\theta + \psi' \\ \omega_2 &= (\Omega+\phi')\sin\psi\cos\theta + \theta'\cos\psi \\ \omega_3 &= (\Omega+\phi')\cos\psi\cos\theta - \theta'\sin\psi \end{aligned}\right\} \qquad (1)$$

where Ω is the angular velocity of the radial line from the gravity center of the earth to the satellite. The Euler equation of the rotational

† Here and throughout the paper, gothic letters denote vector quantities.

‡ Primes denote differentiations with respect to time t.

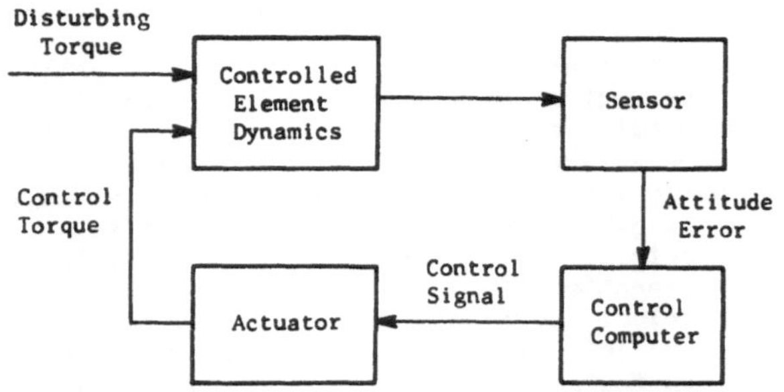

Fig. 1. Schematic diagram of an attitude control system.

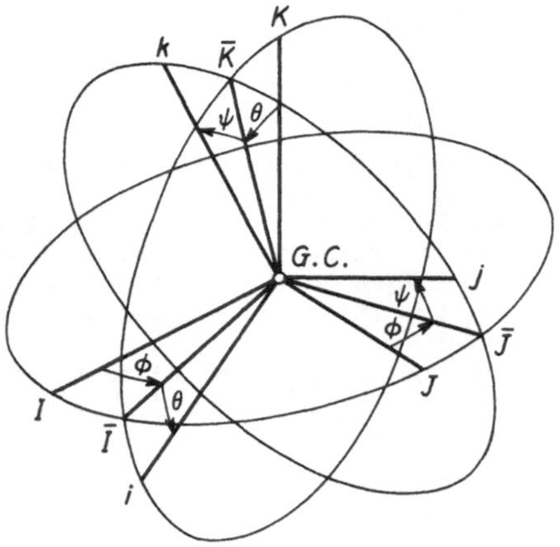

Fig. 2. Reference axes and body-fixed axes.

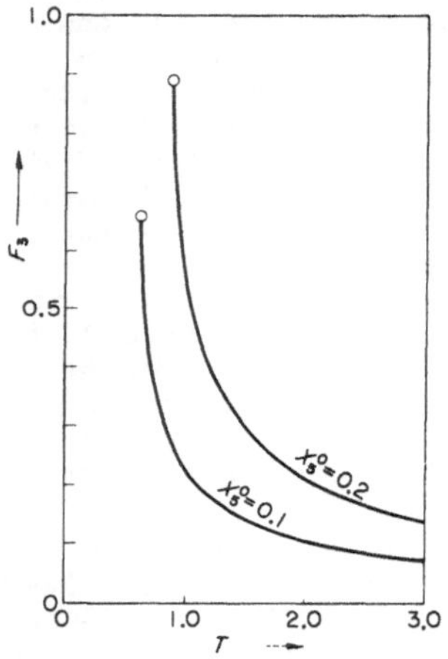

Fig. 3. Influence of the transition time on the fuel consumption ($x_6^0 = 0$).

motion is given by[3]

$$\mathbb{M} = \mathbb{h}' + \omega \times \mathbb{h} \tag{2}$$

where the prime denotes time derivative in the rotating body-fixed frame. In component forms Eq. (2) becomes

$$\left.\begin{aligned} M_1 &= A\omega_1' - (B - C)\,\omega_2\omega_3 \\ M_2 &= B\omega_2' - (C - A)\,\omega_3\omega_1 \\ M_3 &= C\omega_3' - (A - B)\,\omega_1\omega_2 \end{aligned}\right\} \tag{3}$$

Substitution of Eqs. (1) into Eqs. (3) yields a set of the second-order differential equations where angles ϕ, θ, and ψ are dependent variables. Since Eqs. (3) contain nonlinear terms, it is difficult to obtain the exact solution. However, if the deviation angles and their time rates are sufficiently small, we may discard the nonlinear terms. Then, retaining terms linear in ϕ, θ, ψ and letting $A = B = C$ yields

$$\left.\begin{aligned} \psi'' - \Omega\theta' - \Omega'\theta &= M_1/A \\ \theta'' + \Omega\psi' + \Omega'\psi &= M_2/A \\ \phi'' + \Omega' \qquad &= M_3/A \end{aligned}\right\} \tag{4}$$

The first and second equations are mutually related; while the third equation is independent of the first two. In other words, the roll and yaw motions described by ψ and θ are coupled each other; while the pitch motion described by ϕ is decoupled from the other motions.

When an orbit is circular, the angular velocity Ω of the orbital motion is constant. The following two sections are concerned with an investigation of the optimal control of the linear system described by Eqs. (4) with Ω constant. Control of the original nonlinear system (3) is briefly mentioned by showing a particular numerical example. A problem related to an elliptic orbit is studied in the last section.

FUEL MINIMIZATION IN A REACTION GAS-JET SYSTEM

Most of modern gas-jet systems are characterized by the on-off type behavior. In this case the actuator torque is considered as a manipulated variable, or a control. For convenience, we introduce dimensionless quantities defined by

$$\left.\begin{aligned} &x_1 = \psi \quad x_2 = \dot\psi \quad x_3 = \theta \quad x_4 = \dot\theta \\ &x_5 = \phi \quad x_6 = \dot\phi \quad c = t_m\Omega \\ &u_i = M_i/M_0 \quad \tau = t/t_m \quad t_m = \sqrt{A/M_0} \end{aligned}\right\} \tag{5}$$

where M_0 stands for the maximum value of the torque. Here and throughout the paper a dot over a quantity denotes differentiation with respect to the dimensionless time τ.

By use of the quantities in Eqs. (5), Eqs. (4) are rewritten as

$$\left.\begin{aligned} \dot x_1 &= x_2 \qquad \dot x_2 = cx_4 + u_1 \\ \dot x_3 &= x_4 \qquad \dot x_4 = -cx_2 + u_2 \end{aligned}\right\} \tag{6}$$

and

$$\dot x_5 = x_6 \qquad \dot x_6 = u_3 \tag{7}$$

In the minimum-fuel control, the functional to be minimized may be taken as

$$F = \int_0^T (|u_1| + |u_2| + |u_3|)\,d\tau \tag{8}$$

where T is a prescribed transition time. The time T must be larger than or equal to the possible minimum transition time which depends upon an initial angular state.

The optimal controls for the system described by Eqs. (6) and (7) with the consideration of the cost function (8) can be obtained by using the maximum principle. The Hamiltonian function of the problem is given by[1]

$$\begin{aligned} H = {} & p_1x_2 + cp_2x_4 + p_3x_4 - cp_4x_2 + p_5x_6 \\ & + (p_2u_1 - |u_1|) + (p_4u_2 - |u_2|) \\ & + (p_6u_3 - |u_3|) \end{aligned} \tag{9}$$

where the auxiliary variables p's satisfy the set of differential equations

$$\left.\begin{aligned} \dot p_1 &= 0 \qquad \dot p_2 = -p_1 + cp_4 \\ \dot p_3 &= 0 \qquad \dot p_4 = -p_3 - cp_2 \end{aligned}\right\} \tag{10}$$

and

$$\dot p_5 = 0 \qquad \dot p_6 = -p_5 \tag{11}$$

The maximum principle demands that, at every instant of time τ, the H-function must be maximum with respect to all admissible controls u_i ($i = 1, 2, 3$).[1] This condition is satisfied by choosing the optimal controls $\bar u_i$ as

$$\left.\begin{aligned} \bar u_1(\tau) &= \begin{cases} 0 & \text{for } |p_2(\tau)| \leq 1 \\ \operatorname{sgn} p_2(\tau) & \text{for } |p_2(\tau)| > 1 \end{cases} \\ \bar u_2(\tau) &= \begin{cases} 0 & \text{for } |p_4(\tau)| \leq 1 \\ \operatorname{sgn} p_4(\tau) & \text{for } |p_4(\tau)| > 1 \end{cases} \end{aligned}\right\} \tag{12}$$

$$\bar u_3(\tau) = \begin{cases} 0 & \text{for } |p_6(\tau)| \leq 1 \\ \operatorname{sgn} p_6(\tau) & \text{for } |p_6(\tau)| > 1 \end{cases} \tag{13}$$

Therefore the optimal controls must be of a on-off type, assuming only the three values, ±1 or 0.

Solving Eqs. (10) and (11) yields

$$\left.\begin{aligned} p_2(\tau) &= R\sin(c\tau + \theta) + P_2 \\ p_4(\tau) &= R\cos(c\tau + \theta) + P_4 \end{aligned}\right\} \tag{14}$$

$$p_6(\tau) = -P_5\tau + P_6 \qquad (15)$$

where R, Θ, P_2, P_4, P_5, and P_6 are constants to be determined by initial conditions.

Control of the Pitch Motion (System of Two Variables)

In this section we discuss the control of the pitch motion described by Eqs. (7).[4] As we see in Eq. (15), p_6 is linear in time τ. Hence, by virtue of Eqs. (13), the value of $\bar{u}_3$ changes twice at most. Possible modes of switching are listed in Table 1. The time intervals τ_j (j = 1, 2, 3)

Table 1. Mode of switching of $\bar{u}_3(\tau)$ for the minimum-fuel control

Mode	Time interval τ_1	τ_2	τ_3
I	-1	0	1
II	1	0	-1

of Mode I are related to initial values x_5^0 and x_6^0 by

$$\left.\begin{aligned} x_5^0 &= -(\tau_2^2/2 + \tau_2\tau_3 + \tau_3^2) \\ &\quad + (\tau_2 + 2\tau_3)T - T^2/2 \\ x_6^0 &= \tau_1 - \tau_3 \\ \text{and} \quad T &= \tau_1 + \tau_2 + \tau_3 \end{aligned}\right\} \qquad (16)$$

The signs of x_5^0 and x_6^0 should be reversed for Mode II. If the initial velocity is zero, Eqs. (16) give

$$\left.\begin{aligned} \tau_1 &= \tau_3 = [T - (T^2 - 4|x_5^0|)^{1/2}]/2 \\ \tau_2 &= (T^2 - 4|x_5^0|)^{1/2} \end{aligned}\right\} \qquad (17)$$

The amount of consumed fuel may be calculated by using Eqs. (8) and (16). In the case of zero initial velocity the consumed fuel F_3 is given by

$$\begin{aligned} F_3 = \int_0^T |\bar{u}_3| d\tau &= \tau_1 + \tau_3 \\ &= T - (T^2 - 4|x_5^0|)^{1/2} \end{aligned} \qquad (18)$$

The dependence of F_3 on the transition time T is illustrated in Fig. 3.

Control of the Roll and Yaw Motion (System of Four Variables)

We are concerned with the control of the roll and yaw motion described by Eqs. (6). As wee see in Eqs. (14), a representative point on the p_2p_4 plane moves along a circle at the uniform rate c. By virtue of Eqs. (12), the value of $\bar{u}_1$ must be switched at the instant in which the representative point traverses the lines $p_2 = \pm 1$; while $\bar{u}_2$ must be switched when the point passes through the lines $p_4 = \pm 1$. Considering that the normalized angular velocity of the orbital motion c is small and that the transition time T is also not so large because of a slight angular deviation, we may conclude that switching of the values of $\bar{u}_1$ and $\bar{u}_2$ occurs six times at most. Further we may conclude that four switchings are sufficient for most of the initial states; Table 2 shows the possible modes of switching in this case.

Table 2. Mode of switching of $\bar{u}_1(\tau)$ and $\bar{u}_2(\tau)$ for the minimum-fuel control

Mode	Time interval τ_1	τ_2	τ_3	τ_4	τ_5
I.1	-1,-1	-1, 0	0, 0	1, 0	1, 1
I.2	-1,-1	0,-1	0, 0	1, 0	1, 1
I.3	-1,-1	0,-1	0, 0	0, 1	1, 1
I.4	1,-1	0,-1	0, 0	0, 1	1, 1
II.1	1,-1	0,-1	0, 0	0, 1	-1, 1
II.2	1,-1	1, 0	0, 0	0, 1	-1, 1
II.3	1,-1	1, 0	0, 0	-1, 0	-1, 1
II.4	1, 1	1, 0	0, 0	-1, 0	-1, 1
III.1	1, 1	1, 0	0, 0	-1, 0	-1,-1
III.2	1, 1	0, 1	0, 0	-1, 0	-1,-1
III.3	1, 1	0, 1	0, 0	0,-1	-1,-1
III.4	-1, 1	0, 1	0, 0	0,-1	-1,-1
IV.1	-1, 1	0, 1	0, 0	0,-1	1,-1
IV.2	-1, 1	-1, 0	0, 0	0,-1	1,-1
IV.3	-1, 1	-1, 0	0, 0	1, 0	1,-1
IV.4	-1,-1	-1, 0	0, 0	1, 0	1,-1

Equations (6) may be written in a vector form as

$$\dot{\mathfrak{x}} = A\mathfrak{x} + \mathfrak{u} \qquad (19)$$

The time intervals τ_j (j = 1 to 5) in Table 2 are related to an initial value $\mathfrak{x}^0$ by

$$\left.\begin{aligned} \mathfrak{x}^0 = &\int_0^{-\tau_5} X(-T-s)\mathfrak{u}_5 ds + \int_{-\tau_5}^{-\tau_5-\tau_4} X(-T-s)\mathfrak{u}_4 ds \\ &+ \int_{-\tau_5-\tau_4}^{-\tau_5-\tau_4-\tau_3} X(-T-s)\mathfrak{u}_3 ds + \int_{-\tau_5-\tau_4-\tau_3}^{-T+\tau_1} X(-T-s)\mathfrak{u}_2 ds \\ &+ \int_{-T+\tau_1}^{-T} X(-T-s)\mathfrak{u}_1 ds \end{aligned}\right\} \qquad (20)$$

and

$$T = \sum_{j=1}^{5} \tau_j$$

where $X(\tau)$ is the fundamental matrix of the linear part of Eq. (19).[5]

Numerical Example

Let us consider a case where c = 0.1 and T = 1.* The object of control is to make all the state variables x_1 to x_4 zero at $\tau = T = 1$ in the minimum-fuel process. Figure 4 shows the relationship between initial values x_1^0, x_3^0 and the minimum fuel consumption F in the case where $x_2^0 = 0$, $x_4^0 = 0$. From the result of investigation of the time-optimal control,[6] it is known that any initial state located in the region inside the dotted lines can be restored to the origin in an interval of time less than or equal to unity. By making use of Eqs. (20), one may conclude that the square area bordered by the lines $|x_1^0| = 0.25$ and $|x_3^0| = 0.25$ is the region of initial states which can be restored to the origin by the minimum-fuel controls of Table 2. The controls require generally four switchings. This region is divided into sixteen subregions as marked by I.1, I.2,•••, IV.4. If an initial value is prescribed in one of the subregions, say I.1, the optimal control must be switched as indicated in Table 2 by the same number of mode. The regions I to IV are located $\pi/2$-symmetrically about the origin. The narrow areas bordered by the dotted lines and the lines $|x_1^0| = 0.25$ or $|x_3^0| = 0.25$ are the regions of initial states restored to the origin by five or six switchings of the control.

Figure 5 illustrates the iso-τ_j (j = 1 to 5) curves for Mode I on the $x_1^0 x_3^0$ plane ($x_2^0 = 0$, $x_4^0 = 0$). Rotating these curves by $\pi/2$, π, and $3\pi/2$ radians in the counterclockwise direction yields the iso-τ_j curves for Modes II, III, and IV, respectively.

The optimal controls for the linearized system of Eqs. (6) and (7) are not optimal for the nonlinear system of Eqs. (3). An example of the phase-plane trajectories of x_i (i = 1 to 6) is illustrated in Fig. 6 where the initial values are $x_1^0 = 0.103$, $x_3^0 = 0.100$, $x_5^0 = 0.100$, and $x_2^0 = x_4^0 = x_6^0 = 0$. The solid and dotted curves are obtained for the original nonlinear system (3) and the linearized systems (6) and (7), respectively.

ENERGY MINIMIZATION IN A REACTION-WHEEL SYSTEM

The use of a reaction wheel is another popular technique to develop the control torque on a satellite. The reaction wheel is accelerated with respect to the satellite body by an electric motor. We suppose that the driving torque is provided by an armature-controlled, permanent-magnet d-c motor. Three wheels are disposed so that their spin axes are directed along the body axes. The basic equation which relates output torque $\mathcal{M}$ of the motor to the applied voltage $\mathcal{V}$ is given by

$$\mathcal{M} = (K_L/R_a)(\mathcal{V} + \gamma/K_m) \tag{21}$$

where γ is the angular velocity of the wheel relative to the body, R_a is the armature resistance, and K_L and K_m are the torque and the speed constant of the motor, respectively. Denoting the angular velocity of the satellite relative to the space by ω yields approximately

$$\gamma = -A\omega/I_w \tag{22}$$

where A and I_w are the moments of inertia of the satellite and the motor, respectively. It is assumed that $A \gg I_w$. Substituting Eq. (22) into Eq. (21) and using Eqs. (1) and (4) yields the differential equations

$$\left.\begin{aligned} \dot{x}_1 &= x_2 \qquad \dot{x}_2 = -kx_2 + kcx_3 + cx_4 + u_1 \\ \dot{x}_3 &= x_4 \qquad \dot{x}_4 = -kcx_1 - cx_2 - kx_4 + u_2 \end{aligned}\right\} \tag{23}$$

$$\dot{x}_5 = x_6 \qquad \dot{x}_6 = -kx_6 + u_3 \tag{24}$$

where

$$\left.\begin{aligned} &x_1 = \psi \quad x_2 = \dot{\psi} \quad x_3 = \theta \quad x_4 = \dot{\theta} \\ &x_5 = \phi \quad x_6 = \dot{\phi} \quad c = t_m\Omega \quad k = t_m K_L/K_m I_w R_a \\ &u_i = V_i/V_0 \quad \tau = t/t_m \quad t_m = (AR_a/K_L V_0)^{1/2} \\ &V_0 : \text{maximum value of } V_i \end{aligned}\right\} \tag{25}$$

In the derivation of Eqs. (23) and (24), we assume that all the initial velocities of the wheels relative to the satellite are zero. Consequently, the following relations are satisfied:

$$\left.\begin{aligned} x_2^0 - cx_3^0 &= 0 \\ x_4^0 + cx_1^0 &= 0 \\ x_6^0 &= 0 \end{aligned}\right\} \tag{26}$$

---

* By way of an example, when

$\Omega = 1.16 \times 10^{-3}$ rad/sec (90 min/cycle)

$A = 10$ kg-m^2 $\quad M_0 = 10^{-3}$ newton-m

the time constant and the normalized angular velocity become

$t_m = 10^2$ sec $\qquad c = 0.116$

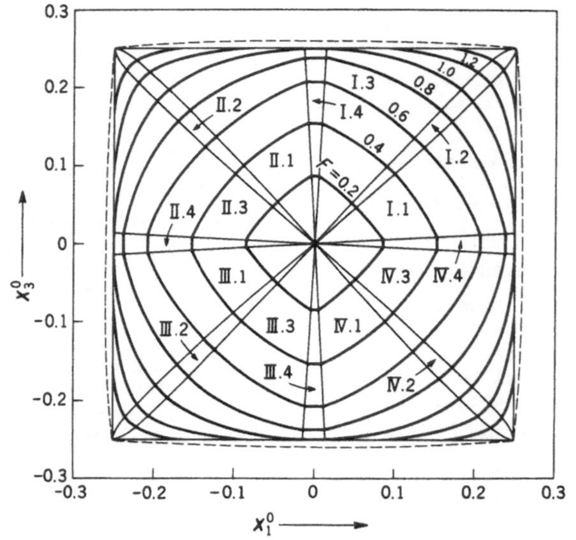

Fig. 4. Iso-F curves on the $x_1^0x_3^0$ plane ($c = 0.1$, $T = 1$, and $x_2^0 = x_4^0 = 0$).

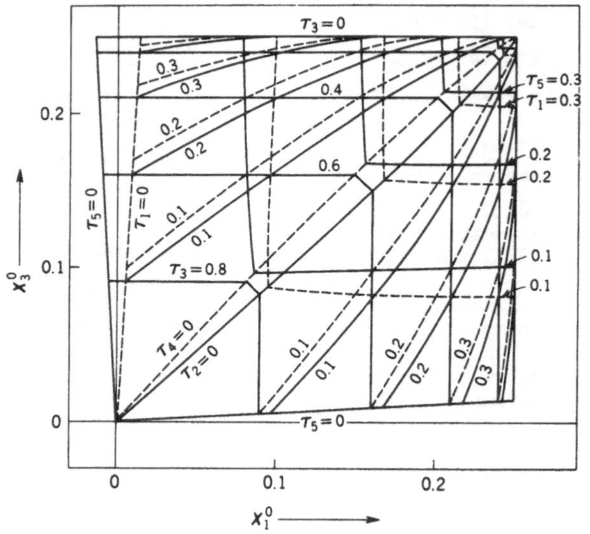

Fig. 5. Iso-τ_j curves for Mode I on the $x_1^0x_3^0$ plane ($c = 0.1$, $T = 1$, and $x_2^0 = x_4^0 = 0$).

The quantity to be minimized in the control is input energy to the motors. The energy is expressed by the dimensionless variables as

$$E = \int_0^T \{u_1[u_1 - k(x_2 - cx_3)] + u_2[u_2 - k(x_4 + cx_1)] + u_3(u_3 - kx_6)\} d\tau \quad (27)$$

where T is a prescribed transition time.

The system of equations for the auxiliary variables p's is given by

$$\left.\begin{aligned} \dot{p}_1 &= kcp_4 - kcu_2 \\ \dot{p}_2 &= -p_1 + kp_2 + cp_4 - ku_1 \\ \dot{p}_3 &= -kcp_2 + kcu_1 \\ \dot{p}_4 &= -cp_2 - p_3 + kp_4 - ku_2 \end{aligned}\right\} \quad (28)$$

$$\left.\begin{aligned} \dot{p}_5 &= 0 \\ \dot{p}_6 &= -p_5 + kp_6 - ku_3 \end{aligned}\right\} \quad (29)$$

Hence, by use of the maximum principle, we obtain the necessary conditions for the optimal controls $\bar{u}_i$:

$$\left.\begin{aligned} \bar{u}_1 &= \begin{cases} q_2 & \text{for} \quad |q_2| \leq 1 \\ \operatorname{sgn} q_2 & \text{for} \quad |q_2| > 1 \end{cases} \\ \bar{u}_2 &= \begin{cases} q_4 & \text{for} \quad |q_4| \leq 1 \\ \operatorname{sgn} q_4 & \text{for} \quad |q_4| > 1 \end{cases} \end{aligned}\right\} \quad (30)$$

$$\bar{u}_3 = \begin{cases} q_6 & \text{for} \quad |q_6| \leq 1 \\ \operatorname{sgn} q_6 & \text{for} \quad |q_6| > 1 \end{cases} \quad (31)$$

where

$$\left.\begin{aligned} q_2 &= [p_2 + k(x_2 - cx_3)]/2 \\ q_4 &= [p_4 + k(x_4 + cx_1)]/2 \\ q_6 &= (p_6 + kx_6)/2 \end{aligned}\right\} \quad (32)$$

The minimum-energy control functions, $\bar{u}_1$, $\bar{u}_2$, and $\bar{u}_3$, vary continuously between and including ± 1. This type of control differs from that of the minimum-fuel control in which the control functions are given by Eqs. (12) and (13).

Control of the Pitch Motion (System of Two Variables)

First, we assume that the optimal control $\bar{u}_3$ does not saturate at any time during the transition. By virtue of Eqs. (31) and (32), the optimal control is given by

$$\bar{u}_3 = (p_6 + kx_6)/2 \quad (33)$$

Substituting Eq. (33) into Eqs. (24) and (29) yields a set of differential equations in x_5, x_6, p_5, and p_6. By solving the equations thus obtained with boundary conditions

$$\left.\begin{aligned} x_5(0) &= x_5^0 & \quad x_6(0) &= 0 \\ x_5(T) &= 0 & \quad x_6(T) &= 0 \end{aligned}\right\} \quad (34)$$

we obtain

$$\left.\begin{aligned} x_5 &= x_5^0[2(\tau/T)^3 - 3(\tau/T)^2 + 1] \\ x_6 &= (6x_5^0/T)[(\tau/T)^2 - \tau/T] \\ p_5 &= -24\, x_5^0/T^3 \\ p_6 &= (6x_5^0/T^2)[k\tau^2/T + (4/kT - 1)k\tau - 2] \end{aligned}\right\} \quad (35)$$

The optimal control does not saturate, i.e., $|\bar{u}_3| < 1$, if the following condition is satisfied:

$$|x_5^0| \leq \begin{cases} \frac{1}{6} T^2 & \text{for} \quad T \leq \frac{2}{k} \\ \frac{2}{3k^2} \frac{kT}{1 + 4/k^2T^2} & \text{for} \quad T > \frac{2}{k} \end{cases} \quad (36)$$

The minimum energy consumption depends on the time T. For the nonsaturating control, the energy E_3 is given by

$$E_3 = 12(x_5^0)^2/T^3 \quad (37)$$

The investigation of the time-optimal control shows that any initial state satisfying the condition

$$|x_5^0| \leq (2/k^2)\cdot\log[\cosh(kT/2)] \equiv (x_5^0)_{max} \quad (38)$$

can be restored to the origin in the interval of time T.[4] Hence we see that, given an initial value satisfying the condition (38) but not satisfying (36), the optimal control saturates in certain intervals of the transition.

Second, we deal with the process in which $\bar{u}_3$ saturates in certain intervals. For convenience, we consider the following two cases separately.

Case 1. $kT \leq 2$

In the time-optimal process, the optimal control is switched from -1 to +1 if $x_5^0 > 0$, and from +1 to -1 if $x_5^0 < 0$. In the nonsaturating minimum-energy process above mentioned, $\bar{u}_3$ increases monotonously for $x_5^0 > 0$ and decreases monotonously for $x_5^0 < 0$. Hence we may infer the modes of switching of $\bar{u}_3$ as listed in Table 3. By substituting the value of $\bar{u}_3$ into Eqs. (24) and (29), we may obtain the optimal trajectories. The time intervals τ_2 and τ_3 are related to an initial value x_5^0 by

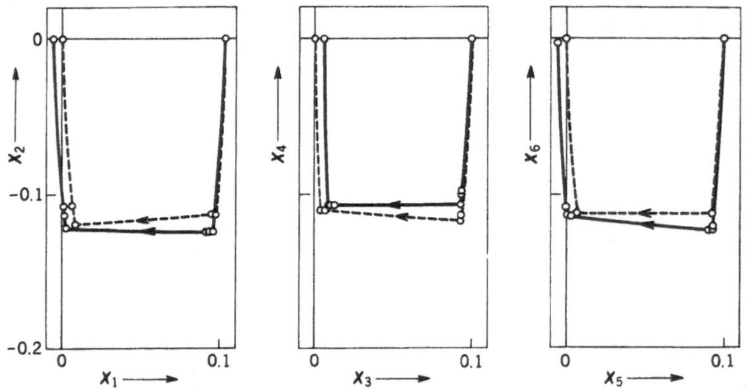

Fig. 6. Trajectories of the state variables $x_1 \sim x_6$ in the minimum-fuel control (c = 0.1 and T = 1).

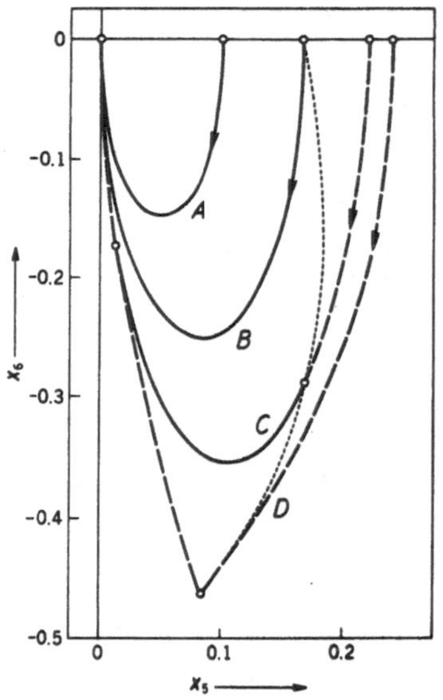

Fig. 7. Trajectories of the state variables x_5 and x_6 in the minimum-energy control (k = 1 and T = 1).

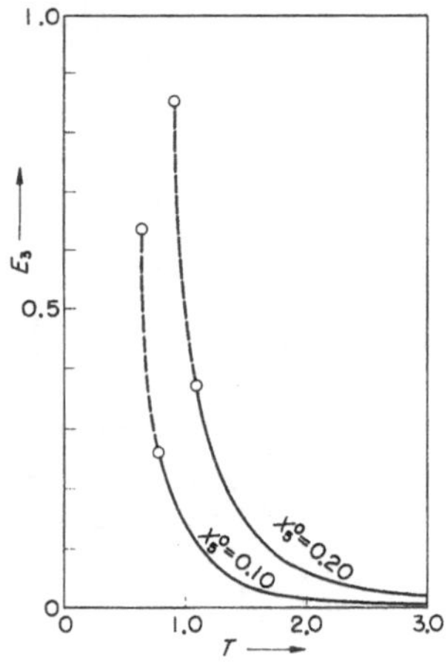

Fig. 8. Influence of the transition time on the energy consumption ($x_6^0 = 0$).

$$|x_5^0| = \frac{1}{6k^2}\{(e^{k\tau_3})[(k\tau_2)^2 + 4k\tau_2 + 6 - 2e^{k(\tau_2 - T)}(k\tau_2 - 3)] + 2k(3T - 4\tau_2 - 6\tau_3) - 12\}$$

and

$$\tau_3 = -\frac{1}{k}\log\{\frac{1}{4}[(k\tau_2 + 2) - (k\tau_2 - 2)e^{k(\tau_2 - T)}]\} \qquad (39)$$

For the control of Mode I.2 or II.2, the minimum energy consumption E_3 is given by

$$E_3 = \frac{e^{k\tau_3}}{12k}[k\tau_2(e^{k(\tau_2 - T)} + 1)^2 e^{k\tau_3} - 12e^{-k(T-\tau_2)} + 12] \qquad (40)$$

where τ_2 and τ_3 are to be determined by Eqs. (39).

Case 2. $kT > 2$

In this case, we encounter a somewhat complicated situation. If x_5^0 takes a value between $(2/3k^2)[kT/(1 + 4/k^2T^2)] \equiv (x_5^0)_1$ and $(1/3k^2)\cdot[e^{(2-kT)} + 3kT - 5] \equiv (x_5^0)_2$, the optimal control $\bar{u}_3$ does not vary monotonously; while, if x_5^0 is between $(x_5^0)_2$ and $(x_5^0)_{max}$, $\bar{u}_3$ varies monotonously like as in Case 1. We may summarize the modes of switching as listed in Table 4.

Table 3. Mode of switching of $\bar{u}_3$ for the partly saturating control ($kT \le 2$)

Mode	Time interval τ_1	τ_2	τ_3	Initial value
I.2	-1	q_6	1	$T^2/6 \le x_5^0 \le (x_5^0)_{max}$
	(monot. increasing)			
II.2	1	q_6	-1	$-T^2/6 \ge x_5^0 \ge -(x_5^0)_{max}$
	(monot. decreasing)			

Table 4. Mode of switching of $\bar{u}_3$ for the partly saturating control ($kT > 2$)

Mode	Time interval τ_1	τ_2	τ_3	Initial value
I.1	q_6	-1	q_6	$(x_5^0)_1 \le x_5^0 \le (x_5^0)_2$
I.2	-1	q_6	1	$(x_5^0)_2 \le x_5^0 \le (x_5^0)_{max}$
	(monot. increasing)			
II.1	q_6	1	q_6	$-(x_5^0)_1 \ge x_5^0 \ge -(x_5^0)_2$
II.2	1	q_6	-1	$-(x_5^0)_2 \ge x_5^0 \ge -(x_5^0)_{max}$
	(monot. decreasing)			

The time intervals τ_2 and τ_3 in Mode I.1 or II.1 are related to an initial value x_5^0 by

$$|x_5^0| = \frac{e^{k\tau_2}[(k\tau_3)^3 - 4] + (k\tau_3)^3 - 12k\tau_3 + 20}{3k^2[(k\tau_3)^2(e^{k\tau_2}) + (k\tau_3 - 2)^2]} + \frac{(T - \tau_3)}{k}$$

and

$$\tau_3 = (1/2k)[k(T - \tau_2) + 2] \qquad (41)$$

The energy consumption E_3 is given by

$$E_3 = \frac{4}{3k[(k\tau_3)^2(e^{k\tau_2}) + (k\tau_3 - 2)^2]^2}\cdot \{6(k\tau_3 - 1)[(k\tau_3 - 3)(e^{2k\tau_2} - 1) + 2(e^{k\tau_2} - 1)^2 - 4] + (e^{k\tau_2} + 1)^2[(k\tau_3)^3 + (k\tau_3 - 2)^3]\} \qquad (42)$$

where τ_2 and τ_3 are determined by Eqs. (41).

The time intervals in Mode I.2 or II.2 are given by Eqs. (39) in Case 1, and E_3 by Eq. (40).

Numerical Example

Case 1. $k = 1$ and $T = 1$ ($kT = 1$)

Figure 7 shows a number of optimal trajectories which start with various initial values x_5^0. The control $\bar{u}_3$ does not saturate on the trajectories shown in solid lines. It saturates on the broken-line trajectories. $\bar{u}_3$ does not saturate in the case of curve A where $x_5^0 = 0.100$. Curve B corresponds to a special case in which $x_5^0 = T^2/6 = 0.167$. Curve C corresponds to a partly saturating control of Mode I.2 in which $x_5^0 = 0.221$ and $\tau_2 = 0.5$. Curve D coincides with the time-optimal trajectory, i.e., $x_5^0 = (x_5^0)_{max} = 0.241$. Since $kT = 1 < 2$, the optimal control of Mode I.1 does not appear. A switching curve is shown dotted in the figure.

The influence of T on the energy E_3 is illustrated in Fig. 8, where the solid lines correspond to nonsaturating control and the broken lines to partly saturating control.

Case 2. $k = 1$ and $T = 3$ ($kT = 3$)

See the optimal trajectories in Fig. 9. Curve A corresponds to a nonsaturating control where $x_5^0 = 1$. Curves B and C correspond to special cases where initial values are $x_5^0 =$

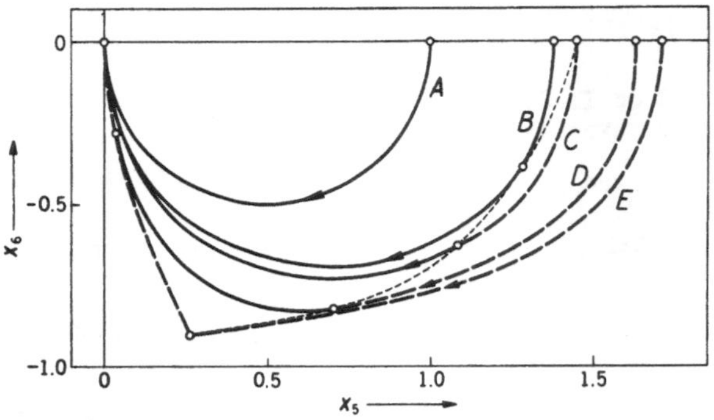

Fig. 9. Trajectories of the state variables x_5 and x_6 in the minimum-energy control (k = 1 and T = 3).

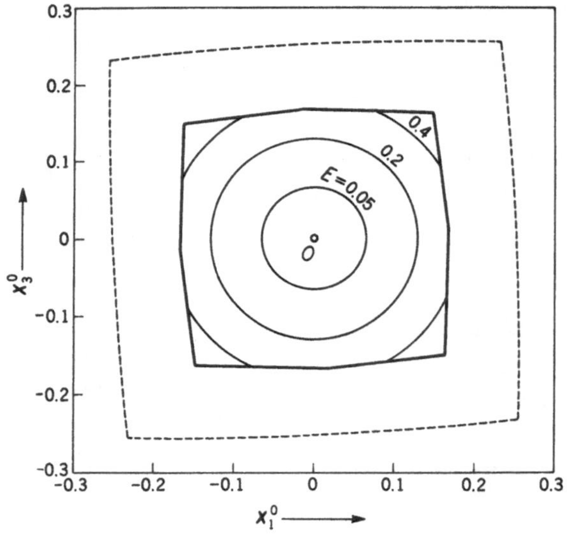

Fig. 10. Region of initial values for the nonsaturating control and iso-E curves on the $x_1^0x_3^0$ plane (c = 0.1, k = 1, and T = 1).

$(x_5^0)_1 = 1.385$ and $x_5^0 = (x_5^0)_2 = 1.456$, respectively. For an initial state existing between those of trajectories B and C, $\bar{u}_3$ follows the switching mode of I.1. Curve D corresponds to a control of Mode I.2 in which $x_5^0 = 1.634$ and $\tau_2 = 1$. Curve E coincides with the time-optimal trajectory where $x_5^0 = (x_5^0)_{max} = 1.711$. The dotted line in Fig. 9 shows the switching curve in this case.

Control of the Roll and Yaw Motion (System of Four Variables)

For the sake of simplicity, we confine our consideration to the nonsaturating control. Also for convenience, the parameter k is set equal to unity.* As shown in Eqs. (30) and (32), the optimal controls are given by

$$\left.\begin{aligned} \bar{u}_1 &= (p_2 + x_2 - cx_3)/2 \\ \bar{u}_2 &= (p_4 + x_4 + cx_1)/2 \end{aligned}\right\} \qquad (43)$$

Substitution of Eqs. (43) into (23) and (28) yields a set of differential equations in x's and p's. By solving the equations with boundary conditions

$$\left.\begin{aligned} & x_2^0 - cx_3^0 = x_4^0 + cx_1^0 = 0 \\ \text{and}\quad & x_1(T) = x_2(T) = x_3(T) = x_4(T) = 0 \end{aligned}\right\} \qquad (44)$$

we obtain the solution

$$\left.\begin{aligned}
x_1 &= -c\beta(\tau-T)^4 - 2(\alpha + cT\beta)(\tau-T)^3 \\
&\quad - T(6\alpha + cT\beta)(\tau-T)^2/2 \\
x_2 &= -4c\beta(\tau-T)^3 - 6(\alpha + cT\beta)(\tau-T)^2 \\
&\quad - T(6\alpha + cT\beta)(\tau-T) \\
x_3 &= c\alpha(\tau-T)^4 - 2(\beta - cT\alpha)(\tau-T)^3 \\
&\quad - T(6\beta - cT\alpha)(\tau-T)^2/2 \\
x_4 &= 4c\alpha(\tau-T)^3 - 6(\beta - cT\alpha)(\tau-T)^2 \\
&\quad - T(6\beta - cT\alpha)(\tau-T) \\
p_1 &= -6[c\beta(\tau-T)^2 + cT\beta(\tau-T) - 4\alpha] \\
p_2 &= -2c\beta(\tau-T)^3 - 3[2\alpha + c(T+4)\beta](\tau-T)^2 \\
&\quad - [6(T+4)\alpha + cT(T+12)\beta](\tau-T) \\
&\quad - 2T(6\alpha + cT\beta) \\
p_3 &= 6[c\alpha(\tau-T)^2 + cT\alpha(\tau-T) + 4\beta] \\
p_4 &= 2c\alpha(\tau-T)^3 - 3[2\beta - c(T+4)\alpha](\tau-T)^2 \\
&\quad - [6(T+4)\beta - cT(T+12)\alpha](\tau-T) \\
&\quad - 2T(6\beta - cT\alpha)
\end{aligned}\right\} \qquad (45)$$

* In the linear system (23), this assumption does not lose the generality of the problem.

The terms of higher order in c than the first are neglected in the above equations. In Eqs. (45) α and β are constants determined by the initial conditions other than those as given by Eqs. (44). The condition for nonsaturating control requires that the parameters α and β satisfy the following inequalities during the transition:

$$\left.\begin{aligned}
&\left|2c\beta\tau^3 + 3[2\alpha - c(T-2)\beta]\tau^2 \right. \\
&\quad \left. - [6(T-2)\alpha - cT(T-6)\beta]\tau - T(6\alpha - cT\beta)\right| \le 1 \\
&\left|2c\alpha\tau^3 - 3[2\beta + c(T-2)\alpha]\tau^2 \right. \\
&\quad \left. + [6(T-2)\beta + cT(T-6)\alpha]\tau + T(6\beta + cT\alpha)\right| \le 1
\end{aligned}\right\} \qquad (46)$$

Numerical Example

Taking the parameters c = 0.1 and T = 0.1 in the conditions (46) gives

$$\left.\begin{aligned}
&|60\alpha + \beta| \le 10 \\
&|60\beta + \alpha| \le 10
\end{aligned}\quad \text{for}\quad \alpha\beta > 0 \qquad
\begin{aligned}
&|60\alpha - \beta| \le 10 \\
&|60\beta - \alpha| \le 10
\end{aligned}\quad \text{for}\quad \alpha\beta < 0\right\} \qquad (47)$$

Equations (45) and the inequalities (47) determine the region of initial states which can be restored to the origin by the nonsaturating controls. Since the initial values x_1^0, x_3^0 and x_2^0, x_4^0 are related by Eqs. (44), an initial state $(x_1^0, x_2^0, x_3^0, x_4^0)$ may be given by prescribing x_1^0 and x_3^0. The area shown bordered by the heavy line in Fig. 10 is the region of x_1^0 and x_3^0 which give the initial states restored to the origin by the nonsaturating controls. The minimum energy consumption E may be calculated by using Eqs. (27), (30), and (45). The iso-E curves in Fig. 10 show the relationship between initial values and the energy consumption.

The broken line shows the boundary of the region of initial states which can be restored in an interval of time less than or equal to unity.[6] Hence, if an initial state is given in the area lying between the heavy line and the broken line, the controls may saturate in certain intervals of the transition.

ATTITUDE OSCILLATION OF A SATELLITE IN AN ELLIPTIC ORBIT (EFFECT OF PARAMETRIC EXCITATION)

For a satellite revolving along an elliptic orbit, angular velocity Ω of the earth-centered frame is approximately given by

$$\Omega = \Omega_0[1 + 2\varepsilon\cos\Omega_0(t - t_p)] \qquad (48)$$

where Ω_0 is the mean value of Ω over one revolution along the orbit, and t_p denotes the time at which the satellite passes through the perigee. Assuming that the eccentricity ε is not so large, terms of order higher than the first in ε are discarded in Eq. (48). Periodical change of Ω produces terms with periodic coefficients in the attitude equations (4). Substituting Eq. (48) into (4) and solving for $M_1 = M_2 = M_3 = 0$, we obtain

$$\left.\begin{aligned}
\psi &= r\sin(\tau+\alpha) + \psi_0 + \varepsilon[-2r\sin(\tau+\alpha) \\
&\quad - \theta_0\sin\tau - 2\psi_0\cos\tau - \psi_0\tau\sin\tau \\
&\quad + \theta_0\tau\cos\tau + r\sin(2\tau+\alpha) + r\sin\alpha + 2\psi_0] \\
\theta &= r\cos(\tau+\alpha) + \theta_0 + \varepsilon[-2r\cos(\tau+\alpha) \\
&\quad + \psi_0\sin\tau - 2\theta_0\cos\tau - \theta_0\tau\sin\tau \\
&\quad - \psi_0\tau\cos\tau + r\cos(2\tau+\alpha) + r\cos\alpha + 2\theta_0] \\
\phi &= \phi_1\tau + \phi_0 - 2\varepsilon(\cos\tau - 1)
\end{aligned}\right\} \quad (49)$$

where r, α, ψ_0, θ_0, ϕ_0, and ϕ_1 are constants of integration and $\tau = \Omega_0(t - t_p)$. The oscillatory terms which grow up indefinitely with time t appear in ψ and θ.

The optimal control of a system with time-varying parameters is a rather complicated problem. If one could assume that $\Omega_0 t$ is small, the problem reduces to the control of a system with time-invariant parameters by replacing Ω with $\Omega_0(1 + 2\varepsilon\cos\Omega_0 t_p)$.

REFERENCES

(1) Pontryagin, L. S.: "The Mathematical Theory of Optimal Processes," John Wiley and Sons, Inc., New York, 1962.

(2) Roberson, R. E.: Attitude Control of Satellites and Space Vehicles, Advances in Space Sciences, Vol. 2, pp. 351-436, Academic Press Inc., New York, 1960.

(3) Thomson, W. T.: "Introduction to Space Dynamics," John Wiley and Sons, Inc., New York, 1961, pp. 111-113.

(4) Flügge-Lotz, I., and H. Marbach: The Optimal Control of Some Attitude Control Systems for Different Performance Criteria, Stanford University Technical Report, No. 131 (1962).

(5) Coddington, E. A., and N. Levinson: "Theory of Ordinary Differential Equations," McGraw-Hill Book Company, Inc., New York, 1955, pp. 62-78.

(6) Nishikawa, Y., C. Hayashi, and N. Sannomiya: Optimal Attitude Control of Orbiting Satellites, Proceedings of the Fifth International Sysmposium on Space Technology and Science, AGNE Corp., Tokyo, 1964.

Ground Stations and Tracking

CONTROL AND GUIDANCE RESEARCH AT NASA

By Charles H. Gould
Chief, Control and Stabilization
Headquarters, National Aeronautics and Space Administration
Washington, D.C., U.S.A.

ABSTRACT

A program of advanced control and guidance research is under way within the National Aeronautics and Space Administration to develop the control and guidance technology required for future space and aeronautical missions. The research objectives are reviewed, and by example, the nature of the efforts is shown.

INTRODUCTION

The National Aeronautics and Space Administration is conducting a program of advanced research and technology development in order to pursue future aeronautical and space missions. As a part of this over-all program, control and guidance technology is developed. It is this control and guidance program that will be discussed

Future missions are difficult to define and no attempt is made to list them here. Required control and guidance technology for these missions is more clear, and several salient points are apparent. Many missions are of the sort typified by communications satellites, with only moderate control and guidance (C&G) performance needed, but with a premium placed on long life, reliability, and low system operating cost. Other missions, such as manned planetary exploration, will require high performance, along with very long life. The NASA C&G program looks to both sorts of requirements.

Underlying all of our C&G work is the requirement for great system reliability, in order to accomplish the mission. This mission accomplishment has aspects of performance, life time, installation cost and operating cost that must be optimized in some sense. Sub-optimal systems, simple reliable sensors, ultra reliable microcircuitry, and reproducible actuators are as necessary as highly precise shaft encoders or exotic computer programs.

Emphasis has been placed recently on design methods, both analytical and synthesis, for manned control of flight. Much use is made of modern control theory in this work, with current stress being on non-linear and stochastic behavior using state space methods. The end goal of this work is to develop a systematic approach to allocation of function and control/display design, in order to produce manned space and aeronautical vehicles that perform the assigned mission reliably.

Another characteristic of C&G requirements is that many related developments must be made available simultaneously for a system to be successful. For example, for a manned spacecraft, it is not enough to consider only the attitude control sub-system, but also the man's controls and displays, the over-all computation capability of the spacecraft, the trajectory, the navigation subsystem, the communications and tracking network, and the designers' tools such as theory and simulation must be considered in an integrated program.

With these ideas in mind, some specific tasks within the C&G program will be described. These tasks are conducted by NASA research scientists at nine NASA Centers, by industry, and by Universities.

EXAMPLES

The cryogenic superconductive gyroscope has been operated in a laboratory environment by industry under contract to Marshall Space Flight Center and by Jet Propulsion Laboratory with drift rates (errors) sufficiently low to demonstrate its feasibility and potential long life. Techniques have been established for measuring and reducing the AC power losses of the spherical rotor, which previously had deterred development progress. Successful demonstrations have been made of all-attitude frictionless levitation systems, stable spin-up and rotation of the rotor, and optical readout devices for detecting rotor orientation. Rotor levitation, used in tests by the Jet Propulsion Laboratory, is shown in Figure 1., and is typical for cryogenic gyroscopes. Actual drift data obtained from tests are also shown. Minute differences in rotors fabricated by the same construction methods, annealing treatments, and balancing techniques account for the wide variance in measured drift rates. The low

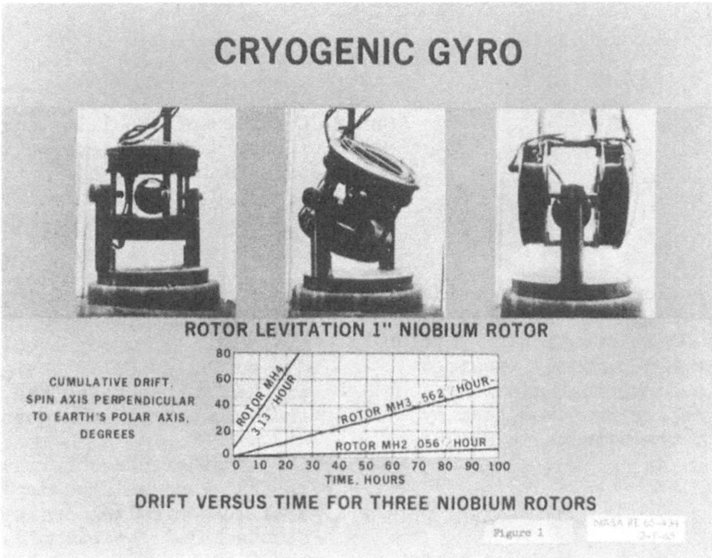

Figure 1

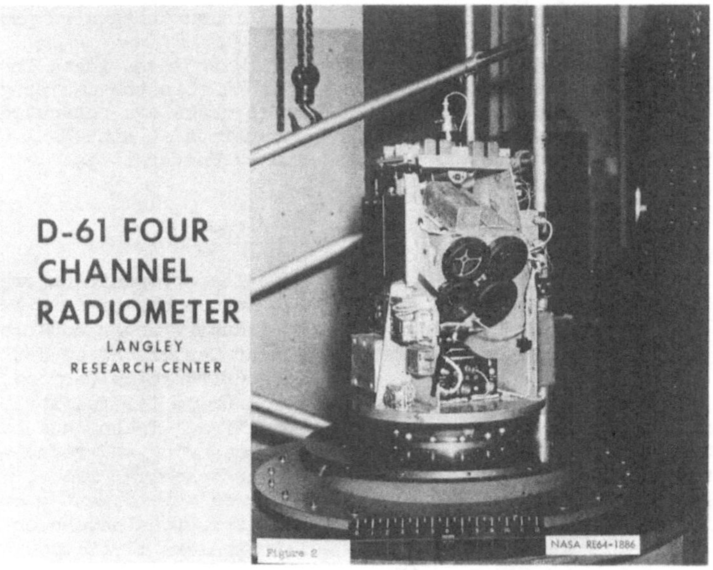

Figure 2

drift rate of .056 degrees/hour for rotor MH2 is significant. Note the linear character of the errors; this provides for accurate prediction and, thus, compensation may be built in. In one test, the data fell on a straight line within ± .0013 degrees/hour. Effort is continuing to improve materials and fabrication techniques and optimize design parameters. The results should provide the basis for a reliable, high-performance gyro for long-term space flight use, especially when cryogenic fuels are carried aboard the spacecraft.

The use of earth or planetary horizons for both guidance and control suffers from a lack of precise knowledge of the exact horizon characteristics[1]. To gather such knowledge, suborbital rocket experiments have been conducted[2], and further experiments are planned in the X-15 aircraft and in the Scanner vehicle. The four-channel radiometer used is shown in Fig. 2. Scanner will investigate the 14-16 and 20-40 micron bands with a vertical resolution of 0.03°. Future plans include long-term orbital experiments to investigate seasonal, topographical and geographical variations of the apparent horizon. Knowledge gained should allow the design of optimized horizon sensors for earth orbital use.

A star tracker currently being developed under contract to Goddard Space Flight Center utilizes advanced concepts of interest for future space applications. The sensor is a Channeltron image dissector, Fig. 3, which gives the same performance as a conventional photomultiplier tube, but is considerably smaller in size. The Channeltron concept substitutes a continuous surface of high resistivity for the multidynode stages of a photomultiplier. It is inherently more rugged and eliminates the voltage dividing network required by the photomultiplier. Also shown is a photograph of the star tracker which is being packaged for possible flight test in an Aerobee rocket. The tracker system uses a unique radial loop scanning pattern that combines high accuracy (15 arc seconds) and a wide field of view (8 degrees) in a single electronic scanning mode. A breadboard tracker has been completed and operated successfully in the laboratory. Development testing will continue.

A laboratory model of an optical radar for spacecraft rendezvous has been developed for Marshall Space Flight Center. The prototype, shown in Fig. 4, uses a combination of two-light sources; a high intensity xenon light and a gallium-arsenide laser. The xenon light acquires a target within a 12-degree field of view at ranges of approximately 17Km and tracks the target to a range of 3 Km. Below 3 Km, the gallium-arsenide diode mode is used. The spatial distribution of the diode light output is also shown. It is to be noted that the diode light output is relatively constant within a cone of 60° solid angle. This characteristic of the transmitter lends itself to simplified and reliable application in a spacecraft in that the sensor may be vehicle mounted without the use of gimbals. It appears possible to eliminate the xenon light by pulsing the gallium-arsenide source at a high rate. The maximum range of the optical radar is limited by the earth background; without earth background the range can be increased considerably. This remains to be investigated.

Computers comprise an integral part of on-board guidance and control systems. Development of magnetic logic computer technology is under way at the Jet Propulsion Laboratory. Magnetic logic computers offer potential advantages of reduced power consumption, increased resistance to temperature and radiation hazards, and improved reliability through the reduction of the number of active components utilized. Apparent limitations are primarily computation cycle time and methods for achieving non-destructive readout of stored information. Construction of a breadboard model was completed in 1964, and laboratory evaluation is currently under way. This computer is a rather slow speed, medium capacity general purpose computer, with emphasis on few active components. During the coming year, portions of the computer will be redesigned to incorporate modifications defined by the evaluation phase and to achieve a model suitable for flight qualification. Future plans include laboratory and flight evaluation of the redesigned model.

In spacecraft stabilized by gravity-gradient forces, very low control torques are encountered, and periods of oscillation of the vehicle are very long[3]. Until recently, components for these vehicles such as bearings, dampers, etc., could not be tested on the ground in a representative environment. The Goddard Space Flight Center has developed a single axis simulator, Fig. 5, which now provides the correct environment. Equipment to be tested is mounted on the table, suspended from the ceiling mount by a metallic ribbon cable having a very low torsional restoring force. To further reduce the restoring force to levels comparable to gravity-gradient torques, twist of the ribbon is detected by an optical device (autocollimotor), and the upper end of the cable is rotated to maintain zero or a small constant maximum twist. This untwisting of the support ribbon causes the periods of oscillation and restoring forces to be comparable

Superior numbers refer to similarly numbered references at the end of this paper.

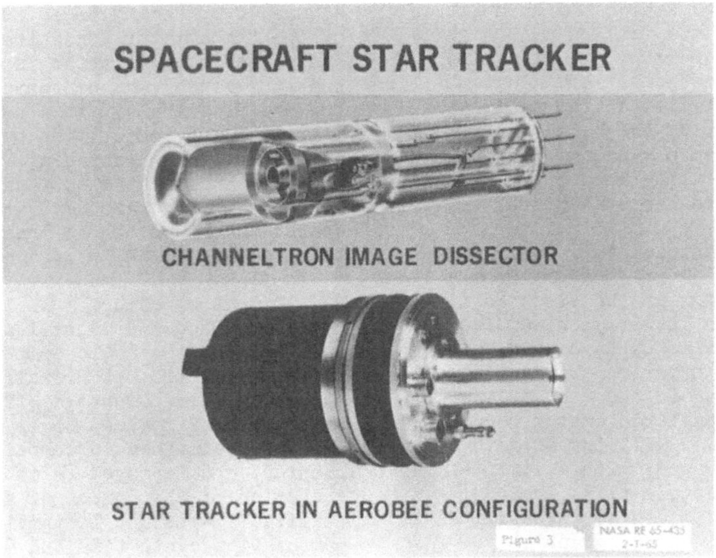

Figure 3

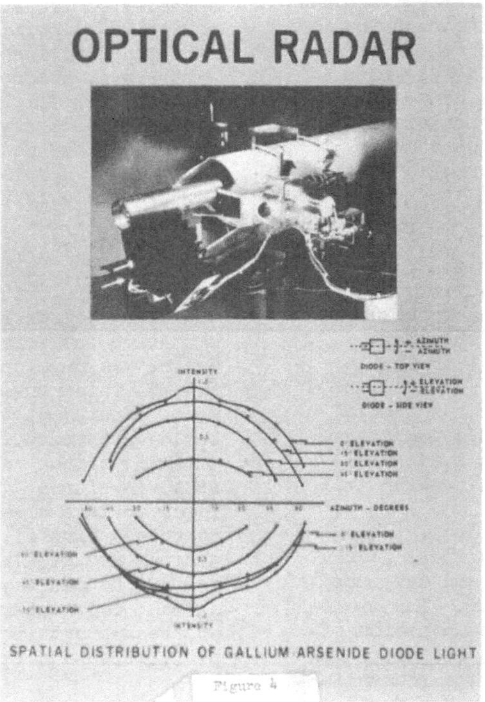

Figure 4

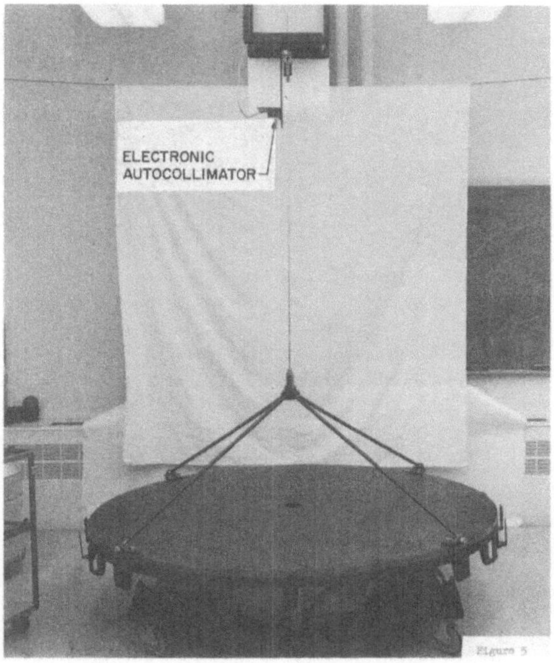

Figure 5

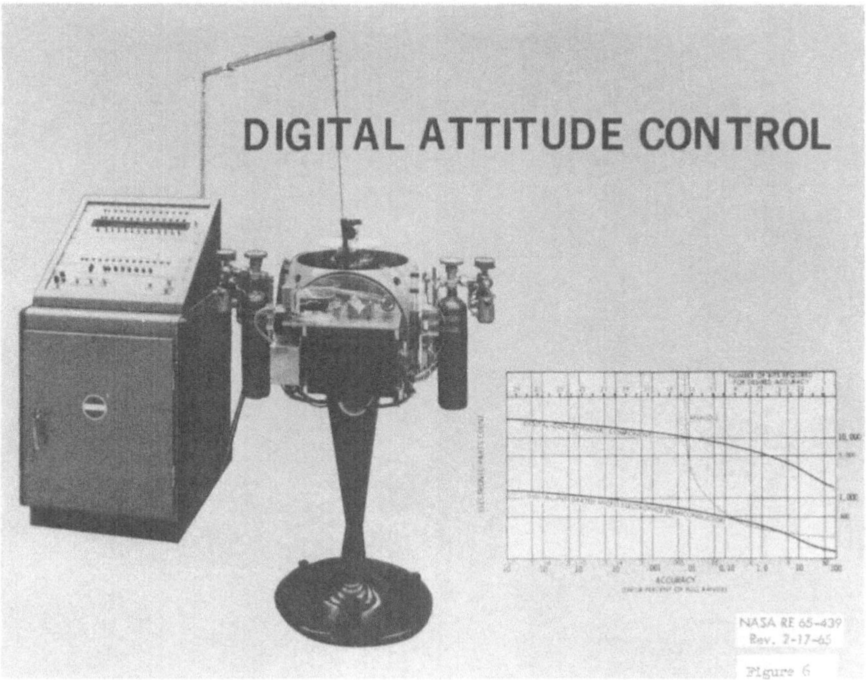

Figure 6

Figure 7

Figure 8

to the levels experienced by a gravity-gradient stabilized vehicle in free space. By varying the inertia of the table through the addition of weights, the oscillatory periods can be controlled to give the two- to four-hour periods representative of the spacecraft period. In the zero torque mode, maximum torque unbalance levels can be similarly controlled to obtain torque levels of 4-8 hundred thousandths of an inch ounce. The simulator performance is several orders of magnitude better (i.e., lower torque unbalance for the zero torque mode, or longer period for the constant maximum torque mode) than has been heretofore possible. This will permit ground testing of gravity gradient satellite bearings and dampers at these torque levels and periods of oscillation experienced in orbit. Dampers and bearings for the Applications Technology Satellite[4] will be tested on this device.

Spacecraft subsystem developments are also required, such as a digital attitude control system with the potential for low drift, high accuracy, and high reliability. The advent of integrated electronic circuits makes this possible. There appears to be an absolute limit to the reliable accuracy of the analog approach as shown in the graph, Fig. 6. As the accuracy requirements approach .01 percent of the operating range, there is a sharp increase in complexity, and therefore a reduction in reliability, an effect not so limiting for lower "parts-count" digital systems. An advanced digital control system study has resulted in the fabrication of a single-axis spacecraft control system which has been successfully demonstrated through airbearing tests. The parts count has been reduced to 5% of that of the original. In addition to the equipment, mathematical models and synthesis methods have been developed. Additional effort is required for three-axis systems development.

Development of future guidance and navigation requirements is dependent on a continuing program of systems and trajectory analysis. Current efforts are: studies of simplified backup manned guidance systems which utilize the astronaut's capability to make navigational fixes and perform guidance corrections; continued attempts to develop closed-form guidance solutions for reducing computer requirements; studies of optimal guidance system configurations for present and future missions;[5] and the development of guidance requirements for lunar surface exploration.

Similarly, control theory and related applied mathematics is necessary for future control systems. The NASA is supporting considerable theoretical work, primarily in Universities, as a part of the nation-wide effort in this field. NASA supported efforts emphasize non-linear analysis and synthesis, optimization, sub-optimal designs, and stochastic and digital control. The theory is applied to problems such as control of large flexible launch vehicles, precise pointing of optical instrumentation, and economical attitude control of spacecraft.

Attitude control of a Manned Orbital Research Laboratory is under study.[6,7] Requirements for stabilization accuracy for a variety of modes of operation have been determined. The problem is complicated by movements of the crew within the MORL. Manual modes of operation for backup, emergency use, and for some primary functions have been studied using the simulator shown in Fig. 7. Preliminary results of these studies indicate that manual control modes can more than accomplish the mission. Very accurate pointing of experiments such as stellar photography require pointing accuracies in excess of those achievable today in the presence of crew movement disturbances. One one-hundred-thousandth of a degree is required here, while achievable accuracies are on the order of one-thousandth of a degree.[8] Future research will concentrate on solving these problems, from both the theoretical and hardware viewpoints. Advances in stochastic and time-varying control theory are needed. Specific components, such as digital momentum exchange devices, are being developed for possible use.

Work at the Langley Research Center has recently contributed directly to the Gemini project. The Rendezvous and Docking Simulation (RDS) has been used for this purpose[9], as well as contributing to the general technology of rendezvous and docking of space vehicles. The RDS is a man-carrying moveable spacecraft capsule arranged so that all of the vehicle motions that would occur during the final stages of rendezvous and docking can be duplicated. Fig. 8 shows the RDS carrying a Gemini capsule in the process of docking with an Agena mockup. Fourteen astronauts have "flown" in the simulator, obtaining training as a by-product of the research work. From the research, came improvements in the Gemini 3-axis hand controller, the Agena visual docking equipment, and improved lights for use during final docking. In addition the evaluation of some actual Gemini hardware was accomplished. This illustrates the close tie that can and must exist between the Research Centers and NASA flight projects.

To obtain data leading to more effective pilots displays, the recent development[10] of a light and accurate eye-movement camera affords a means of obtaining quantitative information that may be used in evaluation of the pilot's display, and ultimately increasing the systems reliability by insuring that the pilot reliably interprets

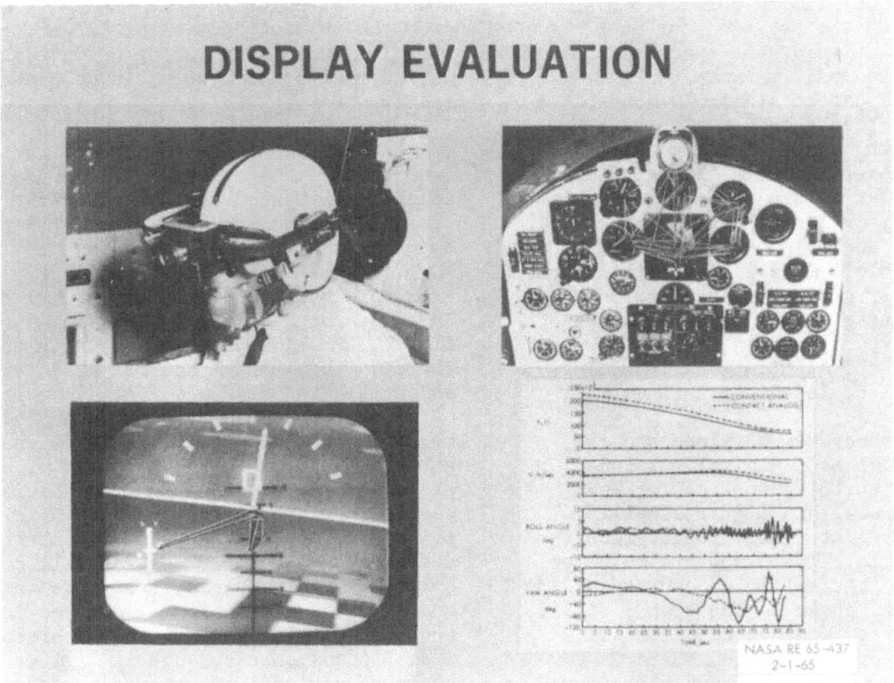

Figure 9

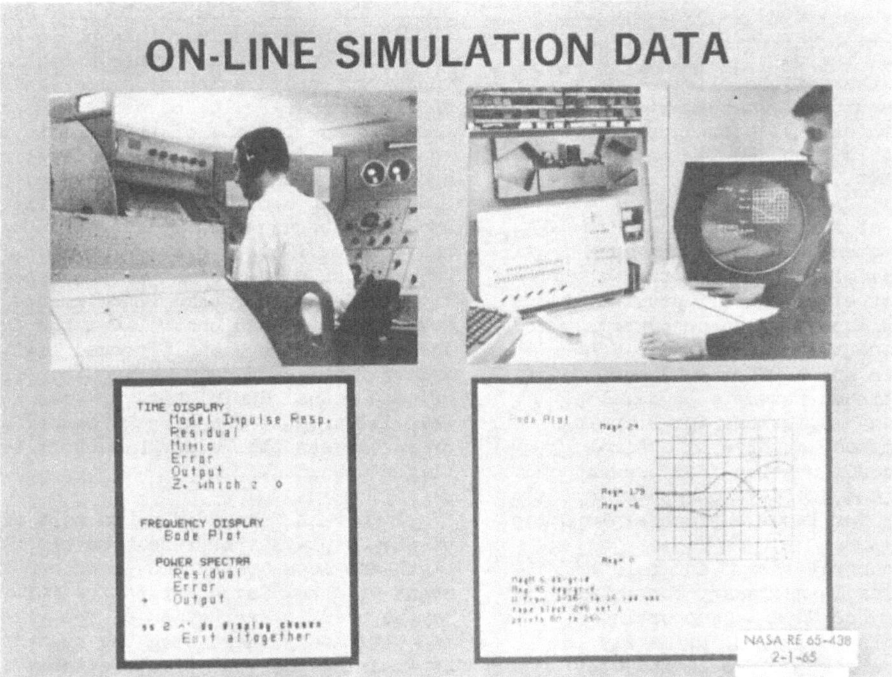

Figure 10

the displayed data. A motion picture camera records both the scene being viewed and the exact fixation point of the eye on that scene, a white spot indicating the pilot's eye position. Fig. 9 shows a camera being worn by a pilot during a simulator flight at the Flight Research Center. On the upper right is shown a conventional X-15 aircraft display panel which provides fixed scales with moving-pointer indicators. The pilot's scan pattern is superimposed, as recorded during the first 40 seconds of a simulation of an X-15 launch. The pilot's task is primarily that of longitudinal control, as evidenced by the high level of concentration on pitch angle and angle of attack. The integrated contact analog display system shown on the left is a computer-driven television system that receives electrical signals from the vehicle's sensors and converts this information into a symbolic picture of the real world moving in six degrees of freedom, corresponding to the real world. The contact analog display lends itself to the superposition of additional information on the screen, such as null indicators or quantitative numerical information. Also depicted is the pilot's scan pattern while using the contact analog system during simulated X-15 launch. A comparison of the scan patterns shows that the pilot is able to obtain the required flight control information from the contact analog display with considerably less scanning. Shown last is a time history taken during a comparative evaluation of these displays for the simulated reentry portion of an X-15 flight. A preliminary evaluation indicates that the pilot's control precision in damping the vehicle's oscillations was improved when he used the contact analog display, verifying the conclusion drawn from the eye-movement camera data.

NASA and NASA contractors effectively employ a wide variety of dynamic flight simulators. Operation of, and data reduction for these remain complicated and expensive. Highly skilled pilots and astronauts are required to operate the simulators in order to obtain valid and relevant experimental results.[11, 12, 13] The efficiency can be increased by providing on-line digital computer data recording and analyzing facilities, which allows the experimenter to concentrate on the critical conditions of the experiment as they occur and by providing him with a more detailed and timely analysis of the data. A recently developed signal analyzer system is illustrated in Fig. 10. On the left is shown a simulator, which houses a pilot "flying" a problem. Also shown is the operating technician. On the right is the experimenter, at the digital computer console, which presents analyzed data on the TV picture to the right of the experimenter. Shown on the lower left is one of the system generated displays, which the experimenter uses to select the results to be viewed and to select any one of as many as ten different methods of analysis. One result that the system computes and displays immediately is shown on the right, a Bode plot of the human pilot's describing function. Illustrated are the pilot's phase lag (Ang) and attenuation (Mag) at various frequencies (W) as the pilot responds to the various components of the signal that he sees during a three-minute run. The graph was obtained in studies of pilot adaptive control characteristics.[14] This technique, soon to be applied on NASA simulators, has saved months in carrying out manual control experimental programs by allowing rapid isolation of the important parameters and conditions of a problem.

CONCLUSIONS

The control and guidance program described by example is varied, highly interdisciplinary, and keyed to future missions of NASA, as well as to a general advance in the state-of-the-art. It represents only a small portion of the C&G efforts in progress within the U.S. aerospace industry. It is intended to contribute to the technology available for future space and aeronautical missions through advanced research.

REFERENCES

(1) R.A. Hanel, W.R. Bandeen, and B.J. Conrath, "The Infrared Horizon of the Planet Earth," NASA Technical Note D-1850, September 1963.

(2) T.B. McKee, Ruth I. Whitman, and C.D. Engle, "Radiometric Observations of the Earth's Horizon from Altitudes between 300 and 600 Kilometers," NASA Technical Note D-2528, December 1964.

(3) Bruce E. Tinling and Vernon K. Merrick, "Exploitation of Inertial Coupling in Passive Gravity-Gradient-Stabilized Satellites," Journal of Spacecraft and Rockets, Vol. 1, No. 4, July-August 1964, p. 381-387.

(4) Donald E. Fink, "GE Gravity Gradient Design in Final Stage," Aviation Week & Space Technology, December 28, 1964, p. 34 & 39.

(5) Gerald L. Smith and Eleanor V. Harper, "Midcourse Guidance Using Radar Tracking and On-Board Observation Data," NASA Technical Note D-2238, April 1964.

(6) Langley Research Center Staff, " A Report on the Research and Technological Problems of Manned Rotating Spacecraft," NASA Technical Note D-1504, August 1962.

(7) Peter R. Kurzhals and Claude R. Keckler, "Spin Dynamics of Manned Space Stations," NASA Technical Report R-155, December 1963.

(8) Jerry R. Havill and Jack W. Ratcliff, "A Twin-Gyro Attitude Control System for Space Vehicles," NASA Technical Note D-2419, August 1964.

(9) J.E. Pennington, H.G. Hatch, Jr., E.R. Long, and J.B. Cobb, "Visual Aspects of a Full-Size Pilot-Controlled Simulation of the Gemini-Agena Docking," NASA Technical Note D-2632, February 1965.

(10) Roger L. Winblade, "Current Research on Advanced Cockpit Display Systems," presented at the 25th Flight Mechanics Panel Meeting of the Advisory Group for Aeronautical Research and Development, Munich, Germany, October 12-14, 1964.

(11) Duane T. McRuer and Ezra S. Krendel, "Dynamic Response of Human Operators," WADC Technical Report 56-524, ASTIA Doc. NR AD-110693, October 1957.

(12) Steven E. Belsley, "Man-Machine System Simulation for Flight Vehicles," IEEE Transactions on Human Factors in Electronics, Vol. HFE-4, No. 1, September 1963, p. 4-14.

(13) David A. Brown, "Langley Simulators Perform Lunar Tasks," Aviation Week & Space Technology, April 13, 1964, p. 54-69.

(14) L.R. Young, D.M. Green, J.I. Elkind, and J.A. Kelly, "The Adaptive Dynamic Response Characteristics of the Human Operator in Simple Manual Control," NASA Technical Note D-2255, April 1964.

DISCUSSION

C. H. Gould

"Control and
Guidance Research at NASA".

Q. What is a breadboard?

A. It is colloquial, referring to an early model of equipment mounted on wood or other material for convenience in design.

Q. What magnetic devices are used in the computer?

A. Ferrite cores.

Q. In what stage is the research in cryogenic gyros?

A. Problems which remain to be solved include A-C losses in materials and means for operating at the temperature of liquid Helium.

Q. How will horizon tracker accuracy improvements be achieved?

A. Lucky breaks, plus the use of best wavelength and knowledge of seasonal variations.

Q. Is the PERT system used by NASA for projects?

A. Yes, on projects.

SPECIAL COMPUTER FOR SIMPLIFIED HIGH PERFORMANCE GROUND ANTENNA STEERING

by Yngvar Lundh
Norwegian Defence
Research Establishment
Kjeller, Norway

ABSTRACT

Description of main principles of antenna steering system utilizing a simple special-purpose digital computer and a precision analog power servo drive. Input to system is 8-channel paper tape with predicted satellite positions in special format; typically 3 feet of tape required for one hour steering. Over-all pointing error $\pm$ 2 minutes of arc excluding prediction errors.

INTRODUCTION

For a ground station which is required to establish radio contact with a satellite via a large ground antenna of high directivity, it is necessary to have some form of automatic program steering. The steering system described in this paper was designed to be as simple as possible both equipment-wise and from an operator's point of view. The antenna is completely program steered by a program on punched paper tape, in a special format. This program contains essentially, the predicted satellite position as a function of time, and is computed in beforehand on the basis of precise knowledge of the satellite orbit.

The over-all pointing accuracy of the steering system is $\pm$ 2 minutes of arc, not including prediction errors. The system referred to was built by the Norwegian Defence Research Establishment, and applied to an 86 foot parabolic radio telescope at Råø in Sweden. The system has been in operation since July 1964, and is operated by the Scandinavian Committee for Satellite Communication.

GENERAL DESCRIPTION OF STEERING SYSTEM

The main parts of the steering system are outlined in Figure 1. A punched paper tape, containing all necessary pointing data and referring to absolute time (UT2), is read by a photoelectric tape reader. This reader is controlled by the special-purpose digital computer, which again refers itself to absolute time by means of a built-in precision clock. The information on tape is given as parameters for second-order curves. These are absolute pointing angles for the two axes of the telescope and the time for which these angles apply; further, the angular velocity of the two axes at that time, and the rate of change of these velocities (accelerations).

One such set of data, giving what we shall refer to as position, velocity, acceleration for the two axes, and time constitute one block of information sufficient for steering the antenna during one interval.

The length of an interval may be variable, and will depend on the actual orbit of the satellite. The actual trajectory as seen from the ground station, has to be approximated piecewise, with a given accuracy by second order curves. Typical interval length will be 2-3 minutes. In the system at Råø, for example, it is occasionally necessary to go a little under one minute for the communication satellites Telstar and Relay. The minimum interval allowed by the system is 1 second. Maximum can be anything up to 24 hours, but it is usually practical to restrict the length of an interval to a few minutes, although the second-order curve fitting might permit longer intervals. The main reason is that of safety. In the case of some mistake or malfunction, it is an advantage to have safe data for a new interval near at hand.

In general, high orbits and trajectories distant from the singularities of the antenna mounting (such as zenith for azimuth-elevation type mounting), tend to permit long intervals.

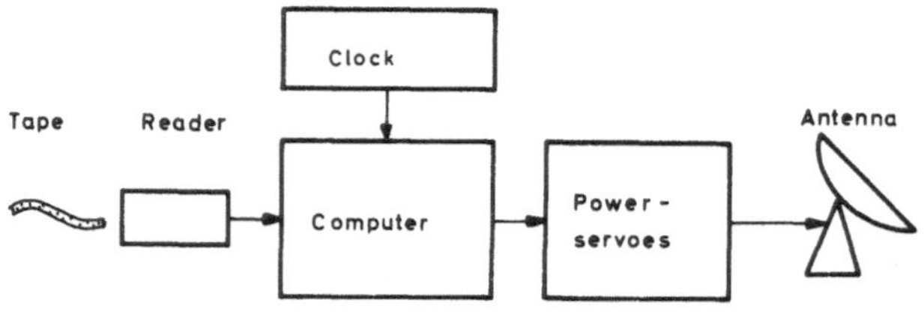

Figure 1 Main blocks of steering system

A block of information is read in from tape immediately before each interval, and data are placed in the appropriate registers in the interpolator part of the computer at the correct instant. The interpolator continuously generates correct angular positions based on the given second order curve parameters.

Outputs from the computer are given in the form of actual shaft angles, or rather, synchro transmitter signals for the two coordinate angles. For each shaft there are two synchros, one coarse and one fine in the speed ratio 1:9.

The power servo has the main task to reproduce the angular positions on the actual antenna shafts. The difficulty of this task will change to some extent, depending on the antenna construction. In the case at Råø, it was rather difficult, since the antenna did not have a radome, and was not constructed for servo control from the outset. The rigidity of the structure and gear drives therefore left quite a lot to be desired. It turned out, however, that it was possible by careful design using "divided reset", to keep the total servo errors within one minute of arc. Power of the output servo motors is 10 kW each.

At the input of the power servo, there is a set of synchro differentials. These may be operated at the station control desk by handwheels, one for each antenna axis, and may be used to manually add corrections to the predicted path. It is also possible in a similar manner to manually add corrections to the time reference given by the clock.

Typical operation of steering is as follows: A few seconds or minutes before the tracking is to begin (allowing sufficient time for the antenna to move to the starting point), the operator places the steering tape in the tape reader, and pushes the "start" button. From then on, the steering goes automatically. Next satellite pass may, if desired, follow on the same tape. No cross reference is necessary in this case (except that no stop code is to be given between passes).

Some redundancy is included in the steering tape to allow a simple parity check on the data. If a reading error is detected, an alarm is given, and entry of the erroneous data into the interpolator is blocked. The interpolation carries on uninterrupted, however, extrapolating on the same second order curve. Meanwhile, facilities are provided for the operator at the control desk to step the tape ahead and check at indicator lights that correct data have again been entered. He can then reset the alarm condition. Thus, a single reading error will usually not interrupt tracking. At critical parts of the trajectory, it could lead to a slight pointing error, which, however, is immediately corrected automatically as soon as correct data again can be applied (next interval).

Preparation of the steering tape is made in beforehand at a general-purpose computer. Based on knowledge of the orbit, or on predicted satellite positions, the computer program fits second order curves to the actual trajectory, selecting appropriate interval lengths at each point. It converts to the coordinate system of the antenna, and punches out the steering tape.

In each block of information on the steering tape, 24 extra bits are included for automatic steering of other equipment at the station. Standard 8-channel paper tape is used. Each block including checking, occupies 24 eight-bit characters.

STEERING COMPUTER

Main computer functions

Main parts of the steering computer, which we call the director are: Updating, Interpolator, Instrument servos and Clock.

The Clock is a set of counters, counting down from a stable quartz oscillator at 100 kHz. It produces a series of various pulse frequencies, which have well defined phase relationships, since they are all derived from the same oscillator. These are used for timing of all operations in the computer. The last counters in the chain are arranged to count seconds, minutes and hours. These therefore present codes for the actual time, which are used by the updating to decide when a new interval is to be started. The time is also displayed at the control desk. To introduce negligible error, the time reference must be accurate within ± 10 ms absolute at all times. This is achieved by a stabilized quartz oscillator. The clock requires correction every second week against radio time standard.

One half second before beginning of an interval, a new block of information is read into the updating unit. At the exact time, the new data replace the old in the interpolator - which carries on continuously.

Data for velocity and acceleration are simply set into appropriate registers in the interpolator. Position data are first compared to the actual position by reading a complete position code (16 bits) from a code disk on each output instrument-servo shaft. This actual position is subtracted from the position command coming from tape, and the resulting difference, one 17 bit number for each of the two axes is set into the appropriate register in the interpolator. The time taken to perform the subtraction is less than that required by the servo to travel one step at maximum speed, thus the correct difference will also appear to arrive in the interpolator at the exact time.

The interpolator works in close cooperation with the instrument servos. Basically, the instrument servo performs the actual interpolation in conjunction with the digital registers in the interpolator. At one point, the velocity is commanded by a certain pulse frequency, each pulse demanding the servo to move one step. The servo feeds back pulses to the interpolator for each step it has moved. This process shall be described in detail a little later.

Thus it is seen that the instrument servos,

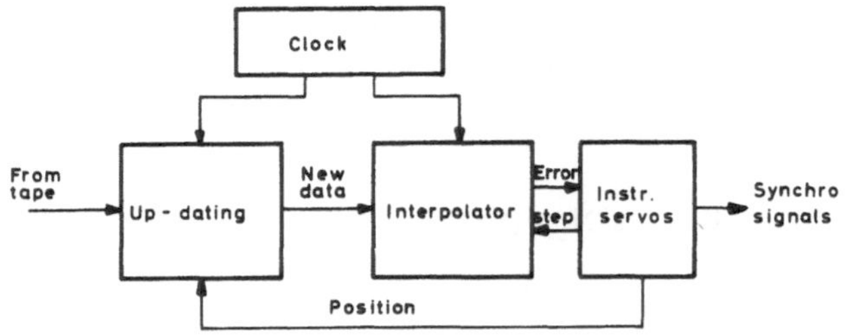

Figure 2 Main parts of "director"

which produce the output shaft angles take part in two servo loops. The inner loop functions as a step servo, and is always operative. The outer loop is in operation only at each updating, i.e. at the beginning of each interval. The feedback in this loop is a reading of the absolute position of the servo shaft. This reading will correct possible errors which have accumulated in the step-servo since the previous updating.

In the step servo errors may accumulate for two reasons. One is the usual for step servos, that of establishing correct starting position after a close down or error introduced by noise. Another is the possibility which one may take advantage of when approximating the trajectory by second order curves: It may be an advantage to let the curve in one interval end at one extreme within the tolerance region, and the next curve start at the other extreme. The outer loop also cancels round-off error, inherent in any method of curve generation.

Interpolator

The interpolation relies on a technique known as "operational digital technique" (1). Basic element is the binary rate multiplier, shown in Figure 3.

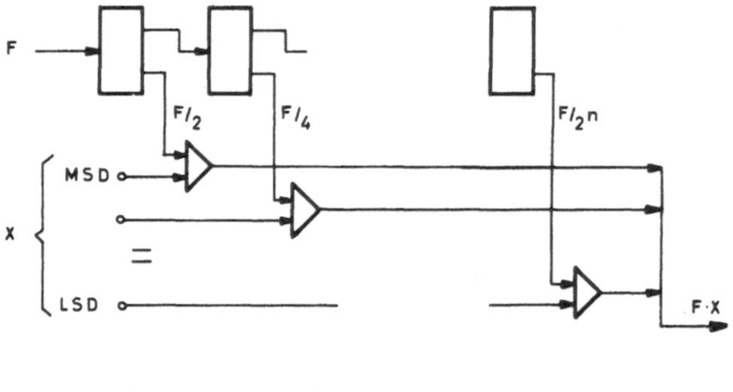

Figure 3 Binary multiplier

The binary multiplier multiplies a pulse frequency F by a binary coded number X, where $0 \leq X < 1$, to produce an output pulse frequency $F \cdot X$. A pulse train of frequency F is divided in a chain of n flip-flops. Each flip-flop generates a pulse frequency $F/2$, $F/4$ .... $F/2^n$. These are derived at non-carry times and pulses therefore do not coincide. Each such binary weighted frequency is gated by the corresponding bit of the binary number X. If a bit is "1", that frequency is let through the and-gate. The frequencies which come through may be added in an or-gate, since the pulses are non-coincident. Resulting frequency is $F \cdot X$. - Note that this is true only on the average, and to be exact, if the averaging time is an integral number of periods

$$T = \frac{1}{F} \cdot 2^n$$

It is, however, easy to arrange scaling such that this multiplication is correct within the required accuracy.

Let us now consider the inner servo loop which performs the actual curve generation or interpolation. The main elements are shown in Figure 4. Velocity command is here shown as

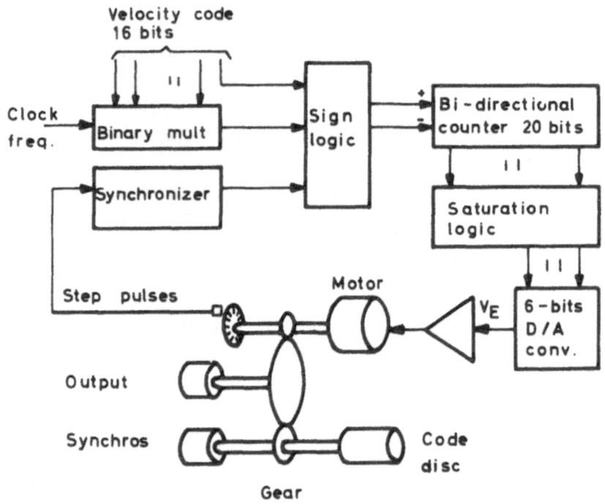

Figure 4 Interpolating step-servo

a binary 16-bit number. A constant reference frequency from the precision clock is multiplied by this number, and the resulting pulse frequency is proportional to the demanded velocity, such that one pulse corresponds to one step by the step detecting wheel on the motor shaft. These and the step-pulses from the pulse-wheel are fed to the bidirectional counter.

The content of the counter is the servo error, i.e. the difference of the actual shaft position from the desired position. This content is

Superior numbers refer to similarly-numbered references at the end of this paper.

decoded to analog form, amplified and used to drive the instrument servo motor. A "sign logic" takes care to direct the command and feedback step pulses to the appropriate inputs of the counter.

To remove the significance of quantization in the step servo, the step-pulse wheel is arranged to give 8 pulses for every step in the 16 bit position code. This comes at the least significant end of the 17 bits of the position difference which is calculated by the updater and set into the bidirectional counter at the beginning of each interval. The counter therefore has 20 bits.

The motor drives, via a precision gear train, the coarse and fine synchro transmitters and the 16 bits multiturn shaft encoder (of the V-scan type) which is used by the updating unit in the outer servo loop.

To obtain change in the velocity during an interval, such as needed for the second order curve generation, another bidirectional counter for "velocity" is used as shown in Figure 5.

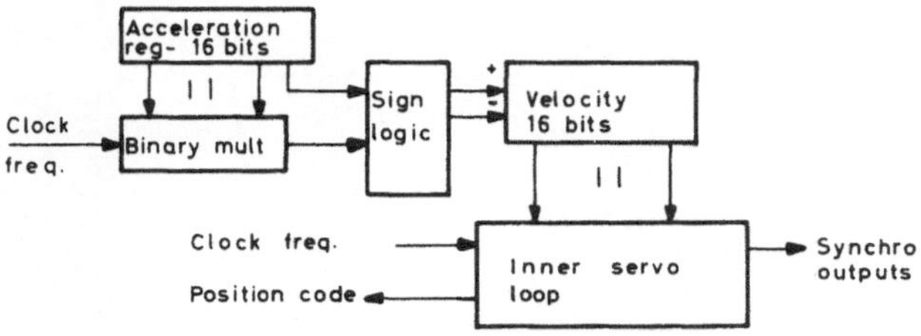

Figure 5 Second order curve generator

A clock frequency is multiplied in a binary multiplier by the code for acceleration, and the resulting frequency of pulses changes the velocity at a constant rate proportional to the acceleration.

There are two identical second-order curve generators and step-servos, one for each antenna axis.

The updating takes place by reading each character from tape into a long shift register for assembly into the six data words for the interpolator. At the instant when the new interval starts, the new data words replace the old ones in the acceleration registers, velocity counters and position error counters. As explained before, the data word set into the error counter is the difference resulting from a subtraction of the position code from the code discs from desired position read in from tape. Subtraction is done simultaneously in two serial subtractors.

The time given in one block of information is, of course the starting time for the next interval.

Control panel

Figure 6 shows a section of the station control desk at Råø. The panel at right comprise the controls for the director and for remote control of power servos.

Included in this panel are in-line displays for time, next updating time, antenna shaft positions and velocities. Further, there are controls for zeroing of synchro differentials (the hand-wheels of which are seen underneath the desk), push-buttons for manual intervention in the case of reading error, and decade switches for clock-time correction.

By a switch on the panel, the director may be set to "joystick" mode, in which the director is independent of tape. In stead the velocities may then be changed by means of a joystick mounted between the handwheels underneath the control desk. The joystick controls gates, which send pulses to the velocity counters. A velocity which is set into the director will, of course, be maintained with the high accuracy of the clock.

In the lower right corner of the control panel there are controls for a special conical scan system. This system obtains a true pointing

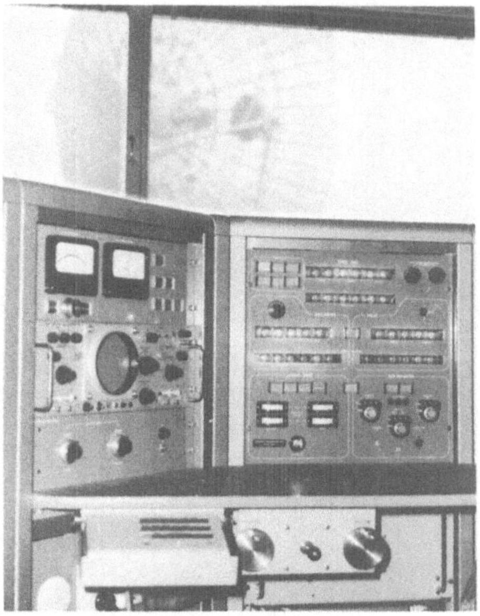

Figure 6 Antenna steering control panel

error by wobbling the subreflector of the antenna, which has a Cassegrain feed system. The pointing error is only used for manual supervision in the system at Råø; not directly in the control. If desired, it should, however, be easy to introduce a suitable pointing error feedback into the system. It could be introduced into the instrument servo or the power servo, depending on antenna characteristics.

REFERENCES

(1) Lundh, Y., Digital Techniques for Small Computations, Jl Brit IRE, Jan 1959

(2) Githens, J.A., Kelly, H.P., Lozier, J.C., Lundstrom, A.A., Antenna Pointing System: Organization and Performance, Bell Syst Tech Jl, July 1963

(3) Marshall, J.E., Coulter, R.J., Binks, J.K, Stevenage, T., Willetts, G.E., Clarke, K.J, Digital Techniques Used in the Steering Apparatus of the G.P.O. Steerable Aerial at Goonhilly Downs, Proceedings of International Conference on Satellite Communication, 22-28 November 1962. Arr. IEE

DISCUSSION

Y. Lundh

"Special Computer for Simplified High Performance Ground Antenna Steering".

Q. Is the antenna self-tracking?

A. No, the steering is programmed. We receive prediction data from NASA which is accurate enough. Our antenna main lobe is $\pm$ 6 minutes of arc.

Q. Do you have a problem in making predictions for motion which takes you near the singularities of the antenna mount?

A. No. We convert ephemeris data to a conic fit before converting to antenna coordinates.

On the System Design of Low-Cost Ground Stations for Satellite Communications.

C.O. Lund, Consultant, Copenhagen.

0. Abstract.

The main design factors for a satellite communications ground station are reviewed and various sections of the stations, where the design is critical, especially the antenna and drive systems, are discussed in more detail with regard to recent results and proposals.

The system design is normalized with respect to the maximum power flux recommended by CCIR, and it is found that present satellite systems do not even approach this maximum level.

In the system design, influences of various sorts, f.ex. influence of heavy precipitation is included.

In a design example, considering only the receiver part of the station, power flux levels comparable to that of early experimental satellites are used and it is concluded, that a Cassegrain type of antenna with a nitrogen cooled parametric amplifier tends to be the most economical for station sites outside tropical regions.

In some tropical regions, additional attenuation due to heavy precipitation will require antenna areas far in excess of those currently used (up to app. 25 meters diameter), if the present rather low power flux from satellites is used.

Then, if the transmitter power and/or antenna gains for the satellites are increased to compensate for this, stations in temperate climates will probably be designed with antenna areas determined by their function as transmitters and/or by interference consideration rather than by their function as a part of a low noise receiver.

1. Introduction.

Satellite Communication is a practical proposition today, with "Early Bird" (HS 303) and several cooperating ground stations providing transatlantic communications.

Preceeding this were some experimental systems like "Telstar", "Relay", and "Syncom".

The major part of the satellite communications ground stations now in operation were planned some years ago and were mainly adapted or designed on a crash-basis, some of them are marvelous examples of engineering, but it is felt that there is room for development of the design towards a practical and economical ground station.

As general basis for design there exists the following:

a. Agreement on frequency bands. This study is based on the use of the 4000 Mc/s band for transmission from satellite to ground and the 6000 Mc/s band for transmission from ground to satellite.

b. CCIR recommendations for maximum power flux at Earth.

Also while the early experimental systems used satellites in low and highly elliptical orbits, most proposals for satellite communications systems favour higher circular orbits, with 6, 8, or 12 hours periods as well as the synchronous orbits with 24 hour period.

The present study will be based on these recommendations and concepts, as well as recent information from the literature. As the most critical parts of a ground station is the low-noise receiving system and antenna steering, the function as a transmitting station will not be discussed.

It has been attempted to collect relevant information on design principles and the various units, - electrónical or mechanical, - from the literature, and these are discussed and evaluated.

Fig. 1

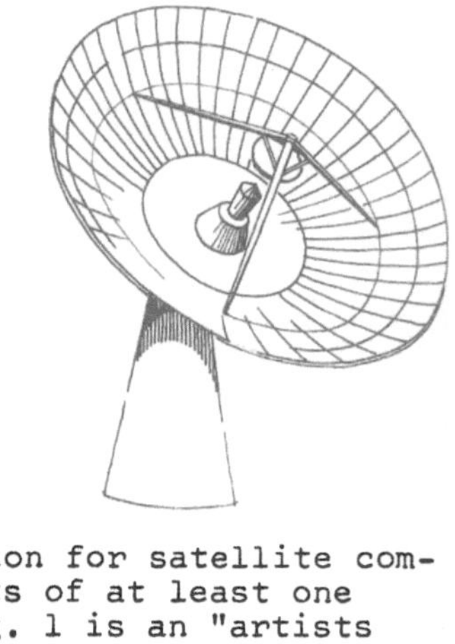

"Artists Concept" of typical ground station for satellite communications.

A ground station for satellite communications consists of at least one antenna system. Fig. 1 is an "artists concept" of a typical design involving a large primary reflector with diameter up to 25 meters, various other parts of the antenna system, and mounted on the movable part are sections of the receiver and transmitter.

The antenna should ideally be supported and driven so that it can be aimed accurately in all positive elevations and azimuths and follow a satellite accurately under all, - or nearly all, - conditions.

It is easily visualized, that a considerable number of engineering disciplines are involved: RF antenna design, microwave receiver design, structural design, design of accurated and rigid gearing for high torques and servo design to name some, where serious interface problems usually exist.

2. Basic Receiving System.

The transmission from a communication satellite to a ground station is considered, see fig. 2. The signal received from the satellite is characterized by the transmitter power, antenna gain, modulation system etc. for the satellite transmitter and the distance to the satellite.

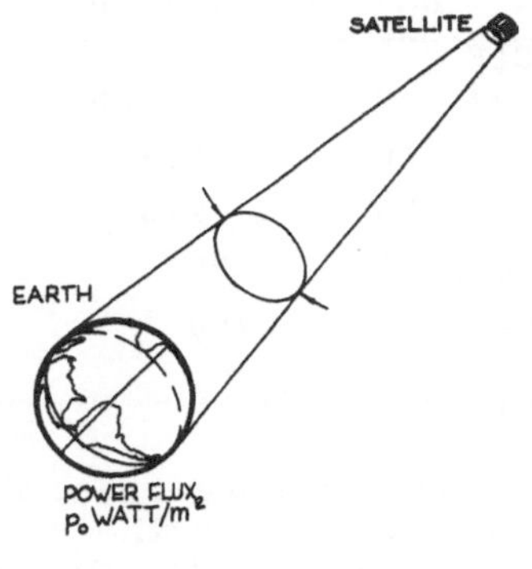

Fig. 2 Satellite-to-ground transmission, schematic.

A simple, - but not in all respects adequate, - way of characterizing the signal level is to assume a certain power flux p_o at ground.

The factors above depend to some extent on the satellite system and some changes can be expected in the future. A limitation on the power flux level at ground is however recommended by CCIR (Recommendation 358, Documents of the Xth Plenary Assembly, vol. IV, pp. 162-163, modified Monte Carlo 1965), this is discussed in appendix 1.

It is recommended by CCIR, that the maximum power flux density set up at Earths surface by emission from a satellite, measured in any 4 kc/s bandwidth, should not exceed -152 dBW/m^2 pr. 4 kc/s for all angles of arrival. The consequence is that there is a higher "legal limit" to the spectral power flux density.

$$10 \log S_o = -188\text{dBW pr.m}^2 \text{ pr.c/s.}$$

It is then proposed to characterize a specific signal from a satellite with $S = S_o/Q$ or, in logarithmic form " q dB below S_o", where $q = 10 \log Q$ for preliminary evaluations.

In satellite systems in operation at present, the values of q range from 29 dB (Relay 1) to 39 dB (HS 303) so, present satellites do not approach the limiting value introduced to prevent interference to terrestrial communications systems working in the same frequency band.

In appendix 1 is outlined a somewhat simplified system calculation, based on the assumption that the transmitted signal can be specified by its transmission bandwidth B_t and signal power spectral flux density S at receiving site.

The signal-to-noise ratio at receiver output (S/N) can then be expressed as a function of the effective antenna area A and the system noise temperature T_s:

$$(S/N) = \frac{S}{k} \cdot \frac{A}{T_s} \qquad (1)$$

where k is Boltzmanns constant.

Introducing $S = S_o/Q$, where

$$S_o = 1.6 \cdot 10^{-19} \text{Watt/m}^2\text{c/s}$$

$$k = 1.4 \cdot 10^{-23} \text{joule/}^o\text{K}$$

$$(S/N) = \frac{S_o}{Q \cdot k} \cdot \frac{A}{T_s}$$

$$= 1.15 \cdot 10^4 \frac{1}{Q} \cdot \frac{A}{T_s} \qquad (2)$$

(A in m^2 and T_s in degrees Kelvin)

or, in logarithmic form:

$$10 \log(S/N) = 10 \log A - 10 \log T_s - q + 40.6\text{dB} \qquad (2a)$$

This is the basic expression used in the following. The aperture area A is proportional to the geometric area A_o:

$$A = \eta A_o$$

where η is the antenna efficiency factor introducing this in equation (2) and rearranging:

$$(S/N)\cdot Q = 1.15\cdot 10^4 A_o \frac{\eta}{T_s} \qquad (2a)$$

The design factors can be split into:

a. The required signal-to-noise ratio (S/N). This is determined by performance specifications, which usually refer to the demodulated signal, then modulation and demodulation methods will also influence the magnitude. The signal-to-noise ratio also includes a margin for performance deterioration, uncertainties in design and various effects of atmospheric disturbances. CCIR has however published a report (see appendix 2) recommending a statistical allowance for increased noise in a hypothetical reference circuit.

b. The factor Q (see appendix 1) depending only on the characteristics of the satellites orbit and equipment.

c. The geometric area of the reflector A_o, this is the only variable with a fairly large range, however, the investment in large areas is a considerable fraction of total investment, then it is of primary significance to optimize.

d. η/T_s, the ratio of antenna efficiency to system noise temperature. As T_s is the sum of various noise temperatures, including the antenna noise temperature T_A, which is not independent of η, the system designers task is mainly to optimixe this ratio η/T_s.

3. Ideal System.

To illustrate a hypothetical design, where the equipment and the atmospheric conditions are ideal, the following is inserted in equation 2a:

a. The factor q as a measure of the level of signal received from the satellite, a range of this factor will illustrate its significance.

b. Area A of the ideal antenna with 100% efficiency i.e. $\eta = 1.0$ and no sidelobes to contribute to system noise temperature.

c. Zero loss ($T_L = 0^oK$) in transmission lines, filters etc. and receiver noise $T_R = 0^oK$.

d. The only noise contribution is noise temperature of the atmosphere T_{AO}, these are referred to 10° antenna -elevation for clear, dry weather $T_{Sr} = T_{AO} = 16^oK$.

This noise temperature is conventionally used for system calculation at 4000 Mc/s, see f.ex. Hogg & Semplak (1961).

e. The performance of the ideal system is referred to (S/N) = 12 dB for a conventional FM detector and, as an alternative (S/N) = 8 dB for the threshold reducing FM demodulator. These values are typical for the threshold of the detection of a wide deviation FM signal. In practice the design values should be somewhat higher as the conventional specifications consider the noise in any telephone channel, see appendix 2.

Rearranging equation 2a:

$$10 \log A_o = 10 \log(S/N) + 10 \log T_s + q + 40 \qquad (2b)$$

from which the required antenna area A_o can be calculated, the table below illustrates the order of magnitude.

	qdB	A_o m^2 (S/N)=12dB	A_o m^2 (S/N)=8dB
(Maximum CCIR recommended)	0	0.025	0.01
	10	0.25	0.1
	20	2.5	1.0
(Relay 1 app.)	30	25	10
(HS 303 app.)	40	250	100

The required antenna areas are in general much less than those in operational use today, even for Relay 1 applications. However, some margin has to be introduced and receiving systems and atmospheric and terrestrial influences are in practice not negligible.

One might speculate, what the consequences will be, if higher transmitter powers and higher antenna gains will be introduced in the satellites, resulting in considerably smaller values of q. According to the table, antenna areas will be negligible, but then the ideal system will be very far from being realistic, this will be discussed in section 9.

4. The Actual System.

The actual system differs from the ideal, especially as the antenna efficiency η is less than 1 and as the system noise temperature T_s is considerably higher than 16°K. These factors can for given parameters of the satellite system only be compensated by increasing the antenna area.

By rearranging equation 2, the geometric antenna area is expressed as:

$$A_o = \frac{k}{S_o}\cdot Q(S/N)\frac{T_s}{\eta} \qquad (2c)$$

k/S_o and Q are fixed as far as the designer of the ground station is concerned. The signal-to-noise ratio required, (S/N), was discussed in section 2, it depends on the modulation and the demodulation system and is only within the control of the designer to the extent, that he can choose to use a conventional FM detector or a threshold reducing detector.

The main factor to be considered is then T_s/η, the ratio of system noise temperature T_s to efficiency η.

The system noise temperature can be expressed as a sum of noise temperatures from various sources:

$$T_S = T_{AO}+T_{AA}+T_{AM}+T_{AS}+T_L+L\cdot T_R \qquad (4)$$

where

T_{AO}: Noise temperature due to space noise and normal clear and dry atmosphere.

T_{AA}: Noise temperature due to precipitation and other atmospheric phenomena.

T_{AM}: Noise temperature due to various losses in and around the antenna system, radomes, loss in spars etc.

T_{AS}: Noise temperature due to antenna sidelobes.

T_L : Noise temperature due to loss in transmission lines, filters etc. $= (L-1)T_o$.

T_o : Standard ambient temperature (= 290°K).

L : Loss in transmission lines, filters etc.

T_R : Receiver noise temperature.

(All noise temperatures are referred to proper terminals).

Rewriting equation (2c) and introducing equation (4) and introducing:

$$K = \frac{k}{S_o}\cdot Q\cdot(S/N) \qquad (5)$$

The geometric area of the reflector is expressed as:

$$A_o = \frac{1}{\eta}\left[KT_{AO}+KT_{AA}+KT_{AM}+KT_{AS}+KT_L+LKT_R\right] \qquad (6)$$

As written, the factors in the brackets all represent effective areas. The first factor, KT_{AO} represents the area of the ideal antenna, as defined in section 2 . The second factor, KT_{AA} is dependent on meteorological statistics at the site.

Both these factors are beyond control of the designers once the site is chosen.

In the following sections, the present status of antenna and low noise receiver design will be reviewed, especially with regard to the factors entering into equation (6) and finally, the optimization will be discussed in section 9.

5. On Antennas.

So far, the antenna has only been considered as a collecting aperture with a certain efficiency and with sidelobes picking up unwanted noise (and possibly interference from unwanted sources).

Various types of antennas have been used or proposed for satellite communications ground stations or similar applications in the past. See f.ex. Hines et al. (1963) for a description of the Andover horn-reflector antenna, Victor (1961) gives a review of large antennas available commercially in 1961.

In this paper, only the Cassegrain type of reflector antenna will be considered in detail as it at present seems to be the most economical. Many of the considerations below are however valid for other types of antennas.

The Cassegrain system has been investigated in some detail, see f.ex. the papers by Hannan (1961), Foldes (1962), Potter (1962), Wilkinson & Appelbaum (1961), Galindo (1964), Morgan (1964), and Visocekas (1964).

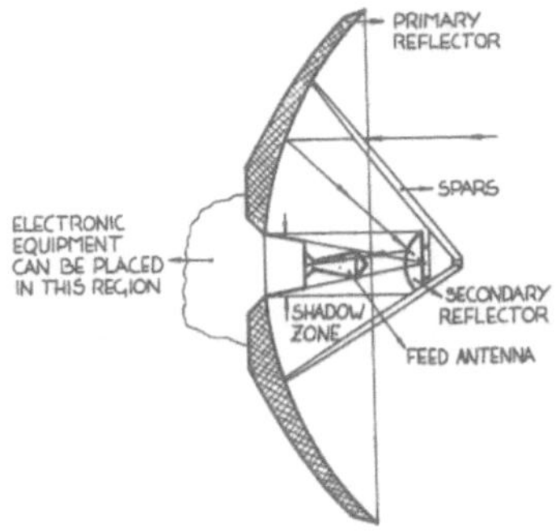

Fig. 3 Cassegrain Antenna System with conical feed horn, schematic.

The main parts of a Cassegrain or folded antenna are (see fig. 3):

a. The primary reflector representing the antenna aperture. It has usually the shape of a paraboloid although other shapes like spherical can be used for an efficient folded antenna system.
The rays entering will be reflected towards the primary focus, but before arriving, they are reflected again by

b. The secondary reflector, which has a much smaller diameter. It is supported on spars and produces a shadow around the center of the main reflector. In this region, the secondary focus and the feed antenna are situated. Behind this, there is room for equipment in a conical cabin. Space for equipment should also be available behind the main reflector.

c. The Feed Antenna is sketched here as a conical horn, but other configurations are used. Hogg & Semplak (1964) have proposed to use a "plane phase-front" type of feed, with a conical horn reflector as a feed antenna, see fig. 4.

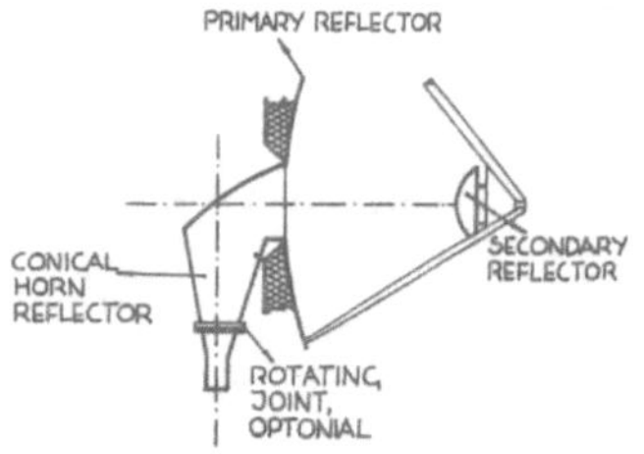

Fig. 4 Cassegrain system, with "plane phase front" type of feed.

The relative merits of these two solutions to the primary antenna are not yet clear although the conical horn reflector for some applications (with elevation-azimuth type of antenna mount) immediately makes a large rotational joint possible so that the receiver front-end can be placed outside the region being tilted. This can be a definite operational advantage.

The antenna system is a costly affair, in most cases, the cost of the antenna is the main investment in a satellite communications ground station, it is even worth while to consider the influence of various degrading factors, the major contributions are:

A. Illumination Efficiency.

This is a problem for the microwave antenna designers, in principle, the parameters of the secondary reflector and the primary horn are the design variables to be optimized, also, the geometry of the primary reflector.
In the literature, there are several studies of this problem, some of the references were given above, further references: Ravenscroft (1962), Viggh (1963), Potter (1963), Jensen (1963), and Potter (1964).
With current techniques, illumination efficiencies around $\eta_i = 0.8$ can be realized, however η_i is not independent of the noise temperature due to side lobes, T_{AS} and some optimizing procedure is necessary in the design.

B. Static Deviations of the Reflector Surface from Ideal Shape.

The relative manufacturing tolerances of the reflecting surface and its supporting structure are quite strict as the design wavelength is app. 7.5 cm (or 5 cm for transmitting frequency) and sizes of the order of 25 meters.

Ruze (1952) has made the original study of deviations of reflector surface on the antenna gain and the sidelobe level.
There are several other studies f. ex. Bracewell (1961) and Dragone & Hogg (1963) of this and related problems.

Experimental confirmations are rare, it is difficult to measure efficiency accurately on large antennas, there has however recently been published some by Ruze (1964) that confirm the original expression for efficiency due to reflector surface deviations:

$$\eta_d = e^{-\left(\frac{4\pi\varepsilon}{\lambda}\right)^2}$$

where ε is the RMS deviation from best -fit paraboloid and λ the wavelength.

This formula is based on the following assumptions:

a. The deviations from best-fit paraboloid are random and distributed in a Gaussian manner.
b. The errors are uniformly distributed over the antenna aperture.
c. The region, over which the deviations are substantially constant, is large compared to a wavelength.
d. The reflector is smooth in wavelength measure.
e. The number of uncorrelated error regions in the aperture is large (at least 5 to 10).

C. Shadow Effects.

On fig. 1 and fig. 3 it can be seen, that both secondary reflector and the spars supporting it will produce a shadow on the primary reflector. These shadows result both in loss in efficiency and increased side lobe levels.

The effect of the sub-reflector has been considered by most authors investigating the Cassegrain systems and the effect of tubular spars has been calculated by Gray (1964), a typical efficiency η_s = 90%.

These values are for solid spars, other types of spars with less shadowing effects can be made so, that an increase η_s is possible.

5.1 On Reflector Panels.

The main, or primary reflector has an area of several hundred square meters, the reflecting area is divided into panels of sizes suitable for shipping, these are attached to a structure to be discussed later.

The main requirements to the reflecting surface is:

a. It must approximate, with very small mechanical tolerance a paraboloid with rotational symmetry. This is usually called the "best fit paraboloid". It is more significant, that the paraboloid has rotational symmetry than it has exactly a specified focal length, - deviation in focal lengths can be compensated by fairly simple adjustments, the effect of small deviations was discussed in the preceeding section.

b. The attenuation for transmission through the paraboloid should be high, at least 70 dB in order to avoid back lobe noise and interference.

c. The geometric shape should be retained with age and under different environmental conditions.

d. The wind resistance should be minimum.

The following types have been used or proposed:

A. Solid reflector panels have been used for several antennas, f.ex. at Jodrell Bank and Goonhilly for antennas without radomes and in the form of honeycomb panels.

The solid sheets have the advantages that the leakage is low, but the disadvantage, that they produce so deep shadows, that distortion due to temperature differences can be a problem.

B. Perforated reflector panels will result in somewhat less wind resistance and shadow effects, however for operation at 6000 Mc/s, the relative opening cannot be very large if a reasonable attenuation is to be retained. Also, perforations can be expected to close up by ice, so the operational and survival wind resistance will not be less than for solid sheets.

C. Grating panels have been proposed by Willoughby (1962), a possible configuration is shown on fig. 5.

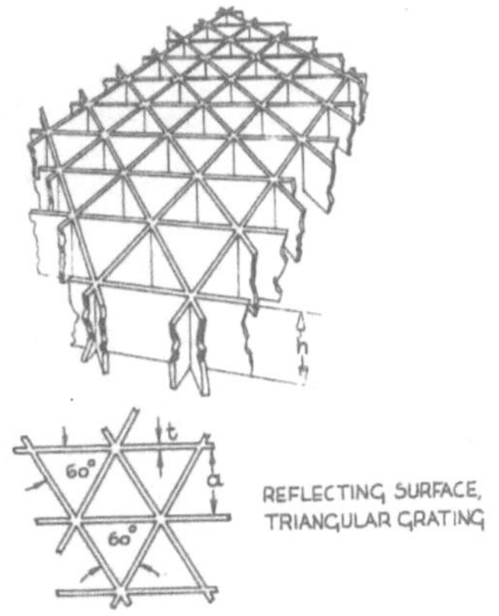

Fig. 5

Reflector Panel, Grating type, schematic.

It is not known, if any large reflectors have been built after this technique, but it has interesting properties.

The attenuation can be quite high for fairly large openings and deep strips, a dimension a = 32 mm can give sufficient attenuation. The aerodynamical properties of such a grating is probably so, that the paraboloid does not act as an efficient aerofoil, this reduces wind forces and especially the usually complicated relationship between orientation and torque due to wind, - a factor of primary significance to the design of the drive system and the servo system.

With deep strips, such a panel can be made self-supporting with sufficient rigidity, without introducing the difficulties of temperature equalization common with sandwich panels. The triangular configuration shown on fig. 5 has the advantage that lateral rigidity is inherent.

5.2 <u>On the Supporting Structure of the Reflector</u>.

The reflecting surface is supported by a grid structure, as rigidity requirements are very high and the loading condition complicated and varying, the design is a major task.

For an illustration of the requirements to accuracy, see fig. 6 where a series of equal-gain reflecting surfaces with increasing values of the RMS surface deviations are shown, calculated for 4000 Mc/s.

For the perfect reflector ($\varepsilon = 0$), D = 20 m. The diameters for $\varepsilon > 0$ are calculated using the equation given by Ruze (see section 5).

From this illustration, it is obvious, that surface deviations up to 2 mm do not result in any large increase in the diameter necessary to preserve the gain of the perfect reflector, whereas deviations in excess of 4 mm result in a significant increase of diameter and consequently antenna area.

2 mm in 20 m ($\varepsilon/D = 10^{-4}$) is comparable to the tolerance of good machine tools and also comparable to the thermal expansion coefficient of aluminium alloy and the relative deformations within the elastic limit of materials.

The main forces on the structure are:

a. Forces due to the weight of the reflector structure, spars secondary reflector etc.

These are changing with reflector orientation, for large reflectors, these forces are comparable to or larger than operational wind loads.

b. Wind loads are complicated and depend very much on the direction of wind relative to the reflector which

EQUAL GAIN REFLECTORS FOR VARIOUS VALUES OF SURFACE DEVIATIONS

THE REFLECTOR SURFACE DEVIATIONS ARE EXAGERATED BY FACTOR 100

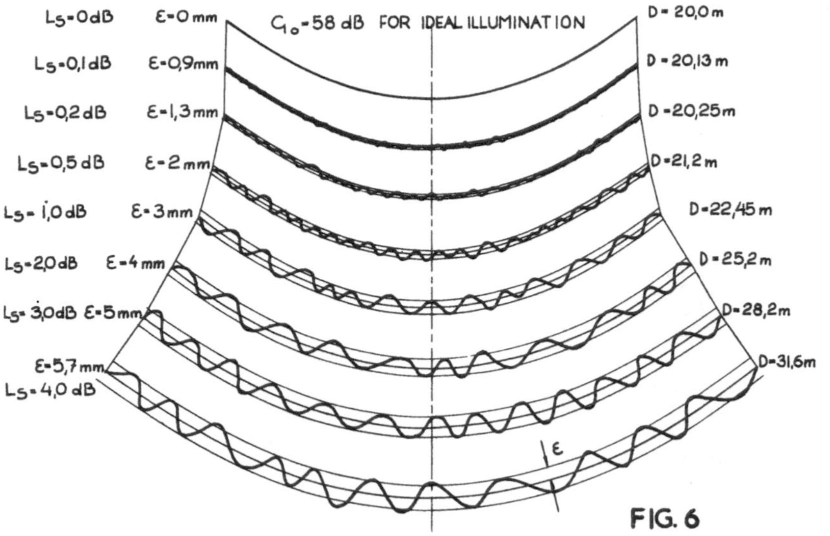

FIG. 6

in the case of solid reflecting surface is a fairly good aerofoil. Hirst & McKee (1964) have reviewed the results of recent calculations and experiments.

c. Acceleration forces have no primary significance as the tracking velocities are low, except when the antenna is slewed.

As stated, the requirements to rigidity are high because the accuracy of the reflector has to be preserved, also, high rigidity is related to a requirement of high mechanical resonance frequencies in the structure imposed by servo design.

A reflector structure can have many modes of resonance, neither of the resonant frequencies must be so low, that they can be excited significantly by the gust spectrum of the wind, also low resonance frequencies have a deteriorating effect on the performance of the servo system for antenna drive. Stallard (1964) states, that structural resonances of 2.2 c/s and 3.5 c/s are the main limitation of one system studied.

5.3 On Antenna Mounts.

The ideal antenna mount allows the reflector to be tracked through any region in the half space without requiring excessive velocities in the driving system, i.e. singular directions must not be present.

This cannot be realized by a two-axis mount, they are however used almost exclusively due to the simplicity of the driving system.

5.3.1 Elevation Azimuth Mount.

Shown on the sketch, fig. 7, is the type mostly used.

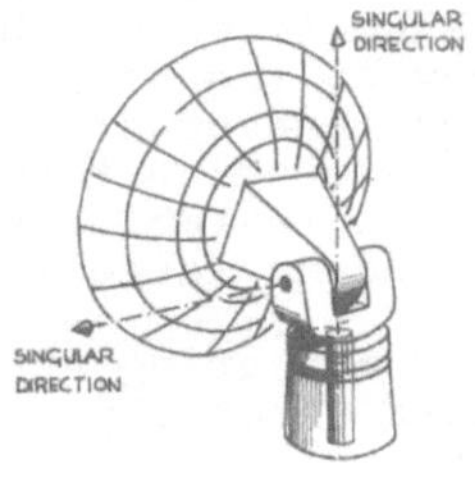

Fig. 7

Elevation-Azimuth Type of Antenna Mount.

Here a fixed singular direction exists at zenith, this will probably not be a disadvantage for a satellite system with aequatorial or near aequatorial orbits for ground stations at higher latitudes, but for satellites in polar orbits or stations near aequator, the fixed singular direction in zenith will prevent continuous tracking.

An advantage of this mount is, that the part rotated in azimuth can have level floors under all orientations, this can be a convenient region for electronic equipment.

A disadvantage can be, that the azimuth rotation in principle should be unlimited. This can be troublesome for cable-connections etc. and requires some programming of the steering for limited rotation.

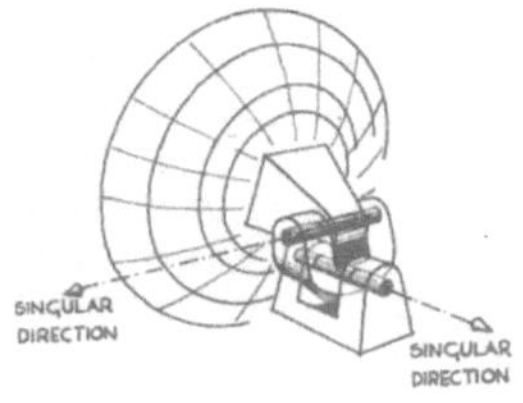

Fig. 8

X-Y Type of Mount.

5.3.2 The X-Y Mount is sketched on fig. 8, it is equivalent to a gimbal suspension. It has one fixed singular direction at horizon, this makes it more universally useful. Satellites are rarely tracked below 10° elevation as the noise and interference level is high for lower elevations.

The angular velocity of the axes is then not much larger than the angular velocity of the satellite in the operational region.

The kinematics of this system have been studied by Rolinski et al. (1962) and Rolinski et al. (1964).

The X-Y Mount has further the advantage, that movements around both axes are limited so that cable connections can be made simple and reliable.

5.3.3 Ball-in-Socket Mount has been proposed for ground stations exclusively designed for cooperation with synchronous (24 hour) satellites, where only very limited movements are needed.

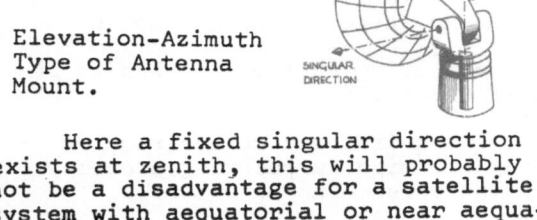

Fig. 9

Ball-in-Socket Type of Antenna Mount.

Fig. 9 illustrates the concept, in practice, kinematic equivalents of the ball arrangement shown will probably be used, and the steering arrangements will have the form of jacks rather than the gearing used in the other drives.

It seems a little early to evaluate this proposed mount configuration.

5.4 On Antenna Noise, - Radomes.

In section 4, various noise contributions were listed:

5.4.1 T_{AO}: Noise temperature due to space noise. This is an unavoidable contribution, for the estimates here, $T_{AO} = 16^oK$ is used.

5.4.2 T_{AA}: Noise temperature due to precipitation and other atmospheric phenomena.

Various studies of this contribution have been published, both expressed as noise temperature and as attenuation, see f.ex. Feldman (1964) and Holzer (1965) who calculate for atmospheric attenuation L_A during precipitation rates not exceeded 1% of the year from meteorological statistics.

The attenuation varies much with geographical position, from L_A = 1 dB at Fairbanks, Alaska, to L_A =6.1dB at Manila, Philipines.

For the present study the computed value for Paris, France, is used, from Holzer (1965), L_A = 1.3 dB, using the relation $T_{AA} = (L_A-1)T_o$, $T_{AA} = 75^oK$, it must however be emphasized, that this value is exceeded less than 1% of all the hours in a year.

Another possible deteriorating atmosphering effect is icing and sleet. Caspar & Sandreczki (1964) have made an interesting study of icing phenomena. This is mostly dependent on geography and it will be assumed, that the ground station is placed in a region where such conditions very rarely occur.

5.4.3 T_{AM}: Noise temperature due to various losses in and around the antenna system.

In an antenna system without radome, T_{AM} is near to negligible, 2^oK to 5^oK has been quoted.

Hogg & Semplak (1964) have made the observation, that spars made from dielectric material increase this temperature by app. 16^oK, metallic structures should then be preferred.

In an antenna system with radome, the attenuation in radome material increases very much during precipitation, values of $T_{AM} = 40^o$ have been quoted for moderate rain, but no consistent information has been found yet.

In this study, radome has not been considered. It is of undoubted value for the antenna driving system and the antenna structure, but the noise performance is not clear and the cost is high. The value of a radome seems very dependent on the meteorology of the site.

5.4.4 T_{AS}: Representing the noise due to antenna sidelobes depends very much on the antenna concept. Hogg & Semplak (1964) have compared horn reflector antennas and various Cassegrain systems. The sidelobes pick up some noise from the surrounding terrain, the reflection coefficient of this has some influence on the apparent temperature so that a site more or less surrounded by sea might be advantageous in that respect.

Such a "free" site will however often have a high interference level compared to a site in a valley.

According to Hogg & Semplak (1964), the sidelobe noise contribution referred to 10° elevation is around $T_{AS} = 5^oK$ for horn reflector antennas and $T_{AS} = 20^oK$ for a typical Cassegrain antenna with "near field feed". There is a tendency for T_{AS} to decrease if a deep primary reflector is used, but T_{AS} is not independent of the illumination efficiency η_i of the reflector.

6. On Mechanical Drive System.

The conventional antenna mount has two-axis movement and drives on both axes (see section 5.3).

Considerable maximum torques in the range 10 to 100 tons-meters or more are excerted by wind so large gears have to be used.

These torques are mainly due to wind pressure and the drive system has to be most rigid, both in order to prevent oscillations excited by wind gusts and also to make a high cut-off frequency of the servo system possible. The resonances are in general of so complex nature, that feed-back damping leads to instability at higher frequencies and no large external damping exsists.

The gearing system has then to be planned carefully with respect to compliance.

The requirements to accuracy of gearing are high as pointing accuracy is in the order minutes of arc.

One way to overcome this, - usually difficult, - problem is to use direct-drive electric motors. Canfield (1959) has described motors for extremely low speeds and large torques, although not in the range usable for antennas larger than app. 8 m diameter.

Direct drive also has the advantage, that there are no problems with backlash in the gears, other methods for the compensation of backlash will be considered in the next section.

7. On Power Servo Systems.

These servo systems drive the antenna so as to follow a satellite in orbit. The input signal is either derived from programmed steering (for one recent system see Lundh (1965)), or from an autotrack system.

The accuracy requirements are usually in the order of one minute of arc and the stiction of the system should be low so that the antenna will follow smoothly a prescribed pattern.

7.1 Angular Velocity.

Early experimental communications satellites had low and highly elliptic orbits resulting in angular velocities of the order 6 minutes of arc/sec and most ground stations now in operation have drive systems able to provide high angular velocities, much in excess of the velocities necessary.

Present plans for satellite communications systems favour almost exclusively considerably higher (larger than 6 hour period) or synchronous orbits.

Seen from a typical ground station a satellite moving in an orbit with 6 hour period will have a maximum angular velocity 1.2 minutes of arc pr. second. For other proposed systems, the velocity is even lower.

It is then an open question, if the antenna drive system should be designed for controlled steering at the rates 1 to 10° pr. second presently used. If not, a considerable reduction in the size of the drive motors is possible. An angular velocity of 1.2 minutes of arc and an extreme torque of 80 tons meters represent a power of app. 0.3 mechanical kilowatts, this is so low, that manual power could replace electrical under emergency conditions.

When the antenna has to be slewed, servo control is in general not necessary and a separate set of motors can be considered.

7.2 On the Servo System.

On fig. 10 is shown a simplified diagram of the servo system. The servo motor drives through a reduction gear, a pinion engaging the bull gear on the antenna axis. On this axis is attached an angle transducer giving the feedback signal to the summing amplifier and the error signal is amplified in the motor amplifier.

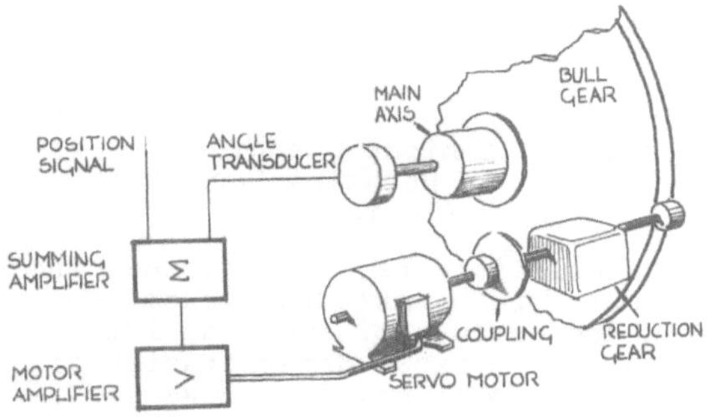

Fig. 10 Servo System, schematic.

Stallard (1964) has reviewed the main design principles for efficient servo drives for large antennas and these will not be discussed here. Only, considering that future satellite systems seem to result in much lower velocities than those used in present designs, so that the maximum power requirements of antenna axes are in the order of fractions of one kilowatt instead of tens of kilowatts, the question is raised, if unconventional concepts for the servo- and drive system should be investigated.

The Scandinavian experimental ground station at Gothenburg, Sweden, has a Cassegrain antenna system and for various reasons tracking information is derived from conical scanning generated by nutating movements of the secondary reflector. Fig. 13 shows a sketch of the arrangement used, the secondary reflector is supported on three hydraulic jacks, these are servo controlled and any scanning pattern within a limited range can be generated.

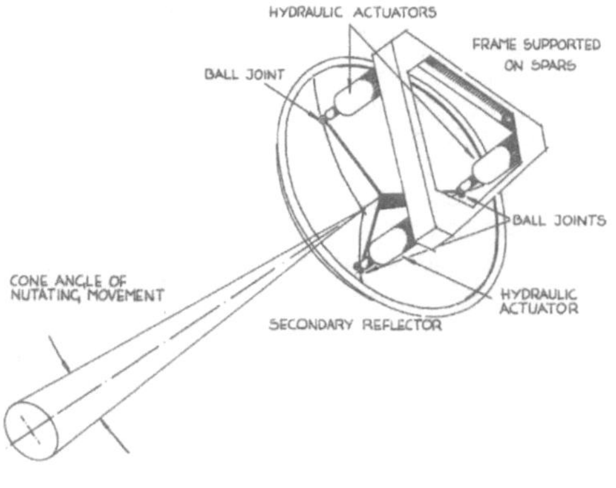

Fig.13 Conical Scanning by Nutating Secondary Reflector.

Calculations have shown, that the main beam of the antenna system can be deflected up to app. two beamwidths without any serious deterioration of antenna performance. The servo system controlling the movement of the secondary reflector can be designed for considerably higher cut-off frequencies than the servo systems driving the reflector system as a whole and it has been proposed to utilize this as a fast "vernier control" for antenna direction so as to relax the requirements to cut-off frequency for the main servo systems.

This proposed system has not been evaluated yet but might have advantageous properties as the accuracy requirements to the main servo system can be relaxed and the vernier servo can take care of deviations due to wind torque.

7.3 On Compliance and Backlash.

In the servo system sketched on fig. 10, compliance and gear system backlash should be reduced as much as possible, - these factors have proved troublesome in many present designs.

The need to reduce mechanical compliance was discussed in section 6, some additional compensation can be introduced by the use of split feed-back.

Backlash can of course be reduced to a minimum by careful manufacture and installation of the gears, but can usually not be preserved due to wear and temperature effects.

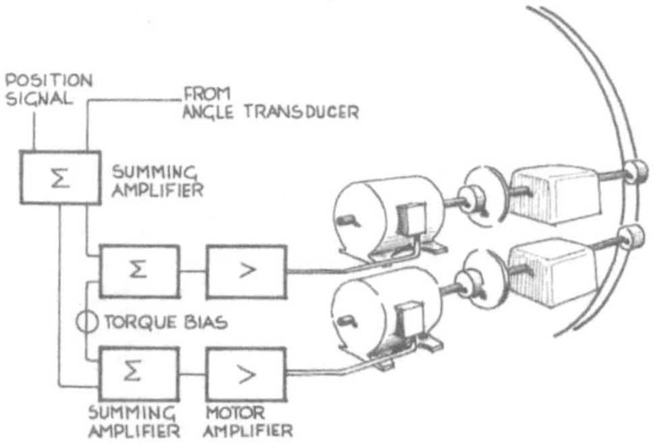

Fig.11 Servo System, with Bias Torque to Eliminate Back Lash.

A common method to control backlash is shown on fig. 11, there are two pinions engaging the bull gear, each driven by its own motor and amplifier so that opposing torques are transferred. These torques must be larger than the operational torques. The opposing torques are here generated at signal level, where compensating networks can be introduced, but other arrangements are possible.

In the case of malfunctions in reductions gears, motors or amplifiers, this arrangement can operate with some performance reduction by removing the torque bias.

The operation of the ground station as a whole is most dependent on the drive system, so it might be advisable to have a third drive system as a spare that can be put into operation at short notice, see fig. 12. With proper design, each drive system can be decoupled mechanically and electrically for repair and maintenance so that the reliability of antenna control can be as high as the electronic systems, where operational spare circuits can be introduced fairly easily.

Fig.12

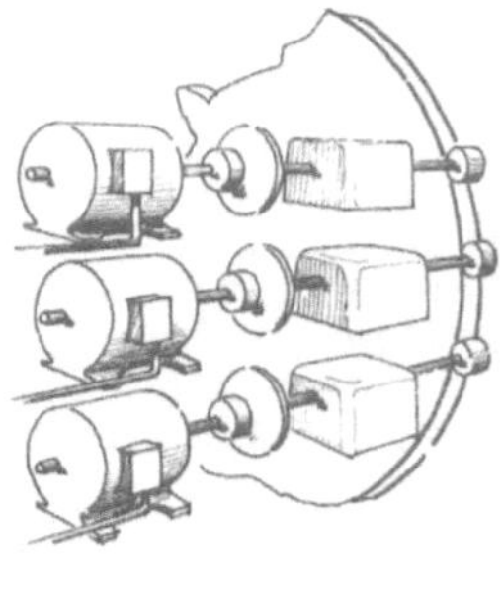

Antenna Drive System, with two Motors and Reduction Gears in Operation and one Set as Spare.

8. On Low Noise Receivers.

The parts of the receiving system to be considered here are:

a. The transmission line from antenna, filters, circulator etc. introducing a loss L and contributing with a noise temperature $T_L = (L-1)T_o$.

b. The actual receiver, contributing with noise temperature T_R.

When Cassegrain antenna systems are used, short transmission lines are possible so that L is very nearly the loss in filters. L = 1.1 corresponding to 0.4 dB attenuation is a conservative figure, - in extreme case, this can be reduced by cooling the filter etc. with liquid nitrogen. The corresponding design value of T_L is app. 30^oK.

A wide range of receiver front ends are possible, representing increasing noise temperatures and decreasing cost and maintenance requirements.

From recent literature, see f.ex. Cuccia (1964), the following typical noise temperatures are derived (for 4000 Mc/s).

A. Maser, cooled by liquid helium $T_R = 10^oK$

B. Parametric Amplifier, cooled by liquid helium $T_R = 20^oK$

C. Parametric Amplifier, cooled by liquid nitrogen $T_R = 40^oK$

D. Parametric Amplifier, room temperature $T_R = 130^oK$

and for comparison

E. Low Noise Crystal Mixer $T_R = 630^oK$

The first four types are two-terminal devices and require a circulator, the ordinary crystal mixer is a four-terminal device and is not so sensitive to balance in the transmission circuit.

Recent development indicates, that cooled four terminal devices, - including transistors, can have rather low noise temperatures, so further development can be expected although it is a little early to make definite predictions.

9. On the Optimization of Design.

In section 4, the main factors for the design of an actual system were included in equation (6) expressing the necessary antenna area for a given signal-to-noise ratio and characteristics of the satellite signal.

In the later sections present obtainable values of the other factors were discussed.

9.1 Design Fundamentals.

Fully expanded, equations (5) and (6) read for the geometric antenna area:

$$A_o = \frac{K}{n_i \cdot n_s \cdot n_d}\left[T_{AO}+T_{AA}+T_{AM}+T_{AS}+T_L+LT_R\right] \quad (6a)$$

where

$$K = k \cdot Q(S/N)/S_o \quad (5)$$

introducing $S_o = -188$ dBW/m²c/s

$$K \approx 10^{-4} Q(S/N)$$

(K has the dimension m^2 pr. ^oK).

In the examples below, the discussion is centered around the following conditions:

a. The station is situated in a temperate climate.

b. The design is based on operation under heavy precipitation, so that $T_{AA} = 75^oK$ and $L_A = 1.3$ dB (see section 5.4.1), all referred to 10° antenna elevation and precipitation rates not exceeding more than 1% of the year.

c. The threshold signal-to-noise ratio $(S/N)_t$ is 8 dB (using threshold -reducing FM detector) and the reference signal-to-noise ratio S/N = 12dB, allowing 1.3 dB for L_A (atmospheric attenuation), 0.7 dB for tracking errors and 2 dB as a general margin for deterioration of performance and allowance for the difference between the statistics of the atmospheric phenomena using 1% and the CCIR noise standards using the concept of 0.2% of any month (see appendix 2).

d. The signal level is characterized by $Q = 10^3$ (see section 2), this corresponds roughly to the spectral power flux density set up by Relay 1 in reference position (see appendix 1).

e. Considering c. and d., K can be calculated (equation (5)):

$$K = 10^{-4} \cdot 10^3 \cdot 16 = 1.6 \text{ m}^2 \text{ pr. } ^oK.$$

9.2 Ideal Antenna and Receiver.

Under this assumption, the efficiency factors $\eta_i = \eta_s = \eta_d = 1$ and the noise temperatures due to the antenna system and receiver, T_{AM}, T_{AS}, and $T_R = 0^oK$. Then only celestial noise T_{AO} and precipitation noise T_{AA} enter; the required antenna area is:

$$A_i = K[T_{AO}+T_{AA}] = 1.6[16+75] = 145 \text{ m}^2 \quad (7)$$

It is easily seen, that in clear weather, when $T_{AA} \approx 0^o$, the signal-to-noise ratio is increased by 1.3(91/16) corresponding to 8.8 dB.

9.3 "Best Available" Antenna and Receiver.

The best available receiver is at present equipped with a Maser amplifier with noise temperature 10^oK with loss corresponding to $L = 1.1$ and $T_L = 30^oK$ (see section 8).

"The best available" antenna is, - with respect to area efficiency, - probably the horn reflector antenna, with $\eta_i = 0.8$, $\eta_s = 1$, and $\eta_d = 1$, sidelobe-noise temperature $T_{AS} = 5^oK$, and $T_{AM} = 5^oK$ (see section 5.4).

For comparison, equation (7) is included in equation (6):

$$A_o = \frac{1}{\eta_i \cdot \eta_s \cdot \eta_d}\left[K(T_{AO}+T_{AA})+K(T_{AM}+T_{AS}+T_L+LT_R)\right]$$

$$= \frac{1}{\eta_i \cdot \eta_s \cdot \eta_d}\left[A_i+K(T_{AM}+T_{AS}+T_L+LT_R)\right] \quad (8)$$

Introducing the values above, the necessary geometric antenna area is:

$$A_1 = \frac{1}{0.8}\left[145+1.6(5+5+30+1.1\cdot 10)\right]$$

$$= \frac{1}{0.8}\left[145+82\right] = 284 \text{ m}^2$$

A_1 is then 139 m^2 larger than A_i, the "clear-weather margin" is now 1.3·227/152 corresponding to 2.5 dB.

9.4 Cassegrain Antenna with Perfect Reflector and Maser Amplifier Receiver.

As outlined in section 5 a Cassegrain antenna has somewhat less efficiency and higher noise temperature due to sidelobes and miscellaneous losses. The values used are:

Illumination efficiency $\eta_i = 0.8$

Shadow effects $\eta_s = 0.9$

(Ideal reflector surface is considered in this case, so $\eta_d = 1.0$)

Sidelobe noise temperature $T_{AS} = 20^oK$

Miscellaneous effects $T_{AM} = 5^oK$

Using equation (8), the necessary geometric antenna area is:

$$A_2 = \frac{1}{0.8\cdot 0.9}\left[145+1.6(5+20+30+1.1+10)\right]$$

$$= \frac{1}{0.72}\left[145+105\right] = 348 \text{ m}^2$$

The clear weather margin is app. 1,8 dB.

The area-increase over the horn reflector is 64 m^2. As horn reflectors are considerably more expensive than Cassegrain antennas, this rather modest area increase will probably not compensate for the cost difference, this can explain the current interest in Cassegrain systems.

9.5 Cassegrain Antenna with Imperfect Reflector, Maser Amplifier Receiver.

In this case, the noise temperatures are the same as above, also the efficiencies η_i and η_s, additionally the efficiency due to reflector surface deviations is introduced by two examples (see section 5):

$\eta_d = 0.9$ corresponding to RMS surface deviation $\varepsilon = 2$ mm

and

$\eta_d = 0.63$ for $\varepsilon = 4$ mm

η_d enters as a simple factor, so for:

$\varepsilon = 2$ mm $A_3 = A_2/0.9 = 388 \text{ m}^2$

$\varepsilon = 4$ mm $A_4 = A_2/0.63 = 555 \text{ m}^2$

As discussed in section 5.2, $\varepsilon = 2$ mm should be realizable with some care, the area "penalty" by going to $\varepsilon = 4$ mm is 167 m^2 or 43% area increase. It is doubtful, if the economic advantages of tolerance relaxations are so great, that this area increase can be justified. So, in the following, $\varepsilon = 2$ mm will be assumed and equation (8) is altered by introducing the data for the Cassegrain reflector:

$$A_o = \frac{1}{\eta_i \cdot \eta_s \cdot \eta_d}\left[A_i + K(T_{AM} + T_{AS})\right]$$

$$+ \frac{1}{\eta_i \cdot \eta_s \cdot \eta_d}(T_L + LT_R)$$

$$= \frac{1}{0.8 \cdot 0.9 \cdot 0.9}\left[145 + 1.6(5+20)\right]$$

$$+ \frac{1.6}{0.65}(T_L + LT_R)$$

$$= 1.55\left[145+40\right] + 2.46(T_L + LT_R)$$

$$= 286 + 2.46(T_L + LT_R)\ m^2 \qquad (9)$$

9.6 "Standard Cassegrain" Antenna with Various Receivers.

In section 8, typical noise temperatures for low noise receivers are listed, the corresponding antenna areas are for the antenna parameters as in section 9.5 and ε = 2 mm.

A. Maser, cooled, as above, $T_R = 10^oK$, $A_3 = 388 m^2$

B. Parametric Amplifier, helium cooled, $T_R = 20^oK$, $A_5 = 414 m^2$

C. Parametric Amplifier, nitrogen cooled, $T_R = 40^oK$, $A_6 = 468 m^2$

D. Parametric Amplifier, room temperature, $T_R = 130^oK$, $A_7 = 711 m^2$

For the value of K used, corresponding roughly to the performance of Relay 1, the necessary increase of antenna area by using cooled parametric amplifiers instead of a maser is quite modest, 26 and 80 m^2. If one considers one square meter antenna area as a monetary unit (roughly equal to US$ 1.000) even 80 m^2 increase of antenna area will probably easily compensate for the complexity of maser and helium-cooling with the present technique.

The difference in antenna area represented by cooled/non-cooled parametric amplifier is 243 m^2, this should easily be compensated by the modest complexity of nitrogen-cooling.

9.7 On Detailed Design.

The antenna performance data used above refer in general to more or less conventional designs and improvements are possible. For example, the shadow effect efficiency η_s, here estimated to 0.9 can probably be increased by proper spar design and improvements in the illumination efficiency η_i seem also possible, then reduction of antenna area by 5 to 10% is conceiveable.

10. Concluding Remarks.

The calculations above tend to show, that a Cassegrain antenna system with 400 to 500 m^2 area (corresponding to 22-25 m diameter) in connection with a cooled parametric amplifier will have sufficient capability to handle satellite transmission with spectral power flux density 30 dB below the maximum limit recommended by CCIR, if the ground station is placed in the temperate zone.

For ground stations in some tropical zones, attenuation due to precipitation is at least 5 dB higher, referring to the same probability (see Holzer (1965)). This cannot be compensated completely by low-noise amplifiers or refined antenna designs and it must be expected, that at least some future communications satellites will have spectral power flux densities at least 5 to 10 dB higher than the value considered in the example in order to provide reliable communication in tropical zones.

In temperate zones, where attenuation due to precipitation is considerably less, the required antenna areas in the example in section 9 will be reduced by a factor 3.2 to 10, when designed as a part of a low-noise receiving system, and if the satellite transmission level is increased by 5 to 10 dB.

If this situation arises, it is most probable, that the antenna system will have to be designed according to interference considerations and for its function as a transmitting antenna rather than as a part of a low-noise receiving system.

For reliable performance with present satellites, extreme low-noise performance is necessary and reflectors with 25 m diameter are nearly marginal. It is to be hoped, that advances in the technology of satellite -borne equipment will make higher signal levels at Earth and consequently smaller ground antennas possible.

If the antenna diameter could be reduced to 10 meters or less, the antenna is not a major capital item in the ground station as a whole and with these sizes together with redundant equipment, it ought to be possible to design a remotely controlled station, attended by a few persons for maintenance.

Much of the material presented above originates from the authors participation in the design of the Scandinavian Earth Station for Satellite Communications Experiments built by the Scandinavian Committee for Satellite Telecommunications, STSK. The persons participating in the design of this station have indirectly contributed considerably, I acknowledge this gratefully.

Especially I will mention Mr. M. Grønlund of the Microwave Laboratory under The Danish Academy of Technical Sciences, Copenhagen, Mr. P. Christensen of the Danish Post Office, Radio Department, Mr. L. Hansson of the Swedish Telecommunications Administration, Radio Department, and Mr. K. Holberg, Mr. E. Evensen, Mr. Y. Lundh of the Norwegian Defense Research Establishment.

Appendix 1

On Basic Equations and Limitations for the Ground Receiving System.

A.1.1 On Signal-to-Noise Ratio.

The following expressions are a simplified model of an actual system, considering the fundamental factors only.

Consider the antenna oriented towards a white noise signal source with an incident signal flux of S Watt pr.m^2 pr. c/s. (The symbol S is used by radioastronomers for the same unit, the unit Jansky for S has been proposed, and is retained here; it should not be confused with the "S" in (S/N).

The received power P_S in an effective aperture area A with a receiver with noise bandwidth B is then:

$$P_S = S \cdot A \cdot B \qquad (A.1)$$

The parameter S is dependent on the Satellite transmitter power P_t and transmitted bandwidth and modulation system as well as distance from the satellite to the ground station.

In the receiver, the total noise referred to the input aperture of the antenna is

$$P_N = k \cdot T_S \cdot B \qquad (A.2)$$

where k is Boltzmanns constant and T_S the system noise temperature.

The signal-to-noise ratio for the system can then be defined as:

$$(S/N) = \frac{P_S}{P_N} = \frac{S \cdot A \cdot B}{k \cdot T_S \cdot B} = \frac{S}{k} \cdot \frac{A}{T_S} \qquad (A.3)$$

A.1.2 On Maximum Power Flux.

Recommendation 358 (Documents of the Xth Plenary Assembly, vol. IV, p. 162-163) concerns maximum levels of power flux density set up at the surface of the Earth for communication satellite systems using wide deviation frequency modulation and sharing frequency bands with line-of-sight radio-relay systems.

Some of these recommendations have been modified in Monte Carlo 1965, so that the maximum power flux density set up at the Earths surface by the emissions from a satellite, measured in any 4 kc/s bandwidth, should not exceed:

-152 dBW/m^2 pr. 4 kc/s for all angles of arrival.

As modified, this is valid for all types of emissions, the specification for maximum power flux density (-130 dBW/m^2)is no longer valid.

The corresponding spectral power flux density S_o (in Watts pr. m^2 pr. c/s) is then:

$$10 \log S_o = -152 - 10 \log B = -152 - 36$$
$$= -188 \text{ dBW/m}^2\text{c/s}$$

This is a convenient reference as it represents the "legal limit" to which actual systems can be referred in "q dB below S_o" or $S = S_o/Q$.

To illustrate:

Seen from a satellite, the apparent area of the Earth is approximately $1.2 \cdot 10^{14} m^2$.

A satellite transmitter with directional antenna illuminating only this area should then radiate a maximum spectral power density:

$$10 \log S_o + 10 \log(1.2 \cdot 10^{14}) = -47 \text{dBW/c/s} \qquad (A.4)$$

As most modulation methods used result in a RF signal with characteristics essentially as band-limited noise, the spectral power density of the transmitter is approximately P_T/B_t, where P_T is the total transmitter power and B_t the bandwidth of the modulating signal.

For B_t = 20 Mc/s, as an example, the maximum effective radiated power of the satellite transmitter is:

$$10 \log P_T = -47+73 = +26 \text{ dBW};$$

$$P_T = 400 \text{ Watt}.$$

This is somewhat more than realized with the present technology.

This calculation assumes that the satellite has a directional antenna limiting radiation to the apparent surface of the Earth. The maximum transmitter power is then independent of the altitude of the satellite.

So far, the antennas of the satellites do not approach that characteristic and present communication satellites have much less power flux, corresponding to high values of the factor q expressing the number of decibels below the "legal limit" S_o.

Relay 1 has been used for many experiments, for this, the nominal values are effective radiated power P_T = 10 Watt and antenna gain G_A = 0 dB. Effective bandwidth B = 25 Mc/s, in a distance d_o = 12.600 km (6.800 nautical miles). The calculated power flux density (corresponding to basic transmission loss L_B = 186.4 dB):

$$10 \log p_1 = -143 \text{ dBW/m}^2$$

and the spectral power flux density:

$$10 \log S = -143-74 = -217 \text{ dBW/m}^2\text{c/s}$$

as

$$10 \log S_o = -188 \text{ dBW/m}^2\text{c/s}$$

$$q = -188-(-217) = +29 \text{ dB}.$$

The COMSAT satellite HS 303 has the same maximum effective radiated power, but 10 dB more transmission loss, for B = 20 Mc/s as in the example above, and q = 39 dB.

Appendix 2

On Noise Standards.

A relevant proposal has been published by CCIR:
Report 208 "Active Communications Satellite Systems for Frequency-Division Multiplex Telephony and Monochrome Television, Form of the basic hypothetic of reference circuit and allowable noise standards; video bandwidth and sound channel for television" (Documents of the Xth Plenary Assembly, vol. IV, pp. 218-221).

A common hypothetical reference circuit is recommended, for both frequency-division multiplex telephony and monochrome television based on a single satellite link.

The allowable noise in the basic reference circuit is specified:

a. "The maximum value 10.000 pW (psophometrically weighted) for the mean noise in any hour in any telephone circuit ...."

b. "The value 80.000 pW (psophometrically weighted) for the one-minute mean value exceeded in less than 0.2% of any month...."

LITERATURE

Bracewell (1961):
Tolerance Theory of Large Antennas.
IRE Trans. on Antennas and Propagation, vol. AP-9, Jan 1961, no.1, pp.49-.

Canfield (1959):
Gearless Drive for Power Servo.
Electrical Manufacturing, pp. 142-, Jan 1959.

Caspar & Sandreczki (1964):
Ice Deposits from the Meteorological Standpoint.
Elektrotechnische Zeitschrift, Ausg.B, vol.16, Dec 18, 1964, no.26, pp. 763-768.

Cuccia (1964):
How to Buy a Low-Noise-Traveling -Wave Tube ?
Microwaves, Sep 1964, pp. 70-.

Dragone & Hogg (1963):
Wide-Angle Radiation Due to Rough Phase Fronts.
The Bell System Technical Journal, vol. 42, Sep 1963, no.5, pp. 2285-.

Feldman (1964):
Estimates of Communication Satellite System Degradation Due to Rain.
AD 609.427, Oct 1964.

Foldes (1962):
The Capabilities of Cassegrainian Microwave Optics Systems for Low Noise Antennas.
Solid State Electronics, vol.4, pp. 319-, Oct 1962.

Galindo (1964):
Design of Dual Reflector Antennas with Arbitrary Phase and Amplitude Distributions.
IEEE Trans. on Antennas and Propagation, vol. AP-12, July 1964, no.4, pp. 403-408.

Gray (1964):
Estimating the Effect of Feed Support Member Blocking on Antenna Gain and Side-Lobe Level.
Microwave Journal, March 1964.

Hannan (1961):
Microwave Antennas Derived from the Cassegrain Telescope.
IRE Trans. on Antennas and Propagation, vol. AP-9, March 1961, no.2, pp. 140-.

Hines (1963):
The Electrical Characteristics of the Conical Horn-Reflector Antenna.
Bell System Technical Journal, vol. 42, pp. 1187-, July 1963.

Hirst & McKee (1964):
Wind Forces on Parabolic Antennas.
Bulletin 602, Andrew Corporation, Chicago, Aug 1964.

Hogg & Semplak (1964):
An Experimental Study of Near-Field Cassegrainian Antennas.
The Bell System Technical Journal, vol. 43, Nov 1964, no.6, pp. 2677-2705.

Hogg & Semplak (1961):
The Effect of Rain and Water Vapour on Sky Noise at Centimeter Wavelengths.
Bell System Technical Journal, vol. 40, pp. 1331-, Sep 1961.

Holzer (1965):
Atmospheric Attenuation in Satellite Communications.
Microwave Journal, vol. 8, pp. 119-, March 1965.

Jensen (1963):
A Low-Noise Multimode Cassegrain Monopulse Feed with Polarization Diversity.
NEREM Record 1963, Session 11: Antenna Feed Systems, pp. 94-96.

Lundh (1965):
Styrningen av radioteleskopet på Råö.
Teknisk Tidskrift, nr. 18, 1965, pp. 479-82.

Morgan (1964):
Some Examples of Generalized Cassegrainian and Gregorian Antennas.
IEEE Trans. on Antennas and Propagation, vol. AP-12, Nov 1964, no.6, pp. 685-91.

Potter (1963):
Aperture Illumination and Gain of a Cassegrainian System.
IEEE Trans. on Antennas and Propagation, vol. AP-11, pp. 373-, May 1963.

Potter (1962):
The Application of the Cassegrainian Principle to Ground Antennas for Space Communications.
IRE Transactions, vol. SET-8, pp. 154-, June 1962.

Potter (1964):
The Design of a Very High Power, Very Low Noise Cassegrain Feed System for a Planetary Radar.
Jet Propulsion Lab.,
Rept. no. 32-653, Aug 24, 1964.

Ravenscroft (1962):
Primary Feeds for a Satellite Communication Aerials.
Internat. Conf. on Satellite Comm.,
22-28 Nov 1962, pp. 154-.

Rolinski et al. (1964):
Satellite Tracking Characteristics of the X-Y Mount for Data Aquisition Antennas.
NASA TN D-1697, June 1964.

Rolinski et al. (1962):
The X-Y Antenna Mount for Data Acquisition from Satellites.
IRE Trans.,
vol. SET-8, pp. 159-, June 1962.

Rule (1963):
Problems of Large Radio Telescopes.
Symposium on Engineering for Major Scientific Programs,
Feb 5-6, 1963, Georgia Inst. of Techn.,
pp. 34-56.

Ruze (1952):
Physical Limitations on Antennas.
Research Laboratory of Electronics,
MIT Tech. Report no. 248, Oct 30, 1952.

Ruze (1964):
Reflector Tolerance Determination by Gain Measurement.
NEREM Record 1964, pp. 166-167.

Stallard (1964):
Servo Problems and Techniques in Large Antennas.
IEEE Trans. on Applications a. Industry,
March 1964, no. 71, pp. 105-114.

Victor (1961):
Ground Equipment for Satellite Communications.
Jet Propulsion Lab.,
Tech. Report no. 32-137 (N 62 10817),
Oct 30, 1961.

Viggh (1963):
Designing for Desired Aperture Illuminations in Cassegrain Antennas.
IEEE Trans. on Antennas and Propag.,
vol. AP-11, March 1963, no.2, pp.198-.

Visocekas (1964):
Noncassegrainian Indirect System for Aerial Illumination.
Proceedings of the IEE,
vol. 111, Dec 1964, no.12, pp. 1969-76.

Wilkinson & Appelbaum (1961):
Cassegrain Systems.
IRE Trans. on Antennas and Propagation,
vol. AP-9, Jan 1961, no.1, pp. 119-.

Willoughby (1962):
Egg Crate Reflector Surface for Large Paraboloids.
Proceedings IRE,
vol. 50, no.12, Dec 1962, pp. 2521-2522.

HOW TO DETERMINE SPACE VEHICLE ORBITS USING AUTOMATED MEASURING GEAR

E. L. Akim, M. D. Kislik, P. E. Él'yasberg, T. M. Eneev

The successful resolution of scientific problems involving artificial celestial bodies (space vehicles) launched into outer space depends in large measure on the accurate determination of the actual orbits of these vehicles.

Orbit determination is a necessity to establish the fact that the vehicle has actually entered a space orbit, to effect a tie-in in intertial space for scientific measurements taken on board the vehicle, to predict the motion and deliver commands controlling the vehicle, and to deliver target destination information via observations and communication. Moreover, the actual orbit traversed by the space vehicle can be utilized in many instances to secure improved information on the forces operating in flight on the vehicle, and consequently on the various geophysical and astronomical parameters governing the magnitude of those forces.

Most of the problems listed above require orbit determination of space vehicles to high order of precision, within very short time spans, and independently of season, time of day, or atmospheric conditions. The observational means applicable in astronomy in the study of the motion of natural celestial bodies fail to meet these requirements. Reliance on astronomical techniques in space-vehicle orbit determination is therefore relegated to an auxiliary plane.

Special-purpose automated measuring gear has had to be devised to facilitate determination of space-vehicle orbits. Mandatory components of this gear include the following items:
a) high-precision radio electronic measuring systems capable of determining the required parameters of the vehicle's motion in space under any and all conditions;
b) means for automatic telemetering of the information;
c) electronic computers capable of mathematically processing the arriving information needed in orbit determination.

Aside from the work of devising the needed technical devices for measurements, transmission and processing of information essential in space-vehicle orbit determination, no less important is the development of special mathematical techniques designed to take the special features of these orbits into account, not to mention the potentialities of electronic computer techniques. These methods differ from the methods employed in astronomy, even though they still naturally hark back to the general laws of celestial mechanics in their fundamentals.

The development of these techniques facilitates the solution of the following major problems:
a) arriving at a suitable set of assumptions under which the mathematical description of the motion in space of the vehicle will occur (determining the form of the right-hand members of the differential equations of motion of the vehicle launched);
b) arriving at a suitable statistical method for estimating the unknown orbit parameters on the basis of redundant information in the measurements;
c) compiling algorithms of orbit determinations which will be optimal in the sense of precision, rapid solution of problems, and reliability in programming the calculations in electronic computers.

It goes without saying that these questions are intimately related and that the distinction drawn is purely arbitrary.

In the selection of the right-hand members of the differential equations of the vehicle's motion, the deciding factor is the shape of the orbit. In the case of low-orbiting artificial earth satellites, a useful rule of thumb is to restrict the problem to the influence of the earth's gravitational field and of atmospheric drag. For satellites having an apogee height exceeding 7000 to 10,000 km, the effect of the lunar and solar gravitational fields will have to be considered as well. If the perigee height of the satellite orbit exceeds 1000 km, then the atmospheric drag force may be excluded from the equations of motion. When the motion of lunar space vehicles is being calculated, it is sufficient to take into account the influence exerted by the earth, moon, and sun, while neglecting any attraction exerted by the planets. The orbits of interplanetary vehicles may usually be described by right-hand members of the differential equations which differ in their structure over different orbit segments, the structure depending on the local intensity of the gravitational fields established by the major bodies of the solar system. If a space vehicle possesses a reasonably small ratio of mass to effective reflecting area, the force contributed by solar radiation pressure should be introduced into the equations of motion. In practically all cases, the space vehicle is regarded as a mass point on the orbital flight segment (i.e., the effect of the rotational motion on the motion of the center of mass will be ignored). It must be borne in mind that the recommendations made on the nature of the right-hand members of the equations of motion of space vehicles of different types cannot be conclusive or definitive. They have to be revised to meet any major alteration of the orbit parameters or vehicle parameters, and to concur with any improvements in our knowledge about outer space and the bodies making up the solar system.

In making a selection of a statistical techniques for estimating the unknown orbit parameters, account must be taken of the nature of the information being processed. The usual assumption is that errors in the a posteriori measuring information are mainly of a random nature, subject to a normal distribution law of zero mathematical expectation and known variance. As a rule, these errors are assumed to be independent. In conjunction with the a posteriori measurement information used in determinations of the parameters of a motion, a priori information on the trajectory is also fitted into the total picture in many instances. A priori information contains expected values of specific functions of the trajectory parameters with their corresponding probabilistic characteristics of the likely errors. This information will result, for example, from processing measurements of earlier segments of the trajectory and from estimates of the spread in the results obtained. It may be used as the totality of several correlation measurements with known a priori statistical characteristic in the form of an error correlation matrix.

The presence of these two groups of measurements in the information being processed leads to the choice of one of the most effective techniques (in the sense of offering minimum dispersion) for estimating the unknown parameters as our statistical technique — the maximum likelihood method. When only uncorrelated measurements are being processed, the maximum likelihood method reduces to the usual method of least squares, an eventuality which immeasurably speeds up the data processing.

The maximum likelihood method employed in order to determine the unknown parameters (Q_i) $i = 1,2,.., m \geq 6)$ from the measurements $\overline{\psi}_1, \overline{\psi}_2, \ldots, \overline{\psi}_n, \ldots, \overline{\psi}_l, \ldots, \overline{\psi}_M$ leads to the necessity of minimizing the functional

$$\phi = \sum_{n,l=1}^{M} P_{nl} \xi_n \xi_l = \sum_{n,l=1}^{M} P_{nl} (\psi_n - \overline{\psi}_n)(\psi_l - \overline{\psi}_l) \quad (1)$$

where P_{nl} are the elements of the "weighted" matrix P expressed in terms of the correlation matrix K_ψ of the measurements by the formula

$$P = K_\psi^{-1} \quad (2)$$

As a result of mimization of the functional (1), we arrive at a system of equations nonlinear in the parameters Q_i:

$$\frac{\partial \phi}{\partial Q_i} = 0 \quad (i = 1, 2, \ldots, m) \quad (3)$$

or (taking into account symmetry of the matrices K_ψ and P)

$$\sum_{n,l=1}^{M} P_{nl} \xi_n \frac{\partial \xi_l}{\partial Q_i} = 0 \quad (i = 1, 2, \ldots, m) \quad (4)$$

The devising of an optimum algorithm for orbit determination involves detailed elaboration of a method for solving system (4). Two methods for solving this problem have found practical applications: Newton's method of generalized tangents and the method of steepest descent.

The solution by Newton's method leads, as we know, to a series of successive approximations. The number of approximations depends on how close the zero-order approximation selected is to the exact solution. At each approximation step in determining the corrections ΔQ to the parameters obtained in the preceding approximation we have to solve a normal system of equations

$$\sum_{j=1}^{m} A_{ij} \Delta Q_j = B_i \quad (i = 1, 2, \ldots, m) \quad (5)$$

where the coefficients A_{ij} and the right-hand members B_i exhibit the form

$$A_{ij} = \sum_{n,l=1}^{M} P_{nl} \frac{\partial \xi_n}{\partial Q_i} \frac{\partial \xi_l}{\partial Q_j}, \quad (6)$$

$$B_i = -\sum_{n,l=1}^{M} P_{nl} \frac{\partial \xi_n}{\partial Q_i} \xi_l \quad (7)$$

The matrix and the right-hand members of systems (5) are formulated with the aid of formulas (6) and (7) in each approximation step for the trajectory to be determined from the parameters of the preceding approximation.

In the absence of any fairly decent zero-order approximation, or when the measurements being processed leave much to be desired in the way of precision, so that Newton's method will not readily yield the solution, the method of steepest descent may fill the gap. This method reduces to a system of differential equations

$$\sum_{j=1}^{m} A_{ij} \frac{dQ_j}{ds} = \frac{B_i}{\sqrt{\phi}} \quad (i = 1, 2, \ldots, m) \quad (8)$$

where the variable s is the parameter along the descent path in the m-dimensional space Q of the variables.

System (8) is numerically integrated by the Euler method. When initial conditions are assigned judiciously, the integration interval will be narrow.

The method of steepest descent substantially improves the reliability of the mathematical data processing techniques in the statistical problem concerning processing of measurements, but is inferior in rapidity of solution to Newton's method. This dictates combining the two methods in attempting to solve the problems, preference being conferred upon the more rapid Newton method. But the method of steepest descent offers the further attraction of substituting reliably when the normal convergence of the process breaks down in the use of Newton's method.

We gather from the foregoing that at each approximation step in the use of Newton's method, or at each integration step in the use of the method of steepest descent, we are obliged to calculate out the values of A_{ij} and B_i from formulas (6) and (7). These formulas include the derivatives $\partial\xi/\partial Q$ and the differences between theoretically predicted and actually observed values of the parameters being measured for each instant of observation and for each measured parameter. Since the total quantity of the measurements involved in the processing may be enormous, while the segment of the flight path from which these measurements were taken may be quite high, reduction of the total data processing time becomes a question of utmost seriousness. This problem is successfully solved by using the most economic techniques in computing derivatives and rated values of the parameters being measured, as well as reduction in the total quantity of approximations (integration steps) during the orbit determining process.

For example, in computing the derivatives, account must be taken of the comparatively mild requirements imposed on accuracy, and the calculations are carried out in almost all cases on the basis of the finite formulas of Keplerian motion. A refinement of the zero-order approximation leading to a reduction in the total number of approximation steps is achieved by resorting to a two-position method of orbit determination, smoothing, and numerical differentiation of the coordinates in order to determine the velocities, etc. Shortening of the time needed for orbit calculations in the parameters Q when the differences ξ are computed becomes possible when an optimum choice is made of the coordinate frame, method, and integration steps, in the integration of the differential equations of motion of the spacecraft. Approaches of this sort have been worked out in large number and are used in the practice of orbit determination.

Summarizing, it should be noted that the solution of the problem of space-vehicle orbit determinations meeting predetermined requirements became possible thanks to the use of sophisticated automatic measuring gear, to telemetering and processing of information on one hand, and thanks to the development of efficient mathematical tools on the other hand.

LITERATURE CITED

1. Sborniki "Iskusstvennye sputniki Zemli." USSR Academy of Sciences: No. 4 (1960); No. 13 (1962); No. 16 (1963).
2. Kosmicheskie issledovaniya. 1, No. 1 (1963); 2, No. 4, No. 5 (1964).

ON SYSTEMATIC ERRORS IN TRAJECTORY DETERMINATION PROBLEMS

by Dr. Alfons J. Claus
Member of the Technical Staff,
Bell Telephone Laboratories, Inc.
Whippany, New Jersey

ABSTRACT

Formulas are presented for the determination of satellite trajectories from observations containing random as well as systematic errors. The case where the systematic errors consist of instrument biases is treated in some detail. It is shown that the accuracy of the computed trajectory becomes highly insensitive to the magnitude of bias errors as soon as the variance of the arithmetic mean of the random errors becomes negligible. The latter conclusion is verified on the basis of numerical computation. A short discussion is devoted to the question of so-called imbred bias errors.

INTRODUCTION

The presence of systematic errors among observations to be used for establishing a satellite orbit invariably has a deteriorating influence on the accuracy of the computed orbital elements. The extent to which the orbit is affected depends obviously on the way the observations are handled. In the vast majority of the present day operational orbit determination methods, an orbit is found which matches the observations according to a weighted least squares fit. Such a procedure extracts the maximum amount of information from the available data only if the latter happen to be uncorrelated, which is hardly ever the case in practice. The reason for the application of a least squares filter is, among others, its inherent simplicity. Although this is undoubtedly a highly desirable characteristic, it seems worthwhile to investigate at what price it is obtained. In particular, the question pertaining to loss of accuracy resulting from ignoring systematic errors tends to be a pertinent one indeed.

It is clear that any acknowledgment of systematic observational errors requires the so-called weighting matrix of the observations to have nondiagonal elements different from zero. Apparently, only the treatment of special cases can be carried through with sufficient detail to be practically useful. However, these cases offer valuable insight and reveal interesting features connected with the use of filters designed for the removal of low frequency errors.

The case that is considered in this paper consists of the determination of the orbit of a near-earth satellite from biased observations from a single tracking station. A brief theoretical discussion, confirmed by numerical computation, shows that the computed orbit becomes completely insensitive to observational bias errors if the variance of the arithmetic mean of the random errors approaches zero. Various curves are given illustrating effects of pass length (smoothing time) and magnitude of the relevant errors on the accuracy of the resulting orbital elements.

The theoretical investigation leads naturally into the issue of "inbred" bias errors and is treated in some detail. Briefly, it is shown that, if the optimum filter ceases to be valid due to singular behavior, an appropriate number of bias errors should be ignored. Although this will undoubtedly cause a loss of accuracy of the established orbit, it preserves the quality of predicted observations provided they are of the same type and pertain to the same station(s) as the ones used for the orbit determination process itself.

FORMULATION AND DISCUSSION

Before we engage in the detailed calculations a few more general remarks may be appropriate. Strictly speaking, they apply only to situations in which a rather substantial amount of observations is processed simultaneously.

Minimum variance estimators, as well as estimators based on the maximum likelihood principle, involve the inversion of the covariance matrix of the observational errors. In one of the simplest applications, namely the least squares method, the inversion is obviously immediate since covariance matrices of all observational errors are assumed to be diagonal. Inversion problems may begin to appear as soon as the latter assumption is removed, thereby recognizing the presence of nonrandom errors.

The reason for this is twofold. First, the size of the covariance matrix could be rather large, the matrix having a number of rows (and columns) equal to the number of observations to be processed. This disadvantage can be somewhat relieved by prefiltering, or reducing observations to so-called normal places. Second, it will be explained later that the more one wishes to enhance the presence of low frequency errors, the more the corresponding covariance matrix approaches the singular case. It seems, then, that a straightforward application of a maximum likelihood (or minimum variance) filter, when simultaneously applied to a large number of observations, should represent a compromise between two possible extremes: high correlation coefficients to account for systematic errors and zero correlation to make the matrix inversion trivial.

We mentioned earlier that prefiltering, perhaps using polynomial filters, can relieve the inversion problem to some extent. Actually, the degree of possible improvement is rather limited. The reason lies in the fact that polynomial filters of reasonably short smoothing time, as must be the case in applications under present discussion, form no effective shield against systematic errors, but may reduce high frequency errors substantially. It follows that the output noise can have quite a narrow bandwidth resulting in a high correlation between the prefiltered observations, thus causing the covariance matrix to be ill-conditioned. Hence, the advantage of having a smaller size matrix to invert is apparently offset by the fact that the new matrix contains substantially higher correlation coefficients than the original one.

The implication of the previous discussion is that any method of orbit determination which calls for the explicit numerical inversion of the covariance matrix of observational errors should be avoided. Apparently, a more fruitful approach to the treatment of systematic errors consists of assuming the errors to originate from random processes such that an analytic inversion of the covariance matrix becomes practically possible. Obviously, this may put quite serious constraints on the error models amenable to analysis. However, it turns out that one of the most important cases in practice, namely errors resulting from an imperfect knowledge of certain constants, belongs to the latter category. Cases of slowly varying errors, such as sometimes encountered in range and range rate systems, can also be handled. In the latter cases, the errors may be assumed to originate from Markov sequences of appropriate order and bandwidth.

Some of the arguments presented above may have appeared rather vague. We will attempt to illustrate them in the case of systematic errors expressible in a functional form involving only unknown parameters, and which was earlier described as being of considerable practical importance.

Assume the error vector $\delta\lambda$ of the observations to consist of a vector ε whose components form a set of statistically independent random variables, and a vector $N\delta\nu$, where $\delta\nu$ is the error vector of the relevant parameters; N is the matrix of sensitivities of the observations with respect to these parameters. Thus, $\delta\lambda = \varepsilon + N\delta\nu$. If Φ, Ψ, Λ represent the covariance matrices of $\delta\lambda$, ε and $\delta\nu$, respectively, it follows that

$$\Phi = \Psi + N\Lambda N'$$

in which the prime denotes transposition.

To simplify the argument, let us consider the case of n observations where each observation is corrupted by a random error with standard deviation σ and an error with standard deviation γ, the latter error being the same for all observations. We then have

$$\Phi = \sigma^2 I + \gamma^2 Q$$

in which I represents an nxn unit matrix and Q is an nxn matrix whose elements are all equal to unity. Let us examine the matrix Φ for small values of σ^2/n.

Clearly, when σ approaches zero, the matrix Φ becomes singular since Q is a matrix of rank one $(n > 1)$. The behavior of Φ for large n is investigated by studying the value Δ_n of the determinant of the correlation matrix,

$$\Delta_n = \frac{(\sigma^2+n\gamma^2)\,\sigma^{2(n-1)}}{(\sigma^2+\gamma^2)^n}$$

It is easily shown that $\lim_{n\to\infty} \Delta_n = 0$. Thus, the smaller the value of σ^2/n, the more troublesome the inversion of the covariance matrix of the observational errors. On the other hand, small values of σ^2/n imply that the prevailing situation is such that the random errors can be "averaged out". Under these circumstances it is reasonable to expect that effects of bias errors in the observations on the computed trajectory are likely to be small if the data are processed in an optimum fashion. Consequently, we are led to the conclusion that situations which look promising for bias removal are indeed the ones where computational difficulties are likely to appear in the numerical inversion of the covariance matrix of the observational errors.

The previous example can also serve as a guide line to illustrate the effects of prefiltering. Suitable polynomial filtering applied to densely spaced data (for instance) should not noticeably affect possible bias errors but will reduce the random noise level considerably.

Hence, although in the expression for Δ_n, n becomes a reasonably small value (the resulting number of "normal places"), σ^2 approaches zero with an increasing amount of raw data.

It becomes quite evident that effective bias removal and ill-conditioned covariance matrices of the observational errors go hand in hand. This is exactly the reason why we insist on an analytic inversion of Φ. To be sure, any inversion of Φ is obviously impossible of $\sigma = 0$ but, even in the latter case, the formulation to be derived will maintain its consistency.

The actual derivation is given elsewhere[2] but is briefly repeated here for completeness. Under appropriate assumptions, a set of orbital elements, represented by the column vector α, is adjusted to the set $\hat{\mu}$ according to the formula[1],[4]

$$\hat{\mu} = \alpha + (C^{-1}+J'\Phi^{-1}J)^{-1}J'\Phi^{-1}\delta\lambda , \qquad (1)$$

in which $\delta\lambda$ is the vector of residuals in the relevant observations, the residuals being computed on the basis of the elements α; C is the covariance matrix of α; Φ is the covariance matrix of the observational errors and J is the matrix of partial derivatives of the observations with respect to the orbital elements. The covariance matrix of the refined elements is

$$\mathrm{cov}(\hat{\mu}) = (C^{-1}+J'\Phi^{-1}J)^{-1} \qquad (2)$$

The analytic inversion of Φ is accomplished by making use of the matrix identity

$$(\Psi+N\Lambda N')^{-1} = \Psi^{-1} - \Psi^{-1}N(\Lambda^{-1}+N'\Psi^{-1}N)^{-1}N'\Psi^{-1} .$$

The matrices $J'\Phi^{-1}J$ and $J'\Phi^{-1}\delta\lambda$ appearing in the formulas (1-2) become

$$J'\Phi^{-1}J = J'\Psi^{-1}J - J'\Psi^{-1}N(\Lambda^{-1}+N'\Psi^{-1}N)^{-1}N'\Psi^{-1}J,$$

$$J'\Phi^{-1}\delta\lambda = J'\Psi^{-1}\delta\lambda - J'\Psi^{-1}N(\Lambda^{-1}+N'\Psi^{-1}N)^{-1}N'\Psi^{-1}\delta\lambda . \qquad (3)$$

We wish to examine the case where the systematic error is predominant. To this effect, let σ^2 approach zero in $\Psi = \sigma^2 I$. This is the case where a straightforward application of the formulas (1-2) is impossible. As mentioned earlier, any method of inverting Φ for $\sigma = 0$ will result in failure, but emphasis must be placed on the fact that analytic inversion of Φ for any nonzero value of σ, however small, leads to a numerical formalism for orbit refinement which is ultimately still valid for $\sigma = 0$. The necessary formulas are easily obtained by substituting $\sigma^2 I$ for Ψ in the expressions (3),

$$J'\Phi^{-1}J = \frac{1}{\sigma^2}\; [J'J-J'N(\sigma^2\Lambda^{-1}+N'N)^{-1}N'J] ,$$

$$J'\Phi^{-1}\delta\lambda = \frac{1}{\sigma^2}\; [J'\delta\lambda-J'N(\sigma^2\Lambda^{-1}+N'N)^{-1}N'\delta\lambda] .$$

Thus,

$$\hat{\mu} = \alpha + [\sigma^2C^{-1}+J'J-J'N(\sigma^2\Lambda^{-1}+N'N)^{-1}N'J]^{-1} \cdot [J'\delta\lambda-J'N(\sigma^2\Lambda^{-1}+N'N)^{-1}N'\delta\lambda] , \qquad (4)$$

and

$$\mathrm{cov}(\hat{\mu}) = \sigma^2\; [\sigma^2C^{-1}+J'J-J'N(\sigma^2\Lambda^{-1}+N'N)^{-1}N'J]^{-1}. \qquad (5)$$

Formulas (4-5) are applicable for any value of σ. In particular they are suitable for the study of cases in which $\sigma = 0$, provided the matrix $J'J-J'N(N'N)^{-1}N'J$ is nonsingular. The case for which the latter condition is not satisfied is quite exceptional and is discussed in some detail in the next section. However, it is assumed throughout this paper that both matrices J'J and N'N are nonsingular.

If the matrices J and N are partitioned according to $J' = \left(J_1' \; J_2' \; \dots \; J_n'\right)$, $N' = \left(N_1' \; N_2' \; \dots \; N_n'\right)$ where J_i and N_i correspond to the i-th observation, formula (5) may be put in the form

$$\mathrm{cov}(\hat{\mu}) = \frac{\sigma^2}{n}\, S^{-1}(\sigma^2,n) \qquad (6)$$

in which

Superior numbers refer to similarly-numbered references at the end of this paper.

$$S(\sigma^2,n) = \frac{\sigma^2}{n} C^{-1} + \frac{1}{n} \sum_{i=1}^{n} J_i' J_i$$

$$- \frac{1}{n} \left(\sum_{i=1}^{n} J_i' N_i \right)$$

$$\cdot \left(\frac{\sigma^2}{n} \Lambda^{-1} + \frac{1}{n} \sum_{i=1}^{n} N_i' N_i \right)^{-1} \left(\sum_{i=1}^{n} N_i' J_i \right).$$

It is readily shown that all elements of the matrix $S(\sigma^2,n)$ remain bounded for either σ approaching zero, or n approaching infinity. Formula (6) then indicates that in the hypothetical case where the random errors are averaged out, that is in the case where σ^2/n is zero, the trajectory will be determined with absolute precision in spite of the fact that bias errors may be present. Let us recall once more that this is exactly the case for which a straightforward application of the formulas (1-2) becomes impossible. It is of some importance to realize that, of course, the previous situation will hardly ever arise in practice, but the fact remains that σ^2/n can become quite small compared to anticipated bias errors, in which case serious numerical difficulties may be encountered during the inversion of the matrix Φ.

In this paper, orbit determination computations are carried out using n slant range measurements and n measurements of azimuth angles and elevation angles from a single tracking station. Systematic errors are assumed to consist of a constant error for range R, a constant error for elevation angle E and an error for azimuth angle A which is inversely proportional to cos E. No prior information is assumed to be known, so that C^{-1} is put equal to a null matrix. Application of the formulas (1-3) yields

$$\hat{\mu} = \alpha + \left[\sum_{i=1}^{n} J_i' W J_i - \left(\sum_{i=1}^{n} J_i' \right) W\Lambda(n\Lambda + W^{-1})^{-1} \left(\sum_{i=1}^{n} J_i \right) \right]^{-1} \cdot$$

$$\left[\sum_{i=1}^{n} J_i' W \delta\lambda_i - \left(\sum_{i=1}^{n} J_i' \right) W\Lambda(n\Lambda + W^{-1})^{-1} \left(\sum_{i=1}^{n} \delta\lambda_i \right) \right], \quad (7)$$

$$\mathrm{cov}(\hat{\mu}) = \left[\sum_{i=1}^{n} J_i' W J_i - \left(\sum_{i=1}^{n} J_i' \right) W\Lambda(n\Lambda + W^{-1})^{-1} \left(\sum_{i=1}^{n} J_i \right) \right]^{-1} \quad (8)$$

in which W^{-1} is the covariance matrix of random errors in a data triplet R,A,E.

If the elements of the matrix Λ are much larger than the ones of W^{-1}/n, that is if the bias errors are predominant, formula (8) becomes

$$\mathrm{cov}(\hat{\mu}) \approx \left[\sum_{i=1}^{n} J_i' W J_i - \frac{1}{n} \left(\sum_{i=1}^{n} J_i' \right) W \left(\sum_{i=1}^{n} J_i \right) \right]^{-1} . \quad (9)$$

Obviously, for any matrix $\Lambda \neq 0$, there exists a sufficiently low noise level such that the above expression for $\mathrm{cov}(\hat{\mu})$ becomes applicable. Thus, for such noise levels, formula (9) provides a measure of the attainable accuracy in the presence of bias errors and, to a first approximation, is seen to be independent of the actual value of these errors.

On the other hand, if $\Lambda = 0$, indicating the fact that no bias errors are present, we obtain

$$\mathrm{cov}(\hat{\mu}) = \left[\sum_{i=1}^{n} J_i' W J_i \right]^{-1} . \quad (10)$$

The formulas (9) and (10) considered together deserve some further attention. Both formulas indicate that, at least for sufficiently small random errors in the case of (9), the accuracy of the established trajectory depends solely on W and n, a statement which is hardly surprising with regard to formula (10). In connection with formula (9), however, the feature to be emphasized is the fact that the mere presence of bias errors, independent of their magnitude, is sufficient to make the formula applicable. The reason for this perhaps somewhat puzzling conclusion must be found in the behavior of the matrix $\Lambda\left(\Lambda + \frac{W^{-1}}{n}\right)^{-1}$ in the vicinity of $\Lambda = W^{-1} = 0$. Indeed, the latter matrix function is discontinuous at $\Lambda = W^{-1} = 0$. If $\Lambda = 0$ (no bias errors), $\Lambda\left(\Lambda + \frac{W^{-1}}{n}\right)^{-1} = 0$ for any $W^{-1} \neq 0$. On the other hand if $W^{-1} = 0$, $\Lambda\left(\Lambda + \frac{W^{-1}}{n}\right)^{-1} = I$ for any $\Lambda \neq 0$ (bias errors present). This peculiarity is further illustrated in the last section by means of the numerical results.

If C_b and C_r represent, respectively, the covariance matrices of the orbital elements in the events that bias errors are and are not present, suitable matrix manipulations on the formulas (9-10) yield

$$C_b = C_r + C_r\left(\sum_{i=1}^{n} J_i'\right) \cdot \left[nW^{-1} - \left(\sum_{i=1}^{n} J_i\right) C_r \left(\sum_{i=1}^{n} J_i'\right)\right]^{-1} \left(\sum_{i=1}^{n} J_i\right) C_r \qquad (11)$$

It can be shown that the matrix $nW^{-1} - \left(\sum_{i=1}^{n} J_i\right) C_r \left(\sum_{i=1}^{n} J_i'\right)$ is positive definite. Thus, it follows from formula (11) that the diagonal elements of C_b, i.e. the variances of the orbital elements in the case bias errors are present, are larger than the diagonal elements of C_r. Hence, with the same noise level, bias errors will indeed deteriorate the trajectory determination but, again, the deterioration is independent of the magnitude of the bias errors provided the noise level is low enough.

A more detailed study would indicate that the filter (7) is more likely to cause some computational difficulties than the usual (weighted) least squares filter. This disadvantage becomes more troublesome either when attempting to account for more sources of bias errors, or when attempting to handle data from short tracking intervals in the event that no prior observations were processed. Of course, the nature of the observations themselves plays an important role.

INBRED BIAS ERRORS

This section is mainly concerned with the study of the accuracy of predicted acquisition quantities for sensors whose earlier data were used for the computation of the trajectory under consideration. It turns out that this topic is very closely related to the fact that the filter (4) may cease to be valid in case the matrix

$$M = J'J - J'N(N'N)^{-1}N'J \qquad (12)$$

is singular. Indeed, if the latter condition prevails, it will be shown that the inability of removing all bias errors does not affect the acquisition accuracy in the hypothetical case of no random errors. It is perhaps worthwhile to remark that, theoretically, the filter (4) never breaks down as long as the random errors, however small, are different from zero, provided $\Lambda^{-1} \neq 0$ and enough observations are processed.

Before proceeding to the detailed discussion, it is found to be convenient to establish a necessary and sufficient condition for M to be singular, which will lend itself to a clear geometrical interpretation.

If there exist two nonzero vectors $\delta\nu$ and $\delta\mu$ such that

$$N\delta\nu = J\delta\mu \qquad (13)$$

it is easily shown that M is singular. Indeed, combination of equations (12) and (13) yields $M\delta\mu = 0$ for nonzero $\delta\mu$.

Conversely, if M is singular, there exists a nonzero vector $\delta\mu$ for which $\delta\mu' M\delta\mu = 0$. Or, with $p = J\delta\mu$,

$$p'[I-N(N'N)^{-1}N']\, p = 0 \qquad (14)$$

Since it can be shown that the matrix $I-N(N'N)^{-1}N'$ is positive semi-definite, it follows that equation (14) is equivalent with the linear system,

$$[I-N(N'N)^{-1}N']\, p = 0 \ .$$

Let n be the number of observations and r the number of bias sources. If the rank of N is r (as assumed earlier), indicated by $\rho(N) = r$, it follows from Sylvester's law[3] that $\rho[I-N(N'N)^{-1}N'] = n - r$. This means that p can be expressed in terms of the r components of a vector $\delta\nu$. It is immediately verified that $p = N\delta\nu$ represents the desired solution. Thus, the two nonzero vectors $\delta\nu$ and $\delta\mu$ satisfy the condition $N\delta\nu = J\delta\mu$.

The situation in connection with the behavior of formulas (4-5) can now be summarized as follows. If conditions (13) cannot be satisfied for nonzero vectors $\delta\nu$ and $\delta\mu$, the formulas are always applicable. They break down if conditions (13) can be satisfied and, either $\Lambda^{-1} = 0$, or no random errors are present, or both. If $\Lambda^{-1} \neq 0$, a nonzero noise level is sufficient to avoid the singular case. Of course, it is tacitly understood that enough observations are processed.

Some interesting conclusions follow quite readily from conditions (13). Let us write the latter in the form

$$(N\ J)\begin{pmatrix}\delta\nu \\ -\delta\mu\end{pmatrix} = 0 \qquad (15)$$

The matrix $(N\ J)$ has n rows and r+6 columns (if we are dealing with six state variables), and thus has a rank less than or equal to $\min(n, r+6)$. Now, the vectors $\delta\nu$ and $\delta\mu$ have, respectively, r and six components. Consequently, if $n < r+6$, the linear system (15) admits a nontrivial solution for $\delta\nu$ and $\delta\mu$. Stated differently, for nonzero $\delta\nu$ and $\delta\mu$ equation (13) does not impose a restriction on N and J unless $n \geq r+6$. It follows that effective bias removal can only be expected in case the number of processed observations equals at least the number of state variables plus the number of bias sources. This conclusion may be somewhat academic since, in practice, one would ordinarily process more observations than the aforementioned minimum.

The conditions implied by the matrix relation (13) lend themselves to a simple geometric interpretation which has a strong intuitive appeal. Consider the hypothetical case in which there are no random errors. If also no bias errors are present, the actual set of orbital elements is obviously capable of producing simulated observations which match the observational data perfectly. Let us now introduce bias errors represented by the components of $\delta\nu$ appearing in equation (13). The observed quantities in this case will differ from the previous ones by an increment $N\delta\nu$, since N is the matrix of the relevant sensitivities. By virtue of equation (13), this increment can be put in the form $J\delta\mu$. It follows that the actual set of orbital elements incremented by $\delta\mu$ corresponds again to a trajectory which matches the (offset) observations perfectly. Consequently, there is no way of detecting whether bias errors were present or not.

The above geometric interpretation of equation (13) leads immediately to the conclusion that the effect of an error in station longitude on the trajectory computed on the basis of observations from the station in question, cannot be eliminated. Another example is found in the case of an equatorial synchronous orbit. Any attempt to account for all bias errors in observations from a sensor located anywhere on earth will result in failure.

The question pertaining to what smoothing formula should be used if the matrix M turns out to be singular is one of primary importance. In particular, we wish to investigate the attainable accuracy of predicted observations for, say, acquisition purposes. For simplicity of explanation, no random errors are assumed. As far as the current discussion is concerned, the latter situation is known to be equivalent to the case where $\Lambda^{-1} = 0$ with random errors present.

First, we briefly examine the case where the matrix M is nonsingular. The error vector $\delta\mu$ in the computed set of orbital elements becomes

$$\delta\mu = M^{-1}[J' - J'N(N'N)^{-1}N']\ N\delta\nu$$

in which $N\delta\nu$ represents the error vector in the observations. Thus, $\delta\mu = 0$ and the orbit can be established without error. Incidentally, this is in agreement with the earlier result that $\mathrm{cov}(\hat{\mu}) = 0$ in the absence of random errors. The vector of residuals in the data is $\delta\lambda = N\delta\nu$. It follows that $\delta\nu = (N'N)^{-1}N'\delta\lambda$. Similarly, the residual in a predicted observation λ_o is $\delta\lambda_o = N_o\delta\nu$. Or, $\delta\lambda_o = N_o(N'N)^{-1}N'\delta\lambda$. Hence, the residual itself is predictable on the basis of the known vector $\delta\lambda$.

It is clear that, in case the matrix M is singular, no attempt should be made to filter out all bias errors. Instead, we must concentrate on the maximum number of such errors which can be handled without causing the smoothing formula to become invalid. We therefore partition the matrix N in the form $N = (U_1 U_2 \ldots U_p V)$, in which the number of columns of V is taken as large as possible, but still leaving the matrix $M_* = J'J - J'V(V'V)^{-1}V'J$ nonsingular; all matrices $U_i\ (i=1,2,\ldots,p)$ consist of one column.

We now consider observations corrupted with errors represented by the components of $N\delta\nu$, where $\delta\nu$ is an arbitrary vector. Let us apply the smoothing formula

$$\delta\mu = M_*^{-1} [J'\delta\lambda - J'V(V'V)^{-1}V'\delta\lambda] .$$

The vector $\delta\nu$ can be arranged in the form

$$\delta\nu = \begin{pmatrix}\delta\nu_u \\ \delta\nu_v\end{pmatrix} \tag{16}$$

in which $\delta\nu_u$ and $\delta\nu_v$ correspond to the matrices $U = (U_1 U_2 \ldots U_p)$ and V respectively.

If $N_i = (U_i V)$, the definition of V implies that all matrices $J'J - J'N_i(N_i'N_i)^{-1}N_i'J$ $(i = 1,2,\ldots,p)$ are singular. It follows then from equation (13) that vectors $(\beta_i \delta\bar{\nu}_i)'$ and $\delta\bar{\mu}_i$ can be found such that

$$U_i\beta_i + V\delta\bar{\nu}_i = J\delta\bar{\mu}_i . \tag{17}$$

Summing the previous expression over all i and putting $\delta\bar{\nu}_v = \sum_{i=1}^{p} \delta\bar{\nu}_i$, $\delta\bar{\mu} = \sum_{i=1}^{p} \delta\bar{\mu}_i$, $\delta\nu_u' = (\beta_1\beta_2 \ldots \beta_p)$, we find

$$U\delta\nu_u + V\delta\bar{\nu}_v = J\delta\bar{\mu} . \tag{18}$$

At this point we wish to emphasize that, in equation (17), all constants β_i are arbitrary. Consequently, the vector $\delta\nu_u$ appearing in equation (18) can be taken equal to $\delta\nu_u$ in (16). Of course, this leaves the vectors $\delta\bar{\nu}_v$ and $\delta\nu_v$ in general different from each other.

The observational error vector $N\delta\nu$ may be put in the form $N\delta\nu = U\delta\nu_u + V\delta\nu_v$ and the errors in the orbital elements become

$$\delta\mu = M_*^{-1} [J' - J'V(V'V)^{-1}V'](U\delta\nu_u + V\delta\nu_v) .$$

Or, using equation (18), $\delta\mu = \delta\bar{\mu}$. Thus, contrary to the case where M is nonsingular, the observations are not capable of establishing the orbit exactly. However, it is still possible to predict observations (for the same sensor as was used to obtain the observations represented by λ) with absolute precision. Indeed, using equation (18), the vector $\delta\lambda$ of observational residuals is expressible as $\delta\lambda = V(\delta\nu_v - \delta\bar{\nu}_v)$. Similarly, for a predicted observation λ_o, we obtain $\delta\lambda_o = V_o(\delta\nu_v - \delta\bar{\nu}_v)$. Finally, combination of the latter two expressions yields $\delta\lambda_o = V_o(V'V)^{-1}V'\delta\lambda$ and $\delta\lambda_o$ can be computed from the known residual vector $\delta\lambda$.

By way of an example we may again consider a single tracker yielding measurements of slant ranges, azimuth angles and elevation angles, of a satellite placed in an equatorial, circular, synchronous orbit. It has already been pointed out that an attempt to remove possible bias errors in all three types of measurements is not desirable. In the present case, a sound policy may consist of ignoring bias errors in ranges and constructing a filter which is optimal with regard to bias errors in azimuth angles and elevation angles.

The previous discussion may have appeared to be rather academic because random errors are always present in practice. Moreover, the developments seem perhaps somewhat theoretical. However, they yield useful guide lines in many practical cases. For instance, we may wish to construct an optimum filter for a certain number of bias sources and subsequently discover that the relevant matrix is highly ill-conditioned. This would indicate that perhaps a number of appropriate bias errors can be ignored without causing a great loss of accuracy in predicted observations for the same sensor, although the accuracy of the established trajectory itself may suffer substantially. Any such deterioration, however, could never be detected without taking observations from an independent source. This argument is in agreement with, for instance, the intuitive notion that a single tracking station may very well acquire on the basis of its own predictions whereas acquisition quantities predicted for other stations could be largely in error. The latter situation has, in fact, been observed in practice.

NUMERICAL RESULTS

The formulas (7-8) have been used to process, first, measurements of ranges, azimuth angles and elevation angles and, second, angular measurements only. All observations belong to the same satellite pass originating from a single tracking station. In order to reduce the number of parameters to a reasonable minimum, the matrices W and Λ appearing in formulas (7-8) are put in the form

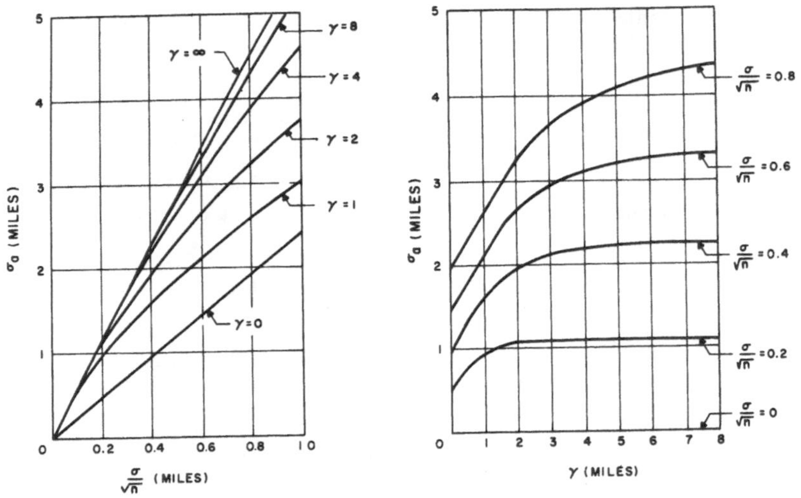

FIG. 1 ANGULAR AND RANGE DATA
COVERING 1 HOUR PASS

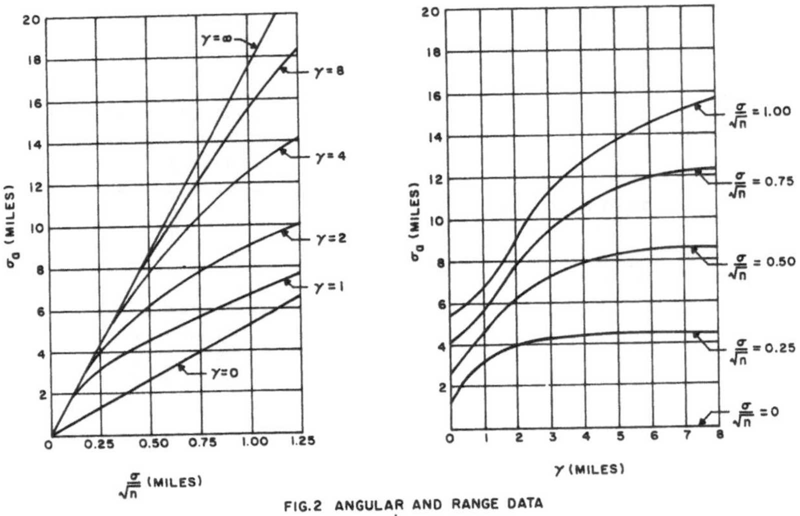

FIG. 2 ANGULAR AND RANGE DATA
COVERING $\frac{1}{2}$ HOUR PASS

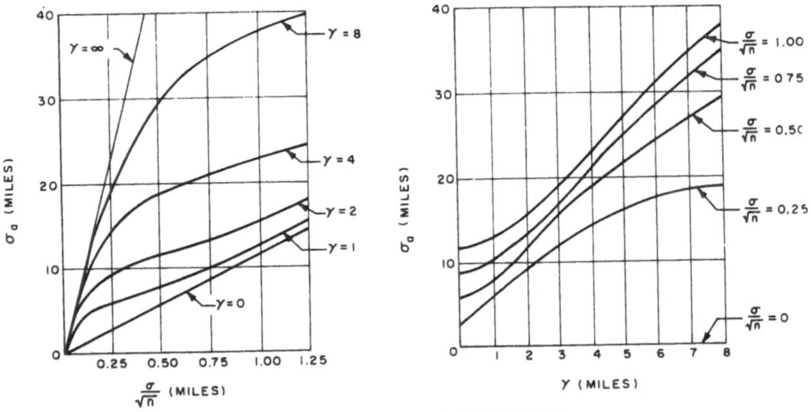

FIG. 3 ANGULAR AND RANGE DATA
COVERING $\frac{1}{4}$ HOUR PASS

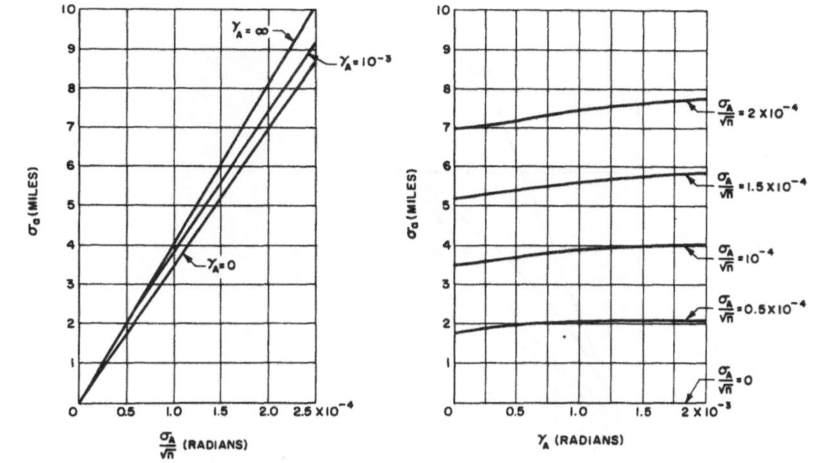

FIG. 4 ANGULAR DATA COVERING
1 HR PASS

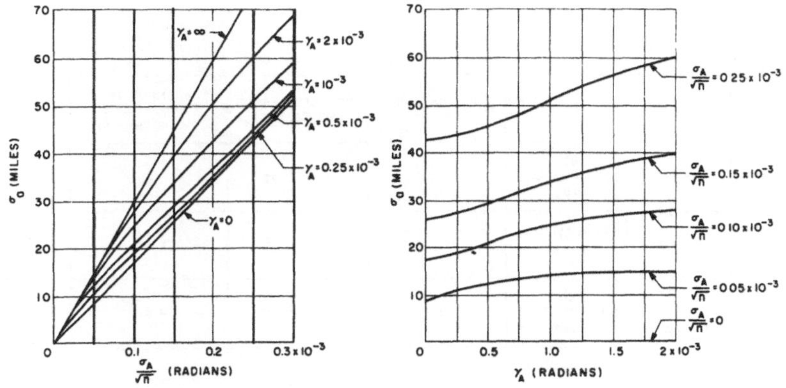

FIG. 5 ANGULAR DATA COVERING
1/2 HOUR PASS

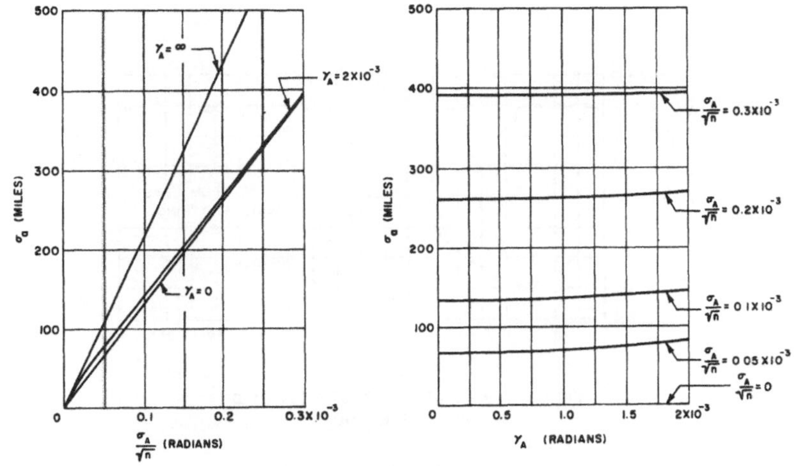

FIG. 6 ANGULAR DATA COVERING
1/4 HOUR PASS

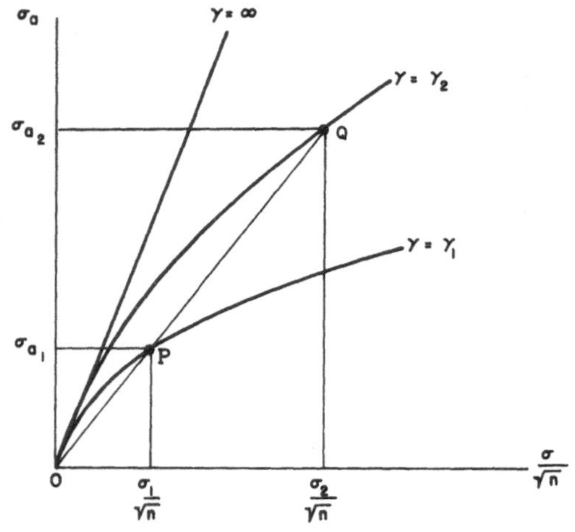

FIG. 7 PROPERTY OF γ CURVES

		POSITION ERROR (MILES)	VELOCITY ERROR (MILES/SEC)	ERROR IN SEMI-MAJOR AXIS (MILES)
LEAST SQUARES FILTER	NO OBSERVATIONAL BIAS	3.14	0.000222	1.41
	OBSERVATIONAL BIAS	162.12	0.011269	37.63
OPTIMUM FILTER	NO OBSERVATIONAL BIAS	3.14	0.000222	1.41
	OBSERVATIONAL BIAS	5.38	0.000347	2.03

FIG. 8 COMPARISON OF LEAST SQUARES FILTER AND OPTIMUM FILTER

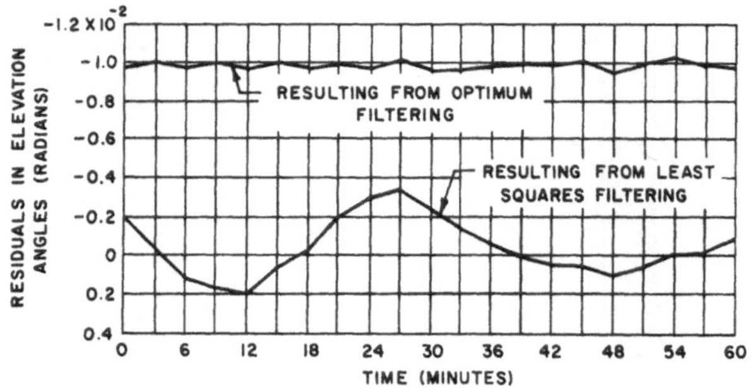

FIG. 9 TYPICAL BEHAVIOR OF RESIDUALS IN THE OBSERVATIONS

$$W = \frac{1}{\sigma^2} W_o, \; \Lambda = \gamma^2 \Lambda_o ,$$

where W_o and Λ_o are fixed matrices. The parameters to be varied in this numerical study are γ and $\sigma/\sqrt{n}$ (n is the number of datapoints). They represent, respectively, a measure of the magnitude of the bias errors and the arithmetic mean of the random errors. In case of the range and angle measurements and for $\sigma = \gamma = 1$ mile, the matrices W_o and Λ_o were selected to make the standard deviation of random and bias errors in the inertial satellite coordinates equal to one mile for a point near the center of the pass. In case of only the angular measurements, the ranges were ignored by assigning them a weight equal to zero. In the latter case the results are presented in terms of the standard deviations γ_A and $\sigma_A/\sqrt{n}$, respectively, of the bias errors and arithmetic mean of the random errors in the angular measurements.

In terms of the above defined parameters, formula (8) can be put in the form[†]

$$\operatorname{cov}(\hat{\mu}) = \frac{\sigma^2}{n} F\left(\frac{\sigma^2}{n\gamma^2}\right), \tag{19}$$

where

$$F\left(\frac{\sigma^2}{n\gamma^2}\right) = \left[\frac{1}{n}\sum_{i=1}^{n} J_i' W_o J_i - \left(\frac{1}{n}\sum_{i=1}^{n} J_i'\right) W_o \Lambda_o \left(\Lambda_o + \frac{\sigma^2}{n\gamma^2} W_o^{-1}\right)^{-1} \left(\frac{1}{n}\sum_{i=1}^{n} J_i\right)\right]^{-1} .$$

Clearly, the adequacy of the established orbit should be judged at least on the basis of the accuracy of all orbital elements. However, in the interest of simplicity we present the results pertaining to only one element, namely the semimajor axis "a". Indeed, the semimajor axis is the most critical element with regard to long range predictions and, as it turned out for the particular geometry of this study, its accuracy can be taken as a reliable measure to judge the accuracy of the computed orbit. If σ_a is the standard deviation of "a", it follows from formula (19) that

$$\sigma_a = \frac{\sigma}{\sqrt{n}} f\left(\frac{\sigma^2}{n\gamma^2}\right), \tag{20}$$

[†]Strictly speaking, only for sufficiently large values of n can F be regarded as a function of the one argument $\sigma^2/n\gamma^2$.

where f represents a scalar function of the argument $\sigma^2/n\gamma^2$. Knowledge of this function enables us to draw families of curves displaying σ_a, either as a function of $\sigma/\sqrt{n}$ for constant γ, or as a function of γ for constant $\sigma/\sqrt{n}$.

The aforementioned curves, shown in Fig. 1 through Fig. 6, enjoy a simple geometric property which is helpful in obtaining additional ones with virtually no supplementary calculations. Let us consider two typical curves giving σ_a as a function of, say, $\sigma/\sqrt{n}$ for two values γ_1, γ_2 of γ (Fig. 7). If σ_1 and σ_2 are values of σ such that $\sigma_1/\gamma_1 = \sigma_2/\gamma_2$, it follows from expression (20) that $\sigma_{a_1}/\sigma_{a_2} = \sigma_1/\sigma_2$, or

$$\frac{OP}{OQ} = \frac{\gamma_1}{\gamma_2} .$$

Thus, if point P describes the γ_1 curve, a second point Q, lying on the radius OP such that the ratio OP/OQ is kept constant, describes another γ curve with a corresponding γ value of γ_1 OQ/OP.

As was demonstrated earlier in this paper, all γ curves corresponding to a nonzero value of γ have the same slope at the origin. This slope is equal to the one of the straight line representing the γ "curve" in the hypothetical case of γ approaching infinity. It is perhaps worth mentioning that the present results can be considered valid only if the observational errors are small enough for the usual linearization processes to be applicable. Hence, except for the feature mentioned above, the "curve" for γ approaching infinity is meaningless.

The graphs giving σ_a as a function of γ for various values of $\sigma/\sqrt{n}$ show clearly the deteriorating influence of bias errors on the accuracy of the computed trajectory. It can be seen that this so-called bias sensitivity is highly dependent on the type of data, the length of the pass, and, to some extent, on the magnitude of the errors themselves. In the case of only angular data covering an interval of 15 minutes, the bias sensitivity is exceedingly small whereas, for example, it becomes quite pronounced with a 15 minute pass of range and angular data. Of course, even in the latter case the results are preferable to the ones obtained by using a least squares filter on the biased data, simply because the present method is optimal for the assumed error model.

As a final part of the evaluation of the present method, the case of the one hour pass with angular data, displaying a limited amount of bias sensitivity, is examined in some detail. In particular, it is interesting to compare the performance of both the optimum method and the method of least squares.

Preliminary numerical computations show that the derived orbit is most vulnerable to bias errors when the ratio of these errors in elevation angles and azimuth angles is approximately equal to -3. Therefore, a trajectory is obtained taking data which, aside from random errors with standard deviation 0.25×10^{-3} radians, contains a bias error of 0.003 radians in azimuth angles and -0.01 radians in elevation angles. The computations are carried out using a least squares filter and, subsequently, the optimum filter presented in this paper. Finally, and as a basis for comparison, the unbiased data are processed according to the least squares method (which happens to be optimum in this case) to yield a third trajectory. Typical results are listed in Fig. 8 and clearly indicate the superiority of the optimum method compared to the method of least squares if observational bias errors are present. For instance, use of the optimum filter reduces position and velocity errors resulting from least squares filtering by a factor of approximately 30 in the case at hand.

The behavior of the residuals in the observations as encountered during the trajectory determination process exhibits an interesting feature. As is expected, the residuals have an arithmetic mean nearly equal to zero when using the least squares method. For instance, this is shown for the elevation angles in Fig. 9. If the orbital elements resulting from application of the least squares method are readjusted by means of the optimum filter, the residuals settle to the ones also indicated in Fig. 9. Hence, the optimum method allows the predominant observational bias error to remain in the residuals rather than forcing a match between the actual and simulated observations. Adding the latter bias error to corresponding predicted observations will greatly improve prediction accuracy.

ACKNOWLEDGMENT

I wish to thank F. T. Geyling for valuable comments on this paper.

REFERENCES

(1) Blackman, R. B., "Methods of Orbit Refinement," Bell System Tech. J., Vol. 43, pp. 885-909, 1964.

(2) Claus, A. J., "Orbit Determination in the Presence of Systematic Errors," Celestial Mechanics and Astrodynamics, Vol. 14, pp. 725-742, 1964.

(3) Gantmacher, F. R., The Theory of Matrices, Vol. I, Chelsea Publishing Company, New York, 1960, pp. 61-66.

(4) Swerling, P., "First Order Error Propagation in a Stage-Wise Smoothing Procedure for Satellite Observations," J. Astronaut. Sci., Vol. 6, pp. 46-52, Autumn, 1959.

DISCUSSION

A. J. Claus

"On Systematic Errors in Trajectory Determination Problems".

Q. Will you compare your method with one in which bias errors are considered as additional unknowns?

A. As the state vector is augmented, the matrix inversion process grows proportionally. With my method only a 6x6 matrix is inverted.

Q. Is your filter suitable only for systematic errors which are constant with time?

A. The filter described in the paper is suitable for systematic errors expressible as a function of time involving two constants. It should be easy to extend it to Markov processes.

Components and Techniques

TRACKING THE PARTIALLY ILLUMINATED EARTH OR MOON*

by Dr. Roger S. Estey
Chief, Radiometry Group
Northrop Space Laboratories
Hawthorne, California

ABSTRACT

The observation of the sightline to the center of the partially illuminated earth or moon is accomplished by a series of measurements by which the circular limb of the body is centered concentrically with the circular field stop in the instrument. Since the limb must be distinguished from the terminator independent knowledge of the sun's direction is required to align the tracker mechanism appropriately. The measurements and calculations are digital and provide servo commands to align the gimballed tracker precisely on the geometrical center of the celestial body.

INTRODUCTION

The navigation of spacecraft within the solar system requires the observation of the line of sight to distant celestial bodies (stars) or to neighboring bodies (the sun, moon or planets). With the exception of the sun these near bodies exhibit phase and consequently an observed line of sight may be ambiguous depending on whether the line of sight terminates at the centroid of luminosity or the geometrical center of the planetary disk. The present paper discusses automatic instrumentation for aligning the optical axis of an instrument in a spacecraft with the geometrical center of a nearby partially illuminated celestial body. Although the concepts and instrumentation to be described are applicable to any planet or satellite, the discussion will be directed to observations of the earth or moon.

At great ranges the angular subtense of the planetary disk is so small that the angular difference in pointing direction between the luminous centroid and the center of the circular disk may be less than the angle error due to other causes and hence negligible. At short range the angular subtense of the body is so large that special optical or mechanical scanning techniques are required. The discussion which follows will emphasize midcourse ranges in which the earth or moon subtends about twenty degrees or less. At the earth-moon distance the moon subtends half a degree and its center can be observed to about 12 seconds by the techniques described here. This accuracy is far greater than the disparity between lines of sight to the luminous centroid or to the geometrical center. Thus in cis-lunar space the partial illumination problem is serious and a solution must be found if intolerable angle errors are to be avoided.

The planet tracking instrument comprises a variable focus telescope, scanning mechanism to sense the planetary disk correctly, gimbals which are articulated about three axes, and a computer for reducing the data and for generating commands to the tracker servo.

CHARACTERISTICS OF THE PLANETARY TARGETS

The trajectory of a spacecraft traveling between the earth and the moon will be used as an example to indicate the nature of the sighting and tracking problem which the instrument must solve. The various considerations which influence sightline observations in cis-lunar space will be discussed under two major headings: the sightline geometry and the properties of the celestial bodies concerned.

Sightline Geometry

The lunar orbit is inclined approximately 5 degrees to the plane of the ecliptic and the intersection of the ecliptic and orbital planes is located at the earth; consequently a sightline to the sun (at 92,000,000 miles) has a maximum deviation from the ecliptic plane of only 0.01 degree, which is negligible for most purposes. The most useful set of coordinates for lunar navigation is a right-handed orthogonal set built up on the first point of Aries on the Y-axis and the north ecliptic pole on the Z-axis. In this set a position point has the coordinates λ, celestial longitude, and β, celestial latitude. The Nautical Almanac, planned for the convenience of terrestrial navigation and earth-based astronomical studies, presents the lunar ephemerides in terms of the north terrestrial pole and equator, and the meridian circle through Greenwich. Conversions between the various coordinate systems are not difficult but may be laborious; consequently, digital computer routines are frequently employed for this purpose.

The angles subtended at the spacecraft by the earth or moon have been computed from the diameters of the bodies considered as spheres and using the values 7926 and 2157 miles respectively. Except for regions in close proximity to either body the subtended angles versus range are shown in Fig. 1 and in Table 1. Lunar trajectory studies have emphasized the need for midcourse corrections at about 90,000 miles and 20,000 miles from the moon. The instrument and technique reported here are adapted to this sort of midcourse navigation.

* Presented at the IFAC Symposium on "Automatic Control in the Peaceful Uses of Space", Stavanger, Norway, June 21-24, 1965.

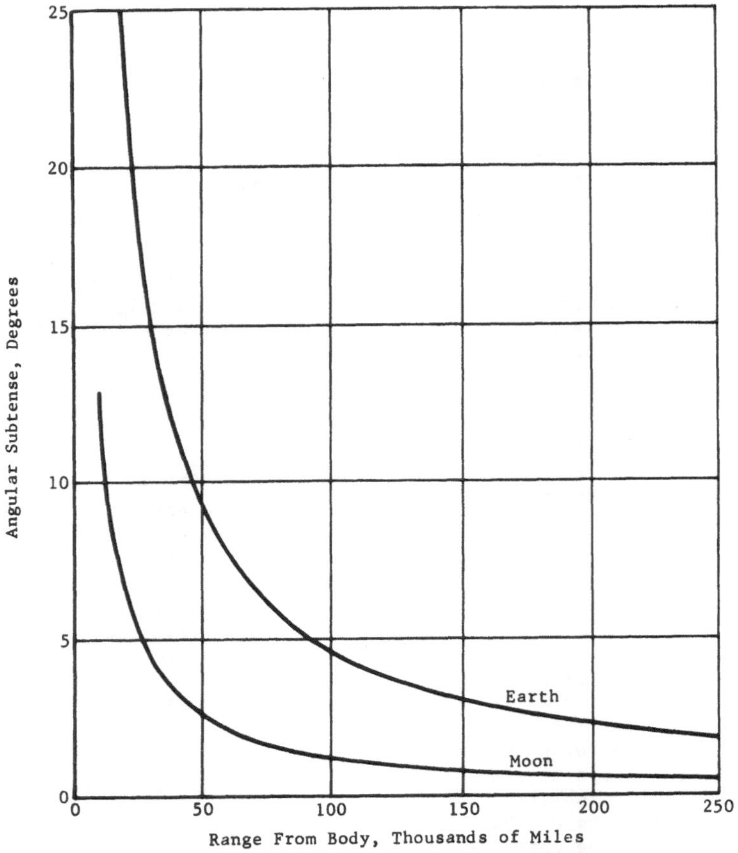

Fig. 1. Angular subtense or earth and moon.

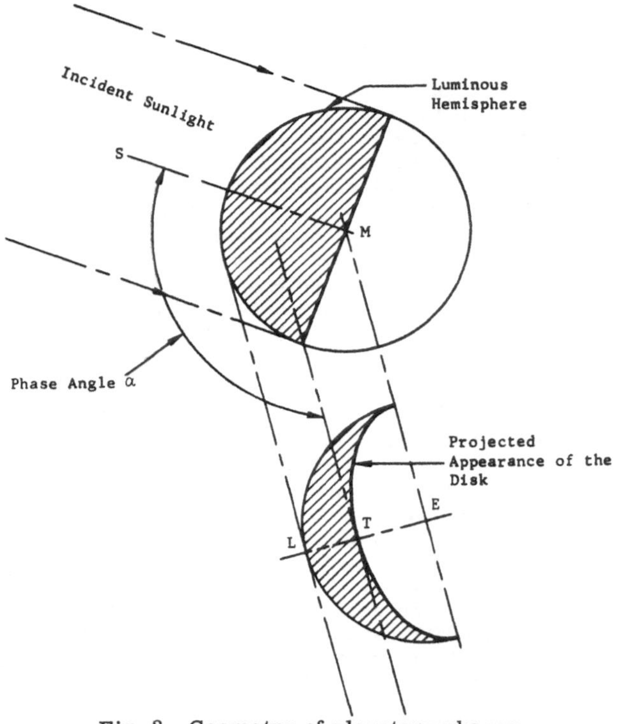

Fig. 2. Geometry of planetary phases.

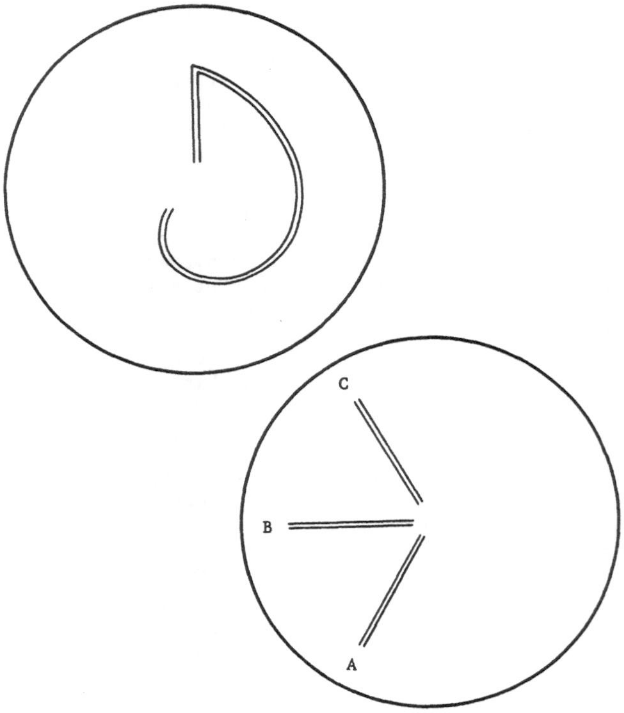

Fig. 3. Spiral and radial slit reticles.

During an earth-moon voyage the lunar tracking problem must be considered to be dynamic in character. The significant angles, angular rates and ranges are time varying over the period of a lunar voyage and typically the rates increase rapidly as the lunar surface is approached. Although an exhaustive study of these variables is not feasible or necessary at this time, limited data are desirable background information for designing and appraising planetary tracking instruments and techniques.

A 42-hour one-way trajectory was available and forms the basis for the following discussion. In the intermediate range slower trajectories or circumlunar trajectories would yield closely similar conclusions. Rates derived from this trajectory are shown in Table 1.

The phase of the earth or moon is identified by the angular relationship between sun, earth or moon, and the observer. The geometry is shown in Fig. 2. Sunlight incident in the direction SM illuminates the left hemisphere of the body as shown. By projection it can be shown that an observer on the line ME will see the partially illuminated body as a crescent disk in which L is the midpoint of the circular limb and T is the midpoint of the elliptical terminator. The phenomenon of phase is described by two terms, "phase" or "phase angle". The former relates to the relative luminous area, sometimes expressed as the relative length of the segment LT. Phase angle is the angle SME denoted by α. The relationship between these concepts is expressed by the equation

$$\text{Phase} = LT/LE = (1 + \cos\alpha)/2.$$

As the illumination changes from new moon, for example, through first quarter to full, the phase changes from zero, through one-half, to unity. The phase angle changes from 180 degrees through 90 degrees to 0 degrees in the same sequence. The tracking technique reported here requires measurements which are rather precisely related to the midpoint of the bright limb of the body. In order to align the tracking instrument the bright limb position angle must be known as an operating parameter. The orientation of the figure of the partially illuminated body is a function of the sun-body-tracker geometry. The discussion which follows relates to the special case in which the moon is observed from a tracking station on earth. The more general cases involving tracking of earth, moon or other planet from a spacecraft are readily inferred. The observation of phase changes from earth is described in the literature. Lunar coordinate values of the position angle of the axis and the position angle of the bright limb are tabulated daily and reported in the American Ephemeris and Nautical Almanac, the section entitled "Ephemerides for Physical Observations". With the exception of narrow intervals at new or full moon where observations are difficult because of reduction in brightness and for other reasons the values are within a few degrees of 90 degrees or 270 degrees. This fortunate circumstance facilitates the alignment of a planet tracker which employs the techniques reported here.

The Moon as a Target

The radiometric characteristics of the earth and moon will be considered separately.

Optical radiation from the moon is a blend of thermal radiation emitted by the body as a consequence of its temperature and of solar radiation reflected from the illuminated hemispherical portion of the lunar surface. The infrared signal derives from the temperature of the lunar surface which represents the balance between the incident radiation from the sun and the outward flow of radiation to space, modified by the effect of the heat capacity, conductivity, and the temperatures of material layers adjacent to the lunar surface. The range of temperature between the center of the fully illuminated disk and the dark limb has been reported by Pettit and Nicholson[1] as 342°K and 120°K, respectively. Studies of the moon discussed by Ehricke[2] and Fielder[3], indicate that the lunar surface is porous, of low heat conductivity, and that the surface temperature responds rapidly to changes in solar irradiation.

The thermal radiation from the warmer portion of the lunar surface is considerable and is peaked at about ten microns. The radiance of the cold surface (the dark limb) is very small and is peaked near twenty-three microns. As the phase

TABLE 1

EARTH-MOON TRAJECTORY PARAMETERS

			Moon		Earth	
Elapsed Time (hr)	Range to Moon (mi)	Range Rate (mph)	Subtended Angle (deg)	Angle Rate (deg/hr)	Subtended Angle (deg)	Angle Rate (deg/hr)
0	2.38×10^5		0.32	0.023		
6	1.83	9.5×10^3	0.68	0.023	8.24	0.84
12	1.48	5.5	0.84	0.030	5.04	0.65
18	1.15	5.3	1.07	0.048	3.69	0.31
24	0.85	5.0	1.46	0.080	2.96	0.20
30	0.55	4.6	2.25	0.20	2.48	0.12
36	0.30	3.7	4.13	0.43	2.18	0.10
42	0.0				1.91	0.09

progresses from full moon through gibbous to crescent, the terminator moves back from the dark circular limb. The thermal gradient is steep right at the dark side of the terminator, but over the rest of the dark area the lower surface temperature is well established. In the shorter wavelength region near seven microns which is readily accessible by photoconductive detectors and where the signal from the warmer surface can be measured effectively, the signal from the colder area (dark limb) is about 3000 times less. In the longer wavelength region accessible only to relatively insensitive bolometers, the signals are less intense but the contrast between bright and dark limbs is reduced to values ranging from about eighty to one down to about twelve to one depending on the experimental conditions. Infrared is not a promising technique for the detection of the dark limb of a planetary body without atmosphere.

Short wave radiation received from the moon is composed almost entirely of reflected sunlight. The contribution of starlight is negligible and the contribution of sunlight reflected from the earth is too small a signal to be useful. Reflected solar radiation has a distribution similar to that of solar energy but is modified by the reflectance of the lunar surface. The useful optical pass band is limited to the interval from about 0.4 to 0.65 microns by the properties of the most useful optical materials and the most sensitive photodetectors available. Within this pass band spectral variations in the reflectance of the terrain are not significant. For the moon (in terms of visible light) the albedo is 0.070 and the average reflectance is 0.106[2,3]. From the position of a spacecraft between the earth and moon, when one body appears full, the other appears new, and vice versa.

Using the terminology familiar to astronomers we will discuss the apparent visual magnitude as the observed illumination at the instrument caused by luminous flux from the moon. These data assume no intervening atmosphere. This illumination, E, expressed for example in lumens/m^2, is related to m_v by the equation

$$-0.4m_v = \log E - \log E_o$$

where m_v and E are the magnitude of and the illuminance produced by the body. E_o has the value 2.43×10^{-6} lumens m^{-2} and is the illuminance produced just outside the earth's atmosphere by a zero-magnitude star.[4] The luminance, B, is expressed in candles/m^2 and depends on the phase and the surface area concerned. The quantity B is independent of the range to the observer, and to the degree that the lunar surface obeys Labert's cosine law, this luminance is independent of the angle of observation.

Photometric information presently available concerning the moon comprises magnitude and phase law data from which either the total flux or the average brightness can be expressed as a function of phase angle. The significant numerical data are as follows:[5]

Star outside earth's atmosphere, $m_v = 0$, illuminance = 2.43×10^{-6} Lux.

Moon at mean opposition, $\alpha = 0$, $m_v = -12.70$, illuminance outside earth's atmosphere, E = 0.292 lumens/m^2 (Lux).

Luminous intensity of full moon, $I = 4.31 \times 10^{16}$ candles.

The luminance of the moon does not vary with range but does vary with phase in a manner which depends on the incident solar radiation and on the luminous projected area. Casual inspection of the full moon indicates, and careful measurements[3] confirm, that the full moon is almost uniformly bright with a narrow zone of greater luminance at the limb. The average luminance of the partial moon falls off with phase as described above, and the cusps decrease in luminance even more rapidly than the central portion. Experiments performed at the Northrop observatory with the tracking equipment described in this paper have disclosed that for the crescent moon, the luminance at points on the limb 60 degrees from the midpoint is only half that at the midpoint. Furthermore, the luminance decreases from limb to terminator in a manner which is more pronounced towards the cusps as compared with measurements made near the center of the bright limb.

The discussion thus far has tacitly assumed the projection of the moon to be a truly circular disk. The actual departures from true circularity are half a per cent of the radius or less. This quantity is trivial unless extreme pointing accuracy is required on a disk subtending a large angle.

The Earth as a Target[6,7]

As in the case with the moon, terrestrial radiation within the limits of the optical spectrum is a blend of emitted thermal radiation and reflected solar radiation. Infrared is emitted from the terrain and from the atmosphere to a degree depending on local temperature. The temperature range from equator to poles which exceeds to a considerable degree the temperature range between day and night controls the signal dynamic range to which an infrared limb sensing instrument must be designed. This temperature range is about -55°F to +110°F. The corresponding range of blackbody radiation intensity, peaking near 9 to 12 microns, is about four to one. However, although the infrared region is effective in sensing the terrestrial limb, either light or dark, it is much less effective in observing bodies without a favorable atmosphere and consequently the instrument which operates in the visual region to sense the bright limb only is considered a preferable instrument because of its universal applicability.

Short wave radiation received from the earth is composed entirely of reflected sunlight. The effect of moon and star light, aurorae, and air glow, is completely undetectable. Reflected solar radiation has a spectral distribution similar to that of the incident solar energy modified by the reflectance of the earth (land or sea) and the cloud layer. Within the useful optical pass band, 0.4 to 0.65 microns, the reflectance varies from 0.03 (ocean) to 0.86 (ice fields). There are moderate variations with wavelength as the unsaturated colors range from blue, through green and yellow to reddish brown. The visual albedo of the earth is 0.36.[8]

Like the case of the moon, the brightness and the total flux from the earth vary with phase angle. The phase laws for the two bodies are substantially complementary, that is the earth is full when the moon is new and vice versa. For the full earth, the visual magnitude as seen from the sun is -3.80 and is as seen from the moon;

Illuminance at moon due to full earth = 12.31 lu/m^2

Luminous intensity of full earth, $I = 182 \times 10^{16}$ candles.

Undoubtedly the luminance of the earth varies with phase in a manner similar to the brightness variation of the moon. In addition the brightness will vary along the terrestrial limb due to the varying reflectance of cloud cover and of terrain. Since the available signal flux is very large, compared to that from the moon for example, the variation in flux due to clouds, variable reflectance of terrain, or phase is rather small and is relatively unimportant.

The discussion thus far has treated the projection of the earth as a truly circular disk, even though partially illuminated. In practice the terrestrial bright limb departs from true circularity due to the typical elevation of the cloud cover and of the terrain and the oblateness of the earth. In trajectories between celestial bodies within our solar system the spacecraft will travel in or close to the plane of the ecliptic. Consequently the effect of oblateness is maximized and the effect of mountainous terrain may be considerable. Typical elevations of these features are listed below in Table 2.

TABLE 2

Earth Features Referred to Mean Radius Sphere
Mean radius = 6,371 km.

Feature	Altitude, km
Polar, sea level	-14
Polar, mountainous	-11
Mid-latitude, mountainous	+10
Mid-latitude, clouds	+10
Equatorial, sea level	+ 7

From the data in the above table it is evident that limb measurements may lead to errors in the estimation of the disk center of about 10 km or 1:1200 of the disk diameter. At shorter ranges the sighting errors due to the non-sphericity of the earth are considerable. The minimum ranges at which the errors will have specified values or less are tabulated below in Table 3.

TABLE 3

Sighting Errors Due to Non-Sphericity of Earth

Minimum Range, mi.	Maximum Sightline Error
2.1×10^3	10 min
2.1×10^4	1 min
1.3×10^5	10 sec
1.3×10^6	1 sec

SURVEY OF TRACKING SCHEMES

Analytically speaking, all methods for identifying and tracking the partially illuminated earth or moon fall into one of two classes. The first, or area-sensing technique bounds the planetary image with a closely matched circular field stop and compares the unvignetted area of the luminous disk with the reference area in the instrument; the lunar flux passing unobstructedly through the instrument is the basic element to be measured. If the circular portion of the planetary image is tangent to a circular aperture in the instrument, then any transverse movement of the planetary image with respect to the aperture will produce a reduction in the common area. Obviously, the sensitive portion of the total area is an annular region bounded by the instrument aperture; the area variations close to the null condition occur in this region. Consequently the balance of the total aperture area can be blanked out since it makes no contribution to the information.

This technique is unsatisfactory because of large differences in angle sensitivity in various azimuths measured with respect to the line of cusps. Furthermore the considerable variation in luminance of different regions of the bright limb militates against this or other photometric technique for establishing the geometrical center of the celestial body.

The second, or radial measurement technique, distinguishes the bright limb from other portions of the lunar contour, and by linear measurements of position locates the lunar image precisely with respect to reference coordinates in the instrument. Since the bright limb is a circular arc, but the terminator is an elliptical arc of variable eccentricity, linear measurements of the limb relative to the instrument axis appear to be an attractive means for determining the circle's center.

The offsets between the limb and the instrument axis can be measured, the center found from the equation of the circle and the tracking servos driven from the sightline errors computed along the gimbal axes. In a second approach emphasized in this report, the data receive minimum processing and are nulled in the servos without solving the circle equation explicitly.

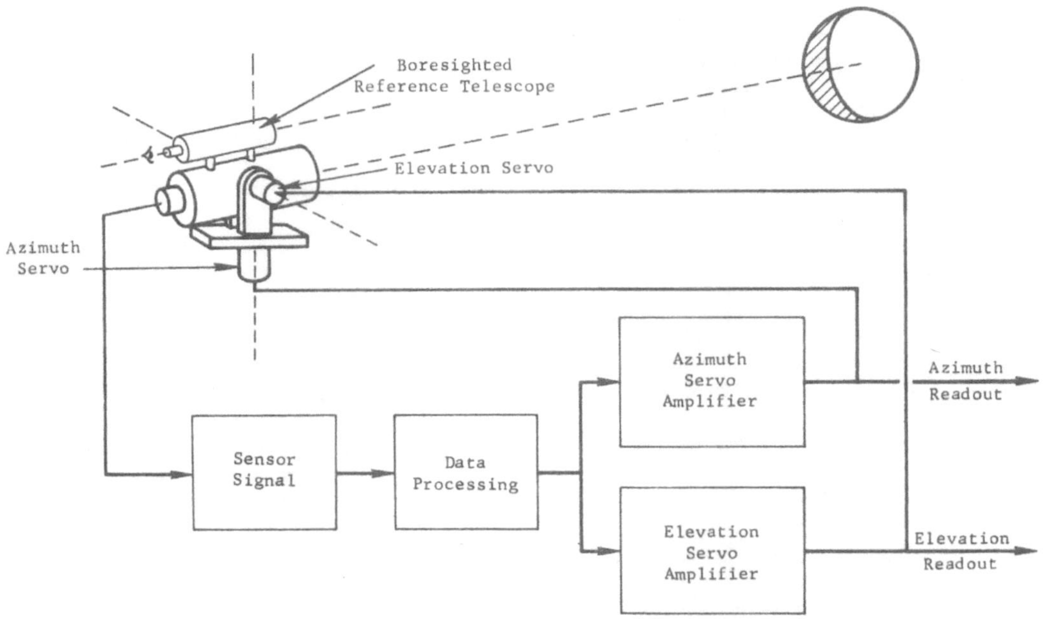

Fig. 4. Moon tracker block diagram.

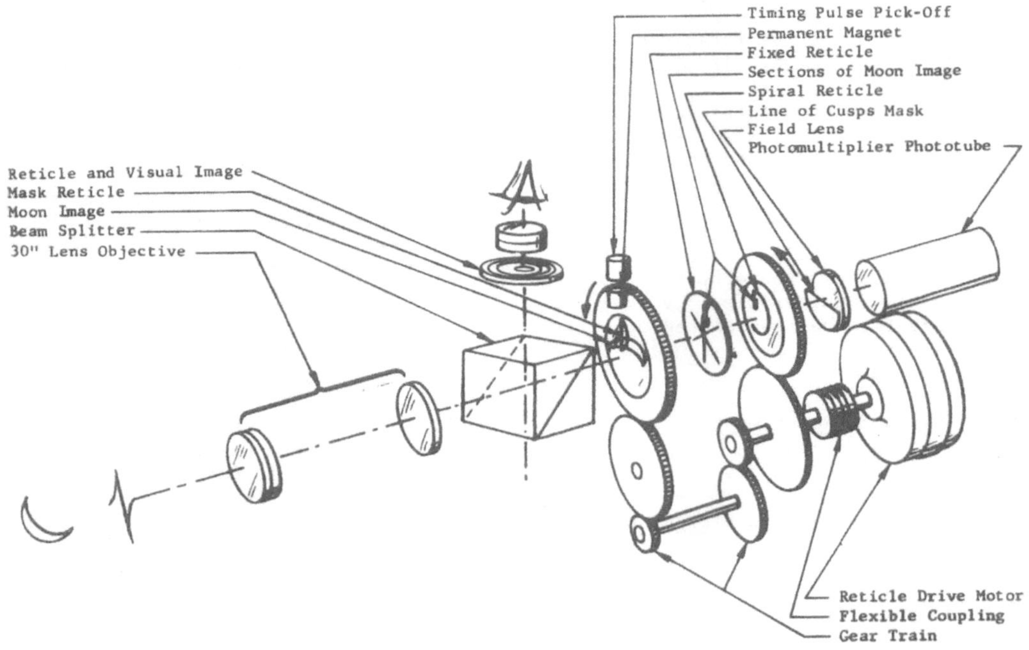

Fig. 5. Opto-mechanical scanner system diagram.

Spiral Reticle Technique

Even a cursory examination of the radial measurements approach suggests the simplification in which radial segments between the lunar image and the circular field diaphragm of the instrument are examined. If these segments are identical in length, the lunar image is concentric with the field diaphragm and the instrument axis coincides with the sightline to the geometrical center of the moon. There is a single restriction: namely, the observations must be limited to the circular limb of the moon.

A many-look approach for processing redundant data was investigated. However the three-look approach described here is more readily reduced to practice. In this technique the radial segments between the rim of the instrument and the moon's limb are determined at three places 60 degrees apart and approximately centered between the cusps. This technique was selected for breadboarding because of its adequacy and simplicity.

Consider an opaque reticle with three transparent radial slots disposed 60 degrees apart and a second opaque reticle with a transparent involute spiral slit as shown in Fig. 3. These reticles are almost in contact in the focal plane containing the planetary image. As the spiral rotates counter-clockwise, the intersection with a fixed slit, B, for example, will travel inward at constant linear velocity along the radius B. Since portions of the spiral intersect slits A and C, the operation with respect to B must be isolated by an appropriate aperture in a separate slowly rotating mask. When the radial slit on the spiral coincides with a radial slit in the fixed reticle, planetary energy anywhere in that part of the field will pass through the two coincident slits and generate a large voltage signal in the photocell which coincides exactly with the start of the linear radial scan.

As the rotation of the spiral proceeds, the scanning aperture moves inward and in due course intersects the bright limb, producing a second abrupt voltage signal. Since the radial scan has constant linear velocity, the time between signals is an exact measure of the distance between the rim of the instrument (and hence the optical axis) and the bright limb of the body. The radial segment distances associated with slits A, B and C can be used to determine the sightline errors ϵ_X and ϵ_Y by the equation

$$2R_B - (R_C + R_A)/2 = \epsilon_X$$

$$0.577(R_C - R_A) = \epsilon_Y$$

where R_A, R_B, and R_C are the radial intervals associated with the slits A, B, and C, respectively. The constant 0.577 is an empirical scaling coefficient which balances the servo.

DESCRIPTION OF A THREE-SPOKE BREADBOARD

A breadboard model emphasizing the spiral scan and the three-look approach to data processing was designed, built and tested to demonstrate the effectiveness of the techniques just discussed by means of lunar observations from earth.

Northrop equipment for sensing and tracking the geometrical center of the partially illuminated moon consists of a gimbaled pointing instrument and a separate electronics and computer package. The system, which is shown as a block diagram in Fig. 4, comprises a) an optical telescope, b) an automatic measuring device to observe the centration of the lunar image in the field of the instrument, c) a computer to calculate gimbal angle errors and generate servo commands, d) gimbal axes and servos, e) a finder telescope for reference, and f) angle readout devices on the gimbals. The optical pointing equipment generates the lunar limb measurements. Programming means combine with a computer to calculate the gimbal errors from equations of the form

$$\epsilon_X = 2R_B - (R_C + R_A)/2$$

$$\epsilon_Y = 0.577\ (R_C - R_A)$$

Design of Tracker Instrument

The moon-sensing part of the breadboard system is an optomechanical instrument mounted in gimbals. The main housing contains the optics, the sensor, the preamplifier, and the motor-driven scanning subsystem. These details are shown in the exploded view, Fig. 5. The telescope objective is a custom designed and built telephoto lens having a 30 in. equivalent focal length, F/20 relative aperture and 2.3 degree angular field. There is a beam splitter in the optical path to permit the use of a separate reticle and eyepiece to form a boresighted telescope which monitors the tracking accuracy when tested against a simulated laboratory target or tested in the observatory against the natural moon. The optical layout includes a simple field lens assembly to image the entrance pupil of the objective into the aperture of the photosensor.

The scan mechanism consists fundamentally of the fixed and radial slits and a rotating mask. The radial slits are 0.34 minutes wide and 60 degrees apart. The spiral slit is 0.17 minutes wide and covers its operating range in 270 degrees at 30 rps. The linear velocity of the aperture formed by the intersection of the radial and spiral slits is approximately 2500 arc-minutes per second of time in the radial direction. Each radial slit is sensed in 25 milliseconds. A shutter rotating at 4.29 rps permits the sensing of only one slit at a time. The rotating parts are driven at the appropriate speeds by a synchronous motor and gear train. When the instrument is aligned with respect to the moon, the three radial slits a, b and c are approximately centered on the bright limb. Since the cusps of the waning and waxing moon face in opposite directions, a second set of slits on the other side of the reticle is provided (not shown in the figures). A supplementary shutter, hand-operated, permits either slit group to be exposed, depending on the moon's orientation at the time.

At one point in the scan system rotation, the radial aperture of the spiral reticle coincides with a fixed slit thus passing an intense pulse of moonlight. As the aperture moves along the radial

Fig. 6. Tracker unit showing partial disassembly.

Fig. 7. Electronics unit.

slit, the image of the lunar limb is reached and a second voltage pulse is produced. The spacing between these pulses is the radial distance between instrument rim and lunar limb at one of the slits a, b, or c. The radiation is sensed by a small 10-stage photomultiplier of commercial type. The signal voltage data from the photomultiplier are processed and error signals for the servos are computed.

The gimbals have a range of ± 10 degrees from their mean position. Each gimbal shaft, mounted on preloaded ball bearings, carries a worm sector of adequate range, which, in cooperation with a single-lead worm, gives a reduction of 432:1. These worm and sector assemblies, developed by Northrop in 1947, are of super-precision quality and have a proven accuracy of 5 arc seconds across the gear mesh. The servo motor and associated train are geared to the worm shaft. The same worm shaft gear drives a ten-turn precision potentiometer through additional gearing such that the 20-degree rotation of the gimbal shaft (limited by stops) corresponds to about 9-1/2 turns of the potentiometer. A potentiometer with a resolution of 0.03%, which is commercially available, will read 20 arc-seconds anywhere in the 20-degree range.

The scanning process was critically examined with respect to various sources of error, including errors in determination of moon radii, Schmitt trigger drift, reticle eccentricity and misalignment, and the error over the tracking range. Since these errors are considered random, they have been combined, and the RMS error due to all of these causes is only 6.8 arc-seconds.

Design of Computer and Data Processor

The various electronic circuits associated with the lunar tracker are, except for the preamplifier, contained in a single electronics cabinet. The functions are discussed in terms of signal shaping and data processing.

The earlier discussion has indicated how two successive pulses of radiant flux from the lunar image identify each of the radial increments from the bright limb to the rim of the instrument's field of view. Each of these pulses is a positive-going step up from the noise. At a selected reference level a Schmitt trigger fires an auxiliary gating circuit in a manner such that a measured train of clock pulses is gated into the computer. The number of clock pulses in the train is a measure of the radial increment R_a, R_b, or R_c as the case may be. Auxiliary circuits with suitable time constants are used to deenergize the trigger circuit except within the interval expected to contain the measuring signal, thus minimizing the exposure to false signals due to noise spikes which extend above the normal level of noise into the region monitored by the Schmitt trigger.

The computer and associated circuitry are responsible for a) extracting the radial distances R_a, R_b, R_c from the signal output of the photo multiplier, and b) computing error signals from the equations

$$\epsilon_X = 2R_b - (R_a + R_c)$$

$$\epsilon_Y = 0.577(R_c - R_a)$$

The tracker computer is a transistorized digital system containing diode logic which performs addition and subtraction and provides storage in a digital-to-analog converter unit. The input signal is a digital series of clock pulses; output is analog voltage proportional to the solutions of the two equations.

Fabrication of Experimental Model Tracker

When fabricated, the experimental tracker was required to demonstrate the effectiveness of the tracking principle by means of a simple, reliable, inexpensive breadboard instrument. As built and delivered the breadboard comprises two units. The tracker unit including finder telescope weighs 82 pounds, and is designed for mounting on a pair of external equatorial axes. The electronics unit comprises signal processors, computer, servo amplifiers, and all necessary power supplies rack-mounted in a cabinet 17 x 22 x 31-1/2 inches. The approximate weight is 150 pounds. Pictures of the tracking and electronics packages are shown in Figs. 6 and 7.

In the fabrication of the opto-mechanical tracker unit techniques appropriate to the economical production of a single precision instrument were employed. Many subassemblies were provided with polished pads to facilitate alignment with autocollimators. Fits were achieved by close-tolerance machining followed by lapping. These procedures avoided costly fixtures and assured accurate and stable alignment. The alignment of the reticle system was critical to provide a quiet gear train and to minimize the air gap between the two reticle elements. The gear train spacing was adjusted to an optimum value and the backlash was removed by compensation with a suitable spring. The alignment of the two conjugate reticle systems, the scanning system and the eyepiece system was within a few seconds tolerance in order not to deteriorate the boresight accuracy. With the alignment just described, if the real or laboratory-simulated lunar image is concentric with the boresight reticle pattern and the tracking error signal outputs indicate zero (i.e., no residual difference exists between the three radial scan line segments), then indeed the tracking system is pointed to the geometrical center of the partially luminous disk. Since the system uses only 120 degrees of the circular portion of the moon's limb to generate the tracking inputs, this analysis is valid for all phases of the moon irrespective of the rest of the luminous contour.

Fabrication and Assembly of Electronics

The electronics associated with the lunar tracker breadboard comprise a) preamplifier and signal shaper, b) data processor and computer, and c) servo amplifiers and readouts. In the fabrication of this design Northrop has elected to use purchased subassemblies and solid-state devices to the greatest extent possible. Furthermore, with

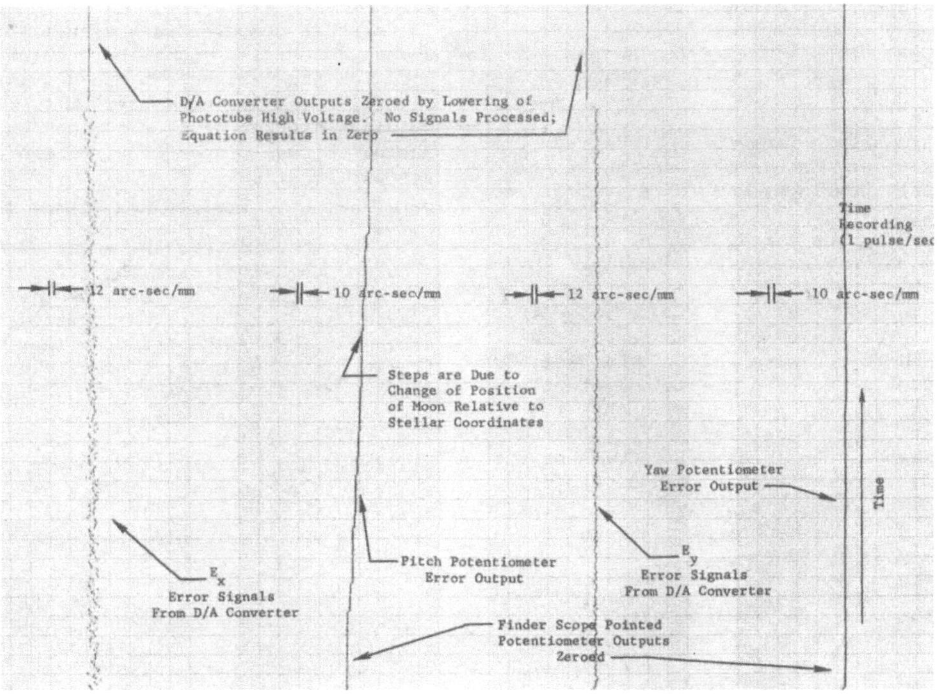

Fig. 8. Typical observatory recording of 3.6-day old moon.

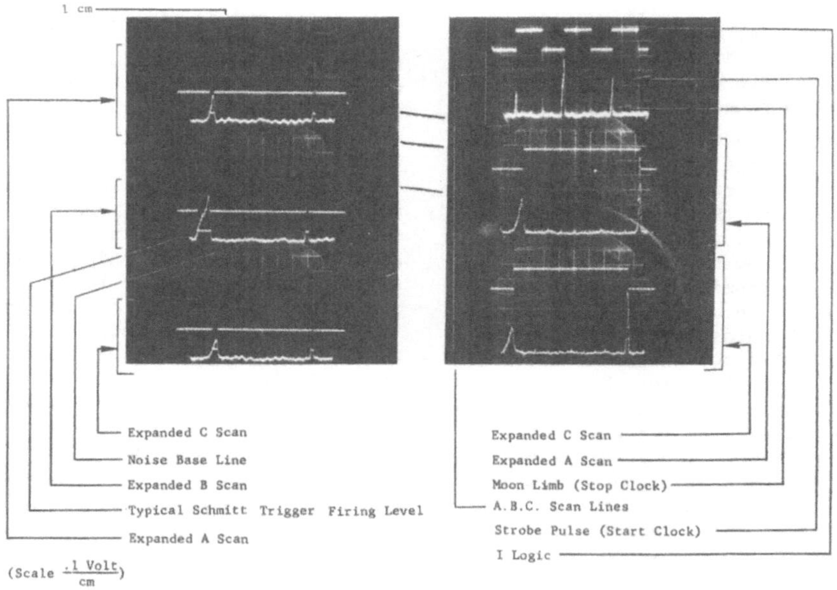

Fig. 9. Typical oscilloscope photographs of 3.6-day old moon taken in the observatory.

the exception of the servo amplifiers and power supply, which are packaged by the respective vendors, all circuits are on cards of similar size and dimensions. This technique has minimized the effort spent in detailed design and component fabrication and checkout, and has correspondingly maximized reliability.

The preamplifier and associated circuits to change voltage levels and match impedances are assembled on a card mounted in the front of the tracker head. The wiring has been brought around the gimbals by flexible leads since the angular movement about each axis is so limited.

The signal shaper and digital computer are assembled on cards and racked into the connector panels, about 400 connections per panel. Appropriate test points are brought out to two sets of 20 pin jacks mounted in strips. All of the cards are inserted from the front and can be checked out through the series of pin jacks located on strips which are accessible from the rear.

The servo amplifiers are vendor-supplied shelf items but adjusted to match the servo motors. Each unit is a multi-stage transistorized amplifier with adjustable gain of one thousand or less. These amplifiers are mounted behind the third panel in the electronics cabinet which is shown in Fig. 7.

Testing Program

The program for testing the experimental model and appraising its performance consisted of two phases: laboratory observation of a simulated moon, and observatory measurements of the natural moon.

Tests were performed in the laboratory using a simulated moon made up of a luminous disk imaged at infinity by means of a large collimator. Diaphragms were provided to simulate changes in range and in phase. Using a proper eyepiece reticle the boresight system was used to align the tracker visually and this alignment supplied a zero reference for studying tracker errors. During some of the tests the simulated lunar image was standardized at 900 foot-Lamberts and 6000 degrees K color temperature to closely approximate the photometric properties of the natural moon.

With the 30-inch objective, simulated lunar images 0.5 and 1.9 degrees in diameter were successfully tracked. When the terminator was roughened to simulate moon craters, the computer failed to function properly. A circuit was designed and built into the computer which prevented these extra on-off signals from introducing erroneous information and a repetition of the tests with the simulated lunar craters showed that the difficulty was corrected, and satisfactory tracking was re-established. The laboratory tests indicated the tracking error to be less than 20 seconds with respect to both reproducibility and stability.

The philosophy of the observatory test program is to compare, by experiment, the true sightline to the natural moon with the sightline generated by the tracker breadboard. In this experiment the instrument was assembled on an equatorial mount having a massive pedestal. The tracker sub-base was aligned with the equatorial adjustments and checked with a sub-base-mounted finder telescope. The assembly was aligned by observing a bright star through finder and boresight telescopes. The reticle assembly was rotated to align the line of cusps. The tracking error is measured from the servo loop. The alignment of the tracking axis with the center of curvature of the bright limb was observed with the boresight telescope. During the waning period of the moon in the latter part of February, 1962, and the waxing period during early March tracking tests were made every clear morning or evening. The maximum brightness of the early morning or evening sky against which tracking was successfully accomplished was approximately 15 foot-Lamberts, although by adjusting the Schmitt firing level to be above the rising sky noise, tracking could have continued through considerably brighter skies.

A typical chart record appears in Fig. 8. Four signal channels and a timing channel are shown with the earlier part of the signal trace at the bottom of the chart. The first signal channel records the X-axis error signal as it leaves the computer. The spread of the data indicates the resolution at that point in the system. The alignment of the data with the steady state line at the top demonstrates correct tracking without bias. The second signal channel records the angle readout. Notice that the servo system has completely smoothed the data. In this experiment the lunar image was centered optically and the declination circle clamped. The hour circle was driven by the sidereal clock. Since no guiding was used to compensate for the movement of the moon in right ascension, the breadboard automatically followed the moon and generated the slanting trace. The third signal records the Y-axis error signal as it leaves the computer. The spread of data indicates the corresponding resolution of Y-axis information. The data out of the computer demonstrate correct tracking except for a 25-second bias, probably caused by imperfect alignment of the line of cusps. The fourth channel records the angle readout on the Y-axis. The data are well smoothed, the bias in channel 3 has disappeared and the final output demonstrates correct tracking.

The wave forms of the signal in the computer are shown in Fig. 9. The oscilloscope pattern serves to monitor the system performance and to measure the signal-to-noise ratio. Traces of the A, B, and C channels are shown separately and the A-B-C sequence is shown as a single trace. The significant features are indicated by the callouts on the figure.

Table 4 summarizes the data derived from experiments in which the natural moon was tracked from the Northrop observatory.

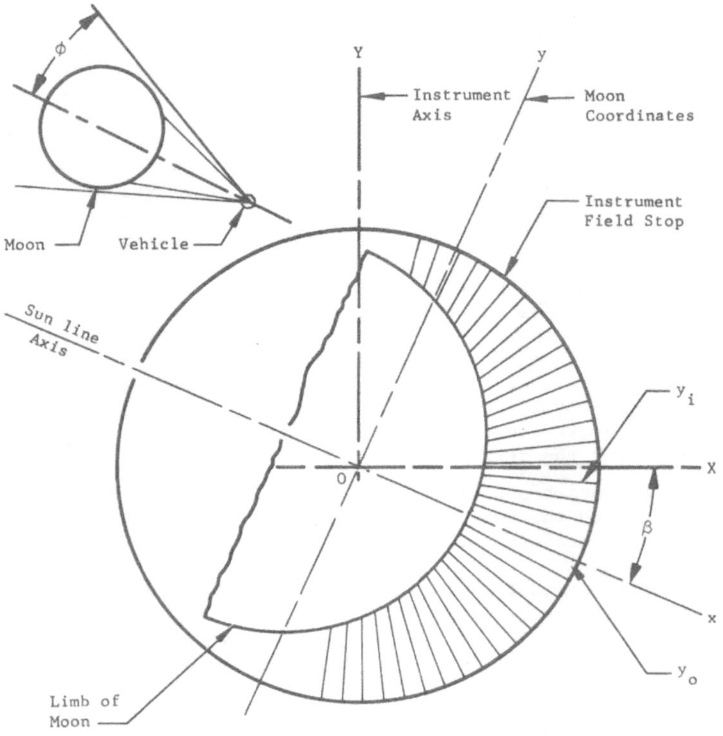

Fig. 10. Geometry of many-look scanner.

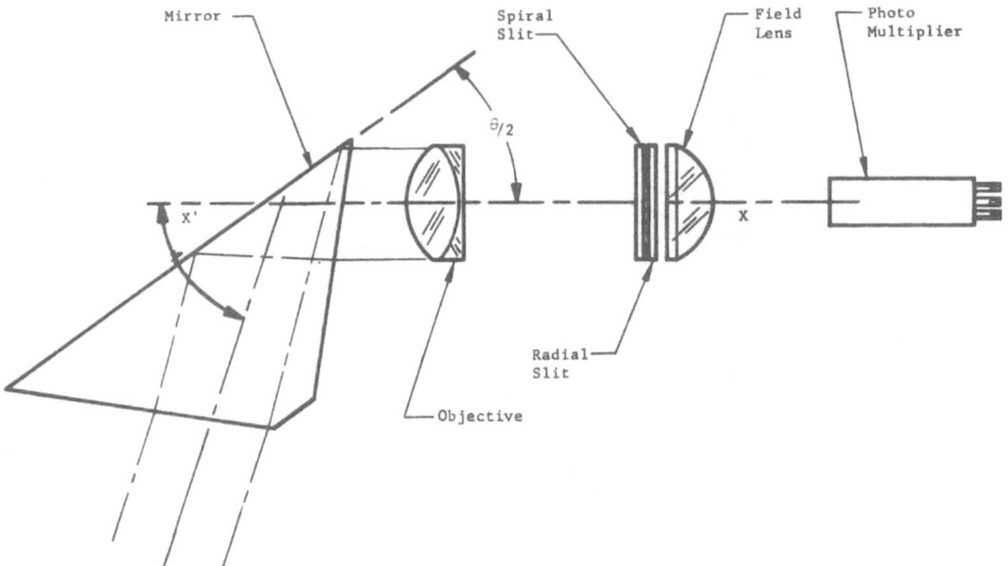

Fig. 11. Optical diagram of many-look scanner.

TABLE 4

SUMMARY OF MOON TRACKING EXPERIMENTS

Date (1962)	Moon Phase & Age (days)	Track or No Track	Signal-to-Noise Ratio (Minimum)	Notes
27 Feb	22 wane	T or NT	11:1	Adjusted line of cusps
28 Feb	23 wane	T	8:1	Early tracking hampered by atmospheric extinction. Later hampered by daylight.
1 Mar	24 wane	T	7.5:8	Stability ±10 arc-sec. Pitch and yaw clock drive off. Temp 35°F.
8 Mar	1.6 wane	NT	Approx. 2:1	Switch to waxing phase
9 Mar	2.6 wane	T(1-3 min)	2:1	Also tracked sun
10 Mar	3.6 wax	T(up to 17 min)	10:1	Max. slew of mount ok Track field is smaller
11 Mar	4.6 wax	T(up to 20 min)	16:1	8 arc-min track field
21 Mar	14.6 wane (near full)	T	170:1	Strongest signals. Start pulse saturates. Intermittent fog.

DESIGN OF A MANY-SPOKE PLANET TRACKER

The validity of the radial measurements approach to planet tracking has been demonstrated by the experimental tracking of the moon with a three-spoke instrument. In the further development of this technique for tracking a partially illuminated body several improvements are desirable. These include observations at frequent intervals along the bright limb, progressive changes in the field-of-view to facilitate tracking in the vicinity of a body where the body subtends a large angle at the instrument. These design improvements are discussed in the following paragraphs.

Logical Design of the Scanning Subsystem

The planetary tracker senses center of the earth or moon by computations based on measurements of the circular portion of the limb contour. These data are obtained from short radial segments of scan spaced in a conical annulus developed by rotating a suitable optical element in the instrument. This instrument design is applicable to any partially illuminated celestial body but as a convenience of expression the discussion will be either general or related to lunar observations.

The geometry of the scanning and sensing scheme is shown in Fig. 10. The insert shows the conical annulus which is 2 degrees wide and defines the outer half cone angle. Consider a lunar orthogonal coordinate system such that the sun lies on the positive extension of the Ox-axis. The circular limb of the moon will be convex to the right. The instrument axis is identified by the center O, or by the concentric field stop circle which is shown. Radial offsets between the limb of the moon and the instrument circle are shown at 5-degree intervals and the magnitude of an offset is symbolized by y_i. The offset on the x-axis (sun line) is identified as y_o. The consecutive measuring points extend symmetrically from y_o at 5-degree intervals. The magnitude of any y_i is a linear measurement in the focal plane and is equivalent to an angular increment at the moon.

Since the measuring stations must be disposed in a definite manner with respect to the orientation of the lunar countour, correlation must be established with the coordinate system fixed in the instrument so that the scanning reticle pattern can be rotated into the correct relationship with the sun line. The instrument coordinate system is identified by the gimbal axes OX and OY. The two systems are related by the angle β.

The scanning equations identify the error in the position of the lunar center with respect to the instrument center. Two somewhat different approaches have been considered. The first uses a least squares approach, has definite defense against random errors but requires digital multiplication by a large number of factors. The second approach, designed to minimize digital multiplication is based on the series expansion of an exact solution. It is not deliberately defensive against random errors, but may be adequately defensive in fact.

Conceptual Design of the Instrument

The optical and mechanical schematic of the many-look tracker is shown in Fig. 11. In the configuration shown two features are of interest. The many-look scanner concept just discussed from a logical point of view is presented. Another design feature, the tiltable head prism, permits the observation of planetary disks over a very wide range of angular subtense. Typical extreme values of half angle are 4-1/2 and 70 degrees. The corresponding ranges to the surface of the moon are 12,600 mi. and 67 mi. To the surface of the earth the ranges for these angles are 46,400 mi. and 257 mi. respectively.

As the figure shows, in order to generate these large sighting angles to the bright limb of a nearby body the device scans a conical field of half angle ϕ by means of a special head prism tilted at the angle $\phi/2$. The sightline is defined by the objective and the optical axis XX'. The prism tilt angle $\phi/2$ is adjustable. In using the instrument the prism is rotated at constant angular velocity about the axis XX'. The detailed scan of the instantaneous field is accomplished by two reticles mounted close together in the focal plane and the diameter of the two reticle patterns defines the instantaneous field of view. The engineering details of this part of the instrument and related logic are discussed below.

When the sightline deviation angle is quite small a first surface head mirror would operate near grazing incidence and would be impractically large. Consequently a Dove prism which is dimensionally compact and is functionally the equivalent of a mirror is substituted. The prism is pivoted at the apex and is large enough to avoid vignetting the lens objective under any condition. This prism is shown in Fig. 11.

As the range from the instrument to the planet changes the angular subtense will change and require progressive adjustment of Dove prism angle.

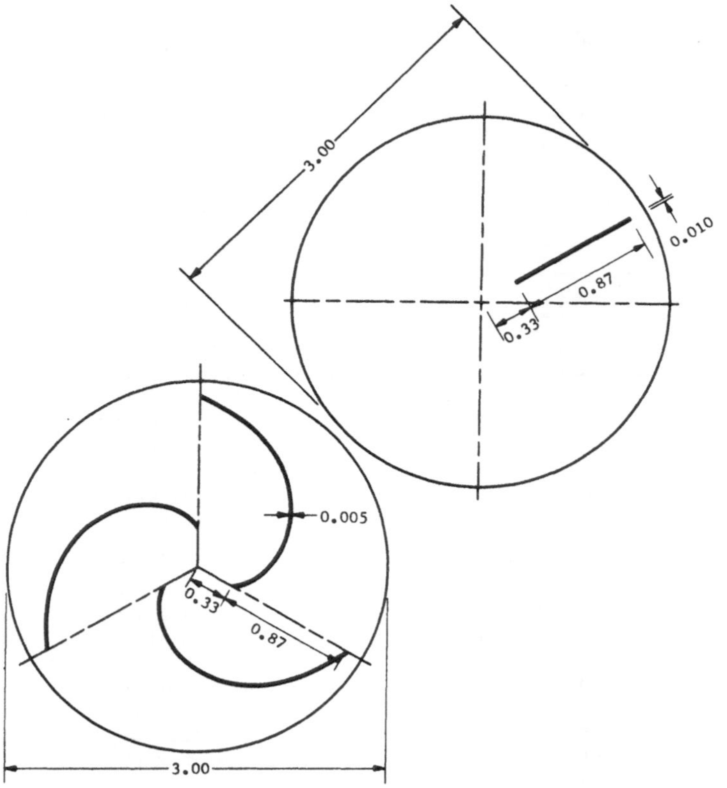

Fig. 12. Design of scanning reticles (dimensions in inches).

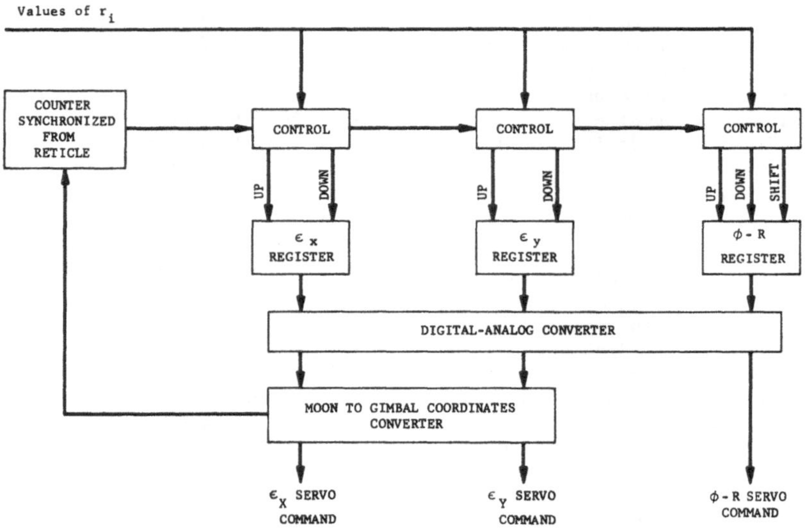

Fig. 13. Block diagram of computing logic.

Typically this slewing rate is about 0.4 deg/sec of time.

The present design uses a telephoto objective with 25 inch focal length. Consequently linear distances in the focal plane correspond to angular distances in object space in the ratio of 135 arc-min/inch. The telephoto design will be as conservative as possible without sacrificing the necessary instrument compactness. The optical specifications for the objective are as follows:

Equivalent focal length	25 inches
Front vertex to focal plane	16-1/2 inch maximum
Clear aperture of front lens	1.35 inches
Relative aperture	F/5.4
Field of View	2-1/2 degrees, half field
Type of achromatization	D, F and G'

The scanning mechanism is based on the same principles discussed earlier, but the equipment is modified to permit many scans along the bright limb.

In the scan routine a small spot sweeps an annulus in a series of equi-spaced radial lines by means of the intersection of a spiral slit and a radial slit mounted in the focal plane. Their dimensions are shown in Fig. 12. The radial slit turns in exact synchronism with the Dove prism and is so phased that the slit remains in the principal plane of the prism. As the Dove prism rotates, the slit sweeps an annulus which is about 2 degrees wide and has a radius dependent on the selection of the Dove prism angle ϕ.

In order to sample the annulus in a series of discrete radial scans, and to avoid excessive reticle speeds, three spirals are used. In a single scanning sequence the radial portion of one of the spiral patterns in the fast reticle lines up for an instant with the radial slit of the slow reticle. As the radial slit elements move out of coincidence the spiral and radial slits intersect in a rectangular aperture which moves at constant linear velocity from the rim toward the center of the field. With three spirals on the reticle, a gear ratio of 23:1 and counter-rotation between the two reticles, a radial line is scanned every 5 degrees and there are 72 discrete looks in a complete horizon circle. The instantaneous field is 1.4 x 0.7 arc min. in size.

Selecting a photomultiplier with S-11 surface as the most desirable sensor, and selecting reasonable values for the other radiometric parameters, the spectral response of the system will extend from 0.38 to 0.68 microns with the peak at about 0.5 microns. This spectral distribution indicates the desirability of achromatizing the objective lens for the wavelengths corresponding to the Fraunhofer lines D, F and G'.

The field lens is a relatively thick plano-convex singlet fabricated to condenser quality. Functionally it accepts the ray bundle defined by the objective and the full length of the radial slit and redirects the flux into the aperture of the sensor. The focal length and other design parameters were selected to effectively project the flux onto the photocathode not withstanding the heavy spherical and chromatic aberrations inherent in so simple an approach.

The data processing function includes those operations relating to the signal in the photocell, the logic and computations, and finally the servo loops which minimize errors in sightline and range angles. Since the bright limb of the moon is a semicircle, symmetrical with respect to the sun-moon line, the related limb measurements naturally refer to a lunar coordinate system using the sun line as one axis of reference. In the course of a tracking exercise this lunar coordinate set is rotated through an angle β with respect to the instrument coordinates identified with the X and Y gimbals. The angle β ranges through 360 degrees depending on the relative position of sun, moon and sighting instrument; the latter being on earth or in a space vehicle as the case may be.

The scanning operation involves 32 looks which are 5 degrees apart and which embrace 160 degrees of arc. The initiation of each look is signalled by a synchronizing pulse generated by a magnetic mark on the reticle rim and the readout head which correlates the moon and the instrument coordinate sets.

The photocell signal results from that portion of the lunar image flux which passes through the intersecting portions of the radial slit and the spiral slit reticles. If the lunar image is within the field of view, a portion will be intercepted by the aperture of the intersecting slits in a manner characteristic of the situation. A considerable portion of the lunar flux will be transmitted by each coincidence of the radial slit elements as the fast reticle pattern rotates past the slow one. Typically the intersection of radial and spiral slits will cross onto the bright lunar image after an interval representative of the spacing between the lunar contour and the rim of the tracker field of view. The spacing between these pulses measures the radial interval between the bright limb and the edge of the instrumental field stop. By auxiliary logic various special cases can be considered such that the partially illuminated body can be tracked under all conditions. The computer logic is shown as a block diagram in Fig. 13. Notice that the outputs include not only two sightline angles, but range also.

CONCLUSION

This paper has discussed two closely related techniques for automatically tracking a partially illuminated body such as a planet or the moon. The sightline to the center is determined from many measurements spaced along the circular bright limb. Means are described by which the change in angular subtense can be controlled. This technique contributes significantly to the methodology by which spacecraft can be navigated between earth, the moon, and the planets.

ACKNOWLEDGEMENT

The information described in this paper resulted from studies performed at Northrop Space Laboratories. The work was supported to a major degree by the Aeronautical Systems Division, Air Force Systems Command, U. S. Air Force. The support of these organizations is appreciated.

REFERENCES

(1) Pettit, E. and Nicholson, S. B., "Lunar Radiations and Temperatures," Astrophysical Journal 71, 102-135, 1930.

(2) Ehricke, K. A., "Space Flight, I. Environment and Celestial Mechanics," New York: D. Van Nostrand, Inc., 1960, p 191.

(3) Fielder, Gilbert, "Structure of the Moon's Surface," New York: Pergamon Press, 1961, pg 47.

(4) Allen, C. W., "Astrophysical Quantities," London: Athlone Press, 1955, pg 24.

(5) Ref. 4 pp 24, 162.

(6) United States Air Force, "Handbook of Geophysics," New York: Macmillan, 1960, Chap. 2.

(7) Ref. 4, various pages.

(8) Kuiper, Gerard P., Ed., "The Earth as a Planet," Chicago: U. of Chicago Press, 1954, pp 733-734.

DISCUSSION

R. S. Estey

"Tracking the Partially Illuminated Earth or Moon".

Q. What was the tracking system used, and what was its accuracy?

A. The tracking system used consisted of a telescope and photomultiplier tube modulated by the reticle system described. The angular accuracy was approximately 10 arc-seconds, using the real moon at any phase greater than 2 days from new.

Q. Could this device be used in a strap-down application, using the spacecraft itself as a gimbal?

A. Yes, but provision for changing field of view due to motion toward the moon would be needed. The device described is a null instrument and requires gimbals in its present form.

THERMAL NOISE OF MEASURING INSTRUMENTS AND LIMITING PRECISION OF SPACESHIP AUTONOMOUS CONTROL SYSTEM

by A. A. Krasovskii, USSR

Autonomous control, based on the use of inertial and astroinertial measuring systems, will play a prominent role in astronautics. Among the multiplicity of factors responsible for errors in inertial and astroinertial measuring systems is thermal noise.

In purely inertial systems, this noise is brought about by the random thermal motion of atoms and molecules in both elastic elements (filaments, springs, etc.) and in gaseous and liquid media surrounding the sensitive elements (gyroscope rotors, moving accelerometer systems), and also by the random thermal motion of charge carriers in electronic components.

In astroinertial systems, we may add to these thermal noise sources the usually dominating thermal noise of radiation detectors.

At any temperature other than absolute zero, thermal noise cannot be completely eliminated. The precision determined by thermal noise will therefore be known as the limiting precision.

This paper is devoted to a study of the limiting precision of inertial measuring systems and to dispersion of spaceship trajectories brought about by thermal noise affecting these systems.

We should start off by stating that the errors in contemporary inertial systems operating under the usual conditions other than space flight conditions are primarily due to design factors and to engineering factors and are many orders of magnitude greater than the errors due to thermal noise. However, a study of the limiting precision of inertial measuring systems becomes a problem of immediate concern in the present period. There are at least two factors which may be held to account for this immediacy:

First, the continual improvement in the design and production technology of inertial systems has been accompanied by a continued reduction and minimization of the errors due to design and engineering factors.

Second, the use of inertial systems in space flight at low thrust or zero thrust sets up favorable conditions for a drastic reduction in the error due to unbalance and friction.

The study of thermal oscillations in inertial systems may be carried out, generally speaking, on the basis of the conventional techniques in the correlation theory of random processes (1, 2). It would be far more convenient, nevertheless, and particularly so in the case of complex systems, to make use of the recently derived general expressions for the matrices of the moments of thermal fluctuations in arbitrary linear passive systems in thermal equilibrium (3).

In this article, the theory is developed in application to open-loop inertial systems whose characteristic feature is the single and double integration of signals from the primary sensors.

GENERAL EXPRESSIONS

Here we consider linear passive systems (i.e. systems containing no amplifiers) subject to intrinsic thermal noise.

The Lagrange equation of the second kind for systems of this type is of the form

$$\frac{d}{dt}\left(\frac{\partial L}{\partial \dot{q}_i}\right) - \frac{\partial L}{\partial q_i} = -\sum_{\nu=1}^{n} r_{i\nu}\dot{q}_\nu + \varphi_i \qquad (i = 1, 2, \dots, n) \tag{1}$$

where

$L = T - V$ is the Lagrangian,

$T = \frac{1}{2}\sum_{i,\nu=1}^{n} m_{i\nu}\dot{q}_i\dot{q}_\nu$ is the kinetic energy,

$V = \frac{1}{2}\sum_{i,\nu=1}^{n} c_{i\nu}q_i q_\nu$ is the potential energy,

$-\sum_{\nu=1}^{n} r_{i\nu}\dot{q}_\nu$ are dissipative (and gyroscopic (4)) forces,

$q_i, \dot{q}_i$ are generalized coordinates and generalized velocities

$\varphi_i = \varphi_i(t)$ are centered random functions corresponding to the intrinsic thermal noise.

The matrices of the coefficients of the quadratic forms for the kinetic and potential energies are symmetrical:

$$\mathbf{m} = \|m_{i\nu}\| = \mathbf{m}^T, \quad \mathbf{C} = \|C_{i\nu}\| = \mathbf{C}^T$$

where the superscript "T" denotes the transpose of the matrix. The matrix of the coefficients $\mathbf{r} = \|r_{i\nu}\|$ may be unsymmetrical, in particular in the presence of gyroscopic forces.

In the expanded form, Eqs. (1) exhibit the form

$$\sum_{\nu=1}^{n}(m_{i\nu}\ddot{q}_\nu + r_{i\nu}\dot{q}_\nu + c_{i\nu}q_\nu) = \varphi_i \qquad (i = 1, 2, \dots, n) \tag{2}$$

The basic theorem, formulated below, and the proof, appear in (3), resting on the following postulate.

If the system is in thermal equilibrium, i.e. the absolute temperatures of all the system components are the same and equal to θ, then the matrix of the termal noise correlation functions φ_i is

$$\mathbf{R}_{\varphi\varphi}(\tau) = \| M[\varphi_i(t)\varphi_\nu(t-\tau)]\| = K\theta(\tau)(\mathbf{r}+\mathbf{r}^T)\times \delta(\tau) \tag{3}$$

where

M - is the symbol for the mathematical expectation,

K - is the Boltzmann constant,

$\delta(\tau)$ - is the delta function.

For mechanical systems with no gyroscopic restraints, as in the case of passive electrical circuits, Eq. (3) is in essence not a postulate, since it may be derived from the expression known as Nyquist formula (5) when the systems are represented in appropriate form.

The theorem in question is stated as follows.

If a passive linear system is in thermal equilibrium and in non-degenerate ($c \neq 0$), then the matrices of the error moments due to thermal noise are satisfied, at steady state, by the equation

$$\mathbf{m}\mathbf{M}_{\dot{q}\dot{q}} = K\theta\mathbf{E}, \quad \mathbf{c}\mathbf{M}_{qq} = K\theta\mathbf{E} \tag{4}$$

and the averages of the kinetic and potential energies are expressed by the formulas

$$M[T] = \tfrac{1}{2}K\theta n, \quad M[V] = \tfrac{1}{2}K\theta n$$

But if $\mathbf{c} = 0$ (degenerate system) then in steady state the expressions

$$\mathbf{m}\mathbf{M}_{\dot{q}\dot{q}} = K\theta\mathbf{E}, \quad \dot{\mathbf{M}}_{qq} = K\theta[\mathbf{r}^{-1} + (\mathbf{r}^{-1})^T] \tag{5}$$

will be valid. Here $\mathbf{M}_{qq}, \mathbf{M}_{\dot{q}\dot{q}}$ are the matrices of the second moments (covariant matrices):

$\mathbf{M}_{qq} = \|M[q_\nu q_\rho]\|$, $\mathbf{M}_{\dot{q}\dot{q}} = \|M[\dot{q}_\nu \dot{q}_\rho]\|$ the unit matrix.

Equations (4, 5) greatly simplify the task of finding the variance and the mutual moments, due to thermal noise, of the coordinates, even in highly complex passive systems.

In particular, the equation $\mathbf{c}\mathbf{M}_{qq} = K\theta\mathbf{E}$ conveys the meaning that the variances and the mutual moments of the coordinates due to thermal noise are numerically equal to the statistical deviations of those coordinates in n experiments in each of which one generalized force φ_i is replaced by a constant magnitude $K\theta$, and all the remaining generalized forces vanish. However, for the study of the limiting precision of inertial systems it is not sufficient to know how to determine the variances and mutual moments of the coordinates of passive systems. The trouble here is that signals from the primary sensors -- which are passive systems -- in inertial systems are usually integrated once and twice. Thus the output variables of the inertial system which characterize the system precision are associated with the coordinates of the primary sensors by single- or double-integration operators, while in the case of closed-loop inertial systems even more complex operators are involved.

Consequently, we have to determine not only the matrices of the moments $\mathbf{M}_{qq}, \mathbf{M}_{\dot{q}\dot{q}}$, but also the matrices of the correlation functions $\mathbf{R}_{qq}, \mathbf{R}_{\dot{q}\dot{q}}$.

It is generally known that the forced component of the motion of a linear system may be represented in the form

$$q_\nu(t) = \sum_{i=1}^{n}\int_0^\infty W_{\nu i}(\tau)\varphi_i(t-\tau)\,d\tau \tag{6}$$

where $W_{\nu i}(t)$ is the weighting function -- the response of the ν-th coordinate to a delta impulse function impressed across the i-th input; the matrix of the weighting functions satisfies the matrix equation

$$\mathbf{m}\ddot{\mathbf{W}} + \mathbf{r}\dot{\mathbf{W}} + \mathbf{c}\mathbf{W} = \delta(t)\mathbf{E} \tag{7}$$

at zero initial conditions. In the case of a non-degenerate system ($\mathbf{c} \neq 0$) the random functions, just as the thermal noise per se $\varphi_i(t)$, are stationary random functions.

We have from Equation (6):

$$M[q_\nu(t)q_\mu(t-\tau)] = \sum_{i,j=1}^{n}\int_0^\infty\int_0^\infty W_{\nu i}(\tau_1)W_{\mu j}(\tau_2)M[\varphi_i(t-\tau_1)\varphi_j(t-\tau-\tau_2)]\,d\tau_1\,d\tau_2$$

or, in matrix form

$$\mathbf{R}_{qq}(\tau) = \int_0^\infty\int_0^\infty \mathbf{W}(\tau_1)\mathbf{R}_{\varphi\varphi}(\tau+\tau_2-\tau_1)\mathbf{W}^T(\tau_2)\,d\tau_1\,d\tau_2$$

And according to Equation (3):

$$\mathbf{R}_{\varphi\varphi}(\tau+\tau_2-\tau_1) = K\theta(\mathbf{r}+\mathbf{r}^T)\delta(\tau+\tau_2-\tau_1)$$

and

$$\mathbf{R}_{qq}(\tau) = K\theta\int_0^\infty \mathbf{W}(\tau+\tau')(\mathbf{r}+\mathbf{r}^T)\mathbf{W}^T(\tau')\,d\tau' \tag{8}$$

where

$$\tau' = \tau_2$$

Similarly

$$\mathbf{R}_{\dot{q}\dot{q}}(\tau) = K\theta\int_0^\infty \dot{\mathbf{W}}(\tau+\tau')(\mathbf{r}+\mathbf{r}^T)\dot{\mathbf{W}}^T(\tau')\,d\tau' \tag{9}$$

In order to find the correlation functions of the coordinates, we have to determine the weighting functions, i.e. in the final analysis we have to solve the characteristic equation

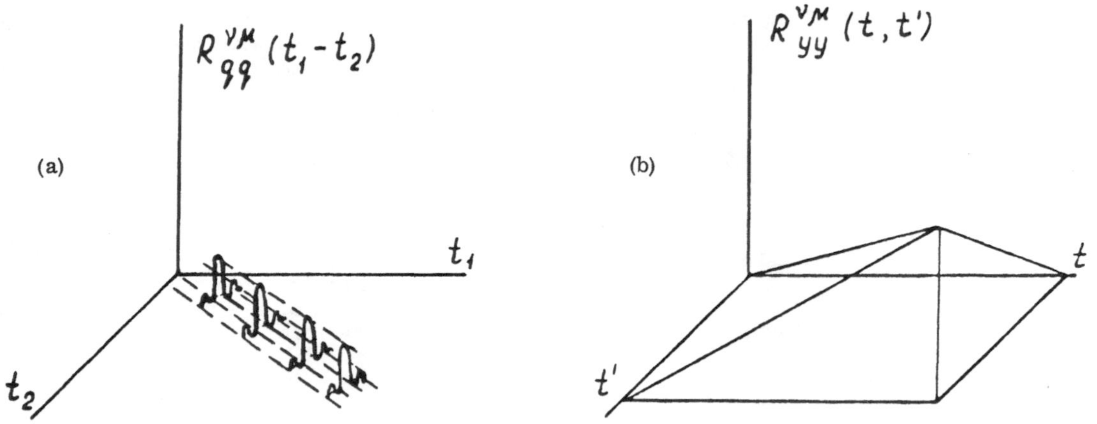

FIG. 1.

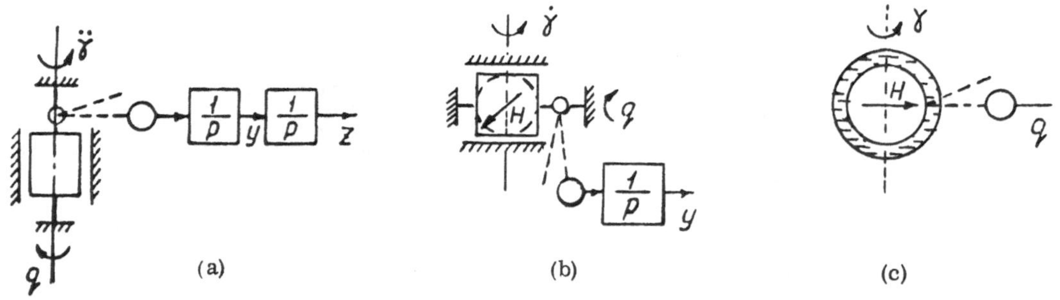

FIG. 2.

$$|\mathbf{m}\lambda^2 + \mathbf{r}\lambda + \mathbf{c}| = 0 \tag{10}$$

However, in order to determine the variances at the outputs of integrators in open-loop inertial systems, a knowledge of the correlation functions per se is not usually required, rather it is sufficient to take the integral estimates of these correlation functions:

$$\int_{-\infty}^{\infty} \mathbf{R}_{qq}(\tau)\,d\tau = 2\int_0^{\infty} \mathbf{R}_{qq}(\tau)\,d\tau$$

Actually, the matrix of the correlation functions of the first quadratures of the coordinates $y_\nu = \int_0^t q_\nu(t_1)\,dt$ is equal to

$$\mathbf{R}_{yy}(t,t') = \int_0^t\int_0^{t'} \mathbf{R}_{qq}(t_1 - t_2)\,dt_1\,dt_2 \tag{11}$$

The correlation functions $R_{qq}^{\nu\mu}$ which are elements of the matrix $\mathbf{R}'_{qq}$, as will become clear from Equation (8), decay almost entirely outside the band having the centerline $t_2 = t_1$ (Fig. 1). The width of this band corresponds to the decay time of the transients in the primary passive system under discussion (2).

If the integration times t, t are much longer than the times of transients in the primary system, and precisely this case is typical of steady-state operation, then the integral

$\int_0^t\int_0^{t'} R_{qq}^{\nu\mu}(t_1-t_2)\,dt_1\,dt_2$ is approximately equal to

$\sqrt{2}\,t\int_{-\infty}^{\infty} R_{qq}^{\nu\mu}(\tau)\,d\tau$ when $t \geq t'$ and is approximately equal to

$\sqrt{2}\,t'\int_{-\infty}^{\infty} R_{qq}^{\nu\mu}(\tau)\,d\tau$ when $t \leq t'$ (cf. Fig. 1b).

In sum

$$\mathbf{R}_{yy}(t,t') \approx \begin{cases} 2\sqrt{2}\,t\int_0^{\infty}\mathbf{R}_{qq}(\tau)\,d\tau & \text{when } t \geq t' \\ 2\sqrt{2}\,t'\int_0^{\infty}\mathbf{R}_{qq}(\tau)\,d\tau & \text{when } t \leq t' \end{cases} \tag{12}$$

The matrix of the correlation functions of the second quadratures

$z_\nu = \int_0^t y_\nu(t_1)\,dt_1$ is equal to

$$\mathbf{R}_{zz}(t,t') = \int_0^t\int_0^{t'} \mathbf{R}_{yy}(t_1,t_2)\,dt_1\,dt_2 \tag{13}$$

According to this formula, and according to formula (12), at $t' = t$ we find

$$\mathbf{M}_{zz}(t) = \mathbf{R}_{zz}(t,t) \approx \frac{2}{3}\sqrt{2}\,t^3\int_0^{\infty}\mathbf{R}_{qq}(\tau)\,d\tau \tag{14}$$

The matrix $\int_0^{\infty}\mathbf{R}_{qq}(\tau)\,d\tau$, as is readily seen, may be determined with ease from formula (9).

Actually, by integrating formula (9) from $-\infty$ to ∞ and carrying out the indicated transformations, we find

$$\int_{-\infty}^{\infty}\mathbf{R}_{qq}(\tau)\,d\tau = \kappa\theta\mathbf{c}^{-1}(\mathbf{r}+\mathbf{r}^T)\mathbf{c}^{-1} \tag{15}$$

And so, we may formulate the following result.

If a passive linear system is in thermal equilibrium and $|\mathbf{c}| \neq 0$ then the matrix of moments of deviations due to thermal noise will become equal to

$$\mathbf{M}_{qq} = \kappa\theta\mathbf{c}^{-1} \tag{16}$$

as a time substantially in excess of the transient time in the system in question elapses. The matrix of the moments (covariant matrix) of the integrals of these same deviations has the expression

$$\mathbf{M}_{yy}(t) \approx \sqrt{2}\,t\,\kappa\theta\mathbf{c}^{-1}(\mathbf{r}+\mathbf{r}^T)\mathbf{c}^{-1}, \tag{17}$$

while the matrix of the moments of the second integrals of these deviations is

$$\mathbf{M}_{zz}(t) \approx \frac{\sqrt{2}}{3}t^3\kappa\theta\mathbf{c}^{-1}(\mathbf{r}+\mathbf{r}^T)\mathbf{c}^{-1} \tag{18}$$

Formulas (17, 18) become exact as the integration time t is increased without bound.

Formulas (16) - (18) enable us to analyze the limiting precision of open-loop inertial systems for nondegenerate cases. For degenerate cases ($\mathbf{c} = 0$), according to formula (5):

$$\mathbf{M}_{qq} = \kappa\theta t\left[\mathbf{r}^{-1} + (\mathbf{r}^{-1})^T\right] \tag{19}$$

By transformations similar to those preceding, we find

$$\mathbf{M}_{yy} \approx \frac{1}{3}\kappa\theta t^3\left[\mathbf{r}^{-1} + (\mathbf{r}^{-1})^T\right] \tag{20}$$

LIMITING PRECISION OF ANGULAR ORIENTATION INERTIAL SYSTEMS

We now compare the limiting precision due solely to thermal noise of three inertial sensors of angular position: an accelerometer of angular accelerations with double integration of the signal (Fig. 2a), of a gyroscopic angular velocity meter with a single integration of the signal (Fig. 2b) and a conventional three-degrees-of-freedom gyroscope (Fig. 2c).

One example of design of a measuring system of the first type may be seen in a cylinder suspended on torsion rods in a liquid-filled or gas-filled chamber and equipped with a torque-free noiseless sensor. The signal from the sensor is integrated twice by means of ideal integrator networks.

If the torsion system of the accelerometer is viewed as a system having one degree of freedom, then the equation taking into account the net moment of inertia will have the form

$$I\ddot{q} + r\dot{q} + cq = -I\ddot{\gamma} + \varphi$$
$$y = \int_0^t q\,dt, \quad z = \int_0^t y\,dt \tag{21}$$

where q is the relative angle of rotation of the torsional system;

$\ddot{\gamma}$ is the angular acceleration about the sensitive axis -- the torsion axis of the system;

I, r, c are the moment of inertia, friction coefficient, and elasticity respectively;

φ is a random function expressing the thermal noise.

Formulas (16) - (18) are of scalar form in this case:

$$\begin{aligned} D_q &= M(q^2) = \frac{K\theta}{c} \\ D_y &= M(y^2) = \frac{2\sqrt{2}\,K\theta r}{c^2}t \\ D_z &= M(z^2) = \frac{2\sqrt{2}\,K\theta r}{3c^2}t^3 \end{aligned} \tag{22}$$

The useful signal at the output as $\ddot{\gamma}$ varies slowly is $z = \frac{1}{c}\gamma$.

Expressing the standard error $\sqrt{D_z}$ in the γ scale, i.e. multiplying it by $\frac{c}{I}$ and introducing a constant characterizing the transient time of the measuring instrument $T_i = 3\ I/r$, we obtain the definitive formula for the standard error of the given inertial measuring system

$$\sigma_\gamma = \sqrt{\frac{2\sqrt{2}\,K\theta}{I}t}\sqrt{\frac{t}{T_i}} \tag{23}$$

The measuring system of the second type (with a two-degrees-of-freedom gyroscope) exists in two variants.

In the first variant, elastic coupling of the moving system to the frame is practically nonexistent, and integration is performed directly in terms of viscous friction. The prototype of this sort of measuring system is the integrating flotation type gyroscope (6).

Under the obvious assumptions, the equation of this sensor exhibits the form

$$I\ddot{q} + r\dot{q} = H\dot{\gamma} + \varphi \tag{24}$$

where H is the angular momentum of the rotor,

q is the rotation angle formed by the cylinder with the rotor.

This is a degenerate system (c = 0) and we should resort to Equation (5):

$$\dot{D}_q = \dot{M}[q^2] = \frac{2K\theta}{r}$$

The useful signal, as $\dot{\gamma}$ varies slowly, is $\frac{H}{r}\gamma$. On reducing the standard error $\sqrt{D_q} = \sqrt{\frac{2K\theta}{r}t}$ to the γ scale, and making use of the notation $3\frac{I}{r} = T_i$, $\frac{H}{I} = f\Omega$ (Ω is the spin angular velocity, $f\Omega$ is the nutation frequency), we have

$$\sigma_\gamma = \frac{1}{f\Omega}\sqrt{\frac{6K\theta}{I}}\sqrt{\frac{t}{T_i}} \tag{25}$$

In the second variant of the measuring system with a two-degrees-of-freedom gyroscope, $c \neq 0$, we have a torque-free noiseless sensor and an ideal integrator.

The prototype of a measuring device for this system is seen in the flotation type gyroscope with torsion-suspended cylinder.

The equations in this case exhibit the form

$$I\ddot{q} + r\dot{q} + cq = -H\dot{\gamma} + \varphi \qquad y = \int_0^t q\,dt$$

Utilizing formula (7), we find

$$\sigma_\gamma = \frac{1}{f\Omega}\sqrt{\frac{6\sqrt{2}\,K\theta}{I}}\sqrt{\frac{t}{T_i}} \tag{26}$$

The angular position pick-off of the third type (Figure 2c) is a free gyro with three degrees of freedom and a gaseous or liquid suspension which by hypothesis brings about no moments other than the random varying moment of molecular thermal origin.

The equations of slight deviations of the gyroscope are of the form

$$\begin{aligned} I\ddot{q}_1 + r_1\dot{q}_1 + H\dot{q}_2 &= \varphi_1 \\ I\ddot{q}_2 + r_2\dot{q}_2 - H\dot{q}_1 &= \varphi_2 \end{aligned} \tag{27}$$

where q_1, q_2 are the rotation angles about the equatorial axis,

I is the equatorial moment of inertia.

For a given two-dimensional degenerate system

$$\mathbf{m} = \begin{Vmatrix} I & 0 \\ 0 & I \end{Vmatrix} \quad \mathbf{r} = \begin{Vmatrix} r_1 & H \\ -H & r_2 \end{Vmatrix} \quad \mathbf{c} = 0$$

$$\mathbf{r}^{-1} = \begin{Vmatrix} \frac{r_2}{r_1 r_2 + H^2} & -\frac{H}{r_1 r_2 + H^2} \\ \frac{H}{r_1 r_2 + H^2} & \frac{r_1}{r_1 r_2 + H^2} \end{Vmatrix}$$

In accord with Equation (5), we find

$$\begin{aligned} \dot{D}_{q_1} &= \dot{M}[q_1^2] = \frac{2K\theta r_2}{r_1 r_2 + H^2} \\ \dot{D}_{q_2} &= \dot{M}[q_2^2] = \frac{2K\theta r_1}{r_1 r_2 + H^2} \end{aligned}$$

Assuming $r_1 = r_2 = r$ and using the notation $3\frac{I}{r} = T_i$ (for the damping time of the nutation), $\frac{H}{I} = f\Omega$, $\sigma_\gamma = \sqrt{D_{q_1}} = \sqrt{D_{q_2}}$, we state

$$\sigma_\gamma = \frac{1}{f\Omega}\sqrt{\frac{6K\theta}{I\left(1 + \frac{9}{f^2\Omega^2 T_i^2}\right)}}\sqrt{\frac{t}{t_i}}$$

Formulas (23), (25), (26), (28) enable us to make a direct comparison of the limiting precision of the systems discussed. Bearing in mind that $f\Omega T_i \gg 1$ as a rule, it should become clear that the expressions of the "fluctuation thermal drift" for a two-degrees-of-freedom integrating gyroscope (25) and a free gyroscope (27) are the same. Formula (26) for the two-degrees-of-freedom gyroscope with external integrator differs solely in the factor $\sqrt[4]{2} \approx 1.19$, i.e. only negligibly. In sum, the limiting precisions of all the gyroscope systems for measuring angular position mentioned here are practically identical under equal conditions (i.e. equal T_i, $f\Omega$, θ, I).

Figure 3 shows graphs of the standard deviations in fluctuation thermal drift for gyroscopic angular position sensors (graph 1) and accelerometer angular position sensors (graph 2). The graphs correspond to formulas (23), (25) for the following data: $\theta = 300^\circ K$, $I = 100$ g cm^2, $T_i = 0.1$ sec, $f\Omega = 1200$ sec^{-1}.

From these formulas and graphs, it becomes clear that the standard errors due to thermal noise increase as $\sqrt{t}$ in the case of gyroscopic measuring instruments and increase as $\sqrt{t^3}$ in the case of accelerometer measuring instruments. Furthermore, even in the case of relatively low t (e.g. $t = 10^{-2}$ h = 36 sec), the limiting precision of an accelerometer type angular position pick-off is several orders of magnitude below the limiting precision of gyroscopic pick-offs.

Remember that all the formulas cited are valid for steady-state operation, i.e. when $t \gg T_i$.

LIMITING PRECISION OF OPEN-LOOP INERTIAL NAVIGATION SYSTEMS

We use the term inertial navigation systems to refer to inertial systems designed to determine the coordinates of the center of mass.

We begin by comparing two simpler one-dimensional systems of this type with absolutely precise alignment of the accelerometers.

The first system consists of an accelerometer sensing linear accelerations and described by the equation

$$m\ddot{q} + r\dot{q} + cq = -m\ddot{x} + \varphi \tag{29}$$

where $\ddot{x}$ is the sensed acceleration; m, r, c are respectively the mass, friction coefficient, and elasticity of the moving accelerator system; φ is a function expressing the effect due to thermal noise; included in the system are two ideal integrators:

$$y = \int_0^t q\,dt \quad , \quad z = \int_0^t y\,dt \tag{30}$$

The second system consists of a rate-integrating accelerometer described by the equation

$$m\ddot{q} + r\dot{q} = -m\ddot{x} + \varphi \tag{31}$$

and a single ideal integrator

$$y = \int_0^t q\,dt$$

Utilizing formulas (16) - (18) for the first system and formula (20) for the second (degenerate) system, we infer that the output errors due to thermal noise are practically the same for the two systems and is expressed by the formulas:

$$\sigma_x = \sqrt{\frac{2\sqrt{2}\,k\theta}{m}}\; t\sqrt{\frac{t}{T_i}} \tag{32}$$

(for the first system),

$$\sigma_x = \sqrt{\frac{2k\theta}{m}}\; t\sqrt{\frac{t}{T_i}} \tag{33}$$

(for the second system). Here σ_x is the root-mean-square error in the determination of the coordinate x and due to thermal noise, while $T_i = 3\frac{m}{r}$ is the characteristic constant of the accelerometer.

The graph corresponding to formula (32) when m = 1 g, $T_i = 0.01$ sec, $\theta = 300^\circ$ K may be seen in Figure 4 (curve 1).

We now determine the errors affecting an inertial system of the first type with account taken of the fluctuation motions of the platform (of the gyroscope) on which the accelerometer is mounted.

Let us take the following model.

An accelerometer described by equation (29) when positioned on a massive base with orientation unaltered is mounted on a gyroscope which is described by equations (27) without the accelerometer. The sensitive axis of the accelerometer is aligned with the spin axis of the gyroscope.

Lagrange equations of the second kind for this system, under constant accelerations, may be expressed in the first-order (linear) approximation by a transformation to the form

(See next page) (34)

where $j_3 = \ddot{x}$ is the sensed acceleration,

j_1, j_2 are "disturbing" accelerations -- components of the total acceleration vectors on the axes orthogonal to the x-direction,

$\varphi_1, \varphi_2, \varphi_3$ are functions accounting for the thermal noise.

The variable $j_3 = \ddot{x}$ is sensed in this system to a precision limited by thermal noise; its effect on the angular orientation (on coordinates q_1, q_2) may be biased out, on that account. Here, as in the entire preceding discussion, we examine the system limiting precision and we need take into account only the perturbing inputs y_1, y_2, y_3.

Bearing in mind j_1, j_2, j_3 = const., we stipulate for this case

$$\mathbf{m} = \begin{Vmatrix} I & 0 & 0 \\ 0 & I & 0 \\ 0 & 0 & m \end{Vmatrix} \quad \mathbf{r} = \begin{Vmatrix} r_1 & H & 0 \\ -H & r_2 & 0 \\ 0 & 0 & r \end{Vmatrix} \quad \mathbf{c} = \begin{Vmatrix} \frac{m^2}{c} j_3^2 & 0 & -mj_2 \\ 0 & \frac{m^2}{c} j_3^2 & mj_1 \\ -mj_2 & mj_1 & c \end{Vmatrix}$$

With the exception of the special case $j_3^2 = j_1^2 + j_2^2$,

$$|\mathbf{c}| = \frac{m^4 j_3^2}{c}\left(j_3^2 - j_2^2 - j_1^2\right) \neq 0$$

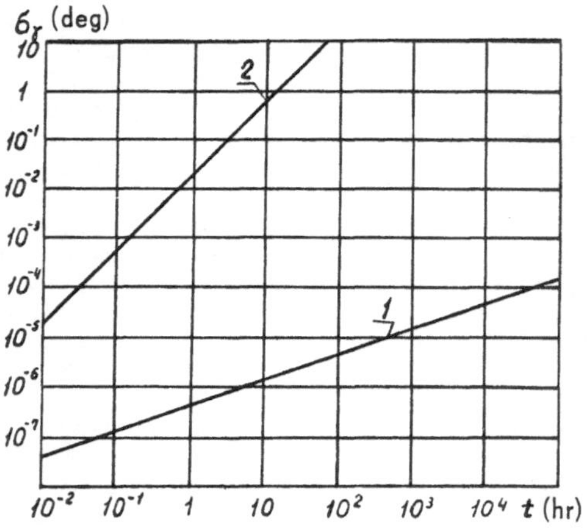

FIG. 3.

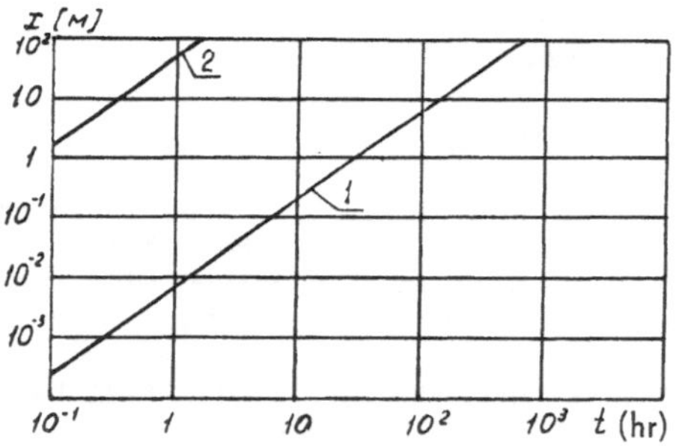

FIG. 4.

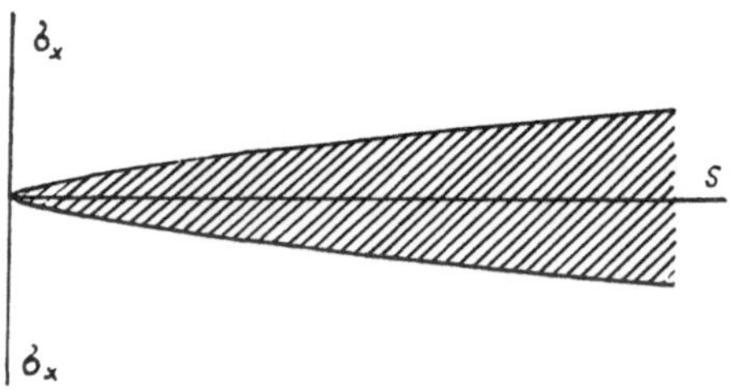

FIG. 5.

The output variable z of the system is expressed by the formula

$$z = \int_0^t y\,dt\ , \quad y = \int_0^t q\,dt$$

On applying formula (18) and expressing the standard error at the output in the scale of the sensed coordinate x, we have

$$\sigma_x = \sqrt{\frac{2\sqrt{2}\,k\theta}{m}\left(\frac{I}{mT_{i_1}}\dot{j}_3^2 + \frac{I}{mT_{i_2}}j_1^2 + \frac{\dot{j}_3^4}{\omega_0^4 T_i}\right)}\ \frac{\omega_0^2 t^{3/2}}{\dot{j}_3^2 - \dot{j}_2^2 - \dot{j}_1^2} \qquad (35)$$

The graph corresponding to this formula at m = 1 g, $T_{i1} = T_{i2} = 0.02$ sec, $T_i = 0.01$ sec, $\omega_0 = 100\ \text{sec}^{-1}$, $I = 2.5 \cdot 10^4$ g cm^2, $j_1 = j_2 = 10$ cm sec^{-2}, $\theta = 300°$K, is depicted in Figure 4 (curve 2).

The juxtaposition of the graphs in Figure 4 reveals that the limiting precision of an inertial system determined with attention given to thermal variations in the accelerometer orientation and to disturbing accelerations will be substantially lower than the limiting precision in the ideal orientation case.

THERMAL NOISE AND DISPERSION OF SPACESHIP TRAJECTORIES IN AUTONOMOUS CONTROL

Errors caused by thermal noise in the primary sensors after single and double integration are slowly varying random functions of the time. Filtering of this type of slowly varying signals in the automatic autonomous control loops of the spaceship is practically out of the question. This is because the bandwith of the automatic control loops is always limited below by the requirements of precision with respect to the large number of external disturbance inputs.

Errors in the inertial system are thus practically entirely handled by the automatic control loops, and are responsible for the corresponding dispersion of the spaceship trajectories.

Clearly, in the light of formulas (32), (33), (35), the generatrices of the trajectory dispersion tubes resulting from thermal noise in the inertial system constitute a parabola of degree 3/4 under the conditions in question and in uniform acceleration (Figure 5).

We note in conclusion that errors in the specification of the initial conditions, errors due to inaccuracies in dealing with the effect of gravitational fields, and miscellaneous other errors for which no exact lower bound can be set, were not taken into account in the study of the system limiting precision.

The estimates of limiting precision which we arrived at are indicative of rather broad potentialities in autonomous navigation of spaceships.

REFERENCES

(1) Laning, J. Halcombe, and Battin, Richard H., Random Processes in Automatic Control, McGraw-Hill Book Company, New York, 1956.

(2) Pugachev, V. S., Theory of Random Functions and Applications to Automatic Control Problems, English translation by Addison-Wesley, Cambridge, Massachusetts, 1965.

(3) Krasovskii, Entropy Variation of Continuous Dynamical Systems, Izvestiya akad. nauk SSSR, Tekhnicheskaya kibernetika, No. 5, 1964.

(4) Merkin, D. R., Gyroscopic Systems, State Technical Press, 1965 (in Russian).

(5) Hygnist, H., Physical Review, 1928, 32, 110.

(6) Smolyanskii, G. A., and Pryadilov, Yu. N., Floated Gyroscopes and Their Applications, National Defense Press (Oborongiz), 1958 (in Russian).

(7) McClure, C. L., Theory of Inertial Guidance, Prentice-Hall, Inc., Englewood Cliffs, New Jersey, 1960.

(8) Fridlender, G. O., Inertial Navigation Systems, Phys.-Math. Press, 1961 (in Russian).

$$I\ddot{q}_1 + r_1\dot{q}_1 + H\dot{q}_2 + \frac{m^2}{c}\dot{j}_3^2 q_1 - mj_2 q = \varphi_1$$

$$I\ddot{q}_2 + r_2\dot{q}_2 - H\dot{q}_1 + \frac{m^2}{c}\dot{j}_3^2 q_2 + mj_1 q = \varphi_2 \qquad (34)$$

$$m\ddot{q} + r\dot{q} + cq - mj_2 q_1 + mj_1 q_2 = -m\ddot{x} + \varphi_3$$

DISCUSSION

A. A. Krasovsky

"Thermal Noise
in Measuring Devices
and the Limiting Precision
of Guidance of a Space Vehicle".

Q. Has there been any effort to correlate this theory with experiment?

A. Only in the case of accelerometers is the sensitivity near that corresponding to thermal noise. Even so, the sensitivity is still too far from the thermal level for experiments.

INFRARED HORIZON SENSORS FOR ATTITUDE DETERMINATION

by Robert W. Astheimer
Barnes Engineering Company
Stamford, Connecticut (U.S.A.)

ABSTRACT

Earth orbiting satellites usually require some means of determining their attitude with respect to the earth, either for control or monitoring purposes. Infrared sensors which detect the sharp thermal discontinuity at the earth-space horizon have been developed for this purpose and are being used in a number of American satellite systems. They may be classed under three general types, namely: conical scan sensors, horizon edge trackers, and radiometric balance sensors. The principles of operation of these general types, and examples of flight hardware are briefly described. The relative advantages and dis-advantages of the different types with respect to system application is discussed. Means for minimizing errors caused by cloud cover and horizon gradients are discussed and some data on flight performance given.

I. INTRODUCTION

It is often necessary to establish the attitude of orbiting and interplanetary vehicles with respect to some external coordinate system. This is usually accomplished by sensing the position of some combination of the sun, stars, or planets. If the object is at a distance such that it appears as a point, the sensor is called a star tracker, while if the planet is close enough so that a definite disc is perceptible, it is usually called a horizon sensor. In the latter case, the direction toward the center of the disc is desired. This paper will consider the problem of determining the horizon position by means of the thermal self-emission of the planets.

In order to sense the position of a planet by its thermal self-emission reliably, and with some degree of accuracy, three properties of the planet are desirable. First, it must be warm enough so that a detectable thermal discontinuity exists between the horizon and the space background. Second, for a good accuracy, this discontinuity or gradient should be very sharp; and lastly, system problems are simplified if the radiance is reasonably uniform over the surface of the planet. The last condition can sometimes be attained fairly satisfactorily by selecting the spectral region of operation.

A plot of the apparent temperature distribution along a diametral scan of the Earth, the Moon, Mars, and Venus is shown in Figure 1 (Ref. 1). A logarithmic radiance scale for the total thermal emission is also shown since it is the radiance rather than the temperature which produces the received signal. Each object is assumed at half phase with the terminator bisecting the disc in order to illustrate the difference in signal level between the dark and sun illuminated sides. This figure is a great over-simplification but is useful for comparing the gross radiant characteristics of the four objects. It must be pointed out that these characteristics may be very different in restricted spectral regions, particularly in the absorption bands of atmospheric constituents; however, restricting the spectral region can only decrease the apparent radiance, although it may also reduce the dynamic range of signal that must be accepted.

A minimum detectable radiance level is shown at 0.1 milliwatts/cm^2 -ster assuming an immersed thermistor detector with a 2-inch diameter collecting aperture, 20% optical efficiency, a 1° x 1° field of view, a 250 cps bandwidth, and a signal 10 times rms noise. These are realistic parameters for a system as will be shown. It is seen that the horizon of all planets is readily detectable with such a system. We conclude then that all these planets have enough thermal emission to make infrared horizon sensors of reasonable aperture practical.

The accuracy with which the horizon can be determined will depend upon the steepness of the horizon gradient which

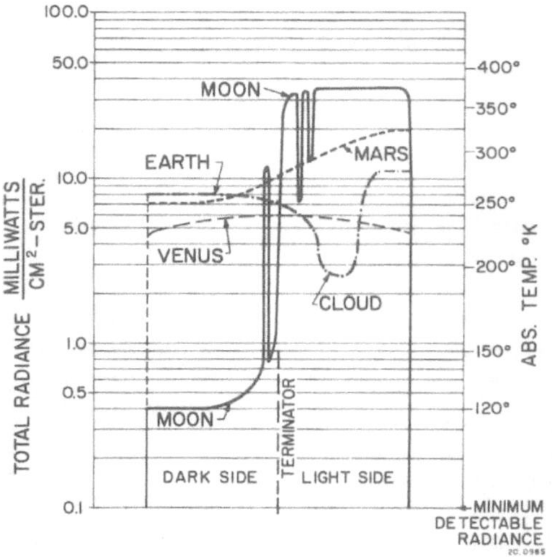

Figure 1 TEMPERATURE PROFILES OF EARTH, MOON, MARS AND VENUS

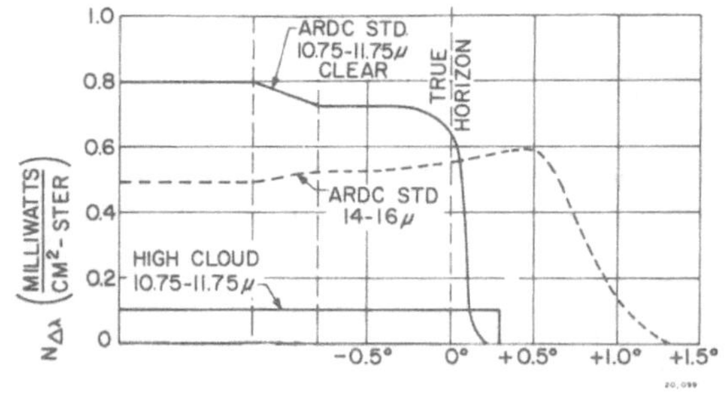

Figure 2 EARTH HORIZON PROFILES IN VARIOUS SPECTRAL REGIONS

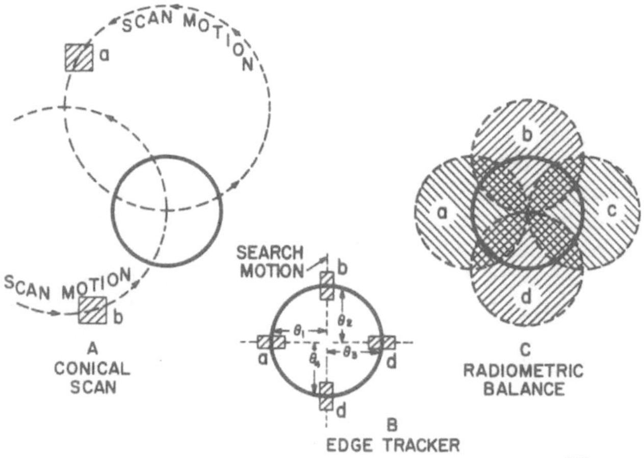

Figure 3 BASIC SENSING TECHNIQUES

is a function of the density and composition of the atmosphere. The moon, possessing no significant atmosphere, has an infinite gradient; and the ultimate accuracy is limited only by terrain irregularities. In a practical system, however, the detector noise would probably be a limiting factor because of the low temperature and emission from the dark edge.

The uncertainty caused by the fuzziness of the horizon discontinuity of a planet with an atmosphere is a function of altitude. At very low altitudes, the error could be quite large, while at high altitudes, the entire subtense of the atmosphere may become negligible. The earth is the only planet on which horizon sensors have been used to date, and considerable data has now been accumulated. The accuracy of these systems has generally been limited by the size of the detector field of view rather than the horizon gradient. A number of theoretical studies (Ref. 2) have been made of the gradient to be expected in different spectral regions, some results of which are shown in Fig. 2. The spectral regions of greatest interest are the CO_2 absorption band at 14 - 16 microns, the water vapor bands extending from 20 - 35 microns (and beyond), and the so-called transparent atmospheric "window" between 8 - 12 microns. It will be seen that the horizon gradient is quite sharp in the atmospheric window, but the range of signal level is greatest here because of cloud cover. Operation in the 14 - 16 micron CO_2 absorption band masks clouds and will present a very uniform signal, but the gradient will be more gradual, extending over almost 1° from an altitude of 400 miles. Accuracy at this altitude is probably limited to about .05 degrees by the indistinctness of the horizon itself.

Much less information is available regarding Mars and Venus. The atmosphere of Mars is fairly tenuous with few clouds and will probably not diffuse the horizon significantly even at low altitudes. Venus has a very dense atmosphere of unknown depth; however, the top of the cloud deck appears to be quite distinct and of uniform temperature. Therefore, a sharp gradient should exist in a spectral region where the clouds are opaque and the atmosphere above transparent.

One of the most troublesome problems in horizon sensing is caused by large variations in temperature or radiance over the surface. Some difficulties of this nature were experienced with early earth horizon sensors because of unexpectedly cold cloud tops associated with large storm areas. Spectral filtering and modifications of the electronic processing systems were required to achieve reliable operation for all weather conditions. The situation is much worse with the moon as indicated in Figure 1. The radiant emission from the sun-baked side is 100 times greater than that from the dark side, and there are additional small "hot" and "cold" areas caused by sun-illuminated crater edges and shadows. These variations in signal strength may not only produce errors due to saturation and time constant effects but could cause the system to confuse the horizon with the terminator, since the terminator discontinuity is 50 times greater than the dark edge horizon discontinuity.

The planet Venus seems to present the easiest object for infrared horizon sensing, but the sharpness of the horizon gradient in various spectral regions, which depends upon the atmospheric structure, is still unknown.

Reflected solar radiation will be super-imposed on the thermal self-emission of the sun-illuminated portions of all the planets. This is in general an unwanted disturbance and can be substantially removed by filtering out wavelengths shorter than 8 microns. Such filtering also greatly reduces the signal produced from direct viewing of the sun which could saturate the system or even damage the detector.

II. DETECTOR CONSIDERATIONS

It has been shown that the range of apparent planetary temperatures to be sensed extends from 120°K to 380°K. Detectors sensitive to such thermal radiation must respond to long wavelengths from 8 to 40 microns. Photoconductive detectors highly sensitive in this region have been developed such as copper or zinc doped germanium (Ref. 3) but these all must be cooled to temperatures in the neighborhood of liquid helium. Reliable cooling to these temperatures for long periods of time is difficult in space systems and the most suitable detectors are the thermal types, particularly thermistor and metal bolometers, and thermocouples. Of these, the thermistor bolometer (Ref. 4) is the most sensitive and has found greatest use to date. These are usually immersed on a germanium lens for increased detectivity.

The thermal detectors operate by virtue of the heating effect when incident radiation is absorbed on the sensitive element. The very small temperature change resulting effects some physical parameter such as the resistance which can be read out electrically. An inherent difficulty in the use of such detectors is the identification or separation of the temperature change caused by the desired radiation from the very much larger ambient temperature variations. For example, in a typical horizon sensor, the change in radiation when the field of view scans from space onto the earth increases the detector element temperature about .001°C. The sensor must be designed so that this temperature differential may be detected in spite of ambient temperature variations 10,000 to 100,000 times greater.

The most common way of separating the desired temperature change from unwanted internal effects is by optical-mechanical scanning or chopping. This modulates the external radiation signal at a relatively high frequency, and spurious signals produced by the slowly changing ambient temperature can be removed by capacitance coupling to the electronic amplifier. It is important to notice that such modulation must be done on the radiation signal before detection and not on the electrical output. Electronic chopping will modulate both external and internal signals and is of no value in this respect.

Thermal detectors which depend upon a resistance change for signal generation, such as thermistor or metal bolometers, are more subject to internal temperature effects than thermocouple types. The reason for this is that in order to sense the resistance change a bias current must be used to convert it to a voltage signal. The desired signal thus appears as a very small change superimposed on the much larger bias voltage. Bridge arrangements of detector elements may be used to buck out the large bias voltage, but it is practically impossible to maintain the degree of balance necessary over a wide ambient temperature range. Optical modulation, of course, solves the problem since the d.c. bias voltage is filtered out, therefore these detectors are seldom used without optical modulation.

Thermocouple detectors (Ref. 5) do not require biasing and are thus free of this problem. The only voltage appearing is the thermoelectric EMF produced by the radiant heating of the junction, and this is not superimposed on a biasing voltage. Unmodulated optical systems are therefore feasible with thermocouple or thermopile detectors. However, even with these detectors, care must be taken in the design to eliminate spurious signals from internal temperature gradients and self-emission of optical parts. Also thermocouple detectors sufficiently rugged for space applications are slower and less sensitive than thermistor detectors.

The elimination of optical chopping and scanning mechanisms is highly desirable for space missions where very long life and low power are necessary. The spurious signals or drifts, inherent in unmodulated systems, make them relatively less accurate than optically scanned systems. Therefore, thermopile detectors are most suitable for low accuracy, long life applications such as antenna pointing.

III. BASIC SENSING TECHNIQUES

The infrared emission characteristics of the planets and suitable detectors for sensing this radiation have been described. Such detectors must be employed in an optical arrangement to provide information from which the coordinates of the center of the planetary disc can be determined to the degree of accuracy desired. A complicating factor is that for most planetary objects the radiant emission will be highly nonuniform over the surface, and for high accuracy, the system must be independent of the radiance level.

A number of systems have been developed for this purpose. These can all be shown to be versions of three general categories which we shall designate as follows:

(1) Wide Angle Scanning Systems.
(2) Edge Tracking Systems.
(3) Radiometric Balance Systems.

The detector field of view and scan pattern associated with these techniques are shown in Figure 3. In the conical scan, which is typical of the wide angle scanning systems, the instantaneous detector field is relatively small and is caused to scan through a large cone whose apex angle may be as much as 180°, although an apex angle between 50° to 120° is more usual. The detector signal generated will be an approximately rectangular waveshape repetitive at the scan frequency. This waveshape is usually limited in some fashion to eliminate amplitude dependence, and then position information is derived by a phase or pulse width comparison technique. Two sensors are used to generate pitch and roll attitude information.

Most horizon sensor systems flown to date have been of the conical scan type because it possesses a number of very significant advantages. It has excellent acquisition capability, attained without additional search modes because of the wide scan angle. The attitude information is derived from time characteristics of an amplitude limited waveshape and is therefore insensitive to radiance variations over the surface of the planet, which as we have seen can be very large in some cases. Another advantage is that, because of the wide field scanned it is certain that some portion of the scan will leave the planet and view space. This provides an absolute zero radiance level against which any portion of any planet must give a positive contrast. Use can be made of this reference in setting limiting levels so as to prevent the system from confusing radiance discontinuities on the planetary surface, such as may be produced by the terminator or clouds, with the true horizon. This can be a serious problem with edge tracking systems. The primary disadvantage of the conical scan sensor is the need for high speed rotating elements which present life and lubrication difficulties in spaceborne applications.

Some sensors have been developed which combine the function of the two scanners by precessing the basic conical scan to produce a rosette or epicyclic pattern (Ref. 6). These may be considered essentially versions of the conical scan system and the same general remarks apply.

The basic concept of the edge tracking system is shown in Figure 3B. A small detector field of view is caused to lock onto the radiance discontinuity at the edge of the planetary disc. This is usually accomplished by oscillating the detector field through a small angle normal to the horizon edge and moving the entire sensor or the oscillating field until a balanced waveshape is obtained. Multiple sensor heads may be used, in which case at least three are necessary, or a single oscillating field may be caused to trace around the edge. In either case, the horizon position is determined by reading out the position of the center of the oscillating field with respect to spacecraft coordinates, i.e., angles θ_1, θ_2, θ_3, and θ_4, in Figure 3B.

For equivalent optics and detectors, this type of system will have a better signal-to-noise ratio than the wide angle scanning types because of narrower electronic bandwidth. It leads to a much more complicated system however because separate search and edge tracking movements must be provided on each sensing head along with precision position readouts.

Figure 4 TIROS HORIZON CROSSING INDICATOR

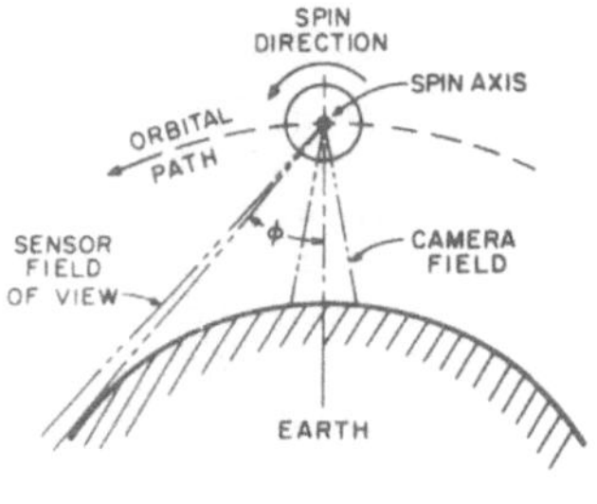

Figure 5A

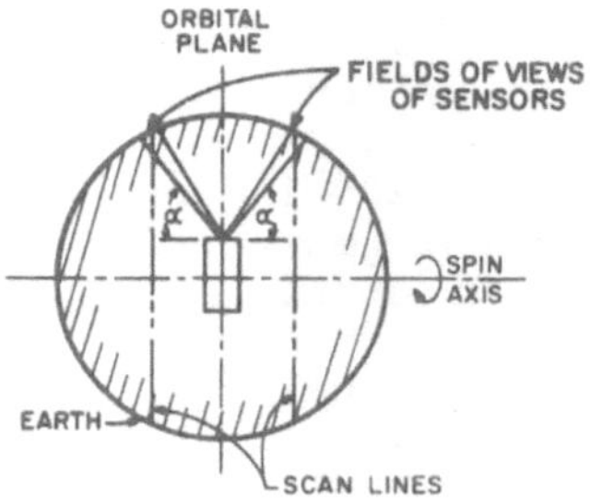

Figure 5B

SPIN STABILIZED SATELLITE GEOMETRY

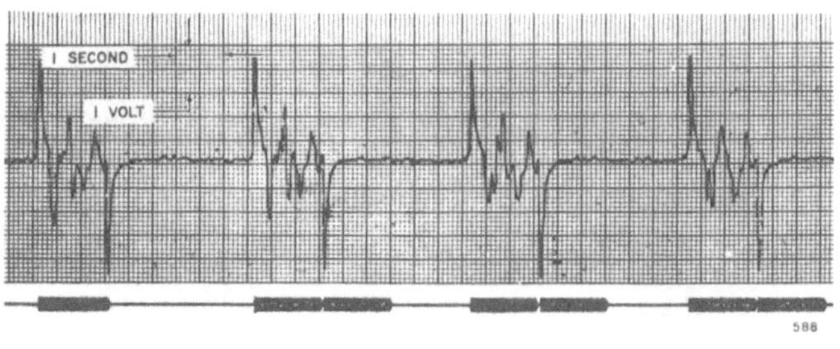

Figure 6 TELEMETERED HORIZON CROSSING INDICATOR RECORDS

A high price is paid for this narrow bandwidth and consequent greater sensitivity. Without the wide angle scan the previously mentioned space reference is lost, and there is no convenient way for the system to sense whether the signal is from the true horizon or some other discontinuity such as a cloud edge or terminator which can be much greater than the horizon signal. Various devices such as auxiliary detectors may be employed to prevent locking on false edges, but they further complicate the system. The edge tracking type of system would appear best suited for application where spectral filtering to an atmospheric absorption band can eliminate radiance variations over the planetary disc. An example for earth horizon sensing would be to filter to the narrow CO_2 absorption band at 15 microns. The increased detectivity of this type system would compensate for the large reduction in signal caused by the spectral filtering.

The former two systems employ optical modulation by mechanical means. It should be pointed out, however, that a stationary array of detectors could be used whose outputs are electronically sampled. For example, in a conical scan system, instead of mechanically causing a small detector field to scan over a wide circle, a stationary array of detector elements can be placed to view the same circle, and the array sequentially sampled electronically. Thermopile detectors are particularly well suited for this technique, and a system of this type is being developed.

The radiometric balance type sensor is a non-scanning system and operates by comparing the radiation received from opposite portions of the planet. Very wide fields of view are used to achieve acquisition and also to average radiance variations over the surface. A typical arrangement of detector fields is shown in Figure 3C. Four wide angle stationary fields are employed designated a, b, c, and d; and attitude information is obtained by the difference in radiant power received from opposite fields, i.e., $P_c - P_a$ and $P_b - P_d$. This is obviously only correct if the planet is uniformly radiant, and therefore this system is primarily suited for use with planets of uniform radiance such as Venus or where only moderately accurate pointing information is necessary.

The great virtue of this system is its extreme simplicity and consequent high reliability. By using thermopile detectors no moving parts are necessary and very long life can be achieved. Non-uniform radiance effects can sometimes be minimized by spectral filtering.

In the remainder of this paper a number of operational sensors employing these principles, which have been developed and used in space missions are described.

IV. "TIROS" HORIZON CROSSING INDICATORS

In a spin stabilized satellite the motion of the vehicle itself can be used to generate a wide angle conical scan. This is the technique used in the TIROS weather satellite series, and permits a very simple sensor design.

A photograph of a TIROS sensor is shown in Figure 4. It consists of a small infrared telescope and germanium immersed thermistor detector having a field of view of 1 degree square with a transistor amplifier. The aperture of the objective lens of the telescope is 16 mm. The entire sensor is approximately 1 X 1 X 12 inches.

As the rotation of the satellite causes the small field of view of the sensor to cross the horizon a steep positive or negative pulse is developed depending upon the direction. These horizon crossing pulses are used for a variety of purposes. The period between successive horizon crossings in the same direction (space to earth) gives the spin rate of the satellite. The ratio of the time between on and off horizon crossings to the spin period is a measure of the inclination of the spin axis to the local vertical. The horizon "on" pulse has been used to trigger other instruments such as cameras at appropriate times. For example in a "wheel" type mode where the spin axis is normal to the orbital plane as shown in

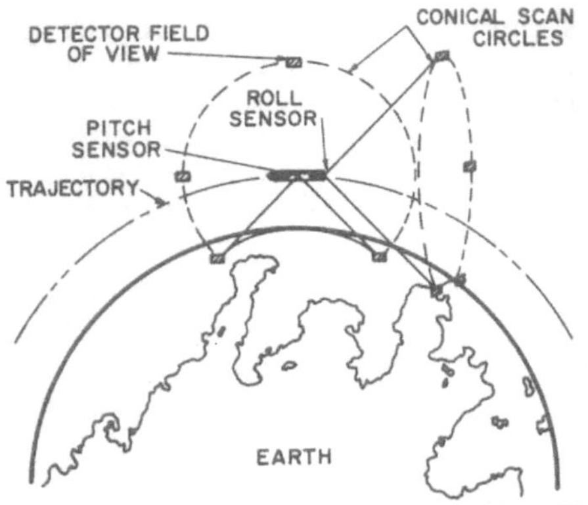

Figure 7 CONICAL SCAN GEOMETRY (PROJECT MERCURY)

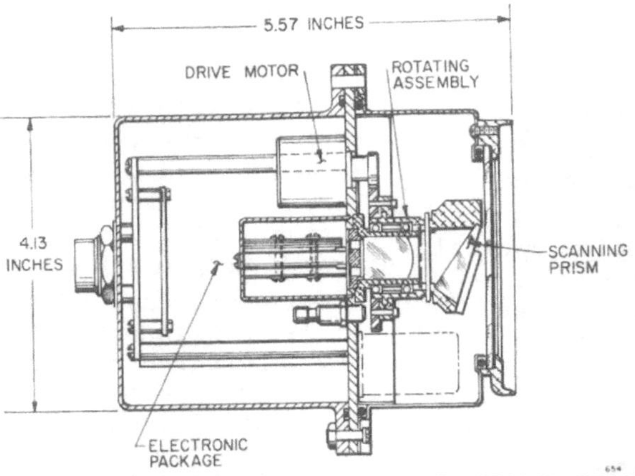

Figure 8 CROSS SECTIONAL VIEW OF MERCURY SCANNER

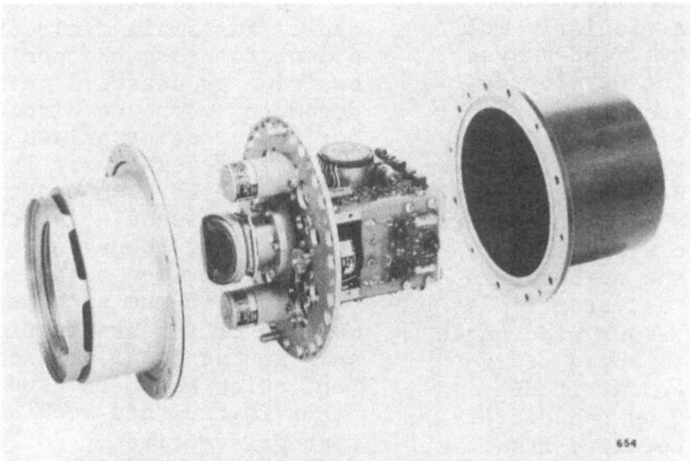

Figure 9 CONICAL SCAN SENSOR

Figure 5a, if a camera and horizon crossing indicator are both pointed radially but displaced angularly by half the earth subtense (ϕ) the horizon pulse can be used to trigger the camera when it is pointed straight down.

Also a pair of horizon crossing indicators (HCI's) whose fields lie in a plane containing the spin axis, at equal opposite angles from it as shown in Figure 5b, can indicate the orientation of the spin axis with respect to the orbital plane. If the spin axis is normal to the orbital plane the two "on" pulses will occur simultaneously. The time delay between the on-pulses will be a measure of the displacement of the spin axis from this position.

Figure 6 shows the telemetered record from a TIROS horizon crossing indicator. The spectral region in which radiation is sensed is 2 - 20 μ. It will be seen that the on and off pulses are clearly recognizable but that many spurious pulses occur during the earth scan because of cold clouds. These can be troublesome if the pulses are to be used for on board automatic control or triggering functions. This problem is usually overcome by making use of only the space to earth "on" pulses and eliminating the rest of the earth signal by a blanking gate which extends slightly beyond the earth to space "off" pulse.

V. "MERCURY" CONICAL SCAN SENSOR

For an attitude controlled vehicle the scan motion must be accomplished internally. A series of very successful sensors have been developed employing a conical scanning principle wherein the projected image of a thermistor detector is caused to scan over a wide circle, or cone in space. Typical of these is the scanner used on the Project Mercury vehicle, the first American manned spacecraft.

The scan geometry is shown in Fig.7. The instantaneous field of view of the detector is 2° X 8° and is caused to scan over a wide circle with a cone apex angle of 110° at 30 revolutions per second. A rectangular waveshape of this frequency is produced by the detector because of the temperature difference between the planet and the cold space background. Another square wave reference signal of the same frequency is generated internally from the rotating scanning mechanism. By phase comparison of the radiation signal with the internal reference the attitude with respect to the conical scan axis can be determined. A pair of such sensors displaced 90° with respect to each other as shown in Fig. 7 can give attitude information about two orthogonal axis, usually designated as pitch and roll.

A cross sectional view of the scanner is shown in Fig. 8. The conical scan is produced by a germanium prism mounted on a hollow barrel which is driven through gearing by a small motor. The internal phase reference signal is developed by a semicircular iron vane on the rotating assembly and a stationary magnetic pick up. The detector is a germanium immersed thermistor detector where the immersion lens also serves as the objective lens of the telescope. The unit is filled with dry nitrogen and hermetically sealed to prevent evaporation of lubricants. A germanium front window is used with a multilayer interference filter blocking all radiation of wavelength shorter than 8 microns. This greatly reduces direct and reflected solar radiation.

A photograph of one of these sensors with the front and rear covers removed is shown in Fig. 9.

In some applications it is inconvenient to mount the roll sensor on the front or aft ends of the space craft where it can have an unobstructed circular scan. An alternate mode of operation is shown in Fig. 10 where the two sensors are pointed 180° apart. In this mode attitude information about the roll axis is obtained by comparing the duration of the earth waveshapes generated by the two sensors while pitch can be determined from either one by the phase comparison method previously described.

The pulse phase or pulse width processing described requires a clean rectangular waveshape of constant amplitude. This is usually produced by

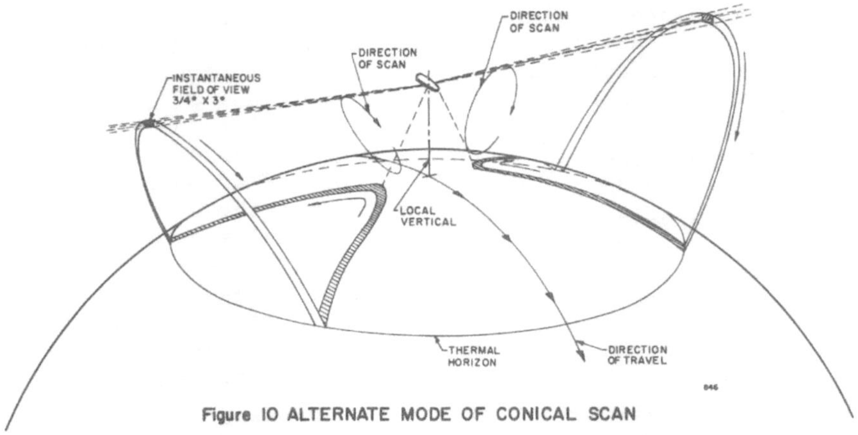

Figure 10 ALTERNATE MODE OF CONICAL SCAN

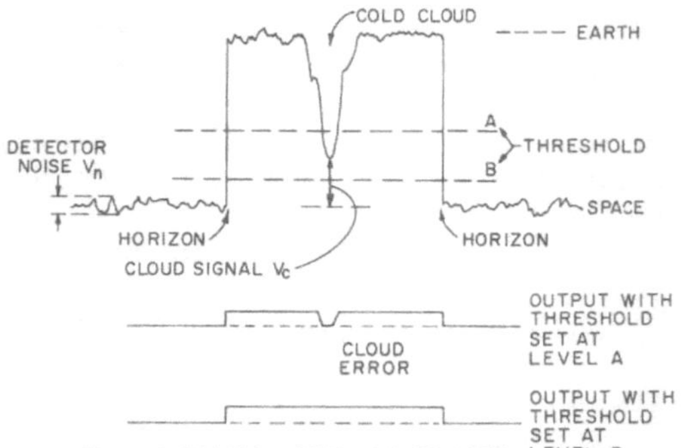

Figure 11 CONICAL SCAN WAVESHAPE

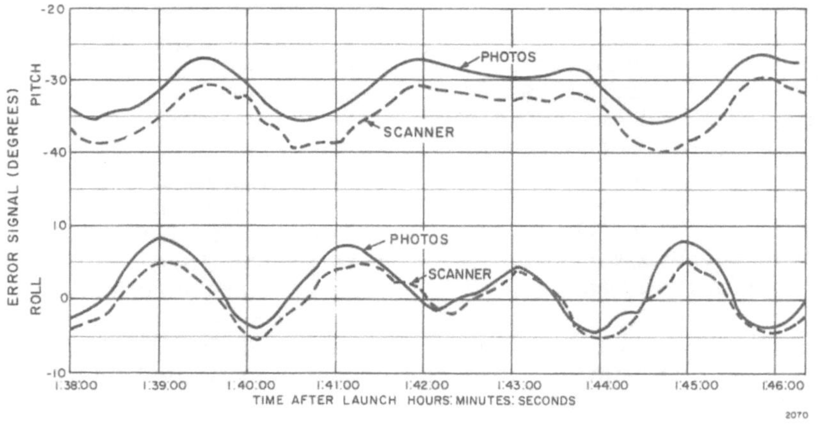

Figure 12 COMPARISON OF HORIZON SCANNER READOUT WITH PERISCOPE PHOTOS.

an on-off threshold circuit as shown in Fig. 11 which gives zero volts if the input signal is less than the threshold and a fixed voltage if it is greater. It is necessary that this threshold be established well above the detector noise level during the space scan but still below the minimum earth signal that may be expected. It is evident from the figure that if a cold cloud causes the earth signal to drop below the threshold, a "gap" will be produced in the processing waveshape which will cause serious errors in the attitude indication.

In the first horizon sensors of this type such errors did occur because the apparent radiation temperature of some cloud tops particularly those associated with large storm areas was considerably colder than expected. This situation was corrected by improving the sensitivity and lowering the threshold level.

The apparent radiation temperature of the earth depends upon the meteorological conditions at the point viewed and the spectral region in which the observation is made. In spectral regions where atmospheric constituents are absorbant the surface radiation will not be received and the signal will consist of emission from the absorbing constituents primarily water vapor and carbon dioxide, which will in general be colder than the surface. In atmospheric windows surface radiation will be received unless there is cloud cover in which case the apparent radiation temperature will be that of the cloud tops. The greatest variation in signal will therefore occur in the atmospheric windows, primarily the 8 - 12 micron region.

Flight data from horizon scanners and radiometers show radiation temperatures ranging between 170°K and 300°K in the 8 - 12 micron window, which produces a variation of ten to one in detector signal. Such large variations in signal level complicate the electronic processing and it is desirable to reduce the dynamic range of signal as much as possible. This can be done by spectral filtering to eliminate atmospheric windows.

The 14.5 - 15.5 micron CO_2 regions absorbtion band appears to be an ideal region in which to operate. (Ref. 7) Carbon dioxide is well mixed in the upper atmosphere and will abscure most clouds. In this spectral region the apparent earth temperature should be fairly uniform at about the temperature of the stratosphere, approx. 220°K. (Ref. 8) Of course operation in such a narrow spectral band considerably reduces the radiation received and requires increased optical gain.

The Mercury capsule contained a wide field periscope which viewed the entire hemisphere below the spacecraft. This periscope had a graticule to indicate the attitude of the spacecraft with respect to the horizon. In one of the early unmanned orbital tests a motion picture camera was installed on the periscope and the pitch and roll outputs of the horizon scanners were also recorded. This provided an excellent independent check on the performance of the horizon scanners. Figure 12 shows a sample of the horizon scanner roll indication with the roll attitude measured from the periscope photographs. A consistant offset of about 2° is evident which was probably due to a boresight error. Applying this correction it was found that the horizon scanner accuracy was about ± 1/2°.

VI. "OGO" EDGE TRACKING SENSOR (Ref. 9)

The horizon sensing system used with the Orbiting Geophysical Observatory (Project OGO) is a good example of an edge tracking system. A cross sectional view of the basic sensor head is shown in Fig. 13. A germanium immersed thermistor bolometer at the focal point of a fixed telescope views the earth by a reflection from a mirror mounted on the rotor of an electro-magnetic actuator called a positor. The latter is essentially a permanent magnet type torquer with a set of auxiliary coils excited at several kilocycles to accurately read out the rotor position in a manner similar to a resolver. It permits the mirror to be positioned at any point over a 45° angle and also super-imposes a small rapid sinusoidal oscillation or "dither" at a frequency

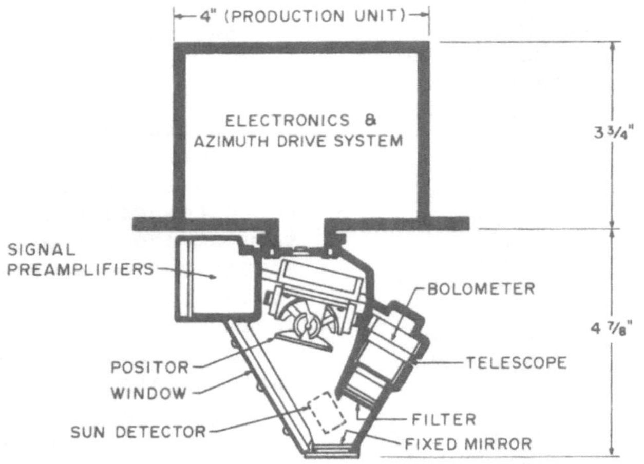

Figure 13 OGO EDGE TRACKING SENSOR

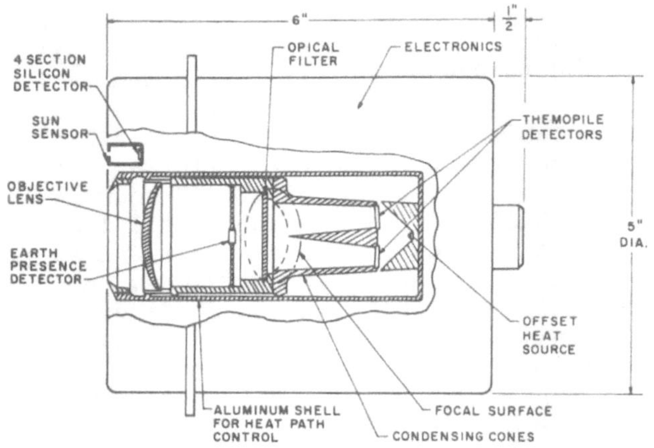

Figure 14 APOLLO ANTENNA POINTING SENSOR

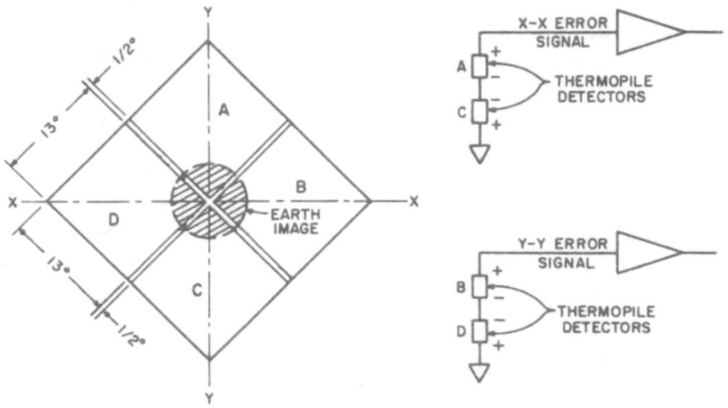

Figure 15
FIELDS OF VIEW OF APOLLO SENSOR

of 30 cps and a peak to peak amplitude of 3°. The rotor is mounted to the frame by a pair of flexure pivots to eliminate lubrication and wear problems. The telescope aperture is 2 cm and the instantaneous field of view of the detector is 1° X 1° which can be positioned over a 90° angle by the Positor.

The sensor has two modes - search and track. In the search mode the instantaneous field of the telescope is slowly scanned over a 90° arc. When the horizon is reached a 30 cps signal will be generated on the detector by the small 30 cps dither oscillation superimposed on the Positor mirror. The position of the horizon within the limits of the dither oscillation amplitude is indicated by the 2nd. harmonic content of the detector signal, which goes to zero when the horizon is centered within the dither oscillation. The Positor mirror is servoed by adjusting the average Positor-mirror angle so that the 2nd. harmonic vanishes. The angle to the horizon is then read out from the average Positor-mirror position by the read out windings which have a precision of ± 0.1°.

The complete system is comprised of four such sensor heads oriented 90° apart. These measure the angles from a reference axis to four points equally spaced around the horizon from which the attitude of the reference axis with respect to the earth is determined. Actually only three heads are required, the fourth giving redundancy in case of a failure. Also if the sun appears within the field of view of any of the four sensor heads that head is removed from the computation without any loss of performance.

Pairs of heads are packaged together in a single casting, the entire system consisting of two such dual tracker head castings and an electronic package. Its total weight is 13.2 lbs. and it consumes 8.5 watts of power (average).

VII. RADIATION BALANCE SENSORS (Ref.10)

A typical radiation balance type of sensor is shown in Fig. 14 which shows an instrument developed for keeping a communications antenna pointed toward the earth. An infrared lens projects an image of the earth onto a curved focal plane which is divided into four adjacent squares by pyramidal condensing light pipes. Each light pipe condenses the radiation in a 15° X 15° field of view onto a thermopile detector. (Ref. 11)

The fields of view delineated by the apertures of the pyramidal light condensors are shown in Fig. 15. Diagonally opposite detectors are connected in series opposition, (A-C) and (B-D), such that when equal radiation is incident on the pair of detectors, their outputs cancel. When a radiation imbalance occurs a positive or negative output signal is produced. The four detectors provide error indications about a pair of orthogonal axis, XX and YY. The instrument is designed to operate at altitudes between 6000 miles and the lunar orbit over which the earth subtense ranges from 35° to 2°.

It will be seen that this type of sensor is extremely simple requiring no scanning mechanisms or other moving parts and consequently has very high reliability and long life. Its accuracy however, depends upon uniformity of radiance over the earths surface since the error indication is derived by the difference in radiation received from opposite sectors of the earth. For this reason optical filtering is employed to restrict the spectral region of sensitivity to the CO_2 & H_2O absorbtion bands between 14 and 25 μ. Theoretical studies and TIROS flight data (Ref. 7 & 8) show the earth radiance to be more uniform in this region.

If the sun should appear within any field, a large error would result. To prevent this an auxiliary array of four silicon sun sensors is provided having identical fields of view with the thermopile infrared detectors. If a sun signal is received by one of these it will automatically disconnect its corresponding thermopile detector and replace its output with the average of the two adjacent thermopile detectors. For example referring to Fig. 15, if the sun should appear in Field A, the A sun sensor would disconnect thermopile A and substitute one half of the summed outputs

Figure 16 THERMOPILE MOSAIC DETECTOR

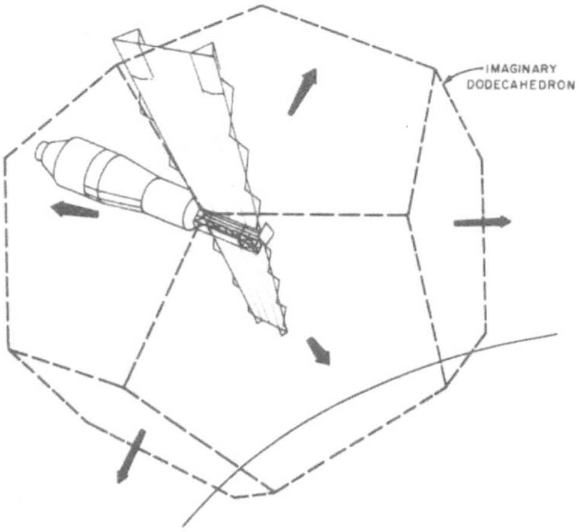

Figure 17 ORIENTATION OF MMC SENSOR FIELDS OF VIEW

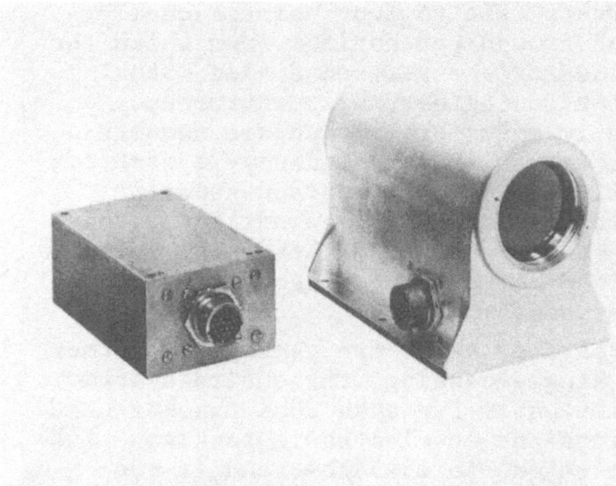

Figure 18 MICROMETEOROID MEASUREMENT CAPSULE SENSOR HEAD

of detectors D and B. At null this will be the same as the signal from detector A without the sun in the field. Field effect transistors are used for switching. To prevent the sun from deactivating two detectors simultaneously, the edges of the pyramidal light condensors are separated from each other by 1/2° (the angular subtense of the sun).

The detectors used with this sensor are vacuum evaporated thermopile mosaics made up of 308 active junctions arranged in a square array 5 mm on a side. A photograph of one of these detectors is shown in Fig. 16.

A unique radiation balance sensing system has been used on the Micrometeoroid Measurement Capsule of Project Pegasus. This is an unstabilized spacecraft which very slowly tumbles in a random manner and the problem was to read out its attitude with respect to the earth upon ground interrogation.

The attitude readout system employed consists of six identical sensor heads, each having two narrow fields of view pointing in opposite directions along the same optic axis. The fields of view are 2° X 2°. On the space vehicle the sensor heads are placed in such a manner that their optical axis are normal to the surfaces of an imaginary dodecahedron encompassing the vehicle as shown in Fig. 17. This produces an angular separation between fields of 63° 26'.

The presence or absence of earth signal within the twelve fields of view is used to establish the orientation of the vehicle relative to the earth. The system is only suitable for use at lower altitudes such that at least one sensor field will always view the earth. When the earth subtense is known and about 130° (alt 625 km) a single interrogation will give an attitude readout accuracy of ± 10° about each axes. By successive interrogations, the vehicle tumbling motion can be determined and the accuracy improved to about ± 1°.

A photograph of one of the six sensor heads is shown in Fig. 18. Two inch diameter germanium objective lenses are mounted at opposite ends of an optical tube focussing radiation onto a pair of thermopile detectors at the center of the tube. The thermopiles are connected in series opposition so that a positive, negative, or zero signal will result depending upon whether the earth is seen in one field, the other, or neither.

In conclusion it may be stated that infrared horizon sensors have, and are, being effectively used on a wide variety of earth orbiting space missions and it is expected that they will find additional use in the future on orbiting missions to the moon and other planets.

REFERENCES

1. Kuiper, G.P. and B.M. Middlehurst (Editors): Planets and Satellites; The Solar System III. University of Chicago Press, 1961.
2. Hanel, R.A., W.R. Bandeen, and B.J. Conrath: "The Infrared Horizon of the Planet Earth", Journal of the Atmospheric Sciences, Vol. 20, No. 2, pp. 73-86, March,1963.
3. Potter, R.F. and W.L. Eisenman: Applied Optics 1, p. 567, 1962.
4. DeWaard, R. and E.M. Wormser: Proc. IRE, Vol. 47, No. 9, pp. 1508-13, September, 1959.
5. Astheimer, R.W. and S. Weiner: "Solid Backed Evaporated Thermopile Radiation Detectors": Applied Optics Vol. 3, p. 493, April 1964.
6. Wormser, E.M. and M.H. Arck: Proceedings of ARS Guidance Control and Navigation Conference, August 7-9, 1961.
7. B.J. Conrath: "Earth Scan Analog Signal Relationships in the TIROS Radiation Experiment and their Application to the Problem of Horizon Sensing", NASA Technical Note D-1341, Washington, D.C.
8. W.R. Bandeen, M. Halev and I. Strange "A Radiation Climatology in the Visible and Infrared from TIROS Meteorological Satellites", NASA Report X-651-64-218 Goddard Space Flight Center Greenbelt, Md.

9. "Infrared Horizon Sensors for Precision Attitude Measurement", ATL-D-1072 published by Advanced Technology Laboratories, Mountain View, California.
10. Falbel, G. and E.A. Kallet: "Infrared Horizon Sensors Having No Moving Parts And Using Evaporated Thermopile Detectors", PROC. IRIS May 1963.
11. Williamson, D.E.: "Cone Channel Condenser Optics", Journal of the Optical Society of America, Vol. 42, No. 10, October 1952.

DISCUSSION

R. W. Astheimer

"Infrared Horizon Sensors for Attitude Determination".

Q. What wave length is most likely to be used for very accurate horizon measurements?

A. We feel the CO_2 band of 14-16 microns is most likely to be suitable for high accuracy. The CO_2 is uniformly distributed in the atmosphere, although there are seasonal variations and variations with latitude.

Q. Are there any investigations underway of horizon sensors based on millimetric or centimetric waves?

A. I know of none.

Q. What scan speeds are used?

A. Conical scanning uses 30 cps.

Q. Why do you use bolometers?

A. They are better for vibration and space environment.

ORGANIZATION OF COMPUTATION AND CONTROL IN THE APOLLO GUIDANCE COMPUTER

by

T. J. Lawton & C. A. Muntz
Staff Members
Instrumentation Laboratory
Massachusetts Institute of Technology
Cambridge, Massachusetts

Abstract

The digital guidance computer is the central control element in the Apollo Guidance and Navigation System. The difficulties in preparing computer programs for the complex Apollo missions are aggravated by the need to keep the computer hardware to a minimum. This paper demonstrates how an effective solution to these problems is achieved with a combination of appropriate programming techniques and computer design.

Introduction

The goal of the Apollo program is to place two men on the moon and return them safely to earth in this decade[1]. An important part of this program is the onboard guidance and navigation system[2]. It is the function of this system to maintain continual knowledge of position and velocity and to use this information to steer the vehicle during non-freefall portions of the mission. The guidance and navigation functions during a lunar landing mission are depicted in Figure 1.

The major components of this system are:

1. The Inertial Measurement Unit
2. The Space Sextant and Telescope
3. The Display and Controls Unit
4. The Digital Guidance Computer

The guidance computer[3] plays the central role in receiving sensor inputs, processing them, and transmitting resulting commands to various control systems within and outside the guidance and navigation system. Figure 2 shows the devices which communicate with the computer.

Because the computer must be carried aboard the spacecraft, the primary design goals are low weight, volume, and power consumption and high reliability. In order to achieve these objectives it was necessary to limit the word length, operating speed, and instruction set (Fig.3)[4]. To offset these apparent limitations, a unique high density memory[5] has been incorporated which provides a large program storage capacity (Figs. 4 and 5).

By appropriate programming techniques, together with an involuntary interrupt feature and unusual input/output techniques, it is possible to achieve the hardware design goals and still meet the system requirements.

Example of Computer Requirements

To illustrate the diversity of requirements imposed on the computer by the guidance and navigation functions, a specific phase of the mission will be examined - control of the vehicle during thrusting. This activity may be considered in three parts: computation of the required guidance and navigation information, communication with the guidance and spacecraft control system, and monitoring the performance of the computer in the overall control loop.

The information flow for this task is depicted in Figure 6. Using present position and velocity, together with a desired terminal condition, the computer calculates an impulsive velocity-to-be-gained and sends corresponding steering signals to the spacecraft control system. This results in vehicle motion which is fed back into the guidance computer via the inertial sensors. The cycle is repeated until the magnitude of the velocity-to-be gained vector is driven to zero[6].

The computation associated with each such iteration begins by updating the vehicle state vector at discrete points in time, using the velocity increments received from the inertial sensors. From the updated state the required impulsive vector velocity change is calculated. A thrust direction is then selected to align this vector and its derivative. The difference between the desired thrust direction and the actual orientation as measured by the inertial system is used to generate steering commands. Each of these computational phases has its own iteration rate: the state vector updating frequency depends on both the integration technique used and the characteristics of the accelerometers; the recalculation of the required velocity is time consuming and can not be performed too often; the rate at which steering commands are updated depends, to a great extent, on the dynamics of the overall guidance loop.

Throughout the powered maneuver, inputs must be received from and outputs delivered to equipment external to the computer. In particular, inputs may consist of:

1. velocity increments
2. gimbal angles
3. guidance and navigation status signals
4. astronaut keyboard commands
5. ground commands

These inputs are basically asynchronous with any internal computer activity and have widely disparate frequencies. Further, it should be noted that commands from ground based control centers are functionally indistinguishable from keyboard inputs initiated by the astronaut.

The outputs from the computer might be:

1. steering commands
2. maintenance of mode and caution lamps
3. digital display updating
4. digital downlink transmission

Performance verification is accomplished by periodically checking the operation of the logic and memory elements of the computer and by incorporating consistency checks of significant computed parameters. The mode and status of the guidance and spacecraft systems must be continually examined for indication of failure while the overall steering loop must be checked for loss of control by examination of gimbal angle inputs. Any malfunction so detected must result in displayed alarms, engine shut-down, or other appropriate action.

Organization of Computer and Programs

A. Input/Output Considerations

The input signals described in the previous section may be divided into three classes:

1. The velocity increments and gimbal angles constitute information of wide band-width and are in the form of pulses representing increments of velocity and increments of angle. To accomodate these inputs, with minimum computer time and interface equipment, a digital differential analyzer type of counter has been incorporated[7]. They enter the computer as pulses which "steal" the next available computer memory cycle so that they may be summed in these counter registers. The only effect of the counter service cycle is to lengthen the program execution time by one memory cycle. The use of counter registers as input buffers premits economical serial interfaces with the inertial measuring devices. The arithmetic registers of the computer are time-shared for the maintenance of these counters, providing, thereby, a significant saving in hardware. Since computer time thus consumed is proportional to the rate of change of the input quantities, wide band-width signals are readily accomodated. Provision is made for several such counter inputs to be exploited for many other purposes, some of which are described below.

2. The system status signals are discrete inputs which may be serviced at a low frequency and are directly coupled with input registers, each corresponding to a particular bit in the register. No immediate response, in the sense of computer speed, is required to changes in state; these inputs need only be sampled periodically by the performance verification programs.

3. The keyboard inputs present a slightly different problem. As soon as one keyed character enters the associated input register of the computer, it must be stored away to prepare for the next. In this way the need for buffer storage is eliminated. To perform this function without frequent scanning of the relevant input register, the entry of a key character causes an involuntary interrupt of the program sequence in prograss. The effect of the interrupt is to cause the next instruction to be selected from a standard location, after sufficient information has been saved to allow resumption of the interrupted program. The ensuing interrupt program may perform short-term processing and then resume the interrupted program via a special instruction. Several interrupt options are provided for similar applications. To minimize interrupt storage requirements and simplify associated programs, only one level of interrupt is allowed. If an interrupt occurs during the execution of another, the second interrupt is inhibited until the completion of the first.

Output commands may be divided into three categories similar to the input signals:

1. Steering signals are the output analog of the input gimbal angles. To utilize the same type of minimum, serial interface, the commands are sent in the form of pulse trains with each pulse representing an angular increment about one of the three spacecraft body axes. The counter feature, previously described for processing inertial sensor inputs, is used to meter out the pulses without requiring program supervision. To initiate such an output signal a program delivers a desired incremental body angle to a counter and then enables an output gate. Every n microseconds a counter service request is generated by the computer clock if such a gate is set. A pulse is sent to the spacecraft control system and the counter contents reduced by one in the following "stolen"

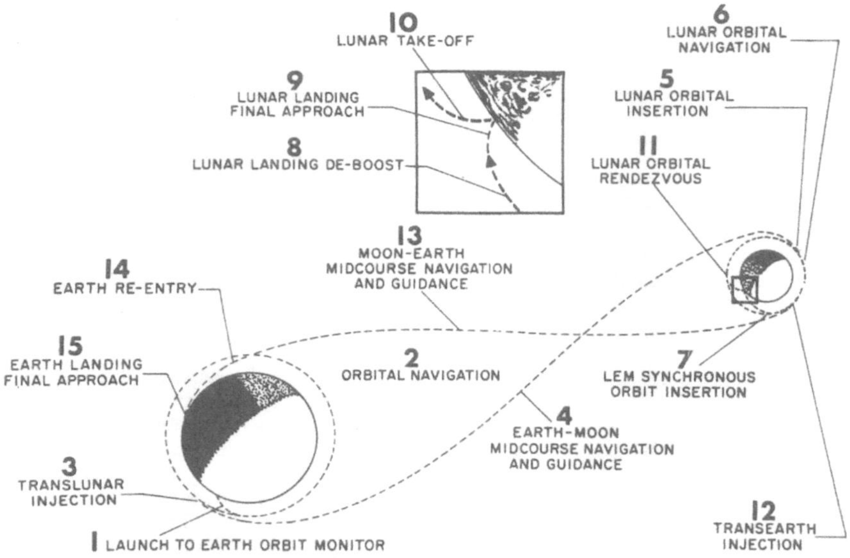

Mission Phase Summary

Fig. 1

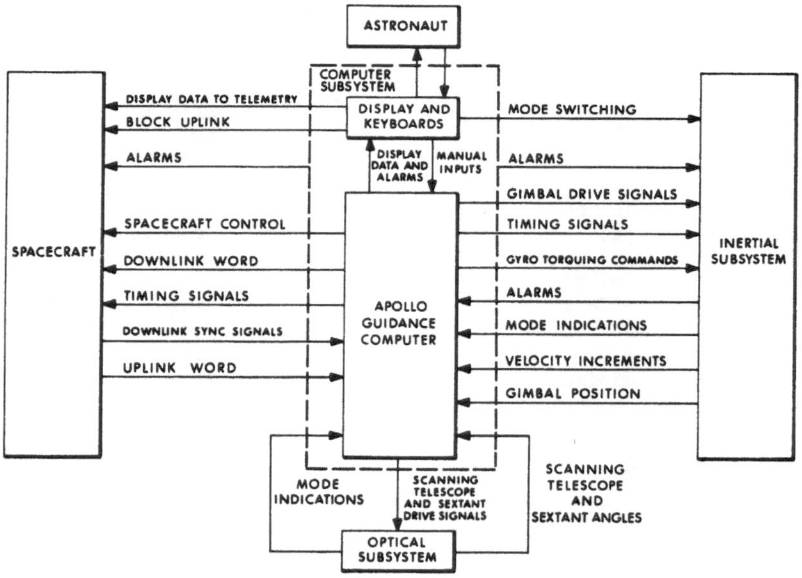

Guidance Computer Interfaces

Fig. 2

Arithmetic - Parallel, Binary, 1's Complement, Fixed Point.

Word Length - 16 bits - 15 bits and parity.

As Data = sign and 14 bits
As Instruction = 3 bit op. code and 12 bit address.

Memory Organization - High Speed Special Registers
Erasable Storage, including Counters-coincident current type.
Program Storage - read-only, core rope type.

Memory Quantity and Addressing - High Speed - order of 10^1, most directly addressable.
Erasable - order of 10^3, all directly addressable.
Program - order of 10^4, expansion capability practically limitless;
2000 directly addressable, remainder addressable
1000 words at a time in conjunction with Bank Register.

Instruction - 11 Total
Arithmetic - Add, Subtract, Multiply, Divide
Information Transmission - Exchange, Transfer to Storage, Clear and Subtract.
Logical - Mask
Decision Making - Count, Compare, and Skip (1 instruction)
Sequence Changing - Transfer Control
Indexing - Index Instruction

Guidance Computer Characteristics

Fig. 3

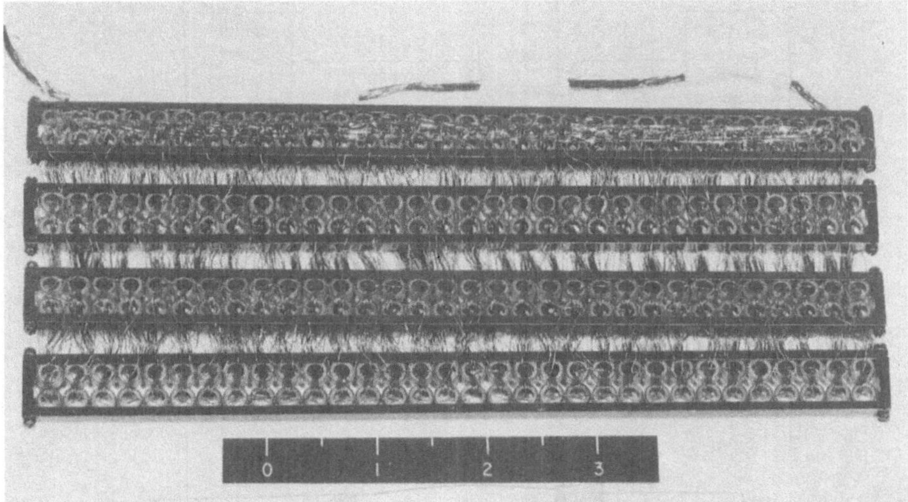

Experimental Core Rope

Fig. 4

Guidance Computer Memory Tray
Fig. 5

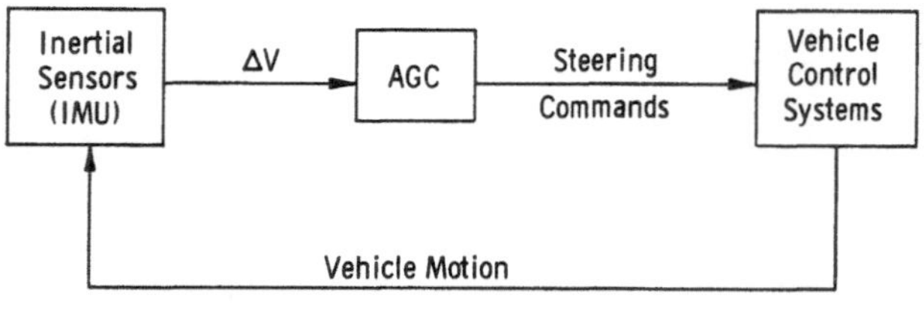

Vehicle Control During Thrusting
Fig. 6

memory cycle. This process is repeated until the required number of pulses has been delivered; i.e., during the counter increment cycle which reduces the counter to zero, the output gate is reset, removing future counter service requests. Note that after initiation of the pulse train, no further program supervision is required. In order that three-axis control may be accomplished with one counter, commands are transmitted serially at a high frequency.

2. Maintenance of mode and caution lamps is a low frequency operation. They are controlled by discretes coupled directly to bits in output registers and may be set as required by the appropriate programs. The process is the reverse analog of the method of handling system status inputs.

3. Transmission of information to the digital display, Figure 7, takes place via a single output register. Essentially, the information in this output register specifies two digits on the display. Since the display lights are not switched by solid state devices, several milliseconds must transpire before a command word may be removed from the output register. To update the entire display, a computer program must place the display information in a buffer table in erasable memory. These characters will be sent out, two at a time, at a constant rate until no further changes in the display are required. This rate is sufficiently low to accomodate the switching time delay without compromising good visual presentation.

Down-telemetry operates on the same basic principle. A single output register is assigned to this task. During transmission, the contents of this register are shifted out serially to the transmitter (the serial to parallel conversion requires no program supervision). An end-of-transmission signal causes an interrupt and the associated program delivers the next word to the output register.

B. Computation Facilities

The wide range of variables, the high accuracy requirements and the general complexity of the Apollo mission dictate the need for a computer with substantial word length, large memory capacity and a flexible and versatile instruction set. Unfortunately, the short word length of the Apollo computer is not compatible with an easy solution to these problems.

The computer instruction word contains 16 bits: a parity bit, three operation code bits, and 12 address bits. Thus, with a direct addressing capability of 4,096 words, the computer, nevertheless, has almost an order of magnitude more memory locations. The problem of generating addresses for all storage locations is successfully handled by means of a special bank register. The first 3,072 computer memory locations carry unique addresses, while the rest of the memory is divided into banks of 1,024 words, each with an individual address determined by a combination of the address bits in an instruction word and the bank register setting.

The operation code set cannot be extended in the same manner as the addressing capability. If only basic language coding is used, the programmer must perform the many tasks required using only eleven simple, single precision instructions. The mathematical and logical complexity of the problem dictates a need for a large and powerful operation code set.

A pseudo-language and a set of routines to accomplish these various ends have been developed. The advantages of a sophisticated algebraic compiler, which would translate equations written in a convenient problem-oriented language into basic machine language, are well known. However, the basic program resulting from such a translation is generally wasteful of storage. As an alternative, an interpretive system was chosen as a suitable compromise between the luxury of the compiler and the tedium of basic language programming.

The interpreter resembles a compiler in that programs are written in a problem-oriented language, but an encoded form of this language is stored directly in the computer memory. The final translation of such a program is done by a group of basic language programs at execution time. This solution requires far less storage than the compiler - generated program but at the expense of increased execution time. The time increase is largely brought about by the need for the translation process. Although the translating program must be carried in memory, it requires only a few hundred words and is certainly a small price to pay for the convenience afforded by the interpreter.

Typically, interpretive systems are based on one pseudo-operation code and one or more addresses placed in the same word. Since an interpreter must have many operation codes to be effective, the short word length of the guidance computer precludes this organization. To provide the desired full-word address and still allow a diversity of operation codes, a form of polish (parenthesis free) notation has been adopted[8]. In this language an equation is written as a number of consecutive operations followed by the required operand addresses. One bit is reserved in each word of the source language to distinguish operation codes from addresses so that list-processing techniques may be employed, thereby providing additional savings in storage.

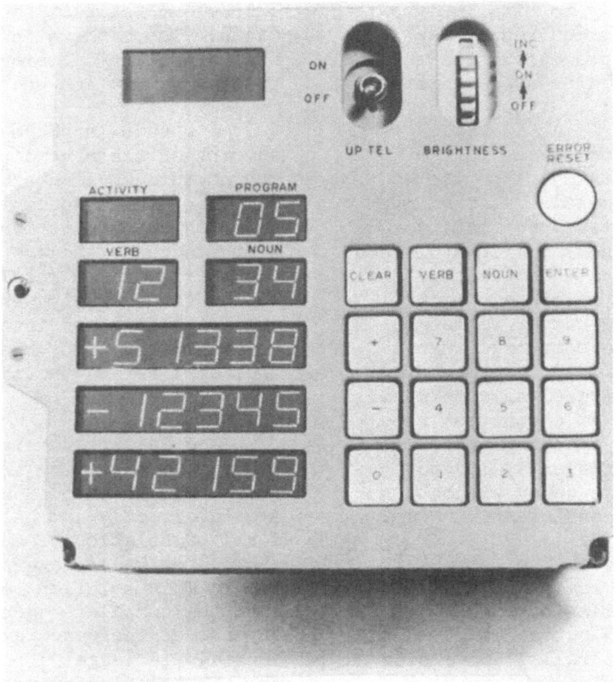

Display and Keyboard
Fig. 7

Two operation codes are packed in each word, yielding 128 possible pseudo-instructions. A full word is available for operand address storage, so that 16,384 registers may be directly addressed.

The bulk of the interpretive instructions is listed in Figure 8 together with a sample interpretive program in Figure 9. An examination of these figures clearly shows the many advantages that have been obtained: double and triple precision arithmetic, explicit vector and matrix operations, and index register addressing.

C. Execution Control

Efficient control by the computer of several different functions occuring at approximately the same time is an inherent requirement of the guidance system. To provide this control, the necessity for some effective technique for initiating these different functions is apparent. Furthermore, with a limited amount of erasable storage available, it is desirable that this storage be time-shared for the various functions.

1. Scheduling

There are , in general, two ways to initiate a program - through an externally generated command or signal, or, internally, on the basis of the elapse of a specific interval of time. The mechanism for the first type of program initiation has already been described in the discussion on keyboard characters. The measurement of elapsed time within the computer is handled by the counter structure. One such counter is fed by a stage of the computer clock divider chain. On receipt of a pulse from the selected stage, the counter is incremented by one during the next memory cycle. When the counter overflows, it causes an interrupt. The program activated by this interrupt then takes whatever action is appropriate and resets the timing counter for the next event in a queue, which can accomodate several such events.

Programs are always initiated when the computer is operating in the interrupted mode. If the program to be initiated is of very short duration (a few milliseconds), its function will be completed in this mode. If, as is normally the case, a longer period of time is required, a request is made of the priority control program to resolve any scheduling conflict and, eventually, allow the program to operate. Clearly, the computer cannot be allowed to be in the interrupt mode for longer than a few milliseconds since, during this time, all other interrupts are inhibited.

2. Dynamic Storage Allocation

When a request is made to the priority control program, one of the first tasks is to perform the erasable memory time-sharing functions. A moderate number of registers form a pool which is time-shared by programs requesting execution via priority control. Requesting programs are assigned a number of registers which are available to the job throughout its execution period but are relinquished when the program terminates. Almost all guidance computer programs operate in this fashion.

3. Priority Control

A program scheduled for execution through priority control may not commence until the program currently in progress terminates or allows itself to be suspended. To resolve scheduling conflicts, each program is assigned a priority number which accompanies the scheduling request. If the newly scheduled program is of higher priority, a flag is set to indicate to the current program that it must temporarily suspend its activity. To assure speedy access to the computer, each program running in this fashion must test the indicator every few tens of milliseconds. Program suspension is, functionally, a programmed interrupt with the required information retained for resumption of the program when it is again of highest priority. If, on the other hand, the new program is of lower priority, it will not gain access to the computer until at least the current program is completed. Approximately ten such programs may be scheduled for execution or in partial stages of completion.

A program may optionally suspend itself with its dynamically allocated storage protected, even if it has highest priority. This may be desirable while awaiting the completion of an input/output event, for example. At the occurrence of the desired event, the program is rescheduled to resume at its original priority.

At the completion of the program the time-shared storage is made available to other programs and the completed program removed from the scheduling list.

Application

The programs required to implement the guidance steering mission phase discussed previously can now be readily organized within the framework that has been constructed. Computations may be coded in a convenient and readable language with adequate precision. The computational

Scalar Arithmetic*

- Add
- Subtract
- Subtract From
- Multiply
- Multiply and Round
- Divide By
- Divide Into
- Shifting
- Normalization

Vector Arithmetic*

- Add
- Subtract
- Subtract From
- Vector Times Scalar
- Dot Product
- Cross Product
- Matrix Pre-multiply
- Matrix Post-multiply
- Unit Operation
- Vector Magnitude
- Vector Complement
- Shifting

Scalar Functions*

- Square Root
- Sine
- Cosine
- Sine^{-1}
- Cosine^{-1}
- Absolute Value
- Complement

Decision Operations*

- Branch Positive
- Branch Negative
- Branch Zero
- Branch if Overflow

Indexing Operations

- Load index directly
- Load index indirectly
- Store index
- Exchange index
- Add to index
- Subtract from index
- Count on index

*Double Precision Throughout

Typical Interpretive Pseudo-Operations

Fig. 8

Problem: Compute $\underline{z} = aM(\underline{x} + \underline{y})$
where a is a scalar and M a 3 x 3 matrix

Program (requires 7 words of storage)		Explanation
VAD MXV VXSC	Operation Codes	1) The first address of an equation is used to load an accumulator; VAD requests a vector load.
X Y M A	Operand Adresses	2) Each op code results in a subroutine call with the corresponding address left in a standard location.
STORE Z	Left-over address used to store result	3) After all op codes have been "executed," the remaining address is used to store the result. Since the result of the last operation is a vector, a vector will be stored in Z.

Sample Interpretive Program

Fig. 9

flow is accomplished using the scheduling and priority control programs. In the remainder of this section, the example problem will be discussed in some detail.

A. Computational Flow

During each navigation cycle a timed interrupt occurs resulting in a request to priority control for integration of the state vector. In addition, steps are taken to ensure that a similar operation will take place at the next cycle time. The calculation then proceeds through to completion, unaffected by higher priority tasks such as a special display request. The guidance law calculation is treated similarly but its priority is higher to prevent changes in the state vector from occurring during the computation cycle. The steering commands are left in buffer registers for servicing by the output control program.

B. Input/Output Control

The output control program has a fast repetition rate but each cycle is completed in the interrupted mode and is, therefore, independent of the priority control scheme. Among the several tasks, performed at integral mutiples of its basic cycle time, are:

1. Status monitoring of all subsystems, comparing the desired and actual states. Any disagreement results in a suitable alarm and other appropriate actions as required.

2. Transmission of the steering commands to the spacecraft control system via the single output channel available. This program acts as a traffic controller determining which spacecraft axis is to be serviced each time.

3. Updating of the digital displays, transmitting two characters at a time from the erasable memory buffer.

As described before, the down telemetry functions independently of these other tasks, being a self contained program.

C. Performance Verification

The requirement that some program must always be in an operational status is a special feature of priority control and is used to advantage. An extensive self test routine is entered whenever no other non-interrupt program is active. Thus, a significant share of the monitoring job is performed during time in which the computer would otherwise be idle. Hence, all machine time is used effectively in one fashion or another to accomplish the overall goal.

Conclusion

The keystone of a space guidance computer is a large memory store for programs. Given this single advantage and the involuntary interrupt, the restrictions of a small operation code, limited input/output facilities, short word length, and addressing difficulties can be cast aside. Of course, a price is paid in operation time and the basic computer design must provide enough speed to counter this problem. Nevertheless, adequate memory and sufficient speed allows incorporation by programming of all those features of a digital computer which are typically accomplished within the circuitry at a substantial price in weight, volume and reliability.

References

1. Sullivan, W., America's Race for the Moon, Random House, New York, 1962.

2. Sears, N. E., "Technical Development Status of Apollo Guidance and Navigation", MIT Instrumentation Laboratory Report R-445, April 1964.

3. Alonso, R. L., and Hopkins, A. L., "The Apollo Guidance Computer", MIT Instrumentation Laboratory Report R-416, August 1963.

4. Alonso, R. L., Blair-Smith, H., and Hopkins, A. L., "Some Aspects of the Logical Design of a Control Computer: A Case Study", IEEE Transactions on Electronic Computers, Vol. EC-12, December 1963.

5. "Core Rope Memory", Computer Design, June 1963.

6. Battin, R. H., "Explicit and Unified Methods of Spacecraft Guidance Applied to a Lunar Mission", Proc. 15th International Astronautical Congress, Warsaw, 1964.

7. Alonso, R. L. and Laning, J. H., Jr., "Design Principles for a General Control Computer", Institute of Aeronautical Sciences, New York, New York.

8. Muntz, C. A., "A List Processing Interpreter for AGC4", MIT Instrumentation Laboratory Report AGC Memo 2, January 1963.

DISCUSSION

T. J. Lawton and C. A. Muntz

"Organization of Computation and Control in the Apollo Guidance Computer".

Q. In the paper you state that you have 3072 direct addresses and sufficient commands for 1000 addresses; the memory is much larger. Will you explain the basis for the constant memory?

A. The basic instruction word has 15 bits. There are 4096 directly addressable words. 3072 of these are unique. The remaining 1000 are addressed in conjunction with a special register to give maximum capacity of about 64000.

Q. From where does the information come for the additional register?

A. Each time an instruction is decoded, if the address is in the range of the 1000 non-unique addresses, then the circuitry interrogates the additional register.

Q. What is the interaction allowed between the operator in the capsule and the computer; the "clear" button doesn't clear the memory, does it?

A. No. The "clear" button clears the last instructions he entered before he enters them in the computer. His instructions are displayed to him and he uses "clear" to remove errors before entering the instruction.

Q. How is the constant memory made?

A. Ferrite cores, electrically non-destructable.

Q. What is the longest command time?

A. Less than 100 microseconds. The memory cycle time is about 10 microseconds.

Q. In what form does the computer receive sensor data?

A. As a pulse train.

Q. What problems does the computer solve?

A. Firing, landing, orbit correction; all problems.

Q. Do you take the point of view of a single computer for all the problems?

A. Yes; I work for Dr. Draper.

Q. Is there only one computer for Apollo?

A. There are two of similar design, one in each manned part of Apollo.

COMPUTER PROGRAMMING FOR CONTROL OF SPACE VEHICLES

by Mr. Claude F. King
President
LOGICON, INC.
Redondo Beach, California

ABSTRACT

Computer programming for control of space vehicles differs from other programming due to the necessity of achieving error-free performance without the opportunity of testing the program in its operational situation. This paper discusses the program content, the procedures for proving the correctness of the content and code of the program and the simulation tools which have been developed and used for that purpose. The primary simulation tool is a general vehicle simulation program that runs on a large scientific computer designed to allow for extensive exercising of the vehicle computer program in all mission phases. The particular features of the simulation program which enhances its utility for this purpose are described.

INTRODUCTION

The programming of computers used in the control of space vehicles is characterized by extensive checking, through simulation, which goes far beyond that necessary for most other applications. The checking is done in order to assure that the hundreds of thousands of bits of information which are entered into the computer prior to each flight represents a set of equations and constants that will cause the system to perform accurately and in a stable and predictable manner in response to the expected measured input data of the system. Of equal importance, is the necessity to include in the checking and simulation procedures, a search for improper behavior under abnormal conditions. Abnormal conditions may influence the system in a way that would be non-catastrophic provided that the computer program properly responded to the situation. Simulation provides a mechanism for detailed evaluation of failure modes. The large number of these potential failure modes make evaluation through analysis impractical.

The procedure which has evolved for the design and testing of programs to be used in space vehicle computers makes use of a large scientific type computer wherein the overall vehicle and its environment are simulated in conjunction with a simulation of the flight computer. For the design phase, the simulation program employs a rudimentary simulation of the flight computer program. However, in the actual testing phase the simulation uses the actual flight program. The flight computer program is thus exercised with simulated missions.

Simulation is used in two phases of the flight program design. These phases are: (1) equation design; and (2) checking of the programmed equations. Because of the differing needs of the two design phases, the most applicable simulation program is one that has the flexibility to provide vehicle subsystem simulation models of varying complexity that can readily be selected depending upon the immediate purpose.

EQUATION DESIGN

The progression of space missions from the simplest to the complex, combined with a progression toward more complicated space vehicles, is characterized by the adoption of an increasing number of the vehicle systems functions into the program of the flight computer. Mission accuracy constraints produced the initial requirement for a flight computer. A digital computer was required in order to perform navigation and guidance computations to the necessary accuracy. Because of the flexibility of the computer and its inherent time-sharing ability, other functions could also be assumed by the vehicle flight computer with little increase in its size and complexity, making the elimination of system functional hardware units possible. For example, time sequencing operations that are provided in some of the simpler space systems by units of electronic equipment can be performed by the computer with little interference to the guidance function. Many other functions may be accomplished by the computer. Among them are, system diagnostic checking, command decoding and execution, telemetry sequencing, radar and optical tracking data processing, engine and control system commanding, payload data buffering, instrument calibration, environmental control, antenna pointing and vehicle attitude stabilization.

This trend toward performing more system functions via the computer program has developed because it aids system design in several ways:

(1) It simplifies the design of very complex systems by divorcing system functional design from electronic circuit design.

(2) It permits system growth and improvement without hardware change.

(3) It eases the interface problems typically associated with system change.

Because of this trend a significant portion of the memory and speed capacity of the computer is devoted to the performance of such functions as sensor compensation, system sequencing and telemetry data processing. These are in addition to the navigation and guidance equations that the computer must solve which have been widely discussed in the literature. These points are important to the understanding of the use of simulation in design of the equations. The set of equations that describe the content of the computer program consists of bits and pieces of many of the major subsystems of the vehicle. The hardware portions of those subsystems are modeled by the simulation program. Equation design consists of analyzing each segment and constructing a mathematical representation. The mathematical representation is then tested by simulation for the adequacy of its performance.

The performance of the flight equations is checked with respect to accuracy, stability, proper behavior of the logic and interaction with other parts of the system. The most practical means which has been devised to verify the accuracy of a set of flight equations is through flight simulations with the intended vehicle and the intended set of equations. This is because a fundamental objective in flight equation design is the achievement of mission trajectories that satisfy all of the imposed mission constraints over as wide a range of perturbations in the system parameters and environmental conditions as is reasonable to expect during the course of a flight. Several simplifications are usually used in the simulations in which the initial accuracy evaluations are performed on an equation set. It is usually expeditious to code the equation sets under consideration in the language of the simulating computer. Then a determination can be made of the basic equation accuracy ignoring the effects that might be introduced by the flight computer. When satisfactory performance of the equation set itself is achieved, the major effects of the flight computer are introduced. These include time delays and shortened word-length. Still later in the development cycle the actual flight computer is simulated with respect to the flight program and the actual flight program is used in the simulation.

There are several reasons for this type of development cycle. First, in the early phases of development it is desirable to study alternate implementations. Keeping the simulation relatively simple saves considerable engineering and computing time. The second reason is to separate out the various effects that may tend to corrupt the accuracy, comparing each level of simulation with previous ones. A second simplification that is used for the initial accuracy investigations comes in the manner in which the loop is closed. That is, in place of trying to achieve an accurate representation of the rotational dynamics of the system, the autopilot is simulated by either a unity attitude or unity rate transfer function. For that model, instantaneous response of the vehicle to the guidance commands is assumed. More sophisticated autopilot transfer functions and rotational dynamics are incorporated into the vehicle simulation later in the development for the final evaluation of accuracy.

Stability is usually investigated in two ways. The overall stability of the attitude control loops is usually studied on the analog computer. For these studies, the guidance effects of quantization, time lags and gains are included. The stability of the steering loops are investigated by analog and/or digital simulation with an appropriate transfer function simulating the vehicle and control system response to the guidance equation steering commands. Combined analog-digital simulations have also been employed to investigate the interactions of the guidance and attitude control systems; however, that order of simulation of hardware complexity is not necessary and perhaps not even the most desirable way to perform the investigagation. That is because there is considerable overlap in well designed independent analog and digital simulations and this overlap region serves as a means of cross checking, with the result that a higher level of confidence is achievable with independent analog and digital simulations than with the combined analog-digital simulation.

Typical logic errors that must be searched for by simulation include incorrect branching because of an unexpected set of logical conditions, improper behavior under limiting conditions, locked loops and undue time lags produced by unexpected branch conditions. The chances for a logic error increases with the complexity of a mission. That is also true when data is presented to the system from a number of different sources (e.g., inertial instruments, optical and radio trackers). A multiplicity of data sources usually requires complex decision making to fully exploit and evaluate the best data available to the system from all of the sources.

Verification of the flight equation's satisfactory interaction with other parts of the total system is similar to the verification concerned with the programmed flight equation logic in that it involves a search for the unexpected. This verification by simulation can be accomplished by exposing the flight equations to many modes of operation and a wide range of operating conditions within the limits of practicality. The exposure of the flight equations, as they interact with other systems, increases the probability of discovering any deleterious behavior that may be present.

CHECKING OF THE PROGRAMMED EQUATIONS

After the equation design is complete, the program for the flight computer is written with the instructions stated symbolically using a mnemonic representation. The program is then converted into computer code via a scientific computer program that converts the symbolic representation into the code of the flight computer. There are various degrees of sophistication to the translating program. However, none of the ones commonly used go to the extent of providing a compiler in the sense of an ALGOL or FORTRAN that provides for higher language programming. The reason for this is that higher language programming increases the difficulty of program checking. Since the emphasis in programming for space vehicle control is on thorough checking, machine language programming has been used almost exclusively.

Once the flight computer code is obtained, extensive simulation is once again used to verify that the programming has been done correctly. This is usually done in two stages. The first stage consists of a subroutine by subroutine and a branch by branch check using both hand calculations and open loop simulations. In the open loop simulation, data is obtained from previously run simulations (that is, those that were used in the equations development) so that comparisons of the simulation outputs can be made with the same inputs being used for both the equations and the program simulations. The second stage consists of the complete closed loop vehicle flight simulation incorporating the coded program in the loop. Once the coded program has been verified in all mission phases (including any ground preflight operating phase) it is then necessary to properly control the coded information in a manner that will preclude any but the simulated program from being entered into the computer prior to launch. The simulation of the preflight program is of importance only if the program is stored with the flight program. When the two programs are intermingled it is possible for an error in the ground program to effect the flight. Means are usually provided to permit the checking of the program after it has been stored in the computer to assure that no error has occurred in the loading process. Where the program instruction code is available to the accumulator this can be provided quite simply by summing the numbers that represent the instruction codes within the computer and checking the total against a previously calculated value for that instruction list.

The simulation program can be brought into use one more time for the evaluation of post-flight data. For this purpose telemetered computer inputs are used to drive a flight simulation on the ground after the flight. The computer outputs then from the simulation can be compared down to the least bit of information to the telemetered computer outputs from the flight.

VEHICLE FLIGHT SIMULATION PROGRAM

The importance and widespread use of the vehicle flight simulation program for the design and checking of the equations and program has been described. Some of the features of this program will be discussed to provide a better understanding of its use. It is helpful for people accustomed to thinking in terms of systems hardware to view a vehicle flight simulation program as a testing tool. This tool is employed as an aid in the design and verification of the flight equations and program in the manner described as follows.

Consider the equations and constants that are entered into the flight computer as another black box in the system. This box must be subjected to the same kind of testing procedures, to verify the design and assure proper system performance, as any hardware box of the system. A vehicle flight simulation program is analagous to the test equipment used to test hardware except that in one case software is used to test system software, and in the other case, hardware is used to test system hardware. There were two basic categories of testing described in the preceding sections. These were:

(1) testing that verifies the proper performance of the equations, this corresponds to the engineering tests of hardware during its development; and

(2) the checking of the program and constants that are to be entered into the flight computer prior to a flight, this corresponds to factory acceptance tests of the finished product in the hardware case.

Criteria used to judge test equipment in such system hardware tests are, by and large, the same as those used in judging a simulation program. These criteria are flexibility, operational simplicity, fast set-up ability, growth potential, low cost of operation and ease of maintenance. A vehicle flight simulation system has been developed and is in use that satisfies the listed criteria. Some of these criteria are satisfied by utilizing the concept of modular software construction. Others are satisfied by the method employed to accept the vehicle flight profile information. In this method each flight profile is defined to be a sequence of trajectory phases and each trajectory phase is initiated and terminated by events. Then, at each event the necessary data and criteria associated with the following trajectory phase is processed and assigned to the module for which it is required.

The functional modules incorporated into this simulation system are: (1) Propulsion, (2) Structure, (3) Aerodynamics, (4) Control, (5) Sensors, (6) Rotational Motion, (7) Translational Motion, (8) Environment and (9) Guidance and Sensor Data Processing.

In addition to the functional modules listed above there are program control modules such as input, output and executive control. Each module performs subfunctions through subservient software units called models. The models to be used for each trajectory phase are selected at event time as specified by the program input. This feature provides for low operating cost since only models of the necessary complexity for each phase are used.

Most model options are apparent, but the models in the Guidance and Sensor Data Processing Module require further discussion. For the simplest case, where the study is of the preliminary performance characteristics of an equation formulation, the guidance and sensor data processing module contains only the equation set under study, programmed in the language of the simulating computer. The models for the control and sensor modules may also be simple for this case, requiring only a unity transfer from input to output for each model. In the next level of complexity, the same equation set as indicated for the simplest case is followed by a computer simulator that simulates computer effects such as computer time delays and quantization of outputs. The model used for the checking of a completed program translates the code from the flight computer program into a set of simulation computer instructions that act on the sensor input data to produce outputs that are identical to the least significant bit and with the same time relationships as that which would be produced by the flight computer for the same inputs. It goes without saying that the differences in computer run time for these three models are considerable.

CONCLUSIONS

The feature of programability that has come into complex systems via the digital computer has produced a mixture of blessings and curses. The blessings accrue from the ability to place a major portion of the system within a large scientific computer where it can be examined by simulating its performance characteristics under conditions that go far beyond those provided for the hardware parts of the system. The curses are due to the fact that a major portion of the system is exposed to continual re-engineering because of the ease of making changes. Since an error in a program change could affect vital parts of the system, program changes should always be followed by detailed simulation as described in the preceding sections.

It is of interest to note, that the ability to make changes through programming has saved the design of more than one complex system that might otherwise have been consigned to the scrap pile, because the complexity had exceeded the ability of the engineers to forsee the problems and difficulties of the design prior to the time that it had been rendered into hardware.

Whether for good or bad, the program of the computer has become a significant portion of many space vehicle systems requiring extensive testing to verify its correctness. This testing by digital computer simulation is an essential part of the vehicle development and flight procedures.

DISCUSSION

C. F. King

"Computer Programming for Control of Space Vehicles".

Q. Do you use the method of modeling for the synthesis of digital computers?

A. Yes.

Q. One of the methods of designing computers is automatic synthesis. Is this method used?

A. Logic simulation is common, although this is not a synthesis method. The design is done by man, and logical synthesis used to check his work.

Q. What is the size of memory needed for modeling?

A. I know of no computer for which we did not wish for more memory after it was built. There seems to be a "Parkinson's Law" for computer memories which states that no matter how large the memory, the program will expand to fill it.

Q. What is the memory size of present computers?

A. The Apollo computer has about 1600 words.

Q. Is the general-purpose computer used for real time simulation with real components? What is the accuracy?

A. The simulations described in this paper take about 3 times real time; there is no limit on accuracy.

Q. What is the ratio of arithmetic to logical problems in modern computers?

A. In the early days most computer work was on arithmetic problems. There has been a great increase in logical operations. On some modern computers logical operations exceed the navigation and guidance equations in their demands on memory.

Q. What is the meaning of "simplicity" in modeling?

A. This refers to the use of only those features necessary to answer the question being studied at the moment.

AUTOMATIC STARTUP FOR NUCLEAR REACTOR ROCKET ENGINES

by Bobby G. Strait - IEEE Member
Staff Member
Los Alamos Scientific Laboratory
of the University of California
Los Alamos, New Mexico, U.S.A.

and Garold L. Hohmann
Supervisor, Control Systems Design and Analysis
Astronuclear Laboratory
Westinghouse Electric Corporation
Pittsburgh, Pennsylvania, U.S.A.

ABSTRACT

The control systems for a nuclear rocket engine will require an automatic technique for changing the reactor from a subcritical, shutdown condition to a sufficiently high power level for the automatic power control system to operate. The startup is achieved by programming the control rods out until the automatic power control system is activated automatically at a preselected reactor power level. The control rod position program provides for high velocities early in the startup and for smoothly decreasing velocities as delayed critical is approached. The control rod program can be adjusted to safely start the reactor in any time within the range of approximately 15 seconds to several minutes. A single range of neutronic instrumentation is used for both the startup and the full-power phase.

The startup technique was successfully tested on Kiwi-B-4D Rover Reactor in May, 1964, at the Nuclear Rocket Development Station, Nevada, U.S.A.

INTRODUCTION

The control system for a nuclear reactor rocket engine will require an automatic technique capable of starting up the reactor while in flight. This is necessary because the nuclear engine will be in the upper stage of the flight vehicle and safety considerations probably will require that the reactor be subcritical during launching. Furthermore, if the flight mission for the nuclear stage includes restarting, it probably will be desirable to completely shut down the reactor and restart as required. Generally, this subcritical startup must be performed quickly in preparation for the programmed full-power phase of the nuclear engine[1].

The reactor startup involves changing the reactor from a subcritical, shutdown condition to a power level which will enable the automatic power control system to operate. During previous Rover reactor testing, the startup was accomplished by the Chief Test Operator (CTO). In general, the procedure was to increase reactivity slowly by manually increasing the setting of the control rod position demand potentiometer. As the CTO manipulated the rods to locate delayed critical, he observed the rod position, neutron multiplication rate, reactor period and reactor power. After obtaining a steady low-power level which was in agreement with the power demand input, the CTO manually switched power control to the automatic power control system. Subsequent power changes were accomplished by changing the power control system demand voltage, which was done by manual manipulations or by an automatic programmer[2,3].

In this paper a technique which accomplishes the reactor startup automatically is presented. Included are data obtained from testing the startup technique on Kiwi-B-4D Rover reactor in May, 1964, at the Nuclear Rocket Development Station (NRDS), Nevada, U.S.A.

THE AUTOMATIC STARTUP TECHNIQUE

The automatic startup technique presented here was designed to operate with the reactor power control system currently being used in ground testing of Rover reactors at NRDS. The power loop is a logarithmic power control system which uses a single five-decade range of nuclear instrumentation[4,5]. A block diagram of the power loop is shown in Fig. 1. A similar power loop probably will be used in flight vehicles using nuclear rocket engines. Power is regulated in Rover reactors by positioning rotary control rods located in the reflector area[6]. These control rods move simultaneously in responding to a common position demand. The full-poison position is 0°, and the maximum positive reactivity position is 180°. Movement from 0° to 180° is considered the out direction.

Superior numbers refer to similarly-numbered references at the end of this paper

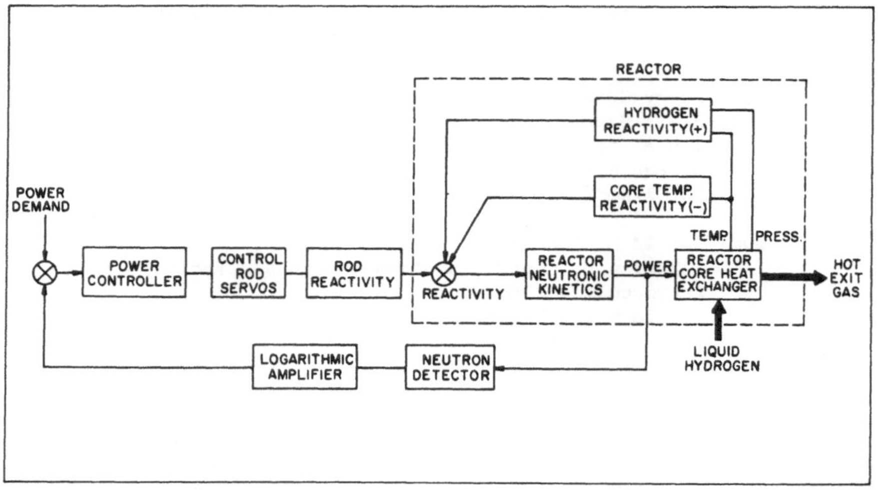

FIGURE 1, POWER CONTROL SYSTEM

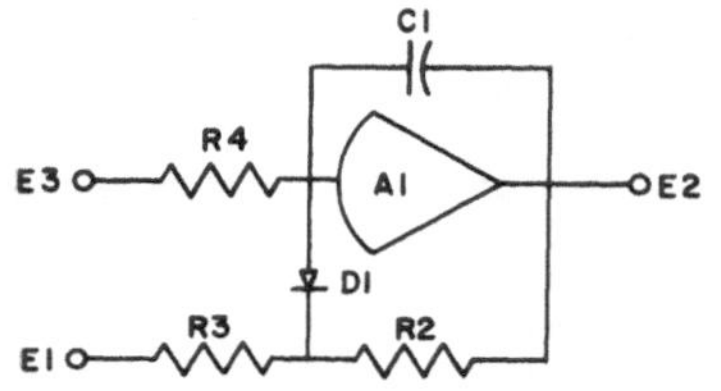

$$E2 = -\left(E1\frac{R2}{R3} + E3\frac{R2}{R4}\right)\left(1 - \epsilon^{-t/R2C1}\right) \qquad \text{for } T_0 \le t \le T_1$$

$$E2 = -E1\frac{R2}{R3} \qquad \text{for } t = T_1$$

$$E2 = -E1\frac{R2}{R3} - E3\frac{t - T_1}{R4C1} \qquad \text{for } t \ge T_1$$

$$T_1 = R2C1 \ln\left(1 - \frac{1}{1 + \frac{E3}{E1}\frac{R3}{R4}}\right)$$

T_0 = initiation of startup

T_1 = end of exponential rise of E2

FIGURE 2, ROD POSITION DEMAND CIRCUIT FOR AUTOMATIC STARTUP

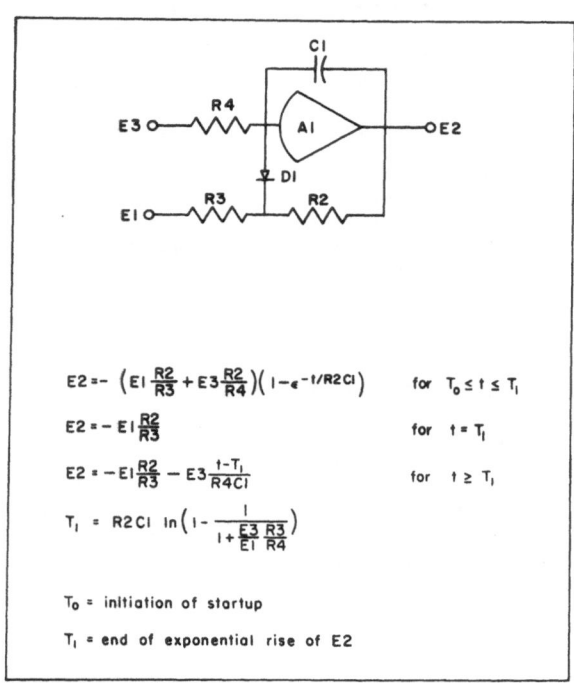

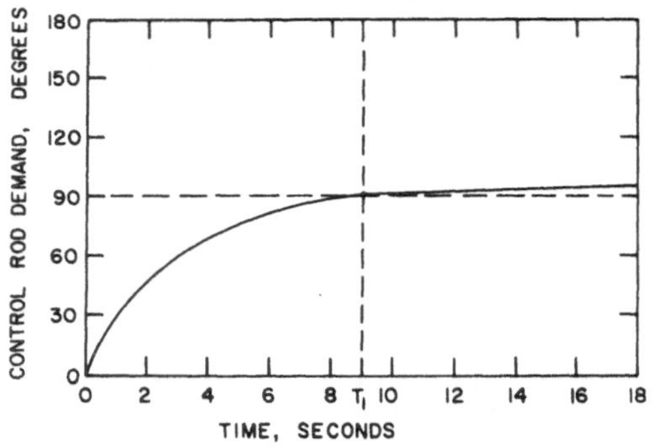

NOTE: THIS IS A NORMAL CONTROL ROD DEMAND SIGNAL FOR A REACTOR WITH DELAYED CRITICAL AT 90°

FIGURE 3, CONTROL ROD POSITION DEMAND

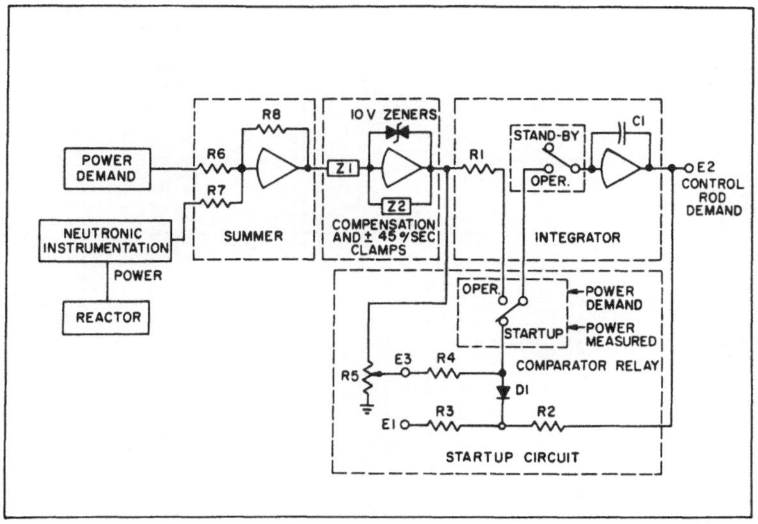

FIGURE 4, POWER CONTROLLER WITH AUTOMATIC STARTUP CIRCUIT

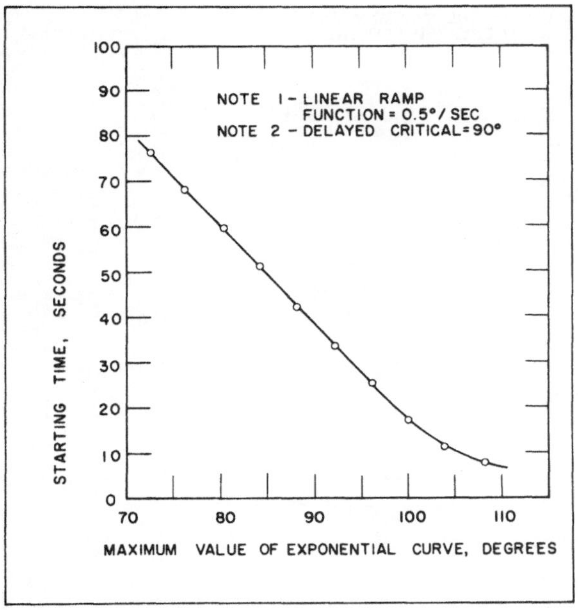

FIGURE 5, TIME FOR AUTOMATIC STARTUP FROM SOURCE LEVEL TO 20 KW FOR VARIOUS MAXIMUM VALUES OF THE EXPONENTIAL CURVE

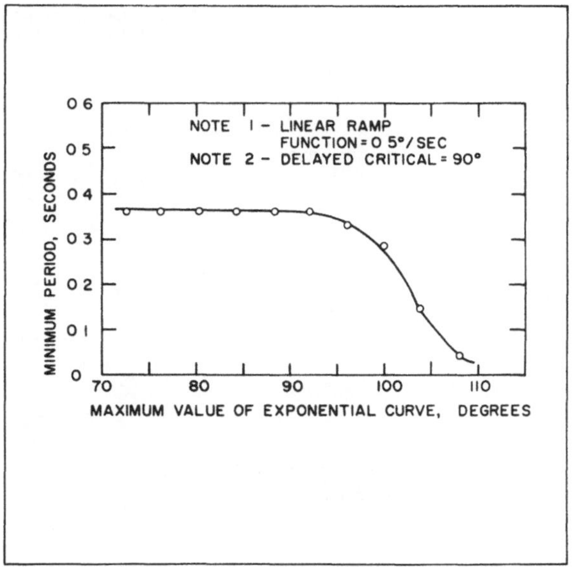

FIGURE 6, MINIMUM REACTOR PERIOD DURING AUTOMATIC STARTUP FROM SOURCE LEVEL TO 20 KW FOR VARIOUS MAXIMUM VALUES OF THE EXPONENTIAL CURVE

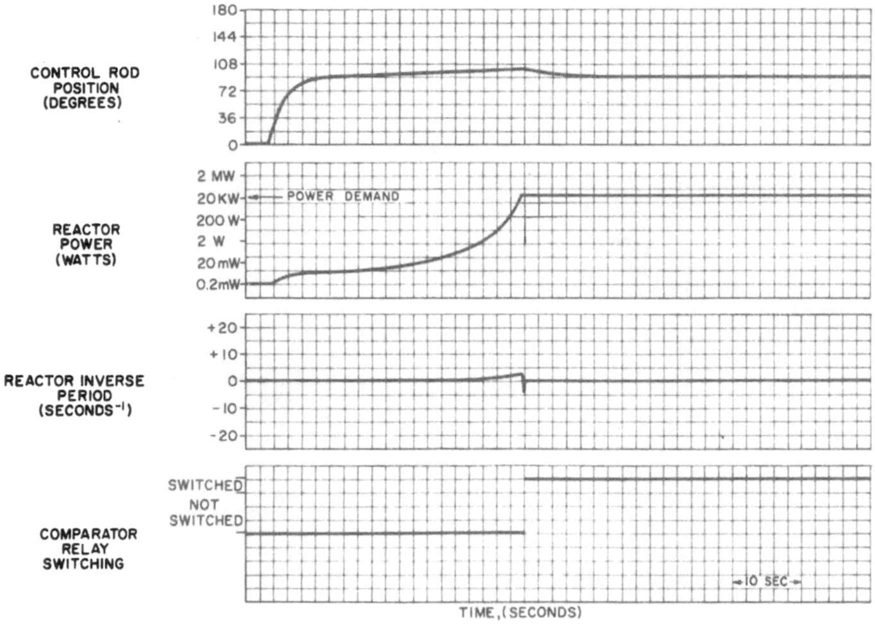

FIGURE 7, AUTOMATIC STARTUP TO 20KW WITH THE MAXIMUM VALUE OF THE EXPONENTIAL SET EQUAL TO DELAYED CRITICAL

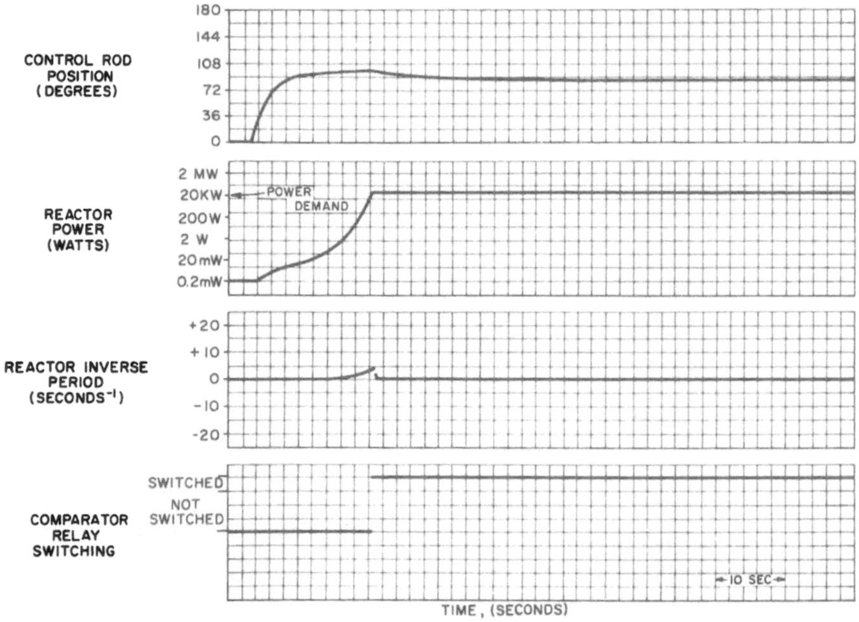

FIGURE 8, AUTOMATIC STARTUP TO 20KW WITH THE MAXIMUM VALUE OF THE EXPONENTIAL SET EQUAL TO DELAYED CRITICAL + 10°

The automatic startup scheme involves a combination of open- and closed-loop power operations. During the early portion of the startup, the power loop is open and the control rods are programmed out. The rods are moved rapidly at first, then slower as delayed critical is approached. They continue moving through delayed critical at a slow rate until the measured power corresponds to the pre-selected power demand. A comparator relay then operates which disconnects the startup circuits and closes the power control loop. Subsequent rod motion is in accordance with the power controller output to keep power at the demanded value. Thus, the reactor is automatically brought from a source power level (about $4 subcritical) through several decades of power to a level slightly above the lower end of the operating range of the logarithmic neutron detector used for full-power operation[7]. The total startup time is adjustable from a few seconds to several minutes.

The choice of the control rod demand program for the automatic startup was based on the reactor power versus control rod position characteristics. The reactor power will rise approximately one decade above source level as the control rods are rotated from zero degrees to the delayed critical position. This allows rapid rotation of the control rods below delayed critical without danger of producing an unsafe period. As delayed critical is approached, speed of the control rods is smoothly decreased to a slow rate to produce the desired power at an acceptable period. The circuit used to generate the rod demand program for the startup is shown in Fig. 2. The amplifier is a conventional high-gain operational amplifier. E2, the rod demand voltage, is positive for the diode connection shown. The conditions prior to initiation of the startup at time T_0 are (1) E1 and E3 are negative DC voltages, (2) E2 is 0, and (3) E1 and E3 are disconnected from the amplifier by switching circuits not shown in Fig. 2. At time T_0, E1 and E3 are switched into the circuit as shown in Fig. 2. E2 then rises exponentially with a two-second time constant toward a rod position demand near delayed critical. This technique gives high velocity early in the startup with continuously decreasing velocity as delayed critical is approached. When E2 reaches the value -E1 R2/R3 at time T_1 diode D1 effectively becomes an open circuit. This changes the amplifier so that E3 is integrated, causing E2 to increase linearly at a rate corresponding to a rod velocity usually less than 0.5°/second. The maximum value toward which the E2 exponential rises is adjusted by setting the values of E1 and R3. The values of E3, R4 and C1 determine the slope of the E2 ramp after time T_1.

Figure 3 is a graph of the control rod demand signal versus time for starting up a reactor that has delayed critical at 90°.

Implementing the automatic startup circuit was simplified by utilizing the integrator of the power controller circuit. A circuit diagram of the power controller with the automatic startup circuitry is shown in Fig. 4. The comparator relay is a single-pole-double-throw latching relay which latches in the "operate" position when switched. The switching occurs when the measured power signal becomes slightly greater than the power demand signal.

The voltage E3, which is integrated to produce the linear ramp function, is determined by the negative clamp of the compensation amplifier and the voltage divider R5. The power demand voltage which is selected prior to startup is sufficient to drive the output of the compensation amplifier to the clamped value. This clamp is used during normal operation of the power controller to limit the rod velocity demand to 45°/sec. The clamped value of the compensation amplifier output voltage is too large to use directly to produce the ramp function because an excessively large value for R4 would be needed. A convenient value was chosen for R4 and the desired ramp rate was obtained by reducing the voltage with divider R5.

RESULTS

The analysis of the automatic startup technique was made on an electronic analog computer using power demand and power controller circuits like those used in the ground testing of Rover reactors at NRDS. A simulation of the reactor heat exchanger was not necessary because the startup operation is restricted to low power levels where heat generation and propellant flow rate are negligible. The reactor neutronic characteristics were simulated with a three delayed group representation[8]. It was assumed that the reactor would start its power increase from a source level of 0.2 mw and the delayed critical rod position was 90°.

Various ramp rates were studied on the analog computer simulation to determine what rate resulted in the best switching transient without making startup time too long. This analysis showed that ramp rates from 0.3°/sec to 0.5°/sec were satisfactory. The analog computer results shown in this paper were made using a ramp rate of 0.5°/sec. The automatic startup tests on Kiwi-B-4D were done with a ramp rate of 0.38°/sec.

The effects of the maximum value of the exponential curve on the startup time and on the minimum reactor period during the startup are shown in Figs. 5 and 6. Setting the maximum value of the exponential curve above the delayed critical rod position gives a faster startup time but the period becomes much shorter. Tests on Kiwi-B-4D included settings at delayed critical and at delayed critical plus 10°.

Figures 7 and 8 present analog computer data for two startups. They were performed using a ramp rate of 0.5°/sec. The power demand was 20 KW. Figure 7 shows the startup with the maximum value of the exponential set to 90°, the delayed critical rod position. The startup time was 38 seconds, the minimum period

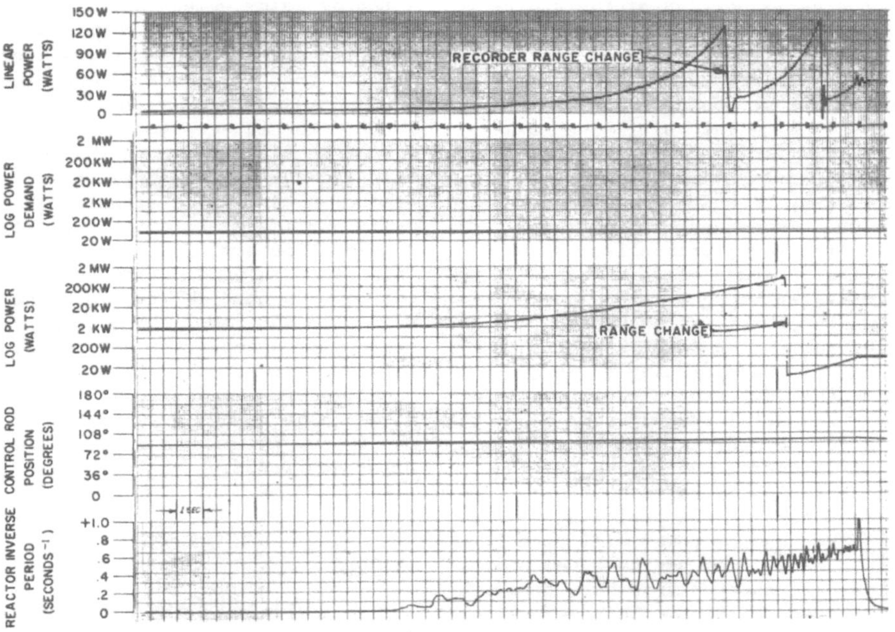

FIGURE 9, AUTOMATIC STARTUP TO 50 W, RAMP RATE 0.38°/sec, EXPONENTIAL DELAYED CRITICAL

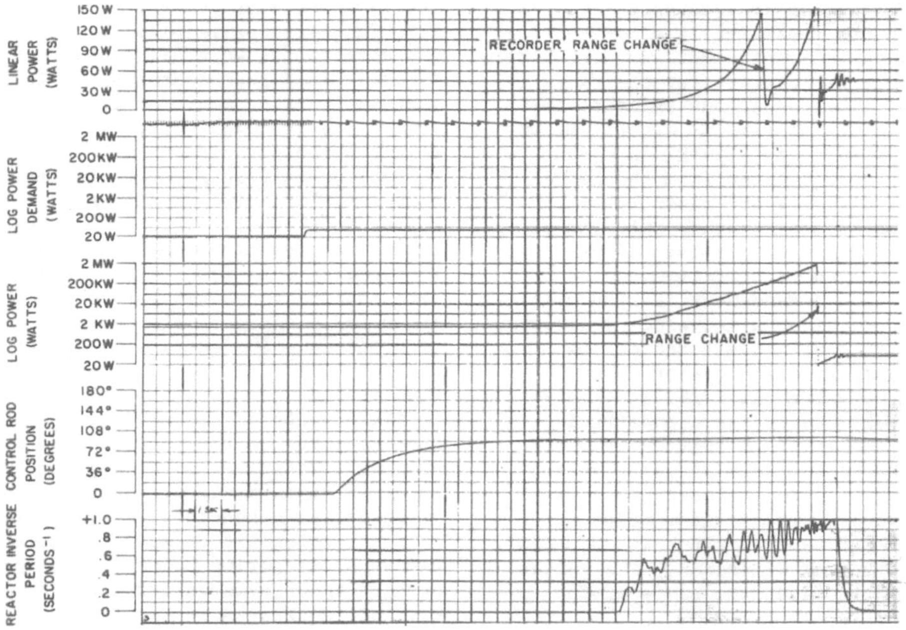

FIGURE 10, AUTOMATIC STARTUP TO 50 W, RAMP RATE 0.38°/sec, EXPONENTIAL DELAYED CRITICAL + 10°

0.364 seconds and the maximum excess reactivity at the switching time was 94¢. The startup in Fig. 8 was done with the maximum value of the exponential set to 100°, delayed critical plus 10°. The startup time was 17 seconds, the minimum period 0.275 seconds, and the excess reactivity at the switching time was 97¢.

The automatic startup was tested on the Kiwi-B-4D Rover reactor on April 30 and May 8, 1964, at NRDS. The automatic startup performed well, and the data were in good agreement with the analog computer predictions. Figures 9 and 10 show results from the automatic startup test conducted on Kiwi-B-4D on May 8, 1964. The power demand for these startups was 50 W, the ramp rate was set to 0.38°/sec, and the source level was less than 0.1 W. Figure 9 shows the last 27 seconds of the startup done with the maximum value of the exponential set to the delayed critical rod position. The startup time was 45 seconds and the minimum period was 1.45 seconds. The maximum value of the exponential was set to delayed critical position plus 10° for the startup shown in Fig. 10. The startup time was 19 seconds and the minimum period was 1.0 second.

As a result of successful testing of the automatic startup technique on Kiwi-B-4D, the startup was accepted for normal use in ground testing nuclear propulsion reactors at NRDS. The final values selected for the exponential maximum value and ramp rate were delayed critical rod position and 0.38°/sec respectively.

CONCLUSIONS

A nuclear rocket engine can be started automatically while in flight using the technique presented in this paper. A single range of neutronic instrumentation can be used for both the startup and the full power phases of the operation. The startup can be accomplished in a few seconds if required. The equipment necessary to implement the startup technique is easily combined with the normal power control system.

REFERENCES

(1) Lt. Col. H. R. Schmidt, "The Nuclear Rocket Engine and Flight Program", IRE Transaction on Nuclear Science, Volume NS-9, No. 1, January, 1962.

(2) G. K. Hess, Jr., et al., "Control Systems for the Kiwi-A Nuclear Reactors", IRE Transaction on Nuclear Science, Volume NS-9, No. 1, January, 1962.

(3) J. Henshall and S. Singer, "Kiwi-B-1A Test Facilities and Procedures", IRE Transaction on Nuclear Science, Volume NS-9, No. 1, January, 1962.

(4) E. A. Brown, "Logarithmic Power Control in Kiwi-A Reactors", IRE Transaction on Nuclear Science, Volume NS-9, No. 1, January, 1962.

(5) E. A. Brown, et al., "Control Systems for a Liquid-Hydrogen-Cooled Kiwi-B Reactor", Proceedings of Nuclear Propulsion Conference, Monterey, TID-7653 (Pt. II), August, 1962.

(6) R. R. Mohler and J. E. Perry, Jr., "Nuclear Rocket Engine Control", Nucleonics, Volume 19, No. 4, April, 1961.

(7) J. McLeod, "Kiwi Ground Test Neutronics Instrumentation", IRE Transactions on Nuclear Science, Volume NS-9, No. 1, January, 1962.

(8) T. E. Springer, "Delayed Neutron Reduced Groups by Least Squares", LA-3082, December 19, 1961.

APOLLO SPACECRAFT CONTROL SYSTEMS

by Robert G. Chilton
Deputy Chief,
Guidance and Control Division
NASA Manned Spacecraft Center
Houston, Texas

ABSTRACT

The overall Apollo spacecraft configuration includes two separate manned spacecraft. They are the command module designed for earth launch and reentry and cislunar flight including lunar orbit, and the lunar excursion module designed for the excursion from lunar orbit to the moon's surface and return. Each spacecraft has its own independent guidance and control system. This paper decribes the Apollo spacecraft control systems including engine control and rocket engine configurations for attitude control. The Apollo guidance systems are not discussed.

INTRODUCTION

The purpose of this paper is to describe the automatic control systems employed for maintaining attitude control over the Apollo spacecraft. The Apollo spacecraft is a modular design and is defined as that part of the Apollo space vehicle which sits atop the Saturn launch vehicle and which is separated from the launch vehicle after injection into a translunar trajectory, in the case of a lunar mission, or into an earth orbit, in the case of an earth-orbital mission. The Apollo space vehicle is defined as the total configuration at launch. Fig. 1 illustrates the Apollo space vehicle, except for most of the Saturn launch vehicle which has been deleted to show more detail of the spacecraft.

The launch escape assembly houses a system of solid-fuel rockets which is employed to separate the command module from the rest of the space vehicle if it is necessary to abort the mission during the atmospheric phase of the launch trajectory. The abort maneuver is passively stabilized, and no control system is employed. In normal missions, the launch escape assembly is jettisoned after the space vehicle has left the atmosphere.

The command module houses the three-man crew during the entire mission, except for the excursion trip from lunar orbit to the moon's surface and return which is accomplished with the lunar excursion module. The command module, which also provides protection to the crew against reentry heating and acceleration, has a system of hypergolic rocket engines for three-axis control during the final earth entry phase of the mission. It also houses the command-service module guidance and control system which includes the sensing devices, electronics, displays, and controls which constitute the automatic and manual control systems for the spacecraft.

The service module houses the service propulsion system and the major support elements for providing environmental control and electrical power for the spacecraft. The service propulsion system includes a gimballed engine which provides the major velocity corrections for changing the trajectory after separation from the launch vehicle, and a system of hypergolic rocket engines similar to those on the command module for providing three-axis attitude control and vernier translational control for the spacecraft.

The lunar excursion module (LEM) which houses a two-man crew separates from the command-service module (CSM) in lunar orbit for the excursion trip to the moon's surface and return. The LEM employs a descent engine, which is staged on the moon's surface, an ascent engine, and a system of hypergolic rocket engines similar to those on the CSM for three-axis attitude control and vernier translational control. The LEM has its own independent guidance and control system including sensors, electronics, displays, and controls.

The Saturn instrument unit is a part of the Saturn launch vehicle and contains the Saturn guidance and control system which controls the space vehicle during all the boost phases of the mission. The guidance system in the command module provides data to the crew for monitoring the performance of the Saturn guidance system. The command module system is also capable of providing the steering signals to the Saturn control system as a back-up to the Saturn guidance system. While coasting in earth parking orbit the crew in the command module can also provide attitude control commands to the Saturn control system. Neither the Saturn control system nor the monitoring and steering interfaces of the command module system with the Saturn system are discussed in this paper.

MISSION PROFILE

A summary of the several phases of the Apollo mission is shown in Fig. 2. The Apollo space vehicle is launched into earth orbit (1) under the control of the Saturn launch vehicle and its separate guidance and control system as described above. The space vehicle coasts for one or more orbits (2) until the correct position for launching into the translunar trajectory is reached. Upon completion of the translunar injection boost phase (3), which is accomplished in the same manner as the earth orbit injection, the CSM separates from the space vehicle, rotates through 180°, and docks with the LEM which is still attached to the last launch-vehicle stage. The CSM and LEM then separate from the launch vehicle and continue the mission alone. The necessity for this procedure can be seen by comparing Fig. 1 which shows the relative position of CSM and LEM at the end of translunar injection, and Fig. 5 which shows their position in readiness for the first midcourse correction maneuver which is to be executed as soon as practical after injection. The CSM service propulsion engine provides the thrust for that maneuver.

A maximum of three midcourse maneuvers (4) is planned prior to reaching the vicinity of the moon where the spacecraft is inserted into a circular orbit around the moon (5). After an appropriate number or orbits, the LEM with a crew of two separates from the CSM and maneuvers into a transfer orbit (6) which will take it near the moon's surface. As the surface is neared the descent engines are started, and the final powered descent and landing (7) is made on the surface of the moon. At the prescribed time the LEM, using the ascent engine, is launched from the moon's surface (8) into a lunar orbit along an intercept trajectory with the CSM. One or more additional maneuvers are performed (9) to bring the LEM into position with the CSM so that docking can be accomplished.

When the crew of three men has once again been assembled in the CSM, it is separated from the LEM and at the appropriate time is injected into a transearth trajectory (10) leaving the LEM in lunar orbit. The necessary midcourse correction maneuvers are made (11) as on the out-bound leg and, finally, the command module is separated from the service module and oriented for entry into the earth's atmosphere (12). During the atmospheric entry phase the command module is stabilized in two axes and maneuvered about the roll axis in order to direct the aerodynamic lift vector of the module in such a way as to achieve the desired landing site on the earth's surface. When the velocity has been reduced to a low value and the module is in essentially vertical fall, a parachute system is deployed to achieve a landing.

OVERALL BLOCK DIAGRAMS

The overall guidance and control block diagram for the CSM is shown in Fig. 3 and that for the LEM is shown in Fig. 4. The early development flights in earth orbit will be accomplished with the CSM Block I configuration shown in Fig. 3a. All subsequent flights of the CSM including lunar missions will fly the Block II configuration shown in Fig. 3b. The similarity between the system concepts of the LEM and CSM is illustrated in Figs. 3b and 4. On the left side of the diagrams are depicted the controls and displays which comprise the interface between the crew and the automatic guidance and control systems. The displays for the CSM and LEM are designed to provide spacecraft attitude and rate information, velocity information, and other data tailored to the particular requirements of each spacecraft. Each spacecraft has a hand-controller for three-axis spacecraft attitude control by command inputs from the crew and a hand-controller for three-axis translational control of the spacecraft. There are several modes of control available to the crew through these hand-controllers including attitude-hold, rate-command, and open-loop acceleration control. No more detailed discussion of the displays and controls will be given in this paper.

On the right side of the block diagram of both CSM and LEM are depicted the engine controls and attitude-control engines which are the primary control elements for accomplishing two general classes of spacecraft maneuvers. These are orientation maneuvers in coasting flight and steering maneuvers in powered flight. In addition, there is a special class of maneuvering flight in the atmosphere in which orientation of the lift vector accomplishes the equivalent of steering the command module to control the reentry trajectory. The engine configurations are the same for CSM Block I and II, but the LEM configuration is somewhat different since the mission is different from that of the CSM. These will be discussed in more detail later in this paper.

The center portion of the block diagrams for CSM and LEM includes the sensors, electronics, and computation necessary to close the overall guidance and control loops. In the CSM Block II and the LEM there are two major systems, a primary guidance and control system and a stabilization and control system which is employed as a secondary channel of control available to the crew. With the back-up attitude reference the stabilization and control system is sufficient to allow the crew, with assistance from ground control systems, to bring the CSM safely back to earth in the event of primary guidance and control system failure. An abort guidance system is required to provide steering signals to the stabilization and control system for the LEM to enable the crew to abort the descent to the moon and steer back to the CSM in case of failure of the primary guidance and control system.

The primary difference between CSM Block I and Block II is that in Block I the two major systems are operated in series and in Block II they are operated in parallel. In the Block I configuration shown in Fig. 3a, the stabilization and control system (SCS) performs the primary control function, accepting steering error commands from the primary guidance system or from the crew controllers and sending signals to the gimbal-drive electronics of the service-propulsion engine and

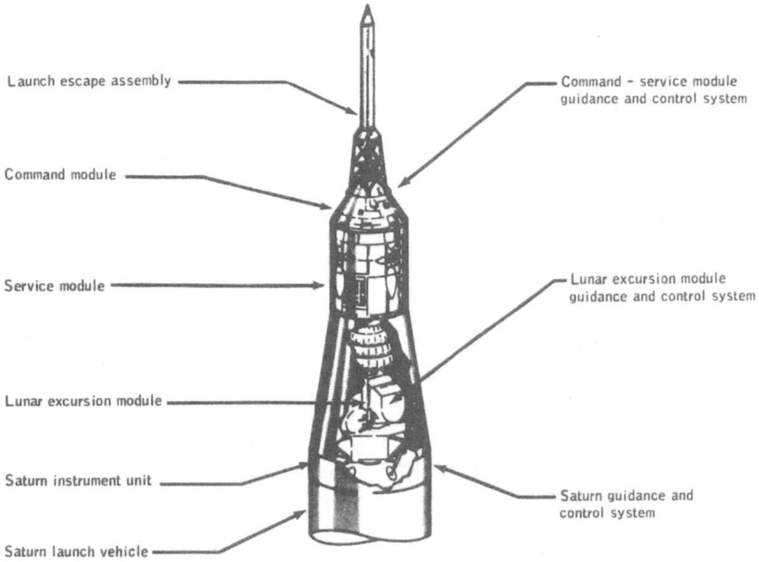

Figure 1 Apollo space vehicle

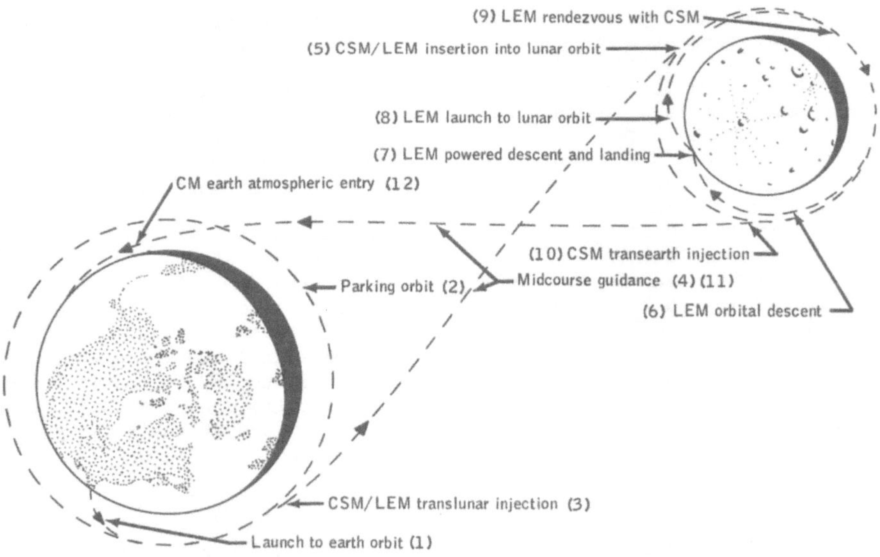

Figure 2 Mission phase summary

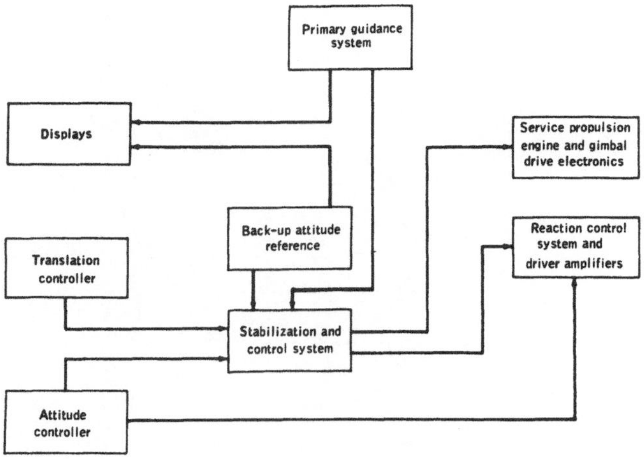

Figure 3. - CSM overall guidance and control block diagram

Part a. Block I configuration

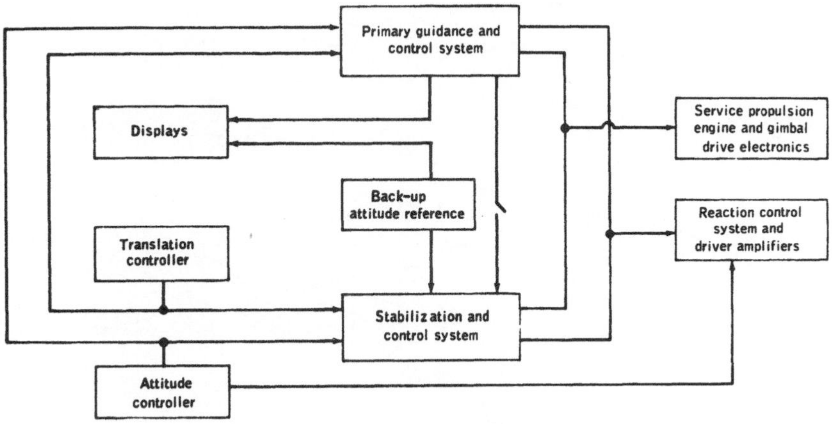

Figure 3. - CSM overall guidance and control block diagram

Part b. Block II configuration

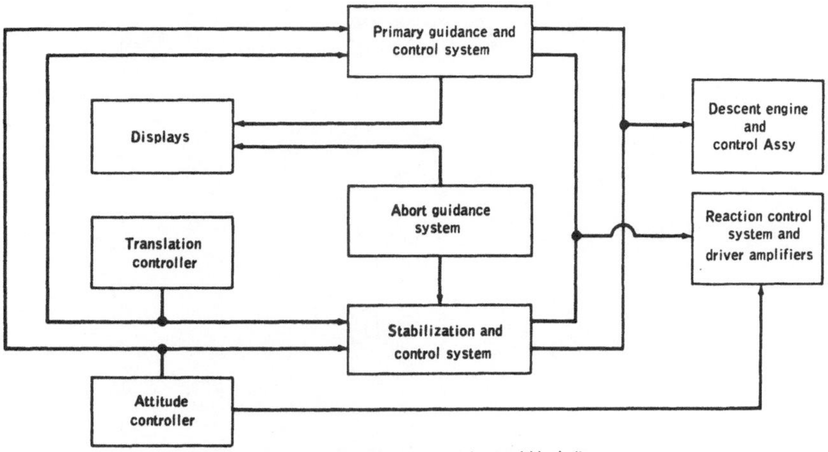

Figure 4. - LEM overall guidance and control block diagram

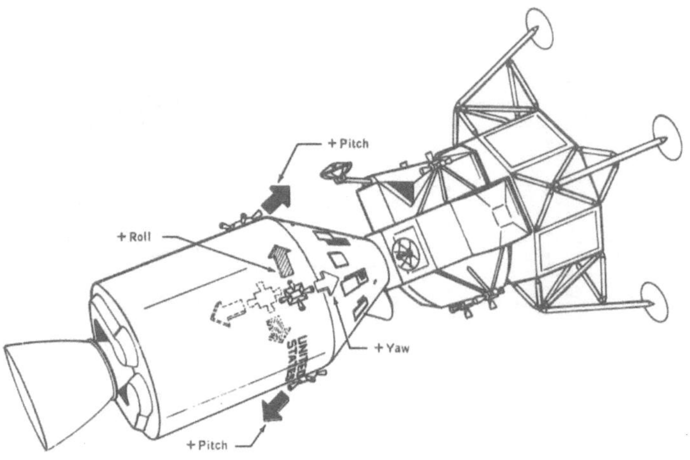

Figure 5 Attitude control of command-service module (with and without LEM)

Part a Coasting flight

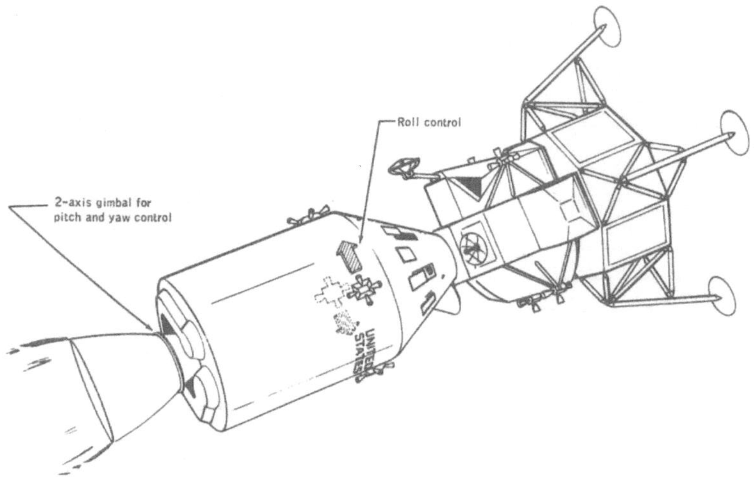

Figure 5 Attitude control of command-service module (with and without LEM)

Part b Service propulsion engine thrusting

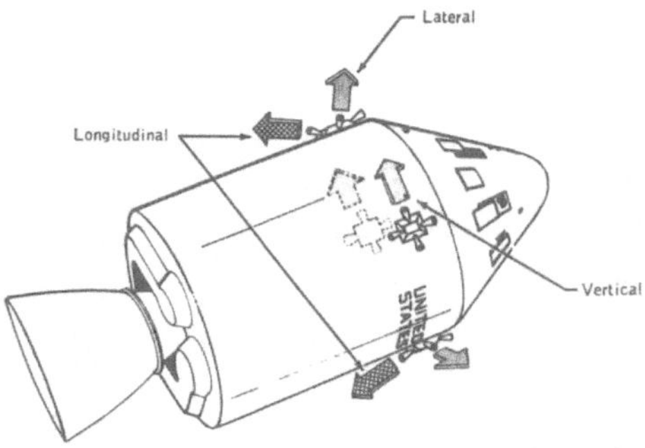

Figure 6 - Translational control of command-service module (with and without LEM)

to the solenoid-control valves of the reaction-control system. In the Block II configuration shown in Fig. 3b, the primary guidance system is expanded by increasing computer capability to include the stabilization and control function. The guidance computer accepts inputs from the crew controllers and sends signals directly to the gimbal-drive electronics for the service-propulsion engine and to the driver-amplifiers for the reaction-control solenoid valves. The SCS becomes a parallel back-up system available to the crew. The link between the primary guidance computer and the SCS retains the option for series operation of the two major systems. The remainder of the paper is concerned with the stabilization and control systems of the CSM and the LEM and with the engine control and attitude-control engine configurations. The primary guidance and control system is not discussed.

CONTROL CONFIGURATIONS

In this section the right-hand portions of the overall block diagrams of Fig. 3 and Fig. 4 are discussed. These are the primary control elements for accomplishing spacecraft maneuvers. They are employed with both the SCS and the primary guidance and control system. Two classes of engines are discussed: the main rocket engines for developing the thrust necessary for accomplishing major maneuvers for changing the spacecraft trajectory, and the small reaction-control engines for developing the control torques necessary for controlling the attitude of the spacecraft. The command module has 12 reaction-control engines. The service module has 1 large rocket engine and 16 reaction-control engines. The LEM has 2 large rocket engines, 1 for descent and 1 for ascent, and 16 reaction-control engines. The reaction-control engines for the LEM and service module are also employed to obtain vernier translational control.

There are some general statements that can be made about the engines before going into particular configurations. All are hypergolic rocket engines designed for multiple starts. They employ hydrazine and nitrogen tetroxide or variations thereof as fuel and oxidizer. All of the reaction-control engines are very similar in thrust level and other design features. Those on the command module require special attention with regard to cooling. The reaction-control engines are solenoid-controlled to provide off-on thrust control. The total of the hypergolic chemical reaction time and the electro-mechanical response time of the solenoid valves and driver is very short; therefore, very short control pulses can be obtained. For this reason very good limit-cycle performance can be achieved in the control loops.

In the remainder of this section the control configurations will be discussed for the command-service module combination both with and without the LEM attached, the command module alone, and the LEM alone.

Command-Service Module Configuration - There are two CSM control configurations, the CSM with LEM attached (Fig. 5) and CSM alone (Fig. 6). As described above, the service module employs 1 large rocket engine for performing the major thrusting maneuvers and 16 small attitude-control engines which make up the reaction control system. The reaction control system is employed for three-axis attitude control, three-axis translational control for docking, ullage for the service propulsion system, and vernier control for midcourse corrections and separation of various stages. As illustrated in Fig. 5 the reaction control system includes four hypergolic rocket engines in a quad arrangement located at four stations around the sides of the service module. Fig. 5a illustrates how the engines are employed for three-axis attitude control in coasting flight. The engines are normally fired in pairs to produce control couples in pitch, yaw, and roll. There is a redundant pair of roll engines. A partially redundant control capability exists for the pitch and yaw axes since the engines can be employed singly with only minor degradation in performance.

Fig. 5b illustrates the operation of the service propulsion engine for performing the major thrusting maneuvers such as lunar orbit insertion, transearth injection, and cislunar midcourse corrections. This engine operates at constant thrust level and is designed for multiple starts. The engine assembly is gimbal-mounted and controlled by electro-mechanical servo actuators to achieve pitch and yaw control while thrusting. The gimbal drive servo consists of a constant speed motor driving a gear train and two magnetic-particle clutches for driving the gimbals in the positive and negative directions. The servo loop incorporates both position and velocity feedback for gimbal position control. There are two complete servo loops for each axis to provide parallel redundancy. Roll control while thrusting is obtained from the service module reaction control system as illustrated in Fig. 5b. Two roll engines are shown operating, but all four roll engines may be operated at the option of the crew. Special provisions will also allow the use of only one roll engine when desired. The need for this occurs when taking navigation sightings under conditions of low inertia. In pitch and yaw a special sequence is followed when a thrust maneuver is initiated. The pitch and yaw reaction control engines are disabled 1 second after ignition of the service propulsion engine to prevent transients during thrust buildup and to conserve fuel. The gimbal servo-actuators are turned off 1 second after engine cutoff to prevent transients during thrust decay.

Fig. 6 illustrates the operation of the reaction control system to achieve three-axis translational control. The engines are operated in pairs in order not to induce disturbing moments.

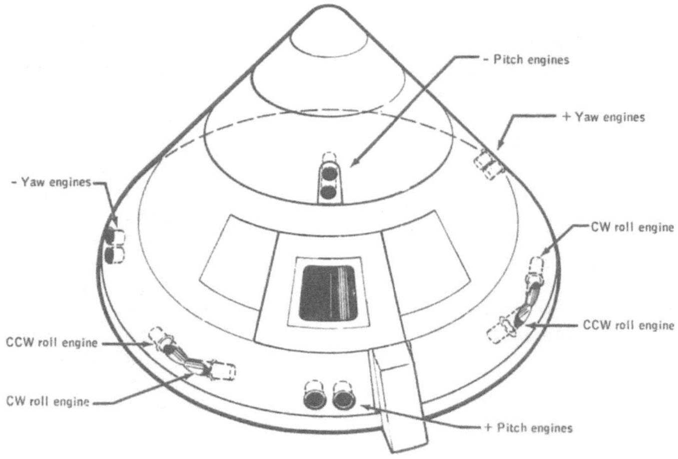

Figure 7 Command module attitude control

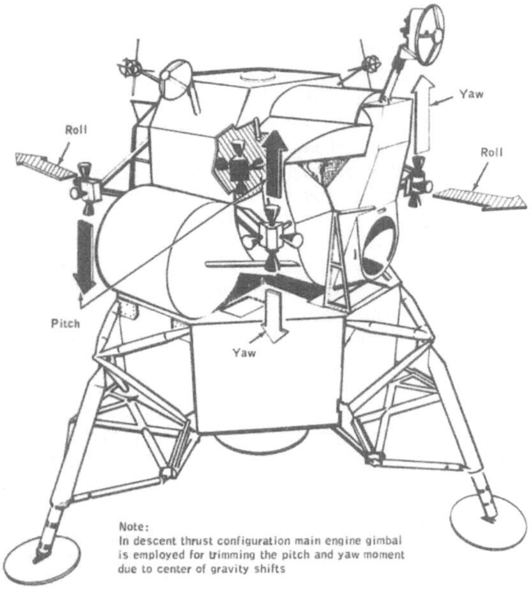

Figure 8 Attitude control of LEM
(descent configuration)

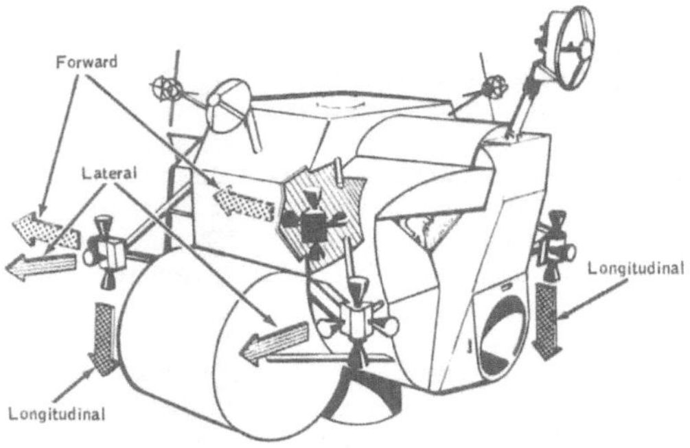

Figure 9 Translation control of LEM
(ascent and docking configuration)

For translation along the longitudinal axis, either two or four engines can be employed at the option of the crew. There in no redundancy afforded for lateral translation except by rolling through 90°.

Command Module Configuration - Fig. 7 illustrates the configuration of reaction control engines required to provide three-axis attitude stabilization and control for the command module. Two completely independent systems are provided for parallel redundancy. The center of gravity of the command module is off-set from the axis of symmetry to cause the reentry body to assume an aerodynamic trim angle with the relative wind. The resulting lift is employed to control the reentry trajectory so that the desired landing site can be reached. This is achieved by controlling the roll angle of the spacecraft during reentry. When the lift is directed downward, the trajectory steepens; when it is directed upward, the trajectory becomes shallow, and right and left roll angle causes the trajectory to curve in that direction.

Because the command module is at a constant angle of attack with respect to the relative wind, one side of the module experiences much more heating and is unsuitable for locating control engines. As shown in Fig. 7 they are all located in the upper half of the conical section of the module. No attempt is made to provide control couples. It is very difficult to position the rocket engines without introducing excessive inter-axis coupling.

LEM Descent Configuration - The LEM descent configuration illustrated in Fig. 8 employs a single hypergolic engine for achieving the necessary thrust for the descent and landing maneuver. The engine is throttle-controlled over a wide range of thrust to allow the capability for hovering above the surface of the moon. Although this engine is gimbal-mounted, as in the case of the service module, the gimbal actuators are very different. The LEM descent-engine gimbal actuators are employed only as a relatively slow-acting trim loop. The main control loop is closed, utilizing the small rocket engines of the LEM reaction control system. The gimbal actuators are screwjacks driven by a reversible constant-speed motor which is operated in an off-on mode.

The LEM reaction control system (Fig. 8) is made up of four sets of engines fixed to the spacecraft with four engines at each location. The engines are located on axes rotated 45° in roll from the spacecraft axes. They are operated as control couples for three-axis attitude control. As can be seen in Fig. 8 two pairs of control couples are available for each axis. At the option of the crew, both pairs may be selected to provide double control authority. The method of providing translational control while in the hovering condition is to tilt the spacecraft by means of the attitude control system. This produces a lateral component of acceleration from the descent engine thrust in the desired direction which is stopped by returning to vertical and reversed by tilting in the opposite direction.

LEM Ascent Configuration - The LEM ascent configuration illustrated in Fig. 9 employs a single rocket engine of constant thrust for achieving the launch ascent. The engine is fixed, and attitude control is obtained by the reaction control system as described for the descent configuration. The control power is sufficient without the necessity for the main engine trim feature. In the ascent configuration, attitude control is obtained as in the descent configuration except that all of the engines pointing upward are disabled. This allows all the downward-firing engines to augment the ascent thrust. The fuel saving for ascent propulsion more than offsets the extra fuel burned as a result of cross-coupling which this introduces. In case of failure of an engine, normal operation is resumed. Fig. 9 illustrates how the same reaction control system is employed for three-axis translational control to achieve final rendezvous and docking with the command module. The operation is essentially the same as that employed for the CSM.

SCS FUNCTIONAL DESCRIPTION

In this section the SCS systems of the CSM and LEM are discussed. The general requirements for both systems are the same. Attitude control is required under conditions of large disturbance torques (during thrusting maneuvers); attitude control is required under conditions of very small disturbance torques (during coasting flight); provisions are required for accurate control (narrow dead band) to meet certain maneuver requirements (for example, LEM attitude control while main engine is thrusting); provisions are required for coarse control (wide dead band) to provide stabilization economically for long periods of time; provisions are required to achieve the ultimate capability of the reaction control engines to deliver a small impulse in order to assure that limit-cycle periods during long duration coasting flights are economical; and provisions are required for control to be exercised by the crew under various conditions of automatic assistance by stabilization loops. However, the configurations of the spacecraft, the engine configurations and the missions are different, and the SCS mechanizations to meet these requirements are different for each spacecraft.

Before describing the individual systems some remarks are in order concerning the design of space stabilization systems employing reaction-control engines which operate in an off-on mode. The characterizing feature of such a system is that a large dynamic range is required. Control for maneuvering and control against relatively

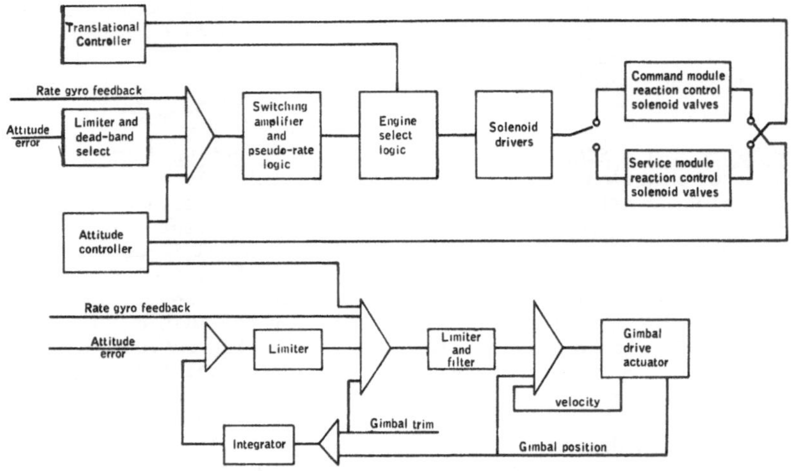

Figure 10. - CSM SCS functional diagram

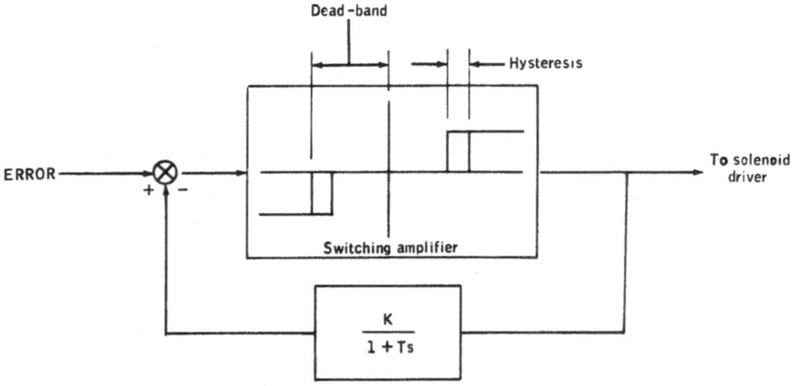

Figure 11 - Switching amplifier and pseudo-rate logic

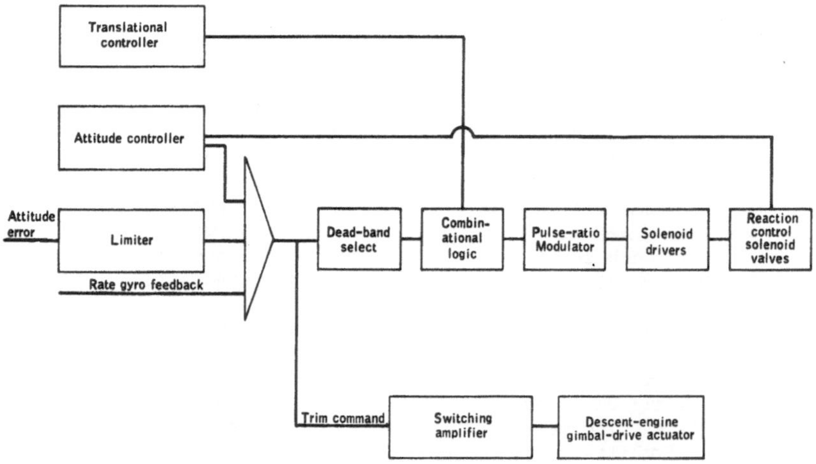

Figure 12 - LEM SCS functional diagram

large disturbance torques require a high gain or effective control power. To provide economical long-term limit-cycle stabilization in the presence of extremely small disturbances requires small vernier control capability. It is desirable, therefore, to employ a control logic which provides off-on control in the classical sense when large control gain is required and which meters out small control impulses to achieve low-rate limit cycles in the presence of small errors. As stated previously the Apollo reaction control engines have a very short response time permitting low impulse operation. Therefore attention to the control logic has large potential benefit to the system operation. Two different principles are employed in the CSM and LEM to achieve essentially the same result. Each transforms the input signal to a series of width and frequency-modulated pulses which drive the solenoid control circuitry. In the CSM the technique is called pseudo-rate logic; in the LEM it is called pulse-ratio modulation. These two techniques will be described in more detail in the paragraphs to follow.

CSM - The SCS functional diagram for the CSM is illustrated in Fig. 10. The control loops for the reaction control engines and for the service propulsion gimbal drive actuators are functionally independent. The latter is called the thrust-vector control loop since it is employed for all major thrusting maneuvers. The primary mode for thrust vector control, as illustrated in the lower half of Fig. 10, has the loop closed to hold a constant attitude in space. A manual gimbal trim capability is provided to aline the engine thrust axis with the estimated position of the spacecraft center-of-gravity prior to initiating thrust. Rate gyro feedback is employed and filtering and gain adjustment are provided to accommodate the large differences in inertia and bending frequencies between the configuration with LEM attached and CSM alone. Gimbal position is integrated and summed with the attitude error in order to reduce the error in pointing the thrust vector which results from lateral shifts of the center-of-gravity. This technique is employed in the interest of simplicity since it is a back-up mode. The primary guidance and control steering loop is not subject to this error. The gimbal-drive actuator servo loop was described in a previous section. By inputs from the attitude controller the crew can control this loop employing visual reference either by direct control of the gimbal-drive servo loop or with assistance by closing the rate gyro feedback loop.

The reaction-control engine loop is illustrated in the upper half of Fig. 10. Attitude error and rate gyro feedback are provided in the same manner as in the thrust-vector control loop. The attitude error is limited in order to limit the maximum maneuver rate in the interest of fuel economy. The output of the attitude controller is limited for the same reason when the crew exercises manual control. The attitude dead band is selectable at either $\pm 5^{\circ}$ or $\pm 1/2^{\circ}$. The switching amplifier and pseudo-rate logic provide an off-on pulse to the engine-select logic in response to the analog error signal input. The principle of operation of pseudo-rate logic is illustrated in Fig. 11. This technique, sometimes called derived-rate increment stabilization[1], has been employed in other space stabilization applications. The on-off output of the switching amplifier controls the thrust of the reaction control engines and, therefore, in an idealized system, is proportional to vehicle angular acceleration. When this signal is fed back through the lag network, the feedback in a short period sense is proportional to angular rate, hence the name pseudo or derived rate. The gain and time constant of the lag network are selected to provide the desired signal for an average spacecraft inertia. When the switching amplifier is closed, the feedback signal builds up until, through the hysteresis loop of the dead band, it is opened again thereby shaping the pulse to the solenoid control valve. The hysteresis loop is set to obtain the desired thrust impulse for limit cycling. During reentry and during manual maneuvers the pseudo-rate feedback is switched out to prevent an overdamped response.

The control pulse enters the engine select logic which also accepts signals from the translation controller. The function of the logic is primarily to provide isolation of the jet driver circuits to prevent undesirable electrical interaction. The solenoid drivers apply a fixed voltage to the engine control solenoid valves. The circuit is designed to suppress inductive spikes at turn-off. The solenoid control valves of both command module and service module reaction control system have a primary and a secondary coil. The primary coil provides the normal driving force. The secondary coils are connected directly to the translation and attitude controllers and are powered from the battery bus.

LEM - The SCS functional diagram for the LEM is illustrated in Fig. 12. The loop is closed in a conventional manner about attitude error and rate gyro feedback. The attitude error is limited for the same reason as stated for the CSM. The loop error is introduced to a separate assembly mounted on the descent stage. This is the descent-engine gimbal trim function which is left with that stage when the LEM is launched from the moon. The trim function was described in an earlier section. As in the case of the CSM, the LEM SCS provides a narrow dead band and a wide dead band. They are $\pm 0.3^{\circ}$ and $\pm 5^{\circ}$, respectively.

The combinational logic for the LEM provides, in addition to the isolation function of the CSM

(1) Superior numbers refer to similarly numbered references at the end of this paper.

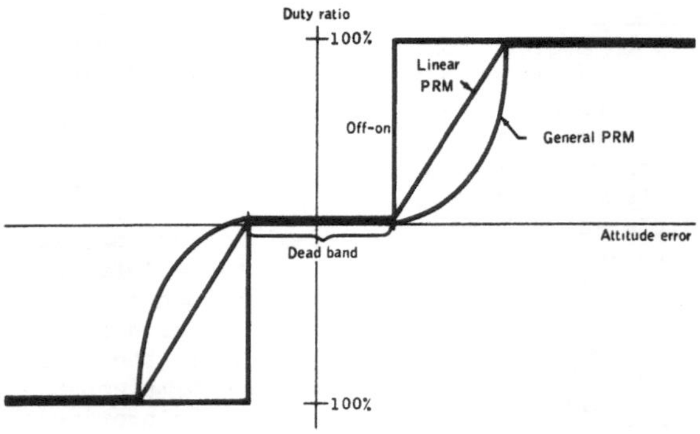

Figure 13. - Switching characteristics of pulse-ratio modulator

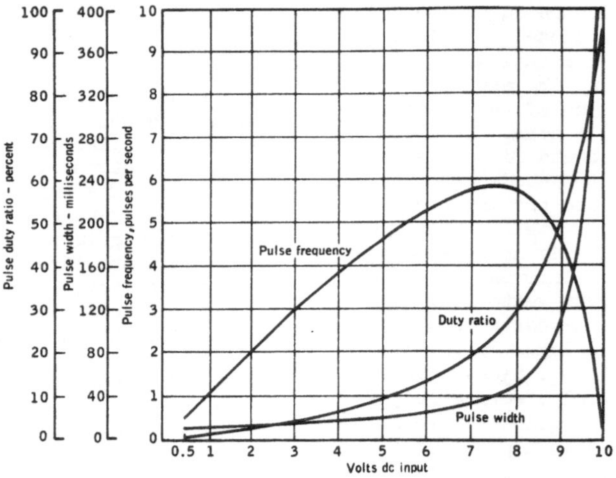

Figure 14. - Performance of pulse ratio modulator

engine logic, the necessary logic circuitry to select the proper combination of engines to be fired to achieve the desired torque. This is required because of the 45° engine configuration described earlier. Each engine is capable of producing torques about two axes. The combinational logic selects the desired engines to perform either rotation or translation about any axis without firing opposing engines. From the combinational logic the signal goes to the pulse-ratio modulator which produces the off-on control signals which drive the solenoid control valves. The solenoid drivers and control valves work in the same fashion as described for the CSM except that the translation controller is not connected to the secondary coils of the solenoid valves. This emergency translation feature is provided in the positive longitudinal direction only and is controlled from a panel switch.

The pulse-ratio modulator (PRM) is so named because the input signal controls the duty ratio of the pulse train[2]. The resulting performance can be similar to that obtained from proportional control. Fig. 13 illustrates the switching characteristics of the PRM in comparison with off-on control. PRM introduces an interim range of values of attitude error between that required to exceed the dead band and that required for full "on" control. The linear PRM characteristic gives a proportional relationship between duty ratio and attitude error. The PRM for the LEM SCS has a non-linear characteristic. PRM performance for LEM is illustrated in Fig. 14. The duty ratio varies nonlinearly with attitude error making the transition from small infrequent pulses to full off-on over a small range of errors. The minimum pulse width is approximately 10 milliseconds. The maximum pulse frequency is less than 6 pulses per second in the interest of reducing the number of cycles for the solenoid valves.

CONCLUSIONS

This paper has described in some detail the engine controls and attitude-control engines which are the primary control elements for accomplishing the maneuvers required of the Apollo spacecraft. The stabilization and control loops for the command-service module and the lunar excursion module have also been described. These loops comprise the back-up method of providing control to the crew for returning safely to earth in the event of failure of the primary guidance and control system.

NOMENCLATURE

CCW	counterclockwise
CSM	command and service module combination
CW	clockwise
LEM	lunar excursion module
K	gain
PRM	pulse-ratio modulation
SCS	stabilization and control system
T	time constant

REFERENCES

(1) Nicklas, J.C. and Vivian, H. C., Derived-Rate Increment Stabilization: Its Application to the Attitude-Control Problem, Jet Propulsion Laboratory, California Institute of Technology, Pasadena, California, Technical Report No. 32-69, July 31, 1961.

(2) Schaefer, R. A., A New Pulse Modulator for Accurate D.C. Amplification with Linear or Nonlinear Devices, IRE Transactions on Instrumentation, Volume I-11, No. 2, Sept. 1962, pp. 34-47.

DISCUSSION

R. G. Chilton

"Apollo Spacecraft Control Systems".

Q. What is the thrust of each of the 16 engines?

A. 100 pounds.

Q. What is the purpose of the logic device used with the engines?

A. It is used to select the best combination of engines for a particular maneuver.

Q. What is the purpose of the "pulse ratio modulator"?

A. The output of this device is a pulse train with a duty ratio (percentage jet on-time) for a constant input which varies with level of input. In the LEM, the variation is nonlinear.

Q. Is duplication of systems used to increase reliability?

A. There are parallel systems. This paper described the second, or backup, system. Mr. Trageser will describe the primary one. There are further redundancies in the details of the system; for example, there are redundant electromechanical actuators for the pitch and yaw gimbals of the service module engine.

Q. What is the task of the pilot during lunar landing?

A. The initial part of descent is automatically controlled for optimal fuel expenditure. At about 10,000 feet and down, the flight deviates from fuel-optimum to provide an attitude allowing the astronaut a view of the landing site. He can correct the automatic system for best landing. Upon reaching the landing site, he may make further lateral corrections from an altitude of a few hundred feet.

Q. Is hovering automatic?

A. The pilots prefer rate command.

Q. What type of engines are used?

A. All engines, large and small, use hydrazine and Nitrogen tetroxide.

Q. Do roll jets act in a direction normal to the axis of the vehicle?

A. No. Because of heating during reentry when an angle of attack is present, all jets are located on top of the vehicle. This makes it more difficult to perform a maneuver because of cross coupling between axes.

Q. What is the reaction time of the small engine?

A. Of the order of 10 milliseconds. The chemical reaction time is negligible with respect to that of the valve.

Q. Is variable thrust used?

A. The LEM descent engine uses variable thrust.

APOLLO SPACECRAFT GUIDANCE SYSTEM

by
Milton B. Trageser and David G. Hoag
Massachusetts Institute of Technology
Cambridge, Massachusetts

ABSTRACT

The guidance and navigation problems inherent in the Apollo mission are discussed. The phenomena to be employed in the solution of these problems are considered. Many of the design features of the equipment which will implement the solutions of these problems are described. The system organizations and the installation configurations for this hardware in both the Apollo Command Module and the Apollo Lunar Excursion Module are presented. In the discussions, elements of the development program and design improvements of Block II over Block I hardware will be revealed.

1. INTRODUCTION

The Primary Guidance and Navigation Problem

The goal of the Apollo Project is to place a human exploration team on the moon and return them safely to earth. The project is well outlined in reference 1. In brief, a spaceship consisting of three modules is launched on a trajectory to the moon with an enormous rocket. The Command Module is designed for atmospheric re-entry and is the home for the three man crew during most of the trip. The Service Module provides maneuver propulsion, power, expendable supplies, etc. The lunar excursion module is the vehicle which actually makes the lunar descent. It carries two of the three-man crew while the other two modules remain in lunar orbit. The Apollo Guidance and Navigation System is the on-board equipment which is used for the determination of the position and velocity of this spacecraft and for the control of its maneuvers. A set of this guidance equipment is contained in both the Command Module and the Lunar Excursion Module. Each set consists of a device for remembering spatial orientation and measuring acceleration, an optical measuring device, displays and controls, equipment to make an interface with a spacecraft control system and indicators, and a central digital data processer.

Organization of Paper

It is the objective of this paper to give a brief description of the Block II Apollo Guidance and Navigation System. Before proceeding, some of the guidelines of the development of this system will be reviewed. Also, the development program which has led to the Block II design will be described briefly. The main differences of Block II compared with Block I are summarized. Section number two reviews the guidance and navigation tasks which must be performed in carrying out an Apollo mission. This material defines the requirements on the hardware. The remaining sections each describe portions of the Block II Guidance and Navigation System.

Design Guidelines

In executing the design of the Apollo Guidance System, there have been many beliefs which one could describe as guidelines for the design. Of the many, four stand out as guidelines which have been particularly strong and consistently present.

The first guideline is that the guidance and navigation system shall be self-sufficient. The system should be designed with the capability of completing the mission with no aid from the ground. Three principles led to this concept at the beginning of the program. The first of these principles is the general appeal of the idea that the crew should be provided with all the means to be the master of its destiny. The second principle is the tendency to supply redundant means of performing each of the

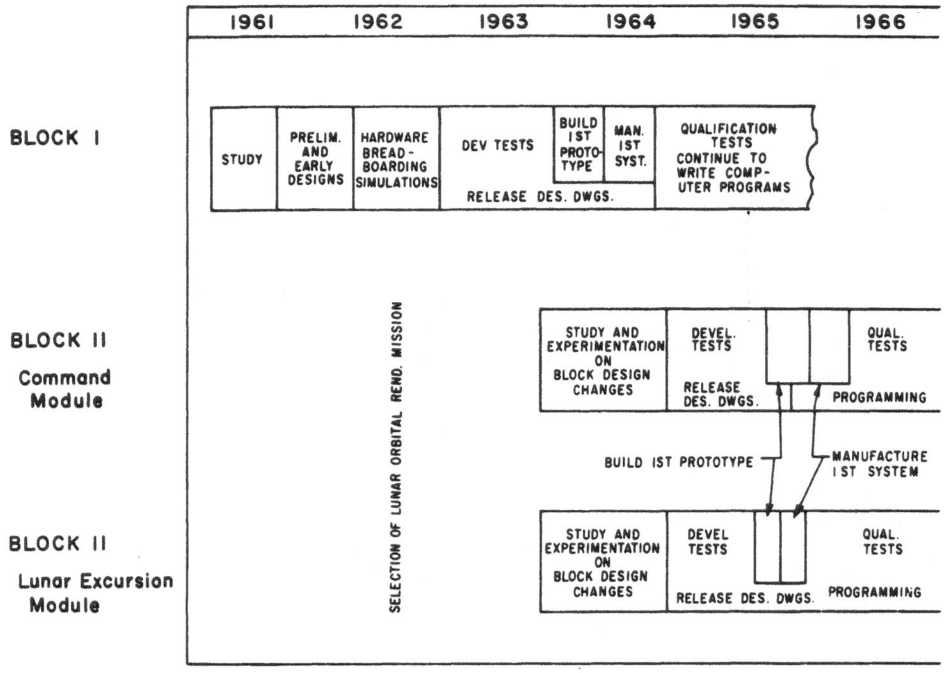

1961
1962
1963
1964
1965
1966
BLOCK I
STUDY
PRELIM. AND EARLY DESIGNS
HARDWARE BREAD-BOARDING SIMULATIONS
DEV TESTS
BUILD 1ST PROTO-TYPE
MAN. 1ST SYST.
QUALIFICATION TESTS CONTINUE TO WRITE COMP-UTER PROGRAMS
RELEASE DES. DWGS.
SELECTION OF LUNAR ORBITAL REND. MISSION
BLOCK II
Command Module
STUDY AND EXPERIMENTATION ON BLOCK DESIGN CHANGES
DEVEL. TESTS
RELEASE DES. DWGS.
QUAL. TESTS
PROGRAMMING
BUILD 1ST PROTOTYPE
MANUFACTURE 1ST SYSTEM
BLOCK II
Lunar Excursion Module
STUDY AND EXPERIMENTATION ON BLOCK DESIGN CHANGES
DEVEL TESTS
RELEASE DES. DWGS.
QUAL. TESTS
PROGRAMMING

Fig. 1.

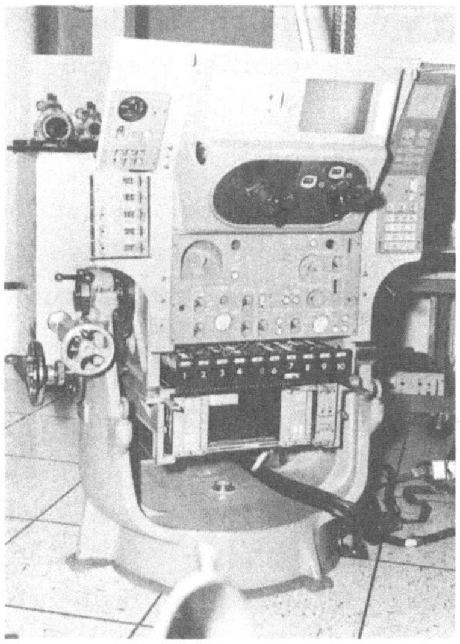

Fig. 2.

functions necessary for crew survival. Thus, although the on-board guidance and navigation system can perform all the guidance and navigational requirements of the mission and the various possible aborts with no aid from the ground, it is planned to have redundant navigational capability in the ground tracking networks. The third principle which makes the guideline of system self-sufficiency attractive is that the effort expended on the Apollo development will advance the state of the art in the manner that will make it of the most general usefulness for various projects and missions that may occur in the future.

System flexibility is a second guideline that was recognized early in the development. During the initial studies it was easy to determine that the Apollo guidance and navigation problem was an extremely complicated one having many phases. In fact, there are many alternative ways of performing the mission. Recognition of this complexity and variety urged great flexibility in system design so that modifications of the trajectories or plans would affect only computer programming. At the same time emphasis was given to eliminating features of the system which appeared to give it flexibility but which really gave little advantage.

The third guideline emphasizes making proper and appropriate use of both man and machine in performing the guidance function.[2,3] The design does not have excessive data displayed to the crew nor does it require the crew to perform dynamic loop closures that are either tedious or high-speed. On the other hand, the man's ability at pattern recognition is employed to establish the space reference by using stars and to make navigational sightings on landmarks. Further, the system provides a good capability for allowing the crew to freely exercise its various options and carry out human decisions efficiently.

The fourth guideline favors using advanced components and techniques in the design but is against overstepping the limits of the state of the art in the design.

Development Program

The development program which led to the Block II design began at the Laboratory in early 1961. It began as a study program with a very modest funding level which was initiated before the American Government decided to pursue the manned lunar exploration program with vigor. During this early study, as shown in figure 1, the basic concepts of the Apollo Guidance and Navigation System were defined and the initial error studies made. This study gave us a good uderstanding of the essentials of the problem as it existed during the middle of 1961. At that time a vigorous design and development program was initiated at the Laboratory.

The first step in this effort was the preparation of a development plan. The projected schedules were very tight. It was determined then that the best method of performing to the tight schedule and yet making the lunar trip with the system of mature design would be to plan on a system configuration change. Therefore, Block I was as quickly as possible defined, designed, built, and tested. This effort proceeded with the realization that hindsight would reveal desirable design improvements. It was decided not to let these improvements retard the progress of the Block I design. Instead, these changes would be saved to make a Block II improved design. During the first two years of the development program the complicated mission became much better defined. There were a number of decisions and some alterations in course. The most notable of these changes occurred in mid 1962. At that time it was determined that the most practical method of making the manned exploration was to use a smaller vehicle designed especially for the landing. This vehicle would make the descent to the moon while the larger portion of the spacecraft stayed in lunar orbit. Prior to this decision to use the Lunar Orbit Rendezvous method, it was envisioned that the whole Command Module which is designed especially for atmospheric re-entry, the service module, and a large landing propulsion module would be landed together on the moon.

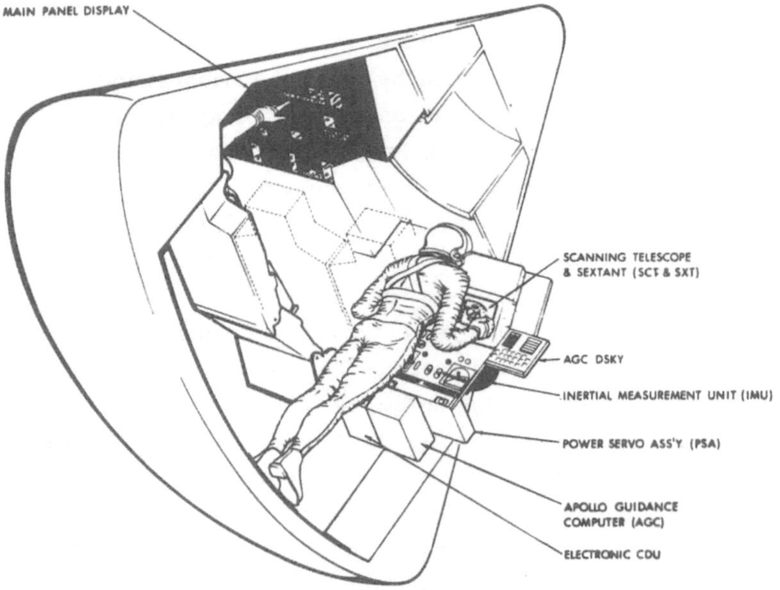
MAIN PANEL DISPLAY
SCANNING TELESCOPE & SEXTANT (SCT & SXT)
AGC DSKY
INERTIAL MEASUREMENT UNIT (IMU)
POWER SERVO ASS'Y (PSA)
APOLLO GUIDANCE COMPUTER (AGC)
ELECTRONIC CDU

Fig. 3.

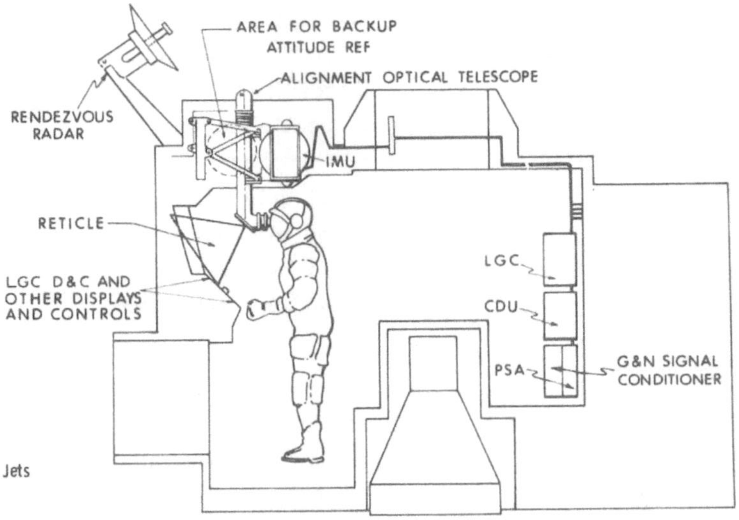
AREA FOR BACKUP ATTITUDE REF
ALIGNMENT OPTICAL TELESCOPE
RENDEZVOUS RADAR
IMU
RETICLE
LGC D&C AND OTHER DISPLAYS AND CONTROLS
LGC
CDU
PSA
G&N SIGNAL CONDITIONER

Fig. 4.

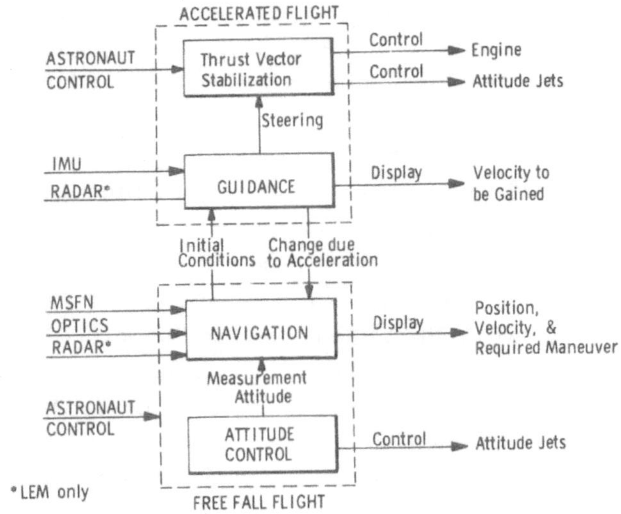
ACCELERATED FLIGHT
ASTRONAUT CONTROL
Thrust Vector Stabilization
Control
Engine
Control
Attitude Jets
Steering
IMU
RADAR*
GUIDANCE
Display
Velocity to be Gained
Initial Conditions
Change due to Acceleration
MSFN
OPTICS
RADAR*
NAVIGATION
Display
Position, Velocity, & Required Maneuver
Measurement Attitude
ASTRONAUT CONTROL
ATTITUDE CONTROL
Control
Attitude Jets
*LEM only
FREE FALL FLIGHT

Fig. 5.

The flexibility inherent in the design of the system to be installed in the command module enabled its application also in the new smaller vehicle, the Lunar Excursion Module.

The Laboratory had the first of its two prototype guidance systems operating before the middle of 1964. The second of these is shown in figure 2. This figure shows the Block I design in the test stand used to check out the system at each field site. This system consists of the same functional parts as the Block II system. Systems of this type will undergo extensive ground testing and some flight tests in Apollo vehicles.

Figure 1 shows that the study and experimentation phases for the Block II design were much shorter than for Block I. This is because the bulk of the guidance and navigation problem had been defined by the execution of the Block I design. Except for this shorter study and experimentation phase, the development cycle for Block II has phases similar to those which were experienced in Block I. At the present stage, mock-ups of the systems exist and operating hardware subsystems exist.

Figures 3 and 4 show the guidance system location in each of the spacecraft. In figure 3 the installation in the Command Module is shown. Figure 4 shows the installation in the Lunar Excursion Module. The elements of these systems are of identical design except for some features that pertain to the details of the installation.

Summary of Block II Changes:

It is intended that this paper shall make the initial presentation of the Block II Apollo Guidance and Navigation System. A discussion of the whole guidance system in a paper of reasonable length will necessarily be somewhat general. Many of the features of Block II have evolved from those of Block I.[4, 5] Of the many changes between the Block I and Block II, only six are substantial. For one, we have been able to incorporate a smaller and lighter instrument for remembering spatial orientation and measuring acceleration. Physically the computer is substantially smaller but substantially more powerful. The third change is the incorporation of the star tracker and horizon photometer in the Block II command module optics to enable the use of the illuminated earth horizon as a navigational phenomenon. The fourth major change is the use of an electronic coupling data unit to provide the interface between the geometrical reference, the computer, and the spacecraft control systems and indicators. Formerly, a less flexible mechanical unit was used. For another, a far reaching packaging change resulted from an intensive effort to moisture-proof all electronic modules and connections against the humid saline environment of the cabin. Sixth, a digital autopilot is incorporated in the Block II guidance system. This gives a better integrated and more efficient overall spacecraft steering and attitude control system in both the Command Module and the Lunar Excursion Module.

SECTION 2.

Guidance and Navigation Task of the Apollo Mission

The Apollo mission consists of a complicated series of intervals of free-fall flight alternated with intervals of powered flight. The tasks of the guidance system during these two types of flight are somewhat different. During periods of free fall, knowledge of the position and velocity of the vehicle is maintained. This activity is referred to as navigation. Figure 5 shows that for navigation, data are used from MSFN (Manned Spaceflight Network for spacecraft tracking), optical measurements between stars and the earth and the moon, and radar. These operations are generally under the control of the astronaut. These data when processed enable the display of position, velocity, and the required maneuver to reach the proposed destination. During these free-flight phases and especially when navigational sensing is in process. attitude control is frequently necessary.

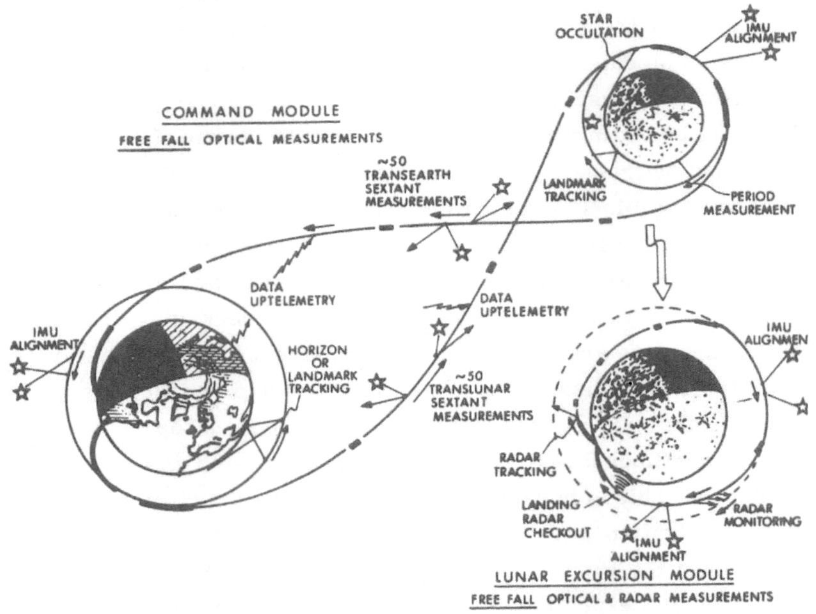

COMMAND MODULE
FREE FALL OPTICAL MEASUREMENTS
STAR OCCULTATION
IMU ALIGNMENT
~50 TRANSEARTH SEXTANT MEASUREMENTS
LANDMARK TRACKING
PERIOD MEASUREMENT
DATA UPTELEMETRY
DATA UPTELEMETRY
IMU ALIGNMENT
HORIZON OR LANDMARK TRACKING
~50 TRANSLUNAR SEXTANT MEASUREMENTS
IMU ALIGNMEN
RADAR TRACKING
LANDING RADAR CHECKOUT
RADAR MONITORING
IMU ALIGNMENT
LUNAR EXCURSION MODULE
FREE FALL OPTICAL & RADAR MEASUREMENTS

Fig. 6.

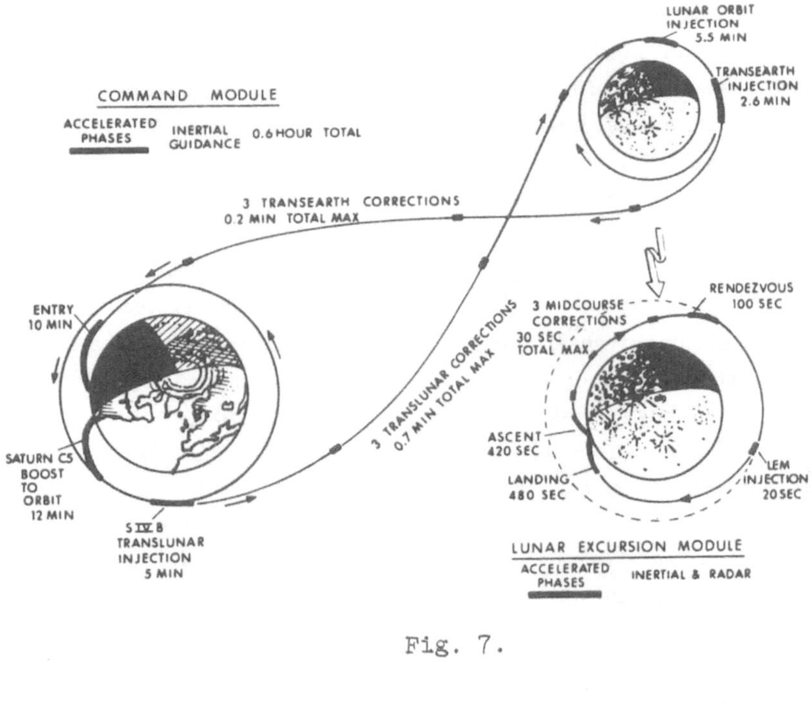

LUNAR ORBIT INJECTION 5.5 MIN
TRANSEARTH INJECTION 2.6 MIN
COMMAND MODULE
ACCELERATED PHASES
INERTIAL GUIDANCE
0.6 HOUR TOTAL
3 TRANSEARTH CORRECTIONS 0.2 MIN TOTAL MAX
ENTRY 10 MIN
3 TRANSLUNAR CORRECTIONS 0.7 MIN TOTAL MAX
3 MIDCOURSE CORRECTIONS 30 SEC TOTAL MAX
RENDEZVOUS 100 SEC
SATURN C5 BOOST TO ORBIT 12 MIN
S IV B TRANSLUNAR INJECTION 5 MIN
ASCENT 420 SEC
LANDING 480 SEC
LEM INJECTION 20 SEC
LUNAR EXCURSION MODULE
ACCELERATED PHASES
INERTIAL & RADAR

Fig. 7.

Guidance occurs during powered-flight phases. Guidance is the measurement of vehicle velocity changes and the control of vehicle attitude to produce the required changes in course and speed. During guidance a spatial reference is necessary. For this purpose a gyroscopically stabilized platform (Inertial Measurement Unit) is employed. This unit also measures acceleration and delivers this information to the guidance computer for processing. Radar is still another sensor used in guidance during the lunar landing phase.

Guidance requires the initial conditions of velocity and position. These initial conditions are generated during the navigation phases. Similarly, velocity changes during thrusting phases are a necessary type of information for maintaining accurate navigation.

Figure 6 shows each of the navigation mission phases and Figure 7 shows each of the guidance phases in a normal mission. These figures will be used in explaining briefly what is to be accomplished by the guidance and navigation system during each of these phases.

Launch to Earth Orbit

During the operation of the launch booster, guidance is provided by a system in the forward section of the booster, during this powered flight, however, the command module guidance system is operating in a guidance mode. During most of the flight it is generating steering signals which can be used by the booster should there be a failure in its guidance system. The command module guidance system data are also useful in showing the crew that the flight is being guided on approximately the right course. It is the objective of this mission phase to achieve an efficient but safe earth orbit.

The Earth Orbit Navigation Phase

The spacecraft and the final stage of the booster remain in earth orbit for one or several periods. Navigational measurements are taken on-board during this orbiting primarily to verify the proper operation of equipment. The optics portion of the command module guidance system is used to better define position and velocity. The on-board data are combined with information sent up from the ground position tracking network. Finally, the gyroscopic reference in the command module guidance system is accurately realigned to the stars, by once more using the optical device.

Translunar Injection Guidance Phase

During this phase the system in the command module is active in supplying the crew with monitoring data and in providing steering signals which are available as a back-up for the booster guidance system. The measurements made of the accelerations during this phase will be used to generate initial conditions for the translunar navigation phase.

Translunar Navigation Phase

The spacecraft separates from the booster and begins a long coasting flight to the moon shortly after translunar injection. The small error in spacecraft velocity at translunar injection would result in a gross error in the arrival conditions of the spacecraft at the moon if left uncorrected. Upon injection, a series of navigation measurements using both the on-board optical system and the ground tracking system is initiated to better determine the position and velocity of the spacecraft. As these measurements proceed, the position and velocity of the spacecraft become more accurately defined. When sufficient accuracy is obtained, a mid-course maneuver is made to improve the accuracy of the course. Three such maneuvers will generally occur on the trip to the moon. Each of these maneuvers has the order of magnitude of several or ten feet per second. The first may occur as early as an hour or two after injection.

LEM Injection Guidance Phase

The first application of the Excursion Module guidance system[6] is to accurately apply a several-hundred-foot-per-second maneuver. The maneuver is intended to produce an excursion module trajectory which has a minimum altitude of ten miles in the vicinity of the designated landing site. The initial conditions for this maneuver are obtained from the navigation phase in the command module. The data are entered into the excursion module guidance system manually by the crew using the computer display and control unit.

Lunar Landing Guidance Phase

At the proper time, as determined by the Excursion Module guidance system, the landing engine is ignited to remove velocity from the LEM. Inertial measurements of acceleration and orientation are used in the early steering of the vehicle. After a fairly low velocity and altitude have been obtained, altitude and velocity data from the landing doppler radar are used in the guidance system to obtain a more accurate set of guidance conditions. Approximately at the same time the programmed destination is displayed to the crew by means of some window markings and the Excursion Module guidance computer display. The guidance objective during this phase is to efficiently use the propellant to rapidly obtain a good landing at the desired location.

Surface Navigation Phase

While the Excursion Module is on the lunar surface, its rendezvous radar is used to track the Command Module to determine more accurately the location of the Command Module orbit relative to the landing site.

Ascent Guidance Phase

The initial conditions obtained while on the surface are used to calculate the rendezvous trajectory in the Excursion Module computer. Inertial measurements are used for the control of steering and rocket cut-off during the ascent. Thrust termination initiates a coasting trajectory which is a collision course with the command module.

Mid-Course Rendezvous Navigation

The Rendezvous Radar on the Excursion Module is used to supply its guidance system with information on the relative positions of the Excursion and the Command Modules. This tracking and range data enable a successively more accurate determination of the relationship between trajectories. As this datum improves, several rendezvous mid-course corrections establish a more accurate collision course. These are applied in the same manner as translunar mid-course corrections.

Rendezvous Guidance Phase

When the vehicles are close together, radar and inertial measurements are again used to steer and control thrust termination to remove most of the hundred or so feet per second of relative velocity between the two vehicles. Final control of the rendezvous at close ranges is strictly a piloted manual function.

Return to Earth

The return trip of the Command and Service modules to the earth is similar to the trip from the earth to the moon. The homeward trip involves a transearth injection guidance phase, coasting mid-course navigation phases, and mid-course correction guidance phases. The object during the last mid-course navigation phase is to determine accurately the initial conditions for re-entry guidance.

Re-Entry Guidance Phase

In appearance the Command Module is a body of symmetry. However, the mass distribution within it is concentrated on one side so that the re-entry body trims asymmetrically, thus providing a modest lift-to-drag ratio of three tenths. During atmospheric entry the guidance system makes inertial measurements of acceleration and orientation and provides control of the re-entry body in roll. The first objective in the exercise of roll control[7] is achieving the correct solution to the altitude problem. Thus, by adjusting the amount of lift in the vertical plane, the desired re-entry range may be established. Trajectories having ranges of several thousand miles may be controlled in this manner. During the later phases of the re-entry this roll control is exercised to adjust the transverse aim and to establish the fine aim in range to bring the Command Module accurately to the designated landing site.

Mission Abort Paths

This section so far has summarized very briefly the main features of a normal lunar exploration mission. The total of such features presents a complex guidance problem having many phases. It does not, however, give the full picture of the flexibility and capability which the guidance system must have. Figure 8 suggests the variety of conditions which may be encountered in various kinds of abort circumstances. This abort logic will not be belabored in this paper other than the suggestion of its complexity given by the rather overwhelming set of options shown in Figure 8.

Section 3.

Inertial Measurement Unit And Its Uses

Function

The function of the Inertial Measurement Unit in the Apollo Guidance and Navigation System is 1) to provide an accurate memory of spatial orientation and 2) to provide an accurate measurement of spacecraft acceleration. Its functions are the same in both the Command Module and the Lunar Excursion Module. A three-gimbal gyroscopically stabilized platform was chosen to implement these functions.

Figure 9 shows a schematic of the chosen arrangement. In the middle of the figure is a stable platform which has mounted on it three gyroscopes represented by the larger size cylinders and three accelerometers represented by the smaller size cylinders. This stable platform is supported by a gimbal system which provides three degrees of freedom between the structure of the spacecraft and the inner member. In both vehicles the outer gimbal axis is aligned approximately along the thrust direction of the vehicle.

Prior to any maneuver the inertial space alignment of the inner axis is chosen. The selected direction is always chosen so that the inner gimbal axis lies approximately normal to a plane that more or less contains the planned maneuver. Since the thrust axis of the vehicle remains approximately in this plane, such a selection of inner axis orientation results in the direction of the outer and inner gimbals being widely separated. It can be seen that if these two axes were parallel or nearly parallel the loss of one degree of gimbal freedom would occur. Thus the inner member could no longer be stabilized. However, in any reasonable maneuver this geometry can always be easily avoided by the proper selection of inner axis orientation as described. Alignment along a preferential axis for each specific mission phase does not represent any penalty. As this portion of the system is unpowered during coasting phases of the mission to conserve electrical energy, it must be realigned prior to use during an accelerated maneuver in any case.

For the stabilization of the inner member, motors are provided on each of the three axes. They provide the necessary torques between pairs of gimbals to accomplish the stabilization. The gyroscopes are only called upon to provide signals. One gyroscope is sensitive to rotations which occur about the inner gimbal axis only. A resolver is used on the inner gimbal axis to operate on the signals of the remaining two gyros to develop a signal to use for the middle-gimbal and outer-axis stabilization motors. With this arrangement the gyro signals are held to very small values. Thus the inner member provides an accurate memory of inertial space orientation since the gyroscopes have very low unbalance drift. Excellent servo performance is maintained even while the inner and

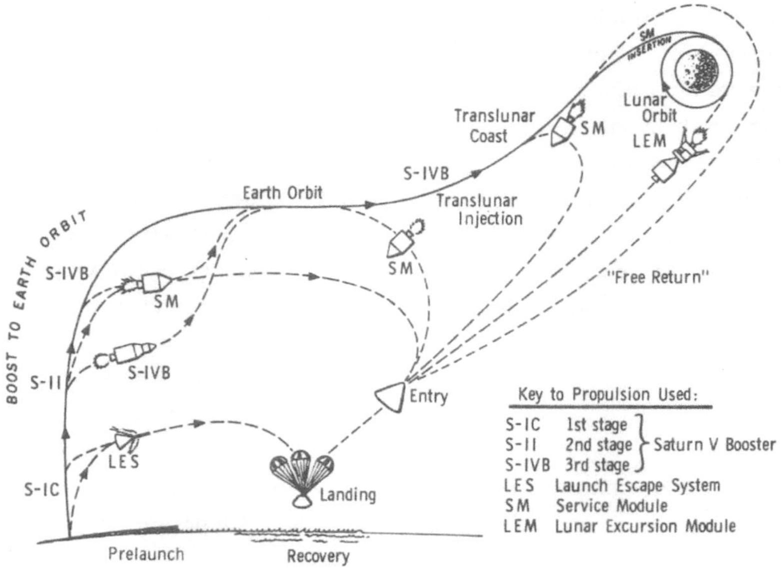
SM INSERTION
Translunar Coast
SM
Lunar Orbit
LEM
S-IVB
Earth Orbit
Translunar Injection
BOOST TO EARTH ORBIT
S-IVB
SM
SM
"Free Return"
S-II
S-IVB
Entry
Key to Propulsion Used:
S-IC 1st stage
S-II 2nd stage
S-IVB 3rd stage
Saturn V Booster
LES Launch Escape System
SM Service Module
LEM Lunar Excursion Module
LES
S-IC
Landing
Prelaunch
Recovery

Fig. 8.

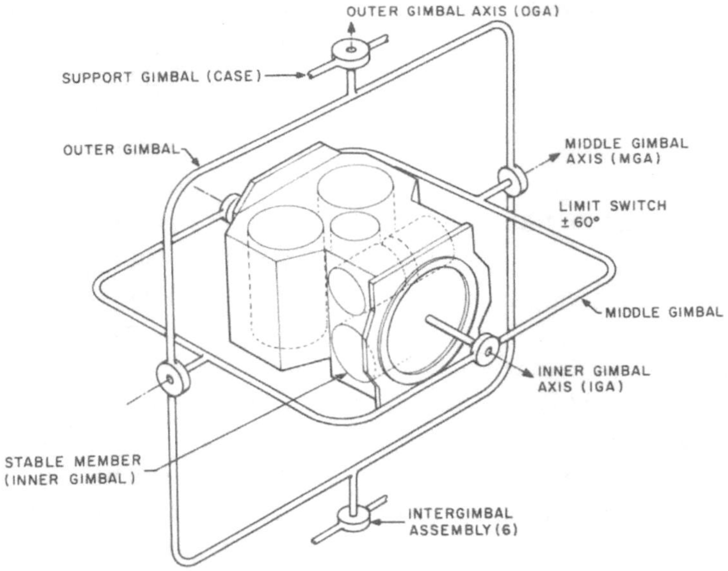
OUTER GIMBAL AXIS (OGA)
SUPPORT GIMBAL (CASE)
OUTER GIMBAL
MIDDLE GIMBAL AXIS (MGA)
LIMIT SWITCH ± 60°
MIDDLE GIMBAL
INNER GIMBAL AXIS (IGA)
STABLE MEMBER (INNER GIMBAL)
INTERGIMBAL ASSEMBLY (6)

Fig. 9.

outer axes are as close together as 15 or 20 degrees. Thus all attitude maneuver capability nearly exists without the added weight, complexity, and dynamic problems associated with a fourth degree of gimbal freedom.

The stable platform is used to provide an inertially fixed coordinate system for the measurement of accelerations. This is accomplished by mounting the three accelerometers on the platform. Each accelerometer is sensitive to acceleration along one direction. Thus the accelerometer set delivers signals to the computer giving the three components of acceleration resolved along the stabilized platform axes.

For the accelerometer coordinate system to have meaning, a method must be provided to accurately establish the orientation of the stabilized platform. The orientation method consists of sensing, transmitting, and acting.

Sensing is accomplished by the use of Command Module and Excursion Module optical devices for determining orientation from the stars. This will be described in the next sections. The optics and the inertial measurement unit are mounted on a common rigid structure so that orientation information obtained from the optics can be transmitted to the inertial unit.

The transmission of this information requires accurately machined gimbals and a precision resolver between each pair of gimbals. Each precision resolver has two windings. A coarse winding generates sine and cosine functions of the respective gimbal angle. The fine windings generate these functions for an angle which equals 16 times the respective gimbal angle. Thus these resolvers provide the means for measuring the geometry within the gimbals. Processing of their data by the computer enables the determination of inner member orientation. Thus transmission of orientation is accomplished.

Action requires accurate and sensitive changes in inner member orientation. To accomplish this each gyroscope has magnetic components for producing a precision torque. These components are excited with pulses directly under the control of the computer so that each pulse produces an increment of angle change.

An operation mode is present so that the converse transmission of geometry is possible. Thus the indication of spacecraft orientation relative to the inner member is provided. This information is used to generate spacecraft steering signals in the guidance computer.

Gyroscope

The three gyroscopes used on the Apollo Block II Guidance and Navigation System are designated 25 Inertial Reference Integrating Gyro Apollo Block II. This gyroscope is a relatively minor modification of Apollo Block I gyro. The designation "25" is associated with a family of inertial components which has a body diameter of 2.5 inches. A cutaway of this gyroscope is shown in figure 10. The drawing shows a spherical inner member which contains a gyro rotor supported on ball bearings. This rotor is driven at 24,000 rpm by 800-cycle-per-second excitation of its hysteresis motor. The motor has an angular momentum of roughly one-half-million dyne-centimeter-seconds.

To minimize disturbing torques the mass of this inner member is buoyantly supported by a dense viscous fluid in the thin gap between the surface of the spherical float and the gyro housing. In addition to the support provided by buoyancy, either of two features provide supplementary suspension.

When operating, a magnetic suspension is used to supply this supplementary support. At each end of the spherical float there is a magnetic device called a ducosyn. The ducosyn has eight magnetic poles and numerous coils. By providing excitation on the proper combination of coils, each ducosyn has the means of translating an end of the gyro float in two transverse degrees of freedom. Furthermore, the ducosyn pole faces are slightly tapered which gives a capability for applying end to end forces on the float. Thus the pair of ducosyns provide small magnetic forces to supplement the

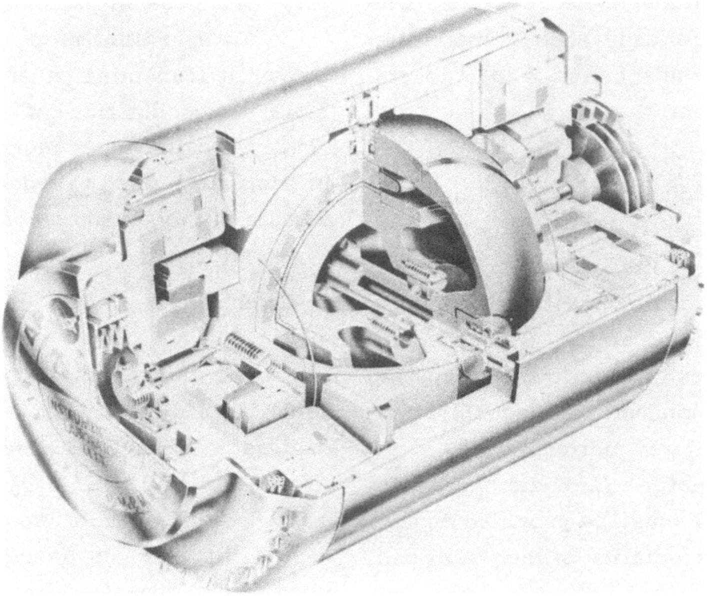

Fig. 10.

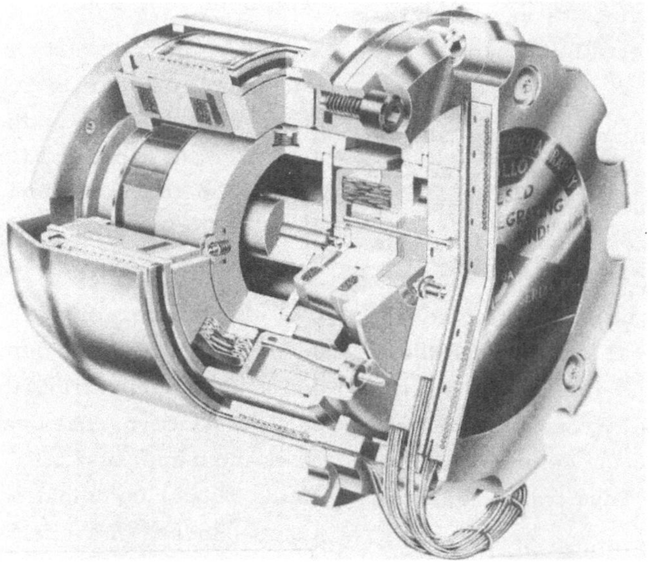

Fig. 11.

buoyancy of the float in the dense fluid. Thus the float is suspended in five degrees of freedom.

The remaining degree of freedom allows the float to rotate when the gyroscope is rotated about a direction transverse to both the rotor and the pivots. Other combinations of coils in the ducosyn sense the small angular excursions. The resulting signals are used in the stabilization servos. Torque to process the gyro for fine alignment of a stable member is also provided by this arrangement of coils.

When not operating, the supplementary support is provided by pivots. The sole purpose of the pivots shown in the cutaway is to keep the float approximately centered during periods of storage.

Ducosyn improvements and the arrangements in the expansion bellows are the primary changes in the Block II over the Block I gyro.

Accelerometer

Three size-16 Pulse Integrating Pendulum units are mounted on the stable member. Figure 11 shows one of these sensors. Again "16" stands for an inertial component family having a 1.6-inch case diameter.

The suspension scheme and internal arrangements of the Pulse Integrating Pendulum unit are generally similar to those of the gyroscope which have just been described. The internal float arrangement of the Pulse Integrating Pendulum is, of course, simpler than that of the gyroscope. An off-center mass distribution is substituted for the rotor assembly. In the accelerometer small angular displacements cause the ducosyn signals used to develop error pulses. These error pulses are applied to the torquing ducosyn as precision pulses of current to restore the Pulse Integrating Pendulum assembly to its reference position. The computer uses these pulses as data representing increments of the velocity change along the respective accelerometer's sensitive direction.

Inertial Unit Assembly

Figure 12 is a photograph of the Inertial Measurement Unit for system 600F, the first Block II prototype system. This Inertial Measurement Unit is designated as "12.5", which is the number of inches corresponding to its spherical diameter. In Block I the corresponding dimension is 14 inches. The Block II design has a correspondingly lighter weight. The change in size and weight resulted largely from improvement in the available state of the art in electromagnetic components, especially the precision resolvers.

In the photograph, the inner member is exposed by the removal of two gimbal hats. The outer of these has been removed from the gimbal case which contains hydroformed cooling passages. Spacecraft Cooling fluid is circulated through these passages.

On the inner gimbal the large shiny disc on the axis houses a direct-drive torque motor, resolvers, and slip rings. The use of direct-drive torque motors enables the achievement of especially excellent servo performance. In this arrangement a permanent magnet stator is mounted in one gimbal. The armature is mounted on the other gimbal without gearing. The armature is excited with the direct current and has an air gap 3.8 inches in diameter. Each axis has an unlimited degree of freedom. The absence of gearing enables the inertias of the gimbals to aid in their own stabilization.

At each end of the middle axis is an assembly containing 40 slip rings each. Visible in the photograph are the ends of the gyro and accelerometer with associated electronics attached. Some of the remaining electronics on the inner member is associated with its temperature control heaters and sensors. The temperature of the inertial components is maintained with a much greater accuracy than one degree fahrenheit.

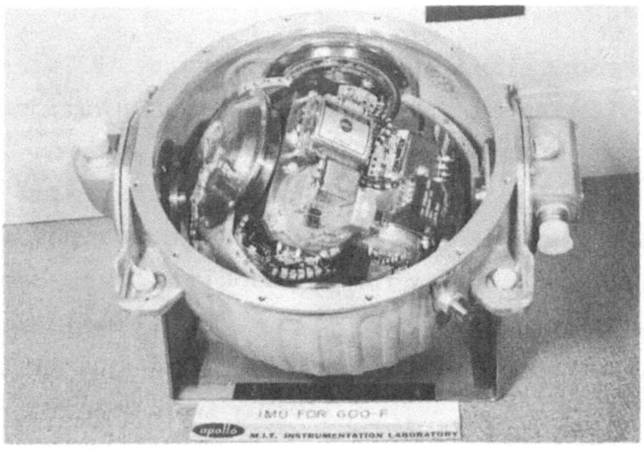

Fig. 12.

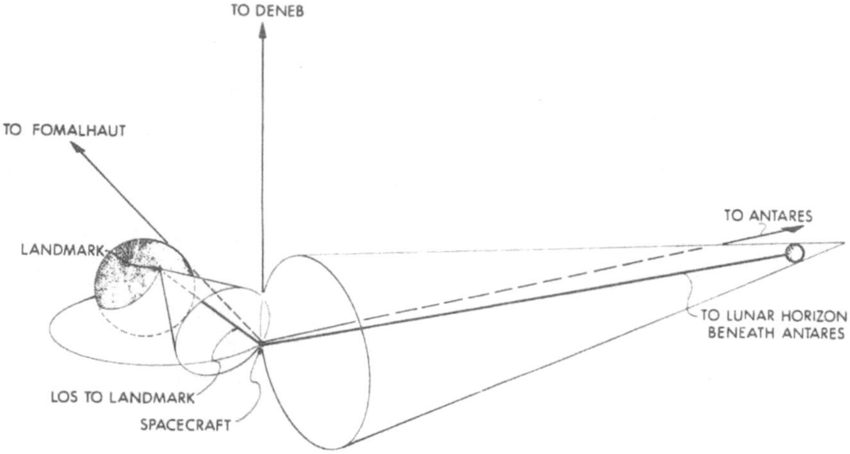

Fig. 13.

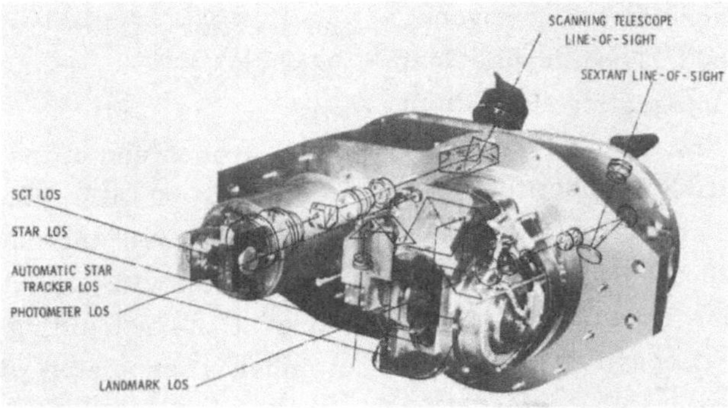

Fig. 14.

SECTION 4.

The Command Module Optics and Its Use

Function:

Unlike other portions of the guidance and navigation system, the design of the Command Module optics is not employed in the system aboard the Excursion Module. The Command Module optics has three functions. The function which determines its characteristics the most is that of navigation. Another function for which it also is well qualified is the measurement of the inertial platform orientation. The Command Module optics are also of use for general viewing since it consists of a pair of articulating telescopes, one of high power and the other of low power.

The principle of navigation which the Command Module optics is designed to implement is illustrated in figure 13. This figure shows a spacecraft on a lunar trip and the directions to some stars. From the spacecraft, an earth landmark can be seen. The cone in the figure having an axis parallel to the direction of the star Fomalhult is a locus of points from which a certain angle can be measured between the direction of the star and the landmark. The precise measurement of such an angle establishes a cone of position in space. A second cone, very flat in the figure, with its axis parallel to the direction of Deneb, intersects the first cone in a line of position. To obtain a measurement which determines the spacecraft position along this line, a third object must be used. This object could be a second landmark or it could be the moon as shown in the figure. In the figure the angle between the moon horizon and the star Antares is employed to obtain the location along the line. Thus, the principle of navigation to be implemented by the Command Module optics is illustrated.

The illustration above assumes the simultaneous measurement of three angles between the stars and two landmarks. Simpler hardware results if time-sequencing single measurements of such angles is done. It is this type of measurement which the Command Module optics is designed to implement. The theory used in the reduction of such data is given in references 8 and 9.

Mechanization

The Command Module optics consists of two telescopes as shown in figure 14. The telescope on the left is the space sextant and the one on the right is the scanning telescope. The sextant is a powerful instrument having a magnification of 28. The scanning telescope has unity power. Both telescopes are fitted with eye pieces which provide eye relief in excess of two inches so that they can be used with space helmets on if necessary. Both head assemblies containing the deflecting optics are each driven by motors and gearing about parallel axes called the "shaft axis."

The scanning telescope has only one line of sight which is deflected normal to the shaft axis by a dove prism which is driven by a second motor and gearing.

The sextant's 1.6-inch objective has in front of it a partially reflecting beam-splitter mirror. Thus the view through the beam splitter is fixed in the spacecraft. The beam splitter enables the deflection of the second telescope line of sight by means of other mirrors including one which is precisely driven. This articulated line of sight is used for seeing stars and thus has been provided the greater efficiency. The body-fixed line of sight used for observing landmarks is given poor efficiency to make up for the difference of brightness of the targets.

The sextant is used during mid-course navigation in a very similar manner to a marine sextant of great power. To make a mid-course navigation sight, the crew first uses the scanning telescope as a finder. The spacecraft is maneuvered to guide the desired earth or lunar landmark onto the shaft axis so that it can be seen with the narrow field of view of the sextant. By applying very small control impulses to the spacecraft it is made to come almost to rest with the landmark in the middle of the sextant field of

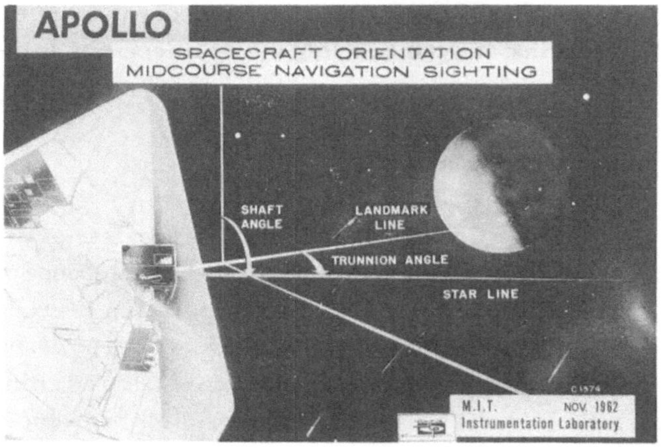

Fig. 15.

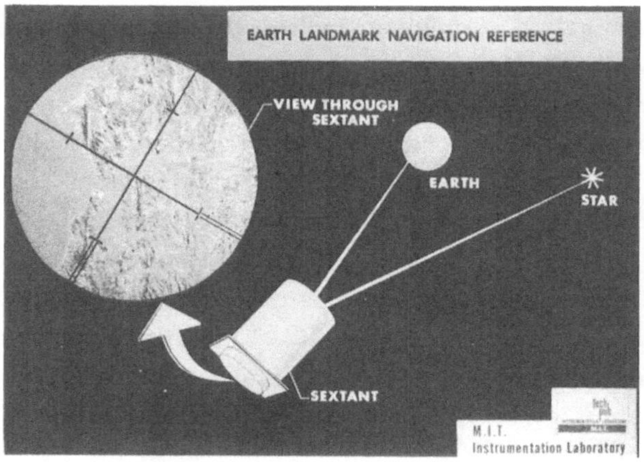

Fig. 16.

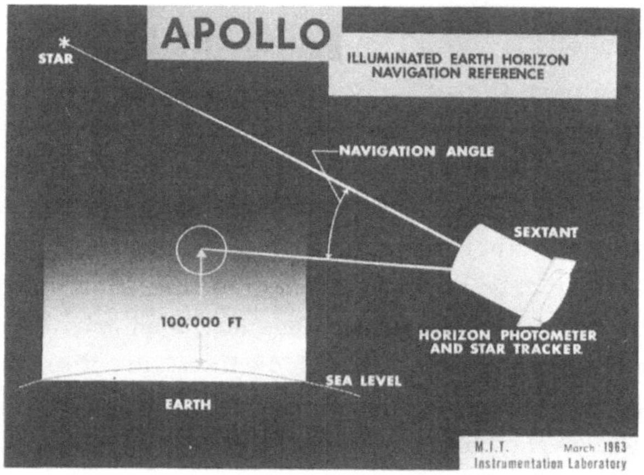

Fig. 17.

Fig. 18.

view. Now a search is made for a suitable star with the scanning telescope. Once acquired with the scanning telescope, the star can be observed through the sextant. The crew now obtains by means of the two degrees-of-freedom of the star line of sight the juxtaposition of the star on the landmark as shown in figure 16. It should also be noted that this juxtaposition need not occur in the center of the field to yield an accurate measurement.

When the measurement geometry is correctly obtained, the observer depresses a button which enables the computer to mark the time and the angle of the accurate trunnion drive of the sextant. This angle is measured by means of a single- and 64-speed precision resolver mounted directly on the trunnion of the drive axis. The accuracy of such measurements has an order of magnitude of ten seconds of arc. Thus, sensitivity to changes of position when midway between the earth and the moon is several miles.

The earth's horizon is also a promising phenomenon for navigation. Its use is illustrated in figure 17. The figure shows an illuminated horizon. Near sea level the atmospheric scattering is so extreme that the appearance of the horizon is cloud white. However, if the horizon is observed through a column of air passing roughly 100,000 feet above sea level, it has the same scattering power as the zenith sky. This results from the column having the same total mass of air as a vertical column. As the altitude of observation of the column changes by 2,500 feet the scattering power should change by 10 percent. Experiments currently under way show promise of good accuracy.

The star tracker and horizon photometer shown on the sextant head in figure 14 are for making such measurements. The body-- fixed horizon photometer line is scanned across the horizon by means of slow rotation of the vehicle. The photometer circuitry provides the computer with a mark at the moment the column being scanned has a certain ratio of brightness to the peak brightness observed near sea level. Since a star is being tracked automatically while the scan occurs, the mark enables the reading of the precision angle.

Figure 18 shows a Block II sextant and scanning telescope. The predominant difference between this and the Block I instrument is the inclusion of the star tracker/photometer. The frame of this addition is shown on the left hand side of the sextant head.

Other Uses

The Command Module optics is well qualified to make navigational measurements in earth or lunar orbit. To do this the Inertial Measurement Unit is referenced to the stars. The scanning telescope is used to observe the bearing of a mapped landmark. By means of the resolvers on the Inertial Unit gimbals and on the scanning telescope drives, these bearings are referenced to the stabilized platform. Several such bearings give sufficient information to derive the characteristics of the spacecraft's orbit.

The inertial platform is aligned to the stars with the Command Module optics. This is accomplished by first approximately aligning the stable platform. Then the direction of a suitable star is tracked with the sextant and marked. The direction of a second star is marked. The computer processes these data to obtain the directions of the two stars in the inertial reference coordinate system. This defines the orientation of the stable platform relative to the stars. Using the capability to process the gyros by torques, the stable platform can now be aligned more accurately to some desired orientation.

SECTION 5.

The Lunar Excursion Module Optics and Its Use

The Optics on the Lunar Excursion Module is very much simpler than that on the Command Module. It consists of a single unit called the optical alignment telescope. The single unit has no mechanical drives but does have two manual degrees of freedom. The sole purpose of this

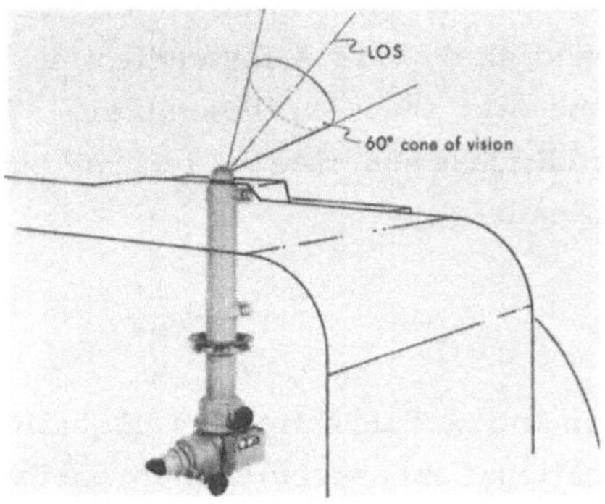

Fig. 19.

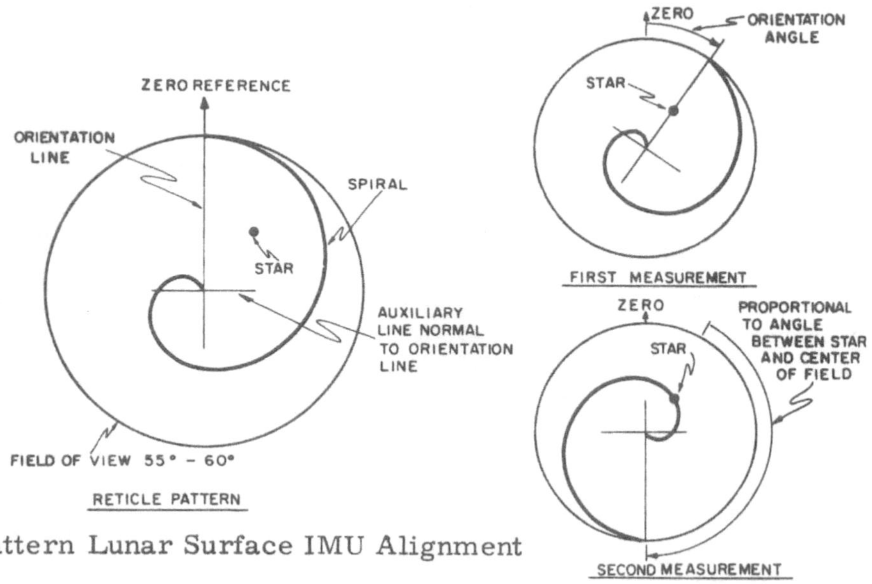

Fig. 20.

AOT Reticle Pattern Lunar Surface IMU Alignment

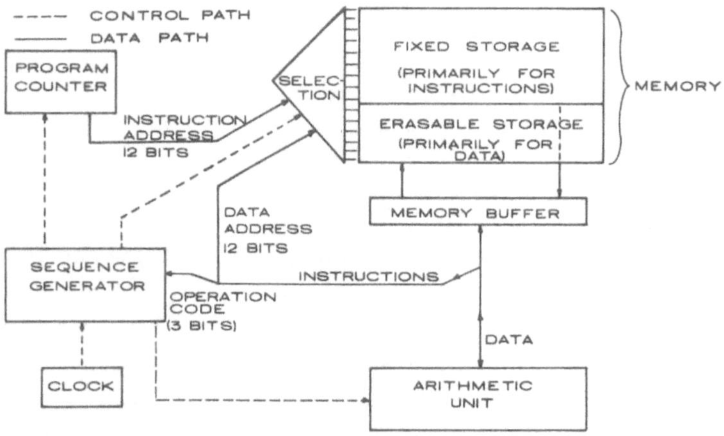

Fig. 21.

TABLE I APOLLO GUIDANCE COMPUTER CHARACTERISTICS-BLOCK II

WORD LENGTH	15 Data Bits plus one Parity Bit
NUMBER SYSTEM	One's Complement
MEMORY CYCLE TIME	11.7 sec
FIXED MEMORY REGISTERS	36,864 Words
ERASABLE MEMORY REGISTERS	2,048 Words
NUMBER OF NORMAL INSTRUCTIONS	34
NUMBER OF INVOLUNTARY INSTRUCTIONS (Interrupt, Increment, etc.)	10
INTERRUPT OPTIONS	10
ADDITION TIME	23.4 sec
MULTIPLICATION TIME	46.8 sec
DOUBLE PRECISION ADDITION TIME	35.1 sec
DOUBLE PRECISION MULTIPLICATION SUBROUTINE TIME	575 sec
INCREMENT TIME	11.7 sec
NUMBER OF COUNTERS	29
POWER CONSUMPTION	100 Watts (AGC + DSKY's)
WEIGHT	58 pounds (computer only)
SIZE	1.0 Cubic Foot (computer only)

TABLE II APOLLO GUIDANCE COMPUTER INTERFACES-BLOCK II

NUMBER OF INPUT DISCRETES	73
NUMBER OF INPUT PULSES	
1 (Serial and Incremental)	33
NUMBER OF D.C. OUTPUT DISCRETES	68
NUMBER OF VARIABLE PULSED OUTPUTS (Serial, Incremental, and Discrete)	43
NUMBER OF FIXED PULSED OUTPUTS	10
NUMBER OF CONNECTOR WIRES	365

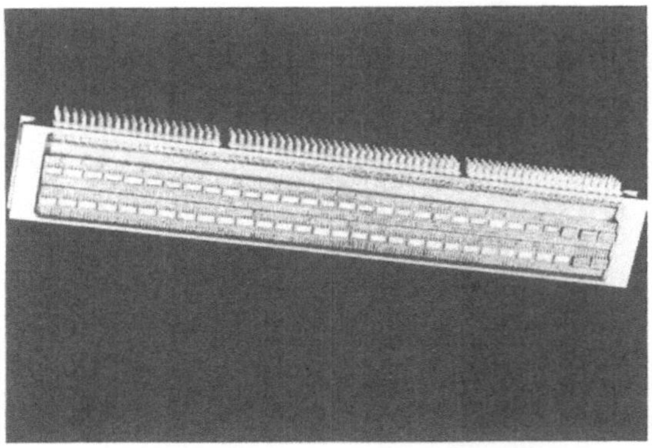

Fig. 22.

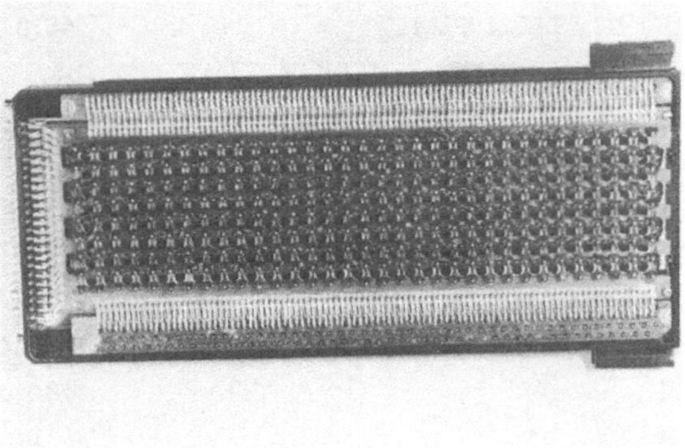

Fig. 23.

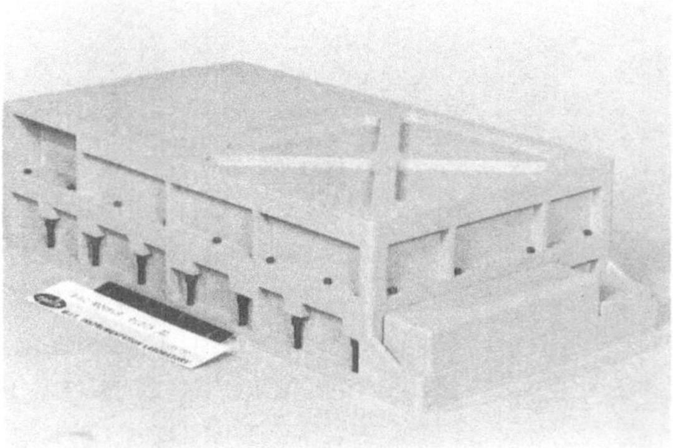

Fig. 24.

telescope is to provide stable-member orientation information to the computer. It does this by means of the crew tracking two stars. The same principles that were discussed for command-module Inertial Measurement Unit alignment apply.

Figure 19 shows a wooden mockup of the Excursion Module Optics. This photograph has been marked up to show the outlines of the LEM vehicle. Also shown is the line of sight of the telescope. As the markings show, this instrument has a single line of sight. This line of sight makes a 45-degree angle with the instrument's tube which is mounted vertically in the Excursion Module. The field of view is 60 degrees. The field of view is aimed upward so that the sunlit moon will not be present to obscure the visibility of the stars being used for prelaunch alignment. The line of sight has several positions. The position is selected by the upper knob on the side of the instrument. The forward position is shown in the drawing. Alternative operating positions are 60 degrees to the left and 60 degrees to the right. These positions enable the crew to see roughly one-half of the overhead sky. In addition to these three operating positions there is a storage position to prevent damage to the Optics from lunar dust during the landing.

In order to make an observation on a star, a suitable detent position is chosen so the star may be seen. The lower knob on the mockup rotates the position of the reticle in the instrument. The reticle has a straight marking and a spiral marking as shown in Figure 20. By using the knob to rotate the reticle the direction of the radius from the center of the field of view to the star is measured. When the alignment is correctly made the observer marks with the buttons. The reticle angle is given to the computer manually by means of a counter and data entry into the computer keyboard. A second measurement must now be made to the same star. To do this the spiral is rotated to overlay the star. Note how the direction of the straight line and the direction of the spiral in the neighborhood of the star are nearly at right angles. Thus the two observations correspond to measuring both degrees of freedom of the bearing of the star.

Similar measurements on a second star are made. The computer has now obtained data sufficient to determine the orientation of the stable platform.

SECTION 6.

Computer

Since another paper in this symposium[10] treats the organization of computation and control in the Apollo Guidance Computer, this section will emphasize the physical nature of the Block II computer rather than the organizational aspect of it. Figure 21 shows a simplified block diagram that is applicable to both the Block I[11, 12] and Block II computers. In this section it will be observed that there is a very extensive fixed memory consisting of core ropes. There is a very limited erasable memory, primarily for data. This consists of a ferrite coincident-current memory plane. There is extensive logic for the performance of arithmetic, sequence generation, computer control, etc.

Tables I and II summarize the Apollo Block II computer characteristics and interfaces. Block II contains twice the erasable memory of Block I. There is an increase of 50 percent in the fixed memory. There are considerable speed increases resulting from extra operation instructions. The interface is considerably expanded to handle additional functions in the control of spacecraft.

In spite of increased capacity and power, the Block II computer is smaller physically. The two reasons for this are shown in figures 22 and 23. Figure 22 is an unpotted computer logic module. The computer contains 24 of these modules. They differ only in the wiring pattern on the matrix located in the middle of the module. On the surface of this module can be seen the active components. Each of the small rectangular elements contains two micrologic nor gates in a flat package. In Block I single nor gates were contained in small

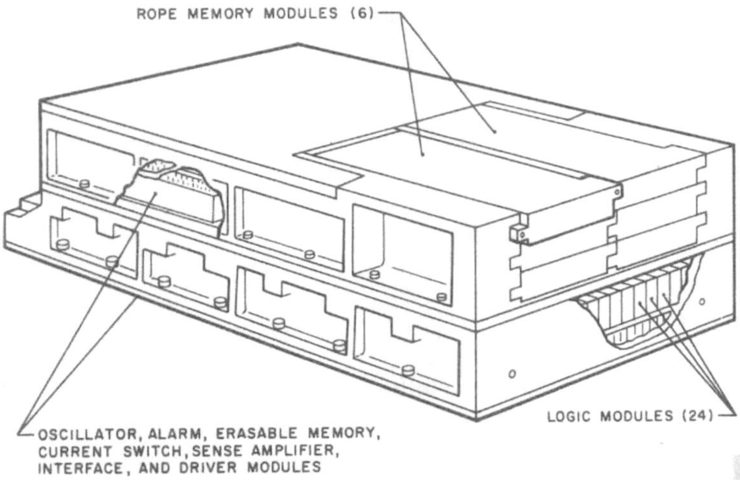

Fig. 25.

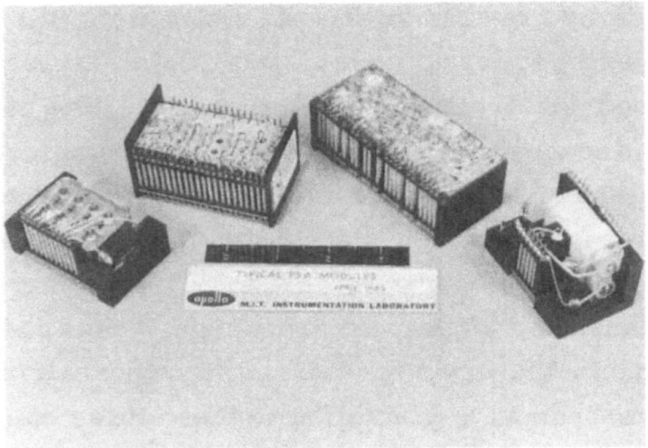

Fig. 26.

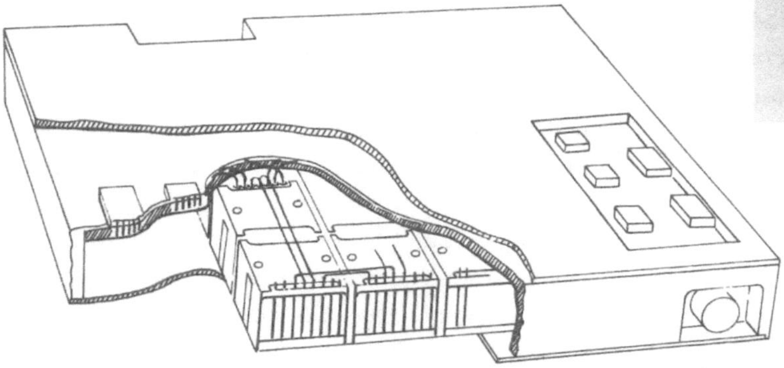

Fig. 27.

Fig. 28.

To-47 transistors cans.[13] The Block II high-density packaging results in a greatly reduced thickness for each logic module. Consequently, a reduction in the overall size of the computer is achieved.

Smaller physical size also results from a higher density in the fixed memory. In Block II, 24 sixteen-bit words are stored per core. An unpotted Block II module is shown in figure 23. In a Block I rope module of of similar volume, there were 16 such words per core. The core rope is an extremely dense technique for storing information. It also possesses considerable reliability advantage since physical destruction of the core or wire is necessary for loss of information.

The mockup of the Block II computer is shown in figure 24. Except for the core rope memory module plugs, the computer is physically sealed against moisture. The internal layout of the computer is shown in figure 25. Basically, it consists of two planes of plug-in modules.

SECTION 7.

Power Servo Assembly Characteristics

The power servo assembly is a collection of electronics in a common package which supports the operation of the Inertial Measurement Unit, the Optics Unit, and other parts of the system with power supplies, servo amplifier, etc. These miscellaneous circuits are packaged in modules of various sizes.

Figure 26 shows several typical Power Servo Assembly modules before potting. The frames of these modules are made of magnesium for light weight and good thermal distribution. These modules plug into a single header which makes the necessary inner modular and system connections. The construction of the Power Servo Assembly is shown in the cutaway of figure 27. It can be seen that the modules of various sizes screw upward into this header in the Command Module. Each one presents the header with a flat area to establish good thermal contact. The header is in turn screwed to the coldplate which is a part of the spacecraft. This coldplate has spacecraft coolant fluid circulated within it, although this feature is not shown in the cutaway.

After the modules are plugged into the header, a cover is placed over the frame to complete the seal of the power servo assembly's modules and plugs. This assembly is installed as a unit onto the spacecraft coldplate. In all there are 44 modules that comprise a set for the Command Module Power Servo Assembly.

The wooden mockup of the Power Servo Assembly for the Command Module is shown in figure 28. The screw holes are for fastening the power servo assembly to its coldplate in the spacecraft. The raised rectangles represent plugs which the Power Servo Assembly presents to the Guidance and Navigation system harness. The cylinders in the foreground represent test connectors which are accessible after spacecraft installation.

SECTION 8.

Control Interfaces

The control interfaces between the guidance and navigation system and the spacecraft consist of signals to and from the computer and of signals to the spacecraft from the Coupling Data Units. In addition, the Coupling Data Units provide interfaces within the Guidance and Navigation system. Before discussing the overall control configuration, the Coupling Data Unit will be briefly examined.

Figure 29 shows a very simplified schematic of an electronic Coupling Data Unit. The first function shown in the schematic is that of converting the mechanical angle data into digital quantities which the computer can process. The schematic shows that each of these mechanical angles is converted to sine and cosine signals by resolvers. In the Coupling Data Unit it is desired to have a solid- state digital counter contain a word which

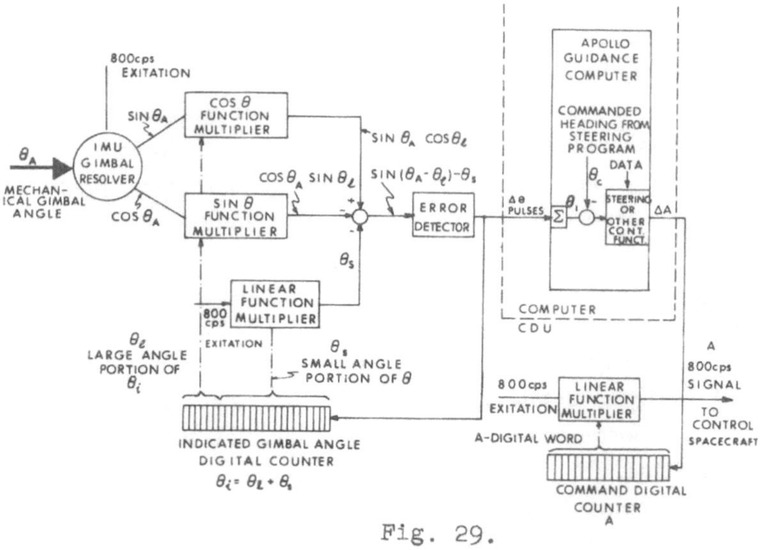

Fig. 29.

Fig. 30.

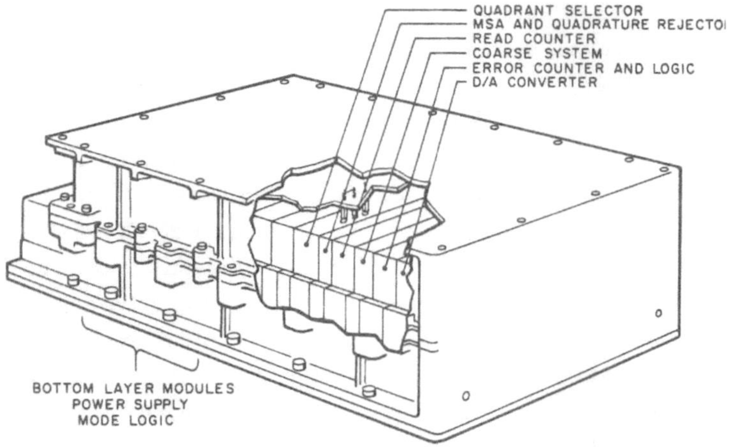

Fig. 31.

represents the value of the mechanical angle. The contents of this counter represent the indicated value of the angle. The most significant four binary digits of this word are used for initial processing of the resolver signals. These four bits θ_1, represent the large part of theta. The least significant bit corresponds to 22-1/2 degrees. These bits are used in the sine and cosine function multipliers. These multipliers consist of solid state switches controlled by the bits through logic to switch appropriate combinations of the sine and cosine of 22-1/2 degrees, 45 degrees, 67-1/2 degrees. . . . . . ., to form a sum of the resolver signal times the sine and cosine, respectively, of the large part of theta.

The less significant figures in the counter indicating theta are summed in a similar linear function multiplier operating on the excitation signal. Examination of the trigonometric forms indicated shows that, to a good approximation, the sum as indicated in the drawing of these signals should be zero or of very small value when the counter correctly indicates the value of theta. When the value of theta is not correctly indicated, the error detector generates pulses which are delivered to the computer to give it knowledge of changes in the value of theta indicated. These pulses are, of course, also used to add or subtract increments from the counter.

The discussion above is extremely simplified and illustrates the principle used in analog-to-digital conversion in the electronic CDU's. In actual fact, each mechanical angle to be encoded has a coarse and a fine resolver. The operations described above are performed on both resolver signals with a nonlinear circuit to select between the coarse and fine resolvers, depending on the size of the error signals.

A second function of the electronic CDU is to generate an analog signal from digital increments delivered for steering by the computer. These increments go into a counter which controls a similar linear function multiplier. An analog 800-cps signal proportional to the argument in the counter is generated. In some cases this is an engine gimbal command. In other cases this represents attitude errors, etc.

A complete discussion of the CDU is complicated because many of the operations are on multiple-speed signals. It is further complicated because the CDU's have within them logic to select a number of modes which are used for various purposes. Examples of modes are Inertial Measurement Unit Coarse Alignment, Zero Optics, Steering, etc.

The wooden mockup shown in Figure 30 illustrates the physical characteristics of the electronic Coupling Data Unit package. In Block II this sealed container has five CDU's plus associated logic and power supplies as shown in the cutaway in figure 31. The packaging is very similar to that found in the computer. There are two layers of modules plugging into a wire-wrapped header, all of which is contained within a seal.

Figure 32 is an overall block diagram of the guidance system operating as a digital autopilot in steering the spacecraft. It can be seen that during powered flight two of the analog signals from the Coupling Data Units go directly to the gimbal engines[14] on the spacecraft. Similarly, during free-fall flight, signals go directly from the computer to the solenoid-actuator amplifiers[14] in the Reaction Control System. Flexibility is the advantage of the electronic Coupling Data Units in Block II over the mechanical units used in Block I. The arrangement gives the computer knowledge of Inertial Measurement Unit gimbal angles while allowing it to generate independent steering commands. These can then bear fairly sophisticated computational relationships to each other.

SECTION 9.

Installation and Displays and Controls

The appearance of the Apollo Guidance and Navigation System installed in the Lower Equipment Bay of the Command Module is shown in figure 33.

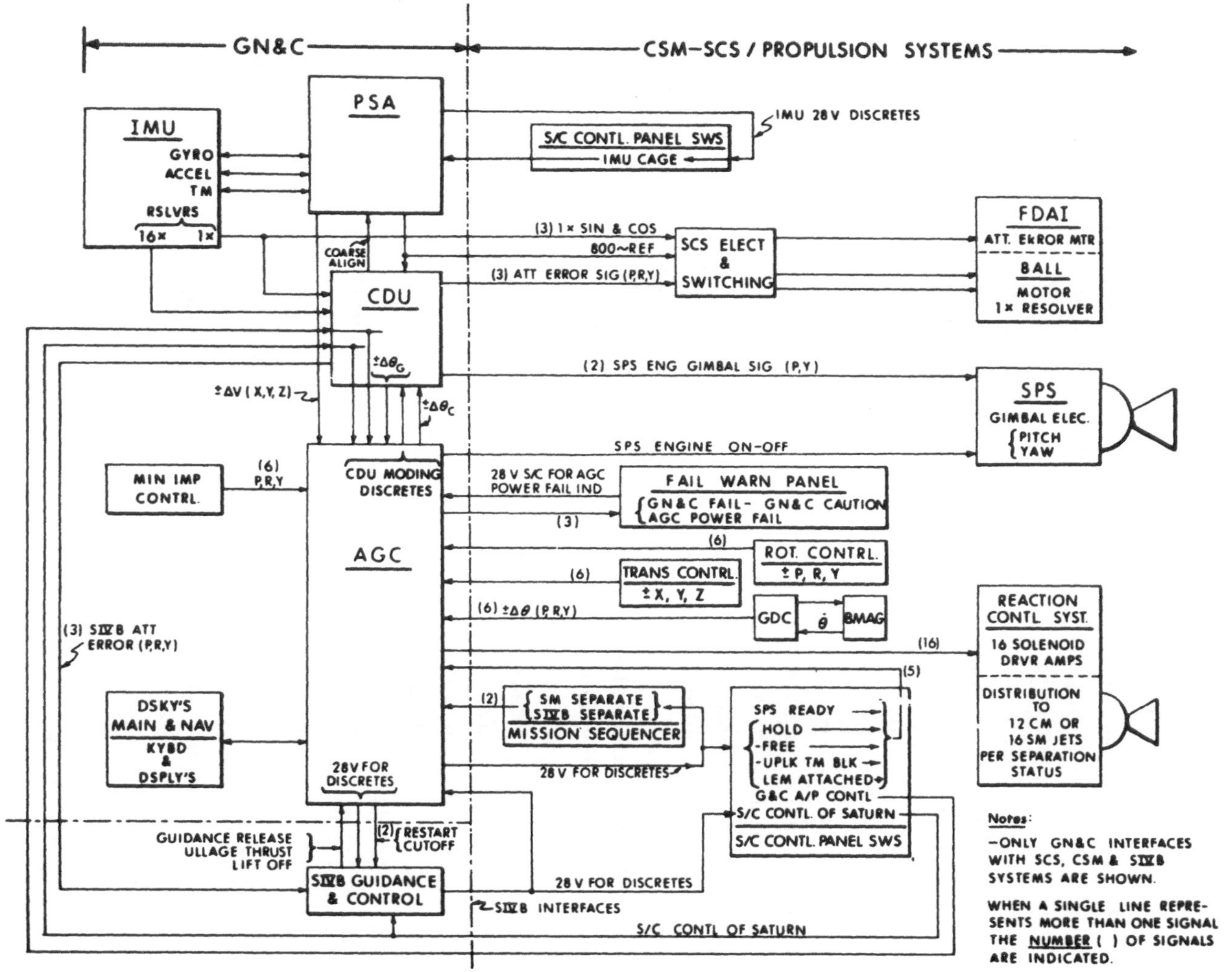
GN&C
CSM–SCS / PROPULSION SYSTEMS
IMU
GYRO
ACCEL
TM
RSLVRS
16x 1x
PSA
COARSE ALIGN
CDU
±Δθ_G
±Δθ_C
±ΔV (X,Y,Z)
IMU 28V DISCRETES
S/C CONTL. PANEL SWS
IMU CAGE
(3) 1x SIN & COS
800~REF
(3) ATT ERROR SIG (P,R,Y)
SCS ELECT & SWITCHING
FDAI
ATT. ERROR MTR
BALL
MOTOR
1x RESOLVER
(2) SPS ENG GIMBAL SIG (P,Y)
SPS
GIMBAL ELEC.
PITCH
YAW
SPS ENGINE ON-OFF
MIN IMP CONTRL.
(6) P,R,Y
CDU MODING DISCRETES
AGC
28 V S/C FOR AGC POWER FAIL IND
FAIL WARN PANEL
GN&C FAIL- GN&C CAUTION
AGC POWER FAIL
(3)
(6)
ROT. CONTRL.
± P, R, Y
TRANS CONTRL.
± X, Y, Z
(6) ±Δθ (P,R,Y)
GDC
BMAG
REACTION CONTL. SYST.
16 SOLENOID DRVR AMPS
DISTRIBUTION TO 12 CM OR 16 SM JETS PER SEPARATION STATUS
(16)
(5)
(3) SIVB ATT ERROR (P,R,Y)
(2)
SM SEPARATE
SIVB SEPARATE
MISSION SEQUENCER
SPS READY
HOLD
-FREE
-UPLK TM BLK
LEM ATTACHED
G&C A/P CONTL
S/C CONTL. OF SATURN
S/C CONTL. PANEL SWS
DSKY'S MAIN & NAV
KYBD & DSPLY'S
28V FOR DISCRETES
28 V FOR DISCRETES
GUIDANCE RELEASE
ULLAGE THRUST
LIFT OFF
(2) RESTART CUTOFF
SIVB GUIDANCE & CONTROL
28 V FOR DISCRETES
SIVB INTERFACES
S/C CONTL OF SATURN
Notes:
-ONLY GN&C INTERFACES WITH SCS, CSM & SIVB SYSTEMS ARE SHOWN.
WHEN A SINGLE LINE REPRESENTS MORE THAN ONE SIGNAL THE NUMBER ( ) OF SIGNALS ARE INDICATED.
SWITCHES = SWS

Fig. 32.

The displays are somewhat abbreviated in Block II compared wth Block I. The control stick on the left is for the Optics. The control stick on the right is for changing spacecraft attitude by means of very small impulses of thrust.[14] This stick provides a sensitive means of steering the landmark line of sight by turning the spacecraft. Above the left-hand control stick are switches which establish the proper mode of Optics system operation. The buttons next to the right-hand stick are for marking the navigational sighting time to the Computer. The numeric panel with keys and display lights to the right is the Computer display and keyboard unit. An identical unit exists on the main panel of the Command Module where the astronauts can reach it during power and re-entry flight. The optics eye pieces are not shown in this mock-up.

The Lunar Excursion Module displays are more abbreviated. An identical Computer display and keyboard is provided on the LEM main panel. Several of the meters, attitude displays, and markings on the Excursion Module window, all display guidance quantities but are not considered a part of the guidance system.

Command Module installation features can also be noted in figure 33. The figure shows the panel installation. The computer and CDU's are the lowest member of the guidance system shown in the photo. They occupy a bay which is displaced to the left of the remaining equipment. Above them is the Power Servo Assembly shown attached to the coldplate with the coolant-carrying flex lines above it. The bellows which isolate the optical unit from structural stress are barely visible near the top of the photograph.

The optics bellows are better visible in figure 34 which shows the behind-the-panel installation. Below the bellows is a spherical member which is the Inertial Measurement Unit. The Inertial Measurement Unit housing is mounted on precise surfaces of the oval-like structure. This oval structure is called the Navigation Base and is mounted with three strain-isolation mounts to spacecraft structure. The Navigation Base also carries the optics unit thus giving the right mechanical connection necessary for Inertial Measurement Unit alignment.

The Excursion Module Guidance and Navigation system installation has two locations. Shown in Figure 35 is the front location. This is outside the vehicle pressure hull and above the crew's heads. In the foreground is the Optical Alignment Telescope. Behind it a toroidal member with the protruding struts is a Navigation Base that connects this Telescope to the spherical Inertial Measurement Unit. Behind the Inertial Measurement Unit is that portion of the electronics which could not easily tolerate a cable run having a length of the order of 20 feet. The remainder of electronics, consisting of most of the PSA, Coupling Data Units, and the computer, is near the very back of the vehicle.

REFERENCES

1. Hugh L. Dryden, Ph.D., "Footprints on the Moon", National Geographic Magazine, Vol. 125, No. 3, March 1964.

2. Dr. C.S. Draper, H.P. Whitaker, and L.R. Young, "The Roles of Men and Instruments in Control and Guidance Systems for Spacecraft," to be published in the Proceedings of the 15th International Astronautical Congress, Warsaw, 1964.

3. J.M. Dahlen and J.L. Nevins, "Navigation for the Apollo Program," MIT Instrumentation Laboratory Report, R-447, May 1964, given at the National Space Meeting, St. Petersburg, Florida, April 30, 1964.

4. M. Kramer, "Apollo Command Module and LEM Guidance Navigation Systems," published in Apollo - A Program Review, SP 257, Society of Automotive Engineers, Inc., 1965.

5. Unpublished MIT Paper by David G. Hoag.

6. N. Sears, "Technical Development Status of Apollo Guidance and Navigation," Paper 64-17, published in Advances in the Astronautical Sciences, pp. 414-450, Vol. 18, 1964.

7. Unpublished MIT Memos by Daniel J. Lickly, Raymond H. Morth, and Bard S. Crawford.

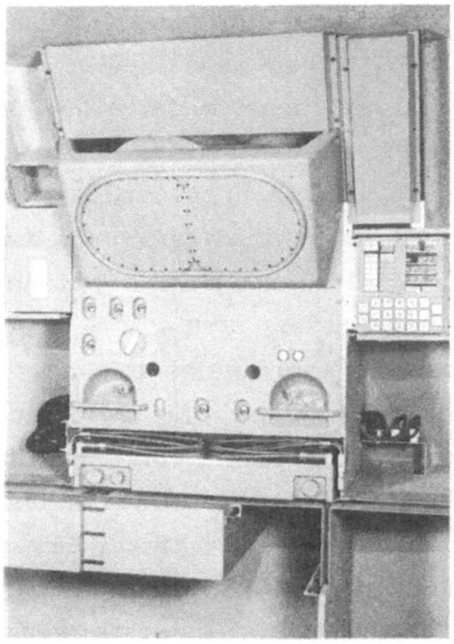

Fig. 33.

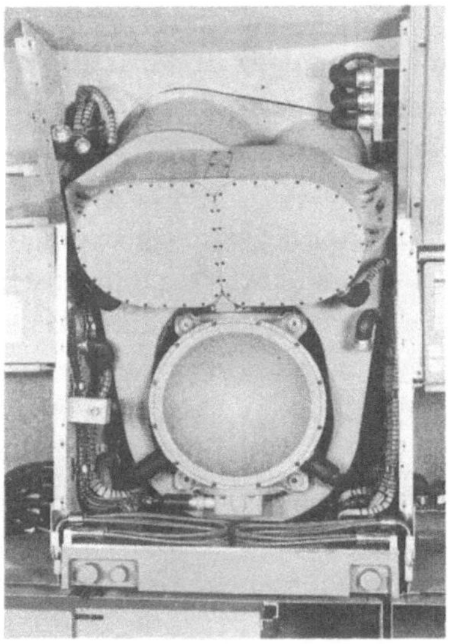

Fig. 34.

Fig. 35.

8. Richard H. Battin, "A Statistical Optimizing Navigation Procedure for Space Flight," MIT Instrumentation Laboratory Report R-341, May 1962, published in the American Rocket Society Journal, November 1962.

9. Richard H. Battin, "Explicit and Unified Methods of Spacecraft Guidance Applied to a Lunar Mission," to be published in the Proceedings of the 15th International Astronautical Congress, Warsaw, 1964.

10. Thomas J. Lawton and Charles A. Muntz, "Organization of Computation and Control in the Apollo Guidance Computer," MIT Instrumentation Laboratory Report E-1758, April 1965, to be given at the IFAC Symposium on "Automatic Control in the Peaceful Uses of Space," Stavanger, Norway, 1965.

11. R.L. Alonso, H. Blair-Smith, and A.L. Hopkins, "Some Aspects of the Logical Design of a Control Computer: A Case Study," Report EC12, No. 5, published in the IEEE Transactions on Electronic Computers, December 1963.

12. R.L. Alonso and A.L. Hopkins, "The Apollo Guidance Computer," MIT Instrumentation Laboratory Report R-416, August 1963, reprinted as Report 6.1, PTGSET Record, published by the Professional Technical Group on Space Electronics and Telemetry of the IEEE, 1963.

13. Jane Partridge, L. David Hanley, and Eldon C. Hall, "Progress Report on Attainable Reliability of Integrated Circuits for Systems Application," November 1964, MIT Instrumentation Laboratory Report E-1679, given at the Symposium on Micro Electronics and Large Systems in Washington D.C., co-sponsored ONR and Univac, November 17, 18, 1964.

14. Robert G. Chilton, NASA Manned Spacecraft Center, "Apollo Spacecraft Control Systems," to be given at the IFAC Symposium on "Automatic Control in the Peaceful Uses of Space," Stavanger, Norway, 1965.

ACKNOWLEDGMENT

IFAC acknowledges the Massachusetts Institute of Technology for assigning the copyright of this paper for publication purposes of the International Federation of Automatic Control.

Entry and Landing

GUIDANCE AND CONTROL IN SUPERCIRCULAR ATMOSPHERE ENTRY

by Rodney C. Wingrove
Research Scientist
NASA, Ames Research Center
Moffett Field, California

ABSTRACT

This paper discusses planetary entry maneuvers which include atmospheric capture for Mars entry velocities up to 12 km/sec; atmospheric capture for Earth entry velocities up to 21 km/sec; skip-out control to either a parking orbit or extended ranges; and terminal range control. Simulator results are compared for both automatic and piloted-guidance systems. The results are presented to illustrate the expected guidance performance as compared with the vehicle's full capabilities. Several factors considered are the control response requirements, the effect of measurement errors and atmosphere uncertainties, and the effect of various display and control techniques.

INTRODUCTION

With the successful entries of manned vehicles from near Earth orbits, attention has been turned to the problems of atmosphere entry for more advanced manned space-flight missions. Manned trips to the planets have been studied in a number of recent investigations.[1-13] Studies have shown that vehicles returning to Earth from these missions will enter the atmosphere at speeds of up to 15 km/sec and perhaps as high as 20 km/sec. Entry velocities at a planet such as Mars are expected to be as high as 12 km/sec. Retrorockets could conceivably reduce these large approach speeds, but the weight of fuel required makes this method of braking impractical, and we are led to aerodynamic braking in the atmosphere as the most realizable solution at this time.

During aerobraking maneuvers, manned vehicles must be able to control aerodynamic lift to satisfy several requirements. Initially on entering the atmosphere, the vehicle must perform a "capture" maneuver to keep from exceeding acceleration limits or skipping back out of the atmosphere. After the capture maneuver and during the supercircular deceleration portion of the flight, the control system must regulate a large negative aerodynamic lift force to counteract the centrifugal force of the trajectory, and thus keep the vehicle within the planetary atmosphere. To maneuver the vehicle to the planet surface or to a parking orbit, the control system must determine the appropriate lift force. Throughout the flight the vehicle heating and acceleration loads must be kept within tolerable limits by the control system. This system must be able to perform adequately even though there are errors in the measuring instruments and uncertainties in the atmosphere characteristics.

In previous studies[14-29] the control of space vehicles entering the Earth's atmosphere from circular or near circular velocities have been analyzed. In this paper these studies are extended to the entry control problems associated with planetary missions. This paper will first consider the basic dynamics of the entry trajectories and how the dynamics of the trajectory variables are related to the guidance and control problem at the extreme entry velocities. Simulator results will be presented that illustrate control in the capture maneuvers, control during skip out to either a parking orbit or to extended ranges, and control to the planet's surface. A comparison will be made of the control problems at both Earth and Mars. The discussion will cover both the use of automatic control and manual back-up control systems.

TRAJECTORY CONSIDERATIONS IN PLANETARY ENTRIES

Trajectory Dynamics

A space vehicle approaching a planet at supercircular velocity must be within a safe entry corridor if it is to be captured within the planet's atmosphere. The entry corridor, illustrated in Fig. 1, is the difference in height of the vacuum perigees of two conic trajectories,[13] the upper trajectory forming the overshoot boundary and the lower trajectory being the undershoot boundary. If the vehicle approaches above the overshoot trajectory, it does not enter the atmosphere sufficiently to be captured; if it approaches below the undershoot trajectory, its total acceleration force will exceed a specified value (usually considered 10g).

After the vehicle has entered the atmosphere, aerodynamic lift control must be correctly applied to insure that the vehicle will not unintentionally skip back out of the atmosphere. In this capture maneuver a negative aerodynamic lift force, centrifugal force, and gravity force must essentially balance along the trajectory

Superior numbers refer to similarly-numbered references at the end of this paper.

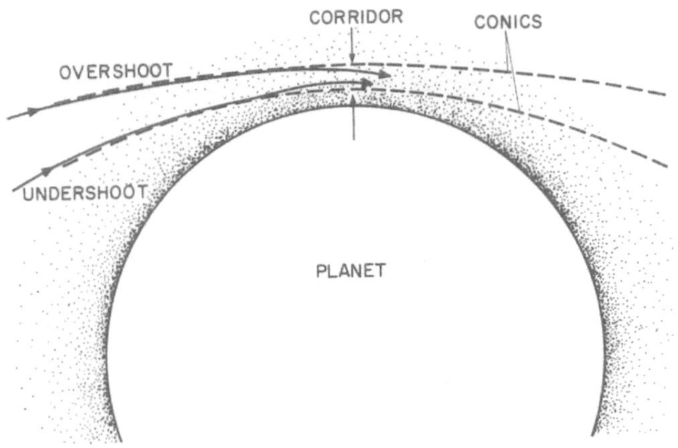

Fig. 1. - Definition of corridor depth.

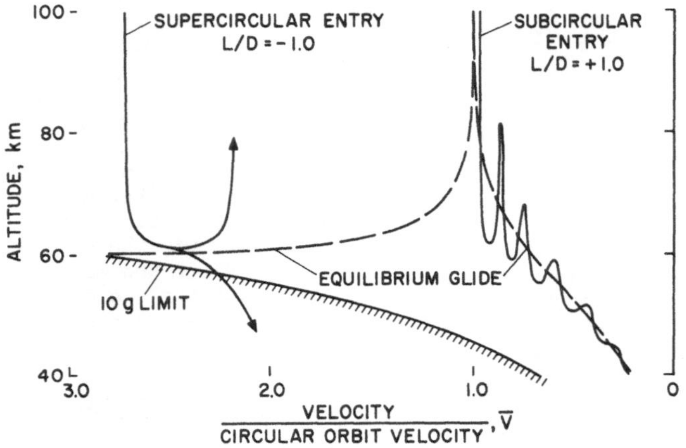

Fig. 2. - Dynamics of constant-trim lifting vehicle in Earth entry.

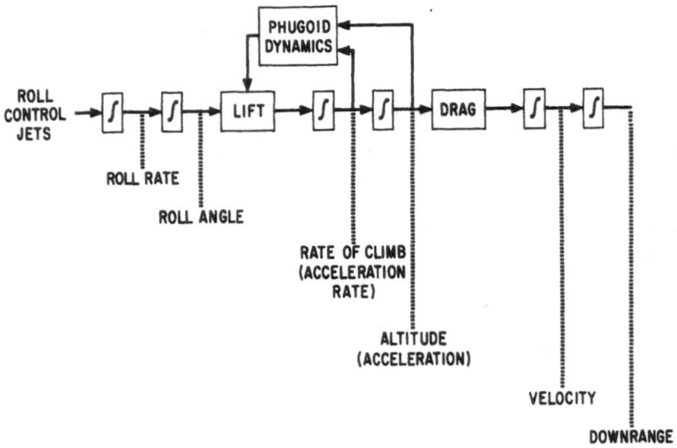

Fig. 3. - Relationship of roll control and trajectory dynamics.

(i.e., equilibrium glide). The stability of vehicle motions near equilibrium flight can be illustrated by the trajectories of a vehicle trimmed at a constant lift/drag ratio. Fig. 2 presents trajectories of a constant L/D vehicle entering the Earth's atmosphere at subcircular and at supercircular entry velocities. The altitude along the trajectories is presented as a function of $\bar{V}$, the ratio of total velocity to local circular velocity. The example trajectories are for a vehicle entering the Earth's atmosphere at an altitude of 100 km where the sensible atmosphere begins. For entry at near circular velocity ($\bar{V}_i = 1$) the vehicle is shown entering with positive lift which allows it to maintain near equilibrium flight. We see these subcircular dynamics are dynamically stable about the equilibrium glide path; that is, the motions are oscillatory and there is a small amount of damping. For the vehicle entering at extreme supercircular velocity ($\bar{V}_i = 2.7$) negative lift is used to maintain near equilibrium flight; however, here we see the dynamics are unstable. One supercircular trajectory is shown to skip back out of the atmosphere, while the other trajectory with just a small change in entry conditions dives deeper into the atmosphere. These uncontrolled dynamics are studied further in detail in references 30 to 32 where they are considered analogous to the classical aircraft phugoid motions.

The data in Fig. 2 show that the phugoid dynamics expected in supercircular entries are unstable and thus illustrate the need for control. This control is critical, and fast response is needed for extreme Earth entry velocity where the time to double amplitude of the uncontrolled dynamics can be on the order of 5 seconds. At Mars the supercircular phugoid dynamics are somewhat more stable than at Earth because the gravitational force is less; therefore less lift is required to hold the vehicle near equilibrium lift.

Now we can point out in Fig. 2 the boundaries within which the vehicle must operate. If the vehicle is near equilibrium flight and is allowed to fly above the supercircular equilibrium glide path, it will skip back out of the atmosphere uncontrolled. The allowable region of flight must be below this equilibrium flight boundary. On the other hand, the vehicle cannot be allowed to dive too far into the atmosphere or the acceleration forces will be greater than 10g. During the supercircular portion of the entry, the vehicle must be flown between the upper skip-out boundary and this lower acceleration boundary.

Relationship of Control and Dynamics

The method of trajectory control most commonly considered for entry at supercircular speeds is to use the roll angle of the vehicle to regulate the vertical component of the lift vector. With this control method, which shall be considered in this paper, it is assumed that the vehicle maintains nearly a constant aerodynamic trim condition about the pitch and yaw axis.

Fig. 3 is a simplified block diagram showing the relationship between vehicle roll control and the trajectory dynamics. This diagram depicts altitude and range variations in only the vertical plane. Crossrange control will be illustrated later in the paper. This figure illustrates the chain of events that result when a roll control moment is applied by the reaction jets. The torque about the roll axis produced by the roll control jets is integrated to produce a roll rate. Integrating the roll rate changes the roll angle. The control of the roll angle is a part of the vehicle "short-period dynamics." The dynamic response of the system which follows a change in roll angle is termed the "long-period" or trajectory dynamics.

The roll angle determines the component of the aerodynamic lift force vector in the vertical plane. The integrated changes in this vertical lift force determine the changes in the vehicle rate of climb. Integrating this vertical velocity determines the changes in the vehicle altitude and consequently the changes in drag acceleration. The subsequent integration of drag acceleration determines changes in horizontal velocity, and a final integration determines the subsequent changes in the downrange.

Fig. 3 illustrates that guiding to a given downrange point is, in effect, controlling a sixth-order function. With an on-board computer and inertial platform, position and velocity information can be accurately determined during entry, and guidance logic equations within the computer can determine the proper roll angle required to capture the vehicle and to reach the destination. Several types of guidance logic equations that can be programmed in the on-board computer have been found to be satisfactory for supercircular entries. These programs include the use of fast-time predicted paths[14,15,21-23,29] as well as stored path information.[16-20,23] In the present paper, automatic control results will be illustrated by the fast-time prediction method outlined in reference 29.

TRAJECTORY CONTROL IN PLANETARY ENTRIES

The results presented were obtained from simulation studies of automatic and manual back-up systems.[27,29] The control problem of capture within the atmospheres of Earth and Mars will be discussed first. Next, skip-out control to either a parking orbit or to extended ranges will be considered, and, finally, the terminal range control to the planet's surface.

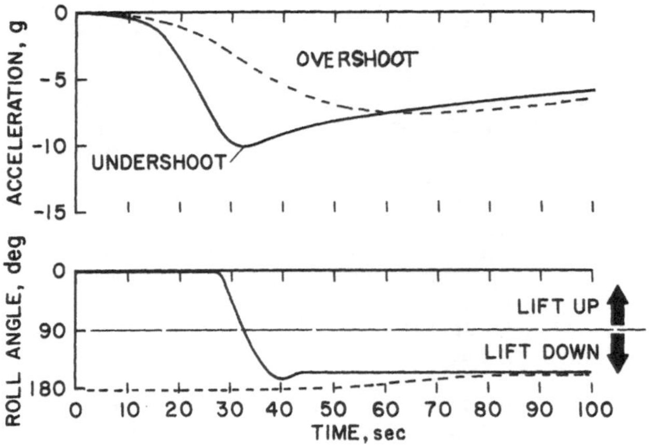

Fig. 4. - Time history of capture maneuvers in Earth entry; maximum L/D = 1, $\overline{V}_i = 2.7$.

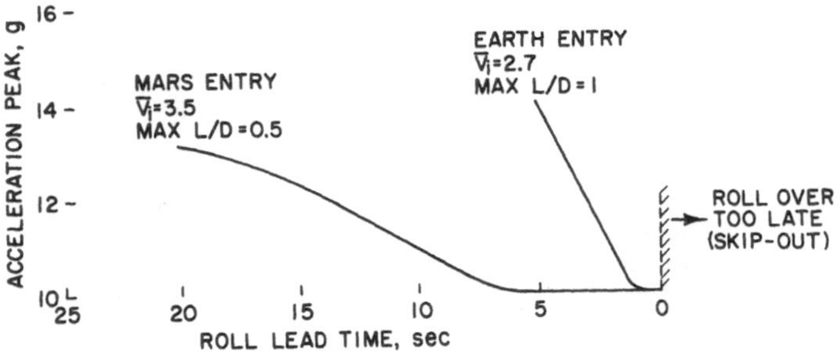

Fig. 5. - Entry at undershoot boundary; maximum roll rate = 20°/sec.

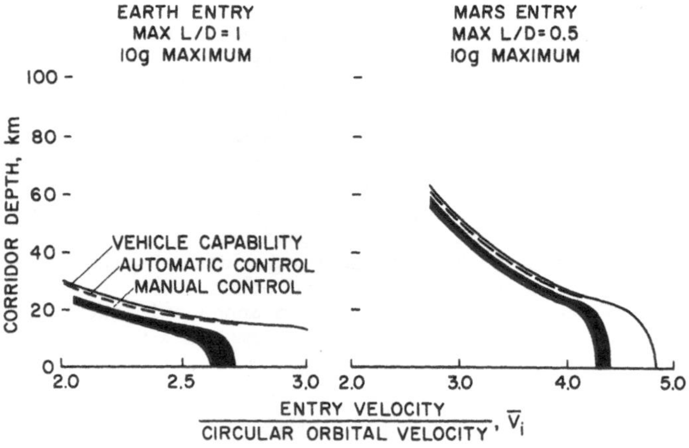

Fig. 6. - Corridor depths as a function of entry velocity.

Capture Control

The control required in capture for the extreme Earth entry velocity is illustrated in Fig. 4 for entries near the overshoot boundary and near the 10g undershoot boundary. These data were obtained using the automatic control method outlined in reference 29. At the overshoot boundary the vehicle lift is held full down throughout the initial maneuver to insure that the vehicle is pulled down into the atmosphere. When the peak acceleration is reached, the roll angle is modulated to stabilize the trajectory about equilibrium and to maneuver onto the desired path. Near the undershoot boundary the full positive lift is held initially to insure that the acceleration will stay within 10g. Near peak acceleration the vehicle must be rolled to that negative lift required to maintain equilibrium and keep the vehicle from skipping back out of the atmosphere. The roll angle is then modulated to stabilize the trajectory about equilibrium and to maneuver onto the desired path.

The proper timing of this roll maneuver for the undershoot boundary case is critical at the higher entry velocities. This is illustrated in Fig. 5 for entries at both Earth and Mars. These data are for representative vehicles with a maximum roll rate of 20°/sec. For the extreme Earth entry velocity there is approximately a 1-second time leeway within which the roll-over maneuver may be initiated. If the roll maneuver is initiated later than this the vehicle will skip back out of the atmosphere. Control applied earlier than this time will cause the vehicle to exceed 10g acceleration. For the extreme entry velocity at Mars, the roll maneuver is less critical and there is about a 6-second leeway within which the roll maneuver may be initiated.

For the capture maneuver in Earth entries it has been found that increasing the maximum roll rate capability of the vehicle from 20°/sec to even an infinite value will allow only about 1/2-second additional leeway for the maneuver. For roll rates less than about 15°/sec there is essentially no time leeway within which the roll maneuver can be performed with the acceleration peak less than 10g. It appears that a maximum roll rate of at least 20°/sec, which is on the order of that for the current Gemini and Apollo vehicles, is adequate to perform the capture maneuvers.

Proper control timing does not appear difficult with automatic control and a high-speed on-board computer.[29] The timing is very critical from the pilot's standpoint, though, if he must perform the entry with only minimal back-up display information. In this case the pilot must "play it safe" and roll a couple of seconds early to insure capture. The corridor capabilities of both the automatic and manual back-up control systems are compared in Fig. 6 with the maximum available corridors. These data are for representative entry vehicles both at Earth and Mars.

For an $L/D = 1$ vehicle entering the Earth's atmosphere at the extreme entry velocity of 21 km/sec ($\overline{V}_i \approx 2.7$) there is about a 15-km corridor. A 15-km corridor is on the order of that required to accommodate midcourse guidance errors and atmosphere uncertainties.[9,33] A vehicle with an $L/D = 0.5$ entering the atmosphere of Mars at a maximum expected velocity of 12 km/sec ($\overline{V}_i \approx 3.5$) will have an available corridor of about 30 km.

The manual control values shown in Fig. 6 were obtained from piloted simulation studies in which the pilot was only given the information that would be available from a back-up roll gyro and a single strapped down accelerometer.[29] The pilot found control to be more difficult in the Earth entry than in the Mars entry. This was due primarily to the critical roll timing required for capture and the more unstable control situation encountered in the Earth's atmosphere as compared to the Mars atmosphere. For the Mars entries the pilot was able to use essentially the full corridor capabilities of the vehicle beyond $\overline{V}_i = 3.5$, the maximum expected entry velocity. For Earth entries the pilot was able to control consistently within a 15-km corridor depth to about $\overline{V}_i = 2.5$. This compares with the full vehicle capability which gives a 15-km corridor at about $\overline{V}_i = 2.7$.

With an automatic system, as outlined in reference 29, vehicles can utilize most of their full corridor. This particular system uses feedback measurements of velocity, acceleration, and altitude rate. Possible errors in measuring the altitude rate have been found to be the most critical. For an Earth entry at $\overline{V}_i = 2.7$ this measurement must be accurate to within ±50 m/sec for the vehicle to utilize 99 percent of the available corridor and within ±90 m/sec to utilize 90 percent of the available corridor. In contrast, for entry to Mars at the extreme entry velocities, $\overline{V}_i = 3.5$, the altitude rate must be known to only about ±120 m/sec to utilize 99 percent of the available corridor.

The day-to-day uncertainties in the Earth's atmosphere do not appear to seriously affect the ability of the guidance systems to utilize the full corridor capabilities of the vehicle. However, we are not at all sure of what variations to expect in the Mars atmosphere. It appears that if the scale height of the atmosphere can be known within about ±25 percent, the vehicle can utilize the full corridor available at the particular time of entry.*

---

*When the acceleration measurement is used in guidance logic, it is only an uncertainty in the scale height of the atmosphere density rather than in reference density level that will affect the capture control.

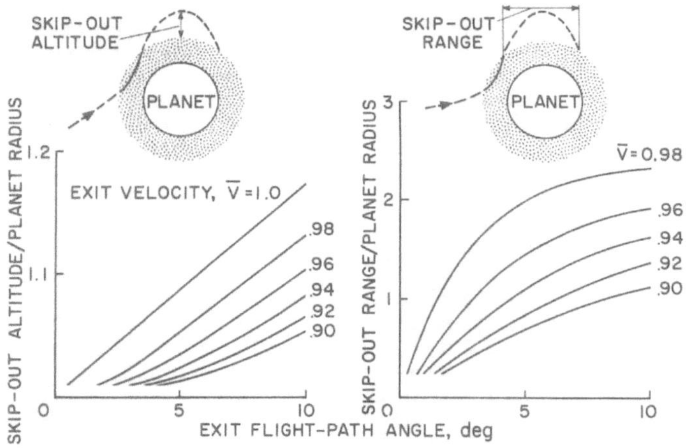

Fig. 7. - The variation of skip-out altitude and skip-out range with exit velocity and exit flight-path angle.

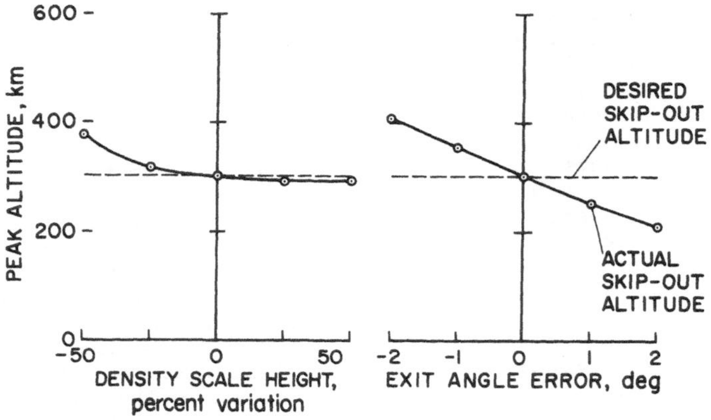

Fig. 8. - Skip-out control at Mars; maximum L/D = 0.5.

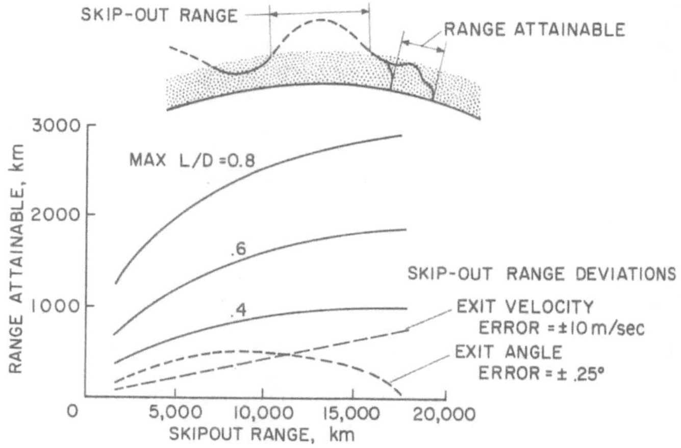

Fig. 9. - Attainable range in the second entry at Earth; exit angle = 5°.

Skip-Out Control

When a vehicle approaches a planet at supercircular velocities it might be controlled to skip out of the atmosphere and into a parking orbit or possibly to extend range. The exit conditions for achieving the desired altitude or range objectives are illustrated in Fig. 7. These data are derived from Keplerian equations of motion for the extra-atmosphere portion of the flight. It can be seen that for the skip out to either a desired altitude or range there is a choice of the combinations of exit angles and exit velocities which will meet these objectives. The choice of exit condition and the manner in which the vehicle must be controlled to arrive at this condition will be discussed briefly. Representative examples will illustrate the skip-out control into a parking orbit about Mars, and the skip-out control to extend the range at Earth return.

Skip-out control to a parking orbit at Mars.- The use of aerodynamic braking to decelerate a spacecraft and establish an orbit about Mars has been considered in a number of studies.[1-12,34-36] The guidance within the atmosphere to reach the parking orbit consists in controlling the vehicle vertically to the desired maximum skip-out altitude and in controlling laterally through the desired plane angle change. A thrusting maneuver is needed then, near the maximum skip-out altitude to circularize the orbit. Only that portion of control within the atmosphere is discussed in this paper.

The choice of exit conditions to control depends on many considerations. These are typically: to minimize thrust required to inject into orbit; to minimize effects of measurement errors; to minimize effects of density uncertainties; to minimize heating within the atmosphere; etc.

In order to minimize the thrust required to inject the vehicle into orbit, the exit should be made at a shallow angle. This implies a near full negative lift at the time of exit. Most of the other considerations, noted however, require that the vehicle be flown to steeper exit angles and exit holding near zero lift. Fig. 8 is included to illustrate control results[29] in such a skip-out maneuver. These data show the deviation of the maximum skip-out altitude actually achieved during the skip-out maneuver from the desired parking orbit altitude as a function of density scale height and exit angle error.

As presented on the left side of Fig. 8 the skip-out error for positive scale height variations is found to be minimal. For negative scale height variations, the maximum skip-out altitude may be much higher than that desired because the density decreases with altitude more than expected in the guidance equations; thus the vehicle exits at a somewhat higher velocity than desired. For uncertainties in the density scale height within ±25 percent there is only a small error in the maximum skip-out altitude.

The right side of Fig. 8 shows that an error in flight-path angle directly affects the maximum skip-out height. For the conditions considered there is an error of about 860 meters in the maximum skip-out altitude for each meter per second of vertical velocity error at exit. An error of about 2000 meters in the maximum skip-out altitude is also found for each meter per second of horizontal velocity error at exit. The final maneuver to accelerate the vehicle into the desired orbit can compensate for some of the errors in the skip-out maneuver.

Skip-out range control in Earth entries.- In the skip-out maneuver in Earth entries the considerations of measurement uncertainties and heating are of primary importance. From Fig. 7 it can be seen that the steeper the exit, the less sensitive the range to changes in exit angle. From the heating standpoint it is desirable to decelerate at the lower altitudes and then pull near full positive lift to extend the range. This also implies an exit at steep flight-path angles. An exit angle of about 5°, close to the maximum exit angle that can be achieved for typical entry configurations, appears to be reasonable for extended range control at Earth.

In the skip-out maneuver for extended range, certain exit errors can be compensated for during the second entry. This is illustrated in Fig. 9 where the attainable range during the second entry is presented for various constant L/D trajectories. For nominal skip-out ranges on the order of 15,000 km, an $L/D = 0.4$ vehicle can compensate for skip-out range deviations on the order of ±500 km and an $L/D = 0.8$ vehicle can compensate for skip-out range deviations on the order of ±1,500 km. Range deviations due to representative measurement errors are compared to these ranging capabilities in Fig. 9. These data illustrate that a vehicle with an L/D capability of less than about 0.4 is marginal in its ability to compensate for these typical skip-out errors. An important tradeoff can be inferred from this discussion; that is, more accurate skip-out control is mandatory with lower L/D vehicles, and less accurate skip-out control can be tolerated with higher L/D vehicles. For a vehicle with an $L/D = 1$, exit angle errors up to $\pm 1.5^{\circ}$ or exit velocity errors up to ±60 m/sec can be tolerated and a satisfactory terminal control maneuver can be performed after the skip out.

Terminal range control in Earth entry.- The discussion in this section applies both to the final range control maneuver after the skip out and to the short range control maneuvers from supercircular velocity in which the vehicle is not required to skip out of the atmosphere. The terminal control requirements are to guide the vehicle near to the desired touchdown point with the vehicle in a position to make the final touchdown.

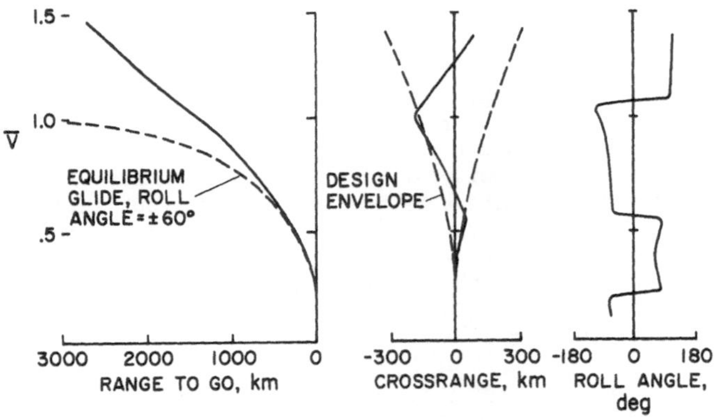

Fig. 10. - Terminal range control with roll modulated vehicle; maximum L/D = 0.5.

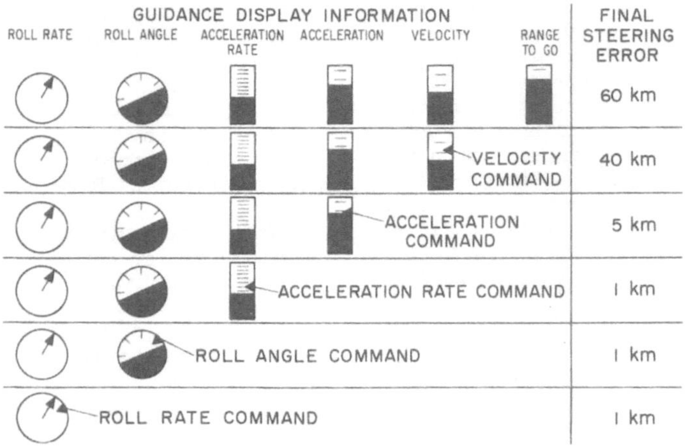

Fig. 11. - Maximum final steering errors for manual control with various levels of guidance display information.

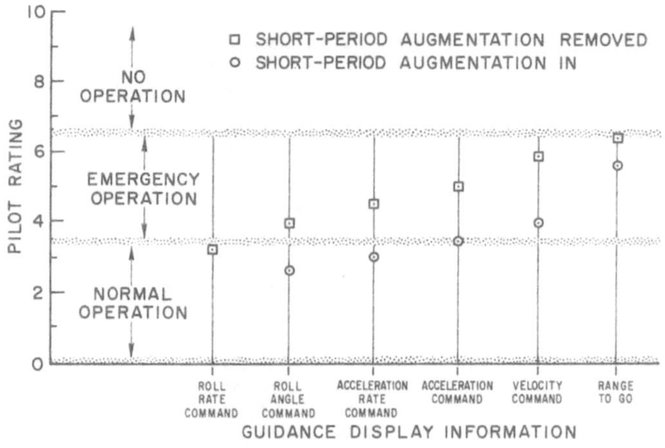

Fig. 12. - Pilot ratings for control with various levels of guidance display information.

A trajectory is illustrated in Fig. 10 for the control of terminal range by a roll-modulated vehicle.[27] The vehicle makes its range maneuver at a velocity of $\bar{V} = 1.4$ and about 3000 km from the desired destination. This is representative of nonskip range control from supercircular velocity. The vehicle is guided toward an equilibrium glide trajectory which terminates at the desired destination. A trajectory with a roll angle of $\pm 60^\circ$, as shown, is near the center of the subcircular downrange capability. A variation in vertical lift force is necessary for the vehicle to fly near this subcircular glide path to its destination. This force determines the magnitude of the roll angle. The sign of the roll angle (i.e., right or left) is determined by the crossrange to the destination. As shown in the figure, the vehicle is allowed to fly to one side until the crossrange exceeds a design envelope at which time the sign of the roll angle is reversed. The design envelope is a converging deadband that represents about one-half of the crossrange capability. The roll-angle history is shown in the figure with three roll reversals corresponding to the crossrange reversal points. With this crossrange control method, the trajectory converges to the destination as shown.

With automatic guidance systems and this type of control technique, the final steering errors are very small, on the order of 1 km. The primary contribution to the over-all final error is the ability of the inertial system to measure the vehicle's position. This navigation error (as opposed to steering error) with present "state-of-the-art" inertial equipment and updating is on the order of 1 km for each 1000 km of range traveled during entry.[37] It should be pointed out that during entry, at altitudes between about 50 and 100 km, a plasma sheath will encompass the vehicle and any updating from the ground may be impossible. Below 50 km guidance information can be relayed to the vehicle to control the final touchdown.

Now the ability of the pilot to perform the terminal range control maneuver is a function of the display information he is given. It was pointed out in Fig. 3 that the control of downrange represents a sixth-order control task. In order for the pilot to control range he must keep in mind six levels of lead information, unless, of course, this information is combined in appropriate display arrangements. Fig. 11 is included to illustrate the maximum expected terminal steering errors associated with various display arrangements that incorporate progressive levels of lead information.[27]

The first arrangement consisted of six separate displays representing measurements of each of the six state variables. Here it is seen that the steering range errors can be as large as 60 km. The range-to-go information in the velocity command display shown next is incorporated (by means of simple guidance logic) into a "velocity-to-fly-to" command. With this display the pilot must interpret five separate levels of information and the maximum error is on the order of 30 km.

In the acceleration command display, shown next, the range and velocity information is incorporated into an "acceleration-to-fly-to" display. With this display the pilot has to interpret four separate levels of information and the maximum terminal steering error is on the order of 5 km.

For either the acceleration rate command, roll angle command, or roll rate command display arrangement, the terminal steering error can be on the order of 1 km. This represents an accuracy as good as that of the fully automatic system.

The pilot's ratings[27] of the various display arrangements with short-period augmentation both in and out are illustrated in Fig. 12. Without stability augmentation the pilot must damp short-period oscillations about the pitch and yaw axes as well as perform the guidance functions about the roll axis.

With short-period stability augmentation in it is seen that display arrangements which give the guidance task to at least a third-order function (i.e., acceleration rate command) are satisfactory for normal operation. Without short-period stability augmentation it is seen that only guidance display arrangements which give a roll rate command are considered satisfactory for normal operation. In this situation the pilot has simply to look at three short-period rate instruments (roll rate, pitch rate, yaw rate) and correct the rates about each axis.

If the on-board computer and precision inertial measuring equipment have not failed during entry and the pilot must control the vehicle (such as with an autopilot failure), it is reasonable to have the pilot fly by the roll rate or roll angle command as would the completely automatic system. If the precision inertial measuring equipment is not operating during entry, however, there is the problem of obtaining navigation measurements. It appears possible, in emergency situations of this type, to use a strapped down roll gyro along with a single strapped down acceleration as measuring devices.[27,28] The gyro (or a view of the outside scene) can indicate to the pilot the horizon roll angle. A display of the acceleration can give relative altitude changes within the atmosphere. With the proper mounting of the accelerometer on the particular vehicle configuration[27] an integration of the accelerometer output can indicate velocity changes, and a second integration can indicate range changes. These types of back-up measurements are expected to give terminal navigation errors on the order of 1 to 3 percent of the entry range as compared to errors on the order of 0.1 percent that are presently expected with more sophisticated inertial measuring units.

CONCLUDING REMARKS

This paper has illustrated several considerations for the guidance and control of space vehicles entering the atmospheres of Earth and Mars.

First, the control at extreme entry velocities requires a large negative aerodynamic force to keep the vehicle within the atmosphere. The uncontrolled dynamics in this situation are highly unstable.

For Earth entry velocities up to 21 km/sec there is approximately a 1-second time interval within which a roll maneuver must be initiated to insure capture without exceeding a 10g limit. When automatic control is used in the capture maneuver, there is essentially no degradation in the usable entry corridor depth. For simple piloted back-up systems, though, successful capture is limited to entry velocity less than about 19.5 km/sec because of the critical timing of the roll maneuver. For entries at Mars, the capture maneuver is less critical, and either the piloted back-up or automatic system can use essentially all of the available corridor depth.

In considering skip-out control to a parking orbit at Mars there is a direct correspondence between system performance and measurement errors. Density scale height uncertainties up to ±25 percent cause essentially no degradation of performance, however.

In the skip-out maneuvers for extended range, certain exit errors can be compensated during the second entry. There is a direct tradeoff between the magnitude of these allowable exit errors and the maximum vehicle L/D.

The final terminal control error with an automatic system is primarily in the navigation error of the initial measuring unit rather than any steering error. With piloted back-up systems, however, both the navigation and steering errors may be sizable. This paper has illustrated the effect of various lead-information displays on the pilot's ability to perform the range control task. It is concluded that displays that give basic velocity and range-to-go information are unsatisfactory for normal operation; whereas displays that include short-period command information are satisfactory.

REFERENCES

(1) Anon., "EMPIRE - A Study of Early Manned Interplanetary Missions," NASA-MSFC Contract 8-5025, Aeronutronic Div. of Ford Motor Co., Pub. U-1951, Dec., 1962.

(2) Himmel, S. C., Dugan, J. F., Jr., Luidens, R. W., and Weber, R. J., "A Study of Manned Nuclear-Rocket Missions to Mars," IAS Paper 61-49.

(3) Anon., "A Study of Early Manned Interplanetary Missions - Final Summary Report," NASA-MSFC Contract NAS 8-5026, General Dynamics/Astronautics, Rep. AOK 63-0001, Jan., 1963.

(4) Anon., "Manned Interplanetary Mission Study - Summary Report - Volume I," NASA-MSFC Contract NAS 8-5024, Lockheed Missiles and Space Co., Rep. 8-32-63, March, 1963.

(5) Jones, A. L., ed., "Manned Mars Landing and Return Mission Study - Final Report," NASA-Ames Research Center, Contract NAS 2-1408, North American Aviation, Inc., Space and Information Div., Rep. SID 64-619, March, 1964.

(6) Sohn, R. L., ed., "Manned Mars Landing and Return Mission," NASA-Ames Research Center, Contract NAS 2-1409, TRW Space Technology Laboratories, Rep. 8572-6011-RV-000, March, 1964.

(7) Anon., "Preliminary Design of a Mars-Mission Earth Reentry Module - Final Report," NASA-MSC Contract NAS 9-1702, Lockheed Missiles and Space Co., Rep. 4-57-69-1, Feb., 1964.

(8) Shopland, D. J., Price, D. A., and Hearne, L. F., "A Configuration for Re-entry From Mars Missions Using Aerobraking," AIAA Paper 64-480.

(9) Wong, T. J., and Anderson, J. L., "A Preliminary Study of Spacecraft for Manned Mars Orbiting and Landing Missions," SAE-ASME National Air Transport and Space Meeting, New York, N. Y., April 27-30, 1964.

(10) Pritchard, Brian E., "Survey of Velocity Requirements and Reentry Flight Mechanics for Manned Mars Missions," AIAA Paper 64-13.

(11) Syvertson, C. A., and Dennis, David H., "Trends in High-Speed Atmospheric Flight," AIAA Paper 64-514.

(12) Sohn, Robert L., "Manned Mars Trips Using Venus Swingby Mode," Proc. AIAA/NASA Third Manned Spaceflight Meeting (Houston, Texas, Nov. 4-6, 1964) pp. 330-338. AIAA Pub. CP-10, 1964.

(13) Chapman, D. R., An Analysis of the Corridor and Guidance Requirements for Supercircular Entry Into Planetary Atmospheres, NASA TR R-55, 1960.

(14) Wingrove, R. C., and Coate, R. E., Piloted Simulator Tests of a Guidance System Which Can Continuously Predict Landing Point of a Low L/D Vehicle During Atmosphere Re-entry, NASA TN D-787, 1961.

(15) Dow, P. C., Jr., Fields, D. P., and Scammell, F. H., "Automatic Re-entry Guidance at Escape Velocity," ARS Preprint 1946-61.

(16) White, Jack A., Foudriat, E. C., and Young, J. W., "Guidance of a Space Vehicle to a Desired Point on the Earth's Surface," Am. Astronaut. Soc. Preprint 61-41.

(17) Foudriat, E. C., and Wingrove, R. C., Guidance and Control During Direct-Descent Parabolic Reentry, NASA TN D-979, 1961.

(18) Young, John W., and Russell, Walter R., Fixed-Base-Simulator Study of Piloted Entries Into the Earth's Atmosphere for a Capsule-Type Vehicle at Parabolic Velocity, NASA TN D-1479, 1962.

(19) Friedenthal, M. J., "Control of Re-entry From Orbit," Transactions of the 7th Symposium on Ballistic Missile and Space Technology (Air Force Systems Command for Aerospace Systems and Aerospace Corp., Los Angeles, Calif., 1962), Vol. II, pp. 33-87.

(20) Wingrove, R. C., A Study of Guidance to Reference Trajectories for Lifting Reentry at Supercircular Velocity, NASA TR R-151, 1963.

(21) Bryant, J. P., and Frank, M. P., "An Automatic Long Range Guidance System for a Vehicle Entering at Parabolic Velocity," IAS Paper 62-87.

(22) Anon., "A Study of Energy Management Techniques for a High Lift Vehicle," General Electric Co., Air Force Systems Command Tech. Doc. Rep. ASD-TDR-62-77, Vols. 1 and 2, June, 1962.

(23) Austin, Robert W., and Ryken, John M., Trajectory Control and Energy Management of Lifting Reentry Vehicles. Vol. 16 of Advances in the Astronautical Sciences, Sept., 1963, pp. 829-866.

(24) Volgenau, E., Boost Glide and Reentry Control, Chapter 11, Guidance and Control of Aerospace Vehicles, McGraw-Hill Book Company, Inc., New York, 1963.

(25) Wingrove, Rodney C., "A Survey of Atmosphere Re-Entry Guidance and Control Methods," IAS Paper 63-86 (AIAA Journal, Vol. 1, No. 9, Sept., 1963, pp. 2019-2029).

(26) Lessing, Henry C., Tunnell, Phillips J., and Coate, Robert E., "Lunar Landing and Long-Range Earth Reentry Guidance by Application of Perturbation Theory," Second Manned Spaceflight Meeting, Dallas, Texas, April 22-24, 1963, pp. 140-150 (AIAA, Dallas, Texas, 1963).

(27) Wingrove, Rodney C., Stinnett, Glen W., and Innis, Robert C., A Study of the Pilot's Ability to Control an Apollo Type Vehicle During Atmosphere Entry, NASA TN D-2467, 1964.

(28) Tannas, Lawrence E., "A Manual Guidance Scheme for Supercircular Entries," Proc. AIAA/NASA Third Manned Spaceflight Meeting (Houston, Texas, Nov. 4-6, 1964) pp. 79-95. AIAA Pub. CP-10, 1964.

(29) Wingrove, Rodney C., "Trajectory Control Problems in the Planetary Entry of Manned Vehicles. Proc. AIAA Entry Technology Conf. (AIAA, Williamsburg and Hampton, Virginia, Oct. 12-14, 1964) pp. 22-33. AIAA CP-9, 1964.

(30) Morth, Raymond, and Speyer, Jason L., "Divergence From Equilibrium Glide Path at Supersatellite Velocities," ARS Journal, Vol. 13, No. 3, March, 1961, pp. 448-450.

(31) Porter, R. F., "The Linearized Long-Period Longitudinal Modes of Aerospace Vehicles in Equilibrium Flight," Air Force Flight Test Center, TN 61-2, Jan., 1961.

(32) Etkin, Bernard, "Longitudinal Dynamics of a Lifting Vehicle in a Circular Orbit," Air Force Office Sci. Res., TN 60-191, Feb., 1960; also, Journal Aerospace Sci., Vol. 28, 1961, pp. 779-788, 832.

(33) Breakwell, John V., Helgustam, Lars F., and Krop, Martin A., "Guidance Phenomena for a Mars Mission," AAS Symposium on Exploration of Mars, Denver, Colorado, June, 1963.

(34) Hanley, Gerald M., and Lyon, Frank J., "The Feasibility of Spacecraft Deceleration by Aerodynamic Braking at the Planet Mars," AIAA Paper 64-479.

(35) Napolin, A. L., and Mendez, J. C., "Target Orbit Selection for Mars Missions Using Aerodynamic Maneuvering," AIAA Paper 64-14.

(36) Finch, Thomas W., "Aerodynamic Braking Trajectories for Planetary Orbit Attainment," AIAA Paper 64-478.

(37) Hansen, Q.Marion, White, John S., and Pang, Albert Y., Study of Inertial Navigation Errors During Reentry to the Earth's Surface, NASA TN D-1772, 1963.

DISCUSSION

R. C. Wingrove

"Guidance and Control
in
Supercircular Atmosphere Entry".

Q. In what range of densities does your system behave well?

A. If we can determine the scale height within $\pm$ 25% we can use all the vehicle capabilities at Mars where the density is not now well known. Simple adaptive methods may make larger variations acceptable.

Q. What about density variations in the earth's atmosphere?

A. These variations can cut the corridor by a mile or two. However, entry guidance systems can operate even with such deviations. In other words, for a given density variation, there is some corridor, and the guidance systems have been able to use all of whatever corridor is available.

Q. What is the pilot's role in reentry? Have you compared automatic and piloted entry? What is the advantage of the piloted system, if any?

A. We use an automatic system for advanced systems such as Apollo. However, there is always a chance, such as occurred in the last Gemini flight (GT4), that the computer will go out. The pilot should be able to provide for his own safety in such a case. Range control is secondary to safety.

In the case of a research vehicle like X-15 complete manual control is used.

Q. Do you use analog simulation?

A. Most runs are done on hybrid-analog machines. We make a few digital runs to check results. The accuracy of the analog machines is about 1 km.

Q. There is great difficulty in performing analog simulation of functions of several variables such as are needed for the aerodynamics. Do you experience this difficulty?

A. Yes, you are right. Approximations to accomodate these terms may limit simulation accuracy.

THE MOTION OF SPACECRAFT IN THE ATMOSPHERE

D. E. Okhotsimskii and N. I. Zolotukhina

This investigation treats the motion of a space vehicle in the atmosphere, and includes an analysis of the possibilites for using a small lifting force to reduce the demands on reentry accuracy, and to diminish the deceleration load on reentry into the atmosphere at the second space velocity.

Landing trajectories are considered which have a locally parabolic velocity at height $y_0 = 100$ km. The reentry angle Θ (the angle between the transversal and the velocity at height y_0) is uniquely determined by the height h of osculating perigee, that is, by the perigee height of an undisturbed orbit in the absence of atmospheric resistance.

The perigee height uniquely characterizes the reentry conditions for a trajectory of given energy. The choice of the perigee height as a basic parameter enables the outer part of the trajectory to be tied in with the descent through the atmosphere in the most natural way. It is convenient to measure the range and time along the descending portion of the trajectory from the point that would have corresponded to perigee passage in undisturbed motion.

The interval in perigee height for which a descent of prescribed range can be effected is called the "width of the perigee corridor." The wider the corridor for reentry, the lower the accuracy required in the approach of the spacecraft to the earth prior to descent.

Control of the lift of the spacecraft permits the deceleration load to be reduced. For a prescribed level of maximum permissible decelerations, the lift control may be used to ensure as wide a reentry corridor as possible.

We shall regard the lift control as providing for a switching operation from the greatest possible positive value to the greatest possible negative value, and inversely. With this discontinuous mode of varying the lift, the problem reduces to making an optimal choice for the number and timing of the reversal operations.

We consider first the case of descent at a constant lift/drag ratio (no lift reversals). We then consider descents with one, two, or more reversals. The manner in which the deceleration varies during the motion and the position of the principal maximum are analyzed. The amounts by which the deceleration and perigee height vary are determined. With some approximation, the standard atmosphere of [1] was used in the calculations.

1. DESCENT ON CONSTANT MODE

The case of descent with a constant value of the lift/drag ratio is investigated in order to determine the interval in perigee height for which a descent is possible, and to determine the decelerations that would then be suffered. Here and subsequently we shall deal with values for the total deceleration, as defined by the formula

$$n = cv^2 \Delta \sqrt{1 + \varkappa^2}$$

where v is the modulus of the velocity, the lift/drag ratio $\varkappa = c_y/c_x$, c_x and c_y are the aerodynamic drag and lift coefficients, $c = 1/2\ (g_0 \rho_0 S/G)\ c_x$ is the ballistic coefficient, g_0 and ρ_0 are the gravitational acceleration and air density at the earth's surface, S is the characteristic area of the vehicle, G is the weight, and $\Delta = \rho/\rho_0$ is the relative density. For these calculations, the value of the ballistic coefficient has been adopted as $c = 1$ km^{-1}.

Suppose that the vehicle has a lift/drag ratio $\bar{\varkappa}$ at its disposal; that is to say, the ratio may assume any value within the interval of control: $-\varkappa \le \varkappa \le \varkappa$. A negative value of the ratio will be regarded as corresponding to a negative value for the lift.

Let h_1 represent the perigee height of a trajectory having the range L for descent at a constant positive lift/drag ratio, and let h_3 be the perigee height of a trajectory having the same range for descent at a constant negative ratio. The interval $h_1 \le h \le h_3$ in perigee height represents the reentry corridor for the prescribed range L and ratio L and ratio $\bar{\varkappa}$. This means that for any height h in the reentry corridor, we may select that ratio within the interval of control which will yield the range L for descent at constant lift/drag ratio. The reentry corridor is thereby specified solely by the condition that the prescribed range be achieved. If supplementary restrictions are introduced, such as heating or deceleration conditions, the corridor may be narrowed.

The deceleration does not vary monotonically along the trajectory. It develops a series of maxima whose values and positions depend on the values adopted for the lift/drag ratio and the perigee height. The value of the principal maximum is of special importance.

Figures 1 and 2 present the results of calculations performed to establish the potentialities of descents at constant lift/drag ratio. In these figures the abscissa is the perigee height, the left-hand ordinate is the range $L = \varphi R$ (φ is the final angular range measured from perigee and R is the radius of the earth), and the right-

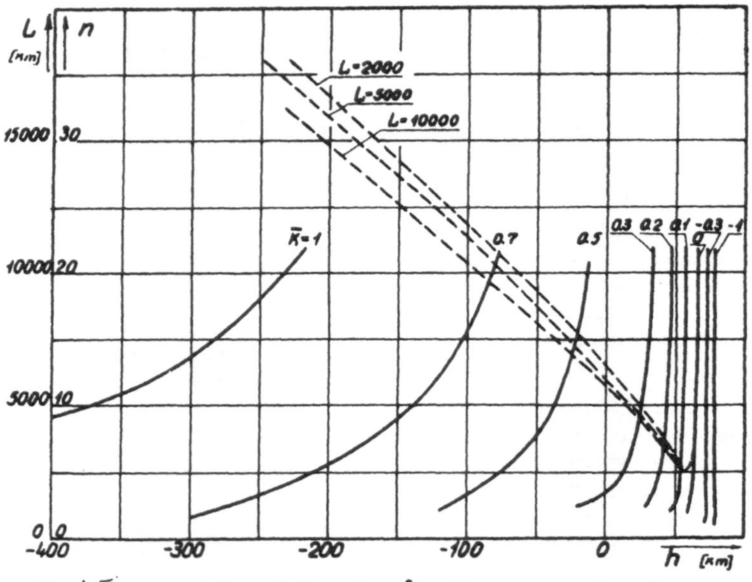

Fig 1. The range L and the deceleration n as a function of the perigee height during re-entry with constant $K = {}^{C_L}/_{C_D}$ ratio

——— The range
– – – Decelerations with the prescribed range.

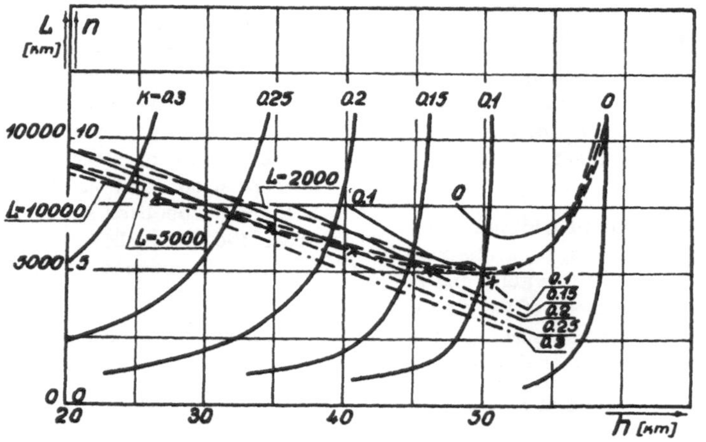

Fig.2. The range L and the deceleration n as a function of the perigee height during re-entry with constant $K = {}^{C_L}/_{C_D}$ ratio ($K \leq 3$).

– – – – – Decelerations for the prescribed range.
—·—·— Decelerations for the prescribed $K = {}^{C_L}/_{C_D}$ ratio.
——— The range.

hand ordinate is the value of the maximum deceleration n. Figure 2 shows on a larger scale the region of perigee heights corresponding to values $\varkappa \leq 0.3$.

The solid curves in the graphs give the dependence of range on perigee height for selected lift/drag ratios. The curves indicate that for descent trajectories of constant $\varkappa$, there is a limiting perigee height $\bar{h}(\varkappa)$ depending on $\varkappa$. Trajectories having perigee heights $h > \bar{h}(\varkappa)$ will not drop to the earth's surface on the first revolution. For trajectories with perigee heights $h < \bar{h}(\varkappa)$, there is a one-to-one correspondence between perigee height and range. As the perigee height decreases, the range decreases.

The families of dashed curves in Figs. 1 and 2 show the maximum deceleration as a function of perigee height for trajectories having selected ranges. In the region of perigee heights of about 50-60 km, these curves display a minimum in the deceleration at $n \approx 5$. The minimal decelerations occur for $\varkappa = 0.1 + 0.15$. The first and second deceleration maxima are close together in this region. For heights $h < 50$ km, the first deceleration maximum is the principal and decisive one, while for heights $h > 60$ km the second maximum is the greater.

Figure 2 also contains a family of curves representing the values of the deceleration at constant lift/drag ratio. These curves consist of two segments separated by a cross. The first segment (thin solid curve) represents the maximum deceleration on landing trajectories. The second segment (dot-dash curve) represents the deceleration encountered during passage through the atmosphere along a path which does not reach the earth on the first revolution. As the figure indicates, the decelerations are relatively mild on the second segments. Similar decelerations can also be achieved on landing trajectories if lift/drag reversal is employed during the descent. We shall return to this point presently. It is noteworthy that according to Fig. 2, the deceleration at given $\varkappa$ varies almost linearly with perigee height for $\varkappa \gtrsim 0.1$.

The maximum deceleration is practically independent of the ballistic coefficient, whose value affects only the height at which maximum deceleration is encountered. If we adopt an exponential law for the variation of density with height, then the displacement of the deceleration maxima in height for a change Δc in the ballistic coefficient may be estimated as

$$\Delta y = H \ln \frac{c_1}{c_0}$$

where H is the scale height of the atmosphere and $c_1 = c_0 + \Delta c$. For example, if the ballistic coefficient increases by a factor of two ($c_1/c_0 = 2$), and if we take a scale height $H = 8$ km, corresponding to deceleration at a height of 50-60 km, we obtain a height displacement $\Delta y = 6$ km in the high-level region. The succeeding calculations have shown estimates of this type to be correct. A change in the ballistic coefficient results in a parallel displacement of the curves in perigee height. The width of the reentry corridor, however, remains invariant.

The graphs presented here show that by considering trajectories with large values of the lift/drag ratio, a reentry cooridor of adequate width can be obtained. However, if the whole corridor is to be utilized the decelerations may reach very high values. The imposing of limits on the deceleration load results in a sharp constriction of the corridor.

Figure 3 illustrates how the width of the reentry corridor depends on the value of the greatest admissible deceleration for descent at constant lift/drag. The abscissa is the value of the admissible deceleration, and the ordinate is the width of the reentry corridor. The solid curves show the reentry-corridor width as a function of the admissible deceleration for selected ranges, while the broken curves correspond to constant values of the lift/drag ratio.

This graph shows the extent to which a decrease in the value of the admissible deceleration n reduces the width of the reentry corridor. For example, if the value of the admissible deceleration is taken as $n = 40$, then for a trajectory with range $L = 5000$ km the width of the reentry corridor would be $\Delta h = 360$ km. In order to obtain this reentry corridor, the vehicle must have a lift/drag $\bar{\varkappa} \geq 1$ at its disposal. If the admissible deceleration is limited to the value $n = 10$, then for the same range the width of the reentry corridor would be $\Delta h \approx 50$ km, and the vehicle would need a lift/drag $\bar{\varkappa} \geq 0.3$. Figure 3 indicates that as the admissible deceleration decreases, there is a corresponding decrease in the value of the lift/drag ratio that must be available in order to achieve a corridor enabling the deceleration to remain within the admissible tolerances.

2. DESCENT WITH LIFT REVERSALS

As mentioned above, a discontinuous mode of lift/drag reversal was adopted to represent the lift control during the descent. Variant descents were considered with different numbers of lift reversals from the maximum positive to the maximum negative value and back. In each case it was assumed that the reversal operation occurs instantaneously; transition effects were neglected.

Two variants were considered with a single reversal: one from negative to positive lift, the other from positive to negative lift.

In the first variant, it was assumed that the reentry of the vehicle into the atmosphere initially proceeds with a lift directed downward ($\varkappa = -\bar{\varkappa}$), so that the aerodynamic force presses the vehicle toward the earth. Then, at time t_1, the direction of the lift changes to a positive sense, and the subsequent descent proceeds with a lift directed upward ($\varkappa = \bar{\varkappa}$). The requirement that a prescribed range be achieved for this part of the descent governs the time t_1 at which the reversal should occur.

In the second case considered for descent with a single lift reversal, it was assumed that the initial reentry into

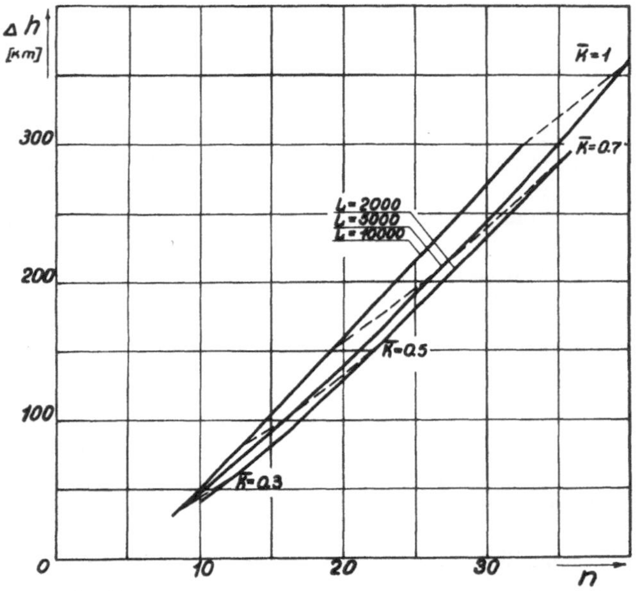

Fig. 3. The corridor width Δh as a function of the value of the permissible deceleration n.
——— Lines L = const.
– – – Lines K̄ = const.

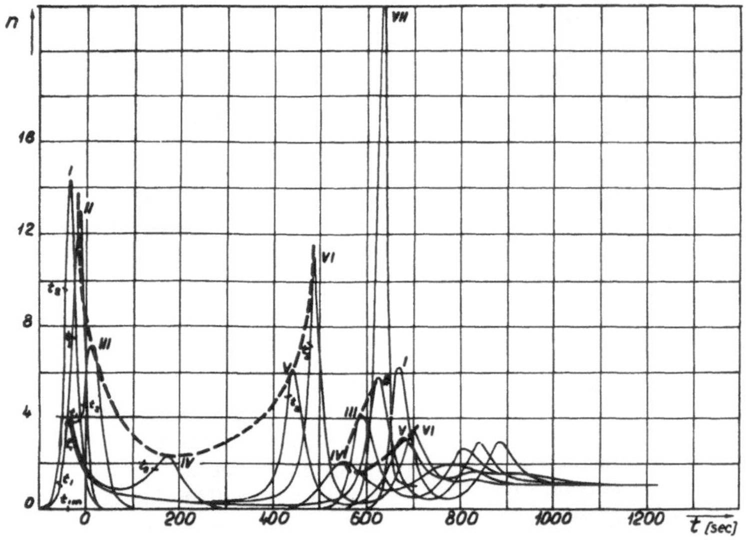

Fig. 4. The deceleration as a function of time.

– – – The envelope of the deceleration maxima.
h = 44 km, κ = 0.5, L = 5000 km.

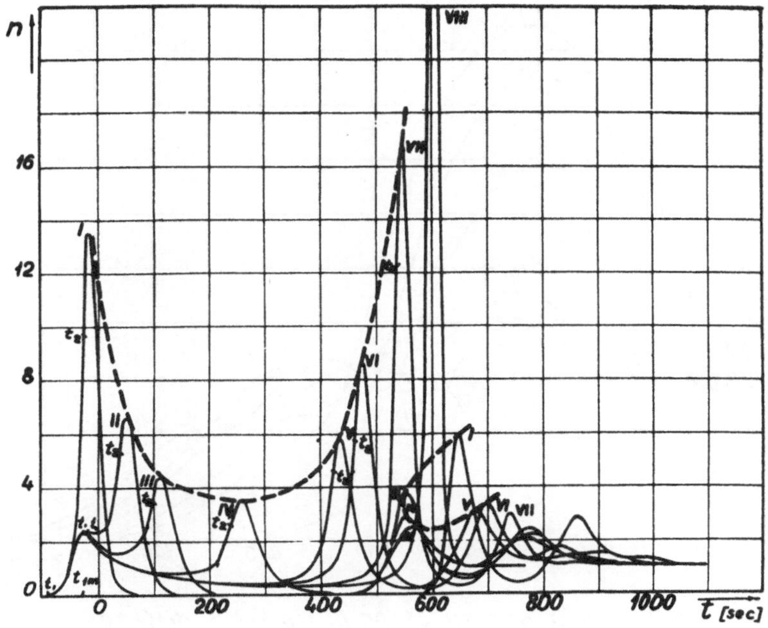

Fig. 5. Decelerations as a function of time for different values t_1.

$h = 54$ km., $K = 0.5$, $L = 5000$ km.

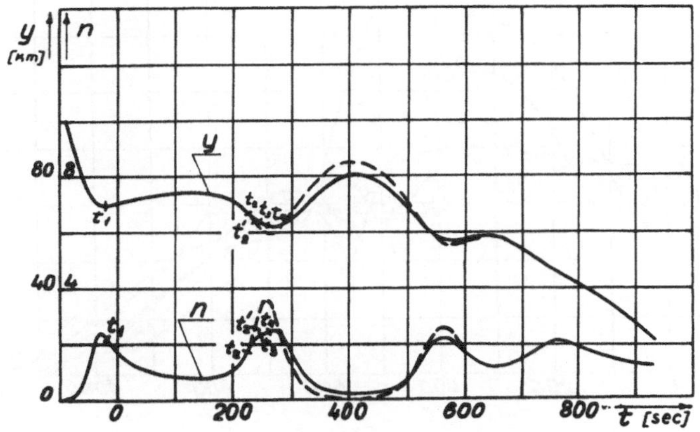

Fig. 6. The deceleration n and the height y as a function of time for the trajectory with the range

$L = 5000$ km, $K = 0.5$, $h = 54$ km.

——— Four switches.
– – – – Two switches.

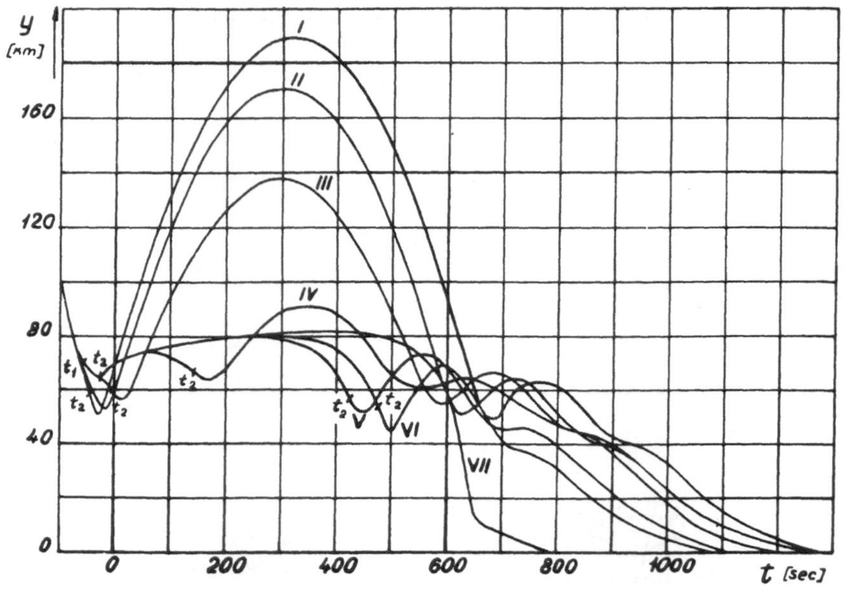

Fig.7. Height as a function of time for different values t_1.

$h = 44$ km, $K = 0.5$, $L = 5000$ km.

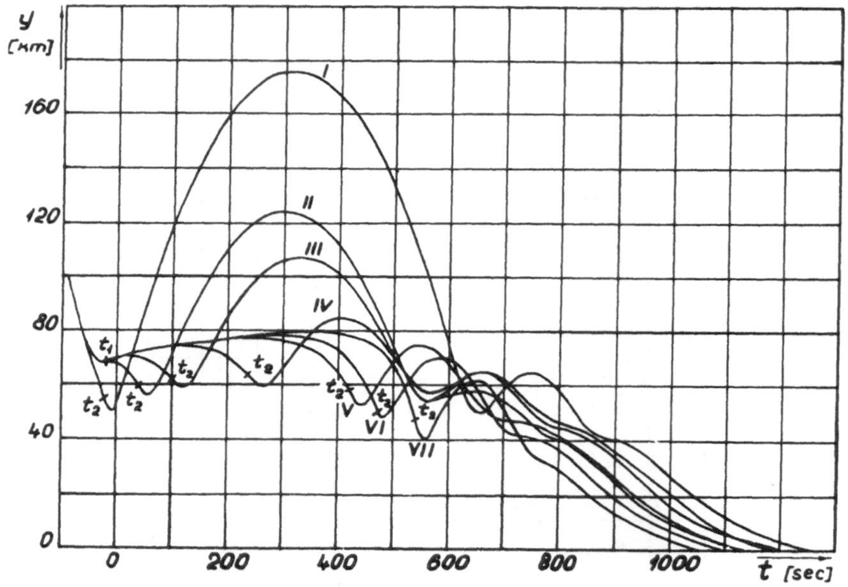

Fig 8. Height as a function of time for different values t_1

$h = 54$ km, $K = 0.5$, $L = 5000$ km.

the atmosphere proceeds with positive lift. At a certain time the lift then instantaneously becomes negative, and remains so until landing. With this mode of descent, very large decelerations arise after the second immersion in the atmosphere, and it is therefore an unreasonable procedure. Nevertheless, the variant is of interest from the standpoint of establishing the time $\bar{t}_1(L)$ at which reversal should occur to yield a given range L. It was found that the time $\bar{t}_1$ is very close to the optimum time for the first reversal in variants with more than one reversal.

The time $\bar{t}_1$ may be called the "controllability limit" for a trajectory of range L, since if $t > \bar{t}_1(L)$ the range L cannot be achieved with any lift reversals (the range would be greater than prescribed, or the vehicle would not land on the earth during the first revolution).

We consider now the case of two changes in the direction of the lift. Let the first part of the descent from height y_0, as far as the first reversal, proceed with positively directed lift ($\varkappa = \bar{\varkappa}$). At time t_1 let the direction of the lift be reversed with the next segment of the trajectory proceeding at negative lift/drag ratio ($\varkappa = -\bar{\varkappa}$). Then at time t_2 the lift is again to become positive, remaining so to the end of the flight.

The physical meaning of this variation in the lift during descent consists in a tendency to diminish the curvature of the trajectory and to prolong the period of deceleration. A protracted deceleration in the upper layers of the atmosphere will cause much of the velocity to be damped out in the upper atmosphere, the time variations in the deceleration to be greatly extended, and the maximum deceleration to be diminished.

The case of two reversals contains two parameters, t_1 and t_2, mutually related by the condition that a given range be obtained. For prescribed values of the lift/drag ratio, range, and perigee height, we obtain a one-parameter family of trajectories. The value of the maximum deceleration differs for the different trajectories of the family. We have to find the trajectory for which the maximum deceleration is least.

For the case of two lift reversals, Figs. 4 and 5 show how the deceleration varies with time for trajectories with range L = 5000 km, lift/drag $\bar{\varkappa} = 0.5$, and the two perigee heights h = 44 km and 54 km respectively. The set of trajectories has one free parameter; we shall take this parameter to be the time t_1 at which the first reversal occurs.

The family of trajectories is bounded by two limiting trajectories. The first limiting trajectory corresponds to descent with only one reversal, from $-\bar{\varkappa}$ to $+\bar{\varkappa}$. It may be regarded as the limiting case where $t_1 = t_0$, with the two-reversal maneuver degenerating to a maneuver with one lift reversal. This trajectory is represented by curve I in Figs. 4 and 5. Another limiting case is a trajectory with a single reversal from $+\bar{\varkappa}$ to $-\bar{\varkappa}$. It may be regarded as the limiting case where the time t_2 of the second reversal satisfies $t_2 = t_{end}$, where t_{end} denotes the concluding moment of the flight. In Figs. 4 and 5, this limit to the family of trajectories is shown by curves VII and VIII respectively. For all trajectories of the family, the time t_1 of the first reversal lies within the interval $t_0 \leq t_1 \leq \bar{t}_1$.

The thick dashed curve in Figs. 4 and 5 joins the maxima of the decelerations for trajectories with different values of t_1. We see from the graphs that both limiting trajectories are very unfavorable with regard to the deceleration.

Consider next how the time dependence of the deceleration develops as the parameter t_1 varies over the region in question. For values of t_1 close to t_0, an increase in t_1 tends to narrow the deceleration maxima. Then, as t_1 passes through a value t_{1m}, representing the time when the first deceleration maximum occurs, an additional maximum develops near the first one (curves II and III in Figs. 4 and 5). As t_1 continues to increase, the isolated first maximum does not change but the value of the second maximum decreases. At the same time all subsequent maxima diminish in value. However, beginning at a certain moment in the immediate neighborhood of the limiting time $\bar{t}_1$, the deceleration begins to increase sharply.

In the first graph presented here (Fig. 4), representing the smaller perigee height, we find that in the cases where the deceleration is minimized, the value of the first maximum is larger than all the subsequent maxima. The time t_1 may be selected in such a way that the subsequent maxima will be as small as possible.

In the second graph (Fig. 5), corresponding to the larger perigee height, we see that for all trajectories close to the optimum one, it is the second deceleration maximum which is greatest.

Let h_2 denote the perigee height of a two-reversal trajectory for which the first deceleration maximum and one subsequent maximum are equal. The corridor may then be divided into two portions with differing types of deceleration minimization.

If the trajectory perigee h falls in the first portion of the reentry corridor, with $h_1 < h < h_2$ (see, for example, Fig. 4), then in order to minimize the deceleration the first reversal should occur after passage through the first maximum, with a judicious choice of the time of reversal from the interval $t_{1m} < t_1 < \bar{t}_1$. The value of the first maximum is well defined and will not be subject to further reduction through use of a more complex reversal maneuver.

The trajectories in Fig. 5 correspond to a perigee height in the second portion of the reentry corridor: $h_2 < h < h_3$. Minimization of the deceleration renders the second deceleration maximum the greatest. By applying a more complicated maneuver with a greater number of reversals, this second maximum may be diminished. As an example of this behavior, Fig. 6 shows trajectories with fourfold reversals. Here the initial trajectory was

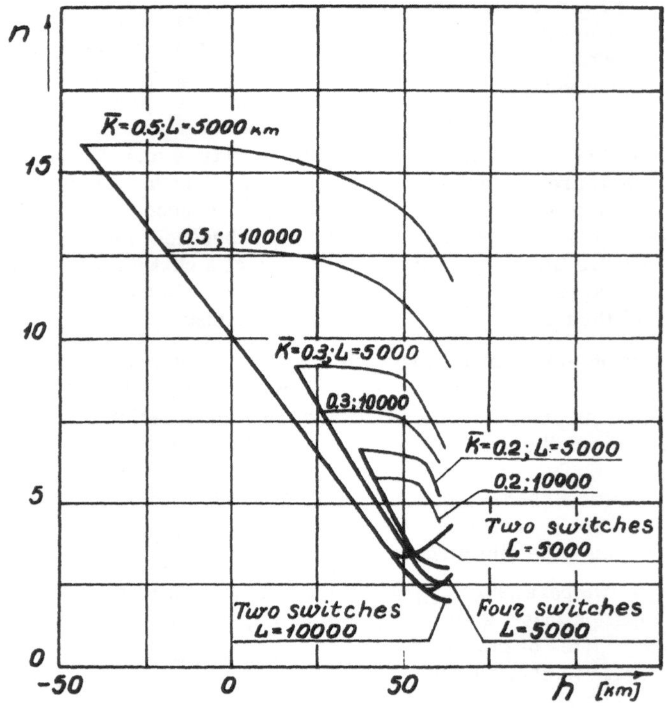

Fig. 9. Boundaries of the region of possible decelerations.

taken to be the optimum trajectory for perigee height h = 54 km, as obtained by minimizing the deceleration in the two-reversal case. The values of the deceleration for this trajectory are represented by curve IV in Fig. 5. A comparison of the two figures shows that by increasing the number of lift reversals to four, one can materially reduce the deceleration.

Figures 7 and 8 illustrate the variation of height with time for the trajectories whose decelerations are shown in Figs. 4 and 5. These graphs show that the first three trajectories rebound strongly, rising to heights y = 150 km or more. All subsequent trajectories, including the trajectory IV that corresponds to optimum deceleration, tend instead to spread out. The minimum height on first immersion increases with increasing t_1. The first trajectory, corresponding to the limiting trajectory I in Fig. 4, penetrates into the atmosphere as far as height y = 50 km. For the optimum trajectories corresponding to the curve IV, the height at first immersion increases to 65-68 km.

The graphs of Figs. 4-8 and the corresponding discussion are intended as examples to characterize the height and deceleration variations for descent trajectories with range L = 5000 km and lift/drag ratio $\bar{\varkappa}$ = 0.5. For trajectories of longer range (say L = 10,000 km) and perigees falling within the second region of the reentry corridor, it is not the second maximum which is the greatest, but the last one, which occurs during the final immersion in the atmosphere.

Just as for the case $\bar{\varkappa}$ = 0.5, the process of minimizing the deceleration was carried out for other values of the lift/drag ratio also. For $\bar{\varkappa} \leq 0.3$, the trajectories differ from those for $\bar{\varkappa} \leq 0.5$ in that the greatest maximum in the second part of the reentry corridor is the final one, while the second maximum can be reduced and made smaller than the first maximum by appropriate selection of the reversal time.

Computations were made for various values of the range and lift/drag ratio. The results are summarized in the next section.

3. LIMITING VALUES OF THE DECELERATION

A summarizing graph, showing the limiting values of the deceleration for descent with lift reversals, appears in Fig. 9. The abscissa is the perigee height, and the ordinate is the value of the maximum deceleration. Each value of the available lift/drag $\bar{\varkappa}$ and range L corresponds to a region in the (h, n) plane.

On each curve bounding a region in the (h, n) plane, the upper left corner corresponds to descent with a constant value of $\bar{\varkappa}$. This corner represents a well defined perigee height h_1, the lower boundary of the corridor in perigee height. For perigee heights less than h_1, the prescribed range cannot achieved by any reversal maneuver using a lift with the given value of $\bar{\varkappa}$.

For perigee heights h > h_1, there is a set of trajectories among which the required range can be attained, and the opportunity arises to diminish the deceleration by means of lift reversals.

On the right-hand side the region is bounded by the upper edge h_3 of the reentry corridor, corresponding approximately to heights $h_3 \approx$ 66-70 km, depending on $\bar{\varkappa}$ and the range. For large $\bar{\varkappa}$ the limiting height h_3 is somewhat greater than for small $\bar{\varkappa}$. Trajectories with perigee heights greater than this cannot descent to the earth's surface on the first revolution by means of aerodynamic forces alone.

The graph of Fig. 9 contains a region of negative perigee height. In this regard, we note that the perigee height h is a characteristic property of the reentry conditions and does not represent the height of closest approach to the earth on the first entry into the dense layers of the atmosphere. For most trajectories, this height is at least 50 km.

The upper boundaries of the regions in Fig. 9 correspond to descent trajectories with a single change of the lift from negative to positive. Each point on a curve represents the maximum deceleration on the corresponding descent trajectory, at a perigee height equal to the value of the abscissa for that point. The perigee height corresponds uniquely to the time of reversal for which the trajectory has the prescribed range. The graph shows that introducing a single change of lift from negative to positive results in a marked reduction of the deceleration only for perigee heights close to h_3.

Decelerations on the upper boundaries of the regions in Fig. 9 represent the greatest limiting values of the deceleration for descent with two reversals, where the time of the first reversal $t_1 \to t_0$, with the two-reversal maneuver degenerating into a single reversal. The points located below the upper boundary of a region correspond to descent with two reversals for a perigee height equal to the abscissa of each point.

The upper boundaries of the regions in Fig. 9 depend on the range. As the range increases, the deceleration decreases and the upper boundaries are displaced downward. In this process the corner point, corresponding to descent with constant lift/drag ratio, moves along the lower left boundary of the region, a common boundary for the entire family of regions having the same $\bar{\varkappa}$ but different ranges.

Of particular interest is the lower boundary of each region, for it represents the minimum values of the deceleration. The lower boundary of a region may be divided into two segments, corresponding to the two portions of the reentry corridor. The first segment, representing the first portion of the reentry corridor, begins at the corner point and is practically linear. In the second portion of the reentry corridor, the lower boundary departs markedly from a straight line.

The first deceleration maximum is the greatest and dominating one in the first portion of the reentry corridor; subsequent maxima do not exceed the first one. For the trajectories corresponding to the first portion, the first reversal occurs at a time after the first deceleration maximum has been passed; thus the first maximum occurs at positive lift. It follows that the deceleration at the first maximum in this case must coincide with the deceleration at the first maximum for the descent trajectory with constant lift/drag ratio equal to the adopted value $\bar{\varkappa}$. This means that the first segment of the lower boundary in Fig. 9 must coincide with the corresponding thin curve in Fig. 2 (becoming a dot-dash curve) for the same $\bar{\varkappa}$.

In the first portion of the reentry corridor, no maneuver utilizing subsequent reversals can reduce the value of the first maximum, which is well defined. From the standpoint of minimizing the maximum deceleration, then, there is no need to employ maneuvers with more than two reversals in this part of the corridor.

The lower boundary in the first portion depends only on the available lift/drag ratio. In the second portion, the lower boundary may depend on both the range and the reversal procedure. The graph of Fig. 9 shows that an increase in the range and an increase in the number of reversals may result in a reduction in the maximum deceleration within the second portion. However, there exist certain limiting values below which the maximum decelerations at the final maximum, and numerically are close to the values of the deceleration upon reentry from descending satellite orbits [2]; we have, approximately, $n = 2$ for $\bar{\varkappa} = 0.5$, $n = 2.5$ for $\bar{\varkappa} = 0.3$, and $n = 3$ for $\bar{\varkappa} = 0.2$.

Figure 9 indicates that for the values $\bar{\varkappa} = 0.2$ and 0.3, the lower boundary in the second portion reaches a limiting minimum position for the ranges 5000 and 10,000 km respectively.

The entire lower boundary is a common one for each family of regions, depending on the value of $\bar{\varkappa}$. The lower boundary itself can be reached by using maneuvers with two reversals.

It also follows from Fig. 9 that for the value $\bar{\varkappa} = 0.5$, the lower boundary depends on both the range and the reversal procedure. For the range $L = 10{,}000$ km, computations have shown that the lower boundary occupies a limiting position for the two-reversal maneuver. For the range $L = 5000$ km, the lower boundary of the region in the second portion lies above the limiting position in the two-reversal case. An increase in the number of reversals to four, as the graph shows, enables the lower boundary to reach its limiting position.

REFERENCES

1. All-Union Standard Atmosphere 4401-64 (Moscow, Standards Press, 1964).
2. Chapman, "An approximate method for investigating the entry of bodies into planetary atmospheres" [Russian translation] (Moscow, Foreign Lit. Press, 1962).

DISCUSSION

D. E. Okhotsimski and N. I. Zolotukhina

"Investigation of
the Motion of
Space Vehicles
in the Atmosphere ".

Q. What is the effect of finite delays in switching?

A. It's effect is not important.

Q. Your studies are of open-loop maneuvers. Have you made closed-loop studies?

A. The closed-loop maneuvers have been considered only to determine principles of operation. Practical results will depend on the devices used to implement the system. One such implementation was considered in the paper by Wingrove.

CERTAIN NONLINEAR LAWS IN THE CONTROL OF A WINGED GLIDE VEHICLE IN TRANSITION FROM A CIRCULAR ORBIT TO A TAKEOFF AND LANDING STRIP

V. S. Bedrov, G. P. Vadichin, A. A. Kondratov,
G. L. Romanov, and V. M. Shalaginov

INTRODUCTION

The fundamental problems encountered in bringing a winged spacecraft down for landing on an airstrip after descent from a circular orbit comprise the following:

1. It is necessary, by the dissipation of energy, to lower the velocity of the vehicle from the first escape velocity to a value approximating the landing speed.
2. It is necessary to reduce the initial variability of the parameters (range, height), which may be as high as several hundred kilometers on entering the denser layers of the atmosphere (70-80 km altitude), to very minimal values on the order of a few hundred meters.
3. The maneuverability of the vehicle is extremely limited in the upper layers of the atmosphere by the insufficiency of aerodynamic forces, in the lower layers by the durability and heating problems. The "corridor" in coordinates (H, V) to which the vehicle must be confined in descent is very narrow.

The present paper is devoted to a consideration of certain nonlinear control laws governing center of mass motion, which permit these difficulties to be surmounted. The problems involved in acquiring the required information are not treated in the paper. It is assumed that suitable on-board or ground-based instrumentation is available for the acquisition and processing of the required information, as well as generation of the control signals to be delivered to the autopilot.

Longitudinal motion is considered in the first part of the paper, the control of lateral motion in the second part.

I. CONTROL OF LONGITUDINAL MOTION

1. Control of Longitudinal Motion and General Considerations Affecting the Choice of Descent Trajectory

We will consider the dependence of the range on the initial conditions and mode of descent. The control of a winged vehicle in the longitudinal plane takes the following form (neglecting the earth's rotation):

$$m\frac{dv}{dt} = -C_x S q - G\sin\theta$$

$$mv\frac{d\theta}{dt} = C_y S q - G(1-\bar{V}^2)\cos\theta$$

$$\frac{dH}{dt} = V\sin\theta \tag{1}$$

$$\frac{dL}{dt} = V\cos\theta$$

where

$$\bar{V}^2 = \frac{V^2}{g^2}, \quad r = R_0 + H, \quad g = \frac{\mu}{r^2}$$

$$R_0 = 6371.2\ \text{km} \quad \mu = 3.986\times10^5\ \text{km}^3/\text{sec}^2$$

The following equation is obtainable from the system (1):

$$K\,dE + (1-\bar{V}^2)\,dL + \frac{V^2}{g}\,d\theta = 0 \tag{2}$$

where $K = C_y/C_x$ is the lift/drag ratio, $E = H + V^2/rg$ is the specific mechanical energy of the vehicle (total energy per unit weight).

From this we have

$$L = \int_{E_K}^{E_H} \frac{K}{1-\bar{V}^2}\,dE + \int_{\theta_K}^{\theta_H} \frac{V^2}{g(1-\bar{V}^2)}\,d\theta \tag{3}$$

where E_b and θ_b are the values of the specific mechanical energy and trajectory angle at the beginning of descent, E_f and θ_f are the values of these parameters at the end of descent.

The first term in Eq. (3) defines the flight range due to the expenditure of mechanical energy, the second is the range due to ballistic glide.

The range with K = const due to a decrease in kinetic energy can be computed from the equation ($r \approx R_0$)

$$L_V = \frac{R_0}{2} K \ln \frac{1-\bar{V}_K^2}{1-\bar{V}_H^2}$$

To estimate the ballistic range we compare it with the range obtained as the result of decreased mechanical energy. We obtain thereby

$$\frac{L_E}{L_\theta} = \frac{g K_{cp}}{V_{cp}^2} \frac{H_H - H_K + \frac{V_H^2 - V_K^2}{2g}}{\theta_H - \theta_K} > \frac{K_{cp}\left(1 - \frac{V_K^2}{V_H^2}\right)}{r(\theta_H - \theta_K)}$$

Calculations based on this equation for a vehicle with high lift-to-drag, on the order of two or three, show that the ballistic range comprises but a small fraction of the total range.

It is convenient in the analysis of L_E to make use of a plane on which are plotted a gridwork of isoenergic lines, lines of constant velocity head $g = \rho V^2/r$ = const, lines of constant temperature or pressure at the critical point, etc. The descent of a space vehicle can also be represented by a definite curve in the coordinates H, V. In order to execute landing of the vehicle at the designated impact point it is necessary to confine the vehicle to motion along a predetermined curve in the coordinates H, V. This problem can be solved for a broad class of curves H(V). The permissible descent curves H(V) must not intersect the same curve E = const twice. The descent curve H(V) must not have sudden discontinuity-type changes, since this would require inadmissibly large overloads in practice. Finally, certain regions in the H, V plane may prove to be forbidden for a certain specific vehicle, either due to very large overloading, excessively high temperatures, or due to a small velocity head. In other words, the curves H(V) must be confined to a definite, permissible landing corridor.

We will assume that the descent curve H(V) has been chosen and that it meets the above requirements. At each point of this curve the quantities g, E, Mach number M, and all aerodynamic characteristics of the vehicle are known. Consequently, at each point we can determine C_y, α, C_x, and $K = C_y/C_x$. This means that we can formulate the function $K/(1-\bar{V}^2)$ in terms of E, thus computing the approximately the range corresponding to descent along the designated curve H(V). Of particular interest is the curve $H_m(V)$ for flight proceeding at maximum lift-drag. If the initial and end points of the descent path lie on the curve $H_m(V)$, the first term of the integral (3), i.e., the magnitude of the energy range L_E, will attain a maximum when integration is performed along the curve $H_m(V)$, since on any other curve $K < K_{max}$. In other words, to attain maximum L_E it is necessary to travel with maximum lift-drag.

If the vehicle has an inertial navigation system or any other means for measuring its altitude and flight velocity, it is then possible to realize descent along any other admissible curve H(V). In particular, we are interested in the curves $H_U(V)$ and $H_L(V)$ for descent along the upper and lower boundaries of the landing corridor, respectively, If the curve $H_m(V)$ lies inside the landing corridor and never intersects the curve $H_U(V)$ and $H_L(V)$, then in flight along the curves $H_U(V)$ and $H_L(V)$ the range is smaller by comparison with the mode of descent along the curve $H_m(V)$.

Based on the above discussion, a series of simple and uniformly effective schemes can be devised for the automatic landing of a vehicle at the impact point.

2. Automatic Control System for Landing a Vehicle in the Intended Impact Area

The automatic control system can be broken down into two main loops (Fig. 1), the internal, or autonomous, loop and the external, or landing-control, loop. The autonomous loop is designed to stabilize the rapid angular motion of the vehicle about its center of mass and to stabilize the variation in height in correspondence with the variation in velocity, i.e., to stabilize the predetermined descent curve H(V). The landing-control loop must generate the required curve H(V) in order to bring the vehicle to the intended impact point. We will consider the autonomous subsystem first. Let us assume that the flight altitude and velocity are known on board the vehicle and, in addition, that there is a device capable of generating a certain function of the form

$$\sigma^{\cdot} = \sigma(H, V^{\cdot})$$

For example, at low velocities the parameter for which we are looking, one that is directly measurable in flight, might be the dynamic pressure, which is equal to the pressure difference between the critical point and the static pressure. The dynamic pressure, of course, depends chiefly on H and M, i.e., on H and V.

It is the task of the autonomous loop to stabilize the basic flight-determinative parameters in correspondence with the signals $\sigma(H, V)$ received from the external loop. The law governing this stabilization will not be considered in the present paper. We simply note that in a number of instances linear stabilization may be assumed, but that in general it is desirable to have nonlinear stabilization functions and, in particular, a self-adaptive autopilot to cope with the large variations in performance of the control units and aerodynamic characteristics due to the large variation of the M number and velocity head. We now turn to the external loop.

Let us suppose that we are able to choose a curve $H_0(V)$ situated between the curve $H_m(V)$ and $H_U(V)$ or between $H_m(V)$ and $H_L(V)$. It would be desirable for this curve to satisfy the condition $\sigma = \sigma_0$ = const, since in this case we are not required to program σ according to V. The choice of the curve H_U or H_L is dictated by special considerations relating to their specific nature and having no particular bearing on the problems in question, for example, the problems of heating or durability. We will assume below that the reference curve $H_0(V)$ is chosen between $H_m(V)$ and $H_L(V)$. We will adopt the curve $H_0(V)$ and corresponding program $\sigma_0(V)$ as our

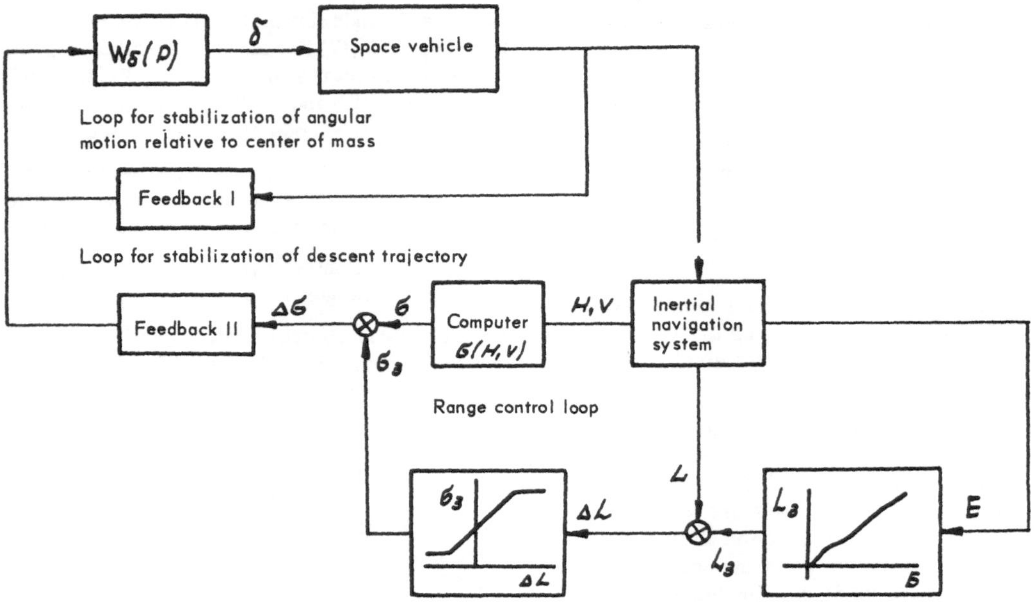

Fig. 1. Block diagram of system for control in longitudinal plane.

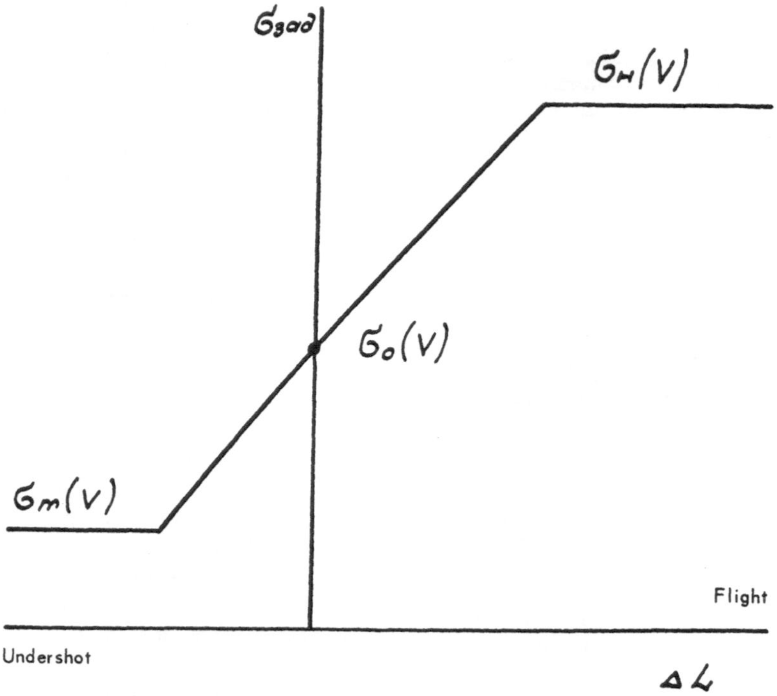

Fig. 2. Dependence of the control signal on the magnitude of the mismatch $\Delta L = L - L_{giv}$ (E).

nominal precomputed curves and descent program. In flight executed according to this program, descent will proceed along the predesignated trajectory and, hence, the range will be known. We will assume, further, that during flight the quantity $E = H + V^2/rg$ is computed on board on the basis of the known H and V. Inasmuch as the function L in terms of E can be simplified for the nominal trajectory, in which case it is denoted by L_{giv}, the control signal σ_{giv} can be generated on the basis of the deviation of the actual range from the prescribed (given) range for the then-existing reserve E, according to the equation

$$\sigma_{giv} = f(\Delta L) = f\left(L - L_{giv}(E)\right)$$

The function $f(\Delta L)$ could have the form shown in Fig. 2. A block diagram of the guidance system is shown in Fig. 1. If on entering the atmosphere the vehicle undershoots the computed distance to the airfield, the flight will be programmed according to $\sigma_m(V)$.

The energy under these conditions will be dissipated more economically than in flight according to the nominal program $\sigma_0(V)$. In the last analysis, at some instant the actual distance to the airfield will become equal to the prescribed value for the vehicle's existing reserve of energy, after which the flight can be continued according to the reference program $\sigma_0(V)$ to such an altitude that the process of leveling out for landing approach can be instituted. Then the velocity at this altitude will be near the nominal value, regardless of the initial deviations in range and velocity vector from the precomputed values, or the effect of external perturbations. An analogous pattern will be observed in the case of overshooting the distance. Flight will proceed at first according to the program $\sigma_L(V)$, then subsequently according to $\sigma_0(V)$.

Calculations performed on a digital computer for the purpose of investigating such an automatic landing system show that the miss distance for a vehicle with $K_{max} \approx 2$ or 3 attains a few kilometers.

It is not too difficult to see that the given automatic landing system is basically almost optimal from the point of view of realizing landing within the maximum permissible range of initial conditions prior to orbital ejection, since in this case the maximum feasible maneuvering capabilities of the vehicle are utilized as fully as possible under given limitations.

3. Ejection of the Vehicle into Reference Trajectories

In general the initial height differs considerably from that required for realization of glide along a prescribed curve $\sigma(H, V)$. At the beginning of descent, therefore, is necessary to restrict the transition process to the prescribed curve.

The elimination of large initial deviations can be accomplished by stabilization of the constant given values of the control parameter. The times of transition from one value of the parameter to another depend on the combinations of initial conditions (H_0, V_0, θ_0, and L_0), i.e., on the initial state vector, and can be computed on digital equipment by means of a special control algorithm.

It is clear that the execution of an optimal transition process in the sense of real time is particularly important with the need for realizing minimum glide range, i.e., in flight according to $\sigma_L(H, V)$.

Possible control parameters would be the initial angle of attack, overload ratio n_y, rudder deflection angle δ_b, etc.

We will consider the simplest case, when the overload n_y is chosen as the control parameter.

The instants for switching of the control parameters can be determined by means of the maximum principle.

Calculations for transition processes demonstrate the possibility of introducing the following simplifications in the system of equations describing the process:

$$V = V_0 = \text{const}$$
$$\sin\theta = 0$$
$$\cos\theta = 1$$
$$g = g_{av} = \text{const}$$

These simplifications reduce the system of equations to one of second order.

It follows from the maximum principle that in this case the optimum trajectory is obtained when the maximum value of the control parameter is given with one sign, while at some instant of time t_1 the sign of the control parameter reverses.

Inasmuch as the load n_y at very large heights is small, we will assume that in the first segment it has a value $n_y = 0$. In the second segment the overload $n_y = n_{ymax}$.

Integration of the simplified system of equations for the time interval in which n_y = const leads to

$$\theta = \frac{g_{av}}{V_0}(n_y + \bar{V}_0^2 - 1)t + \theta_0$$
$$H = g_{av}(n_y + \bar{V}_0^2 - 1)\frac{t^2}{2} + V_0\theta_0 t + H_0 \qquad (4)$$

Eliminating t, we get

$$H = \frac{V_0^2}{2g_{av}(n_y + \bar{V}_0^2 - 1)}\theta^2 + S_1$$

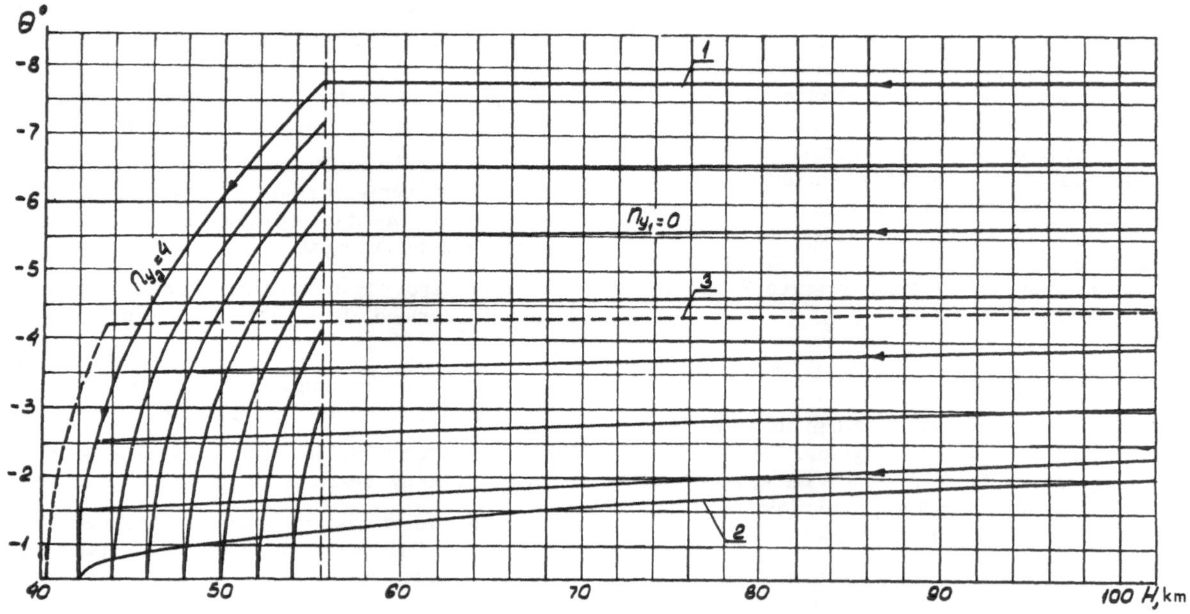

Fig. 3. Phase trajectories of the transition process to the curve σ(H, V).

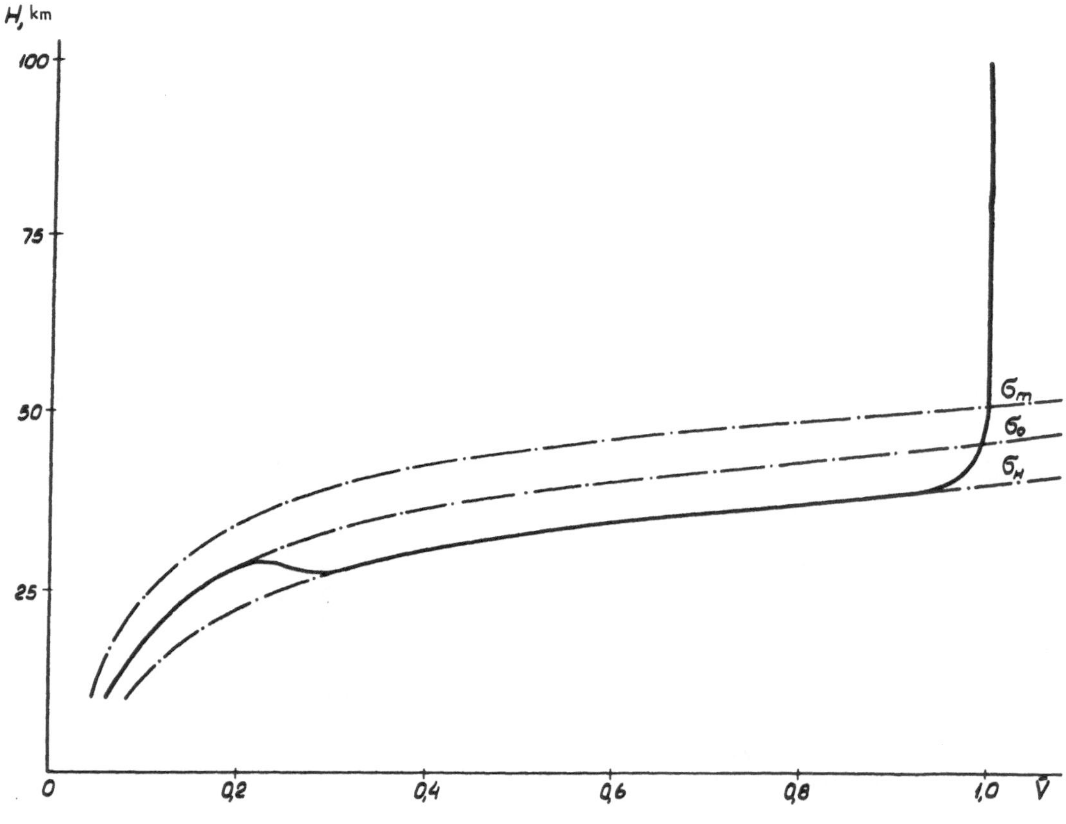

Fig. 4. Trajectory of the vehicle in coordinates H, V.

where

$$S_1 = H_0 - \frac{V_0^2 \theta_0^2}{2 g_{av} (n_y + \bar{V}_0^2 - 1)}$$

Figure 3 shows the phase trajectories corresponding to the transition to the curve σ(H, V) in the case when $n_{y1} = 0$ and $n_{y2} = 4$.

It should be borne in mind that after a certain height an overload $n_{y2} = 4$ can only be attained for $\alpha > \alpha_{max}$. This corresponds to curve 1 in Fig. 3, which is obtained for the maximum permissible angle of entry $|\theta_0|_{max}$. Curve 2 determines the minimum angle of entry θ_{0min} (in this case the time during which the overload n_{y2} is effective is equal to zero).

Solving the system (4), we obtain the necessary effective time of the overload n_{y1} and the total time of the process:

$$t_{1_1} = -\frac{V_0}{g_{av}(n_{y_1} + \bar{V}_0^2 - 1)} \left\{ \theta_0 + \left[\frac{\theta_0^2 (n_y + \bar{V}_0^2 - 1) - \theta_K^2 (n_{y_1} + \bar{V}_0^2 - 1)}{n_{y_2} - n_{y_1}} - \frac{2 g_{av}}{V_0^2} \frac{(H_0 - H_K)(n_{y_1} + \bar{V}_0^2 - 1)(n_{y_2} + \bar{V}_0^2 - 1)}{n_{y_2} - n_{y_1}} \right]^{\frac{1}{2}} \right\}$$

$$t_{\Sigma_1} = \frac{V_0}{g_{av}} \frac{\theta_K - \theta_0}{n_{y_2} - \bar{V}_0^2 - 1} - \frac{n_{y_1} - n_{y_2}}{n_{y_2} + \bar{V}_0^2 - 1} \cdot t_{1_1} \qquad (5)$$

In addition to computation of the times t_{1_1} and t_{Σ_1}, it is necessary to decide upon a curve σ(V, H) guaranteeing the prescribed glide range (see sec. 2).

The exact values of the times t_1 and t_Σ will differ from those computed according to Eq. (5), which are obtained from the simplified equations. The principal error will be introduced by failure of the velocity to remain constant, which becomes especially pronounced when the velocity head is large, i.e., in the second segment of the trajectory.

All of the indicated quantities [t_1, t_Σ, σ(V, H)] must be computed by the method of successive approximations. The calculations show that the curve σ(V, H) is described with the second approximation, the values of t_1 and t_Σ with the fourth.

Figure 3 shows the trajectory obtained by this method of calculation (curve 3).

Figure 4 shows one of the trajectories obtained with this method of control. As is apparent, the process proceeds without overcontrol.

The impossibility of generating this type of process in linear systems is patent; either the process is protracted or it is subject to overcontrol.

The use of an optimal system makes it possible to extend considerably the admissible variations in initial conditions (θ_0 and L_0) by comparison with linear control systems.

II. OPTIMAL CONTROL IN LATERAL MOTION

1. Synthesis of the Optimal Law

The synthesis of optimal lateral control for the landing of a gliding object on an airstrip reduces to the problem of finding a control law that will transfer the object from the initial point with phase coordinates S_0, L_0, ψ (Fig. 5) to a final position with coordinates $S = 0$, $\psi = 0$ in the minimum amount of time. The optimal law in this sense is one that permits the object to be taken from the largest range of initial conditions defined by the maneuverability of the object. Consistent with the existence and uniqueness theorem [1] for optimal control, there exists a function $\gamma = F(L, S, \psi)$ (γ is the banking angle), depending only on the instantaneous phase coordinates, such that it defines all optimal trajectories. The need for seeking the optimal control in the form indicated is dictated by the fact that in the course of its motion the object will be subject to the action of various perturbations which are not known beforehand. In its postulated form, the control law continuously computes the new optimum trajectory, taking into account the new initial conditions, thus compensating the effect of the disturbances.

We assume the following form for the system of differential equations describing the lateral motion of the vehicle:

$$\begin{aligned} \frac{dS}{dt} &= -V \sin \psi \\ \frac{dL}{dt} &= -V \cos \psi \cos \gamma \\ \frac{de}{dt} &= -V \\ \frac{d\psi}{dt} &= -\frac{g}{V} \sin \gamma \end{aligned} \qquad (6)$$

The system (1) was obtained on the assumption of a small angle of inclination θ on the part of the trajectory and small derivative $d\theta/dt$, so that $\sin \theta \approx \theta$, $\cos \theta \approx 1$, and $d\theta/dt \approx 0$; also, the effects of the earth's rotation is neglected.

For our further discussion we assume that the banking angle γ, which determines the available lateral overload, is limited to some value at which the influence of lateral motion on the longitudinal component is small. Without giving the computations, we assert that this can be realized if $-\gamma_0 \le \gamma \le \gamma_0$ ($\gamma_0 = 0.3$), in which case it may be assumed that $\sin \gamma \approx \gamma$, $\cos \gamma \approx 1$. Furthermore, for linearization of the sytem (1) we assume $\sin \psi \approx \psi$, $\cos \psi \approx 1$. We note that the system can be completely integrated without the last assumption, but the results obtained thereby, as the calculations show, differ slightly from those obtained with this assumption.

After all the simplifications have been incorporated, we obtain

$$\frac{dS}{dL} = \psi$$

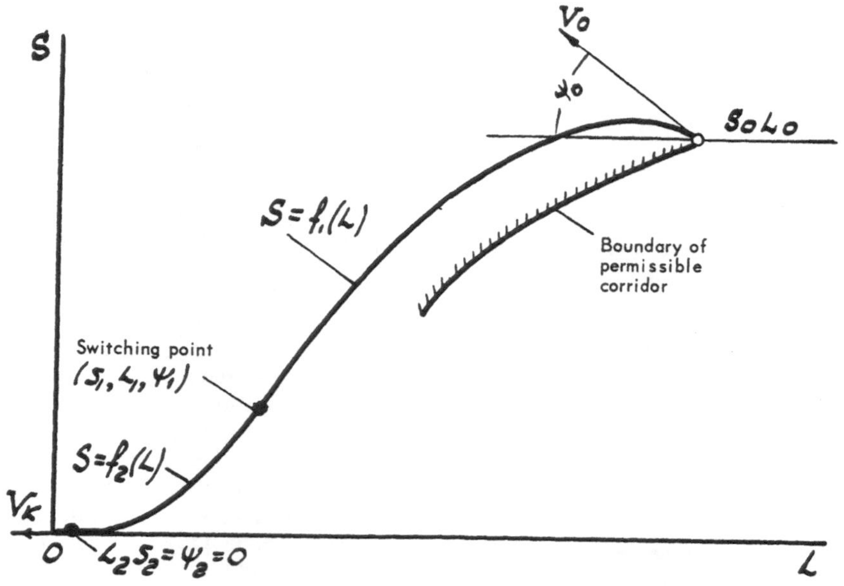

Fig. 5. Phase trajectory in the horizontal plane.

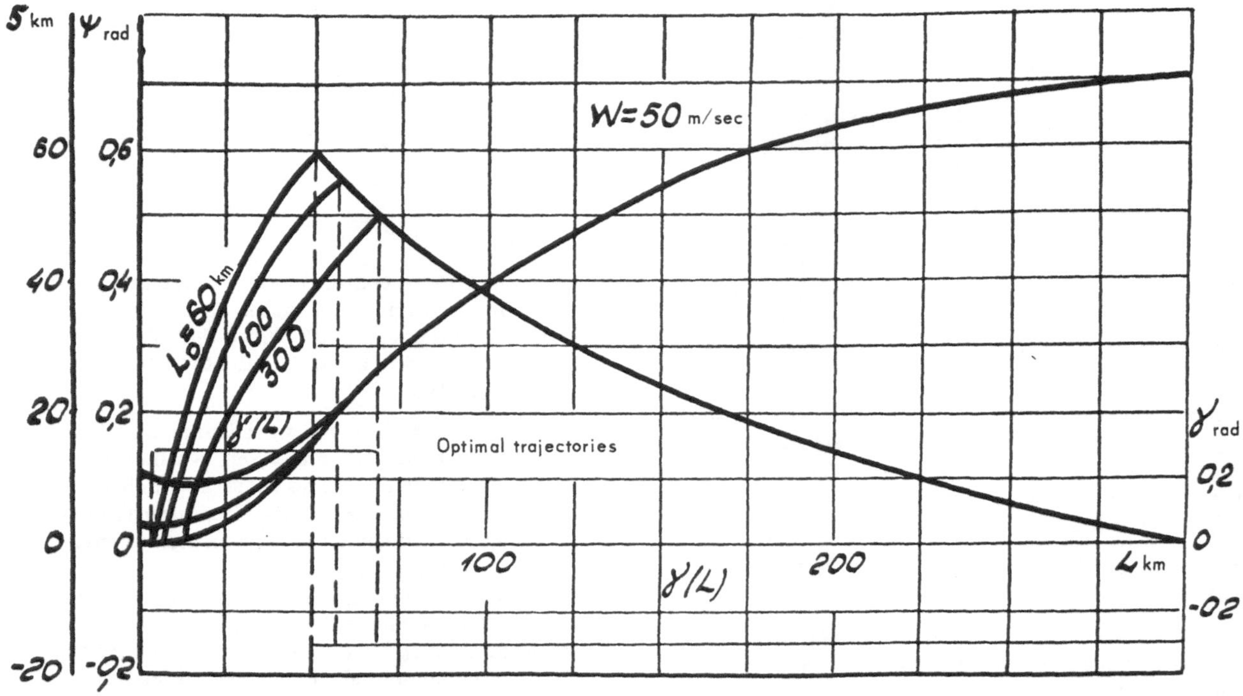

Fig. 6. Trajectories under the action of a steady wind with velocity W = 50 m/sec.

$$\frac{d\psi}{dL} = \frac{g}{V^2}\gamma$$

For the system of equations (2) we will seek the optimal control $\gamma = \gamma(L)$ and the optimum trajectory under the initial conditions $S(L_0) = S_0$, $\psi(L_0) = \psi_0$ for which at some range (the coordinate L_2 is not fixed) the final conditions $S(L_2) = 0$, $\psi(L_2) = 0$ would be satisfied. On the basis of L. S. Pontryagin's maximum principle [1] we find that the function ensuring optimal control in terms of real time response is determined by the equation

$$\gamma(L) = \gamma_0 \operatorname{sgn}(L_1 - L) \tag{8}$$

In this equation L_1 is the range corresponding to the time at which the bank is altered. The optimal control is determined by a piecewise constant function, which assumes the values $\psi = \pm\gamma_0$ and has two intervals of constancy (if there are no external pertubations).

Appearing on the right-hand side of the second equation in the system (2) is the square of the velocity, which is a highly variable quantity. Numerous calculations have shown that a relatively simple analytical expression can be found to describe the range dependence of the velocity, the majority of these functions being contained within the limits of the defined expressions

$$V^2 = K'(L+a) \tag{9'}$$

$$V = K_1'(L+a_1) \tag{9''}$$

Here K^1, K_1^1, a, a_1 are parameters which are functions of the initial glide range.

Let us find a solution of the system (2) defining the trajectory of a glide vehicle from the initial to the final state with a parabolic law relating the velocity to the range, i.e., we wish to solve Eqs. (2), (3), and (4) simultaneously with the following final conditions:

$$\begin{aligned} L = L_0 - \psi = \psi_0 \quad &; \quad S = S_0 \\ L = L_2 - \psi = 0 \quad &; \quad S = 0 \end{aligned} \tag{10}$$

Since the banking angle has two fixed intervals of constancy $\gamma = \gamma_0$ and $\gamma = -\gamma_0$, the trajectory will consist of two segments. Recognizing the stipulation that the solutions must merge at the junction of the two segments, and integrating the system, we obtain

$$L_1 = e^{\psi_0/K}(L_0+a) - a - \Big[e^{2\psi_0/K}(L_0+a)^2 - -\frac{1}{K}e^{\psi_0/K}\big[(L_0+a)^2(\psi_0+K) - S_0(L_0+a)\big]\Big]^{\frac{1}{2}} \tag{11}$$

The coordinates S_0, L_0, ψ_0 can be taken as the instantaneous coordinates at every instant of time, hence switching must be executed at the instant the computed value of the coordinate L_1 becomes equal to the instantaneous value of the coordinate L_0.

Consequently, the control signal delivered to the autopilot input is defined by the formula

$$\gamma_{giv} = \gamma_0 \operatorname{sgn}\phi(S_0, L_0, \psi_0) \tag{11'}$$

where the contron function ϕ is equal to

$$\phi(S_0, L_0, \psi_0) = L_0 - L_1 = L_0 + a - e^{\psi_0/K}(L_0+a) + \\ + \Big[e^{2\psi_0/K}(L_0+a)^2 - \frac{1}{K}e^{\psi_0/K}\big[(L_0+a)^2(\psi_0+K) - \\ - S_0(L+a)\big]\Big]^{\frac{1}{2}} \tag{12}$$

The switching time corresponds to the equation

$$\phi(S_0, L_0, \psi_0) = 0$$

Realization of the control law such that the problem of optimal descent is solved in the form stated can be accomplished in a loop including the following elements:
1) a ground-based or on board instrument providing the necessary primary information (L_{meas}, S_{meas});
2) a computer, which together with the generation of a control signal must process the primary information (smoothing, differentation, noise filtering, etc.);
3) the autopilot, which executes stabilization of the vehicle relative to its center of mass and control in compliance with the command signals.

2. Analysis of Control Quality of External Perturbations

It is essential in the generation of the control signal to ensure high accuracy of control when a variety of disturbances are acting on the system. The most probable perturbations in the lateral motion control of flying craft are crosswind, errors in maintaining the prescribed banking angle, and deviation of the actual variation in velocity from the assumed law used to determine the control function.

The system of equations describing the motion of the vehicle in the field of a steady crosswind directed along the axis can be written in the form

$$\begin{aligned} \frac{dS^*}{dt} &= \psi^* \\ \frac{d\psi^*}{dt} &= \frac{g}{V^2}\left(\gamma + \frac{W}{g}\frac{dV}{dL}\right) \end{aligned} \tag{13}$$

where W is the wind velocity, S^*, ψ^* are the parameters of the motion in the wind field.

It follows from Eq. (13) that the effect of the wind is equivalent to a variation in the prescribed vehicle bank by an amount

$$\Delta\gamma = \frac{W}{g}\frac{dV}{dL}$$

Trajectories in the coordinates S, L and in the phase plane are shown in Fig. 6, providing some notion as to the quality of vehicle control in a steady lateral wind field. It is inferred from the results depicted that the derived control law will ensure a precise trajectory owing to the continuous correction of the switching point. The same result is also obtained with errors in determining the velocity and maintaining the banking angle prescribed by the control law. In every case the motion in the control process is realized with maximum banking angle and, consequently, the maneuvering capabilities of the craft are completely utilized.

Like the quantity $L_1 = L_1(L_0, S_0, \psi_0)$, the other coordinates of the switching point, $S_1 = S_1(L_0, S_0, \psi_0)$, $\psi_1 = \psi_1(L_0, S_0, \psi_0)$ can also be found. The choice of a particular parameter as the determinative one must be made depending on the nature of the initial information and simplicity of the algorithm for the computer equipment. Of major importance is the analytical dependence relating the phase coordinates of the bank switching points. In the case of a linear velocity-range $V = K_1^1(L + a_1)$ the relation between S_1 and ψ_1 is determined by the expression

$$S_1 = K_1 \ln \frac{\frac{\psi_1-\psi_0}{K_1}(L_0+a_1)+1}{\frac{2\psi_1-\psi_0}{K_1}(L_0+a_1)+1} - \frac{\psi_1(L_0+a_1)}{\frac{\psi-\psi_0}{K_1}(L_0+a_1)+1} \quad (14)$$

The control function based on Eq. (14) becomes

$$\Omega(L, \psi, S) = S - S_1(L_0, \psi_0, \psi_1) \quad (15)$$

In a number of practical situations the effect of the initial angle ψ_0 can be neglected, and the exact functions $S_1(\psi, L_0)$ are approximated by straight lines:

$$\phi(S, L, \psi) = S - C(L_0)\psi_0$$

The calculations have shown that in this case high control accuracy is guaranteed both under normal conditions and under the influence of various disturbance factors.

The variable ψ in the control functions $S - C(L_0)\psi$ can be replaced by the derivative $\dot{S}$, since $\dot{S} = V^{(1)}$

LITERATURE CITED

1. L. S. Pontryagin, V. G. Boltyanskii, R. V. Gamkrepidze, and E. F. Mishchenko, The Mathematical Theory of Optimal Processes [in Russian] (Gosizdat Fiziko-Matematicheskoi Literatury, 1961) [in translation: Office of Technical Services, OTS: 62-11751].
2. V. G. Boltyanskii, R. V. Gamkrepidze, E. F. Mishchenko, and L. S. Pontryagin, Maximum Principles in the Theory of Optimal Control Processes, Proc. of the First International Congress of the IFAC, Moscow, 1960 (Izd. Akad. Nauk SSSR, 1961) [English edition: Butterworths, London, 1961].
3. R. Rosenbaum, Longitudinal Range Control for a Lifting Vehicle Entering a Planetary Atmosphere, ARS Paper No. 1911-61 (1961), 34 pages.
4. J. E. Hayes and W. E. Vander Velde, Satellite Landing Control System Using Drag Modulation, ARS Journal, No. 5, 722-729 (1962).

SURVEYOR TERMINAL GUIDANCE*

by R. K. Cheng
Associate Manager, Systems Analysis Laboratory
Surveyor Program
Hughes Aircraft Company
Culver City, California

ABSTRACT

Under the auspices of the Jet Propulsion Laboratory and the National Aeronautics and Space Administration, Hughes Aircraft Company is performing the design and development of the unmanned, lunar softlanding spacecraft system for the Surveyor project. This paper describes the basic concept and implementation of the terminal guidance system for the landing phase.

The terminal descent profile consists of a solid rocket deboost phase with earth-based commands for thrust orientation as well as partial control of ignition. This is followed by a closed-loop vernier guidance phase with throttleable liquid engines. The automatic vernier guidance concept consists of (1) velocity magnitude control in accordance with a velocity versus slant range descent law and (2) gravity turn steering. The differential equations of motion are solved and implementation requirements obtained for anticipated variations in trajectory and site slope conditions. The on-line earth-based guidance program for solid rocket thrust attitude and ignition control is also discussed.

I. INTRODUCTION

The Surveyor is the first U.S. project to softland an unmanned vehicle and its payload on the moon with the primary purpose of discovering those surface and environmental characteristics important to future manned lunar exploration as well as the understanding of the universe. The spacecraft, a model of which is shown in Figure 1, will be launched into a translunar trajectory with a planned midcourse correction to nullify the effects of injection errors such that the approach to the moon is a direct flight to a preselected landing site. The terminal slowdown begins at approximately 60 miles above the lunar surface when a pulsed radar generates a marking signal at a preset range. An earth-commanded time delay ensues and is followed by the ignition of a high thrust, solid propellant engine called the main retro. During the main retro phase, constant attitude is maintained by differentially throttling a set of three liquid vernier engines located symmetrically about the roll or thrust axis. After burnout and staging of the main retro, these same verniers are then used for removal of the remainder of the approach velocity in a closed-loop guidance mechanization involving a three-beam doppler radar for the measurement of vehicle velocity and a fourth beam for measuring slant range to the lunar surface. At a rather low altitude and a correspondingly low velocity, the vernier engines are shut off and the vehicle free falls to the surface. The landing is made with a nominally vertical attitude regardless of the slope condition.

The design of the spacecraft emphasizes simplicity and reliability. In the terminal descent and guidance system for example, the on-board radars are body-fixed, the main retro engine employs no thrust termination device, and the delicate task of softlanding is to be achieved without the use of a digital computer. A prior paper, Reference 1, discussed some of the pertinent design considerations especially concerning approach trajectory and main retro burnout constraints arising due to propulsion and sensing limitations, and their implications with regard to fuel and guidance requirements. In this paper, the guidance mechanization, both in terms of the on-board as well as earth-based portions, is discussed and certain characteristics of interest are obtained through the solution of relatively simple though idealized equations of motion.

II. MECHANIZATION

Ground Computation and Command

Doppler and angle tracking data from earth-based stations will be sent to the Space Flight Operations Facility in Pasadena, California, for

*This work was performed in pursuance of Contract 950056 with the Jet Propulsion Laboratory, California Institute of Technology, under Contract No. NAS 7-100 sponsored by the National Aeronautics and Space Administration.

Superior numbers refer to similarly-numbered references at the end of this paper.

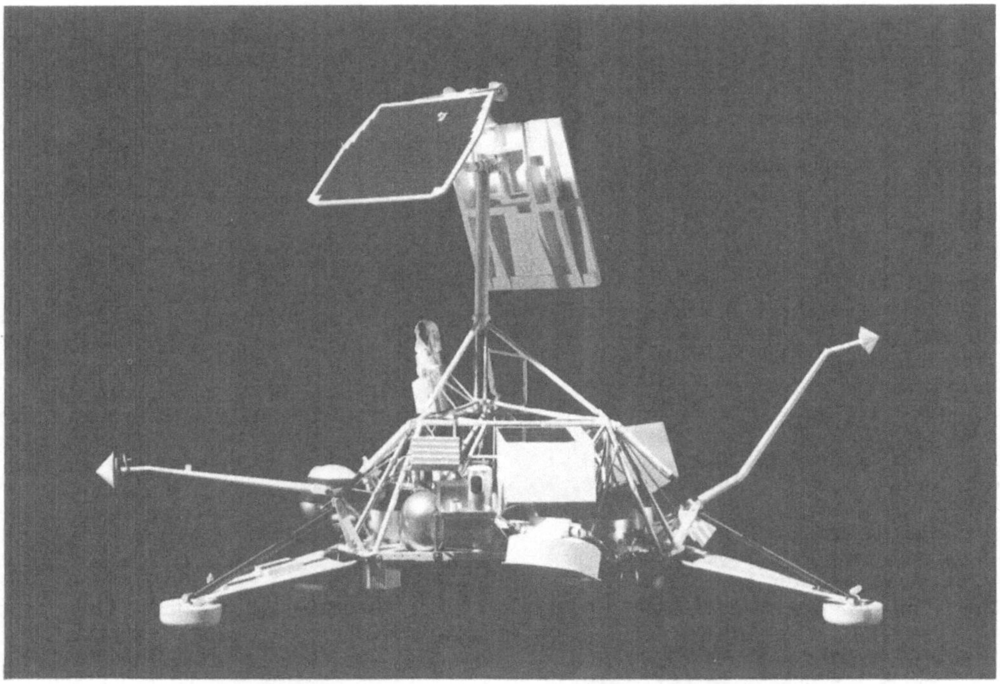

FIGURE 1. MODEL OF SURVEYOR IN LANDED CONFIGURATION.

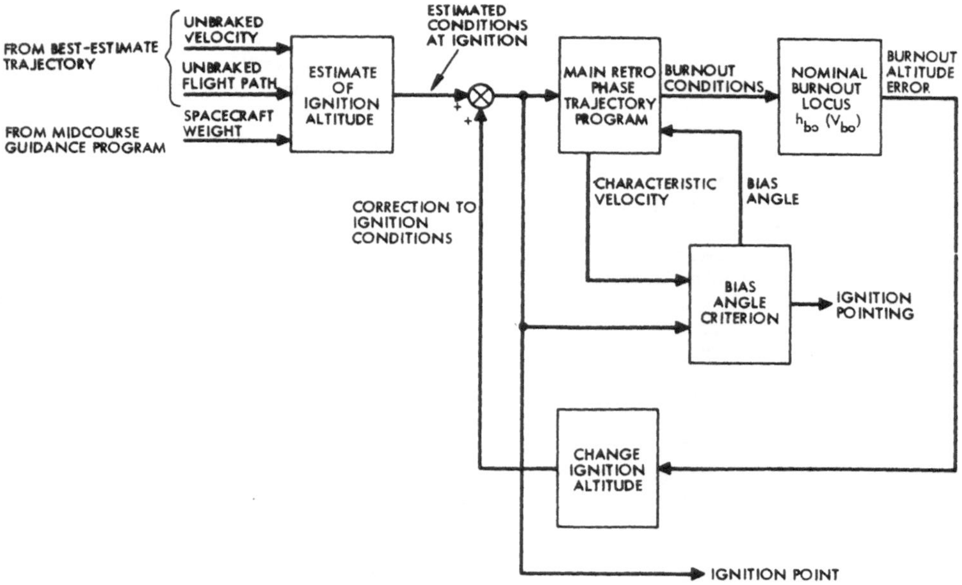

FIGURE 2. TERMINAL GUIDANCE FOR IGNITION CALCULATION

precise orbit determination and guidance computation during the translunar flight. Several hours prior to the vehicle's anticipated impact on the lunar surface, a final orbit determination is made which yields, besides other data, the best estimate of the unbraked impact velocity magnitude, direction, and incidence angle with respect to the lunar surface. This information is fed into a terminal guidance program, the block diagram of which is shown in Figure 2. An iterative procedure based on satisfying a simple relationship between the main retro burnout velocity and altitude yields the nominal ignition altitude as well as the pointing direction of the thrust vector. Given that the pulse radar is preset to generate a marking signal at a given slant range, it is a relatively simple matter then to compute the time delay which must be implemented between the mark and engine ignition. The time delay is telemetered from earth, stored in the flight control programmer of the vehicle, and counted down automatically by the programmer upon reception of the marking signal from the radar.

The other guidance function, that of orienting the vehicle to the proper thrusting attitude, is accomplished by sequentially torquing a set of strapped-down integrating rate gyros through computed angles. The initial orientation of the spacecraft is in this case determined by the vehicle-to-sun and vehicle-to-Canopus directions which are sensed by on-board celestial sensors. Since the thrusting direction in inertial coordinates is obtained from the terminal guidance program, the conversion into spacecraft rotation angles is a straight forward computation. The commands will be sent to and executed by the spacecraft many minutes before the anticipated ignition instant to allow for a period in which television pictures of the surface being approached may be transmitted back to earth for future analysis.

Main Retro Phase

The principal features of the main retro phase are:

1. Constant attitude
2. Open-loop burning - no thrust control
3. Burning to fuel depletion - no thrust termination

The only on-board guidance instrumentation, aside from the strapped-down gyros needed for sensing attitude changes and causing the vernier engines to throttle differentially to counteract such changes, is a crude longitudinal acceleration switch which senses the decay of thrust and initiates a fixed-duration separation sequence.

The conditions at the end of the separation period, also called the burnout conditions, determine to a large degree the satisfactory operation of the subsequent vernier descent phase, assuming that no malfunction occurs. The chief constraints on the altitude as well as the magnitude and direction of the velocity vector arise due to propulsion and terminal sensor limitations as discussed in Reference 1.

The altitude constraint is a function of velocity, the lower limit being further determined by the maximum thrust/weight capability and the upper limit by the amount of propellant remaining. The velocity magnitude is limited approximately to the interval of 150 to 650 ft/sec. The flight path is restricted to a cone of 45° half-cone angle about the local vertical in order to ensure proper incidence of the doppler radar beams to the lunar surface once the gravity turn steering begins. The actual burnout velocity for a given flight even with the effects of the variability in the translunar trajectory and midcourse correction and dispersions in the main retro phase taken into account, still has a very high probability of being within the indicated allowable range. The actual burnout altitude is determined by the earth-commanded setting of the ignition delay and main retro phase dispersions. If there were no dispersions, the burnout altitude would be at a value specified by a function, called the "Nominal Burnout Locus", of the magnitude of the predicted nominal burnout velocity. The locus is designed as a compromise between fuel consumption and clearance above a minimum altitude which can be handled by the limited acceleration capability of the vernier engines, allowing for the random dispersions about any given operating point on the locus. A graph of the burnout locus, dispersion ellipses and constraint contours which illustrate the design problem is shown in Figure 3.

Vernier Phase

The heart of the vernier guidance mechanization is the RADVS (Radar Altimeter Doppler Velocity Sensor) system, the beam geometry of which is shown in Figure 4. There are two split beam antennas placed near the bottom of the vehicle basic framework. Antenna 1 supplies one of the three doppler beams as well as the altitude beam (more appropriately, the slant range along the thrust direction). Antenna 2 supplies the other two doppler beams. The doppler beams are directed at three corners of a square and all are located at an equal angle θ away from the thrust axis. The shift in frequency of the return from that of the transmitted signal is proportional to the spacecraft velocity multiplied by the cosine of the angle between the velocity vector and the beam direction. Referring to Figure 4.

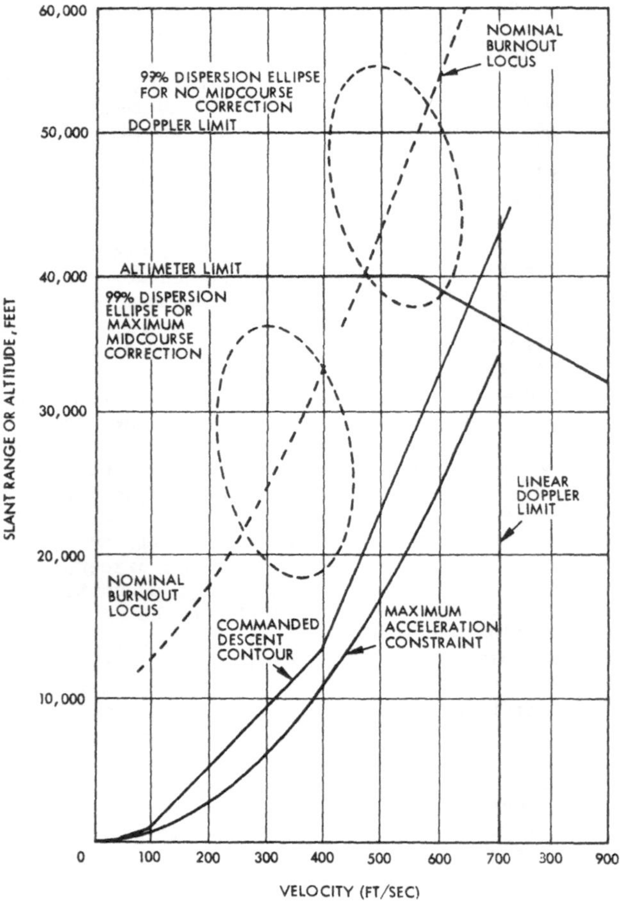

FIGURE 3. VERNIER DESCENT PHASE

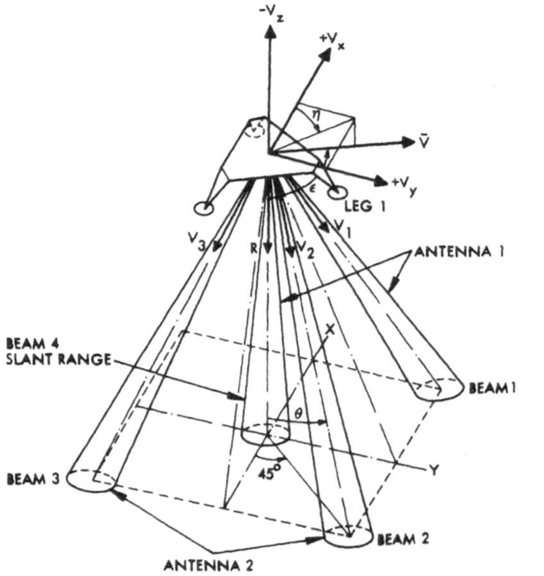

FIGURE 4. RADVS BEAM GEOMETRY

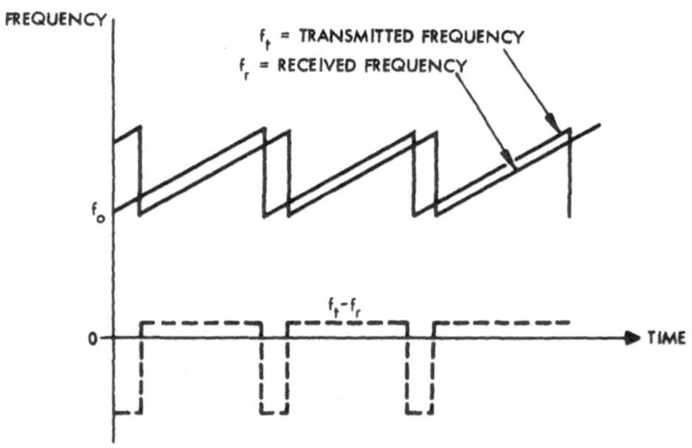

FIGURE 5. ALTIMETER SIGNALS

$$V_1 = \frac{c\Delta f_1}{f_o} = \frac{V_x \sin\theta}{\sqrt{2}} + \frac{V_y \sin\theta}{\sqrt{2}} + V_z \cos\theta \qquad (1)$$

$$V_2 = \frac{c\Delta f_2}{f_o} = -\frac{V_x \sin\theta}{\sqrt{2}} + \frac{V_y \sin\theta}{\sqrt{2}} + V_z \cos\theta \qquad (2)$$

$$V_3 = \frac{c\Delta f_3}{f_o} = -\frac{V_x \sin\theta}{\sqrt{2}} - \frac{V_y \sin\theta}{\sqrt{2}} + V_z \cos\theta \qquad (3)$$

From Equations (1), (2) and (3), the velocity components along the vehicle fixed coordinate system may be solved

$$V_x = \frac{V_1 - V_2}{\sqrt{2}\sin\theta} \qquad (4)$$

$$V_y = \frac{V_2 - V_3}{\sqrt{2}\sin\theta} \qquad (5)$$

$$V_z = \frac{V_1 + V_3}{2\cos\theta} \qquad (6)$$

It may be seen that this particular beam geometry results in the algebraic summing or differencing of only two quantities for each of the three components of velocity along x, y, z. Thus, it permits direct frequency mixing of the return signals in pairs, a distinct simplification in mechanization than in the case where the beams are symmetrically located (at 120° azimuth separation) around the thrust axis.

The altimeter transmits a CW signal modulated in frequency by a triangular wave. The return wave shape is shifted in time by an amount equal to

$$\Delta t = \frac{2R}{c} \qquad (7)$$

If the two frequencies are differenced, the result is a wave represented by the dotted function in Figure 5. The portion above the reference abscissa axis has a height, in the case of zero relative velocity between the radar and the target, proportional to the range along the beam. In the actual situation, the doppler shift adds an additional frequency shift proportional to V_z. This, however, is easily removed in the data processing circuitry since V_z is itself obtainable by mixing two of the three doppler beam frequencies.

The outputs of the RADVS are V_x, V_y, V_z and R, in the form of analog voltages. The first two of these are used to generate pitch and yaw steering commands and the last two for thrust acceleration control.

The flight control system block diagram showing all of the important elements active during the vernier descent phase is shown in Figure 6.

The V_x and V_y signals are used to torque the strapped-down yaw and pitch rate-integrating gyros. The slant range signal R is fed into a function generator which generates a corresponding required velocity V_R. The difference between V_R and V_Z is defined as the velocity error. It is amplified and limited before becoming the thrust acceleration command signal. The output of a longitudinal accelerometer is compared with this signal and the difference is eventually used to raise or lower the thrusts of all three engines by the same amount. Returning to the pitch and yaw channels, the rate errors are integrated by the gyros, then amplified and fed into a mixing network which also accepts the thrust acceleration error signal. The three inputs to the network determine the three outputs which are the thrust commands to the engines. Finally, the vehicle dynamics and geometrical relations complete the feedback to the RADVS and the accelerometer.

III. VERNIER GUIDANCE CHARACTERISTICS

Basic Guidance Concept

The basic concept used in the vernier guidance system involves: (1) gravity turn steering throughout the descent immediately following RADVS acquisition of the velocity vector and (2) a minimum acceleration phase followed by guiding along a nominal descent contour of V vs R for the generation of velocity error which in turn is used to control the thrust acceleration.

The use of gravity turn steering has several outstanding advantages. First, because of the fact that the thrust axis is required to be colinear with the velocity vector which, as was shown earlier, is easily obtained in body coordinates, there is no requirement for the knowledge of another direction. In any other steering scheme, the knowledge of the direction of local vertical is generally necessary, either explicitly or unexplicitly, resulting in more complex instrumentation. Secondly, the gravity turn descent has a very desirable property that, as the velocity approaches zero, the flight path (thus also thrust direction) tends towards the vertical. No violent reorientation maneuver is needed prior to touchdown. Thirdly, a gravity turn with suitable thrust acceleration control is an efficient trajectory fuel wise for a wide range of initial flight path angles ranging from near horizontal to vertical.

In the thrust acceleration control channel, the use of a nominal V vs R descent trajectory permits near maximum utilization of the vernier thrust capability. The transition from burnout to this mode of descent is via a minimum acceleration descent rather than a complete shutoff of the vernier engines because of the necessity of maintaining attitude stabilization for proper sensor operation.

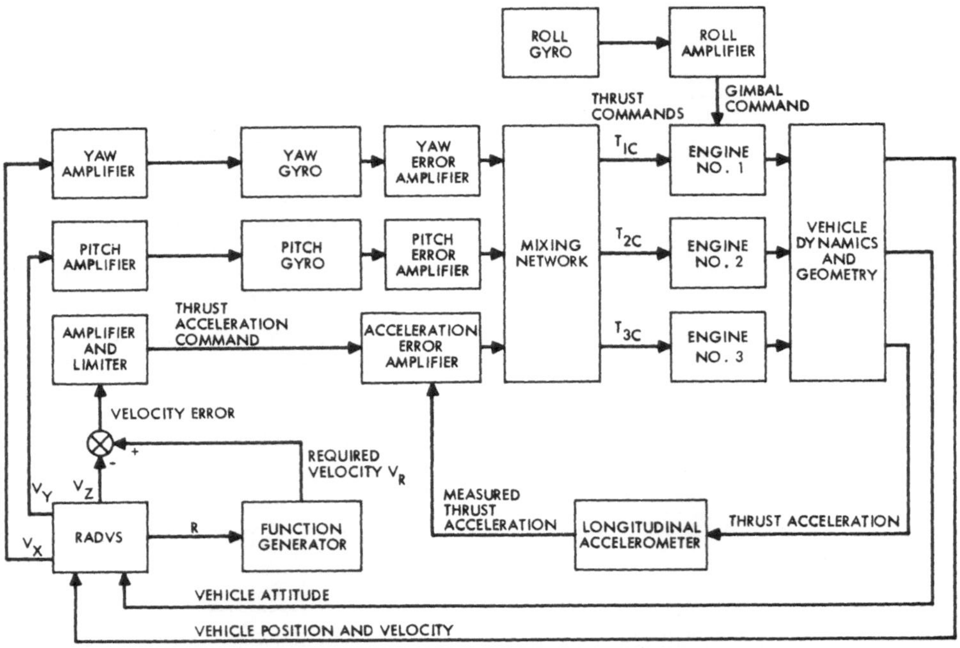

FIGURE 6. FLIGHT CONTROL SYSTEM BLOCK DIAGRAM VERNIER PHASE

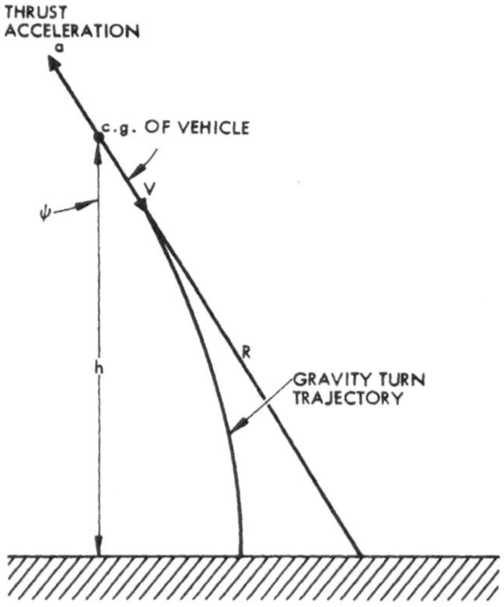

FIGURE 7. GRAVITY TURN GEOMETRY

Doppler and range acquisition by the RADVS is anticipated during the minimum acceleration phase. As long as the measured state of the vehicle, represented as a point in the velocity-range phase plane, is well above the descent contour, the acceleration is constant at a level determined by the minimum thrust capability of the engines as well as the anticipated weight of the spacecraft.

The major portion of the descent contour is an approximation of the idealized parabolic relation:

$$R = \frac{V^2}{2(n - 1)g} \tag{8}$$

which, in the case of vertical descent, requires a constant thrust acceleration equal to ng. Later it will be shown that for a fairly wide range of flight path as well as average surface slope condition, the acceleration requirement does not differ appreciably from ng. This is consistent with the general desire to find a simple guidance law which could cope with a wide range of anticipated conditions while, at the same time, yield nearly maximum performance within the available thrust constraint.

The contour near the origin of the R-V plane deviates from that given by Equation 8. A constant low velocity descent phase is incorporated which serves to absorb altitude errors when the spacecraft velocity reaches this value. Finally, engine shutoff is made when the range beam indicates a specific value (different from zero), in order to avoid possible detrimental effects resulting from the impingement of rocket exhaust on the surface.

Differential Equations

Referring to Figure 7, the basic differential equations for a gravity turn descent with an arbitrary thrust acceleration (not necessarily constant) are

$$\frac{dV}{dt} = \dot{V} = - a + g \cos \psi \tag{9}$$

$$V \frac{d\psi}{dt} = V\dot{\psi} = - g \sin \psi \tag{10}$$

These equations may or may not yield closed form solutions depending on the nature of the accelerations or indirectly, the particular longitudinal guidance law chosen.

Minimum Acceleration Phase

During the minimum acceleration phase which links the operation along the descent contour with main-retro burnout, the behavior is very simply solved by noting that in Equation (9)

$$a = n_{min} g \tag{11}$$

Thus, by dividing (9) by (10) we obtain

$$\frac{1}{V} \frac{dV}{d\psi} = \frac{n_{min} - \cos \psi}{\sin \psi} \tag{12}$$

In the interest of conserving fuel, n_{min} is chosen to be somewhat less than unity. For ψ lying somewhere between zero and 90° it is therefore possible for $\frac{dV}{d\psi}$ to be either positive or negative. If initially $\cos\psi$ is less than n_{min} V tends to decrease with time at first. This trend in most cases however is reversed because the decreasing value of ψ will soon cause $\frac{dV}{dt}$ to change sign.

Integration of Equation (12) results in

$$\frac{V}{V_o} = \left(\frac{\tan \frac{\psi}{2}}{\tan \frac{\psi_o}{2}}\right)^{n_{min}-1} \frac{\sec^2 \frac{\psi}{2}}{\sec^2 \frac{\psi_o}{2}} \tag{13}$$

where the subscript "O" denotes initial values. The factor

$$q\ (\psi) = \left(\tan \frac{\psi}{2}\right)^{n_{min}-1} \sec^2 \frac{\psi}{2} \tag{14}$$

which may be called the "velocity integral" is plotted in Figure 8 for various values of n_{min}. The Surveyor design presently has $n_{min} = 0.9$. As a result q_{min} occurs at $\psi = \cos^{-1} n_{min} = 25.8^\circ$, the point at which $\frac{dV}{dt}$ passes through zero. The phase plane behavior is illustrated in Figure 9 for values of ψ_o on either side of $\cos^{-1} n_{min}$.

Intersection with Descent Contour

The minimum acceleration phase is terminated when the measured slant range reaches a value given by Equation (8). The altitude h_a at this point is a function not only of velocity but also of flight path angle

$$h_a = \frac{V_a^{\ 2}}{2(n - 1)g} \cos \psi_a \tag{15}$$

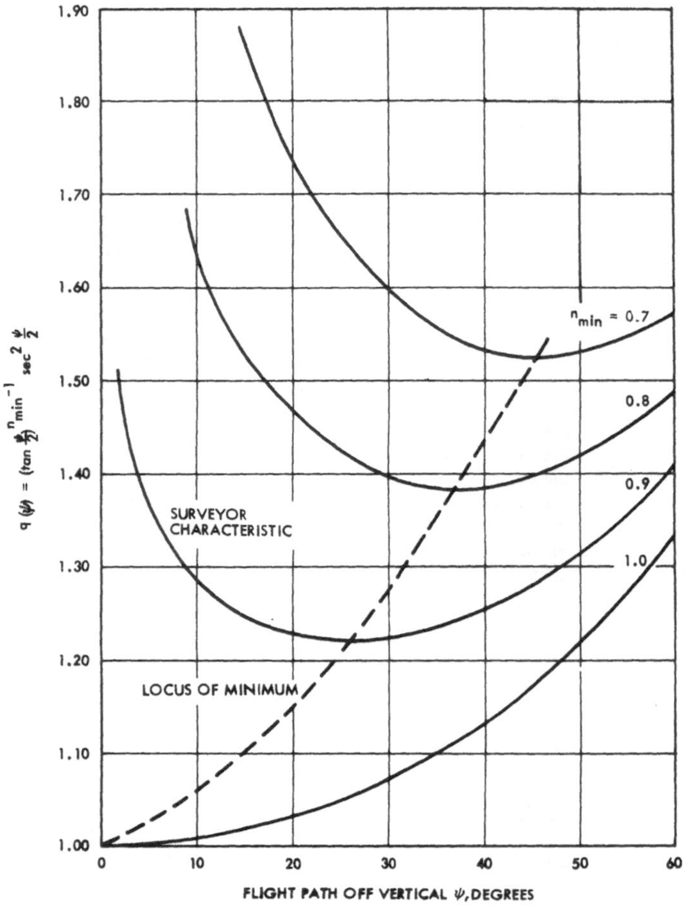

FIGURE 8. VELOCITY INTEGRAL FOR MINIMUM ACCELERATION PHASE

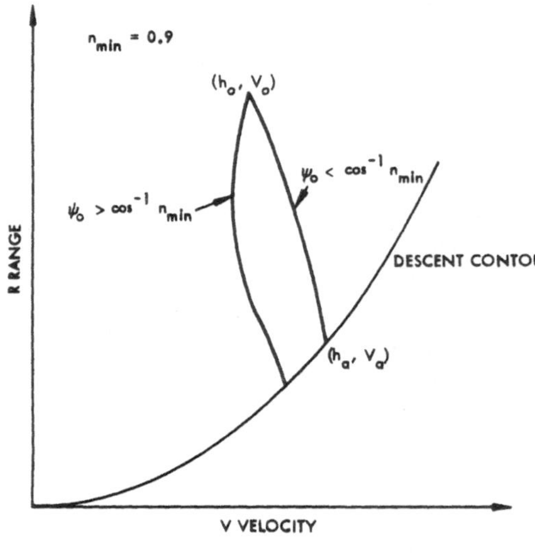

FIGURE 9. TWO POSSIBLE BEHAVIORS DURING MINIMUM ACCELERATION PHASE

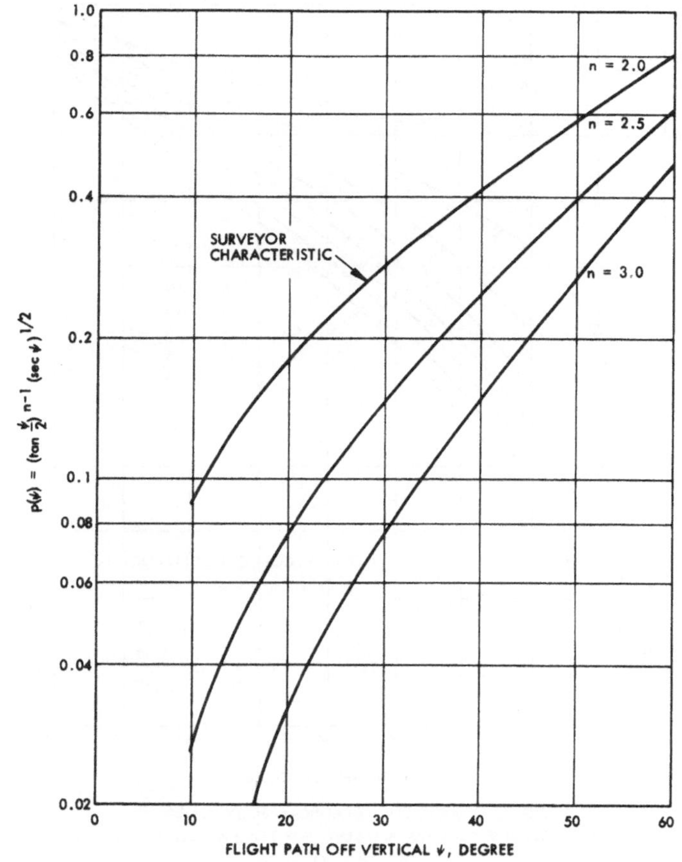

FIGURE 10. VELOCITY INTEGRAL FOR CONSTANT - V^2/R GRAVITY TURN

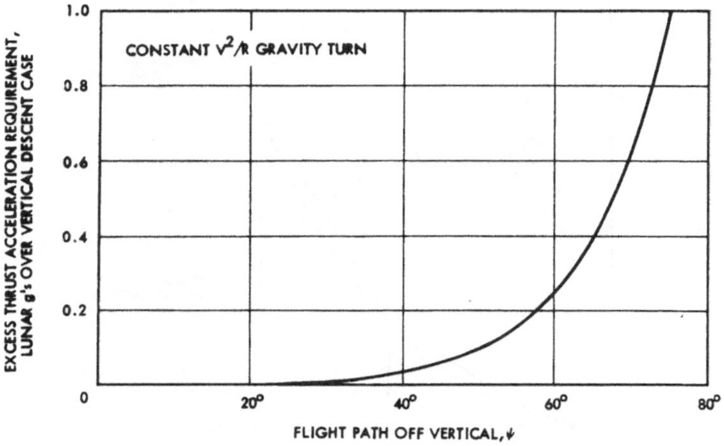

FIGURE 11. EXCESS THRUST ACCELERATION REQUIREMENT OVER VERTICAL DESCENT

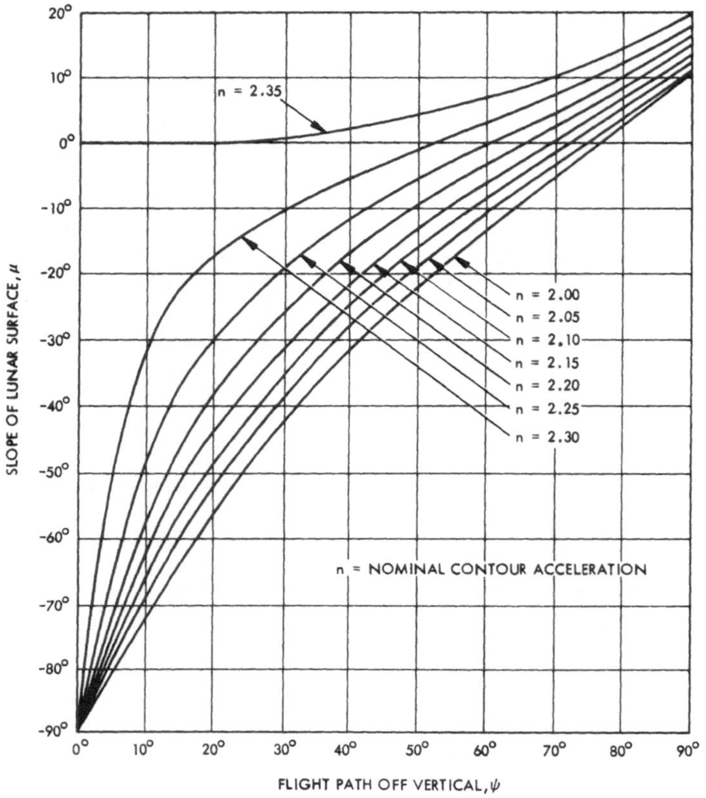

FIGURE 12. SLOPE-FLIGHT PATH CAPABILITY VERSUS NOMINAL CONTOUR ACCELERATION LEVEL FOR 2.35 g ACCELERATION LIMIT (n_{max} = 2.35)

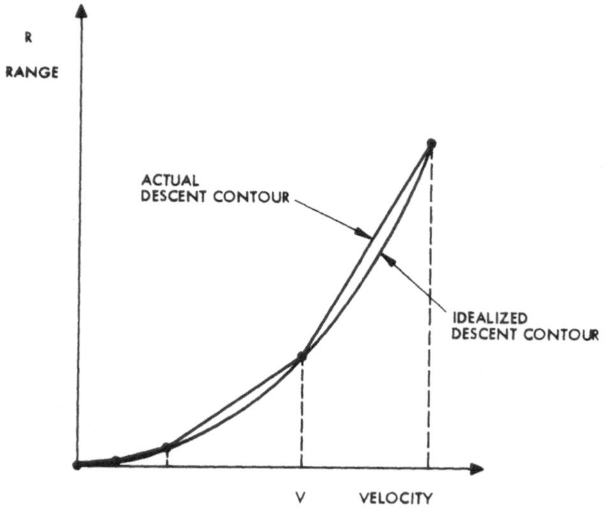

FIGURE 13. 4-SEGMENT APPROXIMATION OF PARABOLIC DESCENT CONTOUR

The exact point of intersection is solvable by the iterative formula

$$
\tan\frac{\psi_a}{2} = \tan\frac{\psi_o}{2}
$$

$$
\times\left[\frac{\dfrac{n-n_{min}}{(1-n_{min})(n-1)} - \left(\tan\dfrac{\psi_a}{2}\right)^4 \dfrac{n_{min}-n+2}{(n-1)(n_{min}+1)}}{\dfrac{2gh_o}{V_o^2}\left(\sec\dfrac{\psi_o}{2}\right)^4 + \dfrac{1}{1-n_{min}} + \dfrac{\left(\tan\dfrac{\psi_o}{2}\right)^4}{1+n_{min}}}\right]^{\frac{1}{2(1-n_{min})}} \tag{16}
$$

which may be used with (13) and (15) for the complete determination of the intersection point. In general, ψ_a is always less than ψ_o due to the effect of the gravity turn steering while V_a may be either higher or lower than V_o. In the case of Surveyor, there is a high probability that V_a will not differ from V_o by more than 50 feet per second or so, due to the fact that n_{min} is so close to being unity.

Idealized Behavior Along Descent Contour

The idealized behavior along the Descent Contour may be found by the following procedure. Noting first from the geometry shown in Figure 7 that

$$
R = h \sec\psi \tag{17}
$$

we differentiate Equation (17) with respect to time,

$$
\dot{R} = \dot{h}\sec\psi + h\,\dot{\psi}\sec\psi\tan\psi \tag{18}
$$

But,

$$
\dot{h} = -V\cos\psi \tag{19}
$$

Also,

$$
h = \frac{V^2}{2\,(n-1)\,g}\cos\psi \tag{20}
$$

Thus

$$
\dot{R} = -V + \frac{V^2\dot{\psi}}{2\,(n-1)\,g}\tan\psi \tag{21}
$$

which, in view of Equation (10), becomes

$$
\dot{R} = -V\left[1 + \frac{1}{2\,(n-1)}\sin\psi\tan\psi\right] \tag{22}
$$

Now, if Equation (8) is differentiated with respect to time

$$
\dot{R} = \frac{V\dot{V}}{(n-1)\,g} \tag{23}
$$

Equations (22) and (23) are combined to eliminate $\dot{R}$

$$
\dot{V} = -\left[(n-1)\,g + \frac{g}{2}\sin\psi\tan\psi\right] \tag{24}
$$

If Equation (24) is divided by Equation (10), the dummy variable time is eliminated and we have the key differential equation relating velocity and flight path angle

$$
\frac{1}{V}\frac{dV}{d\psi} = \frac{n-1}{\sin\psi} + \frac{1}{2}\tan\psi \tag{25}
$$

which may be integrated to yield the general solution

$$
\frac{V}{V_a} = \left(\frac{\tan\frac{\psi}{2}}{\tan\frac{\psi_a}{2}}\right)^{n-1}\left(\frac{\sec\psi}{\sec\psi_a}\right)^{\frac{1}{2}} \tag{26}
$$

a result differing from the constant-acceleration gravity-turn solution only in the exponent of the secant term.

The numerator of the right-hand side of Equation (26) is plotted in Figure 10 for several representative values of n including the case of n = 2, very nearly the Surveyor characteristic. No startling behavior is exhibited here. It is seen that velocity decreases monotonically as ψ becomes smaller.

Of greater interest is the thrust acceleration required to follow the Descent Contour. From Equation (9)

$$
a = g\cos\psi - \dot{V} \tag{27}
$$

But $\dot{V}$ is given by Equation (24), thus

$$
\begin{aligned}
a &= ng + \left(\cos\psi + \frac{1}{2}\sin\psi\tan\psi - 1\right)g \\
&= ng + \frac{(1-\cos\psi)^2}{2\cos\psi}\,g
\end{aligned} \tag{28}
$$

Equation (28) shows that the required thrust acceleration depends on ψ only. The second term in the right-hand side, plotted in Figure 11, is the excess acceleration required above the nominal value of n. The interesting and important property of this term is its negligibly small magnitude compared to g, for ψ ranging from zero up to even as high as 45°. Thus, for all practical purposes, the thrust acceleration required may be regarded as a constant. In the design of the Surveyor, the Descent Contour is therefore made to correspond to a value of thrust acceleration just slightly less than the maximum capability.

MEASURED VELOCITY
ALONG THRUST AXIS
(FROM RADVS)

MEASURED THRUST
ACCELERATION
(FROM ACCELEROMETER)

MEASURED SLANT
RANGE (FROM RADVS)

$\tilde{V}_Z$

REQUIRED
VELOCITY

VELOCITY
ERROR

$\tilde{R}$

$\hat{V}$

$\hat{V}$

K_V
GAIN

LIMITER

$\Delta\hat{a}$

K_a
GAIN

VERNIER ENGINE
THRUST COMMAND

DESCENT CONTOUR
(DIODE FUNCTION GENERATOR)

$\tilde{R}$

$\hat{a}_{MIN}$
(0.9 g)

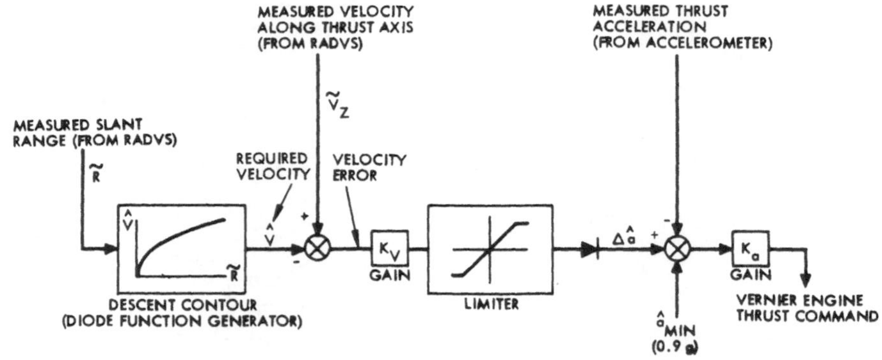

FIGURE 14. THRUST ACCELERATION COMMAND MECHANIZATION

$\Delta\hat{a}$, $\hat{a}_{MIN}$
THRUST ACCELERATION COMMAND

a_{MAX}

$(\Delta\hat{a})_{MAX}$

a_{MIN}

$\dfrac{(\Delta\hat{a})_{MAX}}{K_V}$

0

VELOCITY ERROR, $\tilde{V}_Z - \hat{V}$

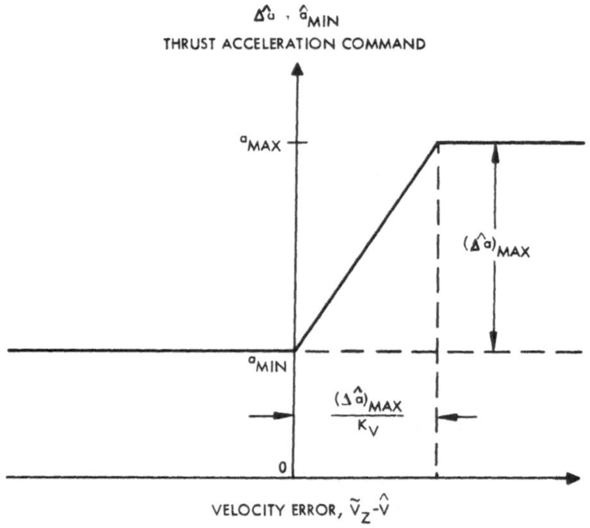

FIGURE 15. THRUST ACCELERATION COMMAND CHARACTERISTIC

POINT	EVENT
A	ACQUISITION OF SEGMENT
B	VELOCITY REQUIRED SWITCHED TO V_C AND INERTIAL HOLD
C	ENGINE CUTOFF
D	TOUCHDOWN

R_O

P_O

FINAL
SEGMENT

ACTUAL TRAJECTORY

SLANT RANGE R

R_b

R_c

0

V_c V_b V_L V_a

V_O

VELOCITY V

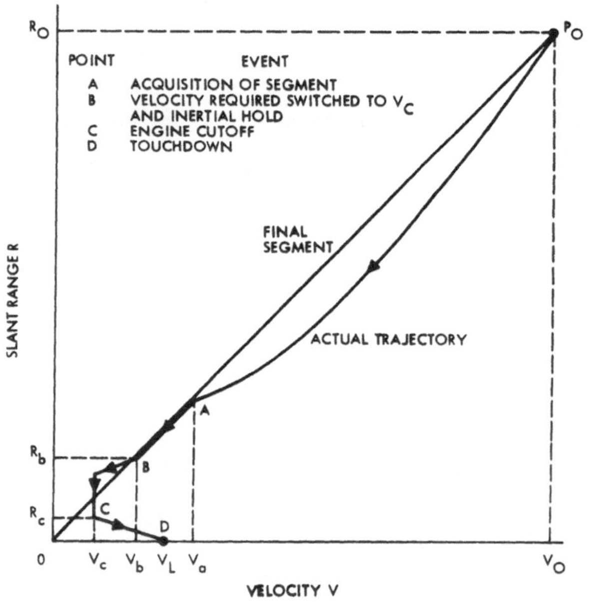

FIGURE 16. TRAJECTORY ALONG FINAL SEGMENT

In the case where a surface slope μ exists in the direction of travel, the acceleration required may be shown to be

$$a = ng + \left[\cos \psi + \frac{1}{2} \sin \psi \tan (\psi - \mu) - 1 \right] g \tag{29}$$

where a positive μ indicates an uphill approach. Thus, the acceleration required tends to be greater for the downhill case where μ is negative. It is not difficult to see from physical interpretation that this must be true. As the gravity turn bends the trajectory towards the vertical, the slant range tends to decrease more rapidly in the downhill than in the uphill situation. Correspondingly in the former case a higher thrust acceleration is demanded of the system.

The choice of n, the nominal descent contour acceleration level, given that there is a maximum capability n_{max}, is determined by a "slope-flight path" capability equation.

$$\tan \mu = \left[\frac{4n_{max}^2 - 1}{4(n_{max}^2 - 1)} \right]$$

$$\times \left[\frac{2(n-1)(1+2n_{max}\cos\psi + \cos^2\psi) - 4(n_{max}^2 - 1)\cos\psi}{4n_{max}^2 - 1 - 2(n-1)(2n_{max} + \cos\psi)} \right] \csc\psi \tag{30}$$

Figure 12 is a plot of μ versus ψ for $n_{max} = 2.35$ and various value of n. For any chosen n, the region above the curve is the allowable operating region. It turns out that the capability curve is a strong function of $n_{max} - n$, the acceleration margin between the maximum capability and the nominal descent acceleration (in the vertical case). In the actual design, allowance must also be made for sensor and electronic errors which may be lumped together as an equivalent dispersion in n. A margin of 0.25 to 0.30 lunar g is found to be required for the Surveyor system.

Another interesting property of the trajectory is the manner the turning rate $\dot{\psi}$ varies as V approaches zero. From Equations (10) and (26),

$$\dot{\psi} = -\frac{g}{V} \sin \psi$$

$$= -\frac{g}{V_a} \sin \psi \left(\frac{\tan \frac{\psi}{2}}{\tan \frac{\psi_a}{2}} \right)^{1-n} \left(\frac{\sec \psi}{\sec \psi_a} \right)^{-\frac{1}{2}} \tag{31}$$

since V and ψ approach zero together,

$$\lim_{V \to 0} \dot{\psi} = \lim_{\psi \to 0} \dot{\psi}$$

$$= \lim_{\psi \to 0} c \, \psi^{2-n} \tag{32}$$

where c is a constant. Thus, there are three possibilities:

1. If $n < 2$, $\dot{\psi}$ approaches zero
2. If $n = 2$, $\dot{\psi}$ approaches a non-zero constant
3. If $n > 2$, $\dot{\psi}$ diverges

The last possibility presents a problem at least in the idealized sense if we attempt to follow the gravity turn to zero velocity. In practice, however, two things can be done to alleviate the situation (a) use of a lower acceleration level near $V = 0$ or, (b) incorporate inertial hold at some finite velocity and accept a small residual horizontal velocity component. Both are implemented in the Surveyor design although not primarily for the above reason.

Straight-Line Approximation of Descent Contour

Equation (8) is not exactly implemented in the spacecraft. Rather, a 4-segment straight line approximation of this equation, shown in Figure 13 and actually furnished by a diode function generator, is used instead. Because of the finite number of segments, there are some deviations in behavior from those characterizing the idealized case, the chief difference being the presence of thrust acceleration saturation periods when the system attempts to follow the Descent Contour. This behavior, discussed in Reference 1, has the beneficial effect of shielding the longitudinal guidance channel from radar noise at least during most of the acceleration-saturated portions. However, towards the bottom end of each segment, close-tracking should and does occur.

Taken as a whole, the straight-line mechanization yields practically the same results with regard to fuel consumption and flight path angle reduction as the idealized case.

The actual thrust acceleration command is derived as shown in Figure 14, with the resultant characteristic of Figure 15. The width of the linear region of operation $\frac{(\Delta\hat{a})_{max}}{K_V}$ is of the order of a few feet per second.

Final Segment and Touchdown

The operation along the final segment to actual touchdown consists of several important features which deserve some discussion.

The final segment design has it passing through the origin in the R - V plane as shown in Figure 16. However, the portion near the origin is not used for guidance, as it is easy to show that following a straight line to the origin is not physically realizable from the fuel requirement standpoint. Instead, a constant-velocity (of nominally V_c = 5 ft/sec) descent subphase is implemented prior to engine shutoff. This also means that altitude dispersions which exist at the beginning of this subphase will produce a negligible effect on the landing velocity. Another discrete change in implementation from the pure gravity-turn mode is the use of inertial attitude hold at a slightly higher velocity V_b of say 10 ft/sec, thus effectively removing the radar inputs to the lateral channel at these very low velocities where radar noise might produce jerky attitude changes. Simultaneously the velocity command is switched to the constant value of V_c.

When the vehicle state reaches P_o the top of the final segment, the flight path angle in general has been reduced to a small enough value (<20°) so that small angle approximations may be used to solve for the behavior from this point onward. The thrust acceleration is saturated at first with a value equal to n_{max} g and the velocity-flight path relationship is approximately

$$\frac{V}{V_o} = \left(\frac{\psi}{\psi_o}\right)^{n_{max} - 1} \tag{33}$$

Also,

$$V_o^2 - V^2 \cong 2\,g\,(n_{max} - 1)\,(R_o - R) \tag{34}$$

At the point where the thrust acceleration comes out of saturation, the trajectory "acquires" the straight line and begins to track it. If b is defined as the slope of the segment, or

$$b = \frac{R}{V} \tag{35}$$

it may be noted from Equation (34) that the acquisition velocity V_a must satisfy the equation

$$V_o^2 - V_a^2 = 2\,g\,(n_{max} - 1)\,(bV_o - bV_a) \tag{36}$$

Solving for V_a

$$V_a = 2\,bg\,(n_{max} - 1) - V_o \tag{37}$$

V_a must in practice be a value somewhat higher than V_b, the point at which the commanded velocity is switched to V_c. This implies that b must be larger than some positive constant

$$b > \frac{V_b + V_o}{2\,g\,(n_{max} - 1)} \tag{38}$$

On the other hand, it may be shown that the fuel requirement from V_a to V_b is

$$\Delta V = V_a - V_b + bg \ln \frac{V_a}{V_b} \tag{39}$$

The gravity loss term is thus directly proportional to b, which must not therefore be excessively large.

The value of b also influences the spacecraft flight path angle at V_b. Since ideally the attitude coincides with the flight path, this angle must be kept small so that the subsequent thrust velocity increment does not give rise to an excessive horizontal velocity component.

During the "tracking" phase from V_a to V_b, the behavior is described by the following differential equation

$$\frac{dV}{d\psi} = \frac{V^2}{bg} \csc \psi + V \tan \psi \tag{40}$$

the solution of which is

$$V \cos \psi = \frac{V_a \cos \psi_a}{1 + \frac{V_a \cos \psi_a}{bg} \ln \left(\frac{\tan \psi_a}{\tan \psi}\right)} \tag{41}$$

For small angles, Equation (41) may be shown to be equivalent to

$$\frac{\psi}{\psi_a} \cong e^{-\left(\frac{V_a - V}{V}\right)\frac{bg}{V_a}} \tag{42}$$

Thus, the value of b needs to be large in order that the above ratio, evaluated at V_b be as small as practically possible.

The above considerations lead to a choice of b which is a compromise between fuel cost on the one hand and final flight path off vertical on the other.

When the state reaches B with a velocity of V_b, the switching of the commanded velocity to a constant value of V_c momentarily causes saturation of thrust acceleration of the system, but within a very short altitude decrement the velocity of V_c is essentially attained. The subsequent descent to point C removes any initial altitude dispersion except for that component which is common to the engine cutoff mechanization.

Engine cutoff is commanded when the RADVS indicates a preset range of R_c. The choice of R_c (equal to 13 feet in the actual mechanization) is a compromise between rocket exhaust interaction and touchdown velocity considerations.

The vehicle lands with a vertical velocity determined primarily by the cutoff altitude and secondarily by the actual velocity during the constant-velocity descent. Both these parameters, aside from their obvious dependence on the nominal settings, are chiefly determined by the RADVS measurement. Likewise, in the lateral channel the horizontal velocity component is primarily caused by three sources:

1. Residual attitude, differing from vertical due to radar noise effect, at V_b.
2. Residual attitude due to terminating gravity turn steering at finite velocity.
3. Thrust to RADVS misalignment.

The noise effect turns out to be the most important for the present mechanization as shown by analyses. Nevertheless, the anticipated horizontal velocities are well within the stability limits as shown by extensive touchdown simulation tests.

IV. CONCLUDING REMARKS

The principal characteristics of the Surveyor Terminal Guidance System, discussed in the preceding sections, have shown that a simple concept with a resulting simple mechanization can be used to solve what might appear at first to be a complex problem. The fact that a number of the characteristics of interest can be deduced in simple, analytical forms is of enormous help in actually specifying numerical performance requirements for the subsystems. It is a firm belief of the author, biased it may be, that a concept resulting in mathematically simple behavior has a definite advantage over those which can only be analyzed numerically on a computer. It is fortunate that such a concept is available for the design of the Surveyor.

Nomenclature

V_1, V_2, V_3	velocity along doppler beams 1, 2 and 3 respectively
$\Delta f_1, \Delta f_2, \Delta f_3$	frequency shifts corresponding to above
c	velocity of light
f_o	transmitted frequency
X, Y, Z	body-fixed cartesian coordinates
V_x, V_y, V_z	velocity along the X, Y, and Z axes
θ	beam to thrust axis angle
R	slant range
V	magnitude of velocity
g	lunar surface gravitational attraction
n	nominal thrust to lunar weight ratio
a	thrust acceleration
ψ	flight path angle, with respect to vertical
n_{min}	thrust to lunar weight ratio during minimum thrust phase
μ	surface slope
n_{max}	maximum thrust to lunar weight ratio
V_c	nominal cut off velocity
R_c	nominal cut off altitude
b	slope of final segment

References:

(1) Cheng, R. K. and Conrad, D. A., Design Considerations for the Surveyor Terminal Descent System. AIAA Paper No. 64-644, presented at the AIAA/ION Astrodynamics Guidance and Control Conference, Los Angeles, California, August 1964.

(2) Cheng, R. K. and Pfeffer, I., Terminal Guidance System for Soft Lunar Landing, Guidance and Control, R. E. Robertson and J. S. Farrior, editors, Academic Press, New York, 1962.

(3) Kriegsman, B. A. and Reiss, M. H., Terminal Guidance and Control Techniques for Soft Lunar Landing. ARS Journal, March 1962.

(4) Green, W. G., Logarithmic Navigation for Precise Guidance of Space Vehicles. IRE Transactions on Aerospace and Navigational Electronics, June 1961.

(5) Citron, S. J., Dunin, S. E. and Meissinger, H. F., Terminal Guidance Technique for Lunar Landing. AIAA Journal, March 1964.

(6) Cheng, R. K., Lunar Terminal Guidance, Chapter 7 of Lunar Missions and Explorations, edited by C. T. Leondes and R. W. Vance, John Wiley and Sons, New York 1964.

DISCUSSION

R. K. Cheng

"Surveyor Terminal Guidance".

Q. What control law for angular position?

A. The thrusting direction of the spacecraft is controlled to be directly opposite to the velocity vector - this is the gravity turn descent mode.

Q. Don't large turning rates result in such a maneuver?

A. In equations (31) and (32) in the paper three possible conditions are described. In particular, if the acceleration level is greater than 2, the turning rate diverges at the end. However, this is true only if we want to guide to zero velocity, so we terminate guidance at a finite velocity. Furthermore, the straight line approximation of the descent contour gives rise to variable thrust acceleration characteristics. If the last segment is chosen with the right slope, the acceleration angular rate will approach zero, even if the control parabola is for an acceleration level greater than 2.

LOW POWER SIGNAL TRANSFORMATION FOR SPACE APPLICATIONS

by

Prof. J. LAGASSE
Director

Dr. I. GUMOWSKI
Associate Professor

Dr. G. GIRALT
Director of Research

Electrical Engineering Research Laboratory
TOULOUSE - FRANCE

ABSTRACT

Exploiting the high input impedance, the reasonable resistance to radiation and the consistant nonlinearity of metal-oxide -semiconductor transistorsfive operational circuits, suitable for space applications, have been designed. These circuits are : a low-level D C current amplifier, a low-level D C voltage amplifier, a wide-band squaring circuit, a power source regulating circuit and a low-level analog to digital converter.

INTRODUCTION

The availability of relatively stable metal-oxide-semiconductor (MOS) transistors has rendered possible the design of new signal shaping circuits and a considerable power-drain reduction of conventional circuits. Even before MOS transistors became commercially available, an application in logic circuits has been reported [1**]. Since logic circuits have already been described [1], this paper will be limited to some operational circuits usually found in measuring devices and in regulating or control systems. Working prototypes will be presented at the Symposium. The components and also the complete units were tested for mechanical resistance to acceleration and vibrational fatigue. Unfortunately no occasion was yet available to test their operation in an actual space environment.

PRINCIPAL PROPERTIES of MOS TRANSISTORS

MOS transistors have eight properties which make them useful in electronic circuits intended for space applications : (a) small size, (b) low weight, (c) mechanical rigidity, (d) relative insensitivity to proton and neutron bombardement, (e) reasonably predictable response to low level gamma-ray irradiation, (f) extremely low power drain, (in the order of microwatts), (g) extremely high input impedance (in the order of 10^{15} ohms), and (h) a consistant and stable nonlinearity. The set of the last four properties is not shared by any other presently known amplifying device.

The study of commercially available MOS transistors has shown, however, that certain complications arise in the design of practical circuits. In particular, it was found that the electrical characteristics of MOS transistors vary widely from unit to unit and that premature failure occurs if no overload protection is provided at the gate terminal (Fig. 1). Furthermore, because of the very pronounced nonlinearity, a nonlinear analysis is required even for moderately small signals. The need for individual attention will probably be reduced with the introduction of improved manufacturing techniques,. If economical mass-production is not considered useful MOS transistor circuits can be designed at present.

EFFECT of PROTON and NEUTRON BOMBARDMENT

Contrary to ordinary transistors, MOS transistors do not derive their useful properties from delicately balanced dislocations and impurities of a very thin crystalline layer. Consequently, the unavoidable dislocations introduced by particle bombardement, have little effect on the MOS transistor conduction channel. The electrical charge introduced by protons and possibly also by other charged

./.

** Superior numbers refer to similary-numbered references at the end of this paper.

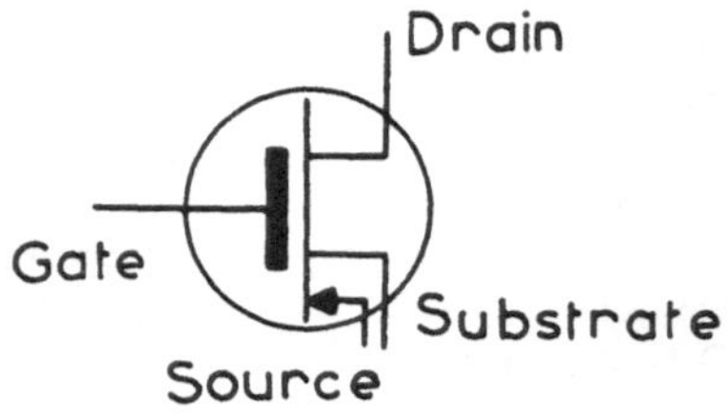

M.O.S. Transistor Symbol

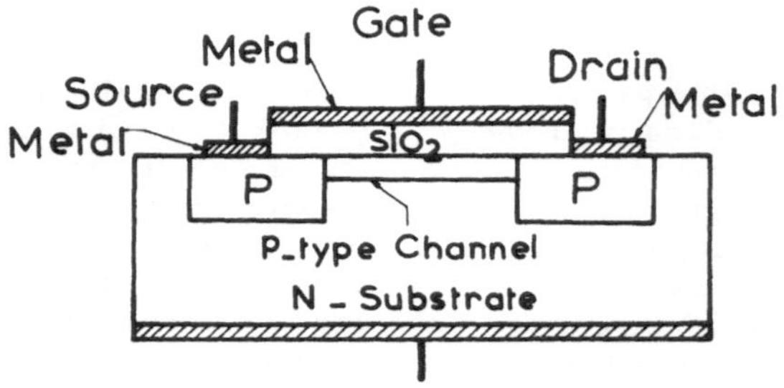

M.O.S. Transistor Cross-sectional view

Fig. 1

particles may build up a field intensity sufficient to produce electrical breakdown of the oxide insulating layer. Otherwise, the oxide layer will retain its insulating function, unless the bombardment is so severe that "macroscopic" physical damage results. Furthermore, protection against protons and other charged particles can be provided by means of electrostatic shielding. Without shielding destruction of the MOS transistor occurs with a dose of about 10^7 RAD, the precise amount depending on the bombardment intensity.

EFFECT of GAMMA RADIATION

Tests carried out in the U.S. [2] have shown that the irradiation of MOS transistor by gamma rays produces a cumulative effect. Consequently, MOS transistorscan be used to monitor the amount of irradiation received by an object. The electrical characteristics will be seriously affected if the intensity of the rays is high and the total dose exceeds about 10^6 RAD. However, if the intensity is relatively low, then an equilibrium state will be reached. For circuit design purposes this equilibrium state amounts practically to the addition of a constant equivalent signal to the gate terminal (Fig. 2 : characteristics without irradiation, Fig. 3 : the same characteristics with irradiation, intensity = 5 . 10^4 RAD).

Within limits, the constant equivalent signal can be either compensated, or the circuit configuration can be so chosen that the overall performance is only slightly dependent on the presence of this constant equivalent signal. The equivalent induced signal is about 20 µV/RAD when the MOS transistor is operating, and 4 µV/RAD when it is not. When the non-destructively irradiated MOS transistor is removed from the irradiation field, its electrical characteristics will recover for all practical purposes their initial state. This recovery takes place with a time constant of about 100 hours.

THE NONLINEAR ELECTRICAL CHARACTERISTICS

According to physical theory [3], the drain current in MOS transistors is given by :

$$(1)\quad Id = k\left[(Vg - \overline{V}g)Vd - \frac{1}{2}Vd^2\right]$$

./.

where Vg, Vd are gate and drain voltages, respectively (Fig. 1), and k, $\overline{V}g$ are constants depending on geometry and materials of the MOS transistors. Tests have shown that (1) is not sufficiently accurate for design purposes. For the MOS transistors used, (1) should be replaced by :

$$(2)\quad Id = \left(Vd + \overline{V}d - \frac{\beta}{Vg - \overline{V}g}\right)\left[(Vg - \overline{V}g)^2 - \alpha(Vg - \overline{V}g)^3\right], \quad \text{for } Vg - \overline{V}g \geq 0$$

$$= 0, \quad \text{for } Vg - \overline{V}g < 0$$

where α, β, $\overline{V}d$ are additional constants. For example, the characteristics shown in Fig. 2 are described with an error less than 2 % by the expression

$$(3)\quad Id = \left[Vd + 78 - \frac{34.5}{Vg - 4.22}\right]\left[(Vg - 4.22)^2 - 0.022\,(Vg - 4.22)^3\right], \quad \mu A,$$

$$5 \leqslant |Vd| \leqslant 20, \qquad 4.22 \leqslant |Vg| \leqslant 10, \quad V.$$

It was found that the constants k, α, β, $\overline{V}d$, $\overline{V}g$ vary widely from one MOS transistor to another, but their values are stable with respect to time. These constants are also practically unaffected by the mode of operation ; in other words, they remain the same in various circuit configurations and do not change with continuous or intermittent operation.

Using (2) it is possible to predict the behaviour of circuits containing MOS transistors. The mathematical analysis is generally tedious but the results obtained are quite reliable.

A LOW-LEVEL CURRENT AMPLIFIER

To accomodate a wide range of current variations, like those likely to occur in the measurement of ionization, the amplifier was designed to have a logarithmic response. Exploiting the very high input impedance of a MOS transistor and the independance of this input impedance of the operating point, the following nominal characteristics were obtained :

- Input current range : 4 decades, from 10^{-12} to 10^{-8} ampers.
- Risetime : less than 50 milliseconds.
- Maximum output voltage : 10 volts.
- Power drain of amplifier : less than 0.1 watt.
- Temperature sensitivity of the log element : 0.01 of the variation of ambiant temperature.

./.

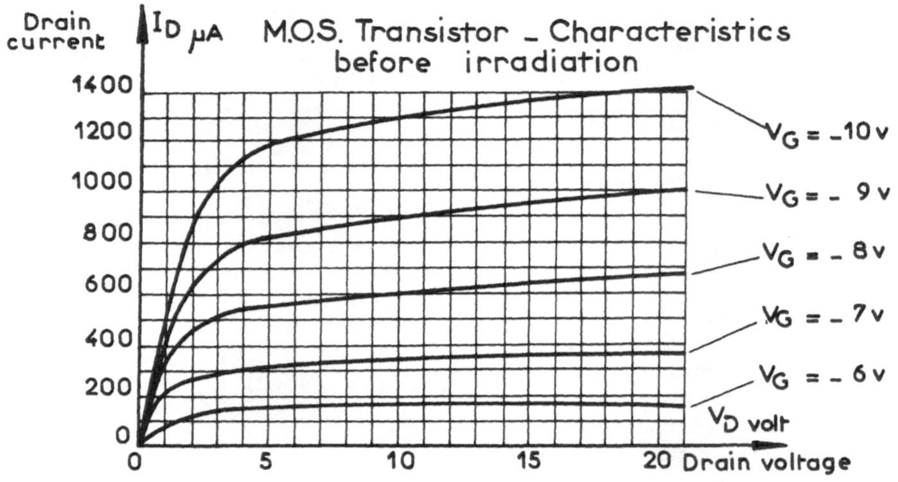

Drain current
I_D μA
M.O.S. Transistor _ Characteristics
before irradiation
1400
1200
1000
800
600
400
200
0
0
5
10
15
20
Drain voltage
V_G = -10 v
V_G = - 9 v
V_G = - 8 v
V_G = - 7 v
V_G = - 6 v
V_D volt

_ Fig. 2 _

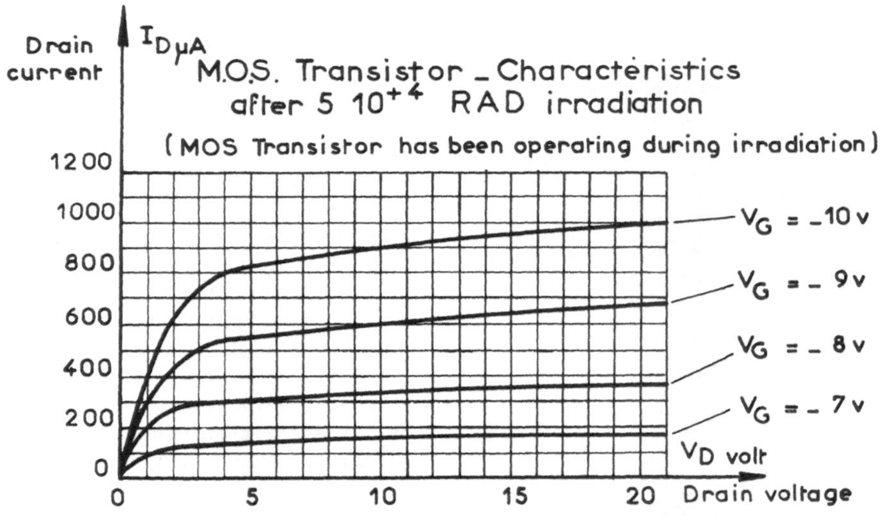

Drain current
I_D μA
M.O.S. Transistor _ Characteristics
after 5 10^{+4} RAD irradiation
(MOS Transistor has been operating during irradiation)
1200
1000
800
600
400
200
0
0
5
10
15
20
Drain voltage
V_G = -10 v
V_G = - 9 v
V_G = - 8 v
V_G = - 7 v
V_D volt

_ Fig. 3 _

Voltage overload protection of the gate terminal is provided by means of a diode network. This network decreases the amplifier input impedance from 10^{15} ohms to about 10^{12} ohms at 323 °K. The only alternative to a MOS transistor would be an electrometer tube, but the latter is mechanically far less reliable.

A successful low-level current amplifier must be highly drift-compensated. Since chopper stabilization is power-wasting, even if mechanical limitations were inexistent, an electronic drift compensation was used. The residual drift and noise level at the input is about 3 % of the minimum input.

The study of drift with respect to time has shown that the long-range drift is less than 5 millivolts during about 200 hours. When a very short but sufficiently strong voltage or current disturbance happens to destroy momentarily the MOS transistor conduction channel a short-range drift is observed. This short-range drift can attain 20 millivolts, but it was found to decrease rapidly and to attain the long-range drift level in about 20 minutes. To diminish the probability of disturbances which are likely to destroy the conduction channel, a protection circuit has been provided. Because of this protection circuit it was necessary to reduce the input impedance of the amplifier from 10^{15} ohms to about 10^{12} ohms.

If the low-current amplifier is designed with high over-all feedback, then it is found that the thermal drift of the input stage is described by a function of temperature which possesses an inflection point. The following formula was deduced from experimental data :

$$\frac{\partial Vg}{\partial T} = \left[5 - 2.5\,(Vg - \overline{Vg})\right] \times 10^{-3} \qquad V/°K$$

Consequently, if the input stage operating point is made to coincide with this inflection point, the resulting thermal drift will be very low. Contrary to ordinary transistors, this thermal drift compensation method does not imply a severe reduction of bandwidth of a MOS transistor. The resulting gain of the MOS transistor is still appreciable (gm = 200 μA/V at Id = 0.1 μA).

A functional diagram of the amplifier is given in (Fig. 4).

./.

A LOW - LEVEL D C VOLTAGE AMPLIFIER

To make the amplifier relatively insensitive to gamma radiation a modulation technique has been used. The key feature of the amplifier consists in the exploitation of the MOS transistor nonlinearity. At low drain-source voltages the MOS transistor can be considered as a variable resistor whose value is controlled by the gate source voltage (Fig. 1). This resistive modulation technique renders the amplifier practically independent of temperature variations. It also compensates the effect of the radiation-induced voltage, provided the latter does not exceed 5 volts. Overload protection is provided in the same manner as in the case of the low current amplifier.

The characteristics are as follows :
- Gain : 100 - Bandwidth : DC to 20 cps.
- Input impedance : 10^8 ohms - Power drain : 0.2 watts.
- Return difference (feedback) : 50 dB.

The bandwidth has been kept very low intentionally because it can be extended up to the megacycle range by the standard addition of an A C high-frequency amplifier.

A functional diagram of the amplifier is given in (Fig. 5).

A WIDE-BAND SQUARING CIRCUIT

From equation (2) it is obvious that for Vd = cte and $|\alpha|$ small the current Id will deviate little from an ideal parabolic segment. The small cubic term can be compensated by two matched MOS transistors operating in parallel, but with opposite signal polarities. Tests have shown that this distortion compensation does indeed take place and that it does not depend critically on the MOS transistor matching. Drift compensation and a reduction of the radiation-induced signal is provided by means of a bootstrap arrangement. A squaring circuit can be used for a direct measurement of RMS voltage and thus, if the load is resistive, for a direct measurement of power.

The squaring circuit has the following characteristics :
- Bandwidth = D C to 250 Kc
- Nominal gain factor k = 0.41 (defined by : (output voltage) = k (input voltage)^2)
- Deviation from square low = less than 0.2 %

./.

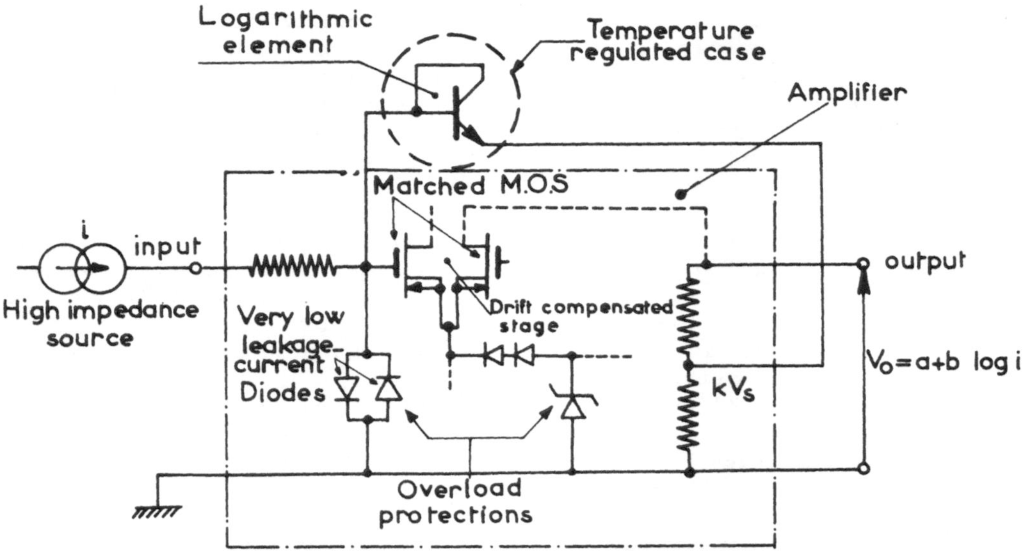

— Low — level D.C. current amplifier —

— Fig. 4 —

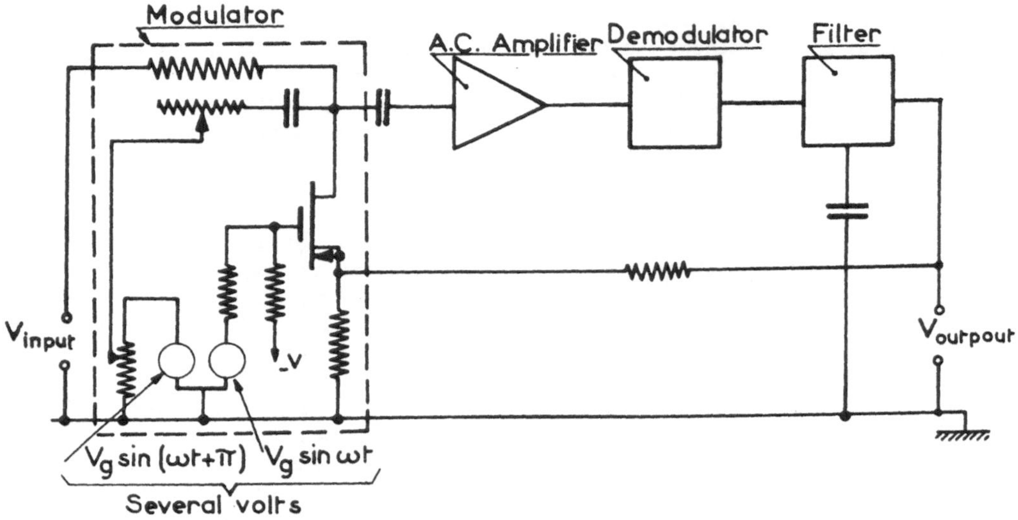

— Low — level D.C. Amplifier —

— Fig. 5 —

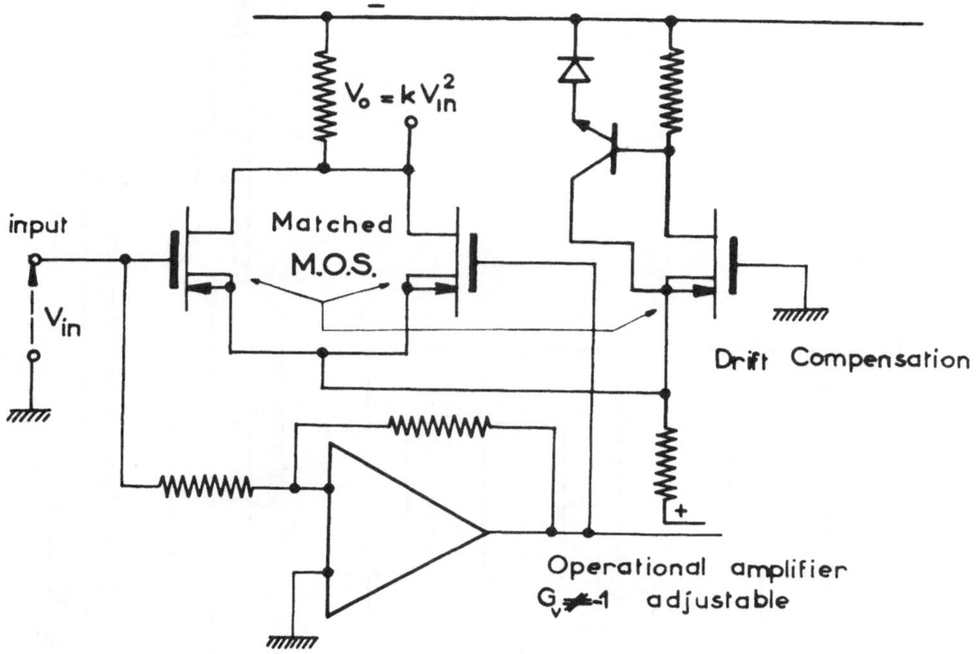

— Wide-band squaring circuit —

— Fig. 6 —

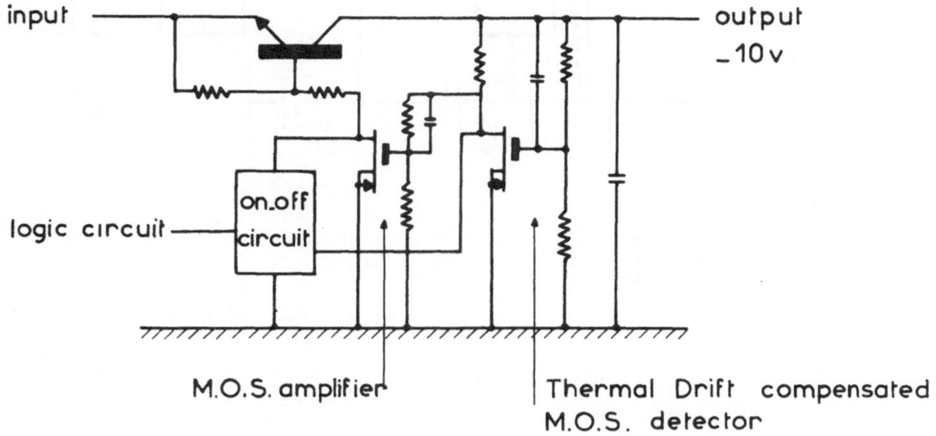

M.O.S. Regulated voltage supply

Fig. 7

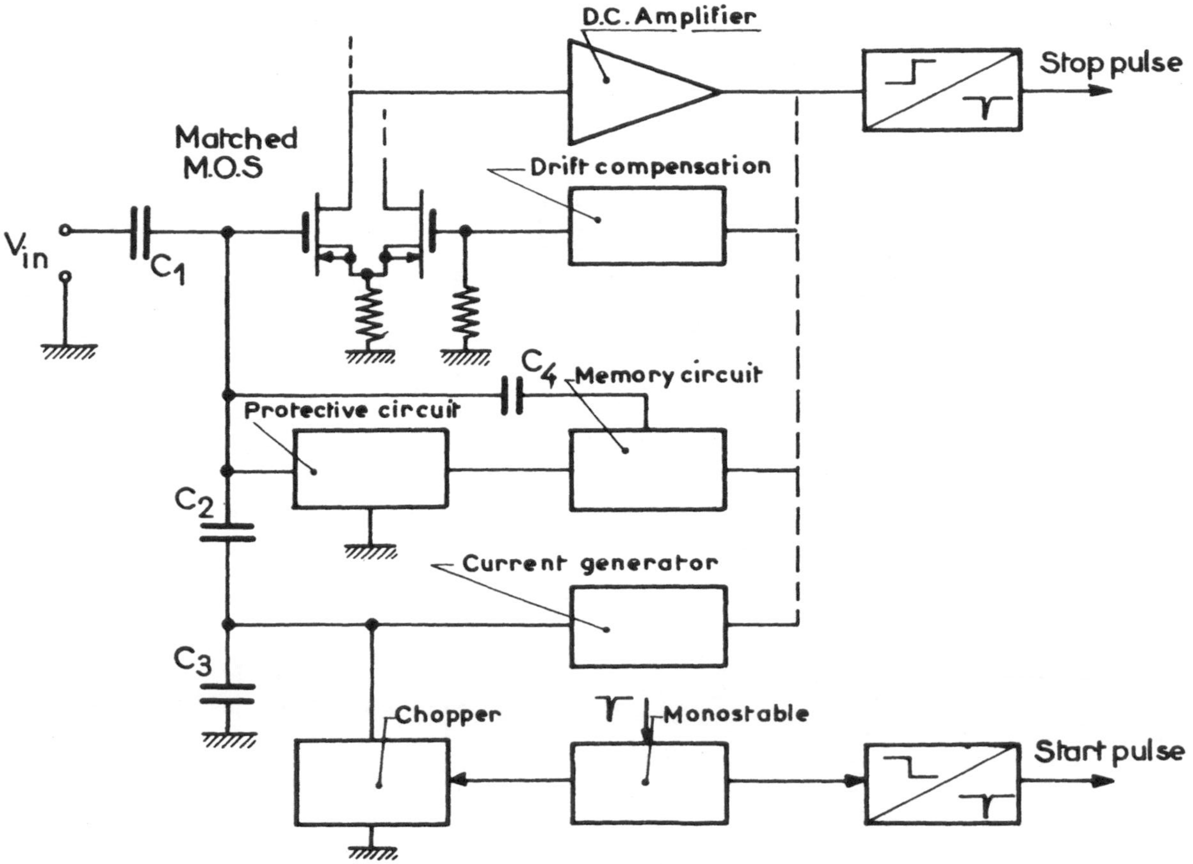

— A low level analog to digital converter —

— Fig. 8 —

- Power drain = 0.1 watts.

The diagram of the squaring circuits is given in (Fig. 6.).

A wide-band multiplying circuit can be obtained by means of two squaring circuits and three linear amplifiers, using the formula :

$$z = 4\,xy = (x+y)^2 - (x-y)^2.$$

The resulting precision and bandwidth will be of the same order of magnitude as that of the amplifying and squaring circuit used. Since this multiplication circuit will operate in four quadrants, it can be used as a controlled gain amplifier with either a positive or negative gain.

POWER SOURCE REGULATING CIRCUITS

If several, possibly variable loads, are to be fed from a single D C power source at different but constant voltage levels, then it is necessary to provide some form of voltage regulating circuits. These regulating circuits should consume the least possible amount of power. A controlled variable series resistance is advantageous when the source-load voltage difference is not too large. Voltage regulators of this type are rendered more efficient by MOS transistors. In fact, since $Id = 0$ for $Vg - \overline{V}g \leqslant 0$ (equation 2), a single MOS transistor is capable of replacing a Zener diode, used as a voltage reference, thus diminishing another power loss.

Unfortunately, no protection against irradiation by gamma rays can be provided, because the radiation-induced signal behaves like if it were added to the cut-off voltage $\overline{V}g$. The constant value of $\overline{V}g$ can therefore no longer be used as a voltage reference.

Without gamma ray irradiation the characteristics are as follows :

- Output voltage : -10 V
- Output Current : 5mA, max.
- Regulation range : - 11.5 V to -20 V
- Stabilization : 1 V input variation produces 2.5 mV output variation.
- Maximum output impedance : 25 ohms.
- A logic circuit operating in on-off conditions sets the output at - 10 V or cuts ./. off the supply.
- Power requirements : Cut-off supply : P≠0

No charge operation, output voltage : - 10 V : P = 0.35 mw.

The circuit diagram appears in (Fig. 7).

A LOW - LEVEL ANALOG TO DIGITAL CONVERTER

Normal analog to digital conversion requires a substantial input signal. If a low-level signal is to be digitalized in this way, it must first be amplified by means of high quality D C amplifier. This high quality D C amplifier can be eliminated when the analog to digital conversion is carried out directly at low level. The high input impedance of MOS transistors permits to achieve this purpose.

It was found that even a single stage MOS transistor amplifier will hold a charge on a small capacitor for a period of over one hour. Consequently, if a constant analog signal is coupled to the input stage of a MOS transistor amplifier through a small capacitor, and the resulting output signal is fed back to the input stage by means of a controlled constant - current generator charging a capacitive voltage divider, then, with a proper choice of polarities, the analog input signal will be converted at the output into essentially a rectangular pulse. The duration of this pulse will be proportional to the amplitude of the input signal[6]. The width of a rectangular pulse is easily digitalized in the usual way.

Since the conversion described can be carried out in periodic cycles, and a few of these cycles can be used for resetting the zero-level of the amplifier, the amplifier will need only moderate drift compensation. Linearity of the amplifier is of course unimportant. The circuit was disigned to digitalize input data available in the pulse-amplitude modulated form.

The characteristics obtained are as follows :

- Input level : 0 - 100 mV
- Sampling time : less than 10μs.
- Conversion factor : 1μsec./mV
- Sampling frequency : 10 Kc
- Precision = 1 %

This analog to digital converter can be used as a memory with non-destructive read- ./.

out, provided the signal is not stored for an excessively long period. An amplitude loss of 1 % will occur in about 30 minutes.

The complete circuit, containing drift compensation, overload protection and partial radiation protection is given in Fig. 8.

ACKNOWLEDGMENT

The following member of the EERL Scientific Staff have contributed technical data for the preparation of this paper :
B. ANDRE, J.P. BABARY, G. BAUZIL, J. BRIAND, J. CLOT, D. ESTEVE, G. FELDMAN, A. LIEGEOIS, R. PRAJOUX, J.C. MARTIN, Dr C. MIRA, D. SEGUIN, Dr Y. SEVELY, J. SIMONNE, J. URGELL.

The manuscript was typed by Miss D. PAUGAM and the illustrations were drawn by E. LAPEYRE - MESTRE.

Some MOS transistors were graciously furnished by S.G.S. FAIRCHILD (Milan) and C.O.S.E.M. (Paris).

REFERENCES

1 F.M. WANLASS, C.T. SAH "Nanowatt logic using field-effect metal-oxide - semiconductor triodes", Digest of Tech. Papers, IRE International Solid State - Conf. 1963.

2 H.L. HUGHES, R.R. GIROUX "Space Radiation affects MOS-FET's" Electronics - December 1964 - pp. 58 - 60

3 C.T. SAH, "Characteristics of MOS transistors", IEEE Transactions on Electron Devices, July 1964 - pp. 324 - 345.

4 Dispositif digital pour la mesure de charges électriques - Brevet C.N.R.S. n° PV 891 487 du 19 mars 1962. J. LAGASSE, G. GIRALT, J. CLOT.

BIBLIOGRAPHIE

- S.R. HOFSTEIN and F.P. NEIMAN - "The Silicon Insulated-Gate Field-Effect Transistor" - Proceedings of the I E E E - Vol. 51 n° 9 - September 1963.

- J.S. MAC DOUGALL - "Some Circuit Applications of Metal over Oxide Silicon Transistors" - S.G.S. FAIRCHILD - Applications Engineering Seminar - January 28 - 1964.

- J. GROSVALET, C. MOTSCH, R. TRIBES - Note de la C.S.F. - "Le Statistor" - Mai 1964

- J. EIMBINDER - "The Field-Effect Transistor : a curiosity comes of Age" - Electronics -November 30, 1964.

- F.P. HEIMAN and S.R. HOFSTEIN -"Metal -Oxide- Semiconductor Field Effect transistors" - Electronics - November 30, 1964.

- J.M. COHEN -"How to measure FET noise" Electronics - November 30 - 1964.

- V. HARRAP, G. PIERSON, M. KUEHLER and B.K. LOVELACE - "Researchers turn to Germanium for a MOS Field - Effect transistor" - Electronics - November 30, 1964

- R.S.C. COTHOLD and F.N. TROFIMENKOFF "Theory and Application of the Field effect transistor" - Proceedings of the I E E -Vol. 111 n° 12 - December 1964.

- P.J. COPPEN - "FET Complementary integrated circuits : aerospace natural" - Electronics - December 1964

- M.H. LIND OLESEN - "Designing against space radiation " - Part 1 - Electronics - December 1964.

- M.H. LIND OLESEN- "Designing against space radiation" - Part 2 - Electronics - January 1965.

./.

DISCUSSION

J. Lagasse,
I. Gumowski and G. Giralt

"Low Power Signal
Transformation
for Space Applications".

Q. For what purpose are the logarithmic amplifiers intended?

A. For applications involving unknown amplitudes of great dynamic range. In this paper, an amplifier with a range of 10,000 to 1 was described. One application is the measurement of ionization currents in an unknown enviornment.

Q. What supply voltage was used for the d-c amplifier, and over what temperature range is it designed to work?

A. $\pm$ 10 volts; -60° C to +40° C.

Q. What is the d-c amplifier bandwidth?

A. It was deliberately limited to 1 kc.

Q. How are the semiconductors effected by neutrons?

A. It depends on neutron energy; for fast neutrons the behavior is stable, but for slow neutrons it is pretty disasterous. We have no space data yet because France has not yet performed space experiments.

Q. What selection is needed to provide transistors with characteristics suitable for your circuits?

A. Selection is done by the manufacturer. We receive transistors which meet our specifications and can only judge selection ratio by price, which is pretty steep.

General Topics

REVIEW PAPER ON THE RESEARCH COMPLETED AT THE COMPUTING CENTER OF THE ACADEMY OF SCIENCES OF THE USSR ON THE THEORY OF OPTIMAL CONTROL FUNCTIONS OF SPACECRAFT

N. N. Moiseev, V. N. Lebedev

ABSTRACT

This paper constitutes a review of the research work completed at the Computing Center of the USSR Academy of Sciences dealing with the theory of optimal orbit transfers for space vehicles powered by low-thrust rocket engines.

Results of the solutions of problems using the maximum principle and dynamic programming techniques are reported.

Systematic research aimed at the development of standard numerical techniques for solving optimal control problems has been in progress for the past several years at the Computing Center of the Academy of Sciences of the USSR. A broad range of problems originating in various fields of science and engineering has been examined. Spacecraft dynamics problems occupy a prominent place in the hierarchy of problems researched.

The purpose of this paper is to furnish a review of the work done on the dynamics and control of spacecraft. The brunt of the attention will be directed to method and technique — to the structure and realization of algorithms.

The research projects described in this review paper are divided under two headings. The first heading includes research on problems admitting of solution by the maximum principle. The second group of problems includes those which do not yield to the maximum principle approach. In the latter case attention will be concentrated on the elaboration of numerical algorithms of the dynamic programming type.

Authors of papers reviewed here include, besides the authors of this review paper, N. Ya. Bagaeva, I. E. Vol'fson, A. N. Zhukov, F. I. Ereshko, I. A. Krylov, B. N. Rumyantsev, V. A. Chebakov, and F. L. Chernous'ko.

The papers included in this review have been published in a variety of journals. The reader will be informed where to find the original papers at appropriate points throughout this review.

I. PROBLEMS SOLVABLE BY THE MAXIMUM PRINCIPLE

1. One of the ways of solving variational problems is to reduce them by the Lagrange-Euler formalism or by the Pontryagin formalism to the solution of boundary problems for ordinary differential equations. Insufficient initial conditions may be determined by Newton's method, consisting of the following. Let $x_{10}, x_{20}, \ldots x_{m0}$ be assumed insufficient initial values for a system of equations; let $x_{11}, x_{21}, \ldots, x_{m1}$ be the specified final values. We now denote as $\bar{x}_{11}, \bar{x}_{21}, \ldots, \bar{x}_{m1}$ the values of the functions $x_1, x_2, \ldots, x_m$ obtained on integration when the assumed values $x_{10}, x_{20}, \ldots, x_{m0}$ are used. We assume that the increments imparted to the variables $\bar{x}_{11}, \bar{x}_{21}, \ldots, \bar{x}_{m1}$ when the assumed initial values are altered are expressed linearly in terms of these variations. The form of these linear expressions may be found after m-tuple integration of the system of equations with the variation of the variables $x_{10}, x_{20}, \ldots, x_{m0}$ taken in succession. Using the linear functions so found, we now make corrections in the assumed initial values, after which the process is repeated. In the case where the iterative process so described is found to diverge, some modifications of the method may be resorted to, the simplest of these being the forced decrease in the increments found for the insufficient initial conditions.

The scheme outlined for solving the boundary problem proves highly effective for many concrete numerical problems (i.e. yields a rapid convergence to the solution) and enables its users to achieve the required accuracy in a small number of iteration steps (5 to 6 steps). The actual process of computation has revealed, however, that the scheme used is by no means so versatile as to answer every application. In some cases the solution has proved highly sensitive to any change in the initial data. Under these conditions, Newton's method no longer makes it possible to arrive at a final result in a small number of iterations.

Below we show the solution of several variational problems in the dynamics of spacecraft powered by low-thrust rocket engines. The assumption entertained in tackling all these problems is that the space vehicle, viewed as a point mass, is moving in a Newtonian field of gravitational attraction; perturbations due to asphericity of the attracting planet, to atmospheric drag, to influences brought to bear by the sun or by neighboring planets, etc., may be disregarded. In all the problems (with the sole exception of the problem of solar wind sailing) the reactive acceleration (thrust to mass ratio of the vehicle) is assumed constant, i.e., any change in the mass of the

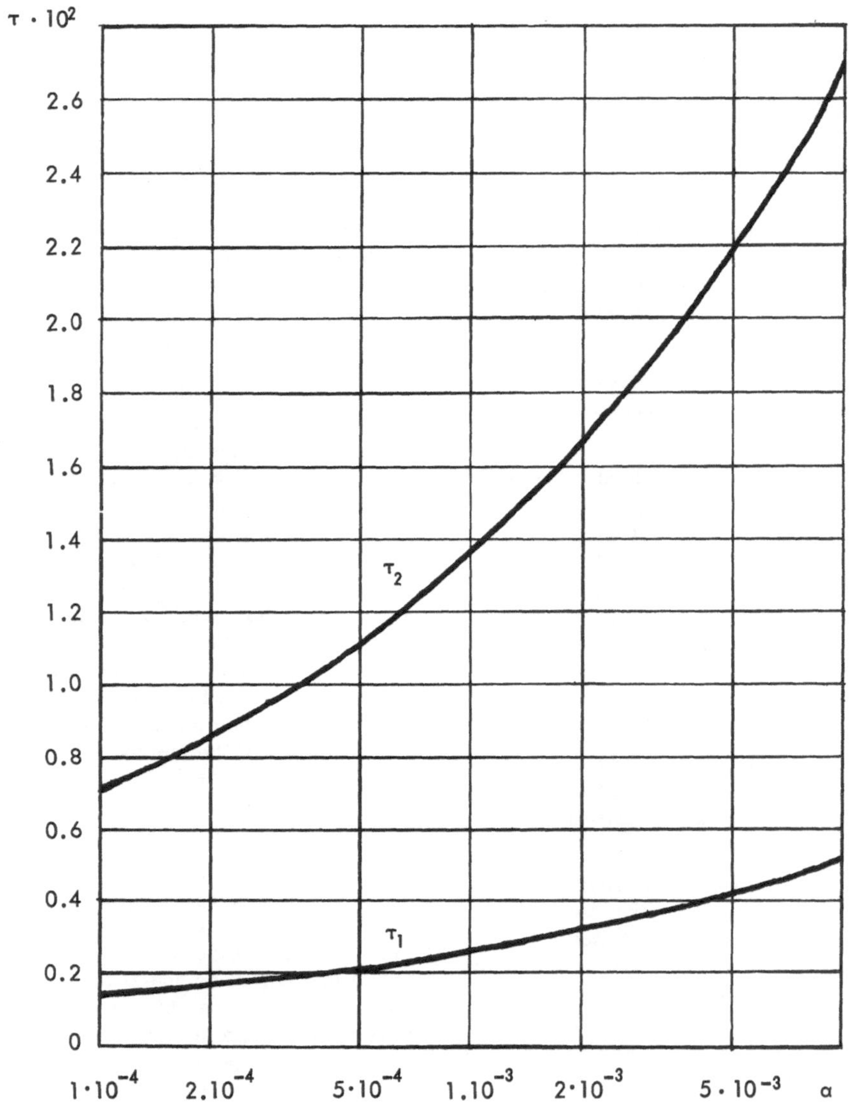
τ · 10²
2.6
2.4
2.2
2.0
1.8
1.6
1.4
1.2
1.0
0.8
0.6
0.4
0.2
0
τ2
τ1
1·10⁻⁴
2.10⁻⁴
5·10⁻⁴
1.10⁻³
2·10⁻³
5 · 10⁻³
α

FIGURE 1

spacecraft is assumed negligible. The control function is consequently only an angle specifying the direction of the thrust vector (or two angles in a nonplanar problem), and the maneuvering time is to be minimized, as a rule.

2. Accelerating a Spacecraft to Parabolic Velocity

The assumption is that the craft is moving in the plane of the initial circular orbit, i.e., we are dealing with a problem in the plane.

The motion executed by the spacecraft is described by the following system of equations:

$$\dot{u} = \frac{v^2}{r} - \frac{1}{r^2} + \alpha\cos\lambda \tag{1.1}$$

$$\dot{v} = -\frac{uv}{r} + \alpha\sin\lambda \tag{1.2}$$

$$\dot{r} = u \tag{1.3}$$

$$\dot{\varphi} = \frac{v}{r} \tag{1.4}$$

where r and φ are polar coordinates, u and v are the radial and transversal components of the velocity, λ is the angle formed by the radius vector of the craft with the thrust vector, α is the reactive acceleration, and dotted variables are differentiated with respect to the time t. All the variables are dimensionless here. The corresponding dimensioned variables marked by asterisks may be obtained from the formulas:

$$r^* = r r_0^*, \quad u^* = u\sqrt{\frac{\mu}{r_0^*}}, \quad v^* = v\sqrt{\frac{\mu}{r_0^*}}$$

$$t^* = t r_0^* \sqrt{\frac{r_0^*}{\mu}}, \quad \alpha^* = \alpha\frac{\mu}{r_0^{*2}}$$

where r_0^* is the radius of the initial circular orbit, μ is the gravitational potential constant of the planet.

Here we require a control function $\lambda(t)$ making it possible to attain parabolic velocity in a minimum time $t = t_k$.

Making use of the L.S. Pontryagin maximum principle, which is explained in the literature [1, 2], we may reduce the search for the optimal control function to the solution of the following boundary problem for a system of ordinary differential equations:

$$\dot{u} = \frac{v^2}{r} - \frac{1}{r^2} + \frac{\alpha p_u}{\sqrt{p_u^2 + p_v^2}} \tag{1.5}$$

$$\dot{v} = -\frac{uv}{r} + \frac{\alpha p_v}{\sqrt{p_u^2 + p_v^2}} \tag{1.6}$$

$$\dot{r} = u \tag{1.7}$$

$$\dot{p}_u = -p_r + \frac{p_v v}{r} \tag{1.8}$$

$$\dot{p}_v = -\frac{2p_u v}{r} + \frac{p_v u}{r} \tag{1.9}$$

$$\dot{p}_r = \frac{p_u v^2}{r^2} - \frac{2p_u}{r^3} - \frac{p_v u v}{r^2} \tag{1.10}$$

$$t = 0, \quad u = 0, \quad v = 1, \quad r = 1, \quad p_v = 1 \tag{1.11}$$

$$t = t_k, \quad r(u^2 + v^2) = 2, \quad p_u = p_r u r^2, \quad p_v = p_r v r^2 \tag{1.12}$$

Here p_u, p_v, p_r, are impulses (Lagrange multipliers).

The control angle function λ is related to those impulses as:

$$\sin\lambda = \frac{p_v}{\sqrt{p_u^2 + p_v^2}}, \quad \cos\lambda = \frac{p_u}{\sqrt{p_u^2 + p_v^2}}$$

The boundary problem is solved by computer, using Newton's method. In cases where the iterative process fails to converge, we used some modifications of the Newton method described elsewhere [3]. Calculations showed that the thrust vector executes oscillating motions about the velocity vector on the optimal trajectory. The amplitude of these oscillations grows with time, and the period of the oscillations is roughly equal to the time of a single complete revolution about the center of attraction. Figure 1 shows plots of the reactive acceleration dependence of the variables

$$\tau_1 = (t_{k1} - t_k)/t_k \quad \& \quad \tau_2 = (t_{k2} - t_k)/t_k$$

where t_k, t_{k1}, and t_{k2} are the optimum acceleration time, the acceleration time when tangential thrust is applied, and the acceleration time when transversal thrust is applied, respectively.

A study of the effect of passive segments of the trajectory on maneuver costs would be of interest. To this end we solved the following variational problem: for the case of a transversally applied thrust it is required to find the number and location of passive segments which will permit the attainment of the parabolic velocity in minimum engine time for a specified total maneuvering time. The mathematical statement of this type of variational problem is discussed in detail in [4].

Some of the results of calculations performed for the case $\alpha = 10^{-2}$ are tabulated:

Total maneuver time	97.0	120.0	140.0	170.0
Number passive segments	5	6	7	8
Power-on time	71.6	67.5	64.7	61.3

The time required to attain parabolic velocity under a constant applied transversal thrust (at $\alpha = 10^{-2}$) is 76.12, and in the case of optimum control without switching off the engine parabolic velocity is attained at a time 74.13. Hence, an insignificant increase in total maneuvering time because of the presence of passive segments results in the transversal thrust becoming cheaper than the optimal thrust.

The first part of the problem (acceleration with no passive segments) has been published in [5].

3. Transfer Between Circular Coplanar Orbits and Rounding of an Elliptical Orbit

The boundary problem to which the search for a control function optimal in rapidity of convergence to solution is reduced differs from the boundary problem stated in formulas (1,5) to (1.12) only in the boundary conditions

$$t=0,\ u=0,\ \dot{v}=1,\ r=1,\ \rho_v=1$$

$$t=t_k,\ r=r_k,\ u=0,\ v=1/\sqrt{r_k}$$

Calculations performed for the case of a transfer from the earth's orbit to orbits around other target planets at $\alpha = 0.1667$ correspond more or less to the dimensioned reactive acceleration 1 mm/sec^2. The time required to arrive at the planets, in days, is: Venus—135, Mars—189, Jupiter—540, Saturn—789, Uranus—1150, Neptune—1480, Pluto—1720. This problem has been published in [6].

We here consider the problem of the transfer of a spacecraft from an elliptic departure orbit of eccentricity e_0 to a circular target orbit of unspecified radius in minimum transfer time.

Recourse to the maximum principle reduces the problem as posed to the solution of the system of equations (1.4), (1.5) to (1.10) with the following boundary conditions:

$$t=t_0,\ u=\sqrt{\frac{1}{1-e_0}}\,e_0\sin\varphi_0,\ v=\sqrt{\frac{1}{1-e_0}}\,(1+e_0\cos\varphi_0),$$

$$r=\frac{1-e_0}{1+e_0\cos\varphi_0},\ \varphi=\varphi_0,\ \rho_v=1;\ t=t_k,\ rv^2=1$$

$$u=0,\ 2\rho_r r=\rho_v v$$

Here e_0 is the eccentricity of the initial orbit, φ_0 is the angle between the direction to the orbit pericenter and the initial radius vector, and the initial distance to the apocenter is taken as the unit distance.

There exists for this problem a locally optimum control which is so selected that the rate of decrease of eccentricity will be minimized.

Calculations performed in the region $3\cdot10^{-2} \leq \alpha \leq 10$, $10^{-2} \leq e_0 \leq 10^{-1}$ revealed that at high accelerations it would be best to initiate the transfer maneuver at the apogee. As the acceleration is decreased the starting point assuring minimum time shifts in the direction opposed to the spacecraft advance. The time required to execute the maneuver with the aid of a locally optimum control function differs only modestly from the time required to execute the optimum maneuver, with the maximum difference 4.9%. The optimal control is close to the locally optimal control. This renders the task of selecting the first-order approximation for solving the boundary problem much easier.

4. Transfer Between Heliocentric Circular Orbits Under Solar Sail Propulsion

The motion of a spacecraft powered by a solar sail is described by the system of equations (1.1) to (1.4), where

$$\alpha = \frac{\alpha_0\cos^2\lambda}{r^2}$$

Here α_0 is the dimensionless acceleration due to the force of radiation pressure on the sail at $\lambda = 0$ and at $r = 1$, where λ is the angle formed by the radius vector with the normal to the surface of the sail.

It is required to solve the system of equations

$$\dot{u} = \frac{v^2}{r} - \frac{1}{r^2} + \frac{\alpha_0\cos^3\lambda}{r^2}$$

$$\dot{v} = -\frac{uv}{r} + \frac{\alpha_0\sin\lambda\cos^2\lambda}{r^2}$$

$$\dot{r} = u$$

$$\dot{\rho}_u = \frac{\rho_v v}{r} - \rho_r$$

$$\dot{\rho}_v = -\frac{2\rho_u v}{r} + \frac{\rho_v u}{r}$$

$$\dot{\rho}_r = \frac{\rho_u v^2}{r^2} - \frac{2\rho_u}{r^3} + \frac{2\alpha_0\rho_u\cos^3\lambda}{r^3} - \frac{\rho_v uv}{r^2} + \frac{2\alpha_0\rho_v\sin\lambda\cos^3\lambda}{r^3}$$

with the boundary conditions:

$$t=0,\ u=0,\ v=1,\ r=1,\ \rho_2=1 \text{ for } r_k>1,\ \rho_2=-1 \text{ for } r_k<1;$$

$$t=t_k,\ r=r_k,\ u=0,\ v=1/\sqrt{r_k}$$

The optimal control is arrived at from the formula

$$\operatorname{tg}\lambda = \frac{\sqrt{9\rho_u^2+8\rho_v^2}-3\rho_u}{4\rho_v},\ \left(-\frac{\pi}{2}\leq\lambda<\frac{\pi}{2}\right)$$

Calculations were performed at $\alpha = 0.3363$, corresponding to the reactive acceleration 2 mm/sec^2. The transfer time in years to target orbits around planets of the solar system are: 0.53 year for Mercury, 0.45 for Venus, 0.88 to Mars, 6.6 years to Jupiter, 17 to Saturn, 49 to Uranus, 96 to Neptune, and 150 years to Pluto.

This problem has been published in [7].

5. Transfer Between Circular Noncoplanar Orbits

Newton's method, which can be invoked for the successful solution of the problems described above, proved far less effective in the search for an optimal control in the sense of rapid convergence to solution for transfers between circular noncoplanar orbits. Solutions were successfully obtained only for cases of high reactive accelerations ($\alpha > 0.1$) and low angles between the planes of the departure and target orbits.

In order to find a possible control function with which this maneuver might be executed successfully at low reactive accelerations, recourse was had to an averaging method described in [8].

Let the reactive acceleration remain constant and be so directed that its radial component will vanish. Then the direction of the thrust vector alignment may be specified by a single variable, viz. the angle ν between the perpendicular to the radius vector in the plane of the instantaneous orbit and the thrust vector ($-\pi \leq \nu < \pi$). If the thrust alignment is changed to the direction symmetric about the plane of the instantaneous orbit at points such that $u = \pm\pi/2$ (where u is the argument of the latitude angle), then the tilt of the orbit will vary monotonically, as is inferred from [9]. We shall assume ν = const. Then the motion of the spacecraft in the plane of the instantaneous orbit will take place in response to a constant transversal accelerations and, as shown in [10], the osculating orbit will remain close to a circular orbit for an indefinite time if the eccentricity of the departure orbit is zero. Hence, this control function can be relied upon to achieve a transfer from a circular departure orbit to another specified close-to-circular target orbit.

The motion of a spacecraft with the control function proposed is described by the following system of averaged equations (the initial equations were taken from [10]):

$$\frac{dr}{dt} = \alpha r\sqrt{r}\cos|\nu|,$$

$$\frac{di}{dt} = \frac{2\alpha}{\pi}\sqrt{r}\sin|\nu|$$

Here i is the angle between the planes of the instantaneous orbit and the departure orbit. In the averaged motion the eccentricity over the polar angle range from zero to $1/\alpha$ is identically zero, and the position of the line of modes remains unaffected.

The resulting averaged equations are also valid in the case where ν is a slowly varying function of the time. Among these functions we may attempt to find a function such that transfer from a point $r = 1$, $i = 0$ to a target point $r = r_k$, $i = i_k$ may be achieved in minimum time. The boundary problem corresponding to the variational problem as posed is solvable analytically. Its solution takes the form:

$$r = \frac{1}{1 - \dfrac{2\left(1 - \dfrac{\cos\frac{\pi i_k}{2}}{\sqrt{r_k}}\right)}{\sqrt{1 - \dfrac{2\cos\frac{\pi i_k}{2}}{\sqrt{r_k}} + \dfrac{1}{r_k}}}\alpha t + (\alpha t)^2}$$

$$i = \frac{2}{\pi}\operatorname{arctg}\frac{\left(\sin\frac{\pi i_k}{2}/\sqrt{r_k}\right)\alpha t}{\sqrt{1 - \dfrac{2\cos\frac{\pi i_k}{2}}{\sqrt{r_k}} + \dfrac{1}{r_k}} - \left(1 - \dfrac{\cos\frac{\pi i_k}{2}}{\sqrt{r_k}}\right)\alpha t}$$

The time dependence of the modulus of angle ν is expressed by the formula

$$|\nu| = \operatorname{arctg}\frac{\dfrac{\sin\frac{\pi i_k}{2}}{\sqrt{r_k}}}{1 - \dfrac{\cos\frac{\pi i_k}{2}}{\sqrt{r_k}} - \sqrt{1 - \dfrac{2\cos\frac{\pi i_k}{2}}{\sqrt{r_k}} + \dfrac{1}{r_k}}\,\alpha t} \qquad (1.13)$$

The time required to execute the maneuver is

$$t_k = \frac{1}{\alpha}\sqrt{1 - \frac{2\cos\frac{\pi i_k}{2}}{\sqrt{r_k}} + \frac{1}{r_k}}$$

We infer from this solution that the time dependence of the radius need not be monotonic. For example, in the case of a transfer between orbits of like radii, the planes of which are perpendicular, the maximum radius will exceed the initial radius by more than nine times. A payoff of 23% in the maneuver time is thereby attained, over the time required for a maneuver involving rotating the plane of the circular orbit by means of a thrust directed normal to the instantaneous orbit plane.

In order to determine the size of the possible error in the parameters of the target orbit introduced by the use of averaged equations, a computer was used to integrate the exact system of equations with the control function (1.13) for the case of a transfer from a circular departure orbit 6671 km in radius to the orbit of a satellite orbiting the earth once daily, the radius of this orbit being 42,240 km. The angle between the planes of these orbits has been agreed upon as 48°. It was shown that when $\alpha^* \leq 2$ mm/sec^2 the eccentricity of the target orbit is less than 0.01, and the error in the tilt of the orbit is less than half an angular degree, while the relative error in the mean radius is less than half a percent.

II. PROBLEMS ATTACKED BY THE METHOD OF DYNAMIC PROGRAMMING

1. In astrodynamics we sometimes encounter a variety of problems which do not yield to the L.S. Pontryagin maximum principle. This is mainly the case in problems containing phase restrictions. The methods of dynamic programming are brought to bear on these problems. Realization of these methods consists in the special organization of variational procedures in state space and in the choice of a simplified phase trajectory. The structure of state space allows us to get by all the difficulties associated with phase restriction requirements in a rather elementary way. The method leads to a converging and stable iterative process. The main difficulty in applying the method involves the construction of an elementary operation, an algorithm, which will place two neighboring points in state space in correspondence with a phase trajectory joining them, a control function with which this mapping can be brought about, and the value of the desired functional on the portion of the trajectory joining the two mentioned points in state space.

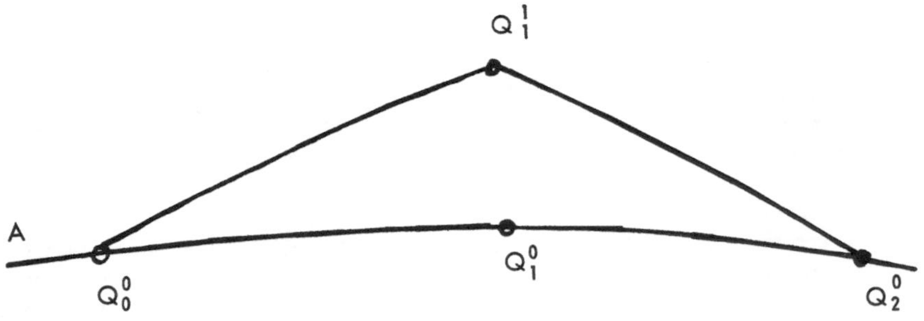

FIGURE 2

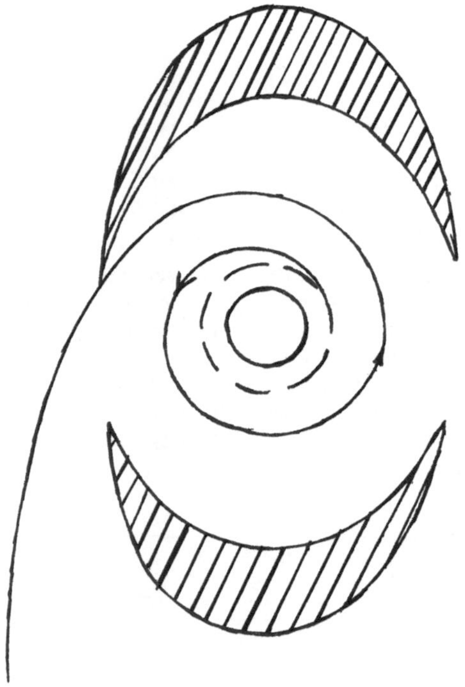

FIGURE 3

Variation of the trajectories and jump techniques are fully standard procedures. Independent programming is required to perform the elementary operation each time.

The scope of this review does not permit sufficient space to present the overall schema of the calculations in great detail. We therefore present only a few explanatory remarks.

Let the functional to be minimized be $J(x, u) = \int_0^T F(x,u)\,dt$ where the control $u \in G_1$, and $x \in G_2$ (G_1 and G_2 are subsets) and furthermore the vector x describing the phase state satisfies the system of equations

$$x' = X(x, u) \tag{2.1}$$

Further, let $x = x^T$ at $t = T$ and $x = x^0$ at $t = 0$. Other restrictions could also be placed on the initial and final states, but we use these only for illustrative purposes.

We denote as τ a step or interval in the variable t and suppose a phase trajectory to be drawn from the point P_0^0 ($x = x^0$, $t = 0$) to each point P_i of the subspace (x, $t = i\tau$), with this phase trajectory yielding the minimum of the functional J over the set of all admissible curves joining point P_0^0 to point P_i. The value of the functional on this trajectory will be designated as $J(x, u, i\tau)$. Now consider any point $P_{i+1}^*[x = x^*, t = (i + 1)\tau]$. The elementary operation, which we shall assume to be known, is the algorithm enabling us to construct a trajectory joining the points $(\tilde{x}, i\tau)$ and $[x^*, (i + 1)\tau]$ for any $\tilde{x}$ and any x^* of G_2, and to compute the value of the functional $J(x, x^*, u, \tau)$. Then, by virtue of the additivity of the functional

$$J(x^*, u, (i+1)\tau) = \min_x \left(J(x, x^*, u, \tau) + J(x, u, i\tau) \right) \tag{2.2}$$

Equation (2.2) may be regarded as the difference analog of the Bellman equation.

On determining $J[x, u, (i + 1)\tau]$ for any $x^* \in G_2$, we go over to the set $P_{i+2}[x, t = (i + 2)\tau]$, and so forth.

This jump is achieved by means of a difference scheme, and the set of mesh nodes is arbitrarily termed the state scale $\{P_j^i\}$ where the subscript is the number of the stratum in t, and the superscript is the number of the node in the j-th stratum. The choice of mesh nodes makes it possible to automatically satisfy all phase restrictions of the inequality type.

Realization of this scheme makes it possible, theoretically in any case, to carry out a search for a global minimum within the framework of the stated scheme, and it should be quite evident that the larger the number of restrictions the smaller will be the state scale and the simpler will be the choice of trajectories to be varied.

But in all problems of dynamics we have to deal with, the volume of state space is so vast that direct realization of this process on medium-capacity computers would be out of the question. We therefore took recourse to some iterative schemes (e.g., cf. [11, 12]). At each step in the iterations a comparison of several phase trajectories only was carried out. The value of this simplification in procedure is enormous: instead of seeking an absolute minimum we are obliged to restrict ourselves to a search for a local minimum. But once embarked on that road we may take the following step. Instead of a bundle of trajectories to be varied we need consider only one trajectory. Schemas of this type have been proposed by I.A. Krylov. These schemas could be termed dynamic programming schemas using global variations. I.A. Krylov and F.L. Chernous'ko have achieved a further simplification of the algorithm, which now uses only variations at a point. The schema has by now been reduced to a very simple form.

Let us have at our disposal some admissible trajectory AB (cf. Fig. 2). At some point Q_1^0 corresponding to some instant of time which we can arbitrarily designate $t_0 + \tau$ (where τ is the step in time) we take the variation of the phase state (point Q_1^1). At that step we compare only the two numbers

$$J(Q_0^0, Q_1^1, \tau) + J(Q_1^1, Q_2^0, \tau)$$

and

$$J(Q_0^0, Q_1^0, \tau) + J(Q_1^0, Q_2^0, \tau)$$

At the next step we again take the variation at a single point, this time at the point Q_2^0, and so forth.

Each of these techniques for improving the trajectories has its own strong points and only a combination of them will successfully furnish us with universality in our numerical schemas.

Let us make one further remark crucial to the practical realization of the schema. Let τ be a step in time. Then for the iterative process to converge there will have to be at some fixed τ a limiting process such that $\|\Delta x\| \to 0$, where Δx is a vector defining the step in phase variables, and only then can the step in τ be narrowed.

2. Structure of the Elementary Operation

The elementary operation can very rarely be materialized explicitly. Some kind of difference schemas is generally invoked to describe the elementary operation. We here cite several examples of difference schemas which will be useful in our further work, restricting them in this case to only schemas of the first order in exactness.

In effect, let there be given a system

$$x_i' = f_i(x_1, \ldots, x_n, u), \quad i = 1, 2, \ldots, n \tag{2.3}$$

The elementary operation now consists in finding a function $u(t)$ which will carry the system from the position x_i^0 to the position x_i^* in a time τ. τ is the step or interval of the difference schema in the independent variable. We shall now construct the controls of a function continuously differentiable in as many orders as required by the problem. Within the framework of the first-order difference schemas, we may assume

$$\Delta x_i = \int_0^\tau f(x_1^0, \ldots, x_n^0, u(t))\, dt \tag{2.4}$$

The vector formula (2.4) here is a system of equations defining u(t). We use the notation

$$f(x_1^0, \ldots, x_n^0, u(t)) = F(t)$$

Then, assuming

$$F(t) = F(0) + F'(0)t + \frac{1}{2}F''(0)t^2 + \ldots \tag{2.5}$$

where

$$F' = \frac{df}{du}u', \quad F'' = \frac{d^2f}{du^2}u'^2 + \frac{df}{du}u'', \quad \text{and so on,}$$

we shall also seek to approximate the function u(t) by the polynomial

$$u = u_0 + c_1 t + c_2 t^2 + \ldots$$

Substitution of this expression into Eqs. (2.5) and (2.4) yields us a system of equations nonlinear in the coefficients c_i. Hence, we still require a standard program for solving the system of nonlinear equations in order to realize this elementary operation.

There exists a whole range of approaches which will aid us in simplifying this problem. If the system of equations (2.3) is reduced to a single equation, then it will suffice for us to assume that at point t = 0 the value of the control of the first $k (k \geq 1)$ derivatives is specified, so that we then obtain a linear equation for the coefficient c_{k+1}.

If n = 2 in Eq. (2.3), we then require that k derivatives $(k \geq 2)$ be specified at t = 0, so that we shall have a system of two linear equations for c_{k+1} and c_{k+2}.

If some of the equations in the system (2.3) contain no control function, then the problem is somewhat simplified by using the method of fractional intervals or fractional steps. For example, let the system (2.3) contain three equations and suppose the control function does not appear in the first of these equations

$$x_1' = f_1(x_1, x_2, x_3)$$

Then

$$x_1^{(1)} = x\left(\frac{\tau}{2}\right) = x_1^{(0)} + \frac{\tau}{2} f_1(x_1^0, x_2^0, x_3^0) \tag{2.6}$$

$$x_1^{(2)} = x(\tau) = x_1^{(0)} + \frac{\tau}{2} f_1(x_1^0, x_2^0, x_3^0) + \frac{\tau}{2} f_1(x_1^{(1)}, x_2^{(1)}, x_3^{(1)}) \tag{2.7}$$

Only the function

$$f_1(x_1^{(1)}, x_2^{(1)}, x_3^{(1)}) = f_1^{(1)} \tag{2.8}$$

is unknown in the last of these equations.

Equation (2.7) will enable us to determine this function.

Formula (2.8) may be represented as

$$f_1^{(0)} + \left(\frac{\partial f_1}{\partial x_1}\right)^0 (x_1^{(1)} - x_1^{(0)}) + \left(\frac{\partial f_1}{\partial x_2}\right)^0 (x_2^{(1)} - x_2^{(0)}) + \\ + \left(\frac{\partial f_1}{\partial x_3}\right)^0 (x_3^{(1)} - x_3^{(0)}) = f_1' \tag{2.9}$$

We specify the control function over this segment in the form

$$u = u_0 + ct$$

where u_0 is given. Then

$$x_2^{(1)} = x_2^{(0)} + f_2^0 \frac{\tau}{2} + \frac{\partial f_2}{\partial u} c \frac{\tau^2}{4},$$
$$x_3^{(1)} = x_3^{(0)} + f_3^0 \frac{\tau}{2} + \frac{\partial f_3}{\partial u} c \frac{\tau^2}{4}$$

Substituting $x_1^{(1)}$ from Eq. (2.6), and $x_2^{(1)}$ and $x_3^{(1)}$ into Eq. (2.9), we obtain an equation for the determination of c.

Remarks. The resulting equation contains terms of different orders. In order for $x_1^{(1)}$ to be determined to an exactness of the order of $0(\tau^2)$ we must use a schema of second-order exactness. And in turn, we have to use the information of the preceding interval just to achieve that much.

Once the value of c has been determined, we proceed to find the approximate value of the control function over the first half-interval. In order to determine this control function over the second half-interval, we have a problem containing only two control functions to work on.

If the control function is limited in absolute value, we shall assume those points unattainable for which the control found as a result of solving the problem exceeds a specified value. In the case of discontinuous controls the problem of constructing the elementary operation becomes still more complicated and at yet it is difficult to talk about any definitive standardization of the procedure. But in concrete problems of astrodynamics the required difference schema can usually be set down successfully. Consider the example

$$\begin{aligned} x' &= u \qquad & u' &= f_1 + \alpha \cos\varphi \\ y' &= v \qquad & v' &= f_2 + \alpha \sin\varphi \end{aligned} \tag{2.10}$$

where f_1 and f_2 are constants. We assume two control functions φ and α, where the function α (t) may assume only the two values 0 and α. The problem consists in seeking a function φ(t) and points at which the engine is switched off in order to realize the jump from the state x^0, y^0, u^0, v^0 to state x^*, y^*, u^*, v^* in a time τ.

We now make use of a first-order finite difference schema: the method of fractional steps:

$$x(\tfrac{\tau}{2}) = x^0 + u^0 \tfrac{\tau}{2}$$

$$y(\tfrac{\tau}{2}) = y^0 + v^0 \tfrac{\tau}{2}$$

Hence

$$x^* - x^{(0)} = \frac{u^0 \tau}{2} = u' \frac{\tau}{2}$$

$$y^* - y^{(0)} = \frac{v^0 \tau}{2} = v' \frac{\tau}{2} \tag{2.11}$$

Equation (2.11) enables us to find $u^{(1)}$ and $v^{(1)}$. But

$$A = u^{(1)} - u^{(0)} - f_1 \frac{\tau}{2} = \alpha \cos\varphi_1 \tau_1$$

$$B = v^{(1)} - v^{(0)} - f_2 \frac{\tau}{2} = \alpha \sin\varphi_1 \tau_1 \tag{2.12}$$

This equation enables us to find:

$$\varphi_1 = \operatorname{arctg} \frac{B}{A}$$

and then to find τ_1 — the operating time of the engine.

With this choice of φ_1 and τ_1 we satisfy the conditions

$$x(\tau) = x^* \quad \varepsilon \quad y(\tau) = y^*$$

After then solving a problem similar to (2.12), we find the values φ_2 and τ_2, which guarantee that $u(\tau) = u^*$ and $v(\tau) = v^*$. If it should turn out that $\tau_0 > \tau/2$ then the point x^*, y^*, u^*, v^* may be assumed unattainable.

Let us stop to consider one further example. Consider the system (2.10). Without placing restraints of any kind on the function α (t), let us consider the variational problem of the jump of a system from the point x^0, y^0, u^0, v^0 to the point x^*, y^*, u^*, v^* in a time τ, so that the functional $J = \int_0^\tau \alpha^2 \, dt$ will be minimized.

Consider this problem within the framework of the maximum principle. The conjugate system will exhibit the following form:

$$p_1' = 0 \;, \quad p_2' = 0 \;, \quad p_3' = -p_1 \;, \quad p_4' = -p_2 \;,$$

and hence

$$p_1 = c_1 \;, \; p_2 = c_2 \;, \; p_3 = c_3 - c_1 t \;, \; p_4 = c_2 - c_4 t \tag{2.13}$$

The control functions α and φ will be found from the condition for the maximum of the hamiltonian H

$$\alpha = \frac{1}{2}\sqrt{p_3^2 + p_4^2} \;, \quad \sin\varphi = \frac{p_4}{\sqrt{p_3^2+p_4^2}} \;, \quad \cos\varphi = \frac{p_3}{\sqrt{p_3^2+p_4^2}} \tag{2.14}$$

Substituting (2.13) and (2.14) into (2.10), we have

$$\begin{aligned} x &= x^0 + u^0 t + \tfrac{1}{2} f_1 t^2 + \tfrac{1}{4} c_1 t^2 - \tfrac{1}{12} c_3 t^3 \\ y &= y^0 + v^0 t + \tfrac{1}{2} f_2 t^2 + \tfrac{1}{4} c_2 t^2 - \tfrac{1}{12} c_4 t^3 \\ u &= u^0 + f_1 t + \tfrac{1}{2} c_1 t - \tfrac{1}{4} c_3 t^2 \\ v &= v^0 + f_2 t + \tfrac{1}{2} c_2 t - \tfrac{1}{4} c_4 t^2 \end{aligned} \tag{2.15}$$

Assuming $t = \tau$ in this system of equations, we come up with four equations for determining the four arbitrary constants c_1, c_2, c_3, and c_4. The elementary operation is thereby reducible to an explicit expression.

Those cases where the elementary operation can be successfully materialized with such ease are of course exceptional. We deduce this fact from two circumstances. First, the type of elementary operation discussed corresponds to the case of an ideally controllable low-thrust engine. This case is of a definite practical interest.

Secondly, this example is interesting from the vantage point of method. It is an example of the application of the maximum principle to the materialization of an elementary operation. This possibility flows from the fact that we satisfied the phase restrictions by our choice of mesh nodes. The jump from the point on the state scale to another point on the state scale takes place within the admissible region.

The scope of this report does not permit us to dwell in greater detail on the methods of realizing the elementary operation. We need only take note here that in those problems in which one admissible trajectory has already been constructed, the structure of the elementary operation may be simplified greatly if the information available on the structure of the possible control function which enabled us to construct that admissible trajectory is used properly. Let us now proceed to enumerate the classes of problems which lend themselves to investigation by the method described.

3. Optimum Orbit Transfers in Radiation Belts

The existence of radiation belts may place certain restrictions on the choice of optimal flight trajectories. A report was presented by N. Ya. Bagaeva and N.N. Moiseev [12] at the 1963 International Astronautics Congress in Paris on procedures for solving two classes of problems in spacecraft dynamics.

The first class of problems involves construction of an optimal trajectory obviating the outer radiation belts (cf. Fig. 3). For example, we suppose that the spacecraft whose motion is described by the system of equations

$$\dot{x} = u \quad , \quad \dot{u} = -f_1 + \alpha \cos\varphi$$
$$\dot{y} = v \quad , \quad \dot{v} = -f_2 + \alpha \sin\varphi \tag{2.16}$$

starting from a certain intermediate orbit, we must attain the second cosmic velocity of escape, but in such a way that the spacecraft trajectory does not pass through the outer radiation belts girdling the earth. In Eqs. (2.16) we have the following notation: f_1 and f_2 are the components of the gravitational field strength

$\left(f_1 = \frac{\mu x}{(x^2+y^2)^{3/2}} , \; f_2 = \frac{\mu y}{(x^2+y^2)^{3/2}}\right)$, α is the engine acceleration. This schematization is frequently resorted to in the dynamics of spacecraft powered by a low-thrust engine. Since no serious complications will be entailed in dealing with the variability of the mass we may content ourselves with this model.

The second class of problems involves the transfer of the space vehicle from a low orbit to a higher one lying in the inner radiation belt, and the major limitation is that the spacecraft may be subjected to a radiation dose not exceeding a certain constant during its jump to the target orbit. This constraint is expressed mathematically as:

$$\int_0^T F(x,y,z)v(x,y,z,t)\,dt < \text{const.} \tag{2.17}$$

Here F is the radiation intensity, and v is the craft absolute speed.

These problems contrast sharply with each other. In the first case the phase restrictions satisfy a simple choice of mesh nodes. In the second case the problem is to compute integral (2.17) along each of the possible trajectories and to discard those trajectories on which the functional (2.17) attains specified values. Since the lower radiation belt is shaped very asymmetrically and the isointensity lines dip steeply in the region of Brazil, condition (2.17) may exert a crucial effect on the choice of optimal trajectory.

Calculations were performed for two kinds of propulsion engine plants: for engines subject to ideal control in flight and for engines developing constant engine thrust. Both these classes of problems were examined with the aid of first-order difference schemes. On the above basis, the functions f_1 and f_2 were replaced by constants in the system (2.16).

In problems with engine subject to ideal control the minimizing functional is $\int_0^T \alpha^2\,dt$, and the orbit transfer time is assumed constant. The state scale is constructed in a 5-dimensional space x, y, u, v, t and our elementary operation is the operation described by formulas (2.13) to (2.15).

The constant engine acceleration problems are problems involving rapidity of solution. The state scale is constructed in the four-dimensional space x, y, u, v. In this case it is more convenient to convert to polar coordinates and to replace system (2.16) by the system

$$\frac{dr}{d\varphi} = \frac{U}{V} r$$
$$\frac{dU}{d\varphi} = F_1 + \frac{\alpha}{V} r\cos\gamma , \quad \frac{dV}{d\varphi} = F_2 + \frac{\alpha}{V} r \sin\gamma \tag{2.18}$$

Here r is the radius, φ is the polar angle, U and V are respectively the radial and transversal velocity components,

$$F_1 = V - \frac{\mu}{Vr} \quad , \quad F_2 = -U$$

μ is Gauss' constant, γ is the control function.

The elementary operation consists in a jump of the system from a point φ^0, r^0, U^0, V^0 to a point $\varphi^0 + \Delta\varphi$, r^*, U^*, V^* where F_1 and F_2 are assumed constant. In this case system (2.18) is seen to belong to the class of systems (2.3) such that n = 3 and one of the equations contains no control function. This elementary operation has been described earlier.

Remarks: The conversion to three-dimensional problems neither leads to any great complication of the programs nor to any appreciable lengthening of computer time, since another control function appears in this instance (the direction of the thrust vector is defined by two angles). And in this case the elementary operation is constructed just as simply.

In the first variant we consider a craft powered by an ideally controllable engine. We stated the elementary operation in full only for the planar case. But for the spatial problem the elementary operation can also be stated in explicit form (a point of which the reader will become convinced on applying the maximum principle).

The second variant deals with a spacecraft powered by a constant-thrust engine that can be switched off at will. It is readily shown that in this case an elementary operation similar to the one described by formulas (2.11) to (2.13) may be used. Actually, in place of the system of equations (2.10) which we had in the planar case we may use the following:

$$x' = u , \quad y' = v , \quad z' = w$$
$$u' = f_1 + \alpha_x , \; v' = f_2 + \alpha_y , \; w = f_3 + \alpha_z$$

Hence

$$\alpha_x^2 + \alpha_y^2 + \alpha_z^2 = \alpha^2$$

On writing down formulas of type (2.12), we immediately realize that

$$\tau_1^2 = \frac{1}{\alpha^2}\left[\left(u^{(1)} - u^{(0)} - f_1 \frac{\tau}{2}\right)^2 + \left(v^{(1)} - v^{(0)} - f_2 \frac{\tau}{2}\right)^2 + \left(w^{(1)} - w^{(0)} - f_3 \frac{\tau}{2}\right)^2\right],$$

$$a_x = \frac{1}{\alpha\tau_1}\left(u^{(1)} - u^{(0)} - f_1 \frac{\tau}{2}\right)$$

and so forth.

4. Interorbital Transfer Problems

Problems of this type are interesting first of all from the standpoint of evaluating the power capabilities in launching a space station from a circumterrestrial orbit to a target orbit around some other planet in the solar system. A special feature of these problems is that the trajectory of the craft must pass within a certain corridor around the planet, and its velocity over that segment of trajectory must also meet certain restrictions. Moreover, it may be required that the craft execute a specified number of revolutions about the planet before embarking on the return trip. These problems are invariably three-dimensional problems, since the orbit planes of the departure planet and target planet do not coincide.

Orbit transfer trajectories to the planets Mars, Venus, and Mercury are practically planar and no phase restrictions other than terminal ones need be considered. In constructing the trajectories of space stations departing for distant planets, we have to cope with the presence of a belt of asteroids forming a toroidal volume. We know very little about the density of matter within this asteroidal belt. The trajectories should avoid this belt, to stay on the safe side. And this requires that the craft execute a spatial maneuver.

Problems of this type have been studied in two variants.

5. Several Problems

The methods described in this part of the paper show great promise. They are practically universal in application, require short computer time, and yield rapid convergence. But there still remain a whole series of questions to be ironed out.

a) Exactness. To date we have been using only schemas of the first order of exactness, and the overall problem of the exactness of calculations of this type has not yet been posed.

b) Stability. A whole series of techniques has been developed "experimentally," and these appear to ensure computational stability. But to date no theoretical investigation of the stability of the difference schemas employed has been carried out.

c) Standardization of the elementary operation. So far only jump schemas have been standardized.

d) The large number of measurements. The problems enumerated thus far involve a selection of optimal trajectories under the assumption of a pointlike spacecraft, with the thrust vector as control function. Actually control is carried out by rotating the entire craft, so that the problem is to learn how to calculate optimal maneuvers within the framework of a more complicated schema.

We could cite a whole series of other important problems facing mathematicians responsible for developing numerical methods of optimizing flight trajectories. But the ones listed already appear to be the most important ones.

SUMMARY

The report demonstrated two approaches to the solution of variational problems in spacecraft control. In the first part of the paper, we discussed problems to which the maximum principle may be applied. This method guarantees a high order of exactness in the calculations. But it consumes much computer time and requires an excellent first approximation. We did not get around to describing the aspect of technique (since it is more or less familiar to those in the field) and dwelt instead mainly on the results of the calculations. All the calculations were carried out to a very high degree of exactness (10^{-6}) and in that sense the results reported are of a "definitive nature."

Dynamic programming methods require far fewer "computer hours." They do not absolutely require an excellent first approximation. But they are inferior to classical methods in exactness. In practice we were working with schemas of only the first order of exactness, which is equivalent to the Euler integration schema. Only now has systematic research begun on difference schemas of higher orders of exactness. In the second part of the paper, therefore, we dealt mainly with technique and enumerated that type of problem for which standard computational procedures have been worked out.

The scope of this review did not allow space for a whole range of ideas of central importance in computational mathematics. In particular, the combination of classical methods and dynamic programming methods appears to offer great promise.

For precisely the same reason we said nothing about methods for constructing first approximations. For that purpose we made use of the method of locally optimal controls which while not optimal in the strict sense (and even not always belonging to the class of admissible controls), nevertheless ensure that the vehicle will end up in the neighborhood of the control target point. These considerations have been developed partially in our earlier paper [13] and published in [5, 6]. A detailed presentation of the arguments would require a special contribution devoted to them.

In conclusion, one further remark. The term "dynamic programming," which we use here to denote methods now being developed, has been employed in rather arbitrary fashion. Only a year ago, Vankorskii successfully proved that the schema outlined in the second part of the report can be used to derive the Bellman equation (when the limiting process is properly organized.) Hence, the methods developed in this paper may be regarded as numerical methods for solving the Bellman equation.

LITERATURE CITED

1. L. S. Pontryagin, V. G. Boltyanskii, R. V. Gamkrelidze, E. F. Mishchenko. The mathematical theory of optimal processes. Phys.-Math. Press, 1961. [In Russian; English translation publ. by J. Wiley, USA, 1962].
2. L. I. Rozonoer. The L.S. Pontryagin maximum principle in the theory of optimal system. Avtomatika i telemekh., 20, No. 10, 11 (1959).
3. V. K. Isaev, V. V. Sonin. One modification of the Newton method of numerical solution of boundary problems. Zhur. vychisl. matemat. i matemat. fiz., 3, No. 6 (1963).
4. G. L. Grodzovskii, Yu. N. Ivanov, V. V. Tokarev. Low-thrust flight mechanics. Inzhener. zhur., 3, No. 3, 4 (1963); 4, No. 1, 2 (1964).
5. V. N. Lebedev. Variational problem of the ascent of a space vehicle from a circular orbit. Zhur. vychisl. matemat. i matemat. fiz., 3, No. 6 (1963).
6. V. N. Lebedev, B. N. Rumyantsev. Variational problem of orbit transfer between two points in a central field. Iskusst. sputniki Zemli, No. 16 (1963).
7. A. N. Zhukov, V. N. Lebedev. Variational problem of orbit transfer between heliocentric circular orbits by solar sail. Kosmich. issled. 2, No. 1 (1964).
8. N. N. Bogolyubov, Yu. A. Mitropol'skii. Asymptotic methods in the theory of nonlinear oscillations. Phys.-Math. Press, 1958 [in Russian].
9. V. F. Illarionov, L. M. Shkadov. Rotation of the plane of a circular satellite orbit. Priklad. matem. i mekh., 26, No. 1 (1962).
10. G. E. Kuzmak, Yu. M. Kopnin. New form of equations of satellite motion and application to the study of near-Keplerian motions. Zhur. vychisl. matemat. i matemat. fiz., 3, No. 4 (1963).
11. N. N. Moiseev. Dynamic programming methods in the theory of optimal controls. Zhur. vychisl. matemat. i matemat. fiz., 3, No. 4 (1963).
12. N. Ya. Bagaeva, N. N. Moiseev. New method for solution of problems in optimal transfer theory. 14th international congress, Paris 1963.
13. V. N. Lebedev. Some problems in optimal transfer theory. 14th international congress, Paris 1963.

Prediction Display, a Way of Easing Man's Job in Integrating Control System

by

R. Bernotat *)

1. Principles and Definitions

Guiding and controlling an aircraft the pilot has to control a special value according to the display of the actual value and command value. **)

The controlled value can be a heading, an altitude, an engine speed, etc. At this action man makes - in general unconsciously - a mental prediction of the future value of the controlled variable. The extrapolation of the movement into the future is necessary for him in order to be able to react in short time and to control systematically.

Which indications does man need for the prediction?

1. Actual value of the variable
2. Derivative information of the variable
3. System-Response
4. Knowledge of the future disturbing functions

Extensive researches have been carried out to the indications 1 and 2.(2,3,4) By this it was proved that man mainly extrapolates from position and speed; acceleration and higher derivatives are hardly considered. The results, however, are not sufficiently significant in order to be able to base on this.

The guidance and control, nowadays, take place - with few exceptions - only by an indication of the actual value. Just in a few cases the first derivative is presented also (rate of climb, turn-indicator). The capability of man to give a prediction according to derivatives of the actual value presented by separate displays is low. He derives these values vastly from the changing of the actual value display. The knowledge of the third indication concerning the systemresponse has to be obtained from a pilot or from an AT-controller by long training. These trainings require the more time the more difficult the system is to control by man. In order to give an effective guiding signal, however, he has to know the consequence in advance.

Furtheron, an indication concerning the expected disturbing functions will also turn considerably to the quality of the prediction. An ignorance of the disturbing direction results in a greater uncertainty of the prediction. In certain cases the prediction time has to be shortened. If on the contrary a prevailing disturbing direction is noted, this influence in the extrapolation of the process can be considered. Examples are air turbulences, of which the effects compensate with time, and upwinds which cause a changing of the flight path.

We realize that a prediction means a considerable mental loading for man - especially with an aircraft or spacecraft manually cumbersome to guide. The question is whether man's job can be eased by supplying a prediction calculated mechanically. Tests have shown that in fact a great facilitation can be attained for the pilot in this way.

By supplying an additional prediction display a higher control quality - with essentially shortened time for study - was obtained.

The reasons for this are:

1. The pilot does not need to know exactly the systemresponse
2. The mechanical prediction is more accurate than the human.

In aviation no displays of this kind - with one exception - have been applied until now. But with the development of new flight techniques - as jet flight and space flight - it may become a necessity at critical systemresponse, if man is furtheron left as operator in the control loop. Before we describe the procedures in details some terms have to be determined. A display which gives an information about the future status of a value is named "prediction display" or "predisplay." The displayed value is the "predicted value."

*) Institute for Guidance, Control and Air Transportation of the Technical University of Berlin

**) Terminology see (1)

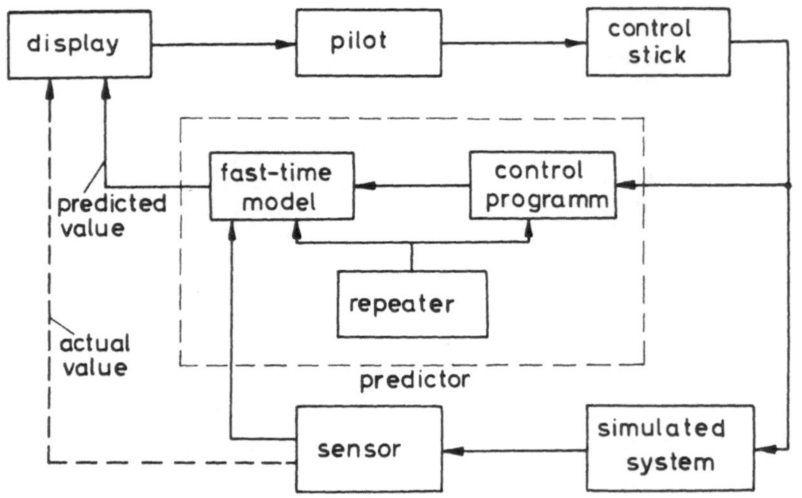

Figure 1 The prediction by a model with accelerated time scale

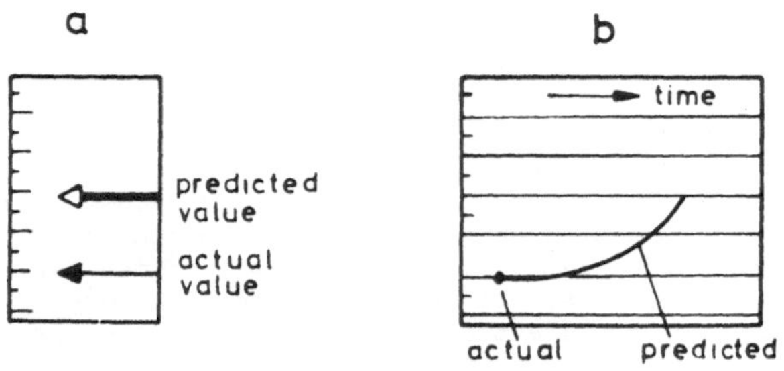

Figure 2 The 1- and 2-dimensional prediction display

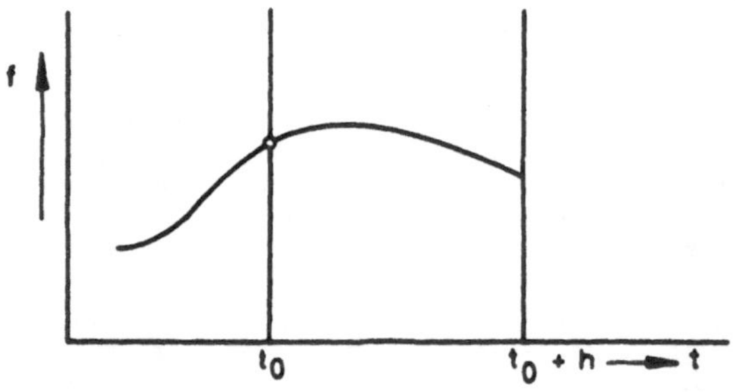

Figure 3 The extrapolating prediction of a movement

Furthermore one has to distinguish for which time into the future the prediction is computed (prediction time).

In the following part we shall exclusively treat the "short-time prediction display" which predicts only a few seconds. It is especially applicable for stabilisation problems.

The "Mean-time prediction display" with a prediction time of minutes is adaptable for distant aiming at a state (landing) or for the avoidance of a state (collision). Such a display gave satisfactory results in the research plane X-15. The pilot received an information by the prediction display which landing field he could reach with the momentary potential energy in the gliding down.

The "long-time prediction display" is a presentation of the future nominal track of a value. It gives to man a general view of the planned total runoff. In general the presented period is great. The indicated value can be the trajectory, the control program or a special parameter. An example is the prediction display of the trajectory of a manned spacecraft orbiting around the earth in the control center on ground.

The display which got known as "quickened display" has not much to do with the prediction display. Indeed derivatives are reduced like the deterministic method described in part 3, but the resulting display has command character in opposition to the prediction display. The pilot will be the performing element which follows the orders of the display (11).

2. The Model-method

We have to distinguish the two aspects. In the first case we have the prediction of a movement, of which the parameters are unknown to the observer, i.e. the prediction of the flightpath of an unknown aircraft. This can be done by the mathematical methods of statistics. The works of Wiener (5,7) and Kolmogoroff (6) have been of great importance in this field.

Of much more interest here is the second case. The systemresponse shall be well known and man is an element of the control loop. If the transfer function is given the prediction can be done by a computer. A special type of predictor is the model. Observing a model with an accelerated time scale you get a prediction of the original system response.

This principle of the model with accelerated time scale and a resulting prediction display has been used with great success in guiding submarines (8,9). Figure 1 shows the block diagram. Input data into the predictor are the actual value of the controlled variable and the human signal, for which the programmer generates an assumed control action during process time. The process in the model is started periodically by the reset device.

The model behaviour is represented by the prediction display. Figure 2a shows a 1-dimensional display with a preselected prediction time. In figure 2b the character of the predicted value, too, is represented above the time. These displays can be simply produced by the cathode-ray tubes.

3. The Extrapolation-method

By the extrapolation-method one renounces the knowledge of the systemresponse in the longer past and starts only from the actual movement. So the method is generally applicable and is not limited for a special control loop. The accuracy can be smaller than during a statistic method. As in any case man extrapolates very rudely it mostly suffices for an unloading. One can expect that the arrangement of the predictor will become more simple.

The prediction in this way is a problem of approximation.

The function f(t) has to be approximated by a serie.

$$f(t) \approx \sum_{n=1}^{N} k_n \cdot s_n(t)$$

k_n weighting factor

s_n linear independent function

At the moment t_o the values of the function are known. The values of the moment $t_o + h$ have to be searched. h represents the prediction time (Figure 3).

From the known series one can only use the Taylor's series because they alone demand the values of the function at the moment $t = t_o$. It is:

$$f(t_o+h) = f(t_o) + \frac{h}{1!} f'(t_o) + \frac{h^2}{2!} f''(t_o) + \frac{h^3}{3!} f'''(t_o) + \ldots..$$

The power series approximate best the moment t_o. Therefore the deviation will generally increase if the prediction time increases, too. Furthermore the accuracy depends on the number of used terms. Those again are limited by the technical possibilities.

For marking we will take the number of terms. The order should be fixed by the power of h.

Prediction of 0th order

$$f(t_0+h) = f(t_0)$$

Suitable to our terminologie $f(t_o)$ is the actual value at $t = t_o$. It is supposed that the actual value remains constant.

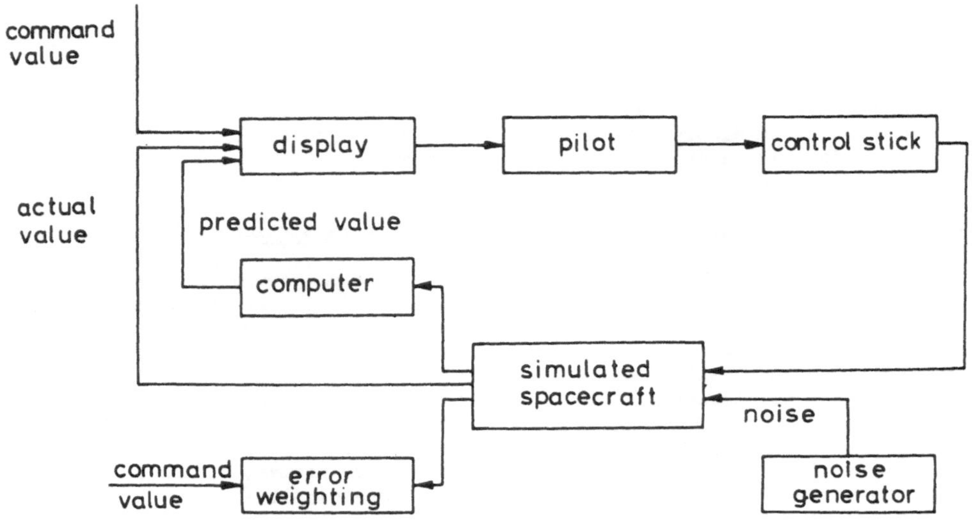

Figure 4 The block-diagram of the simulator

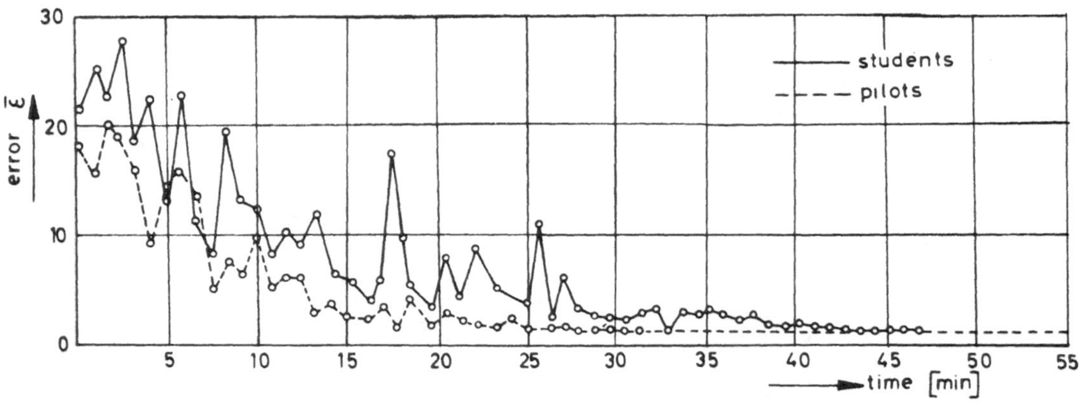

Figure 5 The learning of pilots and students in guiding a 3 times integrated system

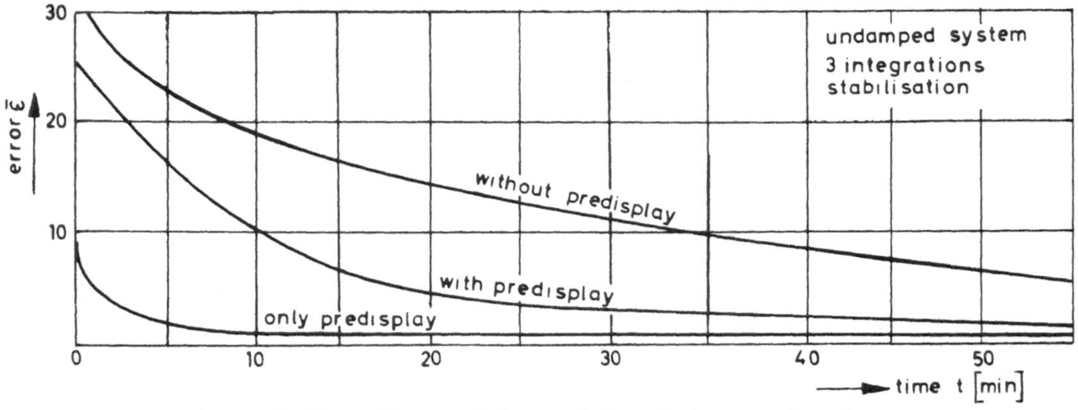

Figure 6 The influence of the prediction display upon learning

Prediction of 1st order

$$f(t_0+h)=f(t_0)+h f'(t_0)$$

The first derivative which is valued in dependence upon the prediction time is added to the actual value. The extrapolation is linear, the elevation is determined by the value of f'.

Prediction of 2nd order

$$f(t_0+h)=f(t_0)+h f'(t_0)+\frac{h^2}{2} f''(t_0)$$

By adding the 2nd derivative the approximation of the function becomes more exactly for t_o. The curve shall bend for $f'' \neq 0$.

Predictions of nth order

The principle can be continued theoretically in the above described manner. For little time of prediction the terms of higher order contribute very few. On the other side the technical expenditure is growing quickly. It remains to examine if predictions of which the order is higher than 2 will ameliorate the method. Until now only a preliminary trial of the prediction display of 1st order is known (10).

4. Some Results of the Extrapolating Prediction

In the Institute for Guidance, Control, and Air Transportation of the Technical University of Berlin some experimental researches have been made in order to examine the availability of the prediction display(12). The aim was to get to know for which sort of control systems the prediction display can be interpreted in the single case.

Figure 4 shows the block diagram. By a display the human pilot controls one degree freedom of the simulated spacecraft. He is sitting in a fixed cockpit. The display is a TV-scope, where - the same as in figure 2a - the actual value of the controlled variable and the predicted value as a horizontal bar against a vertical scale are represented.

The command value will not be represented in particular in this case of stabilisation which is investigated in this report. The pilot only gets the order to hold a certain scale value. The pilot gives his signals upon a control stick, of which the electric output signal goes to an analog computer which simulates the spacecraft response.

Except the guidance signals outer disturbances operate upon the spacecraft. These are generated by a special noise generator.

At first the researches concern an idealized spacecraft of which the response behaviour is determined by the number of integrations. It is:

$$X_A(t)=V\int^{t}\int^{t_1}\ldots\int^{t_{m-1}} X_E(t_m)dt_1\ldots dt_m$$

X_E = input signal of the spacecraft

X_A = output signal of the spacecraft

V = factor of amplification

m = number of integrations

During the guiding of aircraft and spacecraft up to four integrations occur. The investigated systems had 2, 3, and 4 integrations. They were undamped. The stabilisation is done by the pilot.

The actual value of the condition of the spacecraft is indicated directly. In the computer derivatives are formed out of the actual value by differentiating elements and worked up for the deterministic prediction display. The error signal, i.e. the difference between command value and actual value, is weighted and recorded. Further indications for the craft data and for the computation of the error are represented in the appendix.

In order to get some comparable data, a new group of unskilled persons for a fixed parameter value has always to be taken for all experiments which contain the learning phase. For each test 15 persons - pilots and students - were chosen. The pilots stroke by the fact that compared to the students their time for learning was about 50% shorter. The precisions reached after the learning phase were about the same for both groups (figure 5). For the further interpretations the average value of 15 test persons was formed for the skilled state also in mixed groups (see appendix).

For the investigated case of a stabilisation the prediction display brought some perceptible ameliorations . The greater the number of integrations in the spacecraft, the more grew the signification of the advantages. A special advantage represents the great decrease of the time for learning.

From that it can be concluded under the human point of view to the ameliorated controllability of the system. As an example figure 6 shows a 3 times integrated system with and without prediction display.

As the diagram (figure 6) shows the shortest time for learning can be reached by means of the prediction display.

After a longer learning period a precision of the same order can be reached with the three forms of display. But by accounts of the test persons the mental activity is higher during the

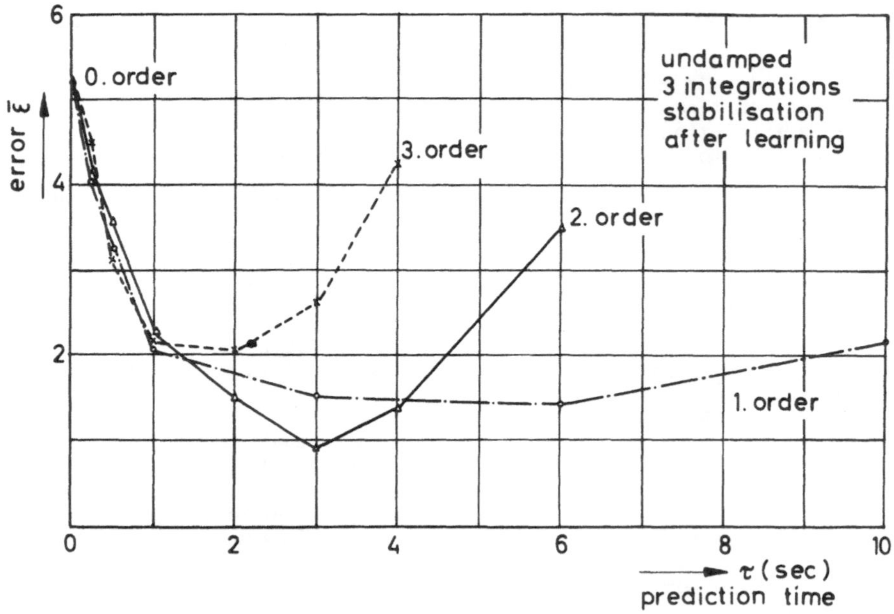

Figure 7: The influence of prediction time and order of prediction upon performance

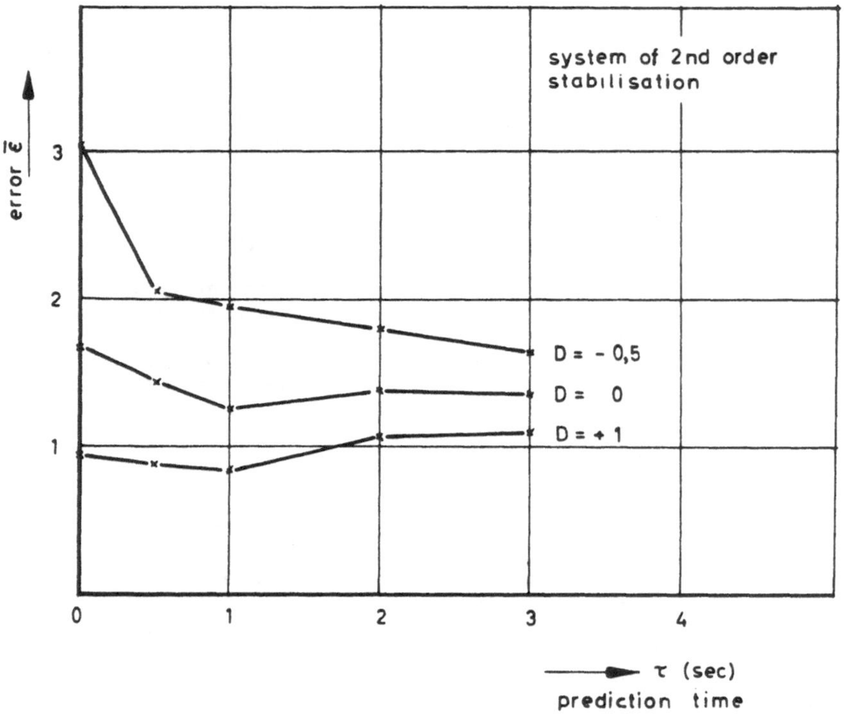

Pict. 8. The influence of the prediction time and the damping factor D upon performance

process with a longer learning phase, even if the number of errors is the same. From this results that during this process an additional mental load or other disturbing functions effect a strong increase of the error. Besides the controllability the mental load of the human pilot should always be considered as a second factor for a valuation. Until now nobody succeeded in finding a standard which could be of use.

The diagram of figure 7 has been taken after finishing the learning phase at a 3 times integrating system. The error has been recorded over the prediction time for several degrees of the prediction display. We see that in the skilled state, too, the prediction display ameliorates the quality of guidance. Without prediction display (order 0) a mean error of 5,3 was reached.

A prediction of 2nd order with the optimal value of the prediction time decreases the error by the factor 5. To be used in practice the prediction display of 1st order is most suitable in this example because on account of the flat minimum it is not necessary to adapt the prediction time when the noise spectrum is changing.

Stabilization of a system of 2nd order by means of a prediction display of 1st order

In order to determine the influence of the prediction display on stabilization problems at present special systems are examined as a function of its parameters.

First we have considered a system of 2nd order which may be described by the following equation:

$$\ddot{x} + D\dot{x} = x_E$$

Once more the errorsignal was examined which should be defined as a function of the prediction time τ with the damping factor D as a parameter.

In the series of experiments 6 persons were tested after having concluded the learning phase with the variable τ and $0 \leq \tau \leq 3$ sec and with the variable D and $D = +1; 0; -0.5$.

The results of this are shown in picture 8.

We want to lay down the characteristics of this test:

1. A prediction display decreases $\bar{\varepsilon}$.
2. The higher the coefficient of difficulty of the system the larger is the improvement ($s.D = -0,5$) by a prediction display.
3. There is an optimum of the prediction time. This is most distinctive in systems with high coefficient of difficulty and it moves with an increasing coefficient of difficulty into the range of higher prediction times.

Furthermore should be indicated that - according to an identical statement of the test persons - the additional prediction display of simple systems involves a higher psychical stress ($s.D=1$). The reason for this is that the test person has to control two displays.

The test results cannot be generalized immediately, because the values resulted from a special system. But as qualitative results they show the possibilities of amelioration which can be reached by means of an adapted display. Recapitulated one can say that the prediction display opens a new way of easing man's job in guiding aircrafts and spacecrafts and simultaneously to have a higher quality of the guidance for the pilot-spacecraft system.

APPENDIX

Spacecraft:

A maximum movement of control stick results by:

$m = 1$ in $x = \pm 25 \quad m\ sec^{-1}$,

$m = 2$ in $x = \pm 17{,}5\ m\ sec^{-2}$,

$m = 3$ in $x = \pm 25 \quad m\ sec^{-3}$.

The controllable span on the scope corresponds to ± 25 m.

Disturbance:

It is necessary that the disturbance which has to be inserted for a research because of the indifferent balance, has a random character and that it is not possible for the pilot to predict it. 10 sine oscillations are summed up of which the frequencies have an odd relation to each other. The frequencies were between 0,05 c/s and 1 c/s. The amplitude was constant for all frequencies.

Error weighting:

The difference d between the actual value and the command value is registrated as a mean quadratic error suitable to the rule:

$$\varepsilon = \sqrt{\frac{1}{T}\int_{t}^{t+T} d^2\,dt} \qquad T = 10 \text{ sec}$$

The average value between the test persons was formed suitable to:

$$\bar{\varepsilon} = \frac{1}{n}\sum_{1}^{n} \varepsilon \qquad n = 15 \text{ test persons}$$

Literature

1) Rößger, E. Bernotat, R.

Zur Systematik der Anzeigen.
Luftfahrttechnik, Raumfahrttechnik, January 1965

2) Sheridan, T. Merel, M. Kreifeldt, J. Ferrel, W.

Some predictive characteristics of the human controller. In: Guidance and Control II, Academic Press, N. Y., London 1964, page 645-663

3) Howell, C. Briggs, E.

Information input and processing variables in man - machine systems. A Review.
Astia-Report No. A.D. 230997

4) Weiner, E.

Motion prediction as a function of target speed and duration of presentation.
J. Appl. Psych. 46. (1962)

5) Wiener, N.

Extrapolation, interpolation and smoothing of stationary time series engineering applications.
The Technology Press of MIT and J. Wiley and Sons, New York 1949

6) Kolmogoroff, A.

Interpolation und Extrapolation von stationären zufälligen Folgen.
Bull. Acad. Sci. USSR. Sev. Math. 5, 3-14 (1941)

7) Wiener, N.

Kybernetik 2te Auflage. Econ Verlag, Düsseldorf, Wien 1963, S. 35 und S. 246

8) Ziebolz, H. Paynter, F.

Possibilities of a two time-scale computing system for control and simulation of dynamic systems.
Proc. Nat'l Electronic Conference 9(1854), page 215 - 223

9) Kelley, C.

Predictor instruments look into the future.
Z. Control Engineering 3 (1962), S. 86 - 90

10) Fogel, L.

Biotechnology: concepts and applications.
page 665 - 668

11) Bernotat, R.

Die Informationsdarstellung als anthropotechnisches Problem der Flugführung.
Bericht Nr. 37 des Instituts für Flugführung und Luftverkehr

12) Kaiser, K.

Untersuchung der Lenkbarkeit von mehrfach integrierenden Flugsystemen in Abhängigkeit von der Art der Voranzeige.
Internal report at the Chair for Guidance, Control and Air Transportation. December 1964.

DISCUSSION

R. Bernotat

"Prediction Display,
A Way of Easing Man's Job
in Integrating Control Systems".

Q. What was the three-integrator system described?

A. It was an unreal hypothetical system used only for studies of the prediction display. We intend in the future to go on to the use of equations describing real systems.

Rendezvous Problem of Space Vehicle with Orbital Station

G. Yu. Dankov and F. A. Mikhailov

Introduction and Formulation of Problem

In recent years the problem of orbital rendezvous has attracted the attention of many specialists. As a result of this, a considerable number of schemes for organizing a rendezvous between a spacecraft and an orbital station have been published. These schemes can be divided into three groups /1, 2, 3/:

I. Direct launching of the spacecraft in the orbital plane of the station.

II. The use of a phasing ellipse or a waiting orbit.

III. Noncoplanar interchange.

In turn, each of the above rendezvous schemes consists of three stages:

I. Action section.

II. Section of coasting flight.

III. Section in which the station orbit is assumed and the maneuvers necessary for accomplishing rendezvous are carried out.

Thus, during the first stage the essential problem is the placing of the spacecraft into the given orbit.

During the second stage, control reduces to the selection of a course for the spacecraft from the condition of collision with the station without the use of propulsion /1/.

During the third stage, the approach velocity is reduced to zero and "berthing" is achieved. This stage is sometimes called "homing".

The present paper deals with the problem of spacecraft guidance during the distant approach stage of rendezvous with an orbital station, information on the rotation of the sighting line being taken from the second derivative of the relative range.

1. Equations of the Relative Motion of the Spacecraft

Let us consider a right-handed orthogonal inertial coordinate system (X, Y, Z) with its origin at the center of mass of the earth. (Figure 1.)

Two points are moving in this system: point I, the spacecraft (X_1, Y_1, Z_1) and point 2, the orbital station (X_2, Y_2, Z_2).

Let us construct a coordinate system (ξ, η, ζ) with origin at the point (X_1, Y_1, Z_1), such that the axes of the two systems were parallel /4/.

Let us consider the motion of the point (X_2, Y_2, Z_2) in system (ξ, η, ζ), i.e., the motion of the point (X_2, Y_2, Z_2) relative to the point (X_1, Y_1, Z_1). The coordinates of the point (X_2, Y_2, Z_2) in the system of relative coordinates are related to the coordinates in the inertial system by the following expressions:

$$\begin{aligned} \xi &= X_2 - X_1, \\ \eta &= Y_2 - Y_1 \\ \zeta &= Z_2 - Z_1 \end{aligned} \qquad (1)$$

Differentiating these expressions twice, we can write the equations of motion in the relative-coordinate system (ξ, η, ζ) as

$$\begin{aligned} \ddot{\xi} &= W_{2X} - W_{1X} \\ \ddot{\eta} &= W_{2Y} - W_{1Y} \qquad (2) \\ \ddot{\zeta} &= W_{2Z} - W_{1Z} \end{aligned}$$

where

$$W_{iX} = \frac{1}{m_i} \sum F_{jX}$$

$$W_{iY} = \frac{1}{m_i} \sum F_{jY}$$

$$W_{iZ} = \frac{1}{m_i} \sum F_{jZ}$$

$$(i = 1, 2)$$

are the equations of motion of points in the inertial coordinate system under the action of the forces given.

It is convenient to study the relative motion in a spherical coordinate system. Let us write down the equations in the spherical coordinate system ($\mathcal{D}$, ε_1, ε_2) (See Figure 2.)

The transformation equations are

$$\begin{aligned} \xi &= \mathcal{D} \cos\varepsilon_1 \cos\varepsilon_2 \\ \eta &= \mathcal{D} \sin\varepsilon_1 \qquad (3) \\ \zeta &= \mathcal{D} \cos\varepsilon_1 \sin\varepsilon_2 \end{aligned}$$

The equations of relative motion in the spherical coordinate system are of the form

$$\ddot{\mathcal{D}} - \mathcal{D}\dot{\varepsilon}_1^2 - \mathcal{D}\cos^2\varepsilon_1 \varepsilon_2^2 = (W_{2X} - W_{1X})\cos\varepsilon_1\cos\varepsilon_2 + (W_{2Y} - W_{4Y})\sin\varepsilon_1 + (W_{2Z} - W_{1Z})\cos\varepsilon_1\sin\varepsilon_2$$

$$\frac{1}{\mathcal{D}}\frac{d}{dt}(\mathcal{D}^2\dot{\varepsilon}_1) + \mathcal{D}\sin\varepsilon_1\cos\varepsilon_1\dot{\varepsilon}_2^2 = -(W_{2X} - W_{1X})\sin\varepsilon_1\cos\varepsilon_2 + (W_{2Y} - W_{1Y})\cos\varepsilon_1 - (W_{2Z} - W_{1Z})\sin\varepsilon_1\sin\varepsilon_2$$

$$\frac{1}{\mathcal{D}\cos\varepsilon_1}\frac{d}{dt}(\mathcal{D}^2\cos^2\varepsilon_1\dot{\varepsilon}_2) = -(W_{2X} - W_{1X})\sin\varepsilon_2 + (W_{2Z} - W_{1Z})\cos\varepsilon_2 \qquad (4)$$

Let us consider in greater detail the following special case of relative motion.

Let us assume that the spacecraft and the orbital station are in free flight. Let us also assume that the relative range is so small, that the influence of the gravitational field on the relative motion can be neglected. Then, equation (2) becomes

$$\ddot{\xi} = \ddot{\eta} = \ddot{\zeta} = 0 \qquad (5)$$

while Equation (4) can be written as

$$\begin{aligned} &\ddot{\mathcal{D}} - \mathcal{D}\dot{\varepsilon}_1^2 - \mathcal{D}\cos^2\varepsilon_1\dot{\varepsilon}_2^2 = 0 \\ &\frac{1}{\mathcal{D}}\frac{d}{dt}(\mathcal{D}^2\dot{\varepsilon}_1) + \mathcal{D}\sin\varepsilon_1\cos\varepsilon_1\dot{\varepsilon}_1^2 = 0 \qquad (6) \\ &\frac{1}{\mathcal{D}\cos\varepsilon_1}\frac{d}{dt}(\mathcal{D}^2\cos^2\varepsilon_1\dot{\varepsilon}_2) = 0 \end{aligned}$$

Integrating Equation (6), we obtain an expression for the relative range

$$\mathcal{D}^2 = \tilde{V}_0^2\left(t + \frac{\mathcal{D}_0\dot{\mathcal{D}}_0}{\tilde{V}_0^2}\right)^2 + \frac{V_{n10}^2 + V_{n20}^2}{\tilde{V}_0^2}\mathcal{D}_0^2 \qquad (7)$$

where

$$V_0^2 = \dot{\mathcal{D}}_0^2 + V_{n10}^2 + V_{n20}^2$$

in which $\tilde{V}_0$ is the relative velocity at zero time,

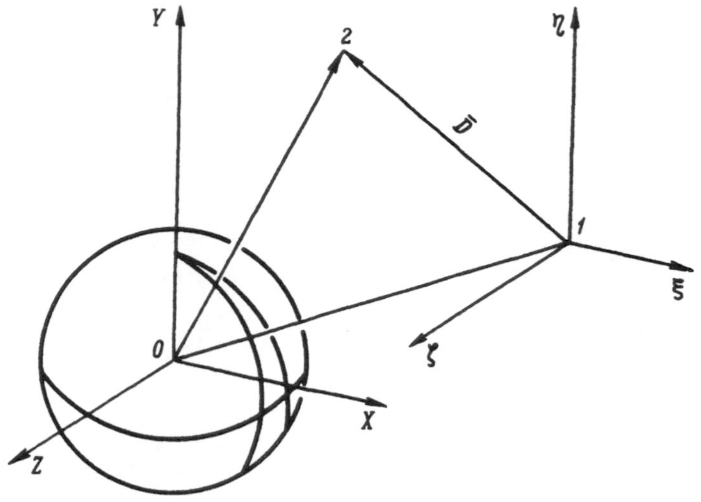

Fig. 1

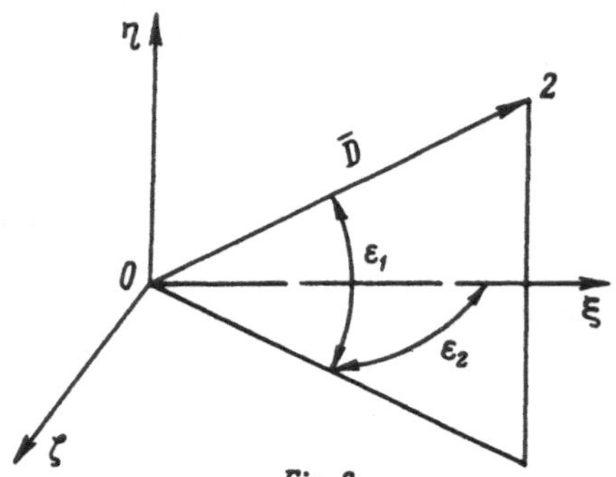

Fig. 2

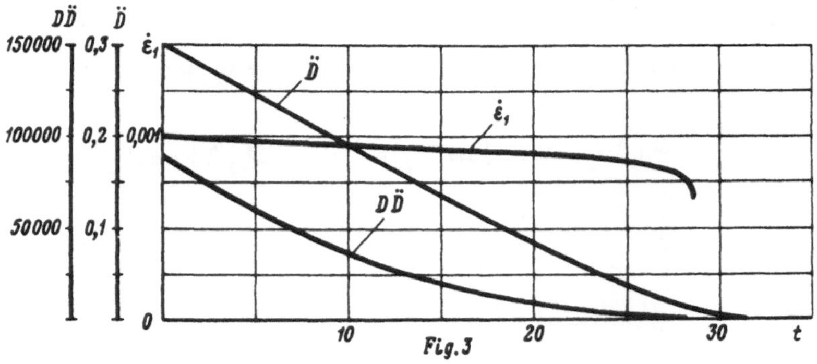

Fig. 3

$\mathcal{D}_0$ and $\dot{\mathcal{D}}_0$ are the relative range and approach velocity at the same time,

V_{n1} and V_{n2} are the components of the relative velocity in the plane perpendicular to the relative range (line of sight), given in terms of the spherical coordinates by

$$V_{n1} = \mathcal{D}\dot{\varepsilon}_1$$
$$V_{n2} = \mathcal{D}\cos\varepsilon_1\dot{\varepsilon}_2$$

V_{n10} and V_{n20} are their values at the initial moment.

One of the first integrals of system (6) is the energy integral

$$\tilde{V}_0^2 = \dot{\mathcal{D}}^2 + V_{n1}^2 + V_{n2}^2 \qquad (8)$$

It follows from formula (7) that if $V_{n10} \neq 0, V_{n20} \neq 0$, then at time $T = \frac{-\mathcal{D}_0\dot{\mathcal{D}}_0}{\tilde{V}_0^2}$ the approaching spacecrafts will attain their closest separation and will then begin to draw farther apart.

The square of this distance is

$$h_{np}^2 = \frac{V_{n10}^2 + V_{n20}^2}{\tilde{V}_0^2}\mathcal{D}_0^2 \qquad (9)$$

2. The Use of the Second Derivative of the Range as Information on the Rotation of the Line of Sight

The model of relative approach in a force-free field has been considered in Section 1. If the spacecraft is in free flight or is moving under the action of a propulsive force which is applied along the normal to the line of sight, then the first equation of system (6) is valid:

$$\ddot{\mathcal{D}} - \mathcal{D}\dot{\varepsilon}_1^2 - \mathcal{D}\cos^2\varepsilon_1\dot{\varepsilon}_2^2 = 0$$

Multiplying this equation by $\mathcal{D}$, we obtain

$$\mathcal{D}\ddot{\mathcal{D}} = \mathcal{D}^2\dot{\varepsilon}_1^2 + \mathcal{D}^2\cos^2\varepsilon_1\dot{\varepsilon}_2^2 \qquad (10)$$

the first terms of Equation (10) represents the square of the velocity component perpendicular to the line of sight.

In fact,

$$V_{n1} = \mathcal{D}\dot{\varepsilon}_1, \quad V_{n2} = \mathcal{D}\cos\varepsilon_1\dot{\varepsilon}_2$$
$$\mathcal{D}\ddot{\mathcal{D}} = V_{n1}^2 + V_{n2}^2 \qquad (11)$$

It follows from Equations (11) that with $\ddot{\mathcal{D}} = 0$, $V_{n1} = V_{n2} = 0$, i.e., the conditions for parallel approach are satisfied. Conversely, with $V_{n1} = V_{n2} = 0$, we find that $\ddot{\mathcal{D}} = 0$.

This, in a force-free field the source of information on the rotation of the line of sight can be taken to be the second derivative of the range instead of the angular velocity.

Let us note one interesting property of the parameter $\ddot{\mathcal{D}}$. As follows from Equations (11), the parameter $\ddot{\mathcal{D}}$ cannot be negative, i.e. $\ddot{\mathcal{D}} \geq 0$.

Let us rewrite Equation (11) in the form

$$\ddot{\mathcal{D}} = \frac{\tilde{V}^2}{\mathcal{D}} - \frac{\dot{\mathcal{D}}^2}{\mathcal{D}}$$

where $\tilde{V}$ is the relative velocity.

It follows from this equation that for a given value of the relative velocity, $\ddot{\mathcal{D}}$ assumes its maximum value when $\dot{\mathcal{D}} = 0$, i.e., at the moment when the distance between the approaching bodies is a minimum, whereas when the relative range is very large, i.e., when $\mathcal{D} \to \infty$, the second derivative of the relative range tends to zero.

The variation of the parameters $\dot{\varepsilon}_1$, $\ddot{\mathcal{D}}$, and $\mathcal{D}\ddot{\mathcal{D}}$, with time is shown in Figure 3.

The first equation of system (6) yields the relation

$$\delta\ddot{\mathcal{D}} = 2V_{n1}\delta\dot{\varepsilon}_1 + 2V_{n2}\cos\varepsilon_1\delta\dot{\varepsilon}_2 \qquad (12)$$

which allows us to compare the accuracies with which the angular-velocity signals and the second derivative of the range can be measured. The partial derivatives

$$\frac{\partial\ddot{\mathcal{D}}}{\partial\dot{\varepsilon}_1} = 2V_{n1} \quad \text{and} \quad \frac{\partial\ddot{\mathcal{D}}}{\partial\dot{\varepsilon}_2} = 2V_{n2}\cos\varepsilon_1$$

are proportional to the components of the relative velocity in the plane perpendicular to the line of sight.

It was pointed out above, that during the distant-approach stage guidance of the spacecraft should be such that the vector of the relative velocity is as close as possible to the line of sight. The usual approach to the solution of this problem is to direct the controlling force along the normal to the line of sight and aim of the control system is to stabilize the line of sight in the inertial space (X, Y, Z), the controlling signal being the angular velocity of the line of sight. If information on the rotation of the line of sight is taken from the second derivative of the range, while the controlling force created by a rocket on the spacecraft is in this case directed along the normal to the line of sight, then the equations of relative motion will become

$$\ddot{\mathcal{D}} - \mathcal{D}\dot{\varepsilon}_1^2 - \mathcal{D}\cos^2\varepsilon_1\,\dot{\varepsilon}_2^2 = 0$$
$$\frac{1}{\mathcal{D}}\frac{d}{dt}\left(\mathcal{D}^2\dot{\varepsilon}_1\right) + \mathcal{D}\sin\varepsilon_1\cos\varepsilon_1\,\dot{\varepsilon}_2^2 = f_1(\ddot{\mathcal{D}}) \qquad (13)$$
$$\frac{1}{\mathcal{D}\cos\varepsilon_1}\frac{d}{dt}\left(\mathcal{D}^2\cos^2\varepsilon_1\,\dot{\varepsilon}_2\right) = f_2(\ddot{\mathcal{D}})$$

where $f_1(\ddot{\mathcal{D}})$ and $f_2(\ddot{\mathcal{D}})$ are the controlling accelerations acting on the spacecraft.

Thus, measuring the second derivative of the relative range, for example, by means of the Doppler effect, and using the information obtained for achieving the required control, we can construct a closed-loop system for controlling the distant-approach stage. It should be noted that one channel (line of sight) is used to obtain information about the relative motion, while two others, orthogonal to the first, are used to control the center of mass of the spacecraft.

The choice of a concrete form for the functions $f_1(\ddot{\mathcal{D}})$ and $f_2(\ddot{\mathcal{D}})$ is a separate problem. In particular, by analogy with the law of proportional navigation, these functions can be represented as

$$f_i(\ddot{\mathcal{D}}) = a_i\,\ddot{\mathcal{D}} \quad (i=1,2) \qquad (14)$$

The parameters $a_i\ (i=1,2)$ must be chosen such that the resultant controlling acceleration acts along the projection of the relative velocity on the plane normal to the line of sight, i.e., the following condition must be satisfied:

$$\frac{a_1}{a_2} = \frac{V_{n1}}{V_{n2}} \qquad (15)$$

In addition, in order to maintain the analogy with the method of proportional navigation, the quantity

$$\sqrt{a_1^2 + a_2^2}$$

should be chosen as a function of $\mathcal{D}$ and $V_n = \sqrt{V_{n1}^2 + V_{n2}^2}$.

Literature Cited:

1. Soule, Peter W., Kidd, Alan T., "Therminal Maneuvers for Satellite Ascent Rendezvous". Preprint ARS 1961, N 206, 28.
2. Straly, W. H., "The Phasing Technique in Rendezvous". ARS Journal, 1962, N 4, p. 620-626.
3. V. Ponomarev, "Orbital Rendezvous". Aviatsiya i Kosmonoavtika, 1963, No. 2, pp. 27-32.
4. Houbolt, J., "Problems and Potentialities of Space Rendezvous". J. Astronautica Acta Vol. VIII, Fasc 5-6, 1961.

DISCUSSION

G. Yu. Dankov and F. A. Mikhailov

"Some Questions of Rendezvous with an Orbital Station".

Q. Are you measuring only doppler?

A. The direction of the axis with respect to the line of sight is also needed. The preliminary estimate can be made from analysis of the orbital flight; the measurement can be made more precise by causing the spacecraft to have small oscillations, using as an optimization condition that the second derivative of the relative distance must be minimum.

Q. Is the control system stable?

A. If the orientation is correct, proportional navigation can be used. This case has been investigated, and nothing new found.

PREDICTING SPACE-CRAFT TRAJECTORIES UNDER DISCRETE CONTROL

V. A. Yaroshevskii

Let the system of equations describing the process of motion under control be of the form

$$\frac{d\bar{x}}{dt} = A\bar{x} + B\bar{\varphi} + C\bar{u} \tag{1}$$

Here $\bar{x} = \begin{pmatrix} x_1 \\ \cdots \\ x_n \end{pmatrix}$ is the vector of deviations of the individual coordinates,

$\bar{\varphi} = \begin{pmatrix} \varphi_1 \\ \cdots \\ \varphi_m \end{pmatrix}$ is the perturbation vector,

$\bar{u} = \begin{pmatrix} u_1 \\ \cdots \\ u_l \end{pmatrix}$ the vector of the "controlling forces,

$A = (a_{ij})$, $B = (b_{ij})$, $C = (c_{ij})$ matrices formed from the variable coefficients $a_{ij}(t)$, $b_{ij}(t)$, $c_{ij}(t)$ of order $n \times n$, $n \times m$, and $n \times l$, respectively. The perturbations $\phi_i(t)$ are Gaussian random functions for which the correlation functions are given

$$M[\varphi_i(t_m)\,\varphi_j(t_n)] = k_{ij}^{(\varphi)}(t_m, t_n)$$

All functions are defined in a time interval $t_0 \leq t \leq T$, the initial values $x_i(t_0)$ being equal to zero.

The random nonzero initial conditions $x_i(t_0) \neq 0$, obeying the normal distribution law, can also be represented by means of functions $\varphi_i(t)$ on the right-hand side of (1) with the help of the generalized delta function $\delta(t-t_0)$.

In the time interval $t_0 \leq t \leq T$, p points

$$t_1, t_2, \ldots, t_p \quad (t_0 < t_1 < \ldots < t_p < T)$$

are specified, at which some of the coordinates x_i are measured,

$$\bar{x}^* = \begin{pmatrix} x_1 \\ \cdots \\ x_q \end{pmatrix} \qquad q \leq n$$

If linear combinations or integrals over the coordinates are measured, then the initial system of equations can be supplemented by several equations in order for us to be able to transform it into the required form.

The measurements are made with errors,

$$\bar{x}_{meas} = \bar{x}^* + \delta\bar{x}$$

which are random normal functions of time with the correlation functions

$$M[\delta x_i(t_m) \cdot \delta x_j(t_n)] = k_{ij}^{(\delta x)}(t_m, t_n)$$

The measurement errors are not correlated with the perturbing forces.

The problem consists in the following: Given the measured values $\bar{x}_{meas}(t_1)$, $\bar{x}_{meas}(t_p)$ to change the controlling forces in such a manner that the probable deviations of the elements of the "final miss" vector are reduced to zero; the latter vector is defined by

$$\bar{z}(T) = \begin{pmatrix} z_1(T) \\ \cdots \\ z_r(T) \end{pmatrix} = D\bar{x}(T)$$

where $D = (d_{ij})$ is a matrix of size $r \times n$ composed of the constant coefficients d_{ij}.

This problem can be subdivided into two parts: firstly, the processing of the results of measurements made at times $t_1, \ldots, t_p$, and, secondly, the design of the logic required to control the variation of the parameters u_i depending on the results of these measurements.

In attempting to solve the first problem, it is natural to make use of the maximum-likelihood principle [1]. We will assume that all the results of measurement, as also the changes in the controlling forces during the preceding time interval are recorded.

On the basis of these data, we have to choose such a change in the controlling forces over the time interval remaining, that the mathematical expectation of the components of the vector $\bar{Z}(T)$ is reduced to zero.

Let us define the matrix of the coefficients of the action of the pulsed control forces $u_j(t)$ on the values of the coordinates $x_i(t_m)$ as

$$E(t_m, t_n) = (e_{ij}(t_m, t_n))$$

a matrix of size $n \times l$, where $e_{ij}(t_m, t_n)$ is the deviation of $x_i(t_m)$ produced by an impulsive force acting on u_j at time t_n, $u_j = \delta(t - t_n)$ (for $t_n > t_m$, $e_{ij} = 0$). In the absence of perturbations

$$\bar{x}(t_m) = \int_{t_0}^{t_m} E(t_m, \tau)\, \bar{u}(\tau)\, d\tau$$

The vector of the measured quantities $\bar{x}^*$ is connected with the controlling forces by analogous relations, apart from the fact that the matrix $E(t_m, t_n)$ is replaced by the "truncated" matrix $E^* \times (t_m, t_n)$ of size $q \times l$.

$$\bar{x}^*(t_m) = \int_{t_0}^{t_m} E^*(t_m, \tau)\, \bar{u}(\tau)\, d\tau$$

(in the absence of perturbations).

The vector of the finite deviations under analogous conditions is defined by the following formula:

$$\bar{z}(T) = \mathcal{D} \int_{t_0}^{T} E(T, \tau)\, \bar{u}(\tau)\, d\tau$$

In order to calculate the coefficients of action e_{ij}, we can use the method of adjoint equations proposed by Bliss [2]. Let the matrix $F(t_m, \tau)$ of size $n \times n$ be the solution of the matrix equation

$$\frac{dF(t_m, \tau)}{d\tau} = -A^T(\tau)\, F(t_m, \tau) \qquad (2)$$

under the following initial conditions:

$$F(t_m, t_m) = J$$

where J is the unit matrix.

We then have,

$$E(t_m, \tau) = F^T(t_m, \tau)\, C(\tau)$$

The matrix $E^*(t_m, \tau)$ is defined in exactly the same manner in terms of the "truncated" matrix $F^*(t_m, \tau)$.

Let us introduce the concept of the expected magnitude of the "miss" $\bar{Z}(T)$ obtained neglecting the effects due to the controlling forces determined from the results of measurements carried out at times $t_1, t_2, \ldots, t_\sigma$ and let us denote this quantity by $\bar{Z}_0^{(\sigma)}(T)$.

Then, the condition for the compensation of the expected miss will be

$$\bar{z}^{(\sigma)}(T) + \mathcal{D} \int_{t_0}^{t_\sigma} E(T, \tau)\, \bar{u}(\tau)\, d\tau + \mathcal{D} \int_{t_\sigma}^{T} E(T, \tau)\, \bar{u}(\tau)\, d\tau = 0 \qquad (3)$$

Here $\bar{u}(t > t_\sigma)$ is the "planned" controlling action over the remaining interval of time.

The concrete law of variation of the controlling action can take any form depending on the peculiarities of the problem being solved, but, in any case, if the control follows the principle of maximum likelihood, expression (3) is a necessary condition.

It is clear that after the next $(\sigma + 1)$-th series of measurements have been carried out, the predicted value $\bar{Z}_0^{(\sigma)}(T)$ is made more accurate and replaced by $\bar{Z}_0^{(\sigma+1)}(T)$, the program to be followed for changing $\bar{u}(t > t_{\sigma+1})$ over the remaining period of time being appropriately changed.

The most difficult problem is the determination of the most probable miss from the results of measurements. To do this, it is first of all necessary to "refine" the results of measurements — to eliminate from them those terms that are due to the effects arising from the controlling action taken over the preceding time interval

$$\bar{x}_0^*(t_m) = \bar{x}^*(t_m) - \int_{t_0}^{t_m} E^*(t_m, \tau)\, \bar{u}(\tau)\, d\tau$$

Thus, the problem has been reduced to the search for the relation connecting the "refined" results of measurement and the expected magnitude of the miss $\bar{Z}_0(T)$.

Let us make use of the formulas given in [1]. Let the vector $\bar{X}$ have an a priori distribution $N(\bar{\nu}, \Sigma)$ — this means that its components are distributed according to the normal law, the vector of the mathematical expectations is $\bar{\nu}$, and the correlation matrix is Σ. We partition the vector $\bar{X}$ into two subvectors,

$$\bar{X} = \begin{pmatrix} \bar{X}^{(1)} \\ \bar{X}^{(2)} \end{pmatrix}$$

Let us assume that the values of the vector $\bar{X}^{(2)}$ have been measured: $\bar{X}^{(2)} = \bar{x}^{(2)}$. Then, the a posteriori probability distribution for the vector $\bar{X}^{(1)}$, the results of measurements being taken into account, can be written as

$$N\left(\bar{\nu}^{(1)} + \Sigma_{12}\Sigma_{22}^{-1}(\bar{x}^{(2)} - \bar{\nu}^{(2)}),\ \Sigma_{11} - \Sigma_{12}\Sigma_{22}^{-1}\Sigma_{21}\right)$$

This means that a posteriori distribution of the vector $\bar{X}^{(1)}$ is a normal one, the mathematical expectation of $\bar{X}^{(1)}$ being the vector $\bar{\nu}^{(1)} + \Sigma_{12}\Sigma_{22}^{-1}(\bar{x}^{(2)} - \bar{\nu}^{(2)})$, and the correlation matrix being the matrix $\Sigma_{11} - \Sigma_{12}\Sigma_{22}^{-1}\Sigma_{21}$. Here $\bar{\nu}^{(1)}$ and $\bar{\nu}^{(2)}$ are subvectors of the mathematical expectations of $\bar{X}^{(1)}$ and $\bar{X}^{(2)}$, while the matrices Σ_{11}, Σ_{12}, Σ_{21}, Σ_{22} are the submatrices defined by

$$\Sigma = \begin{pmatrix} \Sigma_{11} & \Sigma_{12} \\ \Sigma_{21} & \Sigma_{22} \end{pmatrix}$$

Let us apply this result to the case of interest to us. Let us assume that σ series of measurements have been carried out. We now form the vector $\overline{X}$ as follows: the first r components of this vector we take to be the components of the vector $\overline{Z}(T)$ and for the remaining components, we write σ times q of the measured components of the "refined" vector $\overline{x}_{0\,meas}(t_1)$. The first r components we considered to be the subvector $\overline{X}^{(1)}$, the remaining $q\sigma$ components, the subvector $\overline{X}^{(2)}$.

It now remains for us to determine the elements of the matrix Σ.

In order to do this, we express the values $\overline{x}_{0\,meas}(t_m)$ in the following form:

$$\overline{x}_{0\,meas}(t_m) = \int_{t_0}^{t_m} \mathcal{Y}(t_m,\tau)\overline{\varphi}(\tau)d\tau + \delta\overline{x}(t_m)$$

where

$$\mathcal{Y}^*(t_m,\tau) = F^{*T}(t_m,\tau)B(\tau)$$

is $q \times m$ matrix of the weight function and F^* is de-determined from the solution of the adjoint matrix equation (2).

On the basis of this formula, we can show that the $q\sigma \times q\sigma$ correlation matrix Σ_{22}, composed of the correlation functions of the measured parameters, consists of σ^2 submatrices, each of which is defined by

$$M(\overline{x}_{0\,meas}(t_m)\cdot\overline{x}^T_{0\,meas}(t_n)) = \int_{t_0}^{t_m}\mathcal{Y}(t_m,\tau_1)d\tau_1\int_{t_0}^{t_n}K^{(\varphi)}(\tau_1,\tau_2)\mathcal{Y}^{*T}(t_n,\tau_2)d\tau_2 + K^{(\delta x)}(t_m,t_n) \tag{4}$$

and is of size $q \times q$. Here $K^{(\varphi)}(\tau_1, \tau_2)$ denotes the matrix composed of the correlation functions $k_{ij}^{(\varphi)}(\tau,\tau_2)$ and $K^{(\delta x)}(t_m, t_n)$ denotes the matrix composed of the correlation functions $k_{ij}^{(\delta x)}(t_m, t_n)$. If the measurement errors are correlated with the perturbing functions, then formula (4) will contain an additional "cross-term" matrix describing this effect. The elements of the matrices Σ_{11}, Σ_{12}, and $\Sigma_{21} = \Sigma_{12}^T$ are determined in an analogous manner, if the fact that the components of the vector $\overline{Z}(T)$ are linear combinations of the components of $\overline{x}(T)$ is taken into account. It is obvious that terms characterizing the measurement errors do not appear in these matrices.

The $r \times r$ matrix Σ_{11} is defined by the formula

$$M[\overline{z}(T)\cdot\overline{z}^T(T)] = \int_{t_0}^{T}\mathcal{D}\mathcal{Y}(T,\tau_1)d\tau_1\int_{t_0}^{T}K(\tau_1,\tau_2)\mathcal{Y}^T(T,\tau_2)\mathcal{D}^T d\tau_2,$$

where $\mathcal{Y}(t_m,\tau) = F^T(t_m,\tau)B(\tau)$

The $r \times q\sigma$ matrix Σ_{12} consists of σ submatrices of size $r \times q$ of the following form

$$M[\overline{z}(T)\ \overline{x}^T_{0\,meas}(t_m)] = \int_{t_0}^{T}\mathcal{D}\mathcal{Y}(T,\tau_1)d\tau_1\int_{t_0}^{t_m}K^{(\varphi)}(\tau_1,\tau_2)\mathcal{Y}^{*T}(t_m,\tau_2)d\tau_2$$

The matrix Σ_{21} is of size $q\sigma \times r$ and is the transpose of the matrix Σ_{12}.

As a rule, the evaluation of the integrals of the correlation functions is associated with great computational difficulties. Therefore, the problem is frequently simplified through the approximation of random functions (perturbations and measurement errors) by their canonical expansions

$$\varphi_i(t) = \sum_j V_{ij}(t)f_{ij}(t)$$

where $f_{ij}(t)$ are fully defined functions and V_{ij} are random constants (perturbing parameters). If some of the random functions are intercorrelated, then some of these constants are common to several functions. In essence, the problem then reduces to the determination of the most probable values of these constants on the basis of the results of measurements. It should be noted, that the use of this approximation procedure in which a finite number of constants is used can lead to an incorrect estimate of the limiting attainable accuracy of the prediction in a number of cases.

If the control problem is nonlinear or the perturbations are not Gaussian the logic of the control system becomes very much more complicated [5]. Let us consider as an example the simplified case when the perturbations are unambiguously defined by a finite number of random constants $V_1, \ldots, V_m$ and it is required to correct one function of terminal miss with the help of one controlling force.

Let the control system be described by the equation

$$\frac{d\overline{x}}{dt} = \overline{f}(\overline{x}, \overline{V}, u, t)$$

where $\bar{x}$ is an n-dimensional vector of the coordinate deviations.

$\bar{V}$ is an m-dimensional vector of the perturbing parameters, u is the controlling force (a scalar), and $\bar{f}$ an n-dimensional vector function. The principal requirement is the minimization of the modulus of the scalar function of the terminal miss $S(\bar{x}(T))$.

Let us denote the a priori probability distribution for $\bar{V}$ by

$$p_0(V) = P_0(V_1, \ldots, V_m) \tag{5}$$

We assume that $u = 0$ for $t_0 \leq t \leq t_1$. Then, after the measurement of the q components of the vector $\bar{x}$ at $t = t_1$ (we assume that $q < m$ and ignore measurement errors) we will obtain the set of relations

$$\varphi_i(V_1, \ldots, V_m) = C_i \qquad i = 1, \ldots, q \tag{6}$$

On the basis of these relations and (5), we have to determine the probability distribution function of the terminal miss, assuming for definiteness that for $t > t_1$, the planned controlling action is constant.

It is obvious that under these conditions

$$S = S(u, V_1, \ldots, V_m)$$

By solving for V_i in (6), we can transform them to the parametric form

$$V_i = V_i(\alpha_1, \ldots, \alpha_{m-q}, C_1, \ldots, C_q) \qquad i = 1, \ldots, m \tag{7}$$

in which the distribution function for the parameters $\alpha_1, \ldots, \alpha_{m-q}$ in view of (5) and (7) will be written in the form

$$p(\alpha_1, \ldots, \alpha_{m-q}, C_1, C_q) = Cp_0\left[V_1(\alpha_1, \ldots, \alpha_{m-q}, C_1, \ldots, C_q), V_2(\alpha_1, \ldots, \alpha_{m-q}, C_1, \ldots, C_q), \ldots, V_m(\alpha_1, \ldots \alpha_{m-q}, C_1, \ldots, C_q)\right] \tag{8}$$

where C is a normalization factor. Similarly, we have

$$S = S\left[u, V_1(\alpha_1, \ldots, \alpha_{m-q}, C_1, \ldots, C_q), \ldots, V_m(\alpha_1, \ldots, \alpha_{m-q}, C_1, \ldots, C_q)\right] = S(u, \alpha_1, \ldots, \alpha_{m-q}, C_1, \ldots, C_q) \tag{9}$$

Let us assume that the aim of the control system is to minimize the integral

$$J = \int_{-\infty}^{\infty} g(S)\, P(S)\, dS$$

(thus, if $g(S) = S^2$, we will obtain the minimization of the mean square error).

Taking (8) and (9) into account, we can write this condition as

$$J(u, C_1, \ldots, C_q) = \int_{-\infty}^{\infty} d\alpha_1 \int_{-\infty}^{\infty} d\alpha_2 \cdots \int_{-\infty}^{\infty} g\left[S(u, \alpha_1, \ldots, \alpha_{m-q}, C_1, \ldots, C_q)\right] \times p(\alpha_1, \ldots, \alpha_{m-q}, C_1, \ldots, C_q)\, d\alpha_{m-q} = \min$$

from which we can determinte the function $u(C_1, \ldots, C_q)$. The measurements at subsequent times can be analyzed in an analogous manner.

Another criterion for control can be the minimization of the probability for the function S to exceed in absolute magnitude some permissible value $S^* > 0$. In this case, taking (9) into account, we can determine the region U in $\alpha_1, \ldots, \alpha_{m-q}$ space in which

$$|S(u, \alpha_1, \ldots, \alpha_{m-q}, C_1, \ldots, C_q)| \leq S^*$$

and choose U in such a manner that

$$p(|S| < S^*) = \int \cdots \int p(\alpha_1, \ldots, \alpha_{m-q}, C_1, \ldots, C_q)\, d\alpha_1 \ldots d\alpha_{m-q} = \max$$

This expression will also lead to the necessary dependence for $U(C_1, \ldots, C_q)$.

Let us briefly consider the question of the choice of the controlling forces as a function of the results of measurements for the case of a linear control system. Let us rewrite relation (3) which must be satisfied by this program under the condition that the system tends to compensate the most probable value of the miss

$$\bar{z}^{(\sigma)}(T) + \mathcal{B} \int_{t_0}^{t_\sigma} E(T, \tau)\, \bar{u}(\tau)\, d\tau + \mathcal{B} \int_{t_\sigma}^{T} E(T, \tau)\, \bar{u}(\tau)\, d\tau = 0$$

As was shown above, $\bar{Z}_0^{(\sigma)}(T)$ is defined as the product of a matrix by a vector consisting of the "refined" measurement results

$$\bar{z}^{(\sigma)}(T) = \Sigma_{12} \Sigma_{22}^{-1} \bar{x}_{0\Sigma\,meas}(t_\sigma) = \Sigma_{12} \Sigma_{22}^{-1} \left[\bar{x}_{\Sigma\,meas}(t_\sigma) - \int_{t_0}^{t_\sigma} E_\Sigma(t_\sigma, \tau)\, \bar{u}(\tau)\, d\tau\right]$$

Here $\bar{x}_{\Sigma meas}(t_\sigma)$ stands for the vector of all the accumulated results of measurements, i.e., σ vectors $\bar{x}_{meas}(t_i)$, the matrix $E^*_\Sigma(t_\sigma, \tau)$ is composed of σ matrices

$E^*(t_i, \tau)$ ($i = 1, 2, .., \sigma$). In the end, the expression for the determination of the controlling forces for $t \geq t_\sigma$ has the form

$$Q_\sigma \bar{x}_{\Sigma\, meas}(t_\sigma) + \int_{t_0}^{t_\sigma} R_\sigma(\tau)\bar{u}(\tau)\,d\tau + \int_{t_\sigma}^{T} J(\tau)\bar{u}(\tau)\,d\tau = 0$$

where Q_σ, R_σ, J are matrices of size $r \times q\sigma$, $r \times l$, and $r \times l$, respectively.

If l, the number of controlling actions, is not less than r, the number of the components of the miss ($l \geq r$), we can arrange a "plan" for the components of the vector $\bar{u}$ for $t > t_\sigma$ to the constant. For $l = r$, this relation allows us to determine the values of u_i unambiguously, provided that the matrix $\int_{t_\sigma}^{T} J(\tau)d\tau$ is nondegenerate. When $l > r$, there is some arbitrariness in the choice of these values. We can also make use of another rule, specifying in the interval $t_\sigma \div t_{\sigma+1}$ the constant $\bar{u}$ in such a manner that for $t > t_{\sigma+1}$ the proposed controlling action would be zero (it is clear that the results of measurement at $t = t_{\sigma+1}$ will improve the accuracy of the predicted value of $\bar{Z}_0$ and, in general, $\bar{u}$ will not become zero). In this way, the maximum reserve of controlling actions is maintained — in this situation it is less likely that the effectiveness of the control systems proves to be inadequate at the end of the trajectory. In some forms of control systems, the controlling action is an impulse of infinitely small width (see [4] and others). In any case, in the end the dependence of the controlling action on the preceding ones and on the measured values is governed by a matrix which is obtained by the inversion of expression (3). For $l > r$, the inversion requires the introduction of additional $l - r$ conditions. If $l < r$, then to satisfy the conditions for the compensation of the expected miss we have to have a logical system which anticipates several switchings of the controlling actions.

The question may arise whether it may not be possible to avoid the necessity of recording the results of measurements made in the preceding intervals of time without a final loss of prediction accuracy. Let us study this with the help of an example in which $l = r = 1$ (one controlling action and one component of miss). Let u_σ be governed only by the measured quantities $x_1(t_\sigma), \ldots, x_q(t_\sigma)$ and be independent of the results of previous measurements. The final prediction using all of the information is attained after we have all p measurements at times $t_1, t_2, \ldots, t_p$

$$Z_0(T) = \sum_{i=1}^{q}\sum_{j=1}^{p} a_{ij}\gamma_{ij}$$

where γ_{ij} is the measured value of $x_{ij}(t_j)$ minus the term due to the action of the control system during the preceding time interval

$$\gamma_{ij} = x_i(t_j) - \sum_{K=1}^{j-1} C_{ijK} u_K$$

It is assumed that the controlling action is piece-wise constant and is a linear function of the results of the measurement made at the last moment of time (without recording)

$$u_j = u(t_j \leq t \leq t_{j+1}) = \sum_{i=1}^{q} b_{ij} x_i(t_j)$$

As the result of the control system, the terminal miss is wholly or partially compensated, the compensating term being defined by

$$z(T) = -\sum_{j=1}^{p} d_j u_j$$

Let us consider the following problem: to choose the values of b_{ij} in such a manner, that the controlling actions would in the end completely compensate the predicted miss

$$\sum_{i=1}^{q}\sum_{j=1}^{p} a_{ij}\gamma_{ij} - \sum_{j=1}^{p} d_j u_j = 0$$

The solution of this is

$$b_{ij} = \frac{a_{ij}}{\sum_{i=1}^{q}\sum_{K=j+1}^{p} a_{iK} C_{ijK} + d_j} \tag{10}$$

Thus, it would appear that it is possible to achieve perfect guidance even by a system without recording. However, in such a control system, the optimal transfer numbers b_{ij} as a rule increase sharply as $t_j \to T$. Thus $b_{ip} = -a_{ip}/d_R$ but the coefficient a_{ip} as a rule increase as $t_p \to T$ (especially, if the measured coordinate x_i is closely linked with the magnitude of the miss), while the coefficient d_p decreases as $t_p \to T$, inasmuch as the possibility of compensating the miss by means of the controlling actions decrease in the interval $t_p \div T$.

This fact is particularly clearly seen in the case of a continuous tracking system. Using the same notation as above, we can write the expression for the control as

$$u(t) = \sum_{i=1}^{q} b_i(t)\left[\gamma_i(t) - \int_{t_0}^{t} C_i(t,\tau)u(\tau)\,d\tau\right]$$

and the condition for the compensation of the predicted miss as

$$\sum_{i=1}^{q}\int_{t_0}^{T} a_i(\tau)\gamma_i(\tau)\,d\tau - \int_{t_0}^{T} d(\tau)u(\tau)\,d\tau = 0$$

Then, we have

$$b_i(t) = \frac{a_i(t)}{\sum\limits_{i=1} \int_\tau^T a_i(\tau) C_i(\tau, t) d\tau + d(t)} \tag{11}$$

Thus, the problem of finding the optimal values of the transfer coefficients $b_i(t)$ reduces to the problem of finding the best prediction of the miss in the absence of control [$a_i(t)$ are weight functions]. But the numerator in (11) tends to zero as $t \to T$, whereas the functions $a_i(t)$ partly include the impulses of the form $\delta(t - T)$ ([6],[7]) which is unacceptable if the controlling action is limited in magnitude. One of the typical problems in which the choice of the correct strategy for the use of control resources is very important is the problem of the correction of the trajectory of flight outside the atmosphere.

Let us consider the section of the trajectory in which the spacecraft approaches the atmosphere ([8] and others). If we are dealing with the entry of the spacecraft into the atmosphere of a planet, then the principle aim of control is to minimize the deviations of the height of fictitious trajectory perigee from that required for the spacecraft to enter the given "corridor for atmospheric entry" ([9], [10], etc.).

We will assume that in the approach section the spacecraft is wholly within the "sphere of influence" of the given planet [8], so that the motion can be considered to be Kepplerian. It is not difficult to show that for Mars, Earth, or Venus, the radius of the sphere of influence varies from 180 to 100 radii of the planet under consideration (R_0). On this basis, we will assume that the final stage of guidance starts at $\overline{R}_H = R_{init}/R_o = 100$ [8], and that the impact height of the fictitious perigee is small, i.e., that $\overline{R}_p = R_p/R_o \approx 1$

The optimal direction of the correction impulse for most of the trajectory differs little from the direction normal to the radius vector drawn from the center of the planet [8]. In this case, $\frac{\partial R_p}{\partial V_r} = 2(\overline{R} - \frac{1}{\overline{R}}) \frac{R_p}{V_p}$

(by comparison with the optimum

$$\frac{\partial R_p}{\partial V_{opt}} = 2\sqrt{\overline{R}^2 + \frac{1}{\overline{R}} - 2}\; \frac{R_p}{V_p} \quad)$$

when V_p is equal to the second cosmic velocity.

Let us suppose that at certain intervals of time one or more parameters of the trajectory are measured and information about the deviation of the height of fictitious perigee from its nominal value $\Delta\overline{R}_p$ are accumulated. This information may be characterized by the dispersion of the error in the miss, $\sigma^2 = M[(\delta \Delta\overline{R}_p)^2]$, which is independent of the results of measurements and the earlier corrections (errors in the operation of the correcting impulses are neglected). If the correcting impulse at the point $\overline{R} = \overline{R}_i$ compensates the mathematical expectation of the miss and the errors of measurements have a Gaussian distribution, then

$$M(|\Delta V_i|) = \frac{\sqrt{\frac{2}{\pi}} \sqrt{\sigma^2(\overline{R}_{i-1}) - \sigma^2(\overline{R}_i)}}{\frac{\partial R_p}{\partial V_\tau}(\overline{R}_i)} \tag{12}$$

where $\sigma^2(\overline{R}_{i-1})$ and $\sigma^2(\overline{R}_i)$ are the dispersions of the errors in the miss at $\overline{R}_{i-1}$ and $\overline{R}_i$. The probability distribution for any $|\Delta V_i|$ has the shape of a normal curve without its left half. The mathematical expectation of the sum $\Sigma|\Delta V_i|$ characterizing the expenditure of fuel on the correction is

$$M\left(\sum_{i=1}^{K} |\Delta V_i|\right) = \sum_{i=1}^{K} M(|\Delta V_i|) \tag{13}$$

To minimize the total fuel expenditure, it is convenient to make use of dynamic programming techniques [11]. The point of application of the last impulse ($\overline{R} = \overline{R}_k$) is fixed — it is determined by the allowable deviation of the height of fictitious perigee (according to the "three sigma" rule)

$$3\sigma(\overline{R}_k) = \frac{\Delta H_p}{2}$$

where ΔH_p is the width of the corridor of entry into the atmosphere ([9], [10]). Let us express the values of $\partial R_p/\partial V_T(\overline{R}_i)$ in terms of the corresponding $\sigma^2(\overline{R}_i)$. Let us assume that all points at which impulses are to be applied have been chosen except for the last but one. Then this point is determined from the condition

$$M(|\Delta V_{k-1}|) + M(|\Delta V_k|) = min$$

As a result of this, we can express $\sigma^2(\overline{R}_{k-2})$ in terms of $\sigma^2(\overline{R}_{k-1})$ and $\sigma^2(\overline{R}_k)$. Using the analogous procedure, we can express $\sigma^2(\overline{R}_{k-3})$ in terms of $\sigma^2(\overline{R}_{k-2})$ and $\sigma^2(\overline{R}_{k-1})$, i.e., eventually, in terms of $\sigma^2(\overline{R}_{k-1})$ and $\sigma^2(\overline{R}_k)$, etc. Taking various values of $\sigma^2(\overline{R}_{k-1})$, we select from these those that give us $\sigma^2(\overline{R}_{k-m}) = \sigma_0^2$ — the a priori value of the dispersion of the initial deviation $M[(\Delta\overline{R}_{p\,init})^2]$ after a number of steps.

Then, the values of $\sigma^2(\overline{R}_k), \ldots, \sigma^2(\overline{R}_{k-m+1})$ determine the optimum points for the application of m impulses. A comparison of values of $M_{min}(\Sigma|\Delta V_i|)$ various numbers of impulses, gives us the optimum number of impulses.

As a rule, the determination of the guaranteed fuel reserve necessitates the calculation not of the mathematical expectation of the sum of impulses, but a quantity corresponding to a given probability level. If the main contribution to the sum $\Sigma M(1\Delta V_i|)$ comes from one term (for example, either the first or last), then the distribution of the sum is close to the distribution of the dominant term (the right-hand half of the normal distribution). Then, with a probability close to 0.997

$$\sum |\Delta V_i| < 3\sqrt{\frac{\pi}{2}}\, M\left(\sum |\Delta V_i|\right)$$

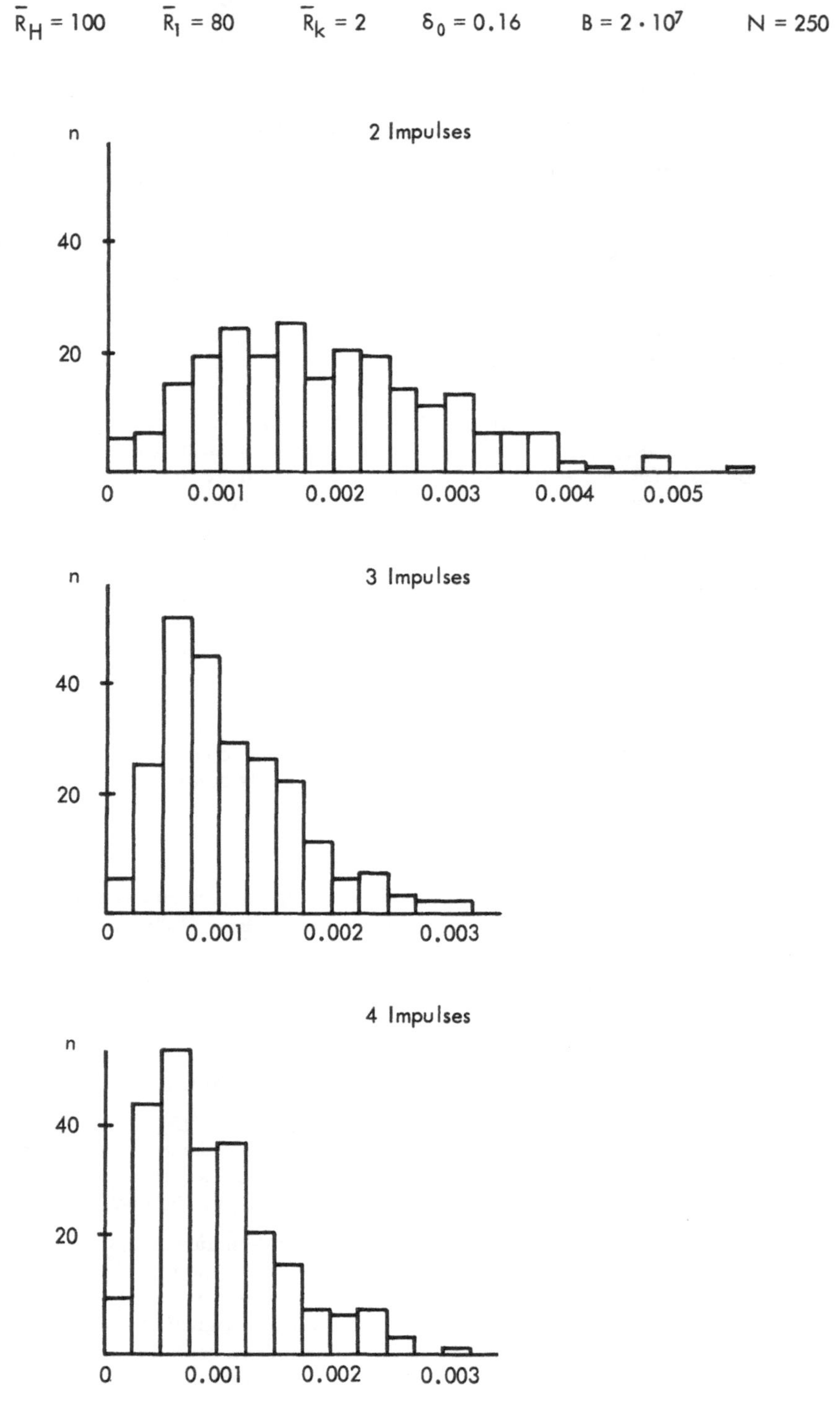

$\bar{R}_H = 100$ $\bar{R}_1 = 80$ $\bar{R}_k = 2$ $\delta_0 = 0.16$ $B = 2 \cdot 10^7$ $N = 250$
2 Impulses
n
40
20
0
0.001
0.002
0.003
0.004
0.005
3 Impulses
n
40
20
0
0.001
0.002
0.003
4 Impulses
n
40
20
0
0.001
0.002
0.003

Fig. 1

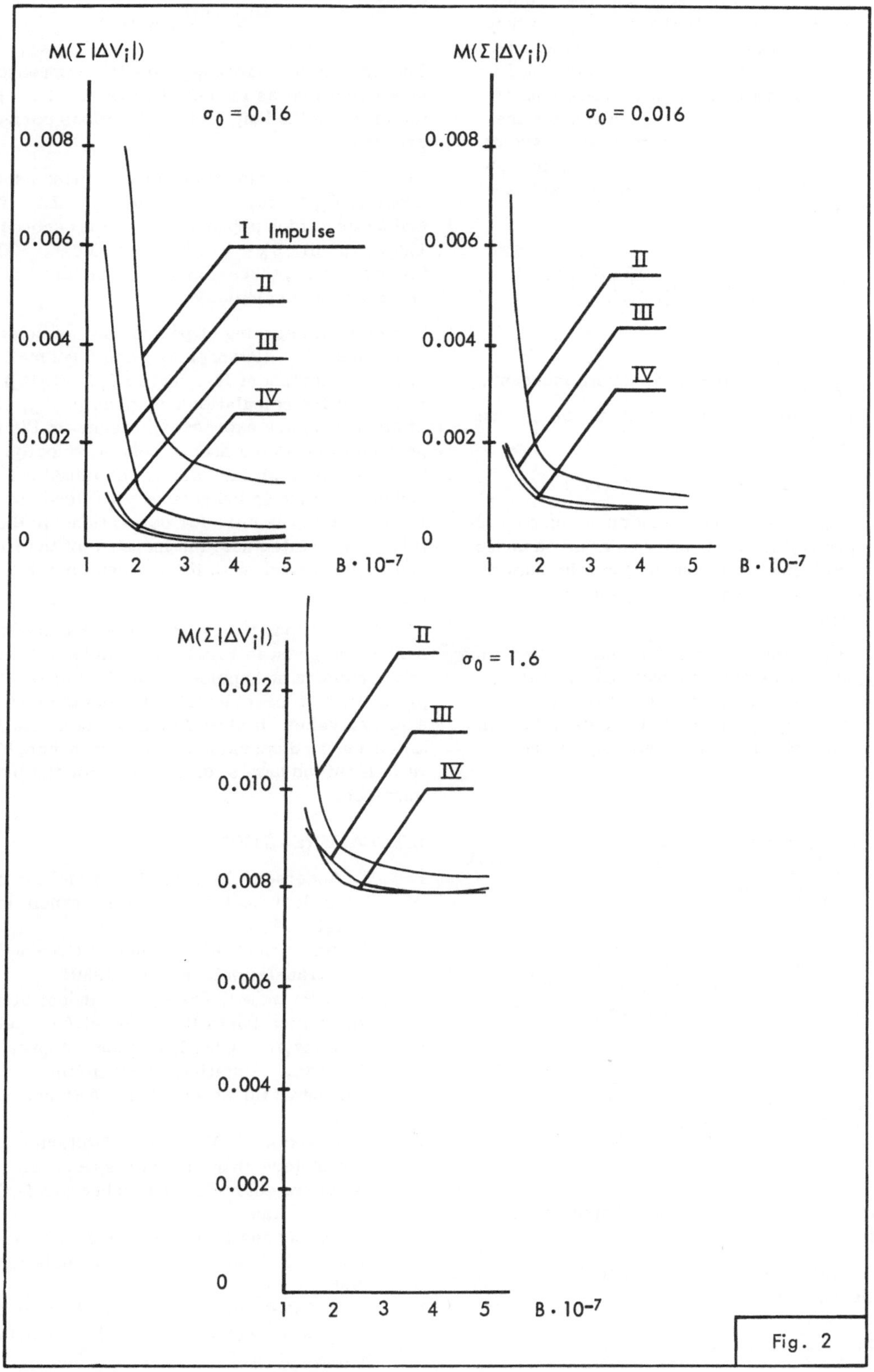
M(Σ|ΔV_i|)
σ_0 = 0.16
0.008
0.006
0.004
0.002
0
I Impulse
II
III
IV
1 2 3 4 5
B · 10^-7
M(Σ|ΔV_i|)
σ_0 = 0.016
0.008
0.006
0.004
0.002
0
II
III
IV
1 2 3 4 5
B · 10^-7
M(Σ|ΔV_i|)
σ_0 = 1.6
0.012
0.010
0.008
0.006
0.004
0.002
0
II
III
IV
1 2 3 4 5
B · 10^-7

Fig. 2

On the other hand, if there are many impulses (more than three), which is the other limiting case, and they are all comparable in magnitude, then they are found to be almost independent of one another. In fact, the first impulse mainly compensates the initial deviations of the parameters, the second, mainly compensates the effects due to the errors of measurement that have accumulated by the time the first impulse is applied, etc. The distribution of the sum of impulses approaches the normal distribution and

$$\sigma^2_{\Sigma|\Delta V_i|} \approx \sum_{i=1}^{K} \sigma^2_{|\Delta V_i|} = \frac{\pi}{2}\sum_{i=1}^{K} M^2(|\Delta V_i|)$$

the quantity $\frac{\sigma_\Sigma |\Delta V_i|}{M(\Sigma|\Delta V_i|)}$ decreasing with increasing number of terms. With a probability close to 0.997, we can consider that

$$\sum|\Delta V_i| < M(\Sigma|\Delta V_i|) + 2.8\,\sigma_{\Sigma|\Delta V_i|}$$

Thus, the minimum of possible expenditures on correction at a given probability level must be attained under approximately the same conditions as the minimum of the mathematical expectation of the expenditure $[M(\Sigma|\Delta V_i|) = \min]$.

If the quantity $\Delta\bar{R}_p$ is measured sufficiently often during the flight (say, at small equal intervals $\Delta\bar{R}$) and the mean square error of an individual measurement is $\tilde{\sigma}(\bar{R})$, then the mean square error in the determination of $\Delta\bar{R}_p$ with the use of all the accumulating information varies as

$$\sigma(R) = \frac{1}{\sqrt{\frac{1}{\sigma_0^2} + \frac{1}{\Delta R}\int_{R}^{R_H} \frac{d\bar{r}}{\tilde{\sigma}^2(\bar{r})}}} \qquad (14)$$

where σ_0 is the a priori mean square deviation of $\Delta\bar{R}_p$. It is obvious that we can reduce the mean square deviation $\sigma_\Delta \bar{R}_p$ to the acceptable value $\Delta H_p/\sigma$ only of

$$\Delta H_p > \frac{\sigma}{\sqrt{\frac{1}{\sigma_0^2} + \frac{1}{\Delta R}\int_{R_k}^{R_H} \frac{d\bar{r}}{\tilde{\sigma}^2(\bar{r})}}}$$

where $\bar{R}_k$ is the last acceptable point of application of an impulse ($\bar{R}_k = 1.2$ [8]).

In order to illustrate the above, let us consider the case $\tilde{\sigma}(\bar{R}) = A\bar{R}$, so that

$$\sigma(\bar{R}) = \frac{1}{\sqrt{\frac{1}{\sigma_0^2} + B\left(\frac{1}{\bar{R}} - \frac{1}{100}\right)}}$$

where

$$B = \frac{1}{A^2\Delta\bar{R}}, \quad \bar{R}_H = 100$$

The hypothetical law approximately expresses the well-known fact that as the destination planet is approached, the error in the height of the fictitious perigee decreases.

The following parameters close to those used in [8] were adopted: $\bar{R}_H = 100$, $\Delta H_p/2 = 10^{-3}$, $B = 1.2 \cdot 10^7$, $2 \cdot 10^7$, and $5 \cdot 10^7$ (which approximately corresponds to $\bar{R}_k$ = 1.2, 2, and 5), $\sigma_0 = 1.6$, 0.16, and 0.016. Values of $|\Delta V_i|$ for one-, two-, three-, and four-impulse correction schemes were calculated.

The optimum spacing of pulses was not selected; they were applied at distances forming a geometrical progression starting at $R_1 = 80$ to $\bar{R}_k$, ([12], [13]). The results of the calculation are given in Fig. 1. The values of $M(\Sigma|\Delta V_i|)$ are expressed in terms of the velocity at perigee equal to the second cosmic velocity. As was to be expected, an increase in the initial scatter (σ_0) and an increase in the accuracy (B) leads to an increase in the fuel expenditure for correction; in the first case, this is associated with the increase of the first impulse and in the second, with the increase in the last impulses [8].

The distributions $\Sigma|\Delta V_i|$ for various numbers of impulses are given in Fig. 2. As can be seen, an increase in the number of impulses leads to a decrease in the probability of large deviations from the mathematically expected value. It should also be noted that in almost all cases the dispersion of the sum of impulses differs little from the sum of dispersions for the individual impulses.

LITERATURE CITED

1. T. Anderson. Introduction to Multi-Dimensional Statistical Analysis [Russian translation]. Fizmatgiz, 1963.
2. Tsyan'-Syue-Sen'. Technical Cybernetics [Russian translation]. Izd. IL, 1956.
3. V. S. Pugachev, Theory of Random Functions and Its Application to Problems of Automatic Control [in Russian]. Gos. Izd. tekhnikoteoret. lit. 1957.
4. R. Battin. A statistical optimizing navigation procedure for space flight. ARS Journal, N11, 1962.
5. E. L. Akim, T. M. Eneev. "Determination of the parameters of motion of a spacecraft from data on its trajectory". Kosmicheskiye issledovaniya, 1, No. 1, 1963.
6. G. Lening and R. Battin, Random Processes in Automatic Control [Russian translation]. Izd. IL. 1958.
7. V. V. Solodovnikov. Statistical Dynamics of Linear Automatic Control Systems [in Russian]. Fizmatgiz, 1960.
8. D. Harry and A. Friedlander. An analysis of errors and requirements of an optical guidance

technique for approach to atmospheric entry with interplanetary vehicles.

9. D. Chapman. An analysis of the corridor and guidance requirements for supercircular entry into planetary atmospheres. NASA JR, NR-55, 1959.
10. V. A. Yaroshevskii. "Approximate calculation of atmospheric-entry trajectory." Kosmicheskiye issledovaniya, 2, Nos. 4 and 5, 1964.
11. R. Bellman. Dynamic Programming [Russian translation] Izd. IL. 1961.
12. T. Breakwell. The spacing of corrective thrusts in interplanetary navigation. AAS Preprint; N 60. 76, 1960.
13. D. Lawden. Optimal programme for correctional maneuvres. Astronautica Acta, N4, 1960.

DISCUSSION

V. A. Yaroshevski

"Predicting the Trajectory of Flight for a Discretely Guided Space Vehicle".

Q. Do you assume that control corrections desired can be realized with perfect accuracy?

A. If the impulses are not quite correct there will be some dispersion. If the errors can be measured after the action of the impulse, the results are not much modified. If, however, the errors are not measured, the problem becomes much more complicated and the method of dynamic programming becomes much more difficult to apply.

ON THE THEORY OF PERTURBATIONAL OPTIMAL GUIDANCE OF SPACE VEHICLES

by J.-P. Mathieu
MBLE-Research Laboratory
Brussels, Belgium

SUMMARY

It is shown first that the classical formula of perturbational optimal guidance requires that the origin of the time axis be set at the instant of injection into orbit whatever the prior trajectory might be. Next, a slightly different procedure is derived in order to avoid the near injection phenomenon. Finally, the computation of the time-to-go-to-injection will receive a physical interpretation.

INTRODUCTION

The hamiltonian formulation of a deterministic optimal guidance problem will be recalled first. A motion is described by N state equations

$$\dot{x}_i = f_i(x_j,\ u_k) \tag{1}$$

involving N state variables $\underline{x}$ and r control variables $\underline{u}$ submitted to c constraints

$$C_i(u_k) \leq 0 \tag{2}$$

The initial state is completely defined by

$$x_i(t_o) = a_i \tag{3}$$

and the final state is given by m conditions only

$$l_i\left[x_j(t_1)\right] = 0 \tag{4}$$

The optimal performance criterion, to minimize, is

$$x_o(t_1) = \int_{t_o}^{t_1} f_o(x_j, u_k)\,dt \tag{5}$$

The hamiltonian of this optimal control problem is

$$H(x_i, z_i, u_k) = \sum_{i=o}^{N} z_i\ f_i(x_j, u_k) \tag{6}$$

where the adjoint variables are defined by

$$\dot{z}_i = - \sum_{j=o}^{N} \frac{\partial f_j}{\partial x_i} z_j \tag{7}$$

and by the final transversality conditions whose resolved form is

$$z_i(t_1) = \sum_{j=1}^{m} \lambda_j \left(\frac{\partial l_j}{\partial x_i}\right)_{t_1} \tag{8}$$

The λ's are arbitrary scalars. The optimal control law $u_k(t)$ meets the maximum condition* [1]

$$H(x_i, z_i, u_k) > H(x_i, z_i, u_k + \Delta u_k) \tag{9}$$

if $z_o \leq 0$ and for any admissible variation of control (2). The maximum condition determines, at any moment, the optimal control as a function of x_i and z_i

$$u_k = E_k(x_j, z_j) \tag{10}$$

that may be introduced into (1) and (7) which, then, turn into

$$\dot{x}_i = f_i^*(x_j, z_j) \tag{11}$$

$$\dot{z}_i = h_i^*(x_j, z_j) \tag{12}$$

These equations constitute, with (3), (4) and (8), a two-side boundary-value problem. From now on, it will be assumed that f_i^* and h_i^* are continuous functions of x_j and z_j. Moreover, x_i, z_i and u_k will be grouped in three vectors $\underline{x}$, $\underline{z}$, $\underline{u}$.

THE GUIDANCE PROBLEM

Optimal guidance sets the problem of finding the control $\underline{u}(t_o)$ yielding the optimal trajectory that starts from any $\underline{x}(t_o)$ and meets terminal conditions (4). A perturbational solution to this problem consists

(i) in establishing a linear relation, valid for small differences,

$$\underline{u}(t_o) - \underline{u}^o = M(t_o)\left[\underline{x}(t_o) - \underline{x}^o\right] \tag{13}$$

where $\underline{x}^o$ and $\underline{u}^o$ correspond to a nominal trajectory;

---

* Superior numbers refer to similarly-numbered references at the end of this paper.

(ii) in establishing how $\underline{x}^o$ and $\underline{u}^o$ should be chosen on the nominal trajectory.
The state perturbation will be denoted by Δx, the control correction by Δu. In the same way, Δz will stand for $\underline{z}(t_o) - \underline{z}^o$.

PERTURBATIONAL GUIDANCE LAW

Knowing a reference trajectory meeting (1) to (9) except (3), $\underline{x}^o$, $\underline{z}^o$ and $\underline{u}^o$, one will, before establishing (13), derive a linear relation between Δz and Δx.

Defining a column-vector $\underline{X}$ with components $(x_1 \ldots x_N,\ z_1 \ldots z_N)$ (11) and (12), (4) and (8) become

$$\dot{\underline{X}} = \underline{F}(\underline{X}) \qquad (14)$$

$$\underline{L}\left[(\underline{X})_{t_1}\right] = 0 \quad \text{(N conditions)} \qquad (15)$$

A small perturbation $\Delta\underline{X}(t) = \underline{X}(t) - \underline{X}^o(t)$ satisfies

$$\Delta\dot{\underline{X}} = \left[\frac{\partial F}{\partial X}\right]\Delta\underline{X} \qquad (16)$$

$$\left[\frac{\partial L}{\partial X}\right]\Delta\underline{X}(t_1) = 0 \qquad (17)$$

where $\left[\frac{\partial F}{\partial X}\right]$ and $\left[\frac{\partial L}{\partial X}\right]$ are Jacobian matrices. A new adjoint vector $\underline{Z}$ (2N components) is now defined by

$$\dot{\underline{Z}} = -\left[\frac{\partial F}{\partial X}\right]^T \underline{Z} \qquad (18)$$

$$\underline{Z}(t_1) = \left[\frac{\partial L}{\partial X}\right]^T \Lambda \qquad (19)$$

where the vector Λ is arbitrary. The scalar product $(\underline{Z}^T\Delta\underline{X})$ is constant. If, moreover, $\underline{Z}(t_1)$ meets (19), $(\underline{Z}^T\Delta\underline{X})_{t_1}$ is null and one gets

$$(\underline{Z}^T\Delta\underline{X})_{t_o} = 0 \qquad (20)$$

If all lines of $\left[\frac{\partial L}{\partial X}\right]$ are linearly independent, one may write for N linearly independent $\underline{Z}$-vectors and at time t_o

$$Z_1^i\,\Delta x_1(t_o) + \ldots + Z_N^i\,\Delta x_N(t_o)$$

$$+\ Z_{N+1}^i\,\Delta z_1(t_o) + \ldots + Z_{2N}^i\,\Delta z_N(t_o) = 0$$

or, in matrix form

$$\Delta\underline{z}(t_o) = P(t_o)\,\Delta\underline{x}(t_o) \qquad (21)$$

The existence of matrix P will be discussed later. The formulae (16) and (17) imply that $\underline{X}$ and $\underline{X}^o$ are compared at the same instant and thus one may define the origin of the time axis when (15) is met, whatever the prior trajectory might be.

According to the **assumption** of continuity of f_i^* and h_i^* and thus of E_k also, one may write

$$\Delta\underline{u}(t_o) = \left[\frac{\partial E}{\partial x}\right]_{t_o}\Delta\underline{x}(t_o) + \left[\frac{\partial E}{\partial z}\right]_{t_o}\Delta\underline{z}(t_o) \qquad (22)$$

where $\left[\frac{\partial E}{\partial x}\right]$ and $\left[\frac{\partial E}{\partial z}\right]$ are Jacobian matrices. The elimination of $\Delta\underline{z}(t_o)$ between (21) and (22) yields

$$\Delta\underline{u}(t_o) = M(t_o)\,\Delta\underline{x}(t_o) \qquad (23)$$

This is a relation of the form (13). Moreover, it has been established how $\underline{x}^o$ and $\underline{u}^o$ should be chosen on the nominal trajectory. But the parameter t_o is generally unknown and a procedure for its determination must be derived in order to solve the guidance problem completely.

PRACTICAL EVALUATION OF GUIDANCE MATRICES

A practical procedure of determining $M(t_o)$ consists in integrating backwards one nominal trajectory and N slightly perturbed reference trajectories meeting (1) to (9) except (3). The identification of (23) on these trajectories yields Nr scalar equations determining the Nr terms of M.

THE NEAR-INJECTION PHENOMENON

The existence of M will be discussed now.

It is convenient to change variables $\underline{x}$ to variables $\underline{y}$ such that the final condition be

$$y_i(t_1) = 0 \qquad 1 \leq i \leq m$$

the N-m other variables y_i being unspecified. Let A be the regular square matrix transforming $\Delta\underline{y}(t_o)$ into $\Delta\underline{x}(t_o)$

$$\Delta\underline{x} = A\,\Delta\underline{y}$$

The guidance formula becomes

$$\Delta\underline{u}(t_o) = R(t_o)\,\Delta\underline{y}(t_o) \qquad (23)$$

with R = MA. When t_o tends to t_1, $\Delta y_i (1 \leq i \leq m)$

tend to zero. Thus, near the injection, the matrix R tends to infinity since its computation by identification involves the inversion of a matrix which is nearly degenerate m times. The matrix $\mathbf{M}$ also tends to infinity when t_o tends to t_1. This gives rise to the near-injection phenomenon which consists in a control instability.

To avoid this phenomenon, the control correction should be computed by applying L'Hospital's rule to (23) which becomes

$$\Delta u_i(t_o) = \sum_{j=1}^{m} R^*_{ij} \Delta \dot{y}_j + \sum_{j=m+1}^{N} R_{ij} \Delta y_j \qquad (24)$$

with

$$R^*_{ij} = \left[\frac{d}{dt} (R_{ij})^{-1} \right]^{-1}$$

The terms R^*_{ij} and R_{ij} may be computed by identification in the same way as explained in the foregoing section.

DETERMINATION OF THE TIME-TO-GO-TO-INJECTION t_o

The problem may be stated as follows : what point t_o of the time axis corresponds to a given state vector $\underline{x}$.

Computing backwards during a time t_o all trajectories meeting (1) to (9) except (3), one determines a surface Σ. The vectors $\underline{\Delta x}(t_o)$ intervening in the guidance formula are located, if they are small enough, in the plane tangent to Σ at the intersection of Σ and $\underline{x}^o$. Consequently, one may write

$$\underline{q}^T(t_o)\underline{\Delta x}(t_o) = 0 \qquad (25)$$

if $\underline{q}(t_o)$ stands for a vector colinear with the normal to Σ.

If T stands for a first approximation of t_o and ΔT for $T-t_o$, one gets

$$\underline{x}(t_o) - \underline{x}^o(t_o) = \underline{x}(t_o) - \underline{x}^o(T) - \Delta T\, \underline{\dot{x}}^o(t_o)$$

and, multiplying leftside by $\underline{q}^T(t_o)$,

$$\underline{q}^T(t_o)\left[\underline{x}(t_o) - \underline{x}^o(T) \right] = \Delta T\, \underline{q}^T(t_o)\, \underline{\dot{x}}(t_o)$$

or, setting

$$\underline{Q} = \frac{\underline{q}}{q^T \underline{\dot{x}}^o}$$

$$\Delta T = \underline{Q}^T(t_o)\left[\underline{x}(t_o) - \underline{x}^o(T) \right] \qquad (26)$$

If $\underline{Q}(T)$ is known, the latter formula suggests an iterative procedure for the determination of t_o. The vector $\underline{Q}(t_o)$ may be determined for any t_o by identification of (26) on the N+1 trajectories used to compute the terms of $\mathbf{M}(t_o)$.

CONCLUSIONS

It has been demonstrated that the origin of the time axis may be set at the instant of injection into orbit whatever the prior trajectory might be. The choice of this time-origin leads to a very simple procedure for determining the guidance matrices, the time-to-go-to-injection and the control correction when the vehicle comes near the injection into orbit.

The theory presented here has been applied to ELDO I launcher guidance during the 44 last seconds of flight. The different parts of the theory have been separately tested with success[2]. The guidance correction at $t_o = -4$ seconds was determined following (24).

REFERENCES

(1) L. Pontrjagin, V. Boltjankii, R. Gamkrelidze, E. Mischenko, The mathematical theory of optimal processes, Interscience Publishers, New York, 1962.

(2) J.-P. Mathieu, "Etude du phénomène d'instabilité à l'injection d'un satellite en orbite", MBLE-Research Laboratory, Report R25, February 1965.

DISCUSSION

J. P. Mathieu

"A Hamiltonian Approach
to an Optimal Guidance Law
of a Space Vehicle".

Q. What quantities are measured?

A. Three components of velocity and three of position; also instantaneous mass.

Q. What kind of computer would be used in this application?

A. It could be very simple. It needs only to perform the multiplication of a 7 x 6 matrix by a vector for guidance in this example. In practice when there are two control variables (angles) a 7 x 2 matrix is used, and the computer needs to store only 14-element matrices. This means 14 polynominals of time, one for each element, must be stored.

Q. When are the matrix elements computed?

A. The matrix can be computed at any time up to 4 seconds before injection when some of the variables become too small for computation. A modified control law using L'Hospital's rule is substituted at that time.

Q. Isn't a large computer needed to compute the matrices?

A. Yes, but this is done before the flight. A small computer is used during the flight for control corrections. The large computer is used before flight.

CLASSIFICATION OF ARTIFICIAL EARTH SATELLITE ORBITAL ORIENTATION SYSTEMS

E. N. Tokar' and V. N. Branets

The general problem of space vehicle orientation implies the alignment of the like axes of some coordinate trihedron Oxyz associated with the vehicle and a given trihedron $Ox_0y_0z_0$.* Breaking down the various types of orientation according to the properties of the absolute motion of the trihedron $Ox_0y_0z_0$, we distinguish the very important class of orientation problems from the practical point of view, when the trihedron $Ox_0y_0z_0$ executes motion in absolute space with a fixed-orientation nonzero angular velocity vector. The present article discusses some of the problems concerning the sensitive elements (sensors) for orientation that can be used to solve the given type of problem.

We postulate a trihedron $Ox_0x_0z_0$ such that

$$p_0 = 0, \quad q_0 = 0, \quad r_0 = -\omega_*(t) < 0$$

where p_0, q_0, r_0 are the projections of the absolute angular velocity vector of the trihedron $Ox_0y_0z_0$ on its axes. As an example, we consider an orbital trihedron connected with the center of mass 0 ot an artificial earth satellite, for which the y_0 axis is the extension of the earth's radius passing through the point 0, z_0 is perpendicular to the orbital plane of the satellite and is oriented to the right with reference to the direction of orbital velocity of the point 0.

Restricting to the case of small vibrations of the axes of Oxyz about the given directions, we will determine the angular coordinates Θ_x, Θ_y, Θ_z of the vehicle as projections on the x, y, z axes of the small rotation vector $\overline{\Theta}$, by means of which the axes of $Ox_0y_0z_0$ are aligned with the like axes of Oxyz. We assert that the functions of the orientation sensors is to determine the variables Θ_x, Θ_y, Θ_z.

We segregate all sensors into two conceptually different types, which we call sensors of the first and second kind. Among the first are devices that determine the quantities Θ_x, Θ_y, Θ_z by utilizing external information received in one form or another as to the spatial orientation of the axes of Oxyz. The second group includes sensors whose operation rests solely on the properties of the absolute motion of the trihedron $Ox_0y_0z_0$. Unlike the first kind, sensors of the second kind can only function as long as the space vehicle is already partially oriented relative to the trihedron $Ox_0y_0z_0$ by virtue of other sensors (sensors of the first kind). On the other hand, these sensors have the significant advantage that they need no external information and hence can always be made in the form of instruments completely isolated from space.

In the ensuing discussion we will consider sensors of the second kind exclusively. We will attempt to find out which of the variables Θ_x, Θ_y, Θ_z can be determined under what conditions using sensors of the second kind in the case when the relation (1) is satisfied, and to present some of the conceivable principles for the operation of such sensors.

Attacking the first problem, we proceed from the fact that any sensor of the second kind relies only on the properties ot the absolute motion of the vehicle for its operation, this motion being determined* by the projections p(t), q(t), r(t) of the absolute angular velocity of the trihedron Oxyz on its axes. In the case of Eq. (1) and $\overline{\Theta} \neq 0$, these projections are equal to the following, correct to small higher-order terms:

$$p = \dot{\theta}_x + \omega_* \theta_y \qquad q = \dot{\theta}_y - \omega_* \theta_x \qquad r = \dot{\theta}_z - \omega_* \tag{2}$$

and for normal orientation of the vehicle ($\overline{\Theta} = 0$):

$$p = 0 \qquad q = 0 \qquad r = -\omega_* \tag{3}$$

We will show that sensors of the second kind cannot be used for the simultaneous stabilization of the vehicle in the coordinates Θ_x and Θ_y. Assuming the contrary, we consider the following motion of the vehicle:

$$\begin{aligned} \theta_x &= C_1 \cos\left[\int_0^t \omega_*(t)dt + C_2\right] \\ \theta_y &= C_1 \sin\left[\int_0^t \omega_*(t)dt + C_2\right] \\ \theta_z &= 0 \end{aligned} \tag{4}$$

(C_1 = const $\neq 0$, C_2 = const). Substituting the relations (4) into Eqs. (2), we obtain values of p, q, r coinciding with (3). Consequently, any sensors of the second kind cannot differentiate the given motion from normal orientation of the vehicle, which proves our assertation.

*We assume that the considered trihedra are right orthogonal.

*Correct to the translational component of the motion, which need not be regarded in the analysis of orientation problems.

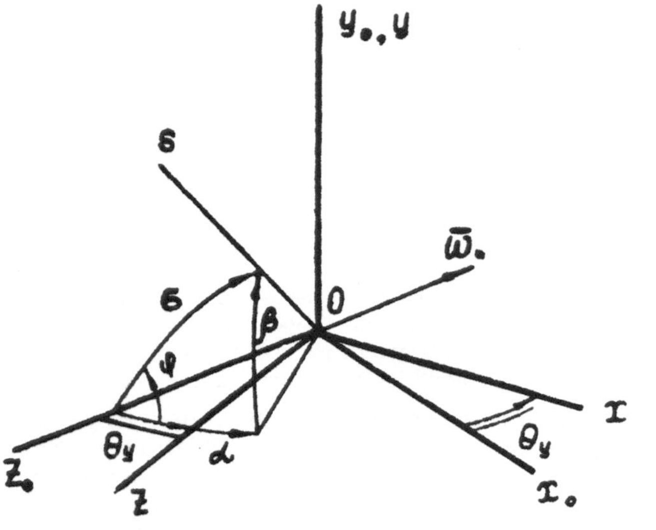

FIGURE 1

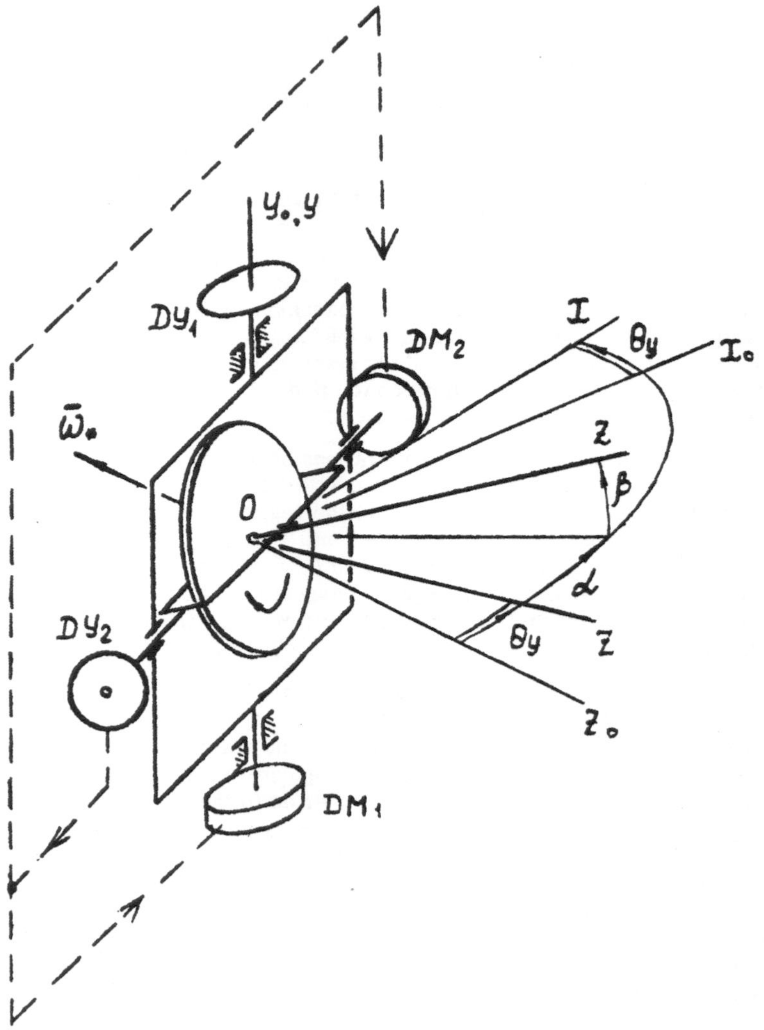

FIGURE 2

Analogously, comparing the case

$$\theta_x = 0 \qquad \theta_y = 0 \qquad \theta_z = \text{const} \neq 0$$

with the case of normal orientation, we are assured that a sensor of the second kind cannot be used to stabilize the vehicle in the coordinate Θ_z under any conditions.

In the remaining cases, i.e., for stabilization of the vehicle in one of the coordinates Θ_x or Θ_y, sensors of the second kind, as seen below, are applicable in principle.

Consequently, sensors of the second kind are suitable only for determining one of the coordinates Θ_x and Θ_y. In both cases the other two coordinates (Θ_y, Θ_z or Θ_z, Θ_x) can only be obtained by means of sensors of the first kind. We note two important special results.

1. Any sensor isolated from outer space and receiving no outside information can only control one channel of the orientation system.

2. In the case of an artificial earth satellite to be oriented in the orbital coordinate system, such a sensor is admissible in the course or pitch channels and is inadmissible in the yaw channel.

We will examine the possible operational principles of a sensor of the second kind.* We will assume, for definiteness, that the task of the sensor is to measure the coordinate Θ_y, and we will consider the condition for its operation to be the following partial orientation of the vehicle:

$$\theta_x = \theta_z = 0 \tag{5}$$

(the errors inherent in the sensors of the first kind and corresponding auxiliary units executing this task will be neglected). The quantity Θ_y determines the angle between the axes x, x_0 or z, z_0 in the plane xx_0zz_0 (Fig. 1).

1. PRINCIPLE FOR MEASUREMENT OF THE PROJECTION OF THE ABSOLUTE ANGULAR VELOCITY OF THE VEHICLE

Given Eqs. (5), the expressions (2) have the form

$$p = \omega_* \theta_y \qquad q = \dot{\theta}_y \qquad r = -\omega_* \tag{6}$$

The quantity Θ_y can be obtained by measuring the projection p. Consequently, the sensor of the second kind could be any type of absolute angular velocity sensor with its sensitive axis parallel to the x axis of the vehicle.

In the given case the source of information for the sensor of the second kind is the linear velocity field of points located on the vehicle (i.e., certain characteristics of this field that arise in the partial orientation of the vehicle relative to the moving reference trihedron $Ox_0y_0z_0$).

2. PRINCIPLE FOR MEASUREMENT OF THE LINEAR ACCELERATIONS OF POINTS LOCATED ON THE VEHICLE

A possible source of information for the sensor of the second kind is the linear acceleration field of points located on the partially oriented space vehicle. We cite an example.

Let us consider two points A_1 and A_2 on the vehicle, having the Cartesian coordinates $(0, 0, l)$ and $(l, 0, 0)$ respectively, in the reference system Oxyz ($l \neq 0$). The projection W_{1x} of the absolute acceleration of the point A_1 on the x axis and the projection W_{2z} of the absolute acceleration of the point A_2 on the z axis in the case of Eqs. (1), (5) are calculated in the form

$$\begin{aligned} W_{1x} &= W_{0x} + (\ddot{\theta}_y - \omega_*^2 \theta_y) l \\ W_{2z} &= W_{0z} - (\ddot{\theta}_y + \omega_*^2 \theta_y) l \end{aligned} \tag{7}$$

from which, combining the two Eqs. (7), we find

$$\theta_y = -\frac{1}{2\omega_*^2 l} (W_{1x} - W_{0x} + W_{2z} - W_{0z}) \tag{8}$$

The quantity Θ_y can be determined if in some way or other the combination of linear acceleration of points on the vehicle, as in Eq. (8), can be measured. In some cases this problem can be solved by means of two single-component accelerometers situated at the points A_1 and A_2, with summation of their output signals.

3. PRINCIPLE OF THE BISTABLE FOUCAULT GYROSCOPE

The characteristics of the absolute motion of a partially oriented space vehicle can be used to determine the coordinate Θ_y by equipping the vehicle with a bistable Foucault gyroscope. If the suspension axis of such a gyroscope is parallel to the Oy, then in the absence of drag torque

$$J\ddot{\varphi} + \Gamma \omega_*(t) \sin \varphi = 0 \tag{9}$$

where J is the moment of inertia of the gyroscope relative to the suspension axis, φ is the angle between the intrinsic kinetic moment vector of the gyroscope $\overline{\Gamma}$ and the negative z_0 axis. The particular solution of Eq. (9) $\varphi = 0$ is stable for a large number of the functions $\omega_*(t)$ that can be encountered in practice. In the case

*Many of the principles considered here have been reflected in the literature with reference to specific problems. See, e.g., [2], [3], [4].

$\varphi = 0$ the axis of the gyroscope rotor is parallel to the z_0 axis, the angle Θ_y is read directly on the suspension axis of the gyroscope.

In the majority of devices that have been built on the bistable Foucault gyro principle, the rigid constraint of one degree of freedom of the gyroscope is replaced by an elastic or otherwise nonrigid constant, in order to reduce the effect of random fluctuations of the base. Following this approach, it is possible to acquire a gyroscopic sensor of the second kind, utilizing a tristable (three-degree of freedom) gyroscope equipped with the correction system. The scheme for such a sensor also follows from the principle presented in section 4 below and is discussed in section 5.

4. PRINCIPLE FOR THE DIRECTIONAL FIXATION OF AN ARBITRARY AXIS IN ABSOLUTE SPACE (KINEMATIC PRINCIPLE)

The quantity Θ_y can be determined by directly measuring the angular coordinates of any fixed-orientation axis in absolute space relative to the vehicle fixed axes, provided the measured data are suitably processed. We will cite one possible means for such processing.

Let a free noncorrected gyroscope mounted on a space vehicle fix an arbitrary axis s in absolute space, where the angles α (t) (angle between Oz and the projections of s on the plane Ozx) and β (t) (angle of elevation of s above the plane Ozx) are measured directly. If $\varphi = \varphi(t) = \int_0^t \omega_*(t)\,dt + \text{const}$ is the angle between the planes sz_0 and z_0x_0, and $\sigma = \text{const}$ $(0 \leq \sigma \leq \pi)$ is the angle between the axes s and z_0 (Fig. 1), then

$$\sin\beta = \sin\sigma \sin\varphi \tag{10}$$

$$\cos(\alpha + \Theta_y)\cos\beta = \cos\sigma \tag{11}$$

The angle Θ_y can be calculated at each instant from the instantaneous values of the angles α (t), β(t) by means of Eq. (11) if the angle σ is known. The latter, in many instances, can be determined by autonomous means on the vehicle by application of Eq. (10). For example, if $\varphi(t) \to \infty$ $(t \to \infty)$, and $|\sin\varphi|$ periodically goes to unity, the quantity $\sin\varphi \geq 0$ is found by recording the peak value of the quantity $|\sin\beta|$.

We note that the result of an autonomous determination of the angle σ can also be used to control this angle, pursuing the objective of reducing it to one of the values

$$1.\quad \sigma = 0$$
$$2.\quad \sigma = \pi/2 \tag{12}$$

which turn out to be the most convenient for determining the angle Θ_y. This problem can be solved by a correction system aligning the axis of the reference gyro with the z_0 axis or the plane x_0y_0.

5. PRINCIPLE OF THE CORRECTED GYROSCOPE, WHICH FIXES THE DIRECTION OF THE z_0 AXIS (GYROSCOPIC ORBIT PRINCIPLE)

Consider an artificial earth satellite moving in a circular orbit with period T, oriented relative to the orbital trihedron $Ox_0y_0z_0$. Consistent with the remark at the end of the preceding section, having mounted a tristable (three-degree of freedom) gyro on the satellite, we pose the problem of constructing a correction system that aligns the gyroscope axis with the z_0 axis. We will show two possible techniques for solving this problem.

a) We set the external axis of the gyroscope Cardan joint parallel to Oy and place on it an angle sensor AS_1 and the moment sensor MS_1, on the internal axis of the Cardan we place an angle sensor AS_2 and moment sensor MS_2 (Fig. 2). Reading the angular coordinates α, β of the gyroscope intrinsic axis z' relative to the axes xyz in the same manner as the angular coordinates of the axis s in the preceding section, we assume

$$-90^\circ < \alpha < 90^\circ \quad -90^\circ < \beta < 90^\circ \tag{13}$$

Let us suppose that the sensor MS_1 is linked to the sensor AS_2 by a frame correction arrangement, which brings the z' into the plane zx; the latter is turned on at equally spaced time intervals $t_k = T/4$ for a fairly short period, sufficient for aligning the z' axis with the plane zx. Taking into account the rotation of the orbital trihedron with constant angular velocity $\omega_* = 2\pi/T$ about the z_0 axis, we realize at once that, regardless of the initial values of α, β within the interval (13), even after the second correction the axis of the gyro z' will be aligned with the z_0 axis, after which the angle Θ_y can be read directly by means of the sensor AS_1.

We see that the correction period need not be exactly T/4, but can have any value as long as it is not a multiple of T/2. In this case the z' axis is also aligned with z_0.

b) We assume that the sensors MS_1 and MS_2 are controlled by signals from the sensor AS_2 and develop moments $M_1 = h\beta$ and $M_2 = k\beta$ ($k = \text{const} > 0$, $h = \text{const} > 0$), which are presumed positive when their vectors for small angles α, β have positive projections on the x and y axes, respectively. With the rotation of the gyro rotor directed as indicated in Fig. 2* and with the supposition of small angles α, β, Θ_y,† we obtain

$$\begin{aligned} &\Gamma\dot{\alpha}' + (\Gamma\omega_* + k)\beta = 0 \\ &\Gamma\dot{\beta} + h\beta - \Gamma\omega_*\alpha' = 0 \end{aligned} \tag{14}$$

where $\Gamma > 0$ is the natural kinetic moment of the gyroscope, $\alpha' = \alpha + \theta_y$, and then from this, the solution

*If the rotation of the rotor is in the opposite direction, the signs of the correction moments must be changed.

†In analyzing the operation of the given correction system in the region of finite angular coordinates, considerable use can be made of the results from [1].

$\alpha' = \beta = 0$ of the system (14), corresponding to the case when the axes z' and z_0 coincide, is asymptotically stable, as can be easily verified.

An artificial earth satellite gyroscopic orientation sensor equipped with a correction system aligning the gyro axis with the axis normal to the orbital plane of the satellite has been proposed by E. N. Tokar' and has been termed a gyroscopic orbit system.

Many other gyroscopic orbit correction systems besides those discussed above are possible (specifically, systems utilizing nonlinear elements, systems whose input represents the time derivatives of the angular coordinates of the gyro, etc.). We note that the gyroscopic orbit can operate on space vehicles moving in elliptical as well as circular orbits, and in some cases in nonclosed orbits.

6. PRINCIPLE OF TWO CORRECTED GYROSCOPES WHICH FIX THE PLANE x_0y_0

At the end of section 4 we pose the problem of a correction system that aligns the axis of a tristable gyroscope with the plane x_0y_0. As shown by analysis, the realization of this concept is also possible. However, unlike the gyroscope which fixes the z_0 axis (gyroscopic orbit), a gyroscope which fixes one axis s_1 in absolute space, this axis lying in the plane x_0y_0, cannot provide a direct reading of the angle Θ_y (it is impossible to read Θ_y at the instants when the y_0 axis coincides with the axis s_1). Two gyroscopes are therefore needed, fixing two axes in absolute space that lie in the plane x_0y_0 and form an angle not equal to 0 or π, i.e., two gyroscopes fixing the orientation of the entire plane x_0y_0 in space.

The gyroscopes must be equipped with a correction system that aligns their axes with the plane x_0y_0 and monitors the mutual orientation of the gyroscope axes in this plane.

The dynamics of a pair of gyroscopes fixing the plane x_0y_0 is similar to the dynamics of the gyroscopic orbit. In particular, they can utilize a correction system for which the equations of small oscillations of the plane containing the gyroscope axes relative to the plane x_0y_0 coincide with Eqs. (14), describing small oscillations of the gyroscopic orbit axis.

7. PRINCIPLE OF DIRECT STABILIZATION OF THE SPACE VEHICLE

The effect of ordered rotation of the trihedron $Ox_0y_0z_0$ in absolute space can be used not only for the design of an instrument to determine the angular coordinate Θ_y of the vehicle, but also for direct stabilization of the vehicle in the plane z_0x_0.

Presupposing, as before, partial orientation of the vehicle in correspondence with Eqs. (5), we assume that the axes Oxyz are the principal central inertial axes of the vehicle, to which correspond the values of the moments of inertia A, B, C. The equation of motion of the vehicle in the plane z_0x_0, taking into account Eqs. (6), has the form

$$B\ddot{\theta}_y + (C-A)\omega_*^2\theta_y = M_y \tag{15}$$

where M_y is the projection of the external moment acting on the vehicle on the y axis. We will restrict to the case $\omega_* = \text{const}$. The solution Θ_y that occurs for $M_y = 0$ will then be stable, provided

$$C > A \tag{16}$$

Providing, by some means or other, damping of the vehicle oscillations (specifically, to accomplish this it is sufficient to introduce a suitably formed control moment M_y), it is also possible to achieve asymptotic stability of the solution $\Theta_y = 0$.

In cases when the geometry of the vehicle and its masses does not permit the inequality (16) to be satisfied, a gyroscope (flywheel) can be mounted on the vehicle with its natural kinetic moment vector Γ oriented in direct opposition to the axis z_0. The equation of small oscillations of the vehicle in the plane z_0x_0 assumes the form

$$B\ddot{\theta}_y + [\Gamma + (C-A)\omega_*]\omega_*\theta_y = M_y$$

The condition for stability of the solution $\Theta_y = 0$ with $M_y = 0$:

$$\Gamma + (C-A)\omega_* > 0$$

is satisfied for any C, A, ω_*, provided Γ is sufficiently large.

We note that damping of the vehicle oscillations in the plane z_0x_0 can be obtained not only as the result of an external moment M_y but also as the result of internal dissipative forces. This can be accomplished by purely passive damping elements, incorporating liquid-filled cavities, specially arranged inertial masses on elastoviscous shock absorbers, etc.

In cases when a powered gyroscope (or several gyroscopes) are used to stabilize the vehicle in the plane z_0x_0, very effective damping can be produced by imparting additional degrees of freedom to the gyroscope relative to the vehicle and constraining these degrees of freedom by dampers. Included among the orientation systems that incorporate this notion, in particular, is a combination gravitational-gyroscopic artificial earth satellite orientation system in an orbital coordinate system, which utilizes gravitational effects to stabilize the satellite in two vertical planes and a special arrangement of powered gyroscopes to stabilize in a third plane and dampen the oscillations of the satellite in all three planes (proposed by E. N. Tokar').

It follows from the above presentation that stabilization of a space vehicle in the coordinate Θ_y by a sensor of the second kind is possible in principle. Assuming partial orientation of the vehicle corresponding to alignment of the x and x_0 axes, one is readily convinced that the problem of vehicle stabilization in the cordinate Θ_x is analogous to the one just treated and can be solved by the same techniques. Consequently, the statement made in the first part of the article may be regarded as proven.

LITERATURE CITED

1. E. N. Tokar', Certain Properties of the Gyrocompass with a 17-Hour Period, Prikladnaya Matematika i Mekhanika, Vo. XXV, No. 3, 1961.
2. W. T. Thomson, Stability of Single Axis Gyros on a Circular Orbit, AIAA Journal, Vol. I, No. 7, July, 1963.
3. H. L. Taylor, Satellite Orientation by Inertial Techniques, Journal of the Aerospace Sciences, Vol. 28, No. 6, 1961.
4. C. T. Leondés and R. E. Roberson, Analysis and Synthesis of Satellite Attitude Control Systems, Journal of the Aerospace Sciences, Vol. 29, No. 12, Dec. 1962.

SYMPOSIUM CLOSING REMARKS

Trageser To conclude the session of which we are co-chairmen, Professor Ulanov will make some comments on behalf of both of us.

Ulanov We are now ending the last session and at the same time the symposium itself. Therefore, we co-chairmen would like to make some concluding remarks. These conclusions we feel will be common to the Soviet and American delegations, as well as the other delegations attending this symposium.

We feel that our first symposium was a success. It was distinguished by a high scientific level and very good organization. We all understood each other and could exchange scientific opinions. It is one of the not very numerous examples of such effective organization.

We would like to express our utmost thanks to the National Norwegian Committee of IFAC and to its president, Mr. Rand; to the Norwegian Council of Scientific and Industrial Research and to its president, Mr. Holberg; to the organizing committee of this symposium in the person of Professor Balchen; the Program Committee in the person of Dr. Aseltine; and to the IFAC Theory Committee in the person of its chairman, Professor Truxal; to the Secretary, Mr. Monrad-Krohn; and to the chairman of the round table discussion, Professor Letov.

Holberg Before we finally close the last session of the First Symposium on Automatic Control in the Peaceful Uses in Space, the organizing committee would like to thank everyone who has contributed to this symposium: the Program Committee; the session chairmen; the authors and speakers, and all the participants. Special thanks are due to the American and Soviet delegations for bringing their interesting films. Also thanks are due to the translators for performing an excellent job under difficult conditions.

I think I can, on behalf of the organizing committee, conclude that the meeting has been reasonably successful from a scientific point of view. I hope that this will not be the last IFAC meeting on space programs, or on any other specialized topic covering theory, applications, and components under one meeting. Thank you all for coming to Norway and I hope that you meet again in the future.